C. Common Units in _____ _____ _____ _____sages

Quantity	Unit	Symbol	Definition
volume	liter‡	L‡	$1 L = 1 dm^3 = 10^{-3} m^3$
mass	metric tonne	t	$1 t = 10^3 kg = 1 Mg$
area	hectare	ha	$1 ha = 1 hm^2 = 10^4 m^2$
pressure	megapascal	MPa	$MPa = 10^6 Pa = 10^3 kPa = 10$ bars
time	minute	min	1 min = 60 s
	hour	h	1 h = 60 min
	day	d	1 d = 24 h
electrical conductivity	siemen per meter	$S\,m^{-1}$	—
water potential	joules per kg	$J\,kg^{-1}$	—
yield	megagrams per ha	$Mg\,ha^{-1}$	(Mg is a metric tonne)

D. Prefixes: To indicate fractions or multiples of the basic or derived units. prefixes are used according to the following list:

Prefix	Multiple	Abbreviation	Multiplication factor
exa	10^{18}	E	1 000 000 000 000 000 000
peta	10^{15}	P	1 000 000 000 000 000
tera	10^{12}	T	1 000 000 000 000
giga	10^{9}	G	1 000 000 000
mega	10^{6}	M	1 000 000
kilo	10^{3}	k	1 000
hecto	10^{2}	h	100
deca (don't use)	10	da	10
deci	10^{-1}	d	0.1
centi	10^{-2}	c	0.01
milli	10^{-3}	m	0.001
micro	10^{-6}	μ (or u)	0.000 001
nano	10^{-9}	n	0.000 000 001
pico	10^{-12}	p	0.000 000 000 001
femto	10^{-15}	f	0.000 000 000 000 001
atto	10^{-18}	a	0.000 000 000 000 000 001

E. Example Relationships for Some Unfamiliar Units

1 siemen equals 1 mho. The boundary for saline vs non-saline soils (4 mmhos/cm) is equivalent to 4 dS·m^{-1} (= 4 dS/m)

1 pascal equals 9.87×10^{-6} atmosphere = 10^{-5} bar.

---

*The kilogram is actually the basic reference unit of mass (weight), even though the gram is the name used with prefixes. One gram equals 0.001 kg.

†The Kelvin temperature scale has the same size of degree as the Celsius scale, but $0°K$ is equal to $0°C - 273.2°$. Therefore, Celsius degrees = $°K + 273.2°$.

‡The liter is now defined as the volume of 1 kilogram of pure water at its maximum density (which is about 4°C) and is equal to 1.000 028 dm^3, although it is often defined in general books as 1 dm^3.

NOTE: A few older terms, which have special convenience, continue to be used by some workers in preference to new **SI** terms. One of these is the use of milliequivalents (meq) instead of moles in cation exchange discussions. The most likely problem will be the confusion of "t" used in past editions with acre to mean U.S. conventional short tons (2000 pounds) per acre. It should refer to metric tonnes (1000 kg) when used with hectare (as 15 t/ha). Using megagrams per hectare (Mg/ha), rather than t/ha, will reduce confusion.

SOILS

In Our Environment

SOILS

In Our Environment
Seventh Edition

Raymond W. Miller

Department of Plant, Soil, and Biometeorology
Utah State University

Roy L. Donahue

Professor Emeritus
Department of Crop and Soil Sciences
Michigan State University

Joyce U. Miller

Editorial Assistant

PRENTICE HALL
ENGLEWOOD CLIFFS, NJ 07632

Library of Congress Cataloging-in-Publication Data

Miller, Raymond W.,
    Soils in our environment / Raymond W. Miller, Roy L. Donahue ;
Joyce U. Miller, editorial assistant.
      p.   cm.
    Rev. ed. of: Soils. 6th. 1990.
    Includes bibliographical references and index.
    ISBN 0-13-095803-4
    1. Soil science. 2. Crops and soils. 3. Soil management.
I. Donahue, Roy Luther. II. Miller, Raymond W.
Soils. III. Title.
S591.M7833 1995
631.4—dc20   94-44096
                                 CIP

Acquisitions Editor: Catherine Rossbach
Production Editor: Terrance Stanton/Carlisle Publishers Services
Production Liaison: Eileen M. O'Sullivan
Director of Manufacturing & Production: Bruce Johnson
Managing Editor: Mary Carnis
Manufacturing Buyer: Ilene Sanford
Editorial Assistant: Craig Campanella
Art Editor: Cynthia Clifford
Cover Designer: Amy Rosen
Cover photograph: X. Richer/Liaison International
Book design: Terri Ellerbach
Formatting/page make-up: Carlisle Communications Ltd.
Printer/Binder: Hamilton Printing Company

Printed in the United States of America

10 9 8 7 6 5 4 3 2 1

ISBN 0-13-095803-4

Prentice-Hall International (UK) Limited, *London*
Prentice-Hall of Australia Pty. Limited, *Sydney*
Prentice-Hall Canada Inc., *Toronto*
Prentice-Hall Hispanoamericana, S.A., *Mexico*
Prentice-Hall of India Private Limited, *New Delhi*
Prentice-Hall of Japan, Inc., *Tokyo*
Simon & Schuster Asia Ptc. Ltd., *Singapore*
Editora Prentice-Hall do Brasil, Ltda., *Rio de Janeiro*

We dedicate this book's seventh edition to the many students and teachers who for the past 39 years have used and have been willing to critique each of the previous six editions and the various textbook products produced by the four authors involved. Your use of this text and the letters sent to us have been sufficient to encourage us in making each new revision. We have attempted to make each edition better and more useful for the user. Our love of the land and wish for its preservation are a great incentive to make each edition better than the one before it. To you who hold stewardship over our lands, we dedicate this book.

Raymond W. Miller
Roy L. Donahue

Contents

4

Soil Physical Properties

5

Soil Colloids and Chemical Properties

6 Soil as a Water Reservoir

7 Organisms and Their Residues

8 Acidic Soils and Their Modification

9   Nitrogen and Phosphorus

10   Potassium, Sulfur, and Micronutrients

11   Salt-Affected Soils and Their Reclamation

12 Diagnosis of Soils and Plants

13 Fertilizer Management

14 Tillage Systems

19   *Soil Surveys, Interpretations, and Land-Use Planning*

Preface

Soils in Our Environment is actually the 7th edition of *Soils: An Introduction to Soils and Plant Growth.* The name change was done (1) to shorten the book title for users and (2) to better represent the book's content, which is the study of soils for all uses, not just for agricultural use. Although the major use of soils is to produce plants, soils are also a basic resource that is extensively used for waste disposal, as a place on which to build our houses and roads, and one of the most extensively involved materials as we attempt to remove wastes from our air and water supplies.

Soil is important to everyone, either directly or indirectly. It is a commodity that we need to protect, to use wisely, to understand its ways, and to realize its value. It is finite; formation of new soil is so slow that we must consider it nonrenewable in our time frame of a few hundred years. Thus, as we often state about the earth's mass—we have all we ever will have—we have all the soil we ever will have. We may alter or relocate it (erosion, hauling, and deposition), we may destroy some soils (wash them into oceans or burn organic soils), and we may "manufacture" new soil by loosening or tilling deep loose deposits and adding fertilizers and organic matter. But, we cannot economically produce "new soils" or wait the several thousands of years for nature to replace or extend the supplies of natural soils we now have available.

Each new edition of a textbook is a challenge. The prior edition had most of the information that the authors thought was essential and important that could be included in its length. In a new edition, the authors must choose what previous material to cut, and what new material is of sufficient importance to be added in replacement of material in the text already. We have considered, also, the needs of you, the readers. The comments of teachers using the textbook and of students in the classes using the text helped to emphasize the strengths and to identify the more obvious deficiencies.

This textbook is intended to be sufficiently rigorous in its detail to provide a strong scientific basis in understanding soils. The necessary chemical information is given to allow persons with minimal chemistry knowledge to learn what is essential to understand the properties of soils. Understanding soils' chemistry is critical to fertility, soil modification, soil stability, drainage, salt control, and erosion control. Knowledge of the characteristics of flow and retention of water in soils is necessary when planning to use soils as a filter of waste waters and for other purposes. Knowledge of the microbial processes and organic material in soils is crucial in wise soil use for plant growth and waste disposal.

This seventh edition rearranges a few chapters. In the first three chapters we discuss the *origin, composition,* and *classification* of soils. In these chapters we generalize about the limitations of soils on the earth and how the earth's soils may be degraded or destroyed. A brief review of the *origin* and *mineral composition* of soils is followed by the *weathering* that forms soils. Although many properties of soils will not be understood as these first three chapters are discussed, there is a benefit in getting the concepts of the general compositions and groupings of soils early in the study of soils, as is described in Chapters 2 and 3. The discussion of soil horizons and of the higher categories of soils in Chapter 2 emphasizes the wide variations existing in soils around the world. *Soil horizons,* both letter horizons and diagnostic horizons, are discussed in this chapter. Chapter 3 details *the soil classification system* used by the United States Soil Conservation Service.

The remainder of the text follows the chapter sequence used in the sixth edition. Two chapters—Soil Organisms and Soil Organic Matter—were combined into a single chapter called Organisms and Their Residues. The final chapter in the sixth edition, Soils Requiring Unusual Management, has been mostly deleted as one way to reduce the length of the text. Some portions of that chapter (acid sulfate soils, organic soils, range hazards) have, in part, been incorporated into the text in other locations, but some portions have been eliminated completely.

The entire text has been edited to remove or rewrite material that seemed to need changes. Yet, many chapters will have the same general overall sequence and organization as was in the sixth edition.

Chapter 4—on the physical properties of soils—has had information on earth's energy cycle added, particularly the influence of water on temperature. Considerable addition was made in Chapter 5 concerning the *soil colloids* and includes a new listing of typical formulas for common clay minerals. The discussion of *chromatographic cation movement* in soils was inserted into this chapter in the section on cation exchange.

Chapter 7, concerning organisms and their residues, is a combination of old Chapters 5 and 6 in the sixth edition and was reduced in length by eliminating some less critical topics. The topic of Low Input Sustainable Agriculture (LISA) is discussed in association with organic farming, and with respect to energy requirements.

Chapters 9 and 10, on the *plant nutrient elements,* were updated and modified to a limited extent. A new nutrient cycle is prepared for nitrogen with additional cycles given for phosphorus, potassium, and sulfur.

Chapter 12, Diagnosis of Soils and Plants, was edited for better reading and shortened a bit by simplifying several tables and rewriting the text. A brief outline of visual nutrient deficiency symptoms is included. Chapter 13, Fertilizer Management, includes a discussion on the use of *variable-rate* (on-the-move change) fertilizer additions. The complex fertilizer calculations used in the sixth edition have been eliminated for simplicity and space conservation, although simple calculations are retained.

Important additions were incorporated in Chapter 15 (Soil Erosion and Sediment Control) to emphasis the current pressures to control erosion. *Best management practices* (BMPs) and their implementation are discussed for all lands, agricultural and otherwise. The *revised universal land use equation* and the need for it are discussed. Chapter 16, Water Resources and Irrigation, is updated. A discussion on water efficiency is enlarged and improved. Cablegation, another method of furrow irrigation, is described.

Chapter 18, Soils and Environmental Pollution, has several additional examples added. Point sources and nonpoint sources of pollution are discussed. Best management practices and their role in today's world are discussed. Heavy metals and other toxic elements as pollutants are more extensively evaluated. The section on air quality was extended.

The final chapter (19), Soil Surveys, Interpretations, and Land-Use Planning, discusses further the problems of evaluating land and classifying it for various uses. Several kinds of classifications, including *Land Evaluation and Soil Assessment* (LESA), are mentioned as methods which are currently used to evaluate land for all kinds of uses. Additional tables of engineering uses of soils for land evaluation are added to the chapter.

The Glossary is extensive. A few items have been removed, many have been more abbreviated, and some additional items have been added. The Index is also detailed and extensive. The glossary and index are long but should justify their space by their help in locating quickly any information the reader is seeking.

Obviously, many additional topics about soils for an introductory text could be included. The limiting problem is the length of the text, and therefore its cost. This text may have more material than is wanted for a single one-quarter or one-semester introductory course in soils. However, selected chapters can be used, and many Details can be omitted to shorten the information the students must learn. It is intended that the textbook will be suitable as a reference text to be kept by the student.

We who work with soil love the study and manipulation of it. We soon recognize that all soils are different. We soon recognize that soils can be manipulated and improved or made even worse. As populations grow, we can see the critical importance of keeping the soils we have in good repair. If those who read this book derive that same feeling about retaining all of our soils for the benefit of future generations, we, the authors, will feel well paid for our efforts.

The authors particularly wish to thank Prentice Hall and Paul Corey, Editor, College Division, and his assistant, Julie Boddorf, for their collective and separate efforts in preparing and distributing the sixth edition. For this seventh edition, we wish to thank Catherine Rossbach, Executive Editor, Prentice Hall Education, Career & Technology; her assistant, Craig Campanella; Eileen O'Sullivan, Production Liaison; and, at Carlisle Publishers Services, Terry Stanton, Production Editor. We acknowledge the many individuals, societies, publishers, and agencies who have granted us permission to quote information and to use photographs and drawings. Particular credit is given for each item where it is used. Thanks are due to our reviewers, particularly to Dr. Thomas Ruehr, who cared enough to spend the time to do careful critiques that saved us many blushes for minor and major mistakes.

Raymond W. Miller
Roy L. Donahue

SOILS
In Our Environment

Soil: Its Composition, Weathering, and Importance

Our entire society rests upon—and is dependent upon—our water, our land, our forests, and our minerals. How we use these resources influences our health, security, economy, and well-being.

—John F. Kennedy

1:1 Preview and Important Facts

PREVIEW

Soils, water, mineral deposits, and forests are some of the resources that are essential for life on earth. Ignorance of soil resources—or the mistaken belief in their endless supply—has caused people to abuse the land, to allow erosion, and to permit other damage to the soil.

Why are soils—those mineral materials that we often consider to be inert—important to plants? Soils provide all but 3 of the 16 or more essential plant nutrients, carbon (C), oxygen (O), and hydrogen (H) being supplied by air and water. Soils support plants by holding down their root masses, although strong winds or saturated soils cause windthrow of some trees. Soils hold enough water to supply plant requirements for several days, sometimes even weeks, between rains or irrigations.

Great differences exist among soils. Each soil consists of a unique combination of numerous kinds of minerals along with organic material (humus)—from dead roots, plant tops, and soil organisms—ranging from 1 to 10 percent in most mineral soils to nearly 100 percent in some organic soils. Some extensively weathered soils consist mostly of low-solubility minerals. Other soils are almost totally new and unaltered, formed in new, unweathered minerals, such as recent volcanic ash and lava flow.

What is soil? **Soil** is the natural covering on most of the earth's land surface. "Soil, in its traditional meaning, is the natural medium for the growth of land plants, whether or not it has developed discernible soil horizons...[In] this sense soil has a thickness that is determined by the depth of rooting of plants."[1] Soils do not cover all of the earth's land. Nonsoil

---

[1] Soil Survey Staff, *Soil Taxonomy: A Basic System of Soil Classification for Making and Interpreting Soil Surveys*, Agriculture Handbook 436, USDA—Soil Conservation Service, Dec. 1975, p. 1.

areas, which will not grow plants, include the ice lands of the polar and high-elevation regions, recent hard lava flows, salt flats, bare rock mountain slopes and ridges, and areas of moving dunes.

IMPORTANT FACTS TO LEARN

1. The definition of soil and some of the many variations in soil compositions
2. The reason soil is important to life on earth
3. The most common elements comprising soil-forming minerals
4. The changes in minerals during weathering and weathering products
5. The general chemical and mineralogical nature of clays
6. The essential plant nutrients and their ion forms supplied from soils
7. The quality of various soils as media in which to grow plants
8. The mechanisms by which soils are degraded or destroyed

1:2 Soil—A Critical Natural Resource

Soils are used to grow most of the world's food and much of its fiber. The world's mushrooming human population imposes increasing pressure on farmers to produce more food each year. The quality of the human diet is determined by the hectares of soil available per person and by the quality and management of various soils to produce foods.

A subsistence diet requires about 180 kg (nearly 400 lb) of grain per person per year; an affluent diet requires about four times as much grain. The production of 180 kg of grain requires an average of about 0.085 hectare (0.21 acre). As shown in Table 1-1, by the year 2000 a worldwide average of 0.62 hectare per person will be available, if all cultivatable (arable) land is cultivated and growing grain. It is true that all arable land isn't currently used. In 1970 only about 37 percent of arable land was used to produce food. This is barely enough land to provide an affluent diet for all people now alive, assuming that the food could be stored and distributed uniformly. In another 40 to 50 years essentially all the world's arable land will be required, at current yields, to provide an affluent diet for the increased number of people in the world. Producing higher yields would help relieve the pressure to use all arable soil. If the world's population continues to increase at the current rate (doubling each 40 to 60 years, or even more quickly), the world will suffer critical food shortages within the next century.

Table 1-1 The Quantity of Arable Land in the World and the World's Population

Continent/ Country	Arable without Irrigation (millions of ha)	Potentially Cultivatable (millions of ha)	Population by A.D. 2000 (millions)	Potentially Cultivatable by A.D. 2000 (ha/person)
Africa	595	995	750	1.33
Asia	530	1,100	4,090	0.27
Australia, New Zealand	115	125	35	3.57
Europe	170	245	580	0.42
North America	450	695	530	1.31
South America	650	715	440	1.63
Old U.S.S.R.	325	355	340	1.04
Total	2,835	4,230	6,765	0.62

Selected data from: R. Revelle, "The Resources Available for Agriculture," *Food for Agriculture,* Scientific American Books, New York, 1976, pp. 113–121.

What will be our chances to provide an "average" affluent diet in the world 50 years from now? As the available arable land is increasingly used for crops, fewer animal products will be produced; these now are often produced on arable land that is not needed to grow crops.

As world petroleum reserves are used up in the next 30 to 50 years, there will be pressures to use more ethanol ("grain" alcohol) for fuel. Although only about 0.34 hectare of land (0.84 acre) is required to supply grain for an affluent diet for a year, about 3.17 hectare (7.83 acres) are required to produce the grain to make ethanol to run a car about 16,000 kilometers (9920 miles) per year, which equals 6.5 km/L (15 miles/gal). Therefore, the use of cropland to produce ethanol is a questionable use of food-producing soils.

Suppose the world's population were to live in the manner of many U.S. families: a family of four, with two cars driven 24,194 km (15,000 miles) per year. Each family, using ethanol for fuel, would require 6.18 ha (about 15.3 acres) per year, an average of 1.54 ha (3.8 acres) per person. On the basis of the world's available arable land, it could only support about 1.11 billion people, but the earth's population was 5.8 billion in 1992. It is obvious that to live a life of affluence, *the earth is already more than five times overpopulated.* The conclusion is simple. Some people live in affluence only because many times more people do not live in such affluence. From the facts available, several conclusions seem evident:

1. The apparent excess of arable land in some countries is excess only because many people live in greater population densities in other countries and eat less affluent diets, some even living precariously close to starvation.

2. As more arable soil is used to produce nonfood products (cotton, grains for alcohol, wood, and paper), there will be a decreasing amount of soil available to produce the necessary food for an affluent diet for many of the world's people.

3. Soil is a finite resource. People must retain all the soil possible by using soil conservation practices and by eliminating erosion of soils. In some countries—for example, (China, India, Nepal, and Peru)—land shortages have already necessitated expensive land terracing (Fig. 1-1).

4. With the present world population, the current yields per hectare of all potentially arable land will not produce enough food to provide everyone in the world with an affluent diet. Any future population growth will only aggravate this imbalance.

5. To avoid the unfavorable effects of an inadequate food supply, average world yields need to be increased. Such increased yields will not be easy to obtain or to sustain.

▣ *1:3* The Mineral Composition of Soils

Most soils are **mineral soils**; only about 0.5 percent of U.S. soils and 0.9 percent of world soils are **organic soils**. Organic soils are formed from plant and animal residues in ponded and/or cold, wet areas where organic-matter decomposition is slow. Some of these soils have been drained in recent decades. Many of these organic soils are referred to as **peat** or **muck** soils. Detail 1-1 gives the criteria for distinguishing mineral soil material from organic soil materials.

The substances from which mineral soils develop are mixtures of literally hundreds of different minerals. Some of the more common of these minerals are shown in Table 1-2. Soils are mostly quartz, feldspars, and dark minerals (biotite, hornblende, augite, epidote,

(a)

(b)

(c)

FIGURE 1-1 Machu Picchu, the Lost City of the Incas, near Cuzco, Peru. Its hidden and remote setting, atop steep cliffs rising 760 m (about 2500 ft) above the valley floor, made it necessary for the inhabitants to become self-sufficient in food production by terracing the slopes and filling them with transported soil. The tall peak in (b), from which photo (a) was taken, also has terraces (note arrow). The terraces, 2–3 m (about 6.5–9.8 ft) wide, have stone retaining walls and extend down steep slopes as shown in (c). (Courtesy of Raymond W. Miller, Utah State University.)

Detail 1-1 Mineral versus Organic Soil Material

Mineral soil material is defined by exclusion from either of these two requirements that define organic soils:

1. Organic soil has more than 20 percent organic carbon (about 35 percent humus) if it is never saturated with water more than a few days.

2. Organic soil has water saturation periods (unless drained by people) and has:

   a. At least 12 percent organic carbon (about 21 percent humus) if the soil has no clay, or

   b. has at least 18 percent organic carbon (about 31 percent humus) if the soil has 60 percent or more clay, or

   c. has an intermediate, proportional amount of organic carbon for intermediate amounts of clay.

Soils with less organic carbon than these amounts are called mineral soils.

Table 1-2 Some Important Minerals in Soils

Name	Formula	Comments
Primary Minerals		
Quartz	SiO_2	Hard; weathers slowly; major material of most sands (see "Secondary Minerals")
Feldspars		Hard; weather slowly or moderately,
Orthoclase	$(K,Na)AlSi_3O_8$[a]	but provide important nutrients and
Plagioclase	$(Ca,Na)Al(Al,Si)Si_2O_8$	clay in the weathered products
Micas		"Glitter" in rocks or wet sands;
Muscovite	$KAl_3Si_3O_{10}(OH)_2$	important source of potassium and
Biotite	$KAl(Mg,Fe)Si_3O_{10}(OH)_2$[a]	clay
Dark minerals (augite, hornblende, biotite mica, others)	$Ca_2(Al,Fe)_4(Mg,Fe)_4Si_6O_{24}$[a]	Include several minerals that weather moderately fast; good clay formers
Apatite	$3Ca_3(PO_4)_2 \cdot CaF_2$[b]	Most common mineral supplying phosphorus
Calcite, dolomite, gypsum	(See "Secondary Minerals")	Can be either primary or secondary
Secondary Minerals		
Calcite	$CaCO_3$	Slightly soluble materials in limestone or
Dolomite	$(Ca,Mg)(CO_3)_2$[a]	dolomite rock common in arid-region soils; calcium or magnesium source
Gypsum	$CaSO_4 \cdot 2H_2O$[b]	A soft, moderately soluble mineral found in arid-region soils
Iron oxides	$Fe_2O_3 \cdot xH_2O$[b,c]	A group of minerals with different amounts of water giving soils their yellow-to-red colors; iron source

Continued.

[a] Two elements in the same set of parentheses means that either or both may be part of the mineral in the crystal site.
[b] The dot within a formula indicates that the compounds on either side of the dot both exist as part of the formula.
[c] The x indicates a variable amount of adsorbed water.

Table 1-2, cont'd Some Important Minerals in Soils

Name	Formula	Comments
Secondary Minerals		
Quartz	SiO₂	Reprecipitated forms such as opal, agate, and petrified wood (see "Primary Minerals")
Clays Kaolinite, montmorillonite Vermiculite, illite	(Complex)	(See Chapter 5)

and others), with various quantities of clay minerals, calcite, gypsum, and sesquioxides (hydrous oxides of iron and aluminum) Detail 1-2).

As an example of the elemental composition of various soils, Detail 1-3 lists the estimated percentages of eight common metal oxides in various rock minerals. **Silica** (SiO_2) and **alumina** (Al_2O_3) predominate, with lesser amounts of the common cations iron, calcium, sodium, potassium, and magnesium. The summation of all these elements (Si, Al, Fe, Ca, Na, K, Mg, O) equals over 90 percent of the mineral volume.

Where are the various plant nutrients in soils located? The minerals listed in the previous paragraph do not contain the macronutrients nitrogen and phosphorous or the micronutrients zinc, copper, manganese, boron, molybdenum, and chloride. In fact, the primary (original) minerals that supply these nutrients, except for nitrogen, occur in small amounts in many rocks, even though the rocks are mostly made up of the minerals discussed previously. Nitrogen is not supplied from minerals, but rather from decomposing humus. The dark-colored minerals (and rocks of those minerals, such as basalt and gabbro) tend to have larger

Detail 1-2 The Importance of Silicon and Oxygen

The earth's crust is almost three-fourths silicon (Si) and oxygen (O). The earth's crust is comprised of

Oxygen (O)	46.6%	Calcium (Ca)	3.6%
Silicon (Si)	27.7%	Sodium (Na)	2.8%
Aluminum (Al)	8.1%	Potassium (K)	2.6%
Iron (Fe)	5.0%	Magnesium (Mg)	2.1%

All other elements make up the remainder, only about 1.5 percent.

Quartz (SiO_2) is the most common form of silicon and oxygen and is an important constituent of soils. Most sands and many silts (smaller than sands) consist of quartz particles. Quartz exists as **coarse crystalline, microcrystalline** (cryptocrystalline), and **amorphous** varieties. Some of these materials are described below.

Coarse crystalline (glassy, greasy, or vitreous luster; hardness 7)

Rock crystal: Colorless, hexagonal crystals

Amethyst: Hexagonal crystals; violet colors from ferric iron

Rose quartz: Without crystals; smoky from free silicon atoms from exposure of the rock to radioactive materials

Milky quartz: Without crystals; milky from minute *fluid* inclusions

Granular varieties: Without crystals; fibrous inclusions in quartz

Microcrystalline

Fibrous varieties: **Chalcedony, agate** (alternating layers of different colored chalcedony), **petrified wood,** and **onyx** (similar to agate but with parallel layers)

Granular varieties: **Flint, chert** (much alike; flint has dark-colored inclusions), and **jasper** (usually colored red by hematite)

Amorphous

Opal of many colors and quality, comprises some parts of petrified wood.

The predominance of eight elements (Si, O, Al, Fe, Ca, Mg, K, and Na) in three common rocks in the earth's crust are shown. Carbon (C) is also found in a significant amount in limestones. Note that the minerals are in their oxide forms.

Element	Igneous rocks (%)	Sandstones (%)	Limestones (%)
SiO_2	59.0	78.0	47.5
Al_2O_3	15.0	5.0	0.8
Fe_2O_3	3.1	1.1	0.5
CaO	5.1	5.5	0.5
MgO	3.5	1.2	7.9
K_2O	3.1	1.3	0.3
Na_2O	3.8	0.5	0.1
CO_2	0.1	5.0	41.5
Total	92.7	97.6	99.1

amounts of the micronutrients than are found in rocks high in both quartz and K-feldspars (granite and rhyolite).

1:3.1 Minerals and Rocks

Minerals are inorganic (nonliving) substances that are homogenous, have a definite composition, and have characteristic physical properties such as shape, color, melting temperature, and hardness. Minerals may be either **primary** (formed from cooling of molten rock) or **secondary** (precipitated or recrystallized as solids from soluble substances). Some important soil minerals are listed in Table 1-2.

Rocks are classified into three major divisions:

1. **Igneous:** cooled molten rock
2. **Sedimentary:** sediments deposited in water and consolidated (made into a hardened mass of rock)
3. **Metamorphic:** igneous or sedimentary rocks changed by heat or pressure (hardened or changed mineral orientations) or chemical solution

When molten magma from under the earth's crust is exposed on the surface or at different depths in the earth, **igneous rocks** are formed from it as it cools. The fastest-cooled, expelled, igneous rocks have a glassy amorphous texture; those less rapidly cooled have small crystals in the rock mass. These are **volcanic** or **extrusive** igneous rocks. Magma near the surface, but not expelled, cools more slowly and forms **plutonic** or **intrusive** igneous rocks comprised of large crystals. Changes in cooling rate during cooling result in the formation of **porphyrys**, igneous rocks having mixed large and small crystal particles. Fig. 1-2 is a simplified classification of some common igneous rocks by crystal size and mineral composition. The relative vertical widths of the mineral portions in the headings represent

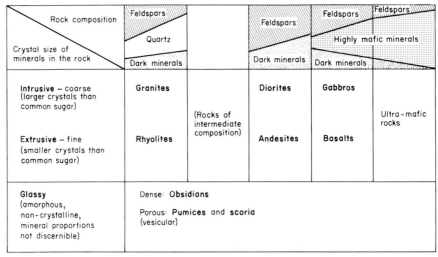

Rock composition / Crystal size of minerals in the rock	Feldspars / Quartz / Dark minerals	(Rocks of intermediate composition)	Feldspars / Dark minerals	Feldspars / Highly mafic minerals / Dark minerals	Feldspars
Intrusive – coarse (larger crystals than common sugar)	Granites		Diorites	Gabbros	Ultra-mafic rocks
Extrusive – fine (smaller crystals than common sugar)	Rhyolites		Andesites	Basalts	
Glassy (amorphous, non-crystalline, mineral proportions not discernible)	Dense: Obsidians / Porous: Pumices and scoria (vesicular)				

FIGURE 1-2 Simplified classification of common igneous rocks. The complete classification includes many other named rocks. The range in proportions of the minerals in each rock group is indicated by the vertical widths of the mineral sections above that column (e.g., granites have few dark minerals and vary in composition from small amounts of quartz to nearly half or more of quartz). *Mafic* refers to high contents of basic cations and lesser contents of silica and alumina portions in minerals.

the range of composition (for example, granites have few dark minerals and vary in composition from small amounts of quartz to nearly equal portions of feldspars and quartz).

Sedimentary rocks were at one time rock and mineral particles or soluble substances that became consolidated or cemented into hard masses. The cementing material makes part of the name for some sedimentary rocks:

- **Calcareous:** for carbonates (lime) (as in calcareous sandstone)
- **Ferruginous:** for iron oxides
- **Siliceous:** for silica (SiO_2)

Sedimentary **conglomerates** and **breccias** are made up of various-sized fragments of rocks cemented together. **Sandstones** are cemented sands (particles 2–0.05mm in diameter) (mostly quartz, but include other minerals) and lesser quantities of smaller particles (smaller than 0.05 mm). **Shales** are consolidated clays and silts (particles 0.05–0.002 mm) with varying amounts of cementation. **Limestones** are calcium carbonate or mixtures of calcium and magnesium carbonates, clays, silts, and sands with more than 50 percent of the mass as carbonates. **Dolomites (dolostones)** are similar to limestones, but have more magnesium carbonate as part of the carbonate mixture. Sedimentary **quartzites** are silica-cemented sands in which the cement is as hard as the sands.

Metamorphic rocks may be as hard as or harder than the igneous, sedimentary, or other metamorphic rock from which they formed, but they weather to produce similar soils. Common metamorphic rocks include:

1. **Gneiss** (pronounced "nice"): from lighter-colored igneous rocks, with minerals segregated and oriented to form light and dark bands (formed mostly from granites, rhyolites, andesites, and other similar minerals)
2. **Schist:** fissile or foliated (appears flaky or layered); comprised of many rocks or minerals, especially micas (the latter are called *mica schists*)

3. **Slate:** hardened shale or siltstone
4. **Quartzite:** recrystallized quartzitic sandstone, formed by heat and pressure; very slow to weather; produces sandy, shallow soils
5. **Marble:** limestone or dolomite hardened enough to polish (easily decomposed; acid-forming gases in the air dissolve in rainfall and slowly destroy marble statues by dissolving them)

Soils that form in similar climates from different kinds of parent rock may differ in nutrients, in depth, in amount of clay, and in color.

1:3.2 Origins of Organic Soils

Organic soils are derived from plants growing in environments that decompose dead residues slowly (Fig. 1-3). Usually, stagnant waters (swamps, marshes, moors, bogs) in temperate climates allow good growth of adapted plants, which die or shed leaves into the water. The decomposition of plant material in stagnant water lacks enough oxygen (is anaerobic) and so is very slow. Over centuries, such accumulations may reach depths of several meters (Fig. 1-4). Examples of these are California's Stockton delta, Florida's everglades, and the many ponded waters in depressions in northern Michigan, Minnesota, and Canada.

Common organic-soil parent materials include many mosses (such as sphagnum), pondweed, cattails, sedges, reeds, grasses, and various "water-loving" deciduous and coniferous shrubs and trees. The term **fibric** means low extent of decomposition; more than two-thirds of the material can be recognized as to its plant part or source.

1:4 Weathering of Soil Minerals

Weathering is the disintegration of **primary** (original) minerals and the reformation of some of those dissolved materials into new, **secondary** minerals. Only when a solid rock mass is

FIGURE 1-3 Organic materials originate in shallow water where over decades generations of plants grow, die, and fall in the water; here they are preserved, even in warm climates, because of the low oxygen content. In this mangrove swamp near Dade County and Monroe County lines in southern Florida is an accumulation of 1.5–2.4 m (about 5–8 ft) of dark brown fibrous (fibric) peat over limestone bedrock.   (Courtesy of Michigan State University.)

FIGURE 1-4   Fibric organic materials (peat) interbedded with silt in the Yukon region of Alaska, where preservation is helped by the cold climate.   (*Source:* U.S. Geological Survey.)

disintegrated to **unconsolidated** material is a soil formed from rock. Water, air, and roots may then move into this loosened material, at which point the material is referred to as *soil*.

When the parent material is a hard rock, such as **granite**, the disintegration to form even a couple of centimeters (an inch or so) of soil can require many hundred to several thousand years. Soils form more rapidly from **sandstones** (cemented sands), **shales** (consolidated clays, silts, and sands), and **limestones** than from granites.

What causes minerals to weather, and what are the products of that weathering? Weathering includes breaking or grinding particles to smaller sizes (**physical weathering**) and the dissolving or chemical alteration of minerals (**chemical weathering**). Particles suspended in flowing water or in blowing winds grind the minerals that the suspended particles hit into smaller particles. Freezing water in cracks in rocks helps split those rocks; temperature changes will weaken some internal bonding, thereby making chemical and physical weathering more effective.

Chemical weathering is usually the most active and effective weathering process. Water in the soil or rock dissolves minerals (**hydrolysis**), softens minerals that absorb the water (**hydration**), and dissolves carbon dioxide. The dissolved carbon dioxide forms carbonic acid, acidifies the water, and increases its solvent action (**carbonation**). Oxygen in the air combines with iron in minerals and forms iron oxides (causing orange and reddish-yellow discoloration). These **oxidation-reduction** reactions increase the rates of disintegration of many dark minerals that have appreciable iron in them. All of these chemical processes will produce in the soil solution many dissolved **ions** from the weathering minerals.

What happens to the dissolved mineral substances? Some dissolved ions are retained in the soil; more-soluble ions are washed away wherever leaching (percolation) water flows, even out of the soil profile. The least soluble materials (iron hydrous oxides and aluminum hydrous oxides) resolidify (precipitate) and remain as part of the soil solids. Soil clays form in this way from these and other low-solubility materials during hundreds of years of weathering. By calculating the soluble mineral materials in river waters, one scientist estimated that the rivers of Europe and North America carried an average of 40 tons of soluble salts dissolved from each square kilometer of watershed. This quantity corresponds to the dissolving and removal of about 1 centimeter of soil depth per thousand years.

The expected products of weathering are indicated by a weathering scheme of *clay-sized particles* (Table 1-3). Soils that still contain gypsum or crystals of sodium salts have been only slightly weathered in their present setting and contain the most easily weathered

Table 1-3 A Weathering Scheme for Clay-Sized Minerals in Soils

Minerals Most Easily Weathered; Present Only in Slightly Weathered Soil		
Typical Clay-size Mineral	*Other Expected Minerals*	*Comments*
1– Gypsum	Halite, other sodium salts	In arid-region soils; only in slightly leached soils
2– Calcite	Dolomite, apatite	In most limestones
3– Olivine-hornblende	Pyroxenes (dark minerals)	Easily weathered primary minerals
4– Biotite mica	Iron-magnesium chlorite	
5– Albite feldspar	Other K-feldspars	
6– Quartz	Cristobolite, other quartz	The most resistant of the primary minerals
Secondary (Reformed) Minerals Produced from Soluble Products of Weathering Primary Minerals		
7– Illite clay	Muscovite, other micas	Least resistant of clays to weathering breakdown
8– Vermiculite clay	Muscovite, other micas	Moderately resistant clay
9– Montmorillonite clay	Beidellite clay	Moderately resistant clay in arid climates
10– Kaolinite clay	Halloysite clay	Moderate-to-high resistance (used for fine "china")
11– Gibbsite [Al (OH)$_3$]	Boehmite, allophane	Resists breakdown; in bauxite (aluminum ores)
12– Hematite [Fe$_2$O$_3$]	Goethite, limonite	Very resistant (in bog iron, sesquioxides)
13– Anatase [TiO$_2$]	Zircon, rutile, corundum	Most resistant minerals
Minerals Most Resistant to Weathering (Insoluble)		

minerals. In their present location these soils have little loss of soluble materials from weathering. Arid regions with little rainfall for leaching have these kinds of soils. In contrast, soils consisting mostly of iron oxides in their clay-sized fraction (as much as 60 to 70 percent of the total material in some wet tropical soils) have been extensively weathered. Most of their primary minerals, including silica, have been washed out of that soil by centuries of weathering and extensive leaching.

Two considerations may modify the patterns shown in Table 1-3. First, if particles are of silt and sand sizes, some quartz and feldspars will still remain in some highly weathered soils, along with the iron and aluminum oxide clay materials. Second, soils formed somewhere else and eroded may have been deposited as *new alluvial deposits.* These new alluvial deposits are starting to form a soil at "time zero," yet the material contains *residues* of highly weathered minerals *(inherited minerals from extensive prior weathering).*

Weathering is a slow process, and clay mineral suites in soils from alluvium and from sedimentary rocks can be **inherited**. Thus, it is usually more important to know the minerals a soil now contains rather than to worry about how it was formed. This "inherited" mineralogy in alluvium (or in rocks such as limestone and sandstone formed from alluvium) must be considered when characterizing any soil being developed from alluvium. Limestones, sandstones, shales, and conglomerate, formed from alluvium in ocean floors, are now uplifted land areas. Soils forming from them may contain clay minerals formed in the weathered soil eons before and in a different climate, before the soil was eroded and formed rocks in an ocean bottom.

The number of elements considered essential for the growth of higher plants now varies from 16 to 20 or more, depending upon the definition of *essentiality*. The authors of this book are aware that Arnon[*] limits essentiality to only those elements that are needed for higher plants to complete all life functions and that the deficiency can be corrected by the application *only of this specific element causing the deficiency*. Other scientists such as Nicholas[†] believe that an element should be considered essential if its addition enhances plant growth even though it merely substitutes for one of the 16 elements that Arnon declares to be essential. For example, because sodium can substitute in plant nutrition for some potassium, and vanadium for some molybdenum, Nicholas would consider both sodium and vanadium as essential, but Arnon would not. On the basis of the criteria used, Arnon specifies 16 elements and Nicholas 20 elements as being essential for the growth of higher plants such as cotton and corn. Three other debated nutrients are nickel (urea transformations), cobalt (N_2 fixation), and silicon. (See Chapter 8 for further information on essential elements for plants.)

[*] D. I. Arnon, "Mineral Nutrition of Plants," *Annual Review of Biochemistry,* **12** (1943), pp. 493–528.
[†] D. J. D. Nicholas, "Minor Mineral Elements," *Annual Review of Plant Physiology,* **12** (1961), pp. 63–90.

1:5 What Soil Provides for Plants

All soils provide the same basic growth factors to plants: **support**, **oxygen** for roots, **adequate temperatures**, **water**, and **nutrients**. But some soils do a better job than others in providing the optimum amounts of these growth factors.

Soils provide support to plants as their roots grow into the soil mantle. Forests on shallow mountainsides, which allow only a foot or so of tree roots, often have catastrophic windthrows of trees. A combination of rain, which softens and lubricates the soils, and high winds will overcome the roots' ability to hold the trees erect.

The water needed by plants is usually retained in the soil reservoir. Most crops need from 300 to 800 mm (11.7 to 31.2 in.) of water during a four or five month growing season. Because most good soils can hold only about 75 to 100 mm (about 3 to 4 in.) of water at one time, the soil must be wetted several times during the growing season by rainfall or irrigation. Some soils are able to hold enough water to supply plants their needed water for 1 to 4 weeks before the soil must be recharged with more water. The size of a soil's water reservoir and the rainfall pattern during the growing season often determines which plants will grow on an unirrigated soil.

Oxygen is supplied to roots from the air in a soil. Plant roots constantly respire during growth; oxygen must be available to roots for respiration. Only a few plants (rice, for example) have mechanisms to transport air (oxygen) *internally* through the aerial part of the plant to its roots. Sandy soils, with few but large pores, are well aerated; clay, with many but small pores, may not have sufficiently rapid gaseous exchange between the atmosphere and deeper soil layers. Roots lacking adequate oxygen replacement in deeper soil layers will not grow very deep or very well. A poorly aerated soil, in effect, will act as a shallow soil. Roots may penetrate only 40 to 60 cm (about 16 to 23 in.) deep in some clayey soils.

Poor porosity of a soil—and thus, poor aeration—reduces the soil's ability to support root growth. Soil structure usually improves soil aeration. Hardpans and compacted sands in deeper layers of the soil may hinder root penetration in those layers.

Plants require 16 nutrients (see Details 1-4 and 1-5). Of these 16 nutrients, 13 are supplied from the soil. Carbon is supplied from carbon dioxide in the air; hydrogen and oxy-

Element Name	Chemical Symbol	Ionic Soil Forms	Comments Concerning Forms of the Element and the Element's Importance in Soils
Aluminum	Al	Al^{3+} $Al(OH)_2^+$	Can be toxic to plants in strongly acid soils; occurs as various hydroxyl forms
Boron*	B	H_3BO_3	A water-soluble plant nutrient in small concentrations
Cadmium	Cd	Cd^{2+}	High atomic weight ("heavy metal"); retained in animals and people and is highly toxic
Calcium*	Ca	Ca^{2+}	Essential plant nutrient; the cation often most prevalent in nonacidic soils
Carbon*	C	HCO_3^- CO_3^{2-}	The basic element of organic substances (mostly made by living organisms); component of carbon dioxide (CO_2)
Chlorine*	Cl	Cl	Occurring in small amounts except when it is a part of soluble salts
Copper*	Cu	Cu^{2+}	May be as Cu^+ (cuprous) in poorly aerated soils
Hydrogen*	H	H^+	A small, active, strongly adsorbed and chemically active ion
Iron*	Fe	Fe^{3+}	Low solubility in most soils; may be as Fe^{2+} in minerals and poorly aerated soil; as iron oxide (Fe_2O_3), it causes the reddish and yellowish coloring in soils
Lead	Pb	Pb^{2+}	Toxic heavy metal; also as PbO_2 in soil
Magnesium*	Mg	Mg^{2+}	Similar in properties and reactions to calcium
Manganese*	Mn	Mn^{2+}	Also as MnO_2 in soil
Mercury	Hg	Hg^{2+}	Toxic heavy metal; also as HgO in soil
Molybdenum*	Mo	MoO_4^{2-}	An essential plant nutrient required in very small amounts
Nickel	Ni	Ni^{2+}	Toxic heavy metal (see Cd)
Nitrogen*	N	NO_3^- NH_4^+	Necessary for proteins; in complex organic forms; both ionic forms are usable by plants
Oxygen*	O	O^{2-} OH^-	As free gaseous form, O_2, it is essential to all respiration
Phosphorous*	P	HPO_4^{2-} $H_2PO_4^-$	Forms many low-solubility phosphates with Ca, Al, Fe, and other heavy metals
Potassium*	K	K^+	Soluble in soils, except mineral forms are very insoluble
Silicon	Si	Si^{4+}	Common in minerals holding oxygens together; sands and quartz are mostly SiO_2
Sodium	Na	Na^+	Not essential nutrient, although it may be for some plants; very soluble; part of "soluble salts"; causes sealing of soil
Sulfur*	S	SO_4^{2-}	Forms S^{2-} (sulfide) form or toxic hydrogen sulfide gas (H_2S) in poorly aerated soil
Zinc*	Zn	Zn^{2+}	Often deficient in calcareous and eroded or leveled soils

*Essential plant nutrient.

gen are derived from water. Nitrogen, the nutrient most often deficient, is released from soil organic materials (humus, plant residues, and animal residues) as they are decomposed. The other 12 essential elements are solubilized from weathering minerals and from surface-adsorbed (exchangeable) cations. Some solubilized elements may form new and different *insoluble* residues with other ions in the soil solution. Each soil's fertility will depend on the soil's amounts and solubilities of these various nutrients.

1:6 Degradation and Destruction of Soils

1:6.1 Degradation of Soil

People used to refer to some soils as "worn out" or as degraded and unproductive. Even the term *destroyed* has been applied to soils buried under deep deposits of flood debris, volcanic ash falls, or wind-blown debris. Burned organic soils and eroded shallow soils certainly can be destroyed.

In the sense that a **degraded soil** has become less productive (less good for growing plants), many soils have been degraded. The worn-out soils of parts of the southeastern United States in the middle 1800s probably suffered deficiencies of one or all of nitrogen, phosphorus, and potassium (or lime) as the yearly crops of cotton and tobacco "mined" these nutrients from the soils faster than they were supplied by fertilizer and weathering. "Resting" the soil for a few years without harvesting crops allows some time for the buildup of soil organic materials (weed residues) to supply nitrogen and an increased accumulation of available phosphorus and potassium from mineral weathering and humus.

What activities degrade soils? Perhaps the best definition of degradation is similar to the definition of a pollutant: *Any substance or action degrades a soil if it makes the soil less usable or desirable for peoples' use.* This definition could include any nonwater changes in the soil—losses or additions—that cause reduced plant growth or make the plant material undesirable for food or other use. Using this definition, reduction in soil humus (the source of nitrogen and much phosphorus and sulfur) would degrade a soil. Cultivation of a virgin humid-area soil without deliberate application of organic materials usually decreases the soil humus over several decades of use to between one-third and one-half of its original amount. This loss of humus is a slow degradation of the soil, which can be partly reversed by good management (adding manure, using green manure crops, using reduced tillage, and using crop rotations).

Soil erosion degrades soils (1) by the loss of soil from areas eroded and (2) sometimes by the deposition of poor soil material on top of good soil. Soil erosion often removes the soil's surface, the most fertile portion of the soil. Deposition of sediment (clays, sands, or rocky materials) on a good soil may lower the quality of that soil (Fig. 1-5).

Soluble salts can hinder plant growth. Soils that were once productive can lose their structure because of high sodium contents in salts and have enough salts to limit water uptake. Nonsalty soils can become salty as irrigation water (with its soluble salts) is applied to them year after year. Some soils do not become salty because there is enough water leaching through them to keep the salts washed out. A soil is not completely destroyed if the salt and sodium can be washed out to help "reclaim" the soil.

What other additions to soils may degrade them? When detrimental materials are added to soils, plants may not grow well or may be poor food. Such chemicals as waste oils, gasoline, and herbicides reduce plant growth. Toxic levels of boron (often accompanying other soluble salts), selenium, heavy metals (from sewage sludge or ore-smelting waste tailings), radioactive fallout elements (strontium, cesium, iodine), and many other materials can degrade the soil. This degradation may cause a direct reduction in plant growth or leave an unacceptable residue in the plant when it is used as a food (heavy metals, pesticides).

FIGURE 1-5 Large plow used for special tillage needs, near Pullman, Washington. Plowing to depths of about 90 cm (about 35 in.) was done to mix impermeable clay layers with coarser textures for better water and root penetration. Such a plow was also used in California to invert good soil back to the surface after it had been covered by alluvial deposits of sand. A plow of about 183 cm (about 6 ft) turned the soil about 120 cm (nearly 4 ft) deep. Such deep tillage is expensive. (*Source*: USDA—Soil Conservation Service; photo by E. E. Rowland.)

1:6.2 Destruction of Soils

The most obvious destructions of soil occur in those shallow soils over rock or hardpan that (1) are organic, such as peat, and are burned away, or (2) are on steep mountain slopes and are washed away or slip away. When these soils are gone, they have been destroyed.

Contaminated soils that can no longer grow plants are essentially destroyed if their reclamation is excessively costly or impractical. Some salty soils in low depressions are not easily drained. To install tile drains and pump water up out of a depression can be impractical. Heavy loadings of some toxic heavy metals (e.g., chromium, cadmium, nickel, selenium, lead, mercury) may be impossible to remove because they do not easily leach from soils. Such contaminated soils are *effectively destroyed,* even though reclamation may be possible sometime in the future.

Some people are so accustomed to getting everything they need from supermarkets that they forget that most foods originate on the land or in the sea. To survive, we must respect those sources and resist destroying or degrading them.

To build may have to be the slow and laborious task of years. To destroy can be the thoughtless act of a single day.

—Winston Churchill

Questions

1. (a) What is soil? (b) Describe the composition of soils. (c) List some of the extremes in composition found in some soils.
2. Name the six most common chemical elements making up soil minerals.
3. Discuss the quantity of arable soil in the world as it relates to adequate future food supplies.
4. What are the possibilities and limits to producing grain alcohol (ethanol) as a fuel source to replace gasoline for internal-combustion engines?
5. (a) What is a mineral? (b) How does a mineral differ from a rock?
6. Differentiate between igneous, sedimentary, and metamorphic rocks.
7. Describe the general composition of these rocks: granite, basalt, limestone, sandstone, gneiss, and quartzite.
8. How is an organic soil different from a mineral soil?
9. The earth's mineral materials change as they weather. What are (a) primary minerals? (b) secondary minerals? (c) some of the more easily weathered minerals? and (d) some minerals that are residual, or resistant to continued weathering?
10. (a) Why is soil important to plants? (b) What roles does soil play in aiding plants to grow?
11. (a) What is an essential nutrient? (b) How many essential plant nutrients are known? List them.
12. (a) What can cause degradation of a soil? (b) What does *degradation* mean?
13. Name the 16 essential plant nutrients, giving their most common ion forms in soil solution.
14. (a) Which essential plant nutrients do not come from soil? (b) Which essential plant nutrients come from air and water? (c) Which essential plant nutrient *other than carbon* comes from soil organic materials but not from soil minerals?
15. Can soil be destroyed? Explain, with some examples.

Soil Formation and Morphology

Soils were not formed in past ages. They are being formed and are ever changing.

—*R. L. Cook and B. G. Ellis*

2:1 Preview and Important Facts

PREVIEW

Soils are formed from hard (solid) rock masses; loose, unconsolidated, transported materials; and organic residues. Originally, even the loose mineral materials were formed by the weathering of rock masses to stones, gravels, sands, silts, clays, and soluble ions. The weathered materials release many soluble ions. Organic soils develop mostly from plants that have fallen into stagnant water where decomposition is slow.

Soil formation comprises two different processes. First, **weathering,** the changes from a consolidated mass (rock) *not capable of growing plants* to the development of an unconsolidated (loose) layer of material that can support plants if climate is suitable and water is available. Second, **soil development,** the changes occuring within the loose material as time passes. Actually, the change from a solid mass to loose soil material and other changes within the soil profile occur simultaneously. **Soil formation** is used to mean both the production of unconsolidated material by weathering processes and soil profile development, which are the changes involved in development of horizons.

Horizons are soil layers that differ from adjacent layers. They tell much about the characteristics of a soil. They include information about hardpans, depth of organic matter accumulation, soil denseness from clay deposition, and the extent of leaching.

Soils have extreme variations. They range from swiftly blowing sand dunes to the deep, wet muds of river deltas and from rocky glacial deposits to the wind-blown plains of the American Midwest. Fence builders in the mountains of the western United States remember painfully the shallow rocky soils and carbonate-cemented hardpans that made digging postholes a blister-raising job. Farm boys soon learn not to plow clayey soils when the fields are wet because the soil becomes a sticky mire that fouls the plow—and tempers. Many northeastern U.S. farmers surround their fields with rock fences built with glacier-strewn stones they have removed from the fields. Tourists in tropical areas can see deep, uniformly red or yellow soil profiles 7 or more meters (23 ft) deep in exposed road cuts. The sands of the Sahara differ greatly from the soils of the jungles of the Congo or Amazon; the sun-baked plains of India contrast sharply with the highly cultivated fields of the Netherlands. Much of the earth's soil cover is far from perfect for its many uses by people.

IMPORTANT FACTS TO LEARN

1. The meaning of *soil development* and *soil formation*
2. The five soil-forming factors and how each influences the soil developed
3. The definition of a *landform* and the various landforms that develop
4. The influence of landform on the soil that develops on it
5. The definition of a *soil horizon*
6. The major *letter horizons* and designated *diagnostic horizons*
7. The information conveyed in diagnostic horizons
8. The relationship between each diagnostic horizon and its equivalent letter horizon
9. One-sentence definitions of these diagnostic horizons: *mollic, argillic, albic, ochric, oxic, spodic, kandic, calcic, duripan*, and *fragipan*

2:2 Soil Formation: Building a Matrix for Living Organisms

What is meant by the term *soil formation?* Soil formation describes a mixture of minerals or hard rock as it changes into loosened material in which plants and other organisms will be able to live and into which air and water can move. To become a soil, the minerals or organic residues must be transformed into separate particles of sand, silt, clay and humus. The soil is a mass of materials with pore space, air, and water that permit root growth. Plants—with a few exceptions—depend on decomposing soil humus to supply their nitrogen and portions of at least 12 other nutrients.

To form soil, the mineral mass weathers to form some clays and accumulate humus. Soil has a depth of several centimeters (an inch) or more, usually a meter (about 40 in.) or more. In this "soil," microorganisms will be active and roots will grow. The cross section of a soil with depth is referred to as a **soil profile.**

The change from minerals or rocks to soil involves changes over centuries of weathering time. The many changes can be summarized as **additions, losses, transformations,** and **translocations.**

Additions are made by water (rainfall, irrigation), nitrogen from bacterial fixation, energy as sunlight, sediment from wind and water, salts, organic residues, and fertilizers, and other substances. **Losses** result from chemicals soluble in soil water, eroded small-sized fractions, nutrients removed in grazed and harvested plants, water losses, carbon losses as carbon dioxide, and denitrification loss of N_2. **Transformations** happen because of the many chemical and biological reactions that decompose organic matter, form insoluble materials from soluble substances, and alter or dissolve some minerals. As water and organisms move within the soil, many substances are **translocated** to different depths—for example, clays to deeper layers, and soluble salts into the groundwater or to the soil surface as water evaporates. Even *localized* sorting is active, such as forming deposits of carbonates or iron oxides in portions of the profile.

2:3 Soil-Forming Factors

Even if all soils were formed through the same weathering processes, they could still differ because of other influences. Five items, called **soil-forming factors,** are primarily responsible for the developed soil:

- **Parent material**
- **Climate** (temperature and precipitation for the most part)
- **Biota** (living organisms and organic residues)
- **Topography** (slope, aspect, and elevation)
- **Time**

2:3.1 Parent Materials and Soil Formation

Parent materials influence soil formation by their different rates of weathering, the nutrients they contain for plant use, and the particle sizes they contain (sandstones = sandy; conglomerates = rocky; shales = clayey). The less developed a soil is, the greater will be the effect of parent material on the properties of the soil. However, even the properties of well-developed soils will be greatly influenced by the parent material. Clay formation is favored by a high percentage of decomposable dark minerals and by less quartz. The results of leaching and many translocations and transformations affected by water movement in soil will be evident even when the soil is well developed. All soils at the lowest category of soil classification (series) are placed into separate series if parent materials are different (Fig. 2-1).

In slightly weathered soils the parent material may be dominant in determining soil properties. Soils from weakly cemented sandstones will be sandy; soils from shales will be shallow and fine-textured. Glacial till, loess, or limestone deposits will be little changed except to accumulate organic material, develop some structure, and perhaps translocate soluble salts and some carbonates. Even moderately to well-weathered soils may not erase the presence of rocks in glacial moraines or greatly change the clays in deltas or the sands in river terraces. It is not surprising that many early soil scientists were also geologists and attributed to parent materials the dominant role in determining the properties of soils.

2:3.2 Climate and Soil Formation

Climate is an increasingly dominant factor in soil formation with increased time, mainly because of the effects of precipitation and temperature. Some direct effects of climate on soil formation include the following:

1. A shallow accumulation or retention of lime (carbonates) in areas having low rainfall occurs because calcium bicarbonates (from dissolving carbon dioxide, minerals, and lime) are not leached if sufficient water is not present. Such soils are usually alkaline.

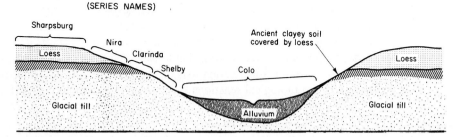

FIGURE 2-1 Cross-section of a landscape in Iowa, illustrating how different times of deposition can result in different soils due to different kinds of parent material, different development times in place, and varying topography. Soils on recent alluvium are the youngest (alluvium = sediment from streams). Eroded loess soils are next youngest (Nira soils). Ancient soil surface (a paleosol now exposed by erosion) is oldest (Clarinda soils). Glacial till now exposed by erosion (Shelby) is next oldest. Stable loess surfaces (Sharpsburg) are more weathered than eroding ones (Nira) (see Fig. 2-5).   (*Source:* Drawn from information in *Soil Survey of Adair County, Iowa.* Soil Conservation Service, 1980, by Raymond W. Miller, Utah State University.)

2. Acidic soils form in humid areas due to intense weathering and leaching out of basic cations (calcium, sodium, magnesium, potassium).
3. Erosion of soils on sloping lands constantly removes developing soil layers.
4. Deposition of soil materials downslope covers developing soils.
5. Weathering, leaching, and erosion are more intense and of longer duration in warm and humid regions like Hawaii, where the soil does not freeze. The reverse is true in cold climates, as in Alaska.

Moisture is important in soil formation. A soil is said to be "developed" when it has detectable layers (horizons), such as of accumulated clays (**B**), organic colloids (**A**), carbonates, or soluble salts that have been moved downward by water. The extent of colloid movement and the depth of deposition are determined partly by the amount and pattern of precipitation, which produce the leaching action.

Climate also influences soil formation indirectly through its action on vegetation. Semiarid climates encourage only scattered shrubs and grasses.[1] Arid climates supply only enough moisture for sparse, short grasses or shrubs, which may not be dense enough to protect the soil against wind and water erosion. Many arid soils exhibit only slight organic matter accumulation and small amounts of profile development.

Soils resulting from year-round hot, humid weathering—such as occurs in some parts of Hawaii on old, stable, and well-drained surfaces—are typically deep (often more than 3 m; 10 ft), reddish in color (due to the presence of oxidized iron), with well-decomposed organic matter, and low in essential elements because of the leaching by the high rainfall.

2:3.3 Biota and Soil Formation

The activity of living plants and animals and the decomposition of their organic wastes and residues (the living environment, the **biota**) have marked influences on soil development. Differences in soils that have resulted primarily from differences in vegetation are especially noticeable in the transition where trees and grasses meet. Minnesota, Illinois, Missouri, Oklahoma, Texas, and most Western states are places where such differences can be observed readily (Fig. 2-2).

Some soils beneath humid forest vegetation may develop many horizons, are leached (washed, eluviated) in the surface layers, and have slowly decomposing organic-matter layers on the surface. In contrast, some grassland soils near the transition zone of forests are rich in well-decomposed organic matter, frequently to depths of 30 cm (1 ft) or more into the mineral soil (Fig. 2-3 and 2-4).

Burrowing animals—such as moles, gophers, prairie dogs, earthworms, ants, and termites—are highly important in soil formation when they exist in large numbers. Soils that harbor many burrowing animals have fewer but deeper horizons because of the constant mixing within the profile, which nullifies the organic colloid and clay movements downward.

Microorganisms help soil development by slowly decomposing organic matter and forming weak acids that dissolve minerals faster than pure water. Some of the first plants to grow on weathering rocks are crustlike **lichens,** which are a beneficial (symbiotic) combination of algae and fungi.

---

[1] J. C. F. Tedrow, J. V. Drew, D. E. Hill, and L. A. Douglas, "Major Genetic Soils of the Arctic Slope of Alaska, " in *Selected Papers in Soil Formation and Classification,* Soil Science Society of America Special Publication Series 1, SSSA, Madison, Wis., 1967, pp. 164–176. *Note:* Although arctic Alaska, receiving about 13 to 26 cm (5–10 in.) of precipitation a year, is climatically a desert, ecologically it is a humid region because of the constantly high relative humidity that condenses as dew at night and thereby adds to the total moisture supply without being recorded as precipitation.

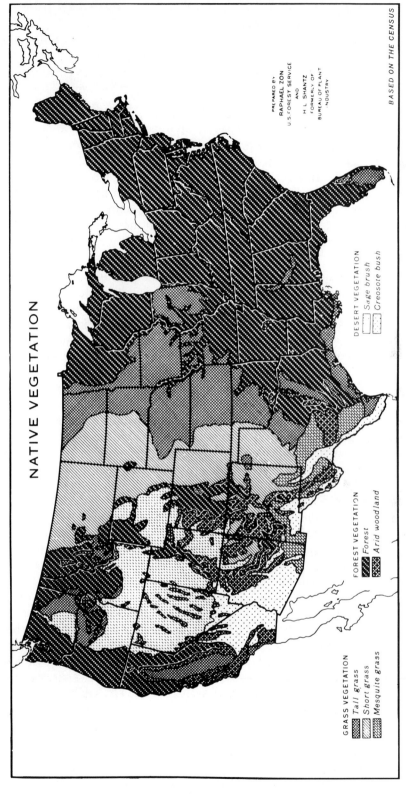

NATIVE VEGETATION

GRASS VEGETATION
Tall grass
Short grass
Mesquite grass

FOREST VEGETATION
Forest
Arid woodland

DESERT VEGETATION
Sage brush
Creosote bush

PREPARED BY:
RAPHAEL ZON
U.S. FOREST SERVICE
AND
H.L. SHANTZ
FORMERLY OF
BUREAU OF PLANT
INDUSTRY

BASED ON THE CENSUS

FIGURE 2-2 Native vegetation exerts a tremendous influence on soil formation, particularly noticeable in the tension zone between forests and grasslands (see Fig. 2-3 and 2-4). (*Source:* USDA.)

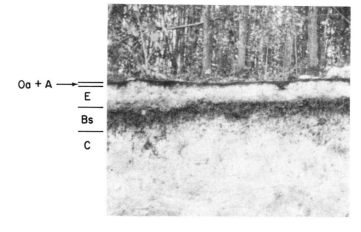

FIGURE 2-3 Under forest vegetation in humid regions, the humus accumulation horizon is more highly leached and there is a well-developed colloid accumulation horizon. Minnesota. (Compare with Fig. 2-4.) (*Source:* USDA.)

Oa + A

E

Bs

C

2:3.4 Topography and Soil Formation

The earth's surface contour is called its **topography** (sometimes called *relief*). Topography influences soil formation primarily through its associated water and temperature relations. Soils within the same general climatic area developing from similar parent material and on steep hillsides typically have thin developing horizons because less water moves down through the profile as a result of rapid surface runoff and because the surface erodes quite rapidly. Similar materials on gently sloping hillsides have more water passing vertically through them than do materials on steeper slopes. The profile on gentle slopes generally is deeper, the vegetation more luxuriant, and the organic-matter level higher than in similar materials on steep topography.

Materials lying in landlocked depressions receive runoff waters from surrounding higher areas. Such conditions favor greater production of vegetation but slower decomposition of dead plant remains because of oxygen deficiency in waterlogged (saturated) soil; this results in soils with large amounts of organic accumulations. If the area above the soil surface is wet for many months of the year, organic (peat or muck) soils develop. If the accumulating waters dissolve salts from surrounding soils, the depression may become a salt marsh with unique tolerant plants growing, or it may develop toxic salt conditions with no plants at all. When soils on the watershed are strongly acid, iron may leach from them and be deposited in depressions to form bog iron (limonite). Alkaline soils on sloping topography in humid regions may result in lime being eroded into depressions and lead to the formation of **marl** (see Chapter 8).

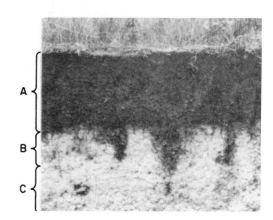

FIGURE 2-4 Under grass vegetation in humid regions, the humus accumulation (**A**) horizon is dark and deep and the colloid accumulation (**B**) is thin or absent. Minnesota. (Compare with Fig. 2-3.) (*Source:* USDA.)

A

B

C

In the northern hemisphere, soils on south- and west-facing slopes receive mroe direct rays of the sun and are therefore warmer and drier than north- and east-facing slopes. In arid climates these drier south- and west-facing slopes are often less productive than soils on north- and east-facing slopes. The reverse slopes are affected in the southern hemisphere. In cold, wet areas, these warmer sites may be the most productive. A higher temperature on south and west slopes results in greater loss of water by evaporation; the net result in regions where water is limiting is often soils with thinner horizons and less vegetative cover than soils on north and east slopes.

2:3.5 Time and Soil Formation

The length of time required for a soil to develop the distinct layers called *genetic horizons* depends upon many interrelated factors of climate, nature of the parent material, the organisms, and topography. Horizons tend to develop most rapidly under warm, humid, forested conditions where there is adequate water to move **colloids** (subvisible-sized clays, humus, and other materials). Acid sandy loams lying on gently sloping topography appear to be the soils most conducive to rapid soil profile development.

Under ideal conditions, a recognizable soil profile may develop within 200 years; under less favorable circumstances, the time may be extended to several thousand years. **Soil development** proceeds at a rate determined by the effects of time plus the intensities of climate and biota (organisms), further modified by the effect of land relief (topography) on which the soil is situated and the kind of parent material from which it is developing (Fig. 2-5).

FIGURE 2-5 Road cut exposing the buried soil from preglacial times that is now part of the currently forming Clarinda soils (see also Fig. 2-1). The ancient clayey subsoil (a paleosol) is very sticky and of low water permeability. Its upper boundary is indicated by the line between arrows. (*Source:* USDA—Soil Conservation Service, State of Iowa.)

Different surfaces of the earth's lands have been exposed for different lengths of time. Some plateau soils have been exposed for hundreds of thousands of years. Glacial till surfaces are more recent but may still be a few hundred thousand years old. More recently, rivers have flooded and covered floodplains and valley bottoms with recent deposition; these land surfaces are only a few years or decades old, but soil development is beginning. Under humid conditions in the central United States, an organic matter–darkened (**A**) horizon and genetic soil structure have developed in surface mine spoils in 5–64 years.[2]

The soils on different-aged surfaces have been forming for different lengths of time. Recent deposits have little soil development; land surfaces exposed thousands of years may have well-developed profiles that are quite different.

2:3.6 Interactions of Soil-Forming Factors

The formation of soil is a diverse and complex process, with effects from the five major factors working in combination. For instance, soil with good drainage (topography or parent material) and mild temperature and high rainfall (climate) will probably support ample plant life (biota) because favorable drainage provides an aerated location for the plant roots. The plants, in turn, decompose, producing carbon dioxide (CO_2), which combines with water from rainfall to form carbonic acid (H^+ and HCO_3^-). Acidity increases the solubility of parent minerals; sodium potassium, calcium, and magnesium are dissolved and leach away in the draining soil water, and the soil becomes ever more acidic. High rainfall also moves some clays and small organic-matter particles deeper into the soil's accumulatin (**B**) horizon.

The same climate (or clayey parent material) but different topography (a low-lying valley or depression instead of a gentle slope) might produce water-logged soil (which has poor drainage). Poor drainage results in stagnant water and nonaerated soil, often with resultant poor plant growth and decreased rates of organic matter decomposition. The accumulating drainage waters also mean the accumulation of dissolved salts. The horizons of these adjacent soils, developed from common factors but with differing topography, will become very different.

Areas of high rainfall and good drainage typically develop acid soils, horizons of organic matter accumulation in the upper layers (**A**), and movement of colloids out of upper horizons and accumulation in the deeper horizons (**B**). Soluble materials are moved deeper or completely below the rooting depth. If the soils have been developing a long time (dozens of centuries), they tend to have high clay content, horizon diffentiation, acidity in wet climates, and salt accumulation in some soils or arid regions. Wet, cool forests become strongly acid, accumulate slowly decomposing organic matter on top of the soil, and have extensive colloid translocation. Wet, warm forests have faster organic-matter decomposition because of warmer temperatures and are extensively weathered and acidic. In drier climates, grass causes deeper layers of dark organic matter accumulation from grass roots and decomposed tops. In semiarid areas, vegetation is sparse, and carbonates often precipitate to form whitish carbonate zones called **lime zones;** accumulated soluble salts are common.

A land body may not be greatly altered during its exposure to the factors of soil development because of many other conditions that retard soil profile development, such as the following:

1. Low rainfall (slow weathering, little soluble material is washed from soil)
2. Low relative humidity (little growth of microorganisms such as algae, fungi, and lichens)
3. High lime or sodium carbonate content of parent material (keeps soil materials less mobile)

---

[2] David Thomas and Ivan Jansen, "Soil Development in Coal Mine Spoils," *Journal of Soil and Water Conservation,* **40** (1985), no. 5, pp.

4. Parent materials that are mostly quartz sands, with few weatherable silts and little clay (slow weathering, few colloids to move)
5. A high percentage of clay (poor aeration, slow water movement)
6. Resistant parent rock materials, such as quartzite (slow weathering)
7. Very steep slopes (erosion removes soil as fast as the **A** horizon develops; low water intake lessens leaching)
8. High water tables (slight leaching, low weathering rate)
9. Cold temperatures (all chemical processes and microbial activity slowed)
10. Constant accumulations of soil material by deposition (continuously new material on which soil development must begin anew)
11. Severe wind or water erosion of soil material (exposes new material to begin afresh to develop a profile) (see Fig. 2-5)
12. Mixing by animals (burrowing) and humans (tillage, digging) minimizes net downward colloid movement
13. Presence of substances toxic to plants, such as excess salts, heavy metals, or excess herbicides

The opposite conditions favor more rapid rates of soil development. For example, material with nearly level topography, not eroding, in an area of high rainfall, developing from easily weathered parent materials, and in warm climates where plants are growing, would develop a differentiated soil profile rapidly.

2:3.7 Ancient Soils: Paleosols

"Normal" soils are those soils developed or developing in the climatic environment now existing. It may be difficult to know whether the climate in which a soil exists has changed within the several thousand years needed for most soils to form. Some soils have been developing for hundreds of thousands of years, and increased development time changes some horizons and increases their degree of development. Climates of many areas have changed during the tens of thousands of years of soil development. In eons past, earth climates, which are now temperate, changed as the glacial and interglacial periods alternated. These previously different climates caused distinctive soils to develop.

At least three terms describe "nonnormal" soils that are generally classified as paleosols: Paleosols, relict soils, and fossil soils. **Paleosols** are *soils formed mostly in previously existing climates* and are now either buried or unburied. Most of these soils have formed during the Quaternary Period (from the beginning of the Ice Age until recently). **Relict soils** are *exposed paleosols that were formed in a previous time under environmental conditions quite different from those currently existing.* These soils may exist as exposed surface soils. **Fossil soils** are *relict soils buried in the geological section of the earth's crust but deeper than present soil development processes.* Although these are relict soils, they are essentially unchanged since burial.

A study of paleosols is used for many purposes, such as to reconstruct ancient history related to climate, crops, and ground cover of buried sites. Such information has helped to verify and even explain conditions existing when catastrophic floods or earthflows buried cities. Radiocarbon dating permits the dating of many sites, even the dates of sequential burials by several catastrophic events.

The most useful *dating method* for buried soils is radioactive carbon dating, based on the fact that radioactive carbon (C^{14}) exists now and has existed in the past in the earth's atmosphere in an exact and constant proportion to the total air carbon dioxide ($C^{14}O_2/CO_2 =$ a constant ratio). Radioactive carbon in the air's CO_2 is produced by *cosmic rays* (protons, alpha particles, other, from space) bombarding atmospheric N_2 molecules forming C^{14} atoms ($_7N^{14} +$ neutron $\longrightarrow$ $_6C^{14} +$ proton off). Any living organism using air CO_2 will incorporate a definite proportion of radioactive CO_2. As long as life continues, the carbon is continuously replaced by new carbon and the *proportion* of radioactive carbon stays constant. As soon as the organism dies, no further continual replacement of carbon occurs. The amount of radioactive carbon in the dead organism decreases as the organism's carbon undergoes radioactive decay. The decay reaction is $_6C^{14}$— decay —> $_7N^{14}$ + beta ray off. Some properties of radioactive carbon are:

Half life (time to lose half its initial activity): 5568 years
Proportion in atmosphere: 1×10^{-12} g of C^{14} per 1 g of total C
Usable time period for dating: 100 to 50,000 years
Sample size needed: varies, 0.1 to 10 g of carbon

Thus, radiocarbon dating of soils or other materials requires a sample of buried carbonaceous materials (wood, bone, coal, oil, carbonate shells, humus, charcoal, or other carbon). Some examples of dates are shown below (see also Detail 7-2):

1. Sediments in Catalina Basin, California (0–0.05 m deep)
    Carbonate fraction   2320 years ± 130 years
    Organic carbon fraction   1970 years ± 150 years
2. Sediments in Catalina Basin, California (4–4.05 m deep)
    Carbonate fraction   23,100 years ± 1000 years
    Organic carbon fraction   18,400 years ± 600 years
3. Linen wrapping enclosing Book of Isaiah included in Dead Sea Scrolls
                                           1917 years ± 200 years
4. Wood and humic acid from a 58 cm-thick soil buried under 0.7 m of leached sand
    and then covered by glacial till
    Wood in buried soil   29,000 years
    Humic acid in buried soil   41,000 years
5. The Provo terrace of the ancient Lake Bonneville, Utah. Tufa at the highest water
    level terrace (west of Salt Lake City)
    Tufa of bedrock outcrops   14,800 years
    Tufa cementing gravel   16,400 years
    Tufa as part of the terrace   11,300 years
    Some geologist believe the dates indicated by the tufa are too young to correctly indicate lake terrace dates.

How would you distinguish between a paleosol and a normal soil? If the soil is buried because of some cataclysmic change (burial by heavy volcanic ash fall, glaciation, or flood alluvium), the soil was obviously formed before burial. The properties (horizons) of any paleosol are different from those of a soil that would develop at that site in the climate currently present or in the time that soil has been developing.

In order to justify calling a soil a paleosol, the scientist must first verify that the buried layer is a *formed* soil, not just deep, unconsolidated earthen deposits. The evidence that the deposit is truly a soil requires evidence of the presence of (1) profile horizons (**O, A, E,** and/or **B**), (2) roots or root channels in place, (3) worm channels that indicate long-time activity of soil organisms, (4) translocated and deposited clays, (5) formed structural peds, particularly those cemented by organic substances, (6) accumulated concretions or translocated lime or gypsum, (7) accumulated plant silica **phytoliths** (plant silica that is formed in various shapes, usually peculiar to the kind of plant), and (8) numerous other features found only in soils as they develop. Some of these other features include horizon sequences, hardpan formation, and fossils of soil organisms or evidence of their extensive activity (earthworm casts). Easily changed soil properties include the diagnostic horizons mollic, gypsic, spodic, and cambic (see Section 2:7). More permanent and irreversible features include oxic, argillic, albic, natric, placic, and petrocalcic horizons (see Section 2:7). Also, plinthite, durinodes, and fragipans are resistant to change because of slow weathering and cementation.

2:4 Landforms and Soil Development

When a soil begins to undergo profile changes because of weathering and leaching, the original material in which the changes are happening is called the **parent material.**

There are numerous possible combinations of parent materials from the sources of rock and mineral mixtures, flood-deposited sediments, glacial deposits, and wind depositions. To organize and describe the properties of these parent materials, they are grouped into the categories of surface rocks, sediments deposited from waters, sediments deposited from winds, sediments deposited from melting ice, and unconsolidated masses accumulated because of gravitational forces. An outline of parent materials is shown in Fig. 2-6.

A mass of parent material often has a distinct shape with characteristic particle sizes. These characteristic mineral or organic masses are **landforms.** (Familiar landforms are

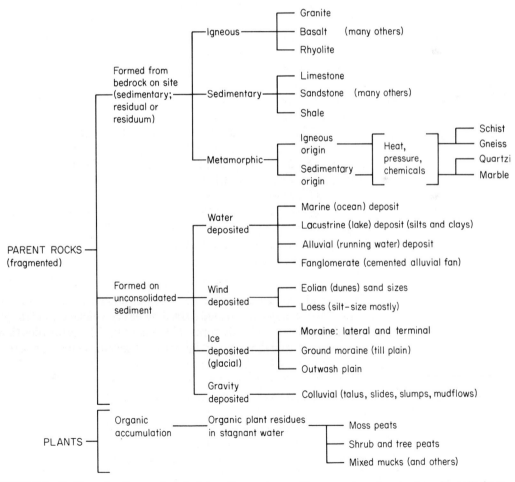

FIGURE 2-6 Outline of soil parent materials with some specific examples of rock given for soils from bedrock. Transported parent materials are by far the most extensive materials on the earth's surface to several feet of depth. (Courtesy of Raymond W. Miller, Utah State University.)

mesas, buttes, plateaus, plains, glacial moraines, terraces, etc.) For example, floodplains tend to be flat and fine-textured, but glacial till remnants are rocky and seldom very smooth-surfaced unless covered by other sediments. Some, but not all, landforms associated with various parent materials are discussed in pertinent sections of this book.

2:4.1 Soils Forming from Rocks

Many soils form in place on hard rocks (sedentary soils). This is a very slow process. In subtropical Zimbabwe, Africa, weathering granite studied at two sites was estimated to form soil at the rate of 11.0 mm (0.43 in.) and 4.1 mm (0.16 in.) per 100 years.[3] Many of the soils forming on rock are shallow because even slight yearly erosion can exceed soil formation rates (Figs. 2-7 and 2-8). Yet on stable slopes, soils can be several meters deep, particularly on porous rocks such as some sandstones, limestones, and volcanic ash.

The landforms of soils forming directly from underlying bedrock are seldom unique. Geologic weathering and erosion tend to produce a leveling effect in the long term, gradually resulting in smooth bedrock slopes with a shallow loose covering of soil. These erosion-formed slopes are called **pediments.** Water cutting into gently sloping pediments or nearly level rock layers can leave large flat **plateaus,** or smaller **mesas,** or tiny **buttes** if the upper rock layers are more resistant to erosion than underlying layers. Erosion undercuts the top layer until it caves in. Nearly vertical **bluffs** and steep **scarps** slope down to the area below (Fig. 2-9).

FIGURE 2-7   This granite rock in Yosemite National Park will require centuries to produce a soil unless slope wash or winds deposit some sediment on the area.   (Courtesy of H. W. Turner, U.S. Geological Survey.)

---

[3] L. B. Owens and J. P. Watson, "Rates of Weathering and Soil Formation on Granite in Rhodesia," *Soil Science Society of America Journal,* **43** (1979), pp. 160–166.

FIGURE 2-8 Lava is an extrusive igneous rock that has cooled rapidly, which does not allow sufficient time for minerals to crystallize (as happens in the formation of basalt or granite, more slowly cooled types or igneous rocks). This is a lava flow in Coconino National Forest, Arizona, where too much erosion or inadequate time and the 380–510 mm (about 15–20 in.) of annual precipitation have not yet been sufficient to develop much soil material. The lone, stunted ponderosa is typical in such environments.   (Courtesy of Michigan State University.)

2:4.2 Materials Deposited from Water

Sediment deposited from flowing water is called **alluvium. Floodplains** are the landforms built by these deposits in the low areas where streams and rivers overflow. Older floodplains now at levels higher than river bottoms, as the waters cut deeper, are called **river terraces.** If soil-laden water is flowing out of mountain canyons, it deposits material as it reaches less steep slopes. These spread out from the canyon mouth to form fan-shaped **alluvial fans.**

If the alluvial fans along the mountain reach each other and coalesce into one continuous slope, the coalesced fans are called **bajadas.** Nearly level **peneplains** can be formed near stream level. If no outlet occurs, a lake in wet climates or a **playa** in more arid climates forms in the low bottoms. The particles of these alluvial deposits are gravelly (2 mm–7.6 cm) where water flows rapidly and more clayey in more level bottoms (floodplains, playas).

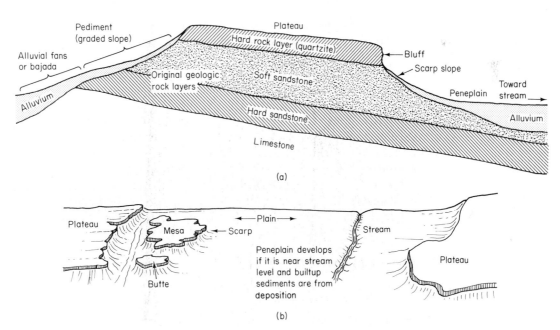

FIGURE 2-9 Some landforms of land areas where soil is forming from bedrock. These are common where the bedrock is sandstones or limestones, particularly horizontally oriented geologic layering. Such landforms typically occur in arid and semiarid regions of the world. (a) is a cross-section of parent materials that may result in the relief view shown in (b).   (Courtesy of Raymond W. Miller, Utah State University.)

FIGURE 2-10 These horizontal lacustrine beds of sands, silts, and clays were formed from previous soils that later were eroded and deposited during glacial times over a period of several thousand years in a freshwater lake in northeastern Washington state. As this area erodes, the materials of different layers are mixed and are deposited somewhere else to begin soil development yet again. (*Source:* U.S. Geological Survey.)

Sedimentation in standing water is mostly in lakes. These are **lacustrine deposits,** lake bottoms (Fig. 2-11). Extensive ancient lakes are now dried up land areas. Large valley areas of lacustrine sediments from ancient glacial Lake Lahontan and Lake Bonneville are now building sites and farmland in Nevada and Utah. Many lacustrine deposits have recent alluvium or wind-carried sediment on top of them. The principal lacustrine deposits in the United States are found bordering the Great Lakes, in northwestern North Dakota and northwestern Minnesota (Red River Valley), northwestern Nevada, and northeastern Washington.

Sediments deposited in oceans or reworked by oceans are **marine sediments.** Many of today's land areas were once marine sediments (salty shales, some dolomites, limestones, and sandstones). Parts of the Great Plains, the Colorado River delta, the Imperial Valley of California, and many coastal areas along the Gulf of Mexico and the Atlantic Ocean have marine sediments. **Beaches** are landforms of marine origin. **Deltas** are formed at the mouths of rivers from waterborne sediment. The Mississippi, Nile, and Amazon deltas (and those of other rivers) are very large land areas, but have shallow water tables.

2:4.3 Materials Deposited by Winds

Wind-transported materials form **eolian deposits.** Carried materials may be coarse and fine sands, silts (smaller than sands), and yet smaller dustlike particles (clays). Small-sized soil materials that were wind-deposited, many following the last glacial period, are known as **loess** (low-ess) (Detail 2-1). Present-day deposits of dominantly silt-sized particles (0.05–0.002 mm diameter) are also called loess.

Loessial soil materials are mostly (60–90 percent) silt and occur mainly in the Mississippi Valley. There are large areas of loess in Kansas, Nebraska, Iowa, Missouri,

Detail 2-1 Particle Sizes and Loess

Soil mineral particle sizes are defined as follows:

Dimensional Units	Size Fraction of the Fine Portion of Soils		
	Sands	Silt	Clay
Inches	0.08–0.002	0.002–0.00008	<0.00008
Millimeters	2–0.05	0.05–0.002	<0.002

When soil contains about 20 percent or less clay and nearly equal amounts of sand and silt in the rest, the soil mixture is called a **loam.** When the words *sandy, silty,* or *clayey* are added as adjectives to the *loam* classification, it indicates a change toward a soil with a higher percentage of sand, silt, or clay, respectively. **Loess,** a German word from *löss,* meaning "pour, dissolve, or loosen," is the term for wind-carried deposits dominantly silt-sized (about 60–90 percent silt). Loess, although forming generally productive soils, is a little unusual in some properties, such as the following:

1. It is quite open and porous.

2. Vertical cuts are more stable than slopes. Grade recommendations are 1 unit distance vertical for each 1/4 unit distance horizontally for 9-m-high (about 30-ft) cuts.* Flatter slopes erode easily by *liquefaction*—supersaturated silt "soup" with water; also known as *mudflows.*

3. When loess is used for earthen dams, problems occur, such as the formation of sinkholes, cavities, and subsurface channel cutting (*piping*).

*Highway Research Board, Highway Research Record 212, Publication 1557, HRB, Washington, D.C., 1968.

Illinois, Indiana, Kentucky, Tennessee, and Mississippi. Extensive deposits also occur in Washington and Idaho (Fig. 2-10). Worldwide, loess deposits of Shaanxi province, China, are said to reach about 300 m (984 ft) deep; other deposits are in the plains of Germany, interior China, and the pampas of Argentina.

Landforms for eolian deposits are **dunes** (sand sizes) and **loess hills** or **plains.** Dunes are further categorized and given names for the sizes and shapes they have.

2:4.4 Materials Deposited from Ice

From perhaps 1 million years ago to as recently as 10,000 years ago, continental ice sheets intermittently occupied the land that is now the northern border area of the conterminous United States. Some geologists claim that we are currently in another such interglacial period. Parts of Alaska, Greenland, Canada, Siberia, Iceland, and the mountains of northern Europe, Switzerland, and Antarctica are now occupied by a mass of ice similar to that which once covered parts of Canada and the northern United States.

The general name for glacial deposits is **glacial till.** Identifying the various landforms of glacial till better describe the parent material. When the ancient ice front melted about as fast as the ice moved, deposits of sediment built up along the melting boundary, resulting in a series of stony hills at the ice front known as **terminal moraines.** Stony ridges also deposited along the outer edges (sides) of the ice mass are known as **lateral moraines** (Fig. 2-12). When the ice front melted faster than it advanced, the glacier shrank and a larger and smoother deposition resulted, known as **ground moraines (till plains)** (Fig. 2-13). Water gushing from a rapidly melting ice mass carried fairly coarse sand and gravel particles and

Loess
layers

Basalt
rock

(a)

(b)

FIGURE 2-11 Loess, wind-blown materials finer than sands, covered many land areas in depths from a few meters to several hundred meters. Sometimes deposition was interrupted by long time periods, even thousands of years, allowing soils to begin to form and then to be covered again. Different deposition layers of Palouse loess on top of Columbia River Basalt near Pullman, Washington. (*Source:* USDA—Soil Conservation Service.)

deposited them in a somewhat gently sloping plain at the outer boundaries of the glacier. These are **outwash plains** (Fig. 2-14).

　　As the glacier bottom melted and tunnels were formed in the ice, sands were deposited along the flow channel, forming **eskers.** When the ice met hard rock hills, it scraped over them, leaving convex mounds (hills) called *drumlins.* Where blocks of ice were left and outwash sediment built up around it, the later melting of the ice left a **kettle** (hole). Unusually large boulders left occasionally on the landscape are called **erratics.**

　　It is not unusual to have quite impermeable rocky ground moraines severely compacted by the ice weight, later covered by outwash sediment, and finally with a top layer of locally

FIGURE 2-12 Summer view of a glacier that is active during current winters in the province of Alberta, Canada, showing a fresh deposit of lateral moraine (as a ridge) in the left foreground as well as directly in front of the glacier on both sides of the U-shaped outlet. (*Source:* U.S. Geological Survey.)

FIGURE 2-13 Glacial ground moraine (till plain) materials are usually stony on the surface (above) as well as throughout the entire deposit (below). (Composite photo courtesy of Julian P. Donahue [above] and U.S. Geological Survey [below].)

FIGURE 2-14 Fresh deposits of coarse sand and gravel from water gushing forth from a glacier (this one in the Gulf region of Alaska) are known as out-wash plains. (*Source:* U.S. Geological Survey.)

sorted sediment from wind or from surface flooding. The gravel and rock hinder tillage, and the compacted till can cause poor drainage and restrict root growth. The thickness of outwash or finer covering over the till is crucial to the use of the land.

2:4.5 Sediments Moved by Gravity

The collective term for all downslope movements caused by gravity of weathered rock debris and sediments is **mass-wasting.** The material moved is called **colluvium.** Some colluvium moves in a relatively dry condition. Rocky accumulations along or at the base of steep slopes are called **talus** (tay′-lus) (*scree; talus cone,* if fan-shaped, Fig. 2-15). Slowly moving slopes caused by expansion and contraction of soil are called **soil creep** (or *rock creep*). Wetting–drying, freezing–thawing, and temperature changes may all aid movement. In lubricated conditions, movement may be a *flow process.* Flow may be from a few centimeters per day (**solifluction**) to rapid **debris flow, mudflow,** or **earth flows** (*slumps,* Fig. 2-16). Some classifications list all mass-wasting flow along discreet shear planes under the general category of **slides.** More-rapid flow is termed **avalanche.** The landform for colluvium from these processes could be specified as **soil creep colluvium, earthflow colluvium,** or **slide colluvium.**

2:5 Development of Soil Horizons

What is a soil horizon, and why do horizons form? **Horizons** are *soil layers that are approximately parallel with the soil's surface.* Each horizon is different from other horizons in the profile. Usually, but not always, the differences between adjacent horizons are quite obvious to the eye (color, structure differences), or the difference can be felt by the fingers (variation in clay contents). The **boundaries** between horizons in a profile range from *indistinct* to *abrupt and clear.* Some boundaries are relatively *smooth* (the horizon is the same thickness in that polypedon); other boundaries have *tonguing* patterns (large vertical changes in the boundary), as is evident in the bottom of the **E** horizon of the Spodosol (color photo 8 in Table 3-2).

Horizons form because of differences in weathering with depth, amounts of humus accumulated, translocations of colloids by water to deeper depths, and losses of colloids from the profile in percolating waters, to name only a few processes. Usually a horizon is separated from others because (1) the horizon has accumulated more humus and is dark-colored, (2) the horizon has had some of its clay and humus moved to greater depths and is a leached horizon, (3) the horizon has accumulated or produced more clay than is in other horizons, and (4) the horizon has accumulated some secondary minerals (calcite, gypsum, silica, or iron oxides), forming horizons such as lime zones or hardpans.

2:6 Letter Horizons

The first scientists to study soils put labels on the different soil layers, using the letters **A, B, C,** and **D.** Since that time, letter-horizon nomenclature has been modified and standardized

FIGURE 2-15 Colluvial material, known as a talus cone, is formed by rocks moving in response to gravity. Glacier National Park, Montana. (*Source:* U.S. Geological Survey.)

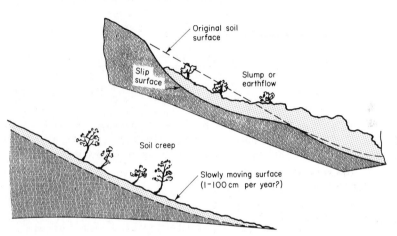

FIGURE 2-16 Effects of gravity on the flow of soil downslope. Rapid movement (a few weeks or less) would be on earthslip or flow, but some areas of soil creep may move only a centimeter or so (fraction of an inch) per month or year. (Courtesy of Raymond W. Miller, Utah State University.)

by the Soil Survey Staff of the U. S. Department of Agriculture's Soil Conservation Service. The latest major "overhaul" of horizon nomenclature was published in May 1981. Because much of the literature currently being published still uses older systems of letters, both the new and the previous system of letter horizons are presented in Table 2-1.

General rules for usage and a few examples follow:

1. Capital letters are used to designate *master horizons,* and some letters have been changed.
2. Lowercase letters are used as suffixes to indicate specific characteristics of the master horizon.
3. Arabic numerals following horizon letters indicate vertical subdivisions within a horizon.
4. Arabic numerals before master horizons indicate discontinuities (used in place of the Roman numerals previously used). *Example:* Loess **A** over limestone **B** would have the **B** written as **2B** to show parent material change (discontinuity).

Examples. An **EB** horizon is a transition horizon between the leached **E** (eluvial) and the accumulation (**B**) horizons. An accumulation of carbonates in a **B** also with clay is a **Btk.**

Table 2-1 Letter Horizons Most Commonly Encountered in Soils

New	Old	Description
		Usually Surface Horizons
Oi, Oe	O1	*Organic* horizon in which most leaves, needles, stems, and other *plant parts* are still *identifiable* (includes recent litter). Usually quite thin—a centimeter or so thick.
Oa,Oe	O2	*Organic* horizon so extensively altered that *identification of the parts of* plant materials *is not* usually *possible.* Can be many centimeters thick.
A	A1	*Mineral horizon darkened by organic matter accumulation.* Under **Oa** horizons it is usually thin; in a cultivated soil it is the surface horizon and may be labeled **Ap.** An **Ap** horizon may be a mixture of several thin horizons, even including part of a shallow **B**.
		Usually Subsurface Horizons
E	A2	A *mineral* horizon lighter colored than an **A** or **Oa** above it or **B** below it. Fine *clays* and minute organic substances have been *washed (eluviated, leached)* out of it by percolating waters. Usually common in high-rainfall areas, especially under forests [see Fig. 2-17(a)]
AB or EB	A3	A transition horizon more like the **A** or **E** above it than like the **B** below it.
BA or BE	B1	A transition horizon more like the **B** below it than like the **A** or **E** above it.
B or Bw	B2	*Layer of illuvial colloids (accumulation)* or evidence of weathering below the "**A**" horizon(s). Small particles that have washed from the **O, A,** or **E** horizons have accumulated because of filtration (lodging) or lack of enough water to move them deeper. Early **B** horizon development stages of soils may have only redder (orange, yellow, brown) colors of weathering caused by the colored iron hydrous oxides. Often higher in clay than the **A**; *always* higher in clay than the **E**. The top of the **B** may start at a depth ranging from about 15 to 50 cm (6–20 in.) below the soil surface.
BC or CB	B3	A transition horizon from **B** to **C** horizons.
C	C	*Unconsolidated material* (unless consolidated *during soil development* by carbonates, silica, gypsum, or other material) below **A** or **B** horizons. Little evidence of profile development.
R	R	Underlying *consolidated* (hard) *rock,* **Cr** for softer material.
		Horizon Subscripts

Subscripts are added to letter horizons for further detail, always as *lowercase letters.* See Glossary Tables G-2 and G-3 for complete list or more detail.

k	ca	A depositional accumulation of calcium and magnesium *carbonates* (lime).
g	g	*Strong gleying,* which is a result of long-time poor aeration, usually because of excess water. Soil colors are grays to pastel blues and greens. *Example:* **Cg.**
h	h	*Deposited (illuvial) humus* from percolating water (**Bh**).
s	ir	*Deposited (illuvial) iron hydrous oxides* (**Bs** or **Bir**).
m	m	*Strong cementation* into hardpans (as by carbonates, gypsum, and silica).
p	p	*Plowed* or other farming disturbance, usually but not limited to **A** horizons.
t	t	*Deposited (illuvial) clay* from horizons above; usually labeled as **Bt.**
x	x	*Fragipan* (hard, silty texture, brittle hardpan).

Source: Adapted by Raymond W. Miller, Utah State University, from (1) USDA—Soil Conservation Service draft of the *Soil Survey Manual,* Chapter 4, pp. 4–39 to 4–50, May 1981; (2) R. L. Guthrie and J. E. Witty, "New Designation for Soil Horizons and Layers and the New *Soil Survey Manual,*" *Soil Science Society of America Journal,* **46** (1982), pp. 443–444.

A parent material with strong gleying and of differing parent material than the horizon above it would be a **2Cg;** the **C** subhorizon below it might be a **2C2g.**

Letter horizons can be subdivided because of differences even within a letter horizon. Thus, **B** could be divided as a **B1** and **B2,** read as "**B** one" and "**B** two." The **C** horizon is divided as **C1, C2, C3,** and so on. A change in parent material (such as loess over limestone) is designated by Arabic numerals 1, 2, and so on. The 1 is understood for the top material and is omitted. Examples of typical profile sequences are listed below:

> A natural grassland: **A1, A2, AB, B1, B2, BC, C1, 2C2, 2R**
> An arid-area soil: **A, B1, B2, C, R**
> Conifer-forested soil: **Oi, Oa, E, Bh, 2Bs, 2C, 3R**
> Plowed prairie soil: **Ap1, Ap2, BA, Bk, C1, C2, R**

All examples use the new horizon terminology. See Fig. 2-17 for two examples of profiles.

■■■ *2:7* **Diagnostic Horizons**

Diagnostic horizons are used to differentiate among soil orders, suborders, great groups, and subgroups. **Diagnostic horizons** that form at the soil surface are called *epipedons;* those forming below the surface, *endopedons.*[4] The most dominant and frequent horizons of U. S. soils are indicated by two asterisks; common but less frequent ones are indicated by one asterisk. The brief definition of each horizon in italics is *not official;* it is given for quickly

FIGURE 2-17 Profiles of soils typical of contrasting conditions of soil formation. The Adams soil (a) formed under forests in humid Vermont and the Dixie soil (b) formed in semiarid southern Utah. Scale is in feet (feet times 30.5 equals cm).   (*Sources:* (a) Vermont Agricultural Experiment Station; (b) USDA—Soil Conservation Service.)

---

[4] *Epipedon,* the surface-developed horizon, is accepted terminology; *endopedon* (from *endo*—in, within), the subsurface horizon, is not established terminology but has been coined by the authors.

orienting the reader. Examples referred to are in the color photos of 10 of the 11 orders in Chapter 3.

2:7.1 Epipedons and Other Horizons Near the Soil Surface

*Albic horizon *A strongly leached E horizon.* Common as a white layer, **E** horizon, near the surface of Spodosols, or in upper Alfisol profiles. A surface or subsurface horizon that is light-colored (>4 in color value, moist; >5 in value, dry) caused by eluviation (leaching out) of coatings of clay and free iron oxides. The light shade is the color of the remaining sand and silt. The clay deposited below it may cause a perched water table.

Anthropic epipedon *A people-made mollic horizon.* A surface horizon formed during use of soil by people for long periods of time as homesites or as sites for growing irrigated crops. Basic cation saturation is high; when not irrigated, the epipedon is dry for 7 years out of 10.

Histic epipedon *An organic surface horizon underlain by mineral soil.* A surface horizon that is saturated with water at some season unless artificially drained, generally between 20 and 30 cm (8 and 12 in.) thick and containing at least 20–30 percent organic matter if not plowed or at least 14–28 percent organic matter if plowed. In each case, the limiting organic matter content depends on the amount of mineral portion that is clay. The lower percentage is used if the horizon has no clay, and the higher percentage is used if the horizon has 60 percent or more clay.

Melanic epipedon *A thick, black, friable horizon formed in volcanic materials.* Usually more than 10% humus content, low bulk density, and at least 30 cm thick. Allophane clays are common with absorbed humus.

**Mollic epipedon *A dark, friable surface horizon, not strongly acidic.* (Found in Mollisols and some Vertisols.) A surface horizon that is dark-colored, contains more than 1 percent organic matter, and is generally more than 25 cm (10 in.) thick unless sandy or shallow to an impermeable layer. It has more than 50 percent basic cation saturation and is not both hard and massive when dry. The Mollic must have Munsell values darker than 3.5 when moist and 5.5 when dry and Munsell chromas of less than 3.5 when moist. For sandy soils, the mollic epipedon may be as shallow as 18 cm (7 in.) or one-third of the depth to a hardpan or bottom of an argillic or natric horizon or to a lime zone, whichever is deeper.

Ochric epipedon *A thin or light-colored surface.* (Common in Aridisols, Entisols, Inceptisols.) A surface horizon that is too light in color (higher value or chroma than mollic epipedon), too low in organic matter, or too thin to be either a mollic or an umbric epipedon.

Plaggen epipedon *A people-caused high-humus horizon.* An anthropic surface layer of soil 50 cm (about 20 in.) or more in thickness that has been produced by long-continued manuring.

*Umbric epipedon *An acidic dark horizon.* (Common as the Ultisol surface.) A surface horizon similar to a mollic epipedon but having less than 50 percent basic cation saturation.

A diagrammatic illustration of relationships among surface-formed diagnostic horizons is shown in Fig. 2-18.

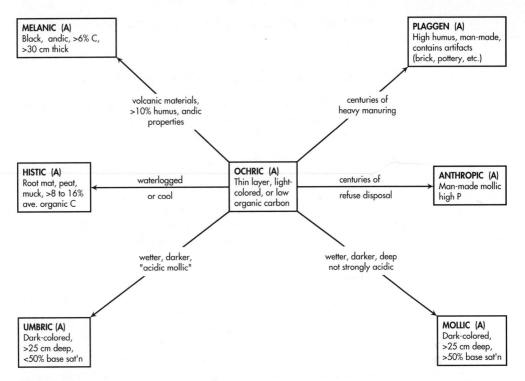

FIGURE 2-18 The relationship among epipedons (surface-formed horizons) using the least-developed, thin, or light-colored ochric as the central concept. (Courtesy of Raymond W. Miller, Utah State University.)

2:7.2 Endopedon ("B") Horizons

These horizons are commonly given letter symbols of **B,** indicating colloid accumulation or weathering changes in a subsurface horizon.

Agric horizon *A tillage-caused clay and humus accumulation horizon.* A subsurface horizon that has formed under a plowed layer by the movement of silt, clay, and humus into voids created by worms, shrink-swell cracks, and capillary pores (not really a heterogenic **B** horizon)

**Argillic horizon *A clay accumulation horizon.* (Common in Alfisols and Ultisols.) A subsurface horizon into which clay has moved. The presence of clay films on ped surfaces and in soil pores is evidence of clay movement. The argillic horizon is at least one-tenth as thick as the sum of all overlying developed horizons unless it is over 15 cm (about 6 in.) thick. An argillic horizon has these clay contents:

Any Part of Overylying (eluvial) Horizon Has:	Argillic Horizon Must Have:
Less than 15% clay	At least 3% more clay than the eluvial horizon
15–40% clay	1.2 times more clay than the eluvial horizon
Over 40% clay	At least 8% more clay than the eluvial horizon

*Cambic horizon *A "color" or weakly developed **B** horizon.* (Common in Inceptisols.) A subsurface horizon that has textures finer than loamy fine sand and in which materials have been altered or removed but not accumulated. Evidences of alteration include the elimination of fine stratifications; changes caused by wetness, such as gray colors and mottling; redistribution of carbonates; yellower or redder colors than in the underlying horizons. A common horizon but weakly developed.

Kandic horizon[5] *An argillic horizon of kaolinite-like clays.* This is a horizon proposed to indicate the accumulation of clays having more cation exchange capacity **(CEC)** than Oxisols and less CEC than montmorillonitic clay layers. The presence of clay films is *not* required. The layer has a CEC of < 16 $cmol_c$/kg of clay and is at least 30 cm (about 1 ft) thick.

*Natric horizon *Like an argillic but with a high exchangeable sodium content.* A subsurface horizon that is a special kind of argillic horizon, containing 15 percent or more of exchangeable sodium, or SAR of the saturation extract is 13 or higher within a 40 cm (15.6 in.) depth.

Oxic horizon *A highly weathered **B** horizon.* (Common in Oxisols.) A subsurface horizon that is a mixture principally of kaolinite, hydrated iron and aluminum oxides, quartz, and other highly insoluble primary minerals, and containing very little water-dispersible clay. It is at least 30 cm (1 ft) thick and has over 15 percent clay of less than 16 $cmol_c$/kg of clay for its CEC.

Sombric horizon *An acidic, humus accumulation, tropical **B** horizon.* A subsurface horizon formed in well-drained mineral soils, consisting of illuvial (leached down) humus. Basic cation saturation is low (less than 50 percent). Restricted to tropical and subtropical regions.

*Spodic horizon *An acidic, cool area, humus and/or sesquioxide accumulation **B** horizon.* (Common in Spodosols.) A subsurface horizon in which amorphous materials consisting of organic matter plus compounds of aluminum and usually iron have accumulated. The spodic is either sandy-textured and has a >2.5-cm-(1-in.)-thick cemented (sesquioxide or humus) layer or has an iron-plus-aluminum content divided by the clay content of 0.2.

A diagrammatic illustration of relationships among diagnostic **B** (colloid accumulation) horizons is shown in Fig. 2-19.

2:7.3 Endopedons from Accumulations of Solubilized Substances

Calcic horizon *A calcium carbonate accumulation horizon.* (Common in Aridisols.) A surface (if exposed by erosion) or subsurface horizon more than 15 cm (6 in.) thick that has more than 15 percent calcium carbonate equivalent, at least 5 percent more carbonates than the **C horizon, and evidence of carbonate translocation (lime-filled cracks, precipitated lime nodules or stalagmites on undersides of rocks and stones).

Gypsic horizon *A gypsum accumulation horizon.* A weakly cemented or noncemented subsurface horizon (or on the surface when eroded severely) that contains a

---

[5] Kaolinite clay is one of the 1 : 1 lattice clays grouped together in a category called *kandites,* which includes kaolinite-like clays including the hydrated clay called halloysite. The **kandic horizon** was proposed by the International Committee on Low Activity Clays (ICOMLAC) in *Agrotechnology Transfer,* **3** (1986), pp. 9–11. This is also described in "Amendments to Soil Taxonomy," *National Soil Taxonomy Handbook,* Issue VIII, April 1986, Part 615.

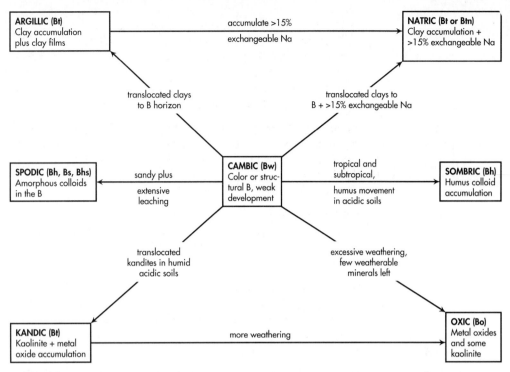

FIGURE 2-19 The **B** diagnostic horizons as they might form from the early-stage horizon, the weakly developed "structural" or "color" **B,** called a *cambic*. (Courtesy of Raymond W. Miller, Utah State University.)

high concentration of gypsum, mostly $CaSO_4 \cdot 2H_2O$. Thickness in cm $\times$ gypsum percentage is equal to or greater than ($\geq$) 150 cm-percent.

Salic horizon *A soluble salt accumulation horizon.* A saline horizon, usually below the surface, at least 15 cm (6 in.) thick that contains at least 2 percent salt. Thickness in cm $\times$ salt percentage $\geq$ 60 cm-percent.

Sulfuric horizon *A horizon high in sulfides.* A surface or subsurface horizon rich in sulfide minerals or high-sulfur organic matter that, when drained, oxidizes to sulfuric acid. Oxidized soil pH is less than 3.5 and is therefore toxic to most plants.

A diagrammatic illustration of relationships among diagnostic horizons formed by leaching or by precipitation of secondary minerals (soluble salts, gypsum, lime, silica, sulfides, and iron and aluminum oxides) is shown in Fig. 2-20.

2:7.4 Hardpan Horizons

*Duripan *A silica-cemented hardpan, usually with some carbonates.* A subsurface horizon that is cemented mostly by silica. Although carbonates may be present, duripans will not slake in water nor in 8 percent hydrochloric (HCl) acid but will disintegrate in hot concentrated KOH solution or alternating acidic and basic solutions. Basic solutions dissolve silica.

Fragipan *A dense, brittle minimal-cemented hardpan.* A natural subsurface horizon with high bulk density relative to **A** and **B** horizons (the solum) above, seemingly cemented (small amounts of silica at contact points?) when dry but showing a

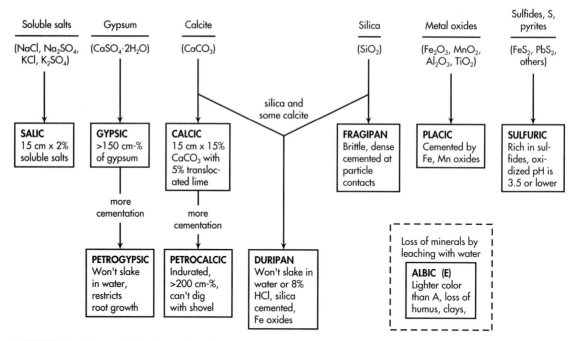

FIGURE 2-20 Diagnostic horizons formed from the precipitation of secondary minerals other than clays and by the loss of minerals by leaching. (Courtesy of Raymond W. Miller, Utah State University.)

moderate-to-weak brittleness when moist. The layer is low in organic matter, mottled, slowly or very slowly permeable to water, and usually shows occasional or frequent bleached cracks, forming polygons. It may be found in profiles of either cultivated or virgin soils but not in calcareous material.

*Petrocalcic horizon *A hard carbonate-cemented hardpan.* An indurated (hardened) subsurface horizon cemented by carbonates and not penetrable by spade or auger. At least 2.5 cm (1 in.) thick and thickness times percentage $CaCO_3$ is ≥ 200 cm-percent.

Petrogypsic horizon *A hard gypsum-cemented hardpan.* A surface or subsurface horizon that is cemented so strongly that dry fragments will not slake in water. Cementation restricts plant root penetration. Fits other requirements for gypsic horizon.

Placic horizon *A very hard iron-cemented hardpan.* A subsurface horizon cemented by iron, iron and manganese, or by iron and organic matter. Forms most readily in both humid tropics and humid cold regions.

2:8 Using Letter and Diagnostic Horizons

Both letter horizons and diagnostic horizons are used. The letter horizons characterize each sequential horizon in relation to each other in that profile and are a part of each profile description. Diagnostic horizons, in contrast, require laboratory data (percentages of lime, salt, gypsum, clay) for their verification. Diagnostic horizons are seldom used in detailed profile descriptions. In the final name given to the soil and in the narration about the soil, some diagnostic horizons may be indicated or mentioned.

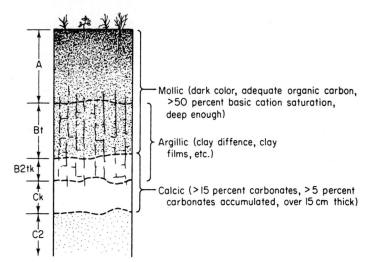

FIGURE 2-21 Example of overlapping diagnostic horizons. Master letter horizons are mutually exclusive, but diagnostic horizons are not. As long as the defined criteria fit, a layer of soil (horizon or part of a horizon) may fit several diagnostic horizons. An example is shown in the diagram.

Although letter horizons cannot overlap, often two or more diagnostic horizons may overlap. For example, a *mollic* horizon (based on color, humus content, base saturation) may include part of all of the *argillic* horizon (based on clay accumulation) in its lower portion (see Fig. 2-21). Many diagnostic horizons, by definition, do exclude all other horizons (for example, ochric, histic, cambic, oxic, spodic, fragipan).

Productivity of young soils reflects parent material; productivity of mature soils reflects soil management.

—R. L. Cook and B. G. Ellis

▮▮ Questions

1. (a) What is a soil profile? (b) What is changed in a soil profile as the soil develops?
2. What is (are) the likely soil texture(s) of soils moderately developed from (a) ground moraine? (b) loess? (c) flat lowland river terraces? (d) glacial moraine outwash? (e) soil creep materials? and (f) dune landforms?
3. List the 16 essential plant nutrients. Give the dominant ion forms in soil solution for the macronutrients N, P, K, Ca, Mg, and S and the nonnutrients Na and Al.
4. (a) Define a *landform*. (b) Define a *soil horizon*.
5. Describe in simple terms (a) an **A** horizon, (b) an **R** horizon, (c) a **B** horizon, and (d) an **E** horizon.
6. Describe some of the characteristics of a petrocalcic horizon.
7. Much amorphous silica in soils is quite soluble in alkaline solution. How can this help explain the presence of duripans below about a meter (about 39 in.) in many arid-region soils?
8. What materials "cement" mineral particles together in various horizons?
9. (a) How do diagnostic horizons differ from letter horizons? (b) Can diagnostic horizons overlap?
10. Briefly define these horizons: (a) mollic, (b) albic, (c) argillic, (d) calcic, (e) ochric, (f) oxic, and (g) duripan.

11. What letter horizon would likely be used for each of these diagnostic horizons: (a) mollic, (b) argillic, (c) calcic, (d) umbric, (e) albic, (f) duripan, (g) ochric, (h) matric, and (i) oxic? Keep in mind, some diagnostic layers may be in more than one letter horizon.

12. Describe, generally, these landforms: (a) plateau, (b) alluvial fan, (c) delta, (d) river terrace, (e) floodplain, (f) aeolian deposits, (g) loessial deposits, (h) ground moraine, and (i) colluvium.

13. Which of the materials in question 12 are likely to be rocky or to have considerable amounts of coarse fragments in them?

14. Write horizon symbols for (a) a **B** horizon with large sesquioxide accumulation, (b) a **B** horizon with high accumulation of smectite clays, (c) a gleyed **C** horizon, (d) a **B** horizon with high carbonate accumulations, (e) a cemented **C** horizon.

15. What is meant by two capital letters together, such as **EB** or **BA,** when referring to horizons? Be general.

16. (a) List the five soil-forming factors. (b) Is erosion one of these factors? (c) How does each factor affect the final soil that is produced?

17. Explain how and why each of these factors influences soil development: (a) aridity, (b) sand with a high percentage of quartz, (c) a high clay content, (d) long cold periods each year, (e) yearly thin depositions of sediment, (f) yearly erosion losses, and (g) profile mixing.

Soil Taxonomy

The ends of scientific classifications are best answered when the objects are formed into groups respecting which a greater number of general propositions can be made, and those propositions more important than could be made respecting any other groups into which the same things could be distributed.

—John Stuart Mill

3:1 Preview and Important Facts

PREVIEW

Soil taxonomy[1] is a classification of soils. It is a grouping of soils into categories based on each soil's morphology (appearance and form).

The first complete U.S. taxonomic classificaiton was published in 1938[2] and modified in 1949.[3] In the 1950s and early 1960s a new system was developed through a series of enlarged publications called *approximations.* In 1961 a comprehensive Seventh approximation was printed and distributed worldwide to soil taxonomists for suggestions. In 1965 the U.S. Soil Conservation Service officially began the use of this system and published it in December 1975.[4] Although this U.S. system has applications worldwide, many countries, among them, France, Canada, China, Brazil, and Russia, use their own systems.

The present U.S. soil classification system (**soil taxonomy**) organizes all soils into 11 **orders,** 54 **suborders,** 238 **great groups,** 1922 **subgroups,** and then **families** and **series.** Each series is subdivided into mapping units, which are called **phases of series;** these mapping units are not a subdivision in the classification system.

---

[1] Taxonomy is derived from the Greek word *taxis,* meaning "arrangement."

[2] M. Baldwin, Charles E. Kellogg, and J. Thorp, "Soil Classification," in *Soils and Men,* USDA Yearbook of Agriculture, U.S. Government Printing Office, Washington, D.C., 1938, pp. 979–1001.

[3] J. Thorp and Guy D. Smith, "Higher Categories of Soil Classification," *Soil Science,* **67** (1949), pp. 117–126.

[4] Soil Survey Staff, *Soil Taxonomy: A Basic System of Soil Classification for Making and Interpreting Soil Surveys,* Agriculture Handbook 436, Soil Conservation Service, USDA, Washington, D.C., 1975, 754 pp.

Order is the most general category in the system. *All soils belong to one of the 11 soil orders.* Five of the soil orders exist in a wide variety of climates: **Histosols** (organic soils), the undeveloped **Entisols,** the slightly developed **Inceptisols** and (volcanic material) **Andisols,** and the swelling-clay **Vertisols.** The other six orders are mostly products of time and of the microclimate in which each developed (Fig. 3-1). **Mollisols** are usually naturally fertile soils, slightly leached, that occur in semiarid to subhumid climates, originally under grasses or broadleaf forests. **Alfisols** are fertile soils in good mositure regimes; they are usually productive nonirrigated lands. **Ultisols** are leached, acidic, and of low-to-moderate fertility, although they are probably the most productive soils because they are formed in areas with long frost-free periods and warm climates. When irrigated, **Aridisols,** the arid-region soils, are often very productive. In contrast, the infertile **Oxisols,** formed in hot, wet tropics, often exist in climates excellent year-round for plant growth. Unfortunately, Oxisols must be left in evergreen forests or carefully fertilized and limed for good yields of cultivated crops. The acidic, sandy **Spodosols,** found in cool climates, are some of the poorest soils for cultivation; they are used for cool-season crops—some grains, potatoes, tree nurseries, and pastures—but they need lime and fertilizer. (See Detail 3-1 and Fig. 3-2 for additional facts about the 11 soil orders.)

The most extensive soils in the United States are Mollisols (over 25 percent). The worldwide distribution of soil orders is Aridisols (19 percent), Alfisols (13 percent), Inceptisols (9 percent), Mollisols (8 percent), and Oxisols (8 percent). The other six orders are present in lesser amounts.

Nearly one-fifth of the world's land area is mountainous and has various amounts of the 11 soil orders so intermixed as to be, on large-scale maps of the United States, uncategorized except as "soils of mountains," "miscellaneous," "salt flats," "rocklands," and "icefields."

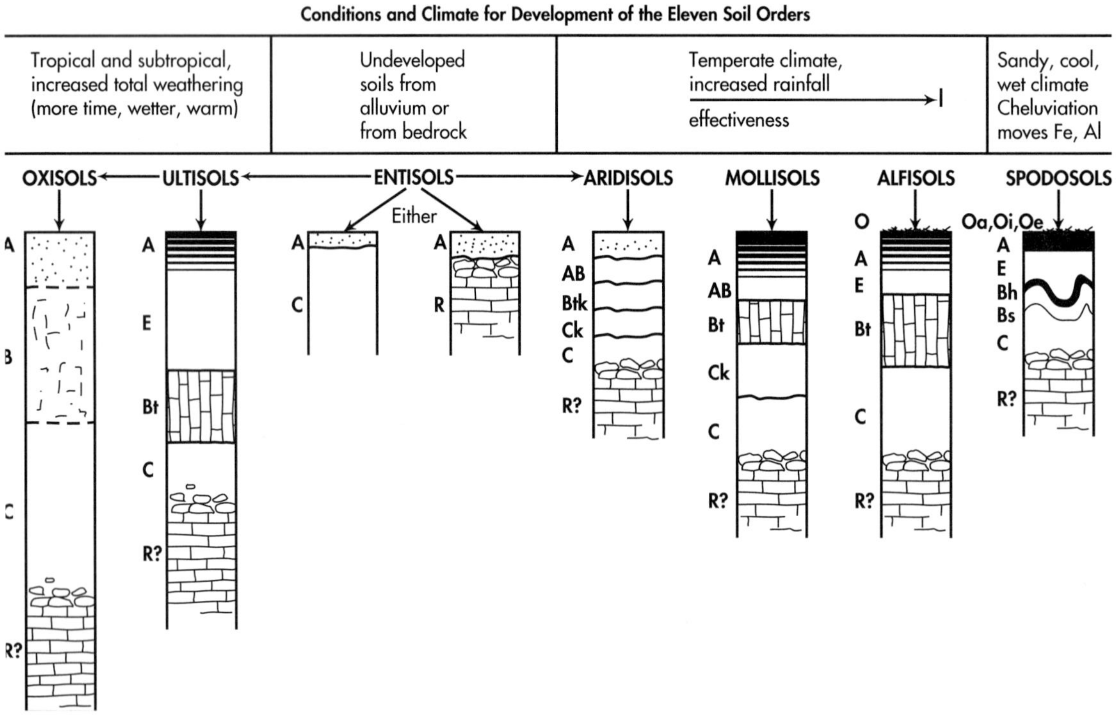

FIGURE 3-1 An outline of 7 of the 11 soil orders suggesting typical climiatic environments and horizon sequences. Question marks indicate horizons that are found in some of these soils but are not found in all soils of that order. (Courtesy of Raymond W. Miller, Utah State University.)

Detail 3-1 Brief Characterizations of the 11 Soil Orders of the United States

Soil Order [a]	General Features
Entisols	Entisols have no profile devlopment except perhaps a shallow marginal **A**.Many recent river floodplains, volcanic ash deposits, unconsolidated deposits with horizons eroded away, and sands are Entisols.
Inceptisols	These soils, especially in humid regions, have weak to moderate horizon development. Horizon development has been retarded because of cold climates, waterlogged soils, or lack of time for stronger development.
Andisols	A tentative soil order. Andisols are soils with over 60 percent volcanic ejecta (ash, cinders, pumice, basalt) with bulk densities below 900 kg/m^3. They have enough weathering to produce dark **A** horizons and early-stage secondary minerals (allophane, imogolite, ferrihydrite clays)[b]. Andisols have high adsorption and immobilization of phosphorus and very high cation exchange capacities.
Histosols	Histosols are organic soils (peats and mucks) consisting of variable depths of accumulated plant remains in bogs, marshes, and swamps.
Aridisols	Aridisols exist in dry climates and some have developed horizons of lime or gypsum accumulations, salty layers, and/or **A** and **Bt** horizons.
Mollisols	Mostly these are grassland, but with some broadleaf forest-covered soils with relatively deep, dark **A** horizons; they may have **B** horizons and lime accumulation.
Vertisols	Vertisols have a high content of clays that swell when wetted. Vertisols require distinct wet and dry seasons to develop because deep wide cracks when the soil is dry are a necessary feature. Usually, Vertisols have only deep self-mixed **A** horizons (top soil falls into cracks seasonally, gradually mixing the soil to the depth of the cracking). These soils exist most in temperate to tropical climates with distinct wet and dry seasons.
Alfisols	Alfisols develop in humid and subhumid climates, have precipitation of 500–1300 mm (about 20–50 in.), and are frequently under forest vegetation. Clay accumulation in a **Bt** horizon and available water much of the growing season are characteristic features. A thick **E** horizon is also common. They are slightly to modearately acid.
Spodosols	Spodosols are typically the sandy, leached soils of cool coniferous forests.Usually, **O** horizons, strongly acidic profiles, and well-leached **E**'s are expected. The most characteristic feature is a **Bh** or **Bs** of accumulated organic material plus iron and aluminum oxides.
Ultisols	Ultisols are strongly acid, extensively weathered soils of tropical and subtropical climates. A thick **E** and clay accumulation in a **Bt** are the most characteristic features.
Oxisols	Oxisols are excessively weathered; few original minerals are left unweathered. Often Oxisols are over 3 m (10 ft) deep, have low fertility, have dominantly iron and aluminum oxide clays, and are acid. Oxisols develop only in tropical and subtriopical climates.

[a] Orders are arranged in approximate sequence from undeveloped soil to increased extent of profile devlopment or increased extent of mineral weathering. The 1981 letter horizons are used.

[b] Allophane is a rapidly formed amorphous aluminosilicate clay; imogolite is slightly more crystallized. Ferrihydrite is rapidly formed amorphous iron hydrous oxide.

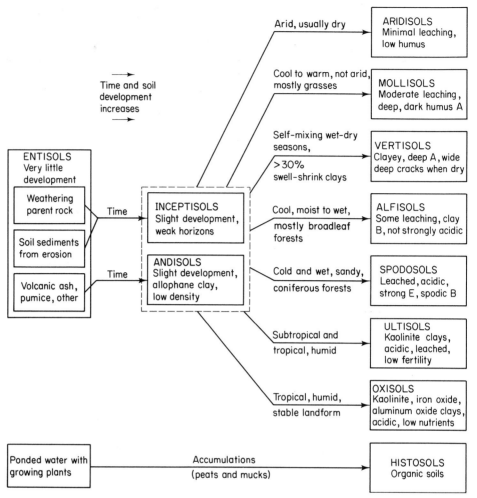

FIGURE 3-2 Schematic outline of the 11 soil orders in the U.S. Soil Classification System, illustrating the general conditions under which each order exists and listing some of the properties of each order. The proposed new order, Andisols, would fit as an early-stage intermediate soil that can develop into other soils   (Courtesy of Raymond W. Miller, Utah State University.)

IMPORTANT FACTS TO KNOW

1. The construction of a soil great group name
2. One-sentence definition of *soil taxonomy, pedon, soil series, soil order, diagnostic horizon,* and *formative element*
3. In general terms, the description of these soil moisture regimes: aquic, udic, ustic, xeric, and aridic
4. The reasons why Entisols and Inceptisols may include very productive soils as well as very nonproductive soils
5. The somewhat unique profile features of many Aridisols and the likely productivity of Aridisols
6. How extensive Mollisols are worldwide and their general fertility
7. The major problems in cropping Histosols, Vertisols, and Spodosols

8. The importance of Alfisols to world agricultural production, especially in past decades
9. The unusual features of Andisols
10. The fertility problems of Ultisols and Oxisols, but the reason why many of these soils are *productive*
11. The composition of plinthite and why it causes problems in cultivation of some Oxisols

3:2 U.S. System of Soil Taxonomy

To select and classify *individual soils,* a minimum soil volume of about 1 to 10 m² (40 to 400 in.²) and as deep as roots grow, called a **pedon** ("ped-don"), is chosen to represent an individual soil. Most *soil individuals* are present in larger than these minimum volumes and are called **polypedons** (many pedons).

The six categories in the U.S. system of soil taxonomy are, in decreasing rank: *order, suborder, great group, subgroup, family,* and *series. Soil phase* is a utilitarian convenience outside the system of soil taxonomy, as explained later.[5]

A given soil, Millville series, from northern Utah would have the following classification:

- *Full name:* **Millville series** of the **coarse silty, carbonatic, mesic family of Typic Haploxerolls**
- *Order name:* **Mollisol;** dark, friable soil
- *Suborder name:* **Xeroll;** deficient water, dry summers
- *Great group name:* **Haploxeroll** ("hap-low-zer-all"); minimum horizon development
- *Subgroup name:* **Typic Haploxeroll;** central concept
- *Family name:* **Typic Haploxeroll, coarse silty, carbonatic, mesic**
- *Series name:* **Millville;** a small town in northern Utah

The categories are a bit difficult to explain, but after a person obtains some experience in soil classifcation, the descriptions will seem logical and clear. The six categories are briefly explained and defined in the following paragraphs. Names are considered as proper nouns and should be capitalized.

Order All soils in the world can be categorized as belonging to one of 11 soil orders. The large number of volcanic soils and their uniqueness caused scientists to separate out of the original ten orders an eleventh soil order called Andisols (Detail 3-1 and Fig. 3-2). When greater detail is known and more soils are studied, new orders (as well as new suborders, new great groups, etc.) may be described.

Suborder Suborders within a soil order are differentiated largely on the basis of additional soil properties and horizons resulting from differences in *soil moisture, soil temperature,* and *dominating effects of chemical or textural features.* Today 54 suborders are recognized; new ones are periodically being considered. Suborders are distinctive to each order and are not interchangeable to other orders.

---

[5] The six categories in the U.S. system of soil taxonomy were inspired by the six categories in the worldwide system of plant and animal taxonomy—namely, *phylum, class, order, family, genus,* and *species.*

Great Group   Soil great groups are subdivisions of suborders. The 238 identified great groups found in the United States have been established largely on the basis of *differentiating soil horizons* and *soil features.* The differentiating soil horizons include those that have accumulated clay, iron, and/or humus and those that have pans (hardened or cemented soil layers) that interfere with water movement or root penetration. The differentiating soil features are the self-mixing (expansion and contraction) properties of clays, soil temperature, and major differences in content of calcium, magnesium, sodium, potasssium, gypsum, and other salts.

It is important to note that the 1938 U.S. classification system included a category called *great soils group,* with names such as *chernozem, podzol, solonetz,* and *laterite.* These are not the same as **great groups** in the current system.

Subgroup   Each soil great group is divided into three kinds of subgroups: one representing the central **(typic)** segment of the soil group; a second, which has properties that intergrade toward other orders, suborders, or other great goups; and a third (extragrade), which has properties that prevent its classification as typic or an intergrade to another soil category. About 1922 subgroups are known in the United States.

Family   Soil families are separated within a subgroup primarily on the basis of *soil properties important to the growth of plants or behavior of soils when used for engineering purposes.* The soil properties used include texture, mineralogy, reaction (pH), average soil temperature, the area's moisture regime, permeability, thickness of horizons, structure, and consistency. More than 5000 families have been identified in the United States.

Series   Each family contains several similar soil series. Each of the more than 15,000 soil series in the United States have narrower ranges of characteristics than the soil family. The name of the soil series has no pedogenic (soil formation) significance but represents a prominent geographic name of a river, town, or area near where the series was first recognized. Soil series are differentiated on the basis of *observable* and *mappable soil characteristics.* Soils in the same series have about the same color, texture, structure, consistency, thickness, and reaction (pH) and have the same *number and arrangement of horizons* and *similar chemical and mineralogical properties.* Soils of a given series must have about the *same kinds, thickness, appearance,* and *properties of horizons.* Examples of series names of mineral soils are

Name	Origin
Jay	*Jay, Oklahoma*
Houston Black	*Houston County, Texas*
Merrimac	*Merrimac River in New Hampshire*
Houghton (peat)	*Houghton Lake, Michigan*

Soil Phase   is used in association with soil series but is not considered as a classification category of the system.[6] **Phases of soil series** are mapping units. Their major use is to delineate soil areas for practical uses, such as farming and municipal or county zoning. Phases have differences in *surface soil texture,* the *solum thickness,* the *percentage slope,* the *stoniness,* the *saltiness,* the *extent of erosion damage,* and other conditions. Mapping units are

---

[6] The word *phase* may also be used in soil taxonomy as a subdivision of a soil order, suborder, great group, subgroup, or family. Its most common use, however, is a subdivision of a series in delineating a soil mapping unit. *Phase of soil series* has replaced *soil type* as a mapping unit.

polypedons that may have *small* portions of other polypedons included in them. Examples of phases are Jay silty clay loam, 0–3 percent slope; Houston Black clay, 2–5 percent slope, eroded; Houghton muck; and Dixie loam, 0–1 percent slope, shallow, rocky.

A soil map of the world (at the suborder level of detail) and based on the U.S. system of soil taxonomy with its original ten orders is given at the end of the chapter in Fig. 3-21. Other systems of soil taxonomy include French, Canadian, Brazilian, and the *USSR.* Utilitarian soil maps of the world, with a common legend in English, French, Spanish, and Russian, have been printed on a scale of 1:5,000,000. They were prepared and published by continents by the Food and Agriculture Organization of the United Nations (FAO) in cooperation with the United Nations Educational, Scientific, and Cultural Organization (UNESCO). As might be expected, computer software with taxonomy keys are being developed.[7] So far, only a limited number of keys have been made available, perhaps partly because the system has been changed since its publication in 1975.

▰ *3:3* **Constructing Taxonomic Names**

To make up suborder and great group names, words were coined from many roots in Latin and Greek, with a few other word roots coming from English, German, French, and Japanese. All names lower than the order (but not the *series* or *family* name) contain a **root** (portion of the order name) as part of the new name; so the soil order to which the soil belongs is always known. The **root** is the syllable preceding a connecting vowel plus *sol*. **Formative elements** are used that suggest properties of the soil (Table 3-1). The fourth most extensive U.S. soil order is the Ultisols, which comprise 12.8 percent of the land surface of the 50 states. An example is given to illustrate the use of formative elements and their origin and meanings in naming the **taxa** (categories). This Ultisol is shown in color photo 9 of Table 3-2.

> **Order Ult***isols* (Latin, *ultimus,* last = well weathered). The root for *Ultisols* is the syllable *ult,* followed by the vowel connecting *i* and the ending *sols.*
>
> **Suborder Ud***ults* (Latin, *udus,* humid = adequate rainfall). Comprises the root preceded by a formative element.
>
> **Great group Pale***udults* (Greek, *paleos,* old = excessive development). The suborder preceded by a formative element.
>
> **Subgroup Aquic** *Paleudults* (Latin, *aqua,* water = overly wet).
>
> **Family Fine-silty over clayey, siliceous, thermic** *Aquic Paleudults* **family.** *Fine-silty* means "a high content of silt and clay"; *over clayey* means "the subsoil is over 35 percent clay"; *siliceous* means ">90 percent silica materials"; *thermic* means "an annual soil temperature between 59 and 72°F (15 and 22°C)."
>
> **Series Sawyer**, named after the city of Sawyer, Oklahoma. The soil occurs in several Southern states.

Throughout the remainder of this chapter, there are many examples of complete soil names.

Although it seems possible to have hundreds of combinations of the formative elements combined with soil order roots, conditions in nature exclude many of them. For example, it is unlikely to find lime in a soil in humid regions, so a *Calciudult* is unlikely to

---

[7] Computer software *Keys to Soil Taxonomy* and a users' manual (1987) are available from the Department of Soil Science, North Carolina State University, P.O. Box 7619, Raleigh, NC 27695-7619. This listing does not suggets an endorsement of the software by the authors or publisher, but is intended as information to interested persons [*Agrotechnology Transfer,* **2** (1986), p. 16].

Table 3-1 Formative Elements Used in Names of Great Groups (and Some Suborders) and the Adjectives Used as Extragrades for All Levels, Particularly Subgroups

Formative Element	Derivation	Mnemonicon	Connotation
		Formative Elements in Names of Great Groups	
Acr	Modified from Gr. *akros*, at tthe end	Acrolith	Extreme weathering
Agr	L. *ager*, field	Agriculture	An agric horizon
Alb	L. *albus*, white	Albino	An albic horizon
And	Modified from *ando* (black soil, Japanese)	Ando	Ando-like, volcanic ash
Aqu	L. *aqua*, water	Aquarium	Aquic conditions
Arg	Modified from argillic horizon; L. *argilla*, white clay	Argillite	An argillic horizon
Bor	Gr. *boreas*, northern	Boreal	Cool
Calc	L. *calcis*, lime	Calcium	A calcic horizon
Camb	L. *cambriare*, to exchange	Change	A cambic horizon
Chrom	Gr. *chroma*, color	Chroma	High chroma, bright color
Cry	Gr. *kryos*, icy cold	Crystal	Cold
Dur	L. *durus*, hard	Durable	A duripan
Dystr, dys	Modified from Gr. *dys*, ill; dystrophic, infertile	Dystrophic	Low basic cation saturation
Eutr, eu	Modified from Gr. *eu*, good; eutrophic, fertile	Eutrophic	High basic cation saturation
Ferr	L. *ferrum*, iron	Ferric	Presence of iron oxide accumulation
Fibr	L. *fibra*, fiber	Fibrous	Least decomposed stage
Fluv	L. *fluvus*, river	Fluvial	Flood plain, alluvium
Frag	Modified from L. *fragilis*, brittle	Fragile	Presence of fragipan
Fragloss	Compound of *fra(g)* and *gloss*		Tongued fragipan
Gibbs	Modified from *gibbsite*	Gibbsite	Presence of gibbsite in sheets or nodules
Gloss	Gr. *glossa*, tongue	Glossary	Tongued, portions of horizon penetrate below it
Gyps	L. *gypsum*, gypsum	Gypsum	Presence of a gypsic horizon
Hal	Gr. *hals*, salt	Halophyte	Salty
Hapl	Gr. *haplous*, simple	Haploid	Minimum horizon development
Hum	L. *humus*, earth	Humus	Presence of humus
Hydr	Gr. *hydor*, water	Hydrophobia	Presence of water
uv	Gr. *louo*, to wash	Ablution	Illuvial
Med	L. *media*, middle	Medium	Of temperate climate
Nadur	Compound of *na(tr)* and *dur*		Sodic duripan
Natr	Modified from *natrium*, sodium		Presence of natric horizon
Ochr	Gr. base of *ochros*, pale	Ocher	Presence of ochric epipedon
Pale	Gr. *paleos*, old	Paleosol	Excessive development, usually very old
Pell	Gr. *pellos*, dusky		Low chroma, dull color
Plac	Gr. base of *plax*, flat stone		Presence of a thin pan
Plagg	Modified from Ger. *plaggen*, sod		Presence of plaggen epipedon
Plinth	Gr. *plinthos*, brick		Presence of plinthite

Psamm	Gr. *psammos*, sand	Psammite	Sand texture
Quartz	Ger. *quarz*, quartz	Quartz	High quartz content
Rhod	Gr. base of *rhodon*, rose	Rhododendron	Dark red color
Sal	L. base of *sal*, salt	Saline	Presence of salic horizon
Sapr	Gr. *sapros*, rotten	Saprophyte	Most decomposed stage
Sider	Gr. *sideros*, iron	Siderite	Presence of free-iron oxides
Sombr	F. *sombre*, dark	Somber	A dark-colored horizon
Sphagn	Gr. *sphagnos*, bog	Sphagnum	Presence of sphagnum, peat moss
Sulf	L. *sulfur*, sulfur	Sulfur	Presence of sulfides or their oxidation products
Torr	L. *torridus*, hot and dry	Torrid	Torric moisture regime
Trop	Modified from G. *tropikos*, of the solstice	Tropical	Humid and continually warm
Ud	L. *udus*, humid	Udometer	Udic moisture regime
Umbr	L. base of *umbra*, shade	Umbrella	Presence of umbric epipedon
Ust	L. base of *ustus*, burnt	Combustion	Ustic moisture regime
Verm	L. base of *vermes*, worm	Vermiform	Worm-worked or mixed by animals
Vitr	L. *vitrum*, glass	Vitreous	Presence of volcanic glass
Xer	Gr. *xeros*, dry	Xerophyte	A xeric moisture regime

Adjectives in Names of Extragrades and Their Meaning

Abruptic	L. *abruptum*, torn off	Abrupt	Abrupt textural change between horizons
Aeric	Gr. *aerios*, air	Aerial	Aeration
Anthropic	Modified from Gr. *anthropos*, man	Anthropology	An Anthropic epipedon
Arenic	L. *arena*, sand	Arenose	Sandy epipedon between 50 cm and 1 m (19.5 and 39.4 in.) thick
Cumulic	L. *cumulus*, heap	Accumulation	Thicker-than-normal epipedon
Epiaquic	Gr. *epi*, over, above, and *acquic*		Surface wetness, standing water
Glossic	Gr. *glossa*, tongue	Glossary	Tongued horizon boundaries
Grossarenic	L. *grossus*, thick, and L. *arena*, sand		Thick sandy epipedon > 1 m (39.4 in.) thick
Hydro	Gr. *hydor*, water	Hydroponics	Presence of water, waterlogged
Leptic	Gr. *leptos*, thin		A thin soil
Limnic	Modified from Gr. *limn*, lake	Limnology	Presence of limnic layer, lake bottom origin
Lithic	Gr. *lithos*, stone	Lithosphere	Presence of a shallow lithic (rock) contact
Pachic	Gr. *pachys*, thick	Pachyderm	Thicker-than-normal epipedon
Paralithic	Gr. *para*, beside, and *lithic*		Presence of a shallow paralithic (hardpan) contact
Pergelic	L. *per*, throughout in time and space, and L. *gelare*, to freeze		Permanently frozen or having permafrost
Petrocalcic	Gr. *petra*, rock, and *calcic* from calcium		Presence of a petrocalcic horizon
Petroferric	Gr. *petra*, rock and L. *ferrum*, iron		Presence of a petroferric contact (ironstone)
Plinthic	Modified from Gr. *plinthos*, brick	Plinthite	Presence of plinthite
Ruptic	L. *ruptum*, broken	Rupture	Intermitten or broken horizons, discontinuous
Superic	L. *superare*, to overtop	Superimpose	Presence of plinthite at the surface
Terric	L. *terra*, earth		A mineral substratum, used with Histosols
Thapto	Gr. *thapto*, buried		A buried soil

Source: Soil Survey Staff, *Soil Taxonomy: A Basic System of Soil Classification for Making and Interpreting Soil Surveys,* Agriculture Handbook 436, USDA, Washington, D.C., 1975, pp. 89–90.

Note: Figure numbers refer to photographs appearing in the color insert between pages 76 and 77.

Table 3-2 Soil Profiles in Color Representing 10 of the 11 Soil Orders: Their Taxonomy and Diagnostic Horizons (Listed Alphabetically)

Figure No.	Order	Suborder	Great Group	Subgroup	Family [a]	Series
1	Alfisols	Boralfs	Cryoboralfs	Typic Cryoboralfs	Loamy, mixed, frigid	Waitville
2	Aridisols	Argids	Haplargids	Typic Haplargids	Fine-loamy, mixed, thermic	Mohave
3	Entisols	Psamments	Ustipsamments	Typic Ustipsamments	Mixed, mesic	Valentine
4	Histosols	Hemists	Cryohemists	Terric Cryohemists	Sphagnic, borustic, euic	Grindstone
5	Inceptisols	Ochrepts	Eurochrepts	Typic Eurochrepts	Coarse-silty, mixed, thermic	Natchez
6	Mollisols	Udolls	Hapludolls	Typic Hapludolls	Fine-loamy, mixed, mesic	Clarion
7	Oxisols	Torroxs	Torroxs	Typic Torroxs	Clayey, kaolinitic isohyperthermic	Molokai
8	Spodosols	Orthods	Haplorthods	Typic Haplorthods	Sandy, mixed, frigid	Kalkaska
9	Ultisols	Udults	Paleudults	Aquic Paleudults	Fine-silty over clayey, siliceous, thermic	Sawyer
10	Vertisols	Usterts	Pellusterts	Udic Pellusterts	Fine, montmorillonitic, thermic	Houston Black

[a] The words given in the table refer to only the family prefix because the family name also includes the subgroup names: loamy, mixed, frigid Typic Cryoboralf.

54

		General Diagnostic Horizons or Differentiating Features		Specific Diagnostic Horizon Indicated by Arrows on Photos	Location of Soil Profile
Figure No.	Phases of Series	Epipedon	Endopedon		
1	Waitville sandy loam	High base saturation	Argillic, ochric, and calcic horizons	Argillic endopedon	Manitoba, Canada
2	Mohave loam, 1–5% slopes	Ochric horizon (light colored)	Calcic, natric, or gypsic horizon	Calcic endopedon	Arizona
3	Valentine fine sand	—	Sandy to a depth of 1 m	Sandy to a depth of 1 m	Nebraska
4	Grindstone peat, deep phase[a]	Histic horizon	—	Histic epipedon	Manitoba, Canada
5	Natchez silt loam 7–40% slopes	Ochric horizon (light colored)	Cambic horizon	Cambric endopedon	Mississippi
6	Clarion loam 0–3% slopes	Mollic horizon (dark colored)	Cambic horizon	Mollic epipedon	Polk County, Iowa
7	Molokai silty clay loam 2–12% slopes	—	Oxic horizon	Oxic endopedon	Molokai, Hawaii
8	Kalkaska sand, 0–40% slopes	—	Spodic horizon	Spodic endopedon	Osceola County, Michigan
9	Sawyer loamy sand, 2–8% slopes	Low base saturation	Argillic (clay) horizon	Argillic endopedon	Dooly County, Georgia
10	Houston Black clay, 1–3% slopes	30% more clay in all horizons, wide and deep cracks, moist chromas of less than 1.5 in upper foot		50–60% clay, moist chroma of 1 in upper 38 in (1 mi)	Travis County, Texas

Courtesy credits: Figures 1, 4, Canada Department of Agriculture; Figures 2, 3, 5, 7, 10, USDA—Soil Conservation Service; Figure 6, Roy W. Simonson; Figure 8, Eugene P. Whiteside; and Figure 9, David A. Lietzke, both at Michigan State University.
[a] When no slope is indicated, it is 0–1 percent.

occur. Many of the formative elements referring to accumulation of soluble salts, gypsum, or lime are unlikely to occur in Ultisols, Oxisols, and Spodosols. Try to think of some other unlikely combinations, as an exercise in looking over the formative elements and learning their meanings. In various tables and figures in the rest of this chapter, many names are given as examples, particularly the ten soils described in Table 3-2.

3:4 Soil Moisture Regimes

The **moisture regimes** represent an attempt to indicate the extent of naturally available water in the soil depth of maximum root proliferation (soil control section). The **soil control section** for moisture regimes is the depth between where 2.5 cm (1 in.) of water would wet that dry soil and where 7.5 cm (2.4 in.) of water would wet it. In clayey soils, this soil control layer is from about 8 to 25 cm (3.1 to 9.8 in.) deep; for sands it is from about 20 to 60 cm (7.8 to 23.4 in.) deep. For greater detail, refer to the Glossary under *Soil moisture regimes*. Many soil suborders have formative elements indicating the moisture regime in which they occur.

Aquic conditions Usually wet with anaerobic saturation for a period long enough to produce visual evidence of poor aeration (mottling and gleying [gray coloring])

Udic Usually adequate water throughout the year

Ustic Deficient water, but most of the water available comes during the summer season

Xeric Deficient water and with a dry cropping season; most of the precipitation comes in the winter time of year

Aridic Very water deficient; long dry periods; short wet periods

Torric Same criteria as aridic, but used at specified locations in the classification system

Because moisture regimes indicate when and how long the soil's "major root zone" is wetted, the precipitation patterns are only part of the cause for the moisture regime. **Aquic conditions** has replaced what was called as *aquic moisture regime* because *aquic conditions* are adaptable to more conditions. The periods of saturation may be only a few days, but *aquic conditions must involve* **reducing conditions.** Aquic conditions include any of: (1) *redoximorphic features* (wetness-caused mottles), (2) *redox concentrations of Fe and Mn,* (3) *redox depletions of Fe and Mn* leaving low-chroma wetness mottles, and (4) *reduced matrix* that changes color when exposed to air (oygen). Extended periods of seepage, the soil aspect (the compass direction), and infiltration all influence the soil's water content.

In Cache Valley, in northern Utah, a sandy ridge could have a dry summer (*xeric* regime), while a poorly drained valley bottom clay could have an *aquic* regime, kept waterlogged by seepage. A higher-elevation mountain slope facing north may have enough coolness, periodic seepage, and snow cover plus light summer rains to have a *udic* regime.

3:5 Terminology for "Family" Groupings

Family groupings are based on (1) mineral particle classes, (2) the soil's mineralogy, (3) soil temperature classes, (4) sometimes the available rooting depth, and (5) less frequently the pH, lime, sand particle coatings, and/or permanent cracks. Only the first three properties (which are the most common) are discussed.

3:5.1 Particle-Size Classes

The intent is to separate soils of quite different textures in the control section into separate families. The **control section** is usually from the top of the **Bt** to a depth of 50 cm (about 20 in.); if no **Bt** exists, the control section is often from 25 to 100 cm (about 10 to 39 in.) (or to rock if less than 100 cm). The seven groups are:

Fragmental Mostly made up of stones, cobbles, gravel, and very coarse sands without enough fine particles to fill voids larger than 1 mm.

Sandy-skeletal Over 35 percent of material is coarser than 2 mm diameter with enough "sand" (see "Sandy," below) to fill voids larger than 1 mm.

Loamy-skeletal As "Sandy-skeletal" except "loam" (see "Loamy," below) fills the voids.

Clayey-skeletal As "Sandy-skeletal" except "clay" (see "Clayey," below) fills the voids.

Sandy All material is sand or loamy sand except "very fine sand" size fraction.

Loamy All material is between "sandy" and "clayey."

Clayey Material is more than 35 percent clay. (*Fine clayey* is 35–60 percent clay; *very fine clayey* is more than 60 percent clay.)

3:5.2 Soil Mineralogy Classes

For most of the root zone soil, terms are used to indicate dominant minerals, such as **carbonatic** (carbonates), **serpentinic** (serpentine minerals), and **siliceous** (over 90 percent silica). The most common term is **mixed,** meaning that no mineral makes up more than 40 percent of the soil. *Clayey* soils with over 50 percent of the clay on one type may be called **montmorillonitic, kaolinitic, illitic, oxic,** and so on, to indicate the dominant clay. *Mixed* is used with clayey soils, also. Other terms may be used for special materials (e.g., **cindery** for over 60 percent of volcanic ash or cinders). The same *control section* is used as for particle-size classes.

3:5.3 Soil Temperature Classes for Family Groupings

Temperature classes are based on *mean annual soil temperature,* the *average summer temperature,* and the *difference between mean summer and mean winter temperatures.* Mean annual soil temperature is determined by measurement for the major root zone (5 to 100 cm) and can be estimated in most of the United States by adding 1°C to the mean annual *air* temperature. A single reading at 10 m (32.8 ft) deep is within one degree Celsius of the annual average value. Also, if the average summer temperature of the top 100 cm is used, the average temperature at different depths is calculated by adding 0.6°C for each 10 cm (4 in.) above 50 cm (20 in.) deep and subtracting the same amount for each 10 cm deeper than 50 cm. The soil temperature classes for temperate region soils are defined in tems of mean annual soil temperature as

- **Pergelic:** 0°C (32°F); permafrost present (unless dry)
- **Cryic:** 0–8°C (32–47°F); summer soil temperature *below* about 15°C
- **Frigid:** 0–8°C (32–47°F); summer soil temperature *above* about 15°C
- **Mesic:** 8–15°C (47–59°F)
- **Thermic:** 15–22°C (59–72°F)
- **Hyperthermic:** 22°C (72°F)

Tropical region soils, characterized by having a difference in temperature between mean summer and mean winter of less than 5°C (9°F), are grouped into the following mean annual soil temperature classes by adding **iso-** in front of the correct temperature name: **isofrigid, isomesic, isothermic,** and **isohyperthermic.** These four *iso* temperatures are common in many tropical climates at different elevation.

3:5.4 Examples of Family Names

The family name contains several words, usually describing texture, mineralogy, and soil temperature. A "correct" sequence for writing family and subgroup names is not established. Because the U.S. Taxonomy system develops from the most specific to the most general, some persons prefer to write the names with the sequence from left to right: *series, family,* and then *subgroup.* The reverse order can be used.

Colby series (Colorado)   The fine-silty, mixed, calcareous, mesic family of Ustic Torriorthents.

Minoa series (New York)   The coarse-loamy, mixed, mesic family of Aquic Dystric Eutrochepts.

Sharpsburg series (Nebraska)   The fine, montmorillonitc, mesic family of Typic Argiudolls.

Houston Black series (Texas)   The fine, montmorillonitic, thermic family of Udic Pellusterts.

Nebish series (Minnesota)   The fine-loamy, mixed, frigid family of Typic Eutroboralfs.

Makaweli series (Hawaii)   The fine, kaolinitic, isohyperthermic family of Oxic Haplustox.

3:6 Soil Orders: An Orientation

The 11 soil orders, which include all the world's soils, are a diverse group. A very generalized outline of these 11 orders in a soil development scheme is shown in Fig. 3-2.

The orders are discussed in the following sections in a development sequence beginning with organic parent materials *(Histosols).* These are followed by *Entisols,* the least developed mineral soils, and then the *Inceptisols* and *Andisols,* having intermediate development. The remaining orders are discussed roughly in order of extent of profile leaching and weathering, beginning with the *Aridisols.* Then follow the soils with increasingly more effective water, *Mollisols* and *Alfisols.* The *Vertisols* tend to occur in many areas where wetting and drying cycles exist, so they are listed after the Alfisols. *Spodosols* are inserted at this point, although they are unique in requiring sandy profiles and strong leaching. The *Ultisols* and *Oxisols* are discussed last because they require very strong weathering of primary materials. These weathering conditions are usually found only in tropical and subtropical climates having very wet cycles and year-round mineral weathering. Weathering may be 300–800 percent greater in these conditions than in many temperate regions.

3:7 Soil Order: Histosols (Greek, *histos,* tissue, organic soils)

Suborders:

Fibrists (Latin *fibra,* fiber): largely undecomposed plant fibers
Folists (Latin *folia,* leaf): mostly leaf mat accumulations

Hemists (Greek, *hemi,* half)[8]: half are recognizable plant fibers
Saprists (Greek, *sapros,* rotten): unrecognizable fibers due to partial decomposition

3:7.1 Properties and Classification of Histosols

Histosols are organic soils. For soils formed in or under water, the distinguishing features for a Histosol is a minimun *organic carbon* content of 12 percent when the mineral portion has no clay, increasing to 18 percent if the mineral portion is 60 percent or more clay. For soils formed on a lithic (shallow to rock) contact, organic carbon content must be 20 percent or more. Organic soils have higher water retention, high cation-exchange capacities, the usual nutrient deficiencies (particularly nitrogen, potassium, and copper), and low bulk densities (as low as $0.1 Mg/m^3$; 6.3 lb/ft^3).

Histosols can form in any climate from the Arctic and Antarctic to the tropics. Humid climates and water ponding are conducive to organic matter accumulation. Formation of Histosols are common in wet, cold areas, such as Alaska, Finland, and Canada, and in wet areas having stagnant marshes and swamps, as in Ireland.

The classification of Histosols is based on both their physical properties (kind of organic material, stage of decomposition, thickness, bulk density, water-holding capacity, temperature, and permeability) and their chemical properties (carbonates, bog iron, sulfates, sulfides, pH, basic cation saturation, and carbon: nitrogen ratio). The extent of decomposition is estimated by the portion of the material recognizable as to source (leaves, plant stem, kind of plant) (Fig. 3-3). In the 1938 U.S. classification system Histosols were known as **Bog** soils.

3:7.2 Management of Histosols

Some management problems of Histosols are unique. When drainage is developed, the soil organic matter decomposes rapidly and drastically reduces the soil volume. Land surfaces have dropped in elevation (**subsided**) as much as 3.7 m (12 ft) in a 40-year period after drainage of the organic soil (Fig. 3–4). The organic matter will burn; setting brush fires, draining fuel from motorized equipment, and smoking pose real hazards. For seed bed preparation, the soil may need to be rolled by 25,000-pound packers to firm it rather than plowed to loosen it. It is difficult, when planting, to get good seed–water contact.

Histosols are used for many valuable crops of vegetables. Costs for drainage and the large amounts of lime usually required, if the soils are acidic, make it economically necessary to grow high-value crops.

In recent years drainage of wetlands, many of which are Histosols, has been discouraged in order to retain these wetlands for bird nesting and food sites, to control runoff and reduce flooding, and for the aesthetic value of preserving these unusual habitats. "Swampbuster" is a recent law that restricts drainage of wetlands not already in crops.

3:7.3 Distribution of Histosols

Histosols occur in large bodies in Florida and Georgia and in numerous areas in the Northeast and Great Lakes states, totaling 0.5 percent of all soils in the United States. Outside the United States the largest contiunous body of Histosols is in Canada south of Hudson Bay; another large area is located in northwestern Canada. Histosols do occur on all continents, but in areas too small to be shown on the soil map of the world. The worldwide extent of Histosols is only 0.9 percent of all soils, and they rank last in total area of all 11 soil orders.

---

[8] Suborder illustrated by Fig. 3-3.

FIGURE 3-3 This *Grindstone Series* is mapped as Grindstone peat (*order:* Histosols; *suborder:* Hemists), deep phase. It has a 30-cm (12-in.) organic-material horizon (organic fiber) made of fibrous, spongy, sphagnum peat moss; is medium acid; and has a 76 percent identifiable fiber content. With further depth, fibers of mosses and herbaceous mixed residues decrease gradually from 54 percent at the 30-cm (12-in.) depth to 45 percent at 121 cm (47 in.). Clay layers exist below 121 cm (>47 in.). Infiltration rate is rapid when saturated. Manitoba, Canada. Scale is 15-cm (6-in.) intervals. Terminology is that of Canada and is not identical to the new U.S. system. (*Source:* Canada Department of Agriculture.)

FIGURE 3-4 Subsidence, the sinking of the land, occurs on organic soils when they are drained and the organic matter rapidly decomposes, drastically reducing the soil volume. The 2.7-m (9-ft)-deep concrete post in the photo was placed in this Terra Ceia (Saprist) soil down to limestone bedrock in 1924, with its top flush with the soil surface. The photo, taken in 1972, shows an average subsidence of about 2.5 cm (1 in.) per year. Florida. (Courtesy of Victor W. Carlisle, University of Florida.)

3:8 Soil Order: Entisols (from recent soils, without pedogenic-developed horizons)

Suborders:

Aquents (Latin, *aqua,* water): overly wet part of the year
Arents (Latin, *arare,* to plow): horizon from plowing
Fluvents (Latin, *fluvius,* river): alluvial deposits

Orthents (Greek, *orthos*, true)[9]: loamy or clayey textures
Psamments (Greek, *Psammos*, sand): sandy profiles

Entisols are soils of slight soil development. No developed (pedogenic) horizons exist except possibly an **Ap** (plowed) or weak **A** from slight organic-matter accumulation. Entisols are identified by the absence of distint pedogenic (naturally developed) horizons. An **ochric epipedon** exists.

The lack of soil development does not imply that Entisols are simple, identical soils. They range from deep sand to stratified river-deposited clays and from recent volcanic-ash deposits (or erosion-exposed surfaces) to dry, arid lake beds.

3:8.1 Properties and Classification of Entisols

The various conditions creating Entisols indicate some of their profile characteristics and their classification. The absence of distinct pedogenic horizons in Entisols may be due, for example, to the

1. Presence of a parent material too inert to develop soil horizons, such as quartz sands. These sand deposits could have existed for centuries with little profile development.

2. Formation of the soil from a parent material that dissolves almost completely with very little residue, such as some rare limestones composed mostly of carbonates.

3. Insufficient time to develop horizons, as in recently deposited volcanic ash, river terrace alluvium, or other recent alluvial deposits.

4. Ecological conditions not conducive to horizon formation, as is true of soil on the moon, in very dry deserts (Fig. 3-5), and in permafrost areas.

FIGURE 3-5 This *Colby Series* is a silt loam that grades from a 1.0-cm (0.4-in.) grayish-brown A into a 10-cm (4-in.) thick AC horizon and finally into a parent material with lime accumulation (Ck) at 20 cm (8 in.) deep. The fairly dry climate where this soil is found does not have enough precipitation to leach lime out of the profile, and the soil pH is moderately alkaline. Infiltration rate is moderate when saturated. The complete taxonomy: *Order*—Entisols; *Suborder*—Orthents; *Great Group*—Torriorthents; *Subgroup*—Ustic Torriorhents; *Family*—Ustic Torriorhents, fine-silty, mixed, calcareous, mesic; *Series*—Colby. Kansas. Torr = very arid climate, orth = typical, ent = Entisols order. Scale is in feet (feet X 30.5 = cm). (Courtesy of Arvad Cline, USDA—Soil Conservation Service.)

---

[9] Suborder illustrated by Fig. 3-5.

5. Occurrence on steep slopes where the rate of surface erosion equals or exceeds the rate of soil profile formation or where deposition "covers" soil too frequently for development.

Entisols are a soil order that includes the former **Azonal** soils and a few **Low Humic Gley** of the 1938 soil classification system.

3:8.2 Management of Entisols

Many recent alluvial deposits may be excellent and productive soils. Much of deposited alluvium is topsoil eroded and deposited over many years, even centuries. The soil texture can vary from gravels, through sands, to silts and clays. Some Entisols may exist on river terraces above flooding levels, while other areas may be subjected to periodic flooding. Well-sorted beach sand deposits, deposited in geologic ages past, may be inert and infertile. Thus, some Entisols may be ancient river terraces of excellent texture, high fertility, deep soils, and highly productive. Poorly drained and flooded low delta areas may be wet, poorly aerated clays (as the Mississippi River delta) and be difficult to cultivate. Recent volcanic-ash deposits (many areas, including the St. Helens blast in this decade) may be erosive because of the high silt and fine-sand contents.

3:8.3 Distribution of Entisols

Entisols are widely distributed in the United States and include river floodplains, rocky soils of mountainous areas, and beach sands. They occupy 8 percent of all U.S. soils. Entisols are on all continents, occupying a total of 8.3 percent of the world land surface and rank sixth in area among the 11 soil orders. Rainfall may be high or low, and vegetation may be forest, savanna (grass, shrubs, and occasional trees), or grass. Precipitation and vegetation are, therefore, not diagnostic of the soil order. Some Entisols are excellent agricultural soils (floodplains, volcanic ash).

3:9 Soil Order: Inceptisols (Latin, *inceptum,* beginning, inception, incipient, pedogenic horizons)

Suborders:

Aquepts (Latin, *aqua,* water): overly wet a part of the year
Ochrepts (Greek base of *ochros,* pale)[10]: light-colored surface
Plaggepts (modified from German, *plaggen,* sod): human-made organic surface
Tropepts (modified from Greek, *tropikos,* of the solstice, tropical): uniform year-round temperature
Umbrepts (Latin, *umbra,* shade): acidic, dark-colored surface

Inceptisols are weakly developed soils. They lack sufficient development to fit into other orders but have more development than Entisols.

---

[10] Suborder illustrated by Fig. 3-6.

3:9.1 Properties and Classification of Inceptisols

Inceptisols are more weathered and developed soils than the Entisols. As the suborder names indicate, some Inceptisols are "young" soils, of relatively recent origin. Other Inceptisols have dark surface horizons, either naturally developed (Umbrepts) or formed by humus additions by people (Plaggepts). The lack of profile and horizon development may be due to several factors, including these four:

1. The deposit may be recent. The weathering processes (clay movement, lime movement, weathering) have been active too short a time to develop strong horizons.

2. The parent material may be resistant to weathering, thereby slowing development. Some quartz sands may be Inceptisols.

3. Erosion may be just fast enough to remove the developing soil before strong horizons can form. Slow, periodic depositions (as in floodplains or by wind erosion) would also be effective in hindering development of a pedogenic horizon.

4. Wetness, cold, or other conditions slowing translocation and weathering in the soil allow Inceptisols to exist longer than in better aerated or warmer soils (Fig. 3-6). Many rice paddy soils and poorly drained soils are Inceptisols.

Inceptisols were officially known in the United States before 1965 as **Brown Forest, Low Humic Gley, Ando,** and **Sol Brun Acide.**

3:9.2 Management of Inceptisols

Inceptisols are as variable as the Entisols. Many are very fertile; others are productive because of excellent climates or textures plus good management. Some Inceptisols may be overly wet or exist in the cold regions of the world (Alaska, Siberia). Many of the soils used

FIGURE 3-6 This *Minoa Series* is a very fine sandy loam with a 25-cm (10-in.)-deep plow layer (**Ap**), dark colored, and medium acid. Yellowish and brown mottles occur at 25-56 cm (10-22 in.) in the granular structure, which indicate both some soil development (**B**) and periodic water tables. Mottles become more distinct with deep (waterlogging is more severe) through the **BC** to parent material (**C**) at 81 cm (32 in.) deep. Infiltration rate is slow when saturated. The complete taxonomy: *Order*—Inceptisols; *Suborder*—Ochrepts; *Great Group*—Eutrochrepts; *Subgroup*—Aquic Dystric Eutrochrepts; *Family*—Aquic Dystric Eutrochrepts, coarse-loamy, mixed, mesic; *Series*—Minoa. New York State. Aqu = excess water much of the time, dystr = low basic cation saturation, eutr = high basic cation saturation, ochr = presence of ochric epipedon, ept = Inceptisols order. Scale is in feet (feet X 30.5 = cm). (*Source:* USDA—Soil Conservation Service.)

for rice production in Asia and the Pacific Islands are large areas of Inceptisols. Waterlogging and soil mixing and puddling of these soils for centuries have hindered profile development. People of many of these areas have tilled the soil, added large amounts of organic wastes and manures as fertilizers, and in other ways produced crops but hindered profile development.

3:9.3 Distribution of Inceptisols

Inceptisols in the United States occur mostly in the Middle Atlantic and Pacific states, accounting for 18.2 percent of all soils in the 50 states. They develop in many climates from the Arctic to the tropics, mostly under trees but sometimes under grasses. On a global basis Inceptisols occupy 8.9 percent of the land surface, third in area of the 11 soil orders. They occur on all continents, but the largest area is in mainland China. Extensive areas also occur on islands of the East Indies, the Northern cold regions, and along the Amazon River basin.

3:10 Soil Order: Andisols (Japanese, *ando: an,* black; *do,* soil; high volcanic ash content)

Suborders:

Aquands (Latin, *aqua,* water): overly wet a period of the year
Cryands (Greek, *kryos,* icy cold): very cold; cryic or pergelic temperatures
Torrands (Latin, *torridus,* hot and dry): torric (dry) moisture regime
Udands (Latin, *udus,* humid): udic (wet) moisture regime
Ustands (Latin, *ustus,* burnt): ustic (dry winter) moisture regime
Vitrands (Latin, *vitrum,* glass): presence of volcanic glass
Xerands (Greek, *xeros,* dry): xeric (dry summer) moisture regime

Andisols are mostly soils weakly to moderately developed. The majority of Andisols form from volcanic ejecta (ash, cinders, pumice, or some basalts). Andisols must have enough soil development to remove them from the Entisols, but not so much as to mask the influence of the unique volcanic parent material. After extensive weathering, Andisols may become soils of other orders. Most Andisols have been developing for less than 5000–10,000 years.

Andisols must have andic soil properties in a cumulative thickness of 35 cm (13.7 in.) or more within the top 60 cm (23.4 in.) of the soil.[11] The *andic properties* include (1) low bulk density, (2) potential for wind erosion, (3) amorphous clays, (4) high macroporosity with rapid drainage, and (5) low soil strength when mechanically disturbed. The **melanic** epipedon (at least 6 percent organic carbon through a 30-cm (depth) was defined for these soils.

3:10.1 Properties and Classification of Andisols

The dominant process in most Andisols is one of rapid weathering. Mineral transformations are only in early stages. The minerals in Andisols form rapidly from the volcanic material and from largely amorphous clays (allophane, imogolite, and ferrihydrite). Imogolite is a slightly crystalline allophane and ferrihydrite is an amorphous iron hydrous oxide. All-humus colloids may be common.

---

[11] R.L. Parfitt, and B. Claydon, "Andisols: The Development of a New Order in Soil Taxonomy," *Geoderma,* **49**. (1991), pp. 181–198.

Andisols may have any diagnostic epipedon. The large number of suborders indicates the great variety of Andisol profiles. These soils exist in the total range of climates from cold to hot and from wet to dry.

The criteria for defining an Andisol are complex but can be approximated by the following items: (1) more than 60 percent by volume of volcaniclastic materials coarser than 2 mm, OR (2) a bulk density of 900 kg/m^3 (lighter than water), OR (3) considerable volcanic-glass content. In addition, Andisols should have (4) a very high phosphate retention (high oxalate-extractable Al and Fe). Finally, (5) the soils will contain mostly "early-stage" minerals, particularly amorphous clays. There will be few crystalline clays (smectites, kaolinite, vermiculite) and minimal translocation of colloids.

Andic properties must exist in at least 35 cm (13.7 in.) of the top 60 cm (23.4 in.) of soil. This allows for minimal mixing with other soil materials in resorting by wind, water, and ice.

3:10.2 Management of Andisols

Many Andisols are among the most productive soils of the world when managed well. The highest measured rice yield (about 10 Mg/ha) was produced on a volcanic-ash soil in the Phillippines. Many volcanic soils of Hawaii are highly productive.

Andisols tend to have large amounts of humus (7–12 percent organic carbon contents in many soils). The amorphous allophane clays have very high cation exchange capacities (often 150 cmol/kg, which is higher than montmorillonite). Unfortunately, these soils also rapidly adsorb and precipitate phophorous. The efficiency of added fertilizer phosphorous is often less than 10 percent, compared to 10–30 percent in most amorphous soils. This phosphorous problem is caused by the high contents of amorphous Al and Fe clays.

Andisols hold large amounts of water. But when they dry, these soils may be difficult to rewet, and the dry soils are often loose and dusty. Erosion is a concern even though the aggregates are often quite resistant to rainfall disintegration.

3:10.3 Distribution of Andisols

Andisols are extensive in many islands of the Pacific Ocean (Hawaii, Indonesia, the Phillippines, Japan, Aleutians). Chile has large areas of Andisols, as does New Zealand, Ecuador, east central Africa, and Central America. Additional areas are found in Spain, France, Italy, the United States, and many other countries.

3:11 Soil Order: Aridisols (Latin, *aridus,* dry, dry more than 6 months a year)

Suborders:

Argids (Latin, *argilla,* white clay)[12]: clay accumulation as argillic or natric horizons
Orthids (Gree, *orthos,* true): typical arid soil profile, without argillic horizon

A long dry period and only short periods of wetness in the upper soil are the dominant soil-forming processes of **Aridisols.** A lack of water reduces leaching of basic cations, retards mineral weathering, and may allow accumulation of soluble salts.

---

[12] Suborder illustrated by Fig. 3-7.

3:11.1 Properties and Classification of Aridisols

The listing of only two suborders for Aridisols should not make you assume that most Aridisols are similar. In temperate areas annual rainfall may be as high as about 300–350 mm (about 12–14 in.) per year. Some of the soils will have carbonates and/or soluble salts throughout the profile. Other soils may have some shallow leaching to 15–40 cm (about 6–16 in.) deep and no salt accumulation. In some soils, layers of carbonate accumulation exist. In some of these soils, deep, thick carbonate and silica-cemented layers occur. These are believed by some to have developed in ancient time during a period with a wetter climate. Some of these argillic horizons may also seem to be too thick or strongly developed for the present dry climates.

The aridity restricts plant growth, resulting in low soil organic-matter contents in the soils (0.5–1.5 percent). The soils may be of almost any texture. Leaching is slight, basic cation saturation is about 100 percent, and primary minerals make up most of the soils forming from parent rock.

Some of the profile features occuring frequently in Aridisols are (1) lime layers, (2) salt or gypsum accumulations, (3) low organic-matter accumulation, and (4) a calcareous profile. Some soils have lime-cemented hardpans (duripans or caliche); other have clay accumulation **Bt** (argillic) horizons (Fig. 3-7). The **Bt** horizons may be very old (other climates) or weathered in place in some soils. Natural vegetative cover commonly is various scattered desert shrubbery, such as sagegbrush, rabbitbrush, mesquite, shadscale, creosote bush, and shortgrasses in hot deserts. Desert regions also occur in *cold* climate areas; the soils are commonly Aridisols and dry Entisols, and the vegetative covers differ from those of *hot* desert areas.

Aridisols include the former great soil groups **Desert, Reddish Desert, Sierozem, Solonchak,** a few **Brown** and **Reddish Brown,** and associated **Solonetz** of the 1938 soil classification system.

3:11.2 Management of Aridisols

Many Aridisols are among the most productive soils when they are irrigated and fertilized. Many of the productive valleys of California, Arizona, Utah, New Mexico, and western

FIGURE 3-7 This *Dixie Series* is a gravelly loam soil used for range with a 15-cm (6-in.) **A** of dark brown granular loam soil. Its pH is 7.2; it has about 20 percent gravel. A clay accumulation (**Bt**) occurs at 23-38 cm (9-15 in.). Between 38 and 91 cm (15 and 35.5 in.), a weakly to strongly lime-cemented hardpan (*calcic* endepedon, *duripan,* or "caliche" and labled as a **Ck** is too hard for roots to penetrate unless it is fractured (see arrow). Below 91 cm (35.5 in.) is soft, massive, calcareous, very gravelly loam of pH 7.9. Infiltration rate is slow when saturated. Scale is in feet (feet X 30.5 = cm). The complete taxonomy: *Order*—Aridisols; *Suborder*—Argids; *Great Group*—Haplargids; *Subgroup*—Xerollic Haplargids; *Family*—Xerollic Haplargids, fine-loamy, mixed, mesic; *Series*—Dixie. Utah. Xer = dry summers, oll = Mollisols-like, hapl = minimum horizon development, arg = argillic horizon, id = Aridisol order. (*Source:* USDA—Soil Conservation Service.)

Texas are mostly Aridisols. The soils' low humus contents make addition of nitrogen essential. In contrast, a lack of leaching allows potassium accumulation. Potassium deficiency is rare, but it is found in sandy soils and a few shallow soils or in soils developed from low-potassium parent material. The lack of leaching does result in soil pH values of about 7–8.5. Deficiencies of zinc and iron, and to a lesser extent manganese and copper, are common in these soils.

Irrigation is essential. If the soils are not already salty, irrigation waters may add enough to develop salty soils. The typically sunny climate makes production very good if water needs are met and correct fertilizers are added.

3:11.3 Distribution of Aridisols

Aridisols in the United States are located primarily in the western Mountain and Pacific states in areas of low rainfall where scattered grasses and desert shrubs dominate the vegetation. They comprise 11.6 percent of all soil orders in the United States. Worldwide, Aridisols rank *first* in area of all 11 soil orders with 18.8 percent (Fig. 3-8). The Sahara Desert is shown on the map of Fig. 3-8 as Aridisols, as are large areas of central and southern Asia, Australia, southern Africa, and southern South America.

3:12 Soil Order: Mollisols (Latin, *mollis,* soft, organic-rich surface horizons)

Suborders:

Albolls (Latin, *albus,* white): have a leached E (albic) horizon
Aquolls (Latin, *aqua,* water): overly wet a period of the year
Borolls (Greek, *boreas,* northern): cold areas
Rendolls (from *Rendzina,* Polish word for noise in plowing dark, limey, clay soils): high lime percentage in parent material
Udolls (Latin, *udus,* humid)[13]: adequate water most of the year
Ustolls (Latin, *ustus,* burnt): dry many months, but some water in summer
Xerolls (Greek, *xeros,* dry): dry summer, some leaching in winter

Mollisols are dark-colored soils of grasslands and some hardwood forests. Their distinguishing feature is a deep, dark-colored, surface horizon (**A**). It is defined as a surface horizon (**mollic epipedon**) that is thick, dark in color, strong in structure (due to the presence of organic matter), and more than 50 percent saturated with basic cations (mostly calcium). In the profile picture the mollic epipedon extends from the surface to a depth of 55 cm (21.5 in.), as shown by arrows on the scale (Fig. 3-9).

3:12.1 Properties and Classification of Mollisols

Mollisols have the largest number of suborders, thus indicating the diverse nature of these soils. Suborder names indicate the wide distribution of Mollisols in cold, temperate and humid, to semiarid climates. Some form on parent material high in calcium carbonate (40 percent or more lime). Others are well leached (Albolls) or exist in wet conditions (Aquolls).

Mollisols were known formerly as **Rendzina, Prairie, Chernozem, Chestnut, Brunizem,** and associated **Solonetz** and **Humic Gley** in the 1938 soil classification system.

---

[13] Suborder illustrated by Fig. 3–9.

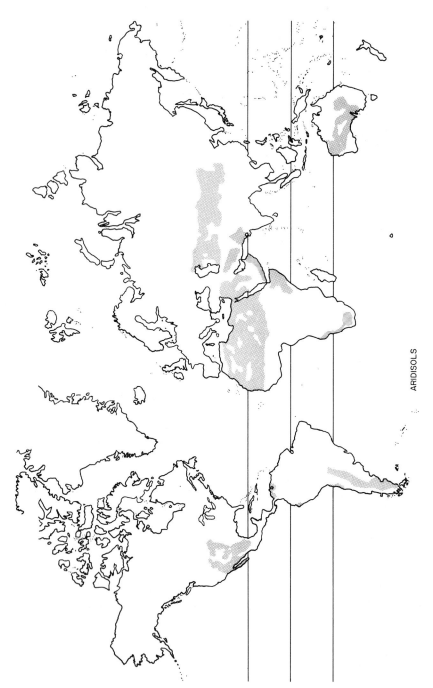

ARIDISOLS

FIGURE 3-8 World distribution of Aridisols. Occupying 18.8 percent of the world land area, Aridisols are the most extensive soil order; they occur in areas of low rainfall. (Courtesy of Raymond W. Miller, Utah State University.)

FIGURE 3-9 This *Sharpsburg Series* is a silty clay clay loam with a dark surface soil to 28 cm (11 in.) (**Alp. A2**), granular and subangular blocky structure, friable, and slightly acid. With further depth the profile grades through a transition zone (**AB**) to a thick clay accumulation layer (**Bt** argillic) from 43 to 112 cm (17-44 in.). This clayey layer is strongly acid, moderately dark, with prismatic structure breaking to blocky structure. Some deep wetness is indicated by a few mottles. The mollic epipedon is about 55 cm (21.5 in.) deep. The parent material begins at 124 cm (48 in.) deep. Infiltration rate is moderate when saturated. The complete taxonomy: *Order*—Mollisols; *Suborder*—Udolls; *Great Group*—Argiudolls; *Subgroup*—Typic Argiudolls; *Family*—Typic Argiudolls, fine, montmorillonitic, mesic; *Series*—Sharpsburg, Iowa, Ud = seldom has drought, moisture well distributed; montmorillonitic = has over 40 percent of clay as montmorillonitic clay. Scale is in feet (feet X 30.5 = cm). (*Source:* USDA—Soil Conservation Service.)

3:12.2 Management of Mollisols

Mollisols, formed under grasses or some broadleaf forests, tend to be some of the most fertile soils. They have higher humus and, thus higher nitrogen (and some other nutrients) than the Aridisols have. Mollisols in the wetter climates (Udolls) do not need irrigation and have been highly productive soils for centuries, even though they may be acidic and need lime additions.

The subhumid areas, such as the Great Plains of the United States, have the well-known *Chernozem* soils (of the old classification) with their black surface soils, often to 60–80 cm (about 23–31 in.) deep. Dryland wheat and sorghum are grown on many of these soils and on even drier Mollisols.

Since about one-fourth of the U.S. soils are Mollisols, it is obvious that there are enormous variations in climates and crops grown on these soils. The drier Mollisols grow dryland grains but are also irrigated for many kinds of crops.

The higher humus content in Mollisols than in Aridisols and the limited leaching, except in the wetter ones, make most Mollisols quite fertile soils without fertilization. Very little if any lime is needed except in wetter climates. Only Alfisols, perhaps, have a higher *natural* fertility. In the present farming practices in the United States, the soils' textures, depths, climates, and lack of inhibiting conditions are the most important for productive crops. When irrigated, limed, and fertilized, they can be very productive soils.

3:12.3 Distribution of Mollisols

The largest contiguous body of Mollisols in North America is in the Great Plains, extending north into Canada and south almost to the Gulf of Mexico. Other areas occur in the intermountain region in the west and northwest. Mollisols are the most extensive of the U.S. soil orders, totaling 25.1 percent of U.S. soils. The largest body of Mollisols in the world is in central Europe, central Asia, and northern China (Fig. 3-10). A second large area is in Argentina and adjoining countries in South America. Mollisols rank fourth and total 8.6 percent of all soils of the world. Precipitation is subhumid to semiarid in regions varying from north temperate to alpine (mountainous) to tropical.

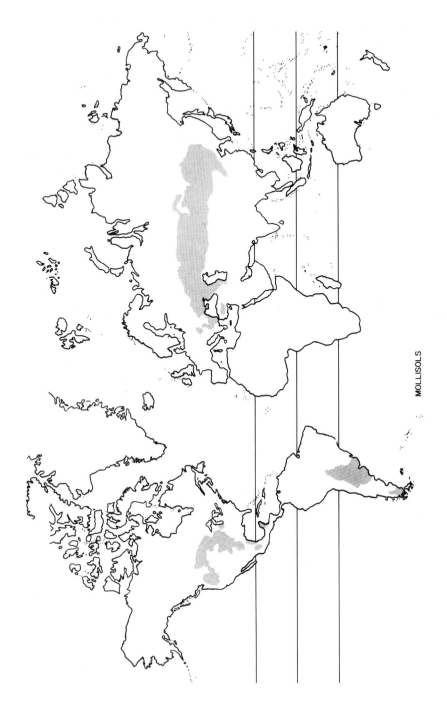

MOLLISOLS

FIGURE 3-10 World distribution of the naturally productive Mollisols (dark grassland soils). Mollisols are the fourth most prevalent soil order, covering 8.6 percent of the world land surface, and the majority are in cultivation. Some of the Mollisols in drier climates can be profitably irrigated. (Courtesy of Raymond W. Miller, Utah State University.)

3:13 Soil Order: Vertisols (origin: Latin, *verto*, turn, self-swallowing clays)

Suborders:

Aquerts (Latin, *aqua,* water) reducing, waterlogged conditions for some time period
Cryerts (Greek, *kryos,* icy cold): cold-area soils
Torrerts (Latin, *torridus,* hot and dry): in tropical areas, quite dry season
Uderts (Latin, *udus,* humid): adequate water most of year
Usterts (Latin, *ustus,* burnt)[14]: dry many months, some water in summer
Xererts (Greek, *xeros,* dry): dry summer, some leaching in winter

Vertisols develop from parent materials high in limestones and marl or from basic rocks, such as basalt. Vertisols expand and contract more than the soils of any other order because of their high swelling-clay contents. The high-swelling, sticky clays make them difficult to cultivate and poor support for roadbeds and buildings.

Vertisols are *self-mixing* soils. When they dry and crack, loosened surface soil falls into or is flushed into the cracks. When the soil again is wetted and swells, the soil in the cracks pushes upward and outward. Over many decades, the surface 30–80 cm (12–31 in.) may be "churned" and mixed, disrupting horizon development within the entire profile. **Gilgai** (wavy soil surface) often develops.

3:13.1 Properties and Classification of Vertisols

By definition, Vertisols have more than 30 percent clay (mostly montmorillonitic), more than 30 $cmol_c$/kg cation exchange capacity, and when dry have wide, deep cracks more than 1 cm (0.4 in.) wide *at a depth of 51 cm (20 in.).* Vertisols occur in humid semiarid climates with noticeable wet and dry cycles. The vegetation is dominantly tall grasses and scattered trees and shrubs.

Vertisols have profiles of deep **A** horizons, often over 1 m (39 in.) deep. Vertisols also show evidence of vertical and angled mass movement during swelling by the presence of wavy soil surfaces (**gilgai** micro relief), **slickensides** (smoothed pressure surfaces on peds), and wedge-shaped (parallelepiped) compound subsoil aggregates tilted at an angle from the horizontal. *Vertical movement* of 3.7 cm (1.4 in.) at a depth of 91 cm (3 ft) due to dislocation during wetting-drying cycles was measured for a Vertisol similar to that in Fig. 3-11.

The light-colored horizon below 91 cm (3 ft) in the Vertisol in Fig. 3-11 contains more than 50 percent calcium carbonate.

Vertisols were known as **Rendzinas** in the 1938 soil classification system and later as **Grumusols.**

3:13.2 Management of Vertisols

Cultivation of Vertisols is difficult. When wet, the soils are very sticky; when dry, they are very hard. Thick, hard soil crusts can inhibit seedling emergence. Plowing when too dry produces large clods and requires enormous power. When wet, the soils have low permeability to water. Often, about the only water the soil absorbs is what immediately fills the dry cracks. Thus, wetting may be shallow, only 40–60 cm (16–23 in.), even with prolonged wetting contact time. Traffic on the soil when it is wet is almost impossible and will compact the soil and leave deep tracks.

---

[14] Suborder illustrated by Fig. 3–11.

FIGURE 3-11 This *Houston Black Series* is a clay with the top 97 cm (38 in.) divide dinto three **A** horizons (a mollic epipedon), which differ from each other in structure mostly; the deeper two horizons differ in color also. Wormcasts, lime and lime concretions, and iron-manganese concretions are found in the profile. The layers at depth of 97-264 cm (38-103 in.) are called **AC** horizons. They have many features of the **A** horizons but are much lighter in color, lower in organic matter, and mottled. Infiltration rate is very slow when saturated. The complete taxonomy: Order—Vertisols; Suborder—Usterts; Great Group—Pellusterts; Subgroup—Udic Pellusterts; Family—Udic Pellusterts, fine, montmorillonitic, thermic; Series—Houston Black. San Marcos, Hays County, Texas. Pell = low chroma, ert = Vertisols order, thermic = annual average temperature between 15 and 22°C (59–72°F). Scale is in feet (feet X 30.5 = cm). (Courtesy of Alan Anderson, USDA—Soil Conservation Service.)

Vertisols generally are quite fertile, with high cation exchange capacities and relatively high humus contents to great depths (>1m). However, aeration of the wetted soil is quite poor. This restricts root growth and penetration. Rooting is often *shallow* on such soils. During drying cycles, cracking soil breaks many roots, retarding plant growth. Infiltration can begin at 10 cm (4 in.)/hr while cracks are being filled, then slow to less than 0.2 cm/hr after the top soil layer wets and swells.

Wetted Vertisols have low support strength. Roads and railroads of Vertisols often become deformed or misaligned by the weight and vibration of traffic during wet soil conditions. Some soil mechanical engineers refer to the *expansive* Vertisol soils as *expensive* soils because of the extra costs in engineering and building costs when using Vertisols. Telephone poles and other vertical poles lean after years of wetting-drying cycles and wind-storms during wet periods (Fig. 3-12).

3:14.3 Distribution of Vertisols

Vertisols occur primarily in central and southeastern Texas and to a lesser extent in Alabama, Mississippi, and California, totaling 1.0 percent of all soil orders in the United States. Outside the United States, large areas of Vertisols occur in India, Australia, and eastern Africa. Ranking ninth in area, Vertisols comprise 1.8 percent of the total land surface of the world.

3:14 Soil Order: Alfisols (coined from *pedalfer; al* for aluminum and *fer* for ferric iron; movement of Al, Fe, and clay into the B horizon)

Suborders:

> **Aqualfs** (Latin, *aqua,* water): seasonally saturated with water
> **Boralfs** (Greek, *boreas,* northern)[15]: cold area

---

[15] Suborder illustrated by Fig. 3–13.

FIGURE 3-12 Cracks in this Vertisol in southern India range from 2 cm (0.8 in.) to over 5 cm (2 in.) wide. Cracking tears plant roots and speeds drying of the soil. (Courtesy Raymond W. Miller, Utah State University.)

Udalfs (Latin, *udus,* humid): adequate water most of year
Ustalfs (Latin, *ustus,* burnt): dry in winter, most of its water in summer
Xeralfs (Greek, *xeros,* dry): dry in summer, some leaching in winter

Alfisols are soils with high enough effective precipitation to move clays downward and form an **argillic** (clay accumulation) horizon. The basic cation saturation percentage is relatively high (>50 percent), and they are usually fertile. These soils usually form under various forests and brush cover.

3:14.1 Properties and Classification of Alfisols

Alfisols occur in many forests receiving marginal amounts of precipitation or in wetter areas where high-lime parent materials retard development of strong acidity. Some of the most notable features of Alfisols are these:

1. Translocated clay in a **Bt** (t = German *ton,* clay) clay accumulation horizon (argillic), shown in Fig. 3-13, occuring 23–74 cm (9–29 in.) in depth (between arrows).

2. A medium-to-high supply of basic cations, such as calcium and magnesium, which is evidence of only *mild leaching.* This is in contrast to Ultisols, which have had severe leaching.

3. Water is available for good plant growth for three or more warm-season months.

Alfisols were formerly classified as **Gray-Brown Podzolic, Noncalcic Brown, Gray Wooded, Degraded Chernozem,** and associated **Half-Bog** and **Planosol** soils in the 1938 classification system.

FIGURE 3-13 This *Nebish Series* is a loam and has a thin 7.6 cm (3 in.) surface layer (**A**) with a 15 cm (6 in.) leached, light-colored layer (**E**) beneath it. Structure is weak; pH is neutral. The clay accumulation layer (**Bt** argillic) is clay loam texture, occurs (between arrows) from the 23- to 74-cm (9- to 29-in.) depth, has blocky structure, and is slightly acid. A transition layer (**BCt**) grades into parent material (**C**) of loam texture at 84 cm (33 in.). The parent material is not strongly leached because it is still moderately basic in pH. Infiltration rate is moderate when saturated. The complete taxonomy: *Order*—Alfisols; *Suborder*—Boralfs; *Great Group*—Eutroboralfs; *Subgroup*—Typic Eutroboralfs; *Family*—Typic Eutroboralfs, fine-loamy, mixed, frigid; *Series*—Nebish. Minnesota. Bor = cold area, frigid = less than 80°C (47°F) average annual soil temperature.   (Courtesy of William M. Johnson, USDA—Soil Conservation Service.)

3:14.2 Management of Alfisols

If relief and climate are favorable, many Alfisols produce well when converted to cropland. Most are leached of lime to at least half a meter (20 in.) or deeper and are slightly to moderately acid in the surface horizon. Leaching can be severe enough to form a leached horizon (**E**). Often the clayey accumulation is not favorable to plant growth, particularly if the surface is eroded, exposing the clay as a surface layer.

Without the assistance of irrigation or fertilization, Alfisols are probably the most naturally productive soils used for crops. They include many of the fertile soils of the corn belt in Indiana, Ohio, Michigan, and Wisconsin, as well as woodland-covered soils in Texas and Colorado. Afisols occupy extensive portions of England, Europe, north central Russia, Ghana, and many other regions. These were prominent agricultural areas in early colonization and developmental periods.

The moderate acidity in the upper soil profile may require addition of lime for the best growth of many crops. Although some of these soils exist in areas with adequate precipitation for good crop produciton without irrigation, long dry periods can occur and supplemental irrigation systems are sometimes used. Addition of fertilizers (nitrogen, phosphorus, and potassium) usually increase yields.

3:14.3 Distribution of Alfisols

As shown on the Soil Map of the United States (Fig. 3-20), Alfisols occur in large bodies in the North Central states and in the Mountain states and comprise 13.5 percent of the total land area of the 50 states. Alfisols occupy 13.2 percent of the land surface of the world (ranking second in area) and occur on all continents (Fig. 3-14). Humid and subhumid climates and tall grasses, savanna, and oak–hickory forests characterize the climate and native vegetation where Alfisols occur.

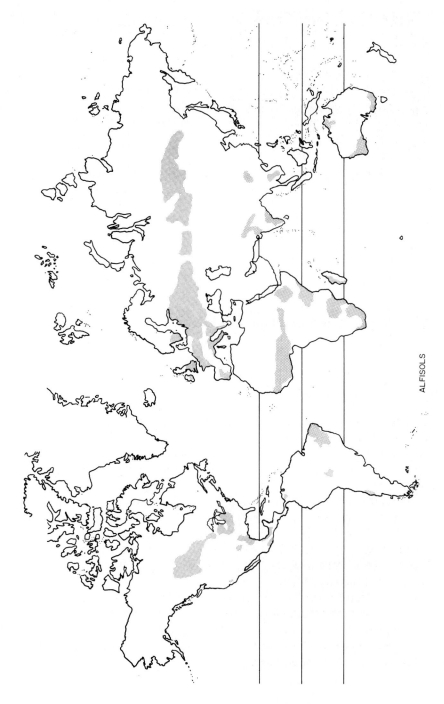

FIGURE 3-14 World distribution of Alfisols. Occupying 13.2 percent of the world land area (second largest soil order), Alfisols occur mostly under forests, especially of the broadleaf type. (Courtesy of Raymond W. Miller, Utah State University.)

ALFISOLS

3:15 Soil Order: Spodosols (Greek, *spodos,* wood ash, gray color of E horizon)

Suborders:

Aquods (Latin, *aqua,* water): seasonally saturated with water
Cryods (Greek, *kryos,* icy cold): very cold soils
Ferrods (Latin, *ferrum,* iron): iron oxide accumulation is extensive
Humods (Latin, *humus,* earth): humus accumulation is extensive
Orthods (Greek, *orthos,* true)[16]: central concept, with humus and iron oxide accumulation

Spodosols, by definition, are of high sand content. The high rainfall and easy leaching produces translocation of humus and/or sesquioxide colloids into a **spodic B** horizon. These soils occur in cold, wet climates, usually under acidic conifer forests or other vegetation that develops acidic soils and produces organic chemicals that mobilize iron, aluminum, and humus colloids.

3:15.1 Properties and Classification of Spodosols

Spodosols have moderately to strongly acidic sandy profiles with an ashy white upper horizon over a dark-brown **B** horizon and yellowish subsoils: The typical profile is (1) black organic litter and humified layer, (2) the white leached **E** layer called **albic** horizon, and (3) a thin deposition **Bs** or **Bh** layer of iron oxides and/or humus colloids (Fig. 3-15). These soils are typical of sandy soils under cool, wet, conifer and deciduous forests. They are usually well leached; basic cation saturation percentage is very low. The unique features of Spodo-

Ft.

FIGURE 3-15 This *Adams Series* is a sandy loam found in northern temperate and subarctic latitudes that has a strongly acid humus-and-roots layer (**Oe** or **Oa**) over a 10-cm (4-in.)-thick, gray, leached layer (**E**, albic). Below the **E** is a very strongly acid, dark-brown humus layer (**Bh**) 5 cm (2 in.) thick. The next 51 cm (20 in.) is brown sand with amorphous iron and aluminum oxide deposits (**Bs** or spodic). Parent material, a strongly acidic light-yellow sand, occurs at 66 cm (26 in.) deep. Infiltration rate is rapid when saturated. The complete taxonomy: *Order*—Spodosols; *Suborder*—Orthods; *Great Group*—Haplorthods; *Subgroup*—Typic Haplorthods; *Family*—Typic Haplorthods, sandy, mixed, frigid; *Series*—Adams. East Middlebury, Vermont. Od = Spodosol order.   (Courtesy of A. R. Midgley, Vermont Agricultural Experiment Station.)

---

[16] Suborder illustrated by Fig. 3-15.

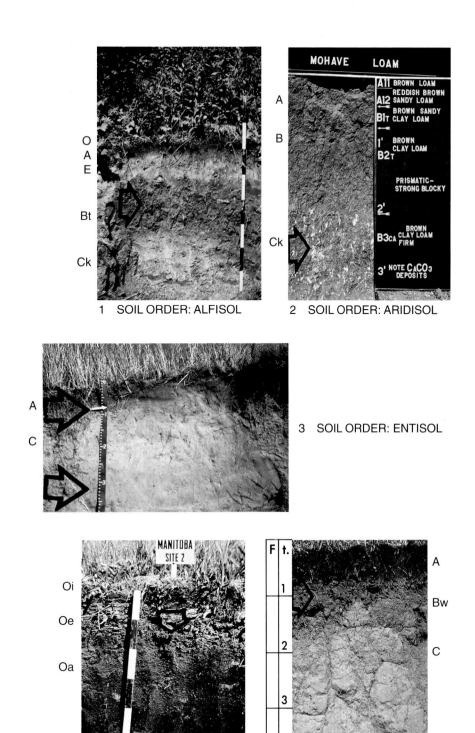

1 SOIL ORDER: ALFISOL

2 SOIL ORDER: ARIDISOL

3 SOIL ORDER: ENTISOL

4 SOIL ORDER: HISTOSOL

5 SOIL ORDER: INCEPTISOL

Note: Corresponding table appears on pages 54–55.

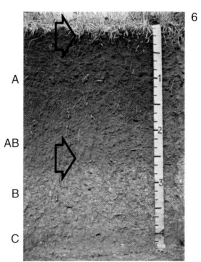

6 SOIL ORDER: MOLLISOL

A

AB

B

C

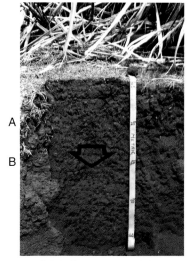

7 SOIL ORDER: OXISOL

A

B

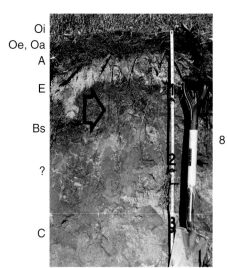

Oi
Oe, Oa
A

E

Bs

?

C

8 SOIL ORDER: SPODOSOL

9 SOIL ORDER: ULTISOL

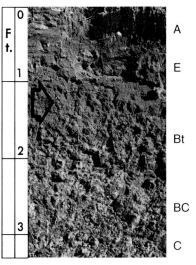

F
t.

A

E

Bt

BC

C

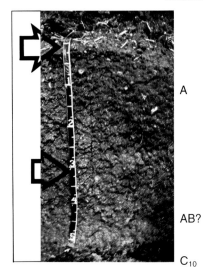

A

AB?

C_{10}

10 SOIL ORDER: VERTISOL

sols are the strongly acidic, leached, white **albic** horizon (**E**) and the sandy **B** horizons, often cemented with humus and noncrystalline aluminum and/or iron, which is a black or brown color (a **spodic** horizon). The albic horizon may be missing (mixed with **Oa** and **B**) in cultivated soils. Clayey soils never develop into Spodosols because of slow infiltration.

There are many variations of Spodosols. A deep 1.8-m (6-ft) iron-pan Spodosol occurs under a pigmy forest only 1.4–3 m (4.6–9.8 ft) tall in coastal California. In contrast, very shallow Spodosols, only 46 cm (18 in.) deep, are found in Alaska. Spodosols were known in the 1938 soil classification system as **Podzol, Groundwater Podzol,** and **Brown Podzolic.**

3:15.2 Management of Spodosols

Cultivation of Spodosols requires considerable fertilization and lime additions to make them less acidic and more productive. Some acidic Spodosols are used without lime additions for blueberry cultivation in coastal North Carolina, but most cultivated Spodosols are used for crops adapted to the cooler north central U.S. climate: grains, potatoes, strawberries, raspberries, pastures, and silage corn.

The typically cold climates of Spodosols restrict their uses to a relatively few crops. Spodosols also typically occur in areas anciently glaciated during the Pleistocene time and often have nonlevel relief and sometimes contain gravel and stones. Relatively small areas are Spodosols. These soils are relatively poor for agricultural crops.

3:15.3 Management of Spodosols

Spodosols exist in large tracts in New England, at high elevations in the Middle Atlantic states, and in the northern part of the Great Lakes states, constituting 4.8 percent of all soil orders in the United States. Between latitude 42°N and 60°N, Spodosols are common in Canada, northern Europe, and northern Asia. They rank eighth and occupy 4.3 percent of the world's land surface.

3:16 Soil Order: Ultisols (Latin, *ultimus,* last, highly leached, clay accumulation in B horizon)

Suborders:

Aquults (Latin, *aqua,* water): seasonally saturated with water
Humults (Latin, *humus,* earth): high-humus surface accumulation
Udults (Latin, *udus,* humid)[17]: adequate water most of year
Ustults (Latin, *ustus,* burnt): dry many months, some water in summer
Xerults (Greek, *xeros,* dry): dry during summer, some leaching in winter

Ultisols are the humid-area soils too low in basic cation saturation (acidic) to be Alfisols or Mollisols but not weathered enough to be Oxisols. Usually Ultisols have developed in humid climates, tropical to subtropical temperatures [average annual temperature more than 8°C (47°F)], and a forest or forest-plus-grass (savanna) vegetation. Many of the "red-clay hills" of the southeastern United States are the exposed **B** horizons of eroded Ultisols.

---

[17] Suborder illustrated by Fig. 3-16.

3:16.1 Properties and Classification of Ultisols

Ultisols often have an **umbric** epipedon (dark-colored, strongly acidic **A**). However, the more common characteristic profile features are a clay accumulation layer (**Bt**), which is moderate to strongly acidic, often a surface horizon dark with humus (**A**), and typically a leached layer (**E**). Intensive weathering forms the clayey **B**, often a reddish soil and with a basic cation saturation < 35 percent. Appreciable quantities of primary minerals (feldspars, micas) are still present and possibly some illite or montmorillonite clays, although kaolinite dominates the clay fraction of most Ultisols.

Ultisols were formerly classified as **Red-Yellow Podzolic, Reddish Brown Lateritic,** and associated **Planosol** and **Half-Bog** in the 1938 soil classification system.

3:16.2 Management of Ultisols

With a high level of management, Ultisols can be some of the world's most *productive* soils. They exist in areas that are frost-free for long periods and also in humid areas with enough rainfall for crops or with adequate water reserves for irrigation (Fig. 3-16). However, their *nutrient reserve,* although better than in Oxisols, is relatively low to moderate. Both fertilization and liming are necessary in continuous cultivation to produce moderate-to-high yields. Optimum yields on these areas require good management of fertilizer and liming alternatives and crop selection. Also, insects are abundant and fungus diseases are prevalent in the warm, humid climate. Southern pines mixed with hardwoods are common as natural cover on these soils; timber production is profitable on many Ultisols.

Ultisols, without added fertilizers, often become "worn-out soils," as happened in the cotton fields in the mid 1800s in the southeastern United States. Humus is quickly decomposed in the year-round warm climate, and soon the supply of nitrogen from humus has de-

FIGURE 3-16 This *Ruston Series* has a moderately acidic, fine sandy loam plow layer 10 cm (4 in.) thick (an **Ap**), a leached (**E**) layer 30 cm (12 in.) thick, and a strongly acidic, 28 cm (11 in.) thick, sandy clay loam (**Bt** or argillic) below the upper arrow. This younger soil above 109 cm (42.5 in.) is developing on an ancient, strongly acidic, deep **E′B′** profile (starting at the lower arrow). The ancient **Bt** lower boundary is at 234 cm (91 in.) deep (below bottom of the picture). The infiltration rate is moderate when saturated. The profile sequence from the soil surface to the depth of the photo is **Ap, E, B1t, B2t, 2E′, 2Bt′.** The complete taxonomy: *Order*—Ultisols; *Suborder*—Udults; *Great Group*—Paleudults; *Subgroup*—Typic Paleudults; *Family*—Typic Paleudults, fine-loamy, siliceous, thermic; *Series*—Ruston. Louisiana. Pale = old, excessive development, ud = seldom has drought, ult = Ultisols order, thermic = annual soil temperature is 15–22°C (59–72°F), siliceous = over 90% of sand and coarse silt are silica minerals, such as quartz and opal. (*Source:* USDA—Soil Conservation Service.)

clined. The high free-iron and aluminum hydrous oxides often keep soluble phosphorus at low values. The extensive leaching in the humid climate reduces plant-available soil potassium, making it commonly deficient in these soils. Because of the favorable climate (long warm periods and high rainfall), Ultisols commonly need heavy fertilization, and they are used for intensive crop production of a wide variety of crops.

3:16.3 Distribution of Ultisols

Ultisols are located mostly in the southern Atlantic states, the eastern South Central states, and in the Pacific states, mostly in subtropical climates. They account for 12.8 percent of the land area of the United States. Central America, South America, western Africa, southeastern Asia, and Australia all have large areas of Ultisols. Ranking seventh in area among the 11 soil orders, Ultisols occupy 5.6 percent of the world's land area.

3:17 Soil Order: Oxisols (French, *oxide,* very highly oxidized throughout profile)

Suborders:

Aquox (Latin, *aqua,* water): seasonally saturated with water
Humox (Latin, *humus,* earth): high humus content in surface soil
Orthox (Greek, *orthos,* true): cental concept
Torrox (Latin, *torridus,* hot and dry): dry for half the year, arid seasons
Ustox (Latin, *ustus,* burnt): dry winter, some water in summer

Oxisols are the most extensively weathered of all the soils. They are typically found on old landforms in humid tropical or subtropical climates. They are usually yellowish to bright red in color and are weathered to several meters in depth.

3:17.1 Properties and Classification of Oxisols

Oxisols have lost much of their silica (because primary minerals have weathered) and are rich in the more residual iron and aluminum hydrous oxide residues, which have very low solubility. Sesquioxide and kaolinite clays dominate.

The unique feature of an Oxisol is the presence of the **oxic endopedon** and often of **plinthite** (Detail 3-2 and Fig. 3-17). Weathering is accelerated by high year-round temperatures and moisture; so the highly weathered Oxisols typically occur in continuously hot and humid tropical and subtropical areas, usually under hardwood forests.

Oxisols have developed on old upland, medium-to-fine textured parent materials that have weathered into crystalline kaolinitic-type clays with a net negative charge and amorphous iron and aluminum oxides with a *net positive charge.* Clay in the *oxic horizon often has a net positive charge* that is mostly a pH-dependent charge.

Weatherable minerals in Oxisols are either absent or present only in trace amounts. There are, therefore, small reserves of basic cations in these soils, mostly those on the limited cation exchange complex, in plant tissue, and in very deep less-weathered soil layers. This fact explains one of the principal causes of the usual rapid decline in crop yields when Oxisols are cleared and cultivated.

There is almost no *translocation* of clay in Oxisols; **Bt** horizon are, therefore, usually faint, diffuse, or absent. Clays in Oxisols are usually not dispersible by shaking in distilled water in contrast to clays in other soil orders. (Oxisol clays will settle in water like fine sand.) High permeability and *low erodibility* are further characteristics of Oxisols. Poor manage-

Oxic horizons, which are diagnostic for Oxisols, are usually unexciting in appearance but may be yellow to bright brownish-red in color. They are so excessively weathered that only a few percentage of primary minerals (feldspars, micas, and ferromagnesium minerals) are left. These horizons consist dominantly of mixtures of materials resistant to weathering and solubility, such as hydrated oxides of iron and aluminum, titanium oxides, a small amount of quartz sands, and kaolinite.

Plinthite (from Greek, *plinthos,* brick) is material high in iron and aluminum oxides found in many Oxisols. Wet anaerobic conditions solubilize reduced iron; drying allows it to solidify as oxidized and low-solubility iron oxides. After many decades plinthite may harden irreversibly to **ironstones,** which are cemented masses of material originally termed **laterite** (from Latin, *later* brick), which has been discarded by the U.S. soil taxonomy system because of its imprecise meaning.

When the softer plinthite is cut into building blocks (bricks), the wetting-drying cycles over several seasons will harden the bricks until they can be used for building (Fig. 3-17). Clearing forests exposes the soil to erosion and more frequent wetting-drying cycles of the exposed plinthite; this speeds its conversion to ironstone. Optimum conditions necessary for *maximum* formation and hardening of plinthite into ironstone are:

1. A fairly level land surface (for reduced erosion) situated at the foot of a seepage slope to accumulate iron in seepage water

2. An adequate supply of soluble iron from incoming seepage or as a residue from weathering

3. Alternating wet and dry seasons of approximately equal duration and sufficient rain continuously during the rainy period to saturate the zone of iron segregation to form temporarily anaerobic conditions in which iron is readily solubilized

Soils with ironstone are usually highly leached and strongly acidic and have low fertility. Nitrogen is the first nutrient to be limiting. Phosphorus is strongly adsorbed to iron oxides and will be a problem to keep available. Zinc, molybdenum, and copper are also usually low. The hardened ironstone layer restricts root elongation, decreases available water during dry periods, and causes poor drainage and excess water during rainy periods.

Ironstone exists as large rock layers or as smaller nodules (gravel) in soil. Some of this gravel is dug out from its pits and used as a surface all-weather material for gravel roads. Recovery of the ironstone (softening it) requires many decades of vegetative cover and rainfall. Complete recovery is unlikely to occur even after centuries of excellent conditions.

---

ment has resulted in large erosion losses. Oxisols were known as **Laterites** in the 1938 soil taxonomy system and **Latosols** from 1950 to 1965.

3:17.2 Management of Oxisols

Oxisols have unique management requirements. Except for the nutrients cycled in organic matter, the soils are very low in nutrients. When the covering native vegetation is burned in shifting cultivation, nutrients from the ash are temporarily added to the soil, but they are used in a year or two by the cultivated crop or lost due to leaching and declining additions of organic matter. Some exposed and eroded soils (but not all Oxisols) may gradually harden and become impossible to cultivate due to formation of ironstone from plinthite. Cultivated crops require careful fertilization for optimum yields. Added phosphorus has low efficiency because it readily forms insoluble iron and aluminum phosphates, which are not very available to plants. With adequate nitrogen, phosphorus, and potassium, the common crops of the

FIGURE 3-17 Hardened plinthite (ironstone, laterite) makes a good building stone, which gets harder with cycles of wetting and drying. Southern India. (Scale in upper photo is feet.) (Photos by Roy L. Donahue.)

area—bananas, sugarcane, coffee, rice, and pineapples—are productive. The soils containing significant amounts of plinthite must be kept covered by vegetation (in timber, coffee, or other tree crops, and/or organic mulches) to hinder drying and irreversible hardening.

Oxisols may be highly productive for carbohydrate and oil crops (both products are mostly carbon, hydrogen, and oxygen; all are derived from air and water rather than from the soil minerals) but are less productive of protein foods (which require large amounts of nitrogen and sulfur). Because Oxisols occur near the equator, where daylight seldom exceeds 12.5 hours daily throughout the year, corn production is less than in temperate regions, such as the U.S. Midwest, where there are 14 to 15 hours of daylight during much of the growing season.

3:17.3 Distribution of Oxisols

The map of the United States located at the end of the chapter (Fig. 3-20) does not include Oxisols in the legend, for their occurrences, which are only in Hawaii, Puerto Rico, and the Virgin Islands, are too small in area to be shown on this map scale. Oxisols occupy 0.01 percent of the soils of the United States but are very extensive in tropical South America and Africa (Figs. 3-18 and 3-21). On a world basis, they rank fifth and total 8.5 percent in area of all soil orders.

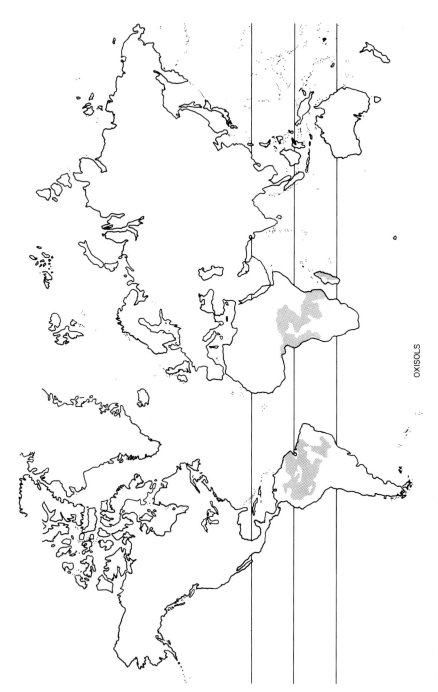

OXISOLS

FIGURE 3-18 World distribution of Oxisols (excessively weathered soils). Although only 8.5 percent of the world land area, they are the fifth most extensive soil order. Oxisols are usually infertile, and some may harden irreversibly because of drying cycles. (Courtesy of Raymond W. Miller, Utah State University.)

3:18 Other Land Areas

Other land areas" in the United States are designated on the soil map as "miscellaneous land types" and total 4.5 percent of all land areas in the 50 states.

When all the percentages given for the 11 soil orders are totaled for the world, they add up to 78.9 percent. The other one-fifth or more of the world land area is so intermixed that it is referred to on the map scale shown at the end of the chapter (Fig. 3-21) only as **Soils in Areas with Mountains** (19.7 percent of total) and **Icefields** plus **Rugged Mountains** (1.6 percent). These areas plus the soil orders total 100.2 percent because of approximating tenths of percentages. Most of the Soils in Areas with Mountains occur in western Canada, intermountain United States, a strip from northern Mexico to Panama, the Andes of western South America, Norway, northeastern Russia, Japan, central Asia, and most of interior China and southeast Asia (Fig. 3-19). These mountain areas contain small valleys of various soil orders typical of the climate and other soil-forming factors in which they occur and with many soil orders on the mountain slopes—all areas too small to be shown separately on a map of this scale.

3:19 Soil Map of the United States

A generalized soil map of the United States, according to patterns of soil orders and suborders, is given in Fig. 3-20. Only the dominant orders and suborders are shown. Each delineation has many inclusions of other kinds of soil. For complete definitions, see *Soil Taxonomy: A Basic System of Soil Classification for Making and Interpreting Soil Surveys,* Agriculture Handbook 436, USDA—Soil Conservation Service, 1975. Approximate equivalents in the modified 1938 soil classification system are indicated in parentheses for each suborder.

Alfisols Soils with gray to brown surface horizons, medium to high base supply, and subsurface horizons of clay
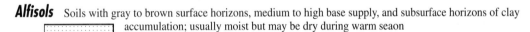
accumulation; usually moist but may be dry during warm seaon

A1   AQUALFS (seasonally saturated with water) gently sloping; general crops if drained, pasture and woodland if undrained (some Low Humic Gley soils and Planosols)

A2   BORALFS (cool or cold) gently sloping; mostly woodland, pasture, and some small grain (Gray Wooded soils)

A2S   BORALFS steep; mostly woodland

A3   UDALFS (temperate or warm, and moist) gently or moderately sloping; mostly farmed, corn, soybeans, small grain, and pasture (Gray Brown Podzolic soils)

A4   USTALFS (warm and intermittently dry for long periods) gently or moderately sloping; range, small grain, and irrigated crops (some Reddish Chestnut and Red Yellow Podzolic soils)

A5S   XERALFS (warm and continuously dry in summer for long periods, moist in winter) gently sloping to steep; mostly range, small grain, and irrigated crops (Noncalcic Brown soils)

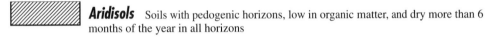 **Aridisols** Soils with pedogenic horizons, low in organic matter, and dry more than 6 months of the year in all horizons

D1   ARGIDS (with horizon of clay accumulation) gently or moderately sloping; mostly range, some irrigated crops (some Desert, Reddish Desert, Reddish Brown, and Brown soils and associated Solonetz soils)

D1S   ARGIDS gently sloping to steep

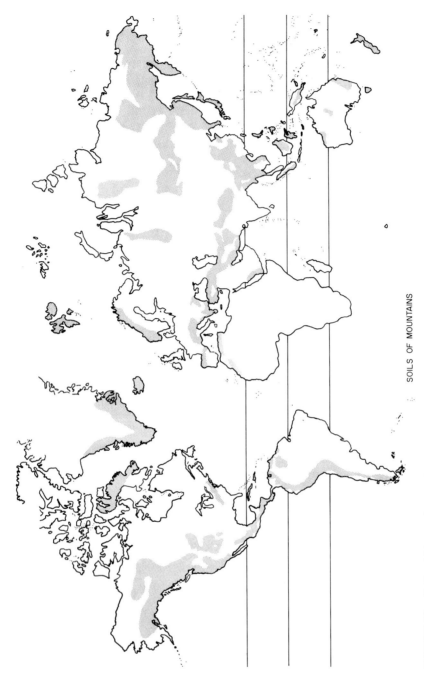

FIGURE 3-19 Soils of Mountains are intermixed regions of mountain slopes and valleys and include many kinds of soils. Nearly 20 percent of the world land area is in such mountainous regions. (Courtesy of Raymond W. Miller, Utah State University.)

SOILS OF MOUNTAINS

FIGURE 3-20 Soil map of the United States. (Courtesy of U.S.D.A. Soil Conservation Service)

Generalized from 1967 Soil Map,
Scale 1:7,500,000, U. S. Atlas

SLOPE CLASSES

Gently sloping = slopes mainly less than 10 percent

Moderately sloping = slopes mainly between 10 and 25 percent

Steep = slopes mainly steeper than 25 percent

Scale 1:17,000,000

D2   ORTHIDS (without horizon of clay accumulation) gently or moderately sloping; mostly range and some irrigated crops (some Desert, Reddish Desert, Sierozem, and Brown soils, and some Calcisols and Solonchak soils)

D2S   ORTHIDS gently sloping to steep

 Entisols Soils without pedogenic horizons

E1   AQUENTS (seasonally saturated with water) gently sloping; some grazing

E2   ORTHENTS (loamy or clayey textures) deep to hard rock; gently to moderately sloping; range or irrigated farming (Regosols)

E3   ORTHENTS shallow to hard rock; gently to moderately sloping; mostly range (Lithosols)

E3S   ORTHENTS shallow to hard rock; steep; mostly range

E4   PSAMMENTS (sand or loamy sand textures) gently to moderately sloping; mostly range in dry climates, woodland, or cropland in humid climates (Regosols)

 Histosols Organic soils

H1   FIBRISTS (fibrous or woody peats, largely undecomposed) mostly wooded or idle (Peats)

H2   SAPRISTS (decomposed mucks) truck crops if drained, idle if undrained (Mucks)

 Inceptisols Soils that are usually moist, with pedogenic horizons of alteration of parent materials but not of accumulation

I1S   ANDEPTS (with amorphous clay or vitric volcanic ash and pumice) gently sloping to steep; mostly woodland; in Hawaii moslty surgarcane, pineapple, and range (Ando soils, some Tundra soils)

I2   AQUEPTS (seasonally saturated with water) gently sloping; if drained, mostly row crops, corn, soybeans, and cotton; if undrained, mostly wooldland or pasture (some Low Humic Gley soils and Alluvial soils)

I2P   AQUEPTS (with continuous or sproadic permafrost) gently sloping to steep; woodland or idle (Tundra soils)

I3   OCHREPTS (with thin or light-colored surface horizons and little organic matter) gently to moderately sloping; moslty pasture, small grain, and hay (Sols Bruns Acides and some Alluvial soils)

I3S   OCHREPTS gently sloping to steep; woodland, pasture, small grains

I4S   UMBREPTS (with thick, dark-colored surface horizons rich in organic matter) moderately sloping to steep; mostly woodland (some Regosols)

 Mollisols Soils with nearly black, organic-rich surface horizons and high basic cation supply

M1   AQUOLLS (seasonally saturated with water) gently sloping; mostly drained and farmed (Humic Gley soils)

M2 BOROLLS (cool and cold) gently or moderately sloping, some steep slopes in Utah; mostly small grain in North Central states, range and woodland in Western states (some Chernozems)

M3 UDOLLS (temperate or warm, and moist) gently or moderately sloping; mostly corn, soybeans, and small grains (some Brunizems)

M4 USTOLLS (intermittently dry for long periods during summer) gently to moderately sloping; mostly wheat and range in western part, wheat and corn or sorghum in eastern part, some irrigated crops (Chestnut soils and some Chernozems and Brown soils)

M4S USTOLLS moderately sloping to steep; mostly range or woodland

M5 XEROLLS (continuously dry in summer for long periods, moist n winter) gently to moderately sloping; mostly wheat, range, and irigated crops (some Brunizems, Chestnut, and Brown soils)

M5S XEROLLS moderately sloping to steep; mostly range

 Spodosols   Soils with accumulations of amorphous materials in subsurface horizons

S1 AQUODS (seasonally saturated with water) gently sloping; mostly range or woodland; where drained in Florida, citrus and special crops (Ground-Water Podzols)

S2 ORTHODS (with subsurface accumulations of iron, aluminum, and organic matter) gently to moderately sloping; woodland, pasture, small grains, special crops (Podzols, Brown Podzolic soils)

S2S ORTHODS steep; mostly woodland

 Ultisols   Soils that are usually moist with horizon of clay accumulation and a low basic cation supply

U1 AQUULTS (seasonally saturated with water) gently sloping; woodland and pasture if undrained, grain and vegetable crops if drained (some Low Humic Gley soils)

U2S HUMULTS (with high or very high organic-matter content) moderately sloping to steep; woodland and pasture if steep, sugar cane and pineapple in Hawaii, truck and seed crops in Western states (some Reddish Brown Lateritic soils)

U3 UDULTS (with low organic-matter content; temperate or warm, and moist) gently to moderately sloping; woodland, pasture, feed crops, tobacco, and cotton (Red Yellow Podzolic soils, some Reddish Brown Lateritic Soils)

U3S UDULTS moderately sloping to steep; woodland, pasture

U4 XERULTS (with low to moderate organic-matter content, continuoulsy dry for long periods in summer) range and woodland (some Reddish Brown Lateritic soils)

 Vertisols   Soils with high content of swelling clays and wide, deep cracks in some seasons

V1 UDERTS (cracks open for only short periods, less than 3 months in a year) gently sloping; cotton, corn, pasture, and some rice (some Grumusols)

V2 USTERTS (cracks open and close twice a year and remain open more than 3 months); general crops, range, and some irrigated crops (some Grumusols)

Andisols   Tentative. Soils mostly developing from volcaniclastics.

Areas   with little soil

X1   Salt flats

X2   Rocklands, ice fields

3:20 Soil Map of the World

The world distribution of soils is shown on the soil map, a folded insert that appears after the Index at the end of the book. The shaded patterns and capital letters are for soil orders; the numbers indicate suborders; lowercase letters indicate combinations of the great groups. The map was compiled and produced by the staff of the USDA—Soil Conservation Service.

A   ***Alfisols***   Soils with subsurface horizons of clay accumulation and medium-to-high basic cation supply; either usually moist or moist for 90 consecutive days during a period when temperature is suitable for plant growth

      A1   BORALFS cool
              A1a—with Histosols, cryic temperature regimes common
              A1b—with Spodosols, cryic temperature regimes

      A2   UDALFS temperate to hot; usually moist
              A2a—with Aqualfs
              A2b—with Aquolls
              A2c—with Hapludults
              A2d—with Ochrepts
              A2e—with Troporthents
              A2f—with Udorthents

      A3   USTALFS temperate to hot; dry more than 90 cumulative days during periods when temperature is suitable for plant growth
              A3a—with Tropepts
              A3b—with Troporthents
              A3c—with Tropudults
              A3d—with Usterts
              A3e—with Ustochrepts
              A3f—with Ustolls
              A3g—with Ustorthents
              A3h—wih Ustox
              A3j—Plinthustalfs with Ustorthents

      A4   XERALFS temperate or warm; moist in winter and dry more than 60 consecutive days in summer
              A4a—with Xerochrepts
              A4b—with Xerorthents
              A4c—with Xerults

D   ***Aridisols***   Soils with pedogenic horizons; usually dry in all horizons and never moist as long as 90 consecutive days during a period when temperature is suitable for plant growth

      D1   ARIDISOLS undifferentiated
              D1a—with Orthents
              D1b—with Psamments
              D1c—with Ustalfs

D2   ARGIDS with horizons of clay accumulation
               D2a—with Fluvents
               D2c—with Torriorthents

E   *Entisols* Soils without pedogenic horizons; either usually wet, usually moist, or usually dry

E1   AQUENTS seasonally or perennially wet
               E1a—Haplaquents with Udifluvents
               E1b–Psammaquents with Haplaquents
               E1c—Tropaquents with Hydraquents

E2   ORTHENTS loamy or clayey textures, many shallow to rock
               E2a—Cryorthents
               E2b—Cryorthents with Orthods
               E2c—Torriorthents with Aridisols
               E2d—Torriorthents with Ustalfs
               E2e—Xerorthents with Xeralfs

E3   PSAMMENTS sand or loamy sand textures
               E3a—with Aridisols
               E3b—with Orthox
               E3c—with Torriorthents
               E3d—with Ustalfs
               E3e—with Ustox
               E3f—with shifting sands
               E3g—Ustipsamments with Ustolls

H   *Histosols* Organic soils

H1   HISTOSOLS undifferentiated
               H1a—with Aquods
               H1b—with Boralfs
               H1c—with Cryaquepts

I   *Inceptisols* Soils with pedogenic horizons of alteration or concentration but without accumulation of translocated material other than carbonates or silica; usually moist, or moist for 90 consecutive days during a period when temperature is suitable for plant growth

I1   ANDEPTS amorphous clay or vitric volcanic ash or pumice
               I1a—Dystrandepts with Ochrepts

I2   AQUEPTS seasonally wet
               I2a—Cryaquepts with Orthents
               I2b—Halaquepts with Salorthids
               I2c—Haplaquepts with Humaquepts
               I2d—Haplaquepts with Ochraqualfs
               I2e—Humaquepts with Psamments
               I2f—Tropaquepts with Hydraquepts
               I2g—Tropaquepts with Plinthaquults
               I2h—Tropaquepts with Tropaquents
               I2j—Tropaquepts with Tropudults

I3   OCHREPTS thin, light-colored surface horizons and little oganic matter
               I3a—Dystrochrepts with Fragiochrepts
               I3b—Dystrochrepts with Orthox
               I3c—Xerochrepts with Xerolls

I4   TROPEPTS continuously warm or hot
I4a—with Ustalfs
I4b—with Tropudults
I4c—with Ustox

I5   UMBREPTS dark-colored surface horizons with medium-to-low basic cation supply 15a—with Aqualfs

M   Mollisols   Soils with nearly black, organic-rich surface horizons and high basic cation supply; either usually moist or usually dry

M1   ALBOLLS light-gray subsurface horizon over slowly permeable horizon, seasonally wet
M1a—with Aquepts

M2   BOROLLS cool or cold
M2a—with Aquolls
M2b—with Orthids
M2c—with Torriorthents

M3   RENDOLLS subsurface horizons have much calcium carbonate but no accumulation of clay
M3a—with Usterts

M4   UDOLLS temperate or warm; usally moist
M4a—with Aquolls
M4b—with Eutrochrepts
M4c—with Humaquepts

M5   USTOLLS temperate to hot; dry more than 90 cumulative days in the year
M5a—with Argialbolls
M5b—with Ustalfs
M5c—with Usterts
M5d—with Ustochrepts

M6   XEROLLS cool to warm; moist in winter and dry more than 60 consecutive days in summer
M6a—with Xerorthents

O   Oxisols   Soils with pedogenic horizons that are mixtures principally of kaolinite, hydrated oxides, and quartz, and are low in weatherable minerals

O1   ORTHOX hot; nearly always moist
O1a—with Plinthaquults
O1b—with Tropudults

O2   USTOX warm or hot; dry for long periods, but moist more than 90 consecutive days in the year
O2a—with Plinthaquults
O2b—with Tropudults
O2c—with Ustalfs

S   Spodosols   Soils with accumulation, of amorphous materials in subsurface horixons; usually moist or wet

S1   SPODOSOLS undifferentiated
S1a—cryic temperature regimes; with Boralfs
S1b—cryic temperature regimes; with Histosols

S2   AQUODS seasonally wet
S2a—Haplaquods with Quartzipsamments

S3   HUMODS with accumulations of organic matter in subsurface horizons
S3a—with Hapludalfs

S4   ORTHODS with accumulations of organic matter, iron, and aluminum in subsurface horizons
S4a—Haplorthods with Boralfs

U   **Ultisols** Soils with subsurface horizons of clay accumulation and low basic cation supply; usually moist, or moist for 90 consecutive days during a period when temperature is suitable for plant growth

U1   AQUULTS seasonally wet
U1a—Ochraquults with Udults
U1b—Plinthaquults with Orthox
U1c—Plinthaquults with Plinthaquox
U1d—Plinthaquults with Tropaquepts

U2   HUMULTS temperate to warm; moist all of year; high content of organic matter
U2a—with Umbrepts

U3   UDULTS temperate to hot; never dry more than 90 cumulative days in the year
U3a—with Andepts
U3b—with Dystrochrepts
U3c—with Udalfs
U3d—Hapludults with Dystrochrepts
U3e—Rhodudults with Udalfs
U3f—Tropudults with Aquults
U3g—Tropudults with Hydraquents
U3h—Tropudults with Orthox
U3j—Tropudults with Tropepts
U3k—Tropudults with Tropudalfs

U4   USTULTS warm or hot; dry more than 90 cumulative days in the year
U4a—with Ustochrepts
U4b—Plinthustults with Ustorthents
U4c—Rhodustults with Ustalfs
U4d—Tropustults with Tropaquepts
U4e—Tropustults with Ustalfs

V   **Vertisols** Soils with high content of swelling clays; deep, wide cracks develop during dry periods

V1   UDERTS usually moist in some part in most years; cracks open less than 90 cumulative days in the year
V1a—with Usterts

V2   USTERTS cracks open more than 90 cumulative days in the year
V2a—with Tropaquepts
V2b—with Tropofluvents
V2c—with Ustalfs

Andisols Tentative. Soils mostly developing from volcani-pyroclastics

X   **Soils in areas with mountains** Soils with various moisture and temperature regimes; many steep slopes, relief and total elevation vary greatly from place to place. Soils vary greatly within short distances and with changes in altitude; vertical zonation common

X1   Cryic great groups of Entisols, Inceptisols, and Spodosols

X2   Boralfs and cryic great groups of Entisols and Inceptisols

X3   Udic great groups of Alfisols, Entisols, Inceptisols, and Ultisols

X4   Ustic great groups of Alfisols, Inceptisols, Mollisols, and Ultisols

X5   Xeric great groups of Alfisols, Entisols, Inceptisols, Mollisols, and Ultisols

X6   Aridisols, torric great groups of Entisols

X7   Ustic and cryic great groups of Alfisols, Entisols, Inceptisols, and Mollisols; ustic great groups of Ultisols; cryic great groups of Spodosols

X8   Aridisols, torric and cryic great groups of Entisols, and cryic great groups of Spodosols and Inceptisols

Z   *Miscellaneous*

Z1   Icefields

Z2   Rugged Mountains—mostly devoid of soil (includes glaciers, permanent snow fields, and, in some places, small areas of soil)

When we try to pick out anything by itself, we find it hitched to everything else in the universe.

—John Muir

Questions

1. Define (1) soil taxonomy, (2) soil classification, (3) a pedon, and (4) a polypedon.
2. Define (1) a "root" of an order, and (2) a formative element.
3. Soils of which order are the most extensively weathered?
4. Which soil order has the "least developed" soils?
5. State which soil order probably fits these landforms: (1) recent deep ash deposits, as from Mt. St. Helens' eruption in this decade, (2) newly exposed ground moraine, (3) deep alluvium in arid Nevada, (4) old landforms of tall grass prairies of the Great Plains, (5) the well-weathered, old, red soils of Georgia (not Oxisols).
6. Why might many Alaskan soils and wet tropical island rice paddy soils be listed as Inceptisols?
7. (a) How are xeric and ustic moisture regimes different? (b) How do udic and aquic differ?
8. (a) How is temperature categorized? (b) What does "iso" mean in these names?
9. (a) Can Histosols have nitrogen deficiency? (b) What are some other problems in using Histosols?
10. (a) Are Entisols good soils, poor soils, or some of both? (b) Can they contain weathered products? Explain. (c) Are Entisols unweathered materials? (d) Describe a non-productive Entisol.
11. (a) What is unique about Vertisols? (b) What special management problems exist with Vertisols in whatever climate they are found?
12. (a) Describe the origin and characteristics of plinthite. (b) How does plinthite become ironstone? (c) How may plinthite complicate cultivation in those Oxisols containing it?

13. (a) Describe "shifting cultivation." (b) Why is it used if it is so low-income and non-stable?
14. To what extent are potassium and lime needed in Oxisols and Ultisols?
15. How can some Inceptisols be (a) very "young" soils? (b) very old soils? (c) Where might some quite productive Inceptisols be found? (d) How do Inceptisols and Andisols differ? Explain.
16. (a) What climatic areas have extensive areas of Aridisols? (b) What are some expected properties in various Aridisols? (c) What do all Aridisols need for high production?
17. Why are Ardisols likely to be high in potassium, calcium, and sulfur?
18. Do any or all of the Mollisols (a) need irrigation? (b) need nitrogen?
19. Briefly define (1) a Mollisol, (2) an Ustoll, (3) an Alboll, (4) a Xeroll, (5) an Aquoll.
20. (a) How good are Mollisols for crops? (b) How should Mollisols be managed? (c) Are Mollisols important soils for the United States? Explain.
21. (a) Explain the reasons Vertisols "invert" themselves. (b) Are drying cycles needed for this "inversion"? (c) Why don't Vertisols have argillic horizons, generally? (d) What is the general definition of *Vertisols?*
22. (a) Why are many Alfisols "the most naturally productive" soils? (b) Why might they be more productive than most Mollisols? (c) To what extent does leaching occur in most Alfisols?
23. Do Alfisols need (a) lime? (b) fertilizers? (c) Where are most Alfisols?
24. (a) Why are Spodosols quite poor soil for growing a variety of crops? (b) Where do most Spodosols occur?
25. (a) How extensively weathered are Ultisols? (b) How inherently fertile are Ultisols? (c) How productive?
26. (a) How important is (1) irrigation, (2) fertilization, and (3) liming in Ultisols? (b) In what climatic areas are most Ultisols? (c) What clays do they have?
27. (a) How do Oxisols differ from Ultisols? (b) What percentage of primary minerals are left in Oxisols? (c) How fertile are Oxisols? (d) In which soil fraction are most of the limiting nutrients found in Oxisols?
28. Locate on the U.S. map the dominant soil orders, area-wise, of (1) the Midwest, (2) the Southeast, (3) the Great Plains, (4) the arid Southwest, and (5) the wet, cold Northeast.
29. Locate on the world map the large areas of (1) Mollisols, (2) Aridisols, (3) Alfisols, and (4) Ultisols.
30. (a) What causes "subsidence" in Histosols? (b) How does subsidence alter management?
31. Why do Histosols usually need drainage?

Soil Physical Properties

The earth, like the body of an animal, is wasted at the same time that it is replaced. It has a state of growth and augmentation; it has another state which is that of diminution and decay.

—James Hutton

4:1 Preview and Important Facts

PREVIEW

The **physical properties** of soils—texture, structure, density, porosity, water content, strength (consistency), temperature, and color—are dominant factors affecting the use of a soil. These properties determine the availability of oxygen in soils, the mobility of water into or through soils, and the ease of root penetration. Soil water, a vital physical property, is described in detail in Chapter 6.

The thousands of different soils exhibit multitudinous differences, such as varying amounts of stones and pebbles, coarse and fine sands, clays, lumps (aggregates) cemented by clays and organic matter, living and dead plant materials, the dark remnants of partly decomposed organic substances (humus), and animal life (ants and earthworms).

The soil, as "seen" by minute organisms such as bacteria and nematodes, is a rough terrain. Sands, silts, and clays are clumped together, with tortuous and irregular open pores of all sizes throughout the soil mass. In undisturbed soil (uncultivated or below tillage depth), pores may be lined with precipitated calcium carbonates, deposited layers of clay, iron oxides, or other materials.[1] **Humus,** the residual, greatly altered organic matter in soils, coats many mineral particles, often "cementing" them together into relatively stable clumps called **aggregates.** Clay particles, smaller than sands and silts, are too tiny to be seen with the visible light microscope. Magnifications of about 25,000 times, using an electron microscope, are required to "see" individual clay particles, some of which are small enough to be attached

---

[1] Soluble substances in soil water can form insoluble materials as water is evaporated and used by plants. Calcium carbonate ($CaCO_3$), iron oxides (Fe_2O_3), silica (SiO_2), and minute particles or organic matter (humus) are some of the common insoluble materials formed in soils or moved into them.

to the surfaces of microscopic-sized organisms such as bacteria. This microscopic soil world is a fascinating one, but still largely unexplored.

Texture is the physical property of particular importance to those using soils. **Texture** indicates the proportions of sands, silts, and clays in each soil. The soil texture controls water contents, water intake rates, aeration, root penetration, and some chemical properties. Soil structure (aggregation) and soil temperature are usually less easily modified. Economically, we cannot often make great changes in the soils' physical properties, but understanding the properties does improve our ability to manage those soils better.

IMPORTANT FACTS TO KNOW

1. The sizes of sands, silts, and clays, both in actual dimensions and in relation to some common objects (pinhead)
2. The meaning of textural classes, such as *loam, clay loam,* and *loamy sand,* in general terms (not exact percentages)
3. The manner of indicating the amounts of large fragments mixed in with sands, silts, and clays
4. The terms used for rock fragments, as size of the fragments increases
5. The meaning of the terms *soil structure* and how the different soil structures are natural processes of soil development
6. The typical bulk density values of "average cultivated soils" and the limits of using bulk density as diagnostic values in studying soils
7. The importance of (a) pore space and (b) the rate of exchange of soil air with atmospheric air
8. The method of measuring soil colors and the importance in knowing a soil's color
9. How soil temperature at different soil depths compares with the air temperature
10. How soil temperature can be altered by various practices, such as wetting or drying, plastic mulches, vegetative mulches, graphite on snow, and the aspect or slope
11. The meaning and use of *growing degree days* and *growing degree hours*
12. The use of "plant hardiness zones" maps

4:2 Soil Texture

Natural soils are comprised of soil particles of varying sizes. The soil particle-size groups, called **soil separates,** are sands (the coarsest), silts, and clays (the smallest). The relative proportions of soil separates in a particular soil determine its **soil texture**.

Texture is an important soil characteristic because it greatly modifies water intake rates (infiltration), water storage in the soil, the ease of tilling the soil, the amount of aeration (vital to root growth), and soil fertility. For instance, a coarse sandy soil is easy to till, has plenty of aeration for good root growth, and is easily wetted, but it also dries rapidly and easily loses plant nutrients, which are drained away in the rapidly lost water. High-clay soils (over 30 percent clay) have very small particles that fit tightly together, leaving little open pore space, which means there is little room for water to flow into the soil. This makes high-clay soils difficult to wet, difficult to drain, and difficult to till.

4:2.1 Sizes of Soil Separates

The U.S. Department of Agriculture (USDA) has established limits of variation for the soil separates and has assigned a name to each size class (Table 4-1). This system has been ap-

Table 4-1 Soil Separates and Their Diameter Ranges

Soil Separate Name	Diameter Range (mm)	Visual Size Comparison of Maximum Size
Very coarse sand	2.0–1.0	House key thickness
Coarse sand	1.0–0.5	Small pinhead
Medium sand	0.5–0.25	Sugar or salt crystals
Fine sand	0.25–0.10	Thickness of book page
Very fine sand	0.10–0.05	Invisible to the eye
Silt	0.05–0.002	Visible under microscope
Clay	<0.002	Most are not visible even with a microscope

proved by the Soil Science Society of America and is the one used in this book. Other particle-size classification systems are used in the United States and throughout the world.

4:2 Soil Textural Classes

Textural names are given to soils based on the relative proportions of each of the three soil separates: sand, silt, and clay. Soils that are preponderantly clay are called **clay** (textural class); those with high silt content are **silt** (textural class); those with a high sand percentage are **sand** (textural class). A soil that does not exhibit the dominant physical properties of any of these three groups (such as a soil with 40 percent sand, 40 percent silt, and 20 percent clay) is called **loam**. Note that loam does not contain equal percentages of sand, silt, and clay. It does, however, exhibit approximately equal *properties* of sand, silt, and clay.

The **textural triangle** (Fig. 4-1) is used to determine the soil textural name after the percentages of sand, silt, and clay are determined from a laboratory analysis.

Because the soil's textural classification includes *only mineral particles* and those of less than 2 mm diameter, the sand plus silt plus clay percentages equal 100 percent. (Note that organic matter is not included.) Knowing the amount of any two fractions automatically fixes the percentage of the third one. In reading the textural triangle, any two particle fractions will locate the textural class at the point where those two intersect.

4:2.3 Particle-Size (Mechanical) Analysis

The procedure used to separate a soil into various size groups from the coarsest sand, through silt, to the finest clay, is **particle-size analysis (mechanical analysis)**. For this purpose, the mineral matter less than 2 mm (0.08 in.) in diameter is considered separately from the larger particles. All rocks, pebbles, roots, and other rubble are removed (and measured) by screening the finer soil parts through a 2-mm sieve before analysis. Humus is removed from the soil sample by destroying it with an oxidizing chemical (such as hydrogen peroxide) before particle-size separation is done. An example of determining soil textural class is given in Calculation 4-1.

The basis of particle-sized separations is **Stoke's law** of settling velocities. *The settling rate of a particle is the net difference between its downward force* (gravity) *against the buoyancy* (resistance to fall) *by surface friction and movement of the water.* It is assumed, because of the complex mathematical restrictions otherwise, that the particles are *smooth spheres.* This is obviously incorrect and is a major error in the calculations. Typical fall rates will be

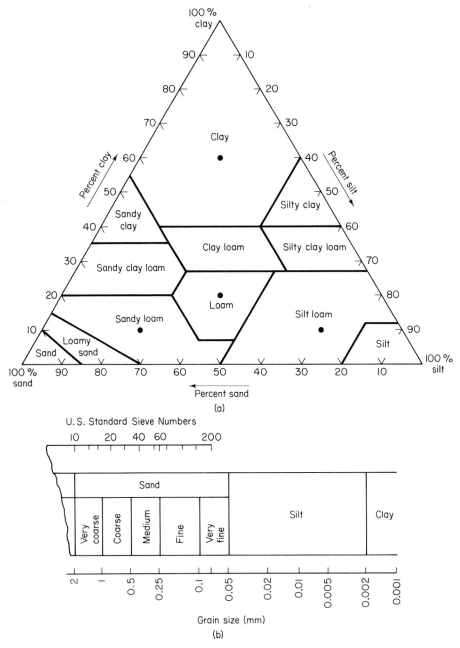

FIGURE 4-1 (a) Graphic guide for soil textural classification of the less-than-2-mm portion. The dots shown in several classes have these percentages of size fractions:

Textural Class	% Sand	% Silt	% Clay
Clay (see center)	20	20	60
Silt loam	20	70	10
Sandy loam	65	25	10
Loam	40	40	20

(b) Separation of sand into various sand sizes and the U.S. standard sieve numbers used for these separations. A 10-mesh sieve has 10 wire divisions per inch in the two directions making the screen.

Problem A sample of soil was screened and had the size separates in material smaller than 2 mm determined by particle-size (mechanical) analysis, with the following results:

$$\text{Sand content (2–0.05 mm diameter)} = 140 \text{ g}$$

$$\text{Silt content (0.05–0.002 mm diameter)} = 38 \text{ g}$$

$$\text{Clay content (< 0.002 mm diameter)} = \underline{22 \text{ g}}$$

$$\text{Total dry soil weight} \quad 200 \text{ g}$$

Determine the textural class name.

Solution Textural names consider only less-than-2-mm portion

$$\frac{140 \text{ g}}{200 \text{ g}} \left| \frac{100}{} \right. = 70\% \text{ sand}$$

$$\frac{38 \text{ g}}{200 \text{ g}} \left| \frac{100}{} \right. = 19\% \text{ silt}$$

$$\frac{22 \text{ g}}{200 \text{ g}} \left| \frac{100}{} \right. = 11\% \text{ clay}$$

1. Using the textural triangle (Fig. 4-1), place the triangle with 100 percent clay at the top (apex) and read across, parallel with the base along the 11 percent line. Keeping this line in mind, turn the triangle so 100 percent silt is now at the top and read across, parallel to the new base of the triangle along the 19 percent line. The 11 percent clay and 19 percent silt lines intersect in *sandy loam.* The percentage sand value could have been used as easily as either clay or silt values, because the lines for all three fractions intersect at the same point. The content of organic matter is ignored. If the soil contains more than 15% (by volume) of particles larger than sand, a "coarse fragment" adjective is added to the textural name (e.g., *gravelly* sandy loam).

2. The correct complete name above is *sandy loam.*

8711 D^2 cm^{-1}sec^{-1}. To calculate fall rates, enter the particle size in centimeters for D and obtain the fall rates in centimeters per second. Some typical values are:

medium sand (0.05 cm)	=	22 cm/sec
fine sand (0.02 cm)	=	3.5 cm/sec
medium silt (0.001 cm)	=	0.087 cm/sec
	=	0.52 cm/min
coarse clay (0.0002 cm)	=	0.00035 cm/sec
	=	0.021 cm/min
	=	1.26 cm/hr
fine clay (0.00002 cm)	=	0.0000035 cm/sec
	=	0.30 cm/day

4:3 Rock Fragments

Mineral particles in soils larger than very coarse sand (> 2.0 mm diameter) are called **rock fragments**. They are classified by shape and size into *rounded* or *flat-shaped* particles.

Table 4-2 Terms for Describing Rock Fragments in Soils

Shape[a] and Size	Noun	Adjective
Rounded, subrounded, angular, or irregular		
0.2–7.6 cm diameter	Gravel[b]	Gravelly
0.2–0.5 cm diameter	Fine gravel	Fine gravelly
0.5–2 cm diameter	Medium gravel	Medium gravelly
2–7.6 cm diameter	Coarse gravel	Coarse gravelly
7.6–25 cm diameter	Cobble	Cobbly
25–60 cm diameter	Stone	Stony
>60 cm diameter	Boulder	Bouldery
Flat		
0.2–15 cm long	Channer	Channery
15–38 cm long	Flagstone	Flaggy
38–60 cm long	Stone	Stony
>60 cm long	Boulder	Bouldery

Source: Revised Soil Survey Manual, Transmittal 430-V, USDA—Soil Conservation Service, June 9, 1981, pp. 4-57 to 4-60, 4-97.

[a] If significant to classification or interpretation, the shape of the fragments is indicated: "angular gravel," "irregular boulders."

[b] A single fragment is called a *pebble.*

Rounded are **gravel** (three sizes), **cobbles, stones,** and **boulders;** *flat* are **channer** (smallest), **flagstone, stone,** and **boulder** (Table 4-2).

The adjective describing rock fragments in soils is used as the first part of the textural class name under these conditions:

- *< 15% by volume.* No mention of rock fragments is used.
- *15–35% by volume.* The dominant kind of rock fragment is used (e.g., "stony loam").
- *35–60% by volume.* The word *very* precedes the name of the dominant kind of rock fragments (e.g., *very cobbly* sandy loam).
- *>60% by volume.* Add the word *extremely* in front of the coarse fragment name (e.g., *extremely gravelly* loam).

Of the 15,000+ soil series described in the United States, about 17% are in soil groupings (*families*) containing 35% or more rock fragments by volume. Although rocks are a hindrance to cultivated crops, rocky soils (<90% rock fragments) are often productive for growing trees.

▬▬ 4:4 Soil Structure

Structure is the arrangement of sands, silts, and clays into stable (cemented) aggregates. **Aggregates** are secondary units or granules composed of many soil particles held together by organic substances, iron oxides, carbonates, clays, and/or silica. In some instances, aggregates are cemented by an abundance of carbonates, iron oxides, or silica. Natural aggregates are called **peds** and vary in their water stability; the word **clod** is used for a coherent mass of soil broken into any shape by artificial means, such as by tillage.

Two terms are often confused with a ped. One is **fragment,** which consists of a piece of a broken ped, and the other is **concretion,** or **nodule,** which is a coherent mass formed within the soil by the precipitation of certain chemicals dissolved in percolating (seepage) waters. Concretions are often small, like shotgun pellets, and they are sometimes referred to as *shot.*

4:4.1 Soil Structural Classes

Soil structural units (peds) are described by three characteristics: **type** (shape), **class** (size), and **grade** (strength of cohesion, Table 4-3). Types of structure describe the ped shape with the terms *angular, blocky, subangular blocky, columnar, granular, platy,* and *prismatic* (Figs. 4-2 and 4-3).

Structure *classes* are the ped sizes such as **very fine, fine, medium, coarse** (or **thick**), and **very coarse** (or **very thick**). Structure *grades* are evaluated by the distinctness, stability, or strength of the peds.

1. **Structureless.** Soils have no noticeable peds. It might be an unconsolidated mass such as noncoherent sand, called **single grain,** or it might be a cohesive mass, such as could occur in some loams or clayey soils, called **massive.**
2. **Structured.** Three structural grades are used:
   (a) *Weak.* Peds are barely distinguishable in part of the *moist* soil; only a few distinct peds can be separated from the soil mass.
   (b) *Moderate.* Peds are visible in place; many can be handled without their breaking.
   (c) *Strong.* Most of the soil mass is visible as peds, most of which can be handled with ease without their breaking.

Soil structure may exist as compound structure in which large peds such as prisms or blocks may further fall apart into smaller blocks or smaller peds

Soil structure influences many important properties of the soil, such as the rate of infiltration of water. Both granular (spheroidal) and single-grain (structureless) soils have rapid infiltration rates; blocky (blocklike) and prismatic soils have moderate rates; and platy and massive soil conditions have slow infiltration rates.

4:4.2 The Genesis of Soil Structure

Structural peds form because of combinations of swelling–shrinking and adhesive substances. As a mass of soil swells (wets or freezes) and then shrinks (dries or thaws), lines of weakness (cracks) are formed. The soil mass between cracks—"cemented" by organic substances, iron oxides, clays, carbonates, and even silica—remains cohesive.

Cracking usually is the minimum necessary to relieve the shrinking stress and results in mostly five- and six-sided shapes. Because the swelling–shrinkage vertically does not require cracks to form (the soil surface just sinks down in shrinkage), the *prismatic* structures develop in early stages. In later development, especially in clayey soils, the cracking horizontally will form *blocky* peds. It is not unusual to have developed prisms consisting of smaller blocky peds (compound structure). These are mostly subsurface peds.

Platy structure requires force to separate horizontal soil layers. Frost heaving, fluctuating water tables, compaction (equipment or animals), and thin layering of different-textured alluvium or lacustrine material can aid the formation of plates.

Granular peds are mineral aggregates "glued" together mostly by organic substances, but so mixed by rodents, earthworms, frost action, and cultivation that all edges are rounded and the peds are small in size. These are limited to surface horizons unless buried by sediment.

Table 4-3 Types and Classes of Soil Structure

	Type (Shape and Arrangement of Peds)					
	Platelike with the Vertical Dimension Greatly Less than the Other Two Dimensions; Faces Mostly Horizontal	Prismlike, with Two Dimensions Limited and Considerably Less than the Vertical: Vertical Faces Well Defined: Vertices Angular		Blocklike, Polyhedronlike, or Spheroidal		
				Blocklike: Blocks or Polyhedrons Formed by the Faces of the Surrounding Peds		Spheroids of Polyhedrons Having Surfaces that Have Slight or No Accomodation to Faces of Surrounding Peds
		Without Rounded Caps	With Rounded Caps	Faces Flattened; Most Vertices Sharply Angular	Rounded and Flattened Faces and Vertices	Relatively Non-porous Peds
Class	Platy	Prismatic	Columnar	(Angular) Blocky	(Subangular) Blocky	Granular
Very fine or very thin	Very thin platy: < 1 mm	Very fine prismatic; < 10 mm	Very fine columnar; < 10 mm	Very fine angular blocky; < 5 mm	Very fine subangular blocky; < 5 mm	Very fine granular; < 1 mm
Fine or thin	Thin platy; 1–2 mm	Fine prismatic; 10–20 mm	Fine columnar; 10–20 mm	Fine angular blocky 5–10 mm	Fine subangular blocky 5–10 mm	Fine granular; 1–2 mm
Medium	Medium platy; 2–5 mm	Medium prismatic; 20–50 mm	Medium columnar; 20–50 mm	Medium angular blocky; 10–20 mm	Medium subangular blocky; 20–20 mm	Medium granular; 2–5 mm
Coarse or thick	Thick platy; 5–10 mm	Coarse prismatic; 50–100 mm	Coarse columnar; 50–100 mm	Coarse angular blocky; 20–50 mm	Coarse subangular blocky; 20–50 mm	Coarse granular; 2–5 mm
Very coarse or very thick	Very thick platy; > 10 mm	Very coarse prismatic; > 100 mm	Very coarse columnar; > 100 mm	Very coarse angular blocky > 50 mm	Very coarse subangular blocky; > 50 mm	Very coarse granular; > 10 mm

Source: Modified from *Revised Soil Survey Manual*, Transmittal 430-V, USDA—Soil Conservation Service, June 9, 1981, pp. 4-70 to 4-81.

FIGURE 4-2 Examples of structure types and crusting. (a) is massive and results from Pleistocene lake layering of different textures; although it appears to be a pedogenic development of platy structure, it is a deposition. (b) and (c) are platy and result from natural soil development: (b) is 1.3 cm (0.5 in.) thick plates of a plowpan in sandy loam; (c) is at 122 cm (4 ft.), caused by a fluctuating water table. (d) and (e) are subangular blocks: (e) has compound structure with the blocks within weak prisms (prism between arrows); (f) is a 10 x 23 cm (4 x 9 in.) prism from a clay loam. (g) shows two prisms 12.5 cm (5 in.) long. (h) shows a dry farm silt loam soil in November, with winter wheat emerging; the soil has lost surface structure and formed a crust with a thin surface layer of massive (structureless) soil (notice upside-down piece of top soil—see arrow. The soil is fairly well structured below the surface crust.   (Courtesy of Raymond W. Miller, Utah State University.)

4:4.3 Deterioration of Aggregates

Increasing exchangeable sodium (Na^+) most often speeds deterioration of structure. The sodium does not effectively neutralize the surface negative charge on soil particles. The result is a repulsion of adjacent soil particles because of similar charges, and destruction of structural peds (dispersion) occurs. The dispersed clays and small organic colloids move with water, lodging in the soil pores and sealing the soil. Soil with too much sodium becomes almost impermeable to water and dries to hard crusts.

4:5 Particle Density and Bulk Density

Density is the mass of an object per unit volume. Approximate densities of some commonly known materials are given as follows on page 104.

Water is a reference for soil density measurements; soil masses are often compared to the density of water. In the metric system, water weighs *one megagram per cubic meter (1 Mg/m³)* or *one gram per cubic centimeter,* which are convenient reference numbers to

FIGURE 4-3 Soil that developed where considerable sodium occurred in the profile. Typically, the top 2.5–5 cm (1–2 in.) is a silty white E horizon (massive leached of colloids) with a B horizon of prismatic or columnar structure at a shallow depth. Often the columns or prisms are coated with dark-colored humus. Massive parent material begins at a shallow depth. In this photograph parent material begins at 20 cm (8 in.) deep. The example occurs in southeast central Utah with precipitation about 30 cm (12 in.) annually. (*Source:* USDA—Soil Conservation Service; photo by John J. Bradshaw.)

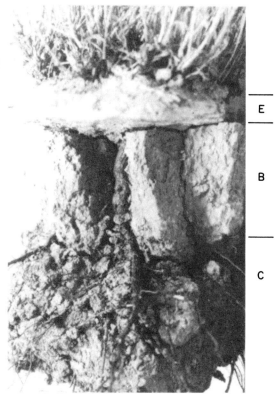

	kg m^{-3}	Mg m^{-3}	g cm^{-3}	lb. ff^{-1}
Water	1000	1.0	1.0	62.4
Pine wood	700	0.7	0.7	44
Loose sand	1600	1.6	1.6	100
Quartz mineral	2600	2.6	2.6	162
Steel, common	7700	7.7	7.7	480
Lead metal	11,300	11.3	11.3	706
Mercury metal	13,500	13.5	13.5	845

remember. Mineral soil densities, except for some volcanic materials, are greater than the density of water; organic soil densities are less than that of water.

Two density measurements—particle density and bulk density—are common for soils. **Particle density** is the density of the solid soil particles only; the measurement does not include water weight or pore (air) space. The dominant soil minerals—quartz, feldspars, micas, and clay minerals—average approximately 2650 kg/m³ (2.65 Mgm⁻³), the "standard" value used in calculations if particle density is not measured. Individual minerals have densities from 2000 kg/m³ for bauxite (aluminum ore) to 5300 kg/m³ for hematite (iron ore) or 7600 for galena (lead ore).

Bulk density, the density of a volume of soil as it exists naturally, includes any air space and organic materials in the soil volume. Because bulk density is calculated for the dried soil, water is not included in the sample weight. The bulk soil volume is assumed not

Table 4-4 Example of Bulk Densities of Soils as Affected by Texture, Compaction, Coarse Particles, and Other Features

Soil Treatment and Identification	Bulk Density (Mg m^{-3})	Bulk Density (kg m^{-3})	Pore Space (%)[a]
Rocky silt loam soil under aspen forest	1.62	1620	40
Loamy sand surface soil	1.5	1500	43
Decomposed peat (low particle density)	0.66	660	65(?)
Cotton field soil			
Tilled surface soil of a cotton field	1.3	1300	51
Trafficked interrows where wheels passed surface	1.67	1670	37
Traffic pan at about 25 cm (10 in.) deep	1.7	1700	36
Undisturbed subsoil below traffic pan, clay loam	1.5	1500	43
Boulder clay in Suffolk, England			
3 years under grass pasture	1.13	1130	57
3 years in barley, no traffic	1.3	1300	51
3 years in barley, normal tractor traffic	1.63	1630	37
Long-term use, Nigeria, sandy loam			
Under bush 15–20 years	1.15	1150	57
3 years corn, last 2 years no-till	1.42	1420	46
5 years corn, conventional tillage	1.51	1510	43
Fragipan soil in Ohio, from glacial till			
Cultivated surface soil, silt loam	1.47	1470	45
Soil layer 30 cm (12 in.) deep, loam	1.65	1650	38
Fragipan, 55 cm (21.5 in.) deep, loam	1.76	1760	34
Beneath fragipan, 125 cm (49 in.) deep, loam	1.85	1850	30
Paddy rice soil, Philippines, clay			
0–15 cm (0–6 in.), average of 9 cores	0.66	660	75
15–30 cm (6–12 in.), average of 9 cores	0.91	910	66
Granfield fine sandy loam, Oklahoma, **Ap**	1.72	1720	35
Lower subsoil	1.80	1800	32
Miami silt loam, Wisconsin, **Ap**	1.28	1280	52
Lower subsoil	1.43	1430	46
Houston clay, Texas, **Ap**	1.24	1240	53
Lower subsoil	1.51	1510	43
Oxisol clay, Brazil, **Ap**	0.95	950	64
Lower subsoil	1.00	1000	62

[a] The percentage pore space was calculated using 2.7 for particle density except for the peat soil, which is estimated.

to have changed by drying; only the water has been removed, leaving empty pores. This is not true for soils with considerable amounts of swelling clays.

A loosened, "fluffed" soil with an increased total pore space will have a smaller weight per unit volume than the same soil after it is compacted. Thus, bulk densities can be used to estimate differences in compaction of a given soil, such as might result after tillage with heavy equipment on wet clay soils.

Unfortunately, because of the wide variety of soil textures and humus contents, bulk density alone is not an indication of soil suitability for plant growth. Soils of different bulk densities, because of different textures, may be equally good for plant growth. Table 4-4 illustrates some of these variations in bulk density that result from compaction, vegetation, and soil use. Figure 4-4 shows the changes in loblolly pine seedlings grown in three different soils with varied bulk densities due to different amounts of compaction. The average soil bulk density of cultivated loam is approximately 1100–1400 kg/m^3. For good plant growth, bulk densities should be below about 1400 kg/m^3 for clays and 1600 kg/m^3 for sands. In greenhouse potting mixtures, the high amount of peat moss, vermiculite, or

FIGURE 4-4 Bulk density in relation to the growth of loblolly pine in Louisiana. Sand, loam, and clay soils were subjected to the following kinds of loosening or pressure, and the resulting soil material was used to grow loblolly pine seedlings: (1) soil loosened, (2) soil undisturbed, (3) soil subjected to static pressure of 3.5 kg/cm, (4) soil subjected to static pressure of 7.0 kg/cm, (5) soil subjected to static pressure of 10.5 kg/cm, (6) soil puddled (dispersed) plus same pressure as in treatment 5. (Courtesy of Louisiana State University.)

perlite used produces low bulk densities, such as 100–400 kg/m³. Recall that all of these density values refer to the weight of *dry* soil.

Bulk density values have various uses. For example, density values are needed to calculate total storage capacity per soil volume. Secondly, soil layers can be evaluated to determine if they are too compacted to allow root penetration or to provide adequate aeration.

Measurements of bulk density commonly involves taking an undisturbed block of soil (clod or soil core), determining its volume, drying it, and weighing it. *Clods* can be coated with paraffin or liquid plastic and dipped into water to measure water displacement, and hence to calculate volume. When soil cores are taken by a metal cylinder, the exact volume is determined by measuring the cylinder volume (Calculation 4-2). The formula for bulk density is:

$$\text{bulk density} = (\rho_B) = \frac{\text{soil mass oven dry}}{\text{soil volume}}$$

Calculation 4-2 Calculating Soil Bulk Density

Problem A metal cylinder pushed into a loam soil is removed from the field and the soil it contains is dried in an oven. The measured data are as follows:

Cylinder height	=	5.0 cm
Cylinder inside diameter	=	4.4 cm
Oven-dried soil weight	=	87.6 g

Calculate the soil's bulk density.

Solution

1. The volume of the soil sample equals the volume of the cylinder. A cylinder's volume equals pi ($\pi = 3.14$) times the radius squared times the cylinder height ($V = \pi r^2 h$):

$$\text{volume} = \pi \left(\frac{\text{diameter}}{2}\right)^2 h$$

$$= (3.14) \left(\frac{4.4 \text{ cm}}{2}\right)^2 (5 \text{ cm}) = 76.0 \text{ cm}^3$$

2. The soil bulk density is soil dry weight divided by the volume of the soil. If measurements were in pounds and cubic feet, the calculations and answer would be in those units. In this example, the measurements are in grams and cubic centimeters. Thus,

$$\text{bulk density} = \frac{\text{soil mass}}{\text{soil volume}} = \left(\frac{87.6 \text{ g}}{76 \text{ cm}^3}\right) \left(\frac{1 \text{ kg}}{1000 \text{ g}}\right) \left(\frac{1{,}000{,}000 \text{ cm}^3}{1 \text{ m}^3}\right)$$

$$= 1150 \text{ kg/m}^3 \ (= 1.15 \text{ g/cm}^3)$$

Some examples of the importance of bulk density to plant growth are given in Detail 4-1.

The average weight of soil for a hectare (or acre) area per unit depth is calculated by multiplying the soil volume by its bulk density. A hectare–15 cm or acre-furrow-slice weight is estimated for bulk densities of about 1300 kg/m³. A hectare–15 cm volume of soil of bulk density 1300 kg/m³ would weigh approximately 2,000,000 kg when oven dry. It can be calculated as follows:

1 hectare = 10,000 m²

1 hectare–15 cm deep = (10,000 m²)(0.15 m) = 1500 m³

1 hectare–15 cm deep of bulk density 1300 kg/m³ weighs 1,950,000 kg

The weight is often approximated to 2,000,000 kg

An acre–7 in. depth is estimated to weigh about 2,000,000 lb. by a similar calculation, but an acre–7 in. depth is slightly deeper than 15 cm. The weight of one hectare–30 cm is estimated to be 4×10^6 kg.

1. Plant roots of field crops are hindered by soils high in bulk density, but in varying degrees. Tolerance to soil compaction is in this order:*

   alfalfa > corn > soybeans > sugar beets > dry edible beans

2. On two silt loam soils in Arkansas, subsoiling decreased soil bulk density by 16% and increased cotton lint yields by 13% in a year with normal rainfall and by 59% during a dry year.[†]

3. Soil bulk densities greater than 1200 kg/m³ (1.2 g/cm²) on medium textured soils in the state of Washington were positively correlated with diseases of beans and peas.[‡]

4. Subsoiling increased the yield of tobacco in North Carolina on soils with a bulk density of greater than 1630 kg/m³ (1.63 g/cm²) and/or a sand content greater than 73%. Responses were greater in dry weather.[§]

---

* A.J.M. Smucker, "Plant Root System Response to Compacted Soils," *Agronomy Abstracts* (1985)
† J.S. McConnell and M.H. Wilkerson, "Soil Compaction Effects on Cotton," Arkansas Farm Research, Mar.–Apr. 1987.
‡ J.M. Kraft, "Compaction/Disease Interaction," *Agronomy Abstracts* (1985).
§ M.J. Vepraskas, G.S. Miner, and G.F. Peedin, "Relationships of Soil Properties and Rainfall to Effects of Subsoiling on Tobacco Yield," *Agronomy Journal,* **79** (1987), pp. 141–146.

4:6 Soil Porosity and Permeability

Pore spaces (also called **voids**) in a soil consist of that portion of the soil volume not occupied by solids, either mineral or organic. Pores in soil are the result of irregular shapes of primary particles and their aggregation; the pushing forces of penetrating roots, worms, and insects; and of expanding gases entrapped by water. Under field conditions, all pore spaces are occupied by air and water. "Tortuous pathways" best describe soil pores. The soil particles have irregular shapes and thus leave the spaces between them very irregular in size, shape, and direction. Sands have large and continuous pores. In contract, clays, although containing more total pore space because of the minute size of each clay particle, have very small pores, which transmit water slowly. Small "bottlenecks" in the pores can fill with water and block air movement through them. Air exchange may be inadequate for plant root growth in some wet, clayey soils. The most rapid water and air movement is in sands and strongly aggregated soils, whose aggregates act like sand grains and pack to form many large pores.

Pores are officially described according to their average diameter in millimeters as follows:

Very fine <0.5 mm	Medium 2–5 mm
Fine 0.5–2 mm	Coarse >5 mm

Water drains by gravitational force from pores larger than about 0.03–0.06 mm. In comparison, root hairs, the smallest plant roots, are between 0.0008 and 0.012 mm in diameter. In soils where most pores are smaller than 0.030 mm, attraction forces in the soil retain water within the fine pores, which results in a waterlogged soil and poor aeration. Thus, *to the growing plant, pore sizes are of more importance than total pore space.* The best balance of water retention (smaller pores) plus adequate air and water movement (larger pores) is in medium-textured soils, such as loams. Aggregation, or the lack of it, can modify this balance of large and small pores, which results from the soil texture (Fig. 4-5). Soluble bicarbonates, silicates, and iron during wet

FIGURE 4-5 A cemented layer can be like semisoft rock. This Upton gravelly loam soil in New Mexico has a very shallow 30-cm (12-in.) surface soil (**A** and **B**, the darker material top foot of depth) over a *hard indurated (lime-and-silica cemented) hard pan called petrocalcic or duripan (caliche)*, the white material below the left 1 ft. marker in the photo. This hardpan cannot be easily dug, even with a pick, and is often nearly impermeable to water movement through it. (*Source: USDA—Soil Conservation Service; photo by Max V. Hodson.*)

periods will move in the soil and precipitate when drying occurs. Over many decades or centuries, pores become filled with the precipitates and often are cemented into hard layers.

The relative amounts of air and water in the pore space fluctuate continuously. During a rain, water drives air from the pores, but as soon as soil water disappears by deep percolation (downward movement), evaporation, and transpiration (evaporation from plant leaf openings), air gradually replaces the water as it is lost from the pore spaces.

The percentage of a given volume of soil occupied by pore space may be calculated from the formula

$$E_p = \% \text{ pore space } = 100\% - \% \text{ solid space}$$

$$E_p = \% \text{ pores space } = 100\% - \left(\frac{\text{bulk density}}{\text{particle density}} \right)(100)$$

or

$$E_p = 100 - \left(\frac{\rho_B}{\rho_p} \right)(100)$$

A sample calculation is given in Calculation 4-3.

Greenhouse and nursery managers are very concerned about pore space and water infiltration, water retention, and drainage. These factors plus costs for transportation have transformed much of the container-grown plant industry into one using mixtures of ground bark, expanded vermiculite, peats, perlite, but only small amounts of heavy materials such

Problem A soil core was taken for determination of bulk density. The measurements were:

Cylinder volume $= 73.6$ cm³ (see Calculation 4-2)

Dry soil weight $= 87.8$ g

"Standard" particle density $= 2650$ kg/m³ (2.65 g/cm³)

Calculate the percentage pore space.

Solution

1. Bulk density $= \dfrac{\text{soil weight}}{\text{soil volume}} = \left(\dfrac{87.8 \text{ g}}{73.6 \text{ cm}^3}\right)\left(\dfrac{1 \text{ kg}}{1000 \text{ kg}}\right)\left(\dfrac{1{,}000{,}000 \text{ cm}^3}{1 \text{ m}^3}\right)$

$= 1190$ kg/m³ [$= 1.19$ g/cm³ $= 1.19$ Mg m⁻³]

2. % pore space $= 100\% - \left(\dfrac{\text{bulk density}}{\text{particle density}}\right)$ (100)

$= 100\% - \left(\dfrac{1.19}{2.65}\right)$ (100)

$= 100\% - 44.9\%$

$= 55.1\%$, the soil's pore space percentage

This soil would be a clay loam or other clayey soil. A sandy soil would have a bulk density closer to 1,400 or 1,500 kg/m³ and a pore space percentage closer to 45–50%.

Examples Several soil textures, bulk densities, and their pore space percentages are shown below for cropped soils:

Soil Texture	Bulk Density (kg/m³)	Pore Space (%)
Gravelly sand	1870	29.4
Coarse loamy sand	1680	36.6
Sandy loam	1510	43.0
Loam	1340	49.4
Clay loam	1260	52.5
Clay	1180	55.5

as sand and soil. Bulk densities of nonsoil mixtures range from 150 to about 450 kg/m³ with most below 300 kg/m³. These materials will hold about 15–25% water against drainage.

4:7 Soil Air

To survive, all living organisms require gaseous exchanges. The most important of these in soils are for *respiration* by plant roots and for organic-matter decomposition by microorganisms. The most desired condition is *well-aerated soil,* a condition in which oxygen exchange between soil air and atmospheric air is rapid. The factors influencing the rate of gaseous exchange include soil pore sizes and continuity, temperature, depth in soil, wetting and drying of soil, and coverings (mulches) on the soil surface.

4:7.1 Composition of Soil Air

Atmospheric air has about the following composition of the gases important in soils:

Dinitrogen (N_2) 79%

Oxygen (O_2) 20.9%

Carbon dioxide (CO_2) 0.03%

Water vapor (relative humidity) 20-90%

The most important of these to plant growth is the oxygen content. Oxygen is required in respiration and organic-material oxidation (decomposition). Because of this, soil air is different from atmospheric air. The differences are as follows:

	Surface Soil %	Subsoils (%)
Higher in carbon dioxide	0.5–6	3–10
Lower in oxygen	20.6–14	18–7
Higher in relative humidity	95–99	98–99.5

As roots and microbes use oxygen, they give off carbon dioxide into the soil air.

4:7.2 Rates of Oxygen Exchange

The **oxygen diffusion rate (ODR)** is the rate at which oxygen in the soil exchanges with oxygen in the atmosphere. Many large soil pores speed air exchange (diffusion); small pores, or pores with "bottleneck" portions filled with water, decrease exchange rates. A depth of 1 m (39 in.) in the soil may have an exchange rate only one-half to one-fourth as fast as in the top few centimeters. Exchange rates of air above 40×10^{-4} g/m^2 per minute seem to be fast enough for most plant roots and microbes. Root growth of some plants ceases when the rate reaches half of this value. Diffusion of CO_2 gas through water is about 10,000 times slower than it is through air-filled pores.

From these data it is easy to visualize reduced root growth occurring in deep clayey subsoils. The small pores of the clays would have a slow ODR. When wetted, water-filled portions of pores would block oxygen diffusion further. Because of this problem, some clayey soils are, in effect, *shallow soils,* as far as plant root growth is concerned.

4:7.3 The Oxidation-Reduction Potential (Eh or Redox Potential)

In most soils, plants grow best in *oxidized* (aerated) soil (rice and marsh plants are exceptions). The major reason is the need for oxygen. Oxygen is the *primary acceptor of electrons* released in the oxidation of carbon in organic materials. If the oxygen concentration is too low, certain elements or ions become the electron acceptors and are, in turn, reduced. The major elements involved in this process and comments about them are given in Table 4-5. Mostly undesirable things happen when the ODR is too low and the "redox value" drops: Toxic organic acids form, nitrate is denitrified and nitrogen is lost to the atmosphere, available sulfate is reduced to sulfide, and most plant roots cannot respire. Thus, plant root growth slows or stops.

4:7.4 Aeration and Plant Growth

All plants need oxygen for respiration. Respiration is necessary for life and growth. Most plants need the oxygen to be in the soil pores where roots are growing. However, a few

Table 4-5 Oxygen and the Ions in Soil Commonly Acting as Electron Acceptors and Their Ion Forms That Are Produced (When Oxygen Concentrations Are Low)

Form in Oxidized Soil	Form in Reduced Soil	Approximate Eh Where Change Occurs (volts)[a]	Comments about the Solubility of the Reduced Form Compared to Oxidized Form
O_2	H_2O	0.38 to 0.32	Much more soluble
NO_3^-	N_2, N_2O	0.28 to 0.22	Lost to atmosphere
MnO_2	Mn^{2+}	0.28 to 0.22	Much more soluble
Fe_2O_3	Fe^{2+}	0.18 to 0.15	Much more soluble
SO_4^{2-}	S^{2-}	−0.12 to −0.18	Less soluble
CO_2	CH_4	−0.2 to −0.28	Lost to atmosphere

Source: Selected and modified data of W.H. Patrick, Jr., and C.N. Reddy, "Chemical Changes in Rice Soils," in *Soils and Rice,* International Rice Research Institute, Los Baños, Philippines, 1978, pp. 361–379.

[a] Positive voltage values indicate oxidized soils. A typical soil is borderline and may be poorly aerated if the Eh is smaller than about 0.25 wet or 0.30 when at good moisture for plant growth. The values change with soil pH and location in the profile or even within a soil ped.

plants, such as rice, can move oxygen *internally* from their tops, which are in the atmosphere, to their roots. Airflow is through internal large-diameter pores. By these means, they are able to supply themselves oxygen, even when growing in stagnant waters.

Plants obtain energy from the sun, store it in chemical bonds, and release and use the energy as they break these bonds. Splitting a glucose sugar (6-carbon sugar) into 2-carbon or 3-carbon fragments (fermentation) is common in **anaerobic glycolysis**. In the presence of oxygen, **glycolysis plus respiration** (conversion of glucose carbon to carbon dioxide) releases and makes available much more energy, about 19 times more than anaerobic breakdown (308 kcal vs. 16.2 kcal stored into ATP bonds). The energy that is needed to form new ATP energy bonds is released by the anaerobic glycolysis (47.3 kcal) and aerobic glycolysis plus respiration (686 kcal.) Thus, much less energy flow exists in anaerobic conditions; anaerobic decomposition rates of organic matter are much slower than are rates of aerobic decomposition.

Reduced aeration causing deficient oxygen concentrations can be caused by various factors. Some of these are (1) waterlogging (ponding), (2) soil compaction of loams and clays, (3) very high contents of clay that swell when wet, thereby closing large pores, and (4) organic-matter decomposition by soil microorganisms that uses oxygen in soils already low in ODR.

Some practices—deep ripping, drainage of excess water, and incorporation of "loosening" organic residues—are partly designed to aid soil aeration. Inadequate aeration should be expected in these conditions:

1. Poorly drained soils
2. Soils of high clay content shortly after rainfall or irrigation
3. Deep subsoils in clayey soils, especially if wet
4. Highly compacted soils of fine texture
5. Deeper portions of clayey soils having no structure (massive)

4:8 Soil Strength

Soil strength is the degree of resistance of a soil mass to crushing or breaking when force is applied. A noncemented soil mass is evaluated under two moisture conditions: air-dry and field capacity water content. When a soil mass is cemented, soil strength is evaluated when air dry and again when wet after it has soaked in water for 1 hour. The descriptive terms are presented in Table 4-6.

4:9 Soil Color

For centuries people have recognized that wearing white clothes in hot climates in cooler than wearing dark-colored clothes because more heat is absorbed by dark colors. Similarly, dark soils absorb more heat than do light-colored ones. Some black coal mining wastes and dark-colored oil-shale residues reach temperatures of 65.6–70.0°C (150–158°F), which are lethal to many plants that could otherwise grow in those soils. Although black soils having high humus content absorb more heat than light-colored soils, they also frequently hold more water. The water requires a relatively larger amount of heat than the soil minerals to raise its temperature; water also requires considerable heat to evaporate it. The net result is that many dark soils are *not* warmer than adjacent lighter-colored soils because of the temperature-modifying effects of the soil moisture; in fact, they may be cooler except for a centimeter or so at the *dry* surface.

Table 4-6 Strength Classes of Soil

Field Method and Condition of Failure of Specimen	Units (newtons)[2]	Consistency Term When:		
		Air Dry	Field Capacity	Cemented Material
No specimen can be obtained		Loose	Loose	
Specimen crushes or breaks when very slight force applied by thumb and forefinger	<8 N[a]	Soft	Very friable	
Specimen crushes or breaks when slight force applied by thumb and forefinger	8–20 N	Slightly hard	Friable	
Specimen crushes or breaks when moderate force applied by thumb and forefinger	20–40 N	Slightly hard	Firm	Weakly cemented[b]
Specimen crushes or breaks only when strong force applied by thumb and forefinger	40–80 N	Hard	Very firm	Weakly cemented[b]
Specimen cannot be crushed or broken by thumb and forefinger but can be by squeezing slowly between hands	80–160 N	Very hard	Extremely firm	Weakly cemented[b]
Specimen cannot be crushed or broken in hands but can be broken or crushed underfoot by person weighing 80 kg applying weight slowly	160–800 N	Extremely hard	Extremely hard	Weakly cemented[b]
Specimen cannot be crushed underfoot but can be crushed or broken by blow of 3 J[c]	800 N = lower limit 3 J = upper limit			Strongly cemented
Specimen cannot be broken or crushed by blow of 3 J[c]	> 3 J			Indurated

Source: Revised Soil Survey Manual, Transmittal 430-V, USDA—Soil Conservation Service, June 1981, p. 4–100.

[a] One newton (N) = 1 kg–m/s[2].

[b] If applicable.

[c] A weight of 1 kg dropped a distance of 10 cm = 1 joule (J). A geologist's hammer weighing 1.36 kg dropped a distance of 0.22 m exerts an impact energy of 3 J.

Soil color indicates many soil features. A change in soil color from the adjacent soils indicates a difference in the soil's mineral origin (parent material) or in soil development. White colors are common when salts or carbonate (lime) deposits exist in the soil. Spots of different color (**mottles**), usually rust colored, indicate a soil that has periods of inadequate aeration each year. Bluish, grayish, and greenish subsoils (**gleying**), with or without mottles, indicate longer periods each year of waterlogged conditions and inadequate aeration. Usually, a soil horizon with a chroma ≤ 2 indicates a low level of aeration.

Within geographic regions, darker colors usually indicate higher organic-matter contents. However, between contrasting climatic conditions, color is not a good indicator or organic matter content. Soils with high concentrations of dark-colored minerals may be dark colored. Some partially decomposed humus is darker in some environments than in others.

Soil color determination is standardized and determined by the comparison of the soil color to **Munsell color charts.** These color charts are similar to books of color chips found in paint stores where many gradations of different color groups are shown (Figs. 4-6 and 4-7) there are color charts for plant foliage and for other uses. Soil color notation is divided into three parts:

Hue The dominant spectral or "rainbow" color (red, yellow, blue, and green)
Value The relative blackness or whiteness, the amount of reflected light
Chroma Purity of the "color" (chroma number increases and the color is more
brilliant as grayness decreases)

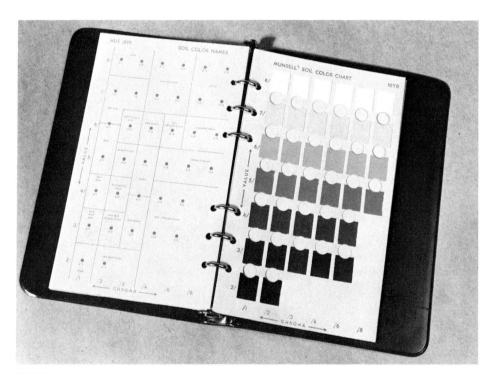

FIGURE 4-6 Sample page from the Munsell soil color charts, showing the hue 10YR. Color values appear vertically on the chart, and chroma (brilliance) is horizontal. Official color names at left are assigned to the specific color chips shown on the right page. Example: 10YR 4/3 is a dark brown. (Courtesy Munsell Color, 2441 N. Calvert St., Baltimore, MD 21218.)

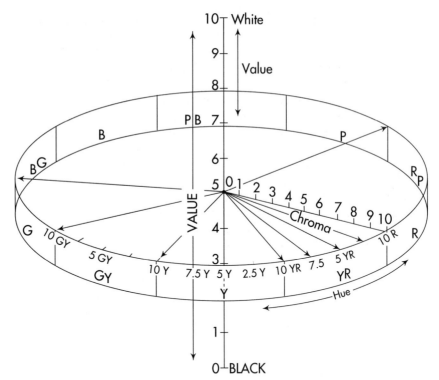

FIGURE 4-7 llustration of the Munsell color scheme, showing the hue terms and their subdivisions into 2.5, 5, 7.5, and 10 units. Chroma subdivisions range from 0 (neutral) to 10 (a very select portion of wavelengths, called *brilliance*). The value term ranges from 0 (black, no reflected light) to 10 (white, maximum light reflected).   (Courtesy of Raymond W. Miller, Utah State University.)

4:10 Soil Temperature

Soil temperatures vary from perpetual frost (**permafrost**), at a shallow depth in many soils in frigid Alaska, to warm tropical Hawaii, where daytime temperatures of the bare soil surface seldom fall below 40°C (104°F) on sunny days.

4:10.1 Relationship of Soil and Air Temperatures

Heat is energy. Heat energy is lost from the earth to space and is replaced by energy from the sun. The temperature of the earth and its atmosphere is the *net* (R_N) of *incoming shortwave solar radiation* (R_S) minus the losses into space of (1) *reflected shortwave radiation* (albedo, R_R) and (2) *longwave far-infrared radiation* (emissivity, R_L).

$$R_N = R_S - R_R$$

Figure 4-8 illustrates the more complex energy flow, including *heat absorbence by the ground* (G), *heat absorbed by the air* (H), the *latent heat* (LE) used mostly to evaporate water, and *longwave radiation emitted from the earth* (RL), which is lost as degraded heat energy. The more complete relationship is:

$$R_N = R_S - R_R = G + H + LE = RL$$

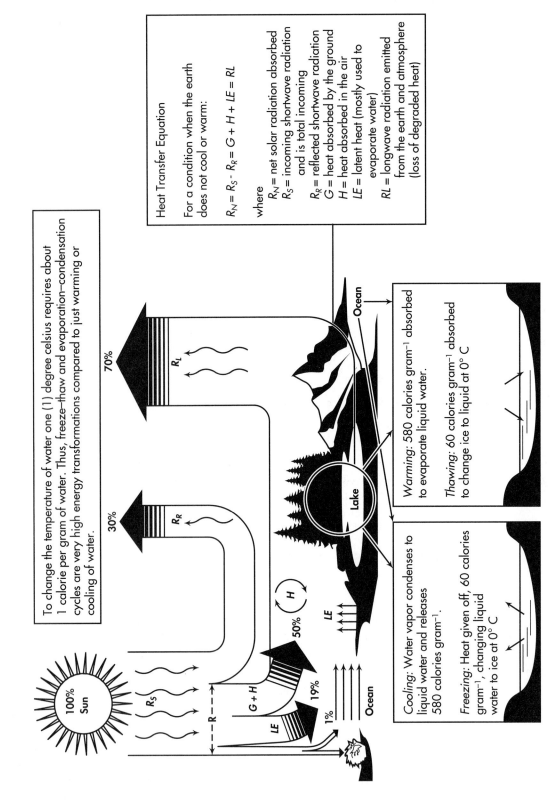

FIGURE 4-8 An illustration of short-wave incoming energy (sun) and energy transformations in and from the earth and its atmosphere, emphasizing the energy losses and gains modified by changes in phases of water. (Raymond W. Miller, Utah State University)

Heat Transfer Equation

For a condition when the earth does not cool or warm:

$$R_N = R_S - R_R = G + H + LE = RL$$

where

R_N = net solar radiation absorbed
R_S = incoming shortwave radiation and is total incoming
R_R = reflected shortwave radiation
G = heat absorbed by the ground
H = heat absorbed in the air
LE = latent heat (mostly used to evaporate water)
RL = longwave radiation emitted from the earth and atmosphere (loss of degraded heat)

To change the temperature of water one (1) degree celsius requires about 1 calorie per gram of water. Thus, freeze–thaw and evaporation–condensation cycles are very high energy transformations compared to just warming or cooling of water.

Warming: 580 calories gram⁻¹ absorbed to evaporate liquid water.

Thawing: 60 calories gram⁻¹ absorbed to change ice to liquid at 0° C

Cooling: Water vapor condenses to liquid water and releases 580 calories gram⁻¹.

Freezing: Heat given off, 60 calories gram⁻¹, changing liquid water to ice at 0° C

100% Sun

R_S

R

$G + H$

LE

19%

1%

Ocean

50%

H

LE

Lake

Ocean

30%

R_R

70%

R_L

Detail 4-2 Energy (Heat) Budgets

The earth receives energy (short-wave) from the sun and loses energy (long-wave and heat) to space. The overall earth's balance is zero, but at any one location on the earth that location may have a net energy gain (spring warming) or net energy loss (fall cooling). A typical yearly average for the earth might be as shown below:

Gain and loss[a]	Sun's Short-wave	Degraded Long-wave	Net
Space energy budget (= no change in earth's temperature)			0
Atmospheric energy budget			0
Emission by water vapor to space	——	−38	
Emission by liquid water (clouds) to space	——	−26	
Absorbed sun energy by water, dust, ozone	+16	——	
Absorbed sun energy in clouds	+3	——	
Net heat from earth (see last 4 items below)	——	+45	
Surface energy budget			0
Incoming solar radiation to ground	+51	——	
Emission of energy by earth to space	——	−6	
Emission of energy by earth to atmosphere	——	−118	
Absorption by earth from water and CO_2	——	+103	
Sensible heat to atmosphere	——	−7	
Latent heat transfer (evaporation) to atmosphere	——	−23	

[a] Average long-term (year-round) average of 100% of incoming sun's radiation.

The net heat absorbed by the earth equals the heat lost as far-infrared radiation. If this were not so, the earth would be cooling or heating up gradually year after year.

Water has great influence on temperatures of soils and of the atmosphere (Fig. 4-8). Heating, cooling, and evaporation of water exert large effects on the heat energy of an area. The evaporation process absorbs relatively large amounts of heat; condensation, likewise, releases large quantities of heat back to the system.

Detail 4-2 presents relative numerical values for the energy balance in the atmosphere and in the earth's surface. *Positive values are absorption of heat; negative values are losses of heat from that system.* Temperatures of an area of the earth will be greatly modified by the water contents in that area.

The "standard" soil temperature is measured at a depth of 50 cm (19.5 in.) or at the rock or hardpan contact if the soil is not 50 cm deep. The **average annual soil temperature** (***AST***) can be measured at a depth of 6 meters (20 ft.), although the value at about 3 meters (10 feet) deep is usually quite close. The AST can be approximated by adding about 1°C (about 2°F) to the mean annual *air* temperature. This rule-of-thumb is not as accurate in arid, sunny regions as it is in other areas.

Soil temperatures resulting from the sun's radiation change with depth and with the time of day. For example, the maximum daily temperature at deep soil depths is delayed, even by several hours, after the time when the air temperature reaches a maximum. Conversely, on cool nights, the deep soil layers do not cool as fast as surface layers because of the insulating effect of the overlying soil. Heat flow is slower in soil than in the atmosphere. The deeper the soil layer, the longer it takes a temperature change to reach it and thus the less will be the ac-

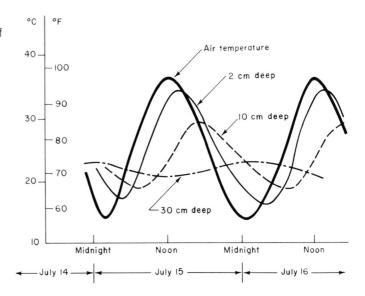

FIGURE 4-9 Temperature fluctuations with the depth of soil in northern Utah, illustrating the reduced amplitude of change with deeper soil depth and the delay in time when maximums and minimums are reached. (Courtesy of Raymond W. Miller, Utah State University.)

tual temperature fluctuation from day to day or week to week (Fig. 4-9). Daily temperature fluctuations seldom affect the soil deeper than about 30–40 cm (12–16 in.). Below about 1 m (3.3 ft.) the soil changes slowly from season to season. The mean summer and mean winter temperatures at 1 m deep seldom differ by more than 5°C (9°F) in the subtropics; differences increase several degrees in temperate regions.

4:10.2 Factors Influencing Soil Temperature

The temperature a soil attains depends on (1) how much heat reaches the soil surface (heat supply) and (2) what happens to that heat within the soil (dissipation of heat). The heat supply to the soil surface from external sources is reduced by organic-soil coverings, which act as insulators. All plastic mulches make the soil hotter when the sun is shining, but transparent plastic allows greater warming than opaque plastic. A sun angle less than perpendicular to the soil surface also supplies less heat per unit of soil area (Fig. 4-10). North-facing slopes in the northern hemisphere will receive less heat than will other directional slopes during

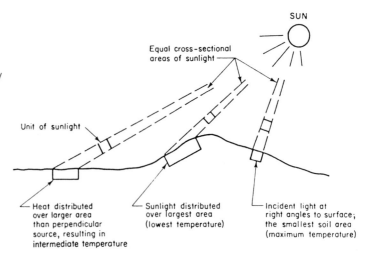

FIGURE 4-10 Effect of soil slope, aspect (direction of slope), and the sun's position on the heat supplied to soil. (Courtesy of Raymond W. Miller, Utah State University.)

winter months, when the sun is low in the southern sky. (The reverse slopes have these properties in the southern hemisphere.) Rock gardens, or nonlevel lands, sloped to be at right angles to the springtime sun's rays will receive more heat per soil area and will warm faster than flat (oblique-to-the-sun) surfaces. Light-colored soils will reflect more heat than will darker soils.

Some heat that is absorbed by soil is lost in evaporating soil water. If a soil's water content is high, much more heat is needed for temperature changes because the heat capacity ("heat reservoir") of the water is 3 to 5 times more than for soil minerals. In a dark soil where the color is caused by large amounts of humus, the larger amounts of water held by the humus may offset the increased heat absorption due to the dark color. Although compacted soils have better heat conductivity than loosened soil, they also have more material to heat per volume of soil. Thus, a rock garden sloped south or west with sandy soil (which holds less water) will warm fast in the spring after warming begins.

4:10.3 Living with the Existing Temperatures

Generally, existing temperatures have just been "lived with" and agricultural activities modified to fit them. Temperature restrictions can be mitigated usually only by expensive measures. Acknowledging the limitations and planning for them is wise soil and crop management. Some examples follow:

1. To get maximum germination and growth of seeds, soil temperatures must be correct; 4–10°C (40–50°F) for wheat and peas; 10–29°C (50–85°F) for corn; 16–21°C (60–70°F) for potatoes; about 27°C (80°F) or above for sorghums and melons. Optimum emergence temperatures for other plants: 8–11°C (46–52°F) for cabbage and spinach; 11–18°C (52–64°F) for beets and cauliflower; and 18–25°C (64–72°F) for asparagus, carrots, celery, endive, lettuce, onion, radish, and tomato. Direct planting of onion seeds in cool spring soil (a recent change from transplants; done because of labor costs) produces late plants. When pregerminated seeds were planted, they emerged in 1.7–7.3°C soil within 7 days; regular seeds required 30 days. Maturity of onions from pregerminated seeds was 10–12 days earlier. Geranium seeds have most rapid germination at about 27°C (80°F).

2. When applying anhydrous ammonia in the fall, it is best to wait until the soil at a depth of 10 cm (4 in.) is 10°C (50°F) or less. Below this temperature nitrification of ammonium to nitrate is slow, and therefore leaching losses of nitrate will be minimal.

3. Freezing and thawing of bare, saturated, fine-textured soils in cold areas, such as the intermountain West and the northern United States, may cause heaving and then death of shallow-rooted crops.

4. Alternate freezing and thawing under conditions of moderate soil moisture improves the structure of cloddy soils, but this process with excess moisture destroys structure. Freezing and thawing of wet loessial and other silty slopes causes erosion as mudflows.

4:10.4 Modifying Temperature Effects

As technology enables us to modify the environment, we become less satisfied with tolerating existing undesirable climates. Conquering temperature, except in greenhouses, is not currently possible, but the effects in the field can be modified in limited ways such as these.

1. The use of clear-plastic surface covers in cold areas or seasons increases soil temperatures and permits successful plant growth during the cold period (Fig. 4-11). The use of

FIGURE 4-11 These clear-plastic strips are trapping the rays of the sun and warming the root environment, thus making the growth of this corn possible in Alaska. (*Source:* Lee Allen, D.H. Dinkel, and Arthur L. Brundage, University of Alaska Institute of Agricultural Sciences, *Agroborealis,* **1,** no. 2, Sept. 1969.)

clear-plastic mulches in Iowa and Alaska will mature sweet corn 4 to 8 days earlier than without it. In Orange County in southern California, plastic mulch speeds the growth of an early strawberry crop worth $15 million, produced on only 891 ha (2200 acres).

2. A black polyethylene mulch on the soil in a pineapple plantation in Hawaii was responsible for a 50% increase in growth of pineapple. This growth increase was attributed to an increase in soil temperature of 1.5°C (2.7°F) during winter and not to an increase in soil moisture. Black mulches retain moisture and control weed growth (no light) but transmit less heat rays through the soil than do clear plastic mulches.

3. In February, early snow removal, using fly ash or graphite, has reduced winter wheat losses caused by snow mold in northern Utah and Idaho.[2,3] Snow-mold damage is increased with the length of time snow covers the land in spring. Sometimes snow mold kills 50–70% of the grain; yields are reduced or replanting of spring grains is required. *Graphite* is applied at rates of 13–20 kg/ha (at $0.45/kg) by airplane or snow-cat. Commercial applicators charge $15/ha. Graphite can be applied in liquid nitrogen and/or herbicide solutions. Treated snow usually melts 2–3 weeks earlier than without graphite. In Utah, small grain breeding nursery stock has been secured by yearly use to minimize snow-mold losses. It is being used on mountain pastures in April to speed plant growth on grazing lands (Fig. 4-12).

---

[2] T. A. Tindall, "Darkening Agents in a UAN solution: Influence on Snow Mold and Yield of Winter Wheat," *Journal of Fertilizer Issues,* **3** (1986) no.4, pp.129–132.

[3] T. A. Tindall and S. A. Dewey, "Graphite-Nitrogen Suspensions with Selected Herbicides Applied to Snow Cover in Management of Winter Wheat," *Soil Science,* **144** (1987), pp. 218–223.

FIGURE 4-12 Distributing graphite on snow in Utah in early spring to remove snow cover. A fungus, called snow mold, becomes very active under the snow as the soil thaws and warms. Early snow removal can save a wheat crop. (a): Spreading. (b): Cleared areas (black strips) show where snow melted on research fields because of ash cover. (Courtesy of Dr. Terry Tindall, Utah State University.)

(a)

(b)

4. Mulches left on the soil through the winter act as insulation and retard warming in the spring. Incorporating straw stubble into the soil in the fall rather than leaving it on the surface as a mulch resulted in about 40% more growth of early corn in Minnesota.

5. Early fruit bloom following a prolonged warm period is often followed by a freeze that can eliminate a year's production of apples or other fruits. Cooling by sprinkling has maintained fruit buds in a longer dormant condition until flowering is safe. Spring sprinkling of lettuce to prevent frost damage is also a common practice.

6. Specially contoured and planted rows can modify low spring temperature effects by protecting the young plants with soil and positioning them to receive more direct sunlight (greater warming), as in many rock gardens.

4:10.5 Using Temperature: Growing Degree Days

Most plants require a certain amount of growth energy before they reach certain growth stages, such as seed germination, flowering, or fruit maturity. This needed energy is a summation of the temperature multiplied by the time units to that date. In some instances the number of daylight hours may also be important.

Predictions of when to plant, when to harvest, and when to expect to apply chemicals to control insect infestations now can be made more accurately (and scientifically) by utilizing calculations of the heat energy required to reach these states. Unfortunately, the systems known at the present time are few.

The energy requirements are expressed as the numbers of hours at air temperatures above the minimum base temperature (*TB*) and are called **growing degree hours (GDH)** Less exact summing may calculate the heat energy as **growing degree days (GDD)** (Fig. 4-13). Nature does not provide days neatly packaged with the same constant temperatures; the heat supplied during a hot summer day is much greater than that supplied during a cool spring one. In some manner these changes must be averaged in order to calculate values important in plant physiology.

The method of averaging hourly or daily temperatures is shown in Fig. 4-13. The dashed line *TB-TU-TC* is a typical growth curve as influenced by temperature. To simplify calculations, the curve may be drawn as straight lines *AB-BC-CD*. The "corn model" uses this simplification. No growth occurs below *A* (10°C or 50°F for corn), growth increases as temperature increases up to *B* (30°C or 86°F for corn), and growth remains at the maximum constant rate above temperature *B* to temperature *C* (about 44°C or 111°F). The simplified model using the *A-B-C-D* lines of Fig. 4-13 averages the daily temperature from the day's maximum and minimum temperatures (T_{max}, T_{min}).

Calculating GDD for corn uses the day's average. This average is calculated with a temperature minimum of no less than 10°C (50°F) and a maximum no greater than 30°C (86°F), regardless of how far the actual measurements exceed these limits. It is assumed that

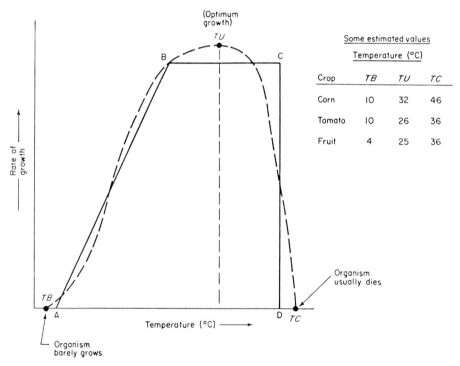

FIGURE 4-13 Influence of temperature on the growth of organisms (dashed line). Equations are used to approximate the curve; the straight lines *AB-BC-CD* are examples of simplifying the curve into easy-to-use models. (Courtesy of Raymond W. Miller, Utah State University.)

Problem Calculate the growing degree days for these two days using the equation

$$GDD = \frac{T_{max} + T_{min}}{2} - 10°C = \frac{T_{max} + T_{min}}{2} - 50°F$$

Day	Maximu m Temperature	Minimum Temperature
1	23°C (73.4°F)	9°C (48.2°F)
2	32°C (89.6°F)	13°C (55.4°F)

Solution Substituting numbers in the equation, day 1[a] is

$$GDD = \frac{23 + 10}{2} - 10 = 6.5 \text{ GDD}$$

For day 2[b],

$$GDD = \frac{30 + 13}{2} - 10 = 11.5 \text{ GDD}$$

The two days furnish a total of 18 GDD.

[a] In day 1, the minimum temperature was actually 9°C, so the least value for minimum in the equation, 10, was used.
[b] In day 2, the maximum temperature exceeds the maximum of 30°C used in the equation, so 30 is used rather than 32.

plant growth is not benefited beyond these temperature extremes. Calculation 4-4 presents a calculation. The equation for GDD is:

$$\mathbf{GDD} = \left[\frac{T_{max} + T_{min}}{2} - 10°C \right]_{daily}$$

Energy requirements for plant or animal development can be determined under controlled laboratory conditions and these values used in predicting planting times, harvesting times, or suitable plant varieties for particular areas where average annual growing days have been measured (Table 4-7). Most values have been in °F units.

The equations are calculated for each hour of the day, and the development stages are given in growing degree hours. With such detail it has been possible to predict growth stages within 1 or 2 days of actual development measured or observed.

4:10.6 Using Temperature: Average Annual Minimum Temperatures

Roots of most plants will not start growing until the soil temperature around them is 4.4°C (40°F) or higher. By contrast, their tops require 10°C (50°F) or higher to grow. Therefore, early planting of dormant perennials encourages roots to become well established before leaves require water and nutrients from roots growing in the soil.

Table 4-7 Average Annual Growing Degree Days for Various States and the Average Required Growing Degree Days for Selected Corn Varieties (Based on Fahrenheit Temperature)

Location	Average Annual GDD[a]	Corn Variety	Required GDD
Itasca County, Minnesota	1175		
East Lansing, Michigan	2122	Northrup King A08	1950–2050
Logan, Utah	2275	DeKalb XL 311	2050–2150
Ames, Iowa	2465	Northrup King Px20	2150–2300
Monmouth, Illinois	2651	Pioneer 3780	2300–2400
Columbia, Missouri	2844	Northrup King Px610	2400–2500
Manhattan, Kansas	2938	Northrup King Px74	2500–2650
St. George, Utah	3035		

Sources: R. H. Shaw, "Growing-Degree Units for Corn in the North Central Region," North Central Regional Research Publication 229 and Iowa State Experiment Station Research Bulletin 581, Ames, Iowa, 1975, pp. 795–807. R. E. Neild and M. W. Seeley, "Growing Degree Days Predictions for Corn and Sorghum Development and Some Applications to Crop Production in Nebraska," Nebraska Agricultural Experiemnt Station Research Bulleting 280, 1977, pp. 1–11. Personal communication with Dr. DeVere McAllister, former Extension Specialist on corn production, Utah State University.
[a] Between average frost-free dates, 100 GDD may require 4 or 5 summer days or 10–20 cooler fall days.

Plant tolerance to cold is extremely variable. Among the most tolerant are species of willow and trembling aspen that grow north of the Arctic Circle, where temperatures may reach −46°C (−50°F). By contrast, banana, coconut, and pineapple will not tolerate freezing (0°C, 32°F).

A generalized minimum temperature map, designated as "zones of plant hardiness," has been developed especially to help people in selecting the temperature tolerance of species new to an area (Fig. 4-14). All first-class nurseries and seed houses use this map in their catalogs.

4:11 Other Soil Physical Properties

Additional soil physical properties include plasticity, stickiness, smeariness, and fluidity. **Plasticity** is the degree to which soil is permanently deformed, without rupture, by a force applied continuously in any direction. Plasticity is determined in the field by rolling wet soil between the hands until a 3-mm (⅛-in.) cylinder is formed. With continuous rolling while it dries, this 3-mm "wire" will break when the soil has dried to approximately the *plastic limit.*

Stickiness means the property of wet soil to adhere to another object.

Smeariness is a field term for clays that are thixotropic. **Thixotropy** is the property exhibited by certain fine-textured clay gels of being "solid" upon standing but turning liquid when shaken or mixed. Many swelling clays used in well-drilling muds to lubricate the drill and lift cuttings to the surface are thixotropic.

Fluidity refers to nonthixotropic clay soils that flow under hand pressure.

The light in the world comes principally from two sources—the sun, and the student's lamp.

—Bovee

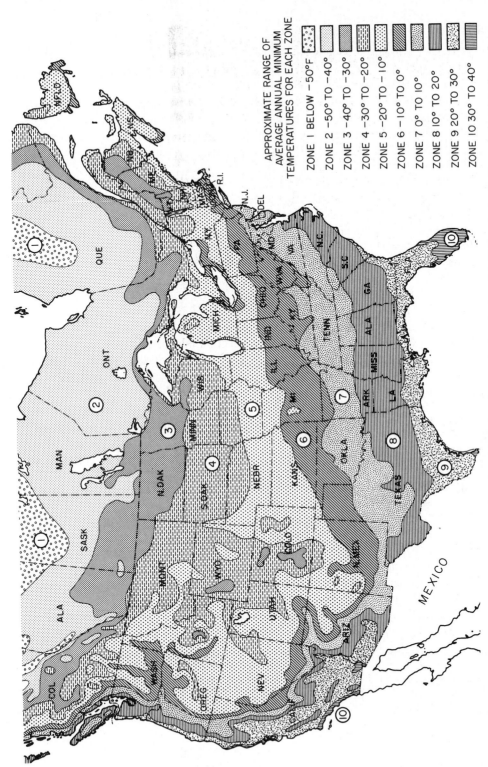

FIGURE 4-14 Mean minimum temperature ranges for the coldest months are designated on this U.S. map as "zones of plant hardiness." Most nurseries reproduce this map in their catalogs and, correspondingly, indicate the zone(s) of hardiness of each plant offered for sale. (*Source:* Agricultural Research Service Miscellaneous Publication 814, USDA, Washington, D.C., 1965.)

APPROXIMATE RANGE OF
AVERAGE ANNUAL MINIMUM
TEMPERATURES FOR EACH ZONE

ZONE 1 BELOW $-50°F$
ZONE 2 $-50°$ TO $-40°$
ZONE 3 $-40°$ TO $-30°$
ZONE 4 $-30°$ TO $-20°$
ZONE 5 $-20°$ TO $-10°$
ZONE 6 $-10°$ TO $0°$
ZONE 7 $0°$ TO $10°$
ZONE 8 $10°$ TO $20°$
ZONE 9 $20°$ TO $30°$
ZONE 10 $30°$ TO $40°$

Questions

1. Using the textural triangle, determine the texture for these three soils: (1) 10% sand, 15% clay, and 75% silt; (2) 41% clay, 40% silt, and 19% sand; and (3) 16% clay, 28% silt, and 56% sand.
2. State how the following are altered, in general, by increased clay content; (1) water infiltration, (2) pore size, (3) total pore space, and (4) bulk density.
3. In considerable detail (giving particle sizes, for example), state what each of these names tells about that soil: (1) a gravelly loam, (2) a very cobbly loamy sand, and (3) an extremely stony sandy loam.
4. Draw to actual size a ped of (1) a coarse columnar structure and (2) a coarse granular structure. Do not memorize exact dimensions, but have an idea of sizes.
5. (a) Which usually has the higher bulk density, a loamy sand or a clay loam? (b) Which of these soils has the greater total pore space?
6. (a) How does soil air differ from atmospheric air? (b) What are the critical gases in the soil air for root growth?
7. The revised *Soil Survey Manual* uses the term *soil strength*. What is measured to indicate soil strength?
8. Of these colors—5YR 2/2, 5YR 6/2, 5YR 3/6, and 5Y 6/2—state which is (1) the blackest, (2) the least reddish, and (3) the most brilliant colored.
9. What are some of the practical uses of soil color information?
10. If daily temperature of the air changes 30°F from its coldest to warmest, what amount of daily temperature change might be expected in soils at a depth of (1) 2 cm, (2) 10 cm, (3) 30 cm, and (4) 60 cm?
11. State how each of these affects the warming or the cooling of soils: (1) transparent-plastic mulches, (2) opaque-plastic mulches, (3) organic mulches, (4) water content in the soil, and (5) soil color.
12. List three examples of how soil temperature has been modified to improve crop production, other than using greenhouses or growth chambers.
13. Give an example of the growing degree days (GDD) equation and explain several of its practical uses.
14. What is meant by "plant hardiness zones"? Why are these maps used?
15. Explain how and why large bodies of water modify nearby land area.
16. How is energy lost from the earth so that incoming sun's energy does not gradually heat the earth to higher and higher temperatures?
17. How are freezing–thawing cycles major controls on heat flow in the immediate vicinity (within a few centimeters or feet)?
18. Explain the general observation that in Northern land areas (northern U.S.), organic mulches (plant residues) left on soils (a) cause the soils to warm more slowly in the spring and (b) to be generally a "cold soil" in the spring.
19. By what mechanisms do soils lose heat to the surrounding atmosphere?
20. In tropical countries with similar year-round temperatures, it is often said that the "rainy season" is their "winter." Explain.

Soil Colloids and Chemical Properties

Clay minerals form by fixation of free silica onto a layered hydroxide that is in the process of development. Suitable conditions, therefore, are necessary so that the cation or cations can precipitate with hexacoordination in the form of layered hydroxides.

—*G. Millot*

5:1 Preview and Important Facts

PREVIEW

The chemical properties of soils are determined by the colloids of soils. What is a colloid? What are the colloids in soils? A **colloid** is any solid substance whose particles are very small; thus, its surface properties are relatively more important than its mass (weight). Most colloids are smaller than a few *micrometers* (microns) in diameter. In Greek, colloid meant *glue*. The small particles, with their enormous surface-to-surface contact, would "stick things together."

Because of their large surface-to-mass ratio, soil colloids settle slowly from suspension. The predominant soil colloids are soil **clays** and **humus.** Humus is a mixture of residues left after partial decay of organic substances in and on top of soils. Chemical properties of soils are more influenced by the clays and humus than by equal weights of the larger silt and sand particles.

The chemical properties of soils include mineral solubility, nutrient availability, soil reaction (pH), cation exchange, and buffering action. Clays have negatively charged sites in their lattices and attract and hold positively charged ions (cations) at the clay surface. The quantity of these cations that can be held, or exchanged, by a given amount of soil is the **cation exchange capacity** of that soil. The exchangeable cations are not easily removed by leaching action until they are replaced (exchanged) by other cations. Plants often excrete hydrogen ions (H^+) to the soil solution, which can replace nutrient cations from their exchange sites, thus allowing the nutrients to move to the plant root and be absorbed.

If the adsorbed ions on the colloids are mostly $Al(OH)_2^+$ and some H^+, the soils are acidic. The pH is a measure of acidity or alkalinity: **soil pH** ranges from about pH 3.5 (extremely acidic) to pH 7 (neutral) to about pH 11 (extremely alkaline [or basic]). Plants grow well between pH 5.5

and pH 8.5. Strongly acidic soils are undesirable because soluble aluminum and manganese can reach toxic levels and microbial activity is greatly reduced. Strongly alkaline (basic) soils have lower micronutrient availability, except for boron, chloride, and molybdenum; iron, zinc, manganese, and macronutrient phosphorus may be deficient. Colloids dominate the soil's physical and chemical properties more than their percentage by weight would suggest.

IMPORTANT FACTS TO KNOW

1. The definition, identity, and properties of "soil colloids"
2. The contrasting properties of montmorillonite, kaolinite, and metal oxides and the climates in which each becomes the dominant clay
3. The origins and general properties of "clay minerals"
4. The nature of the "active groups" in humus colloids and the variation in organic colloid composition
5. The causes and action of cation exchange and the soil ions mostly involved in cation exchange in soils
6. The meaning of soil pH, soil acidity, and the ions causing acidity
7. Some general amounts of calcium and potassium held on the cation exchange sites per hectare–30 cm deep
8. The effect of the soil cation exchange capacity (CEC) on the amount of lime needed and on the available potassium, magnesium, and calcium held in soils
9. The meaning of "basic cation saturation percentage" (BCSP)

5:2 Soil Clays

Clays, the active mineral portion of soils, are colloidal, and most clays are crystalline. The term *clay* has three meanings in soil usage: (1) it is a particle-size fraction composed of any mineral particles less than 2 micrometers (microns)[1] in effective diameter, (2) it is a name for a group of minerals of specific composition, and (3) it is a soil textural class. Many materials of the clay-size fraction—such as gypsum, carbonates, or quartz—are small enough to be classified as clay on the basis of size but are not clay minerals. In contrast, some clay mineral particles may reach sizes of 4 or 5 μm, double the upper size limit of the clay-size fraction. Clay, as discussed in this chapter, includes only *clay minerals* of specific crystalline or amorphous composition and of any size up to several micrometers (Fig. 5-1).

5:2.1 The Origin of Clays

Prior to the x-ray study of mineral compositions, clays were incorrectly thought to be just smaller particles of primary minerals, such as small particles of quartz, feldspars, micas, hornblende, or augite. Now clay minerals are known to have specific compositions that are not very similar to the primary minerals, except micas. **Clay minerals** are mostly newly formed crystals re-formed from the soluble products of the primary minerals and are *secondary* minerals. Laboratory syntheses of clays have proven that the kind of clay formed is determined by the proportions of the different ions in the solution during formation. Removal of some of the soluble products by leaching will reduce clay formation rates and alter the kind of clay formed. Soils of regions that have hot, moist climates, but are not leached excessively because of poor drainage, have large amounts of the primary minerals dissolved, which then crystallize to clays. These hot, humid, tropical soils tend to be high in percentage clay, even to depths of 5–20 m (16.4–65.5 ft). Other nearby soils may have large portions of

---

[1] One micrometer (μm), or micron, equals 0.001 mm; 1.0 mm equals 0.04 in. Thus, 25,400 μm equals 1.0 in.

FIGURE 5-1 Microscopic evidence that primary minerals and secondary (clay) minerals are crystalline is portrayed here. The regular shapes indicate definite crystal patterns. The large crystals are mica, a primary mineral (1), and the small crystals are kaolinite, a secondary clay mineral (arrow). Mexico. (Magnification 26,300 times.)   (Courtesy of R. L. Sloane, University of Arizona.)

the weathered primary minerals flushed away. Some clays apparently form from slight alteration (selective solubility and rebuilding) of some primary minerals, particularly from the micas (biotite and muscovite), to form vermiculite and hydrous mica.

The composition of clays in a given soil can be a complex mixture of clays from various origins. Clays have been forming for thousands of centuries in some soils. Erosion moves some of those clays and redeposits them as alluvium in oceans and as sediments on land. Thus, a soil may have clays from several origins:

1. **Inherited clays,** deposited in sediments that are now forming a soil. Perhaps the clays were formed in a different climate eons of time ago.

2. **Modified clays,** changed by further weathering (degradation) of moved and deposited clays.

3. **Transformed clays,** (aggradation) accumulated from highly weathered clays deposited in alluvium that now provide soluble silica and other clay-forming materials.

4. **Neoformed clays,** new clays formed entirely from crystallization of clays from soluble ions in solution from the weathering minerals of the developing soil.

5:2.2 The Nature of Clays

Most clays are crystalline; they have definite, repeating arrangements of atoms of which they are composed. The majority of clays are made up of planes of oxygen atoms with silicon and aluminum atoms holding the oxygens together by **ionic bonding,** which is the attraction of positively and negatively charged atoms (Fig. 5-2).

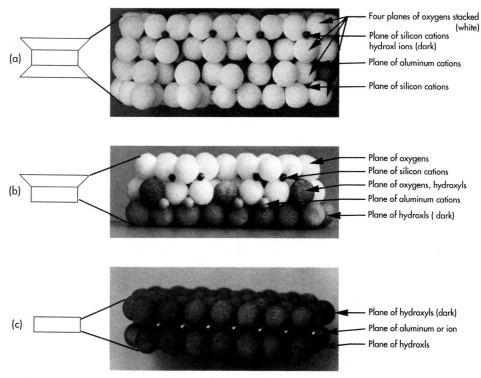

FIGURE 5-2 Representation of the basic orientation of atoms in clay minerals. The unit lattice of (a), typified by montmorillonite, has four planes of mostly oxygen ions with positively charged silicon (Si^{4+}) or of aluminum (Al^{3+}), mostly in the small spaces between oxygen ions, holding the oxygen ions together. Where only three oxygen ion planes occur (b), as in kaolinite, one plane is hydroxyls (OH^-, dark-colored gray). Many of the aluminum and iron clays (sesquioxides) have only planes of hydroxyls (c) or a mixture with oxygen ions and hydroxyl ions, which are about the same size. Clay minerals are mostly oxygen ions (70–85% by volume). In addition to aluminum and silica, the ions of iron, zinc, magnesium, and potassium occur in the structural lattice of some clays (See Fig. 5-5). (Courtesy of Raymond W. Miller, Utah State University.)

Three or four planes of oxygen atoms with intervening silicon and aluminum ions (or other cations depending upon the clay) make up a **layer.** One clay particle is composed of many layers stacked like a deck of cards. A clay particle is called a *micelle.* A few clays have the oxygen and other atoms regularly oriented only in small sections and are called *amorphous* materials. Structural details are given in Detail 5-1.

Clays have a net negative charge, which will attract and hold positively charged ions (**cations**).[2] Common soil cations include potassium (K^+), sodium (Na^+), ammonium (NH_4^+) calcium (Ca^{2+}) magnesium (Mg^{2+}), hydrogen (H^+), aluminum dihydroxide ($Al[OH]_2^+$) and aluminum (Al^{3+}). The amounts of these positive cations held by clays vary with the kind of clay. Plant roots can use some exchangeable cations as nutrients. Leaching by rainfall over several years removes exchangeable basic cations replaced by H^+ or $Al(OH)_2^+$ ions. Exchangeable cations must be replaced by *other cations as they are removed* (exchanged).

---

[2] Atoms react with other atoms to form compounds through interactions of their electrons. The tendency for an element to attract or to give up electrons is indicated by "+" or "–" symbols. If the element has lost or lacks some electrons, it will have a "+ charge" indicating the number of electrons in which it is deficient. The positively charged ions are attracted to negatively charged ions or sites. Any atom or group of atoms with "charge" is an *ion.* Positively charged ions are cations (pronounced cat'-eye-ons); negatively charged ions are anions (an'-eye-ons).

When a cation such as Si^{4+} or Al^{3+} is surrounded in a "close fit" by oxygens and has planes fitted on the exposed surfaces, the names of the geometric shapes are used to indicate the spacial relationships. Thus, the Si^{4+} surrounded by 4 oxygens forms a tetrahedron (4 sides when planes cover the surfaces) and the unit has **tetrahedral** coordination. The 6 oxygens needed to fit snugly around the larger Al^{3+} forms an octahedron (8 sided) and the Al^{3+} is in **octahedral** coordination.

Si hidden in the interstice

Al hidden in larger interstice

(a)

In Fig. 5-2 the planes of oxygen held together by Si^{4+} are tetrahedrally oriented and are referred to as a **(silica) tetrahedral sheet.** One of these planes of oxygen plus a second plane of oxygens (or hydroxyls) are held together by Al^{3+}, are octahedrally oriented, and are referred to as the **(alumina) octahedral sheet.** One silica sheet per one alumina sheet is a 1:1 lattice. With 2 silica sheets per 1 alumina sheet, the lattice is a 2:1 type. Vermiculite and chlorite can be referred to as 2:1:1 or 2:2 lattice types.

For a different look at the structure, an expanded view of the "structure A" of Fig. 5-2 is shown below.

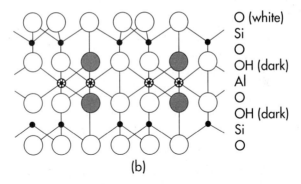

O (white)
Si
O
OH (dark)
Al
O
OH (dark)
Si
O

(b)

Other cations, weathering from minerals or hydrogen from the dissolving of carbon dioxide in water, are able to *exchange* for ions already on the exchange sites. (**Acidic cations** are mostly H^+, the hydroxy aluminum ions and aluminum Al^{3+} ions).

The nomenclature of clays is complex, and no single complete system is agreed upon by clay mineralogists. Generally, the outline in Table 5-1 illustrates the kinds of criteria used to group clays; clays are indicated for some of the categories. The actual system is much more detailed and extensive than shown in Table 5-1. *Kinds* of octahedral ions (Al vs. Mg) is indicated by *dioctahedral* (2 Al for each 3 possible sites) and *trioctahedral* (3 Mg for each 3 possible sites). See Detail 5-2 for chemical formulas.

The common clays and some of their properties are listed in Table 5-2. The different clay structures are shown diagramatically in Fig. 5-3.

Table 5-1 Brief Outline of Clay Minerals

1. **Amorphous** (nonoriented, small crystal units)
   a. Alumino-silicates: **allophane, imogolite**
   b. Iron hydrous oxides: **ferrihydrites**
   c. Silica: **opal, glass**

2. **Crystalline Layer Silicates**
   a. *Kandite group,* 1:1 lattice (one tetrahedral, one octahedral sheet per layer)
      i. Alumino-silicate: **kaolinite**
      ii. Magnesium silicate: **serpentine**
      iii. Water layers between kaolinite sheets**: halloysite**
   b. *Smectite group,* 2:1 lattice (two tetrahedral, one octahedral sheet per layer)
      i. No isomorphous substitution
         a. Alumino-silicate: **pyrophyllite**
         b. Magnesium-silicate: **talc**
      ii. Various isomorphous substitutions (swelling)
         a. Alumino-silicate: **montmorillonite**
         b. Magnesium-silicate: **saponite**
   c. *Hydrous mica group,* 2:1 lattice, K between layers
      i. Alumino-silicate: **illite** or **hydrous mica**
      ii. Swelling mica-like clays: **vermiculite**
   d. *Chlorite group,* 2:2 or 2:1:1 lattice (two tetrahedral and two octahedral sheets of different composition)
      i. Many **chlorites,** mostly nonswelling clays

3. **Crystalline Chain Silicates**
   a. Alumino-magnesium silicates, fibrous: **attapulgite**

4. **Crystalline Iron, Aluminum, and Titanium Oxides**
   a. *Metal oxides of iron:* **goethite** [FeOOH]
   b. *Metal oxides of aluminum:* **gibbsite** [Al (OH)$_3$]
      boehmite [AlOOH]
   c. *Metal oxide of titanium:* **anatase** [TiO$_2$]

Table 5-2 Common Groups of Soil Clays, Their Relative Swelling in Water, and Environments in Which They Are Likely to Be the Predominant Clay in the Soil

Clay Groups and Prominent Elements	Relative Swelling When Wetted	Relative Stickiness	Environment Where the Clay Group Is Predominant
Smectite[a] (O, Si, Al)	High	High	Arid to humid soils having limited leaching
Hydrous mica (illite) (O, Si, Al, K)	Low	Low	Subhumid and cool areas, parent rock containing micas
Vermiculite (O, Si, Al, Mg)	High	Moderate	Subhumid to humid soils high in micas
Chlorite (O, Si, Al, K, Mg, Fe)	None	None	Clays formed in marine sediments, now uplifted and exposed to weathering
Kaolinite (O, Al, Si)	Almost none	Slight	Moist, warm to hot, subhumid and humid leached soils
Sesquioxides (metal oxides) (O, Fe, Al)	None	None	Wet, hot, excessively weathered, old soils of tropics
Amorphous (allophane) (O, Al, Si)	None	Slight	In rapidly weathering, young volcanic ash

[a]*Smectite* is a group name for several clays; the best known is montmorillonite.

Detail 5-2 Composition of Clays

Except for a few amorphous clays (*allophane, imogolite, ferrihydrite,* others), soil clays are crystalline, with compositions that can be described by chemical formulas, as is shown in the following. In the formulas, the *M* shown in brackets at the end of some formulas represents the charges of *adsorbing cations* that are needed to neutralize charges caused by isomorphous substitution. The subscripts *x, y,* and *z* indicate variable amounts of substitution, with each letter indicating a particular numerical fraction less than 1.0.

Kaolinite	$Al_2 (Si_4O_{10}) (OH)_8$
Pyrophyllite	$Al_2 (Si_4O_{10}) (OH)_2$
Muscovite	$KAl_2 (AlSi_3O_{10}) (OH)_2$
Illite	$K(Al_{2-x-y}Fe_xMg_y) O_{10} (Si_{4-z}Al_z) (OH)_2$ $\cdot [x+y+z = M]$
Talc	$Mg_3 (Si_4O_{10}) (OH)_2$

Montmorillonite	$(Al_{2-x}Mg_x) (Si_{4-y}Al_y) O_{10}(OH)_2$ $\cdot [x+y = M]$
Saponite	$Mg_6(Si_{8-x}Al_x)O_{20}(OH)_4$
Chlorite	$Mg_3(Si_{4-x}Al_x)O_{10}(OH)_8(Mg_{6-x-y}Al_xFe_y)$
Vermiculite	$(Mg_{3-x-y}Al_xFe_y) (Si_{4-z}Al_zO_{10}) (OH)_2$ (4.5 HOH)
Sesquioxides	$Fe_4O_{6-x}(OH)_x$ and $Al_4O_{6-x}(OH)_x$

These formulas illustrate the dominance of oxygen in minerals and its bonding to aluminum and silicon. Common "substituting" ions include Al^{3+} for Si^{4+}, and Fe^{3+}, Fe^{2+}, Zn^{2+}, and Mg^{2+} for Al^{3+}. Particular clays will vary from the "example" compositions shown above. Clays not listed above include Allophane, an amorphous aluminum silicate; imogolite, a microcrystalline allophane; and ferrihydrites, amorphous iron hydrous oxides.

5:2.3 The Charge on Clays

Why do clays have a charge? The charge in clays comes from ionizable hydrogen ions and from isomorphous substitution. **Ionizable hydrogen ions** are hydrogens from hydroxyl ions on clay surfaces. The –Al—OH or –Si—OH portion of the clay ionizes the H and leaves an unneutralized negative charge on the oxygen (–Al—O⁻ or –Si—O⁻). The extent of ionized hydrogen depends on solution pH; more ionization occurs in more-alkaline (basic) solutions.

Isomorphous substitution, the second source of charge on clay particles, is the substitution of one ion for another of similar size and often with lower positive valence. In clay structures, certain ions fit into certain mineral lattice sites because of their convenient size and charge (see Table 5-3). Only during clay formation from soluble materials can other ions of similar *size* and *charge,* if they are present, occupy some of those sites. The Si^{4+} dominates in tetrahedral sites, and Al^{3+} dominates in octahedral sites. The Al^{3+} can replace some Si^{4+} cations, and some Al^{3+} is replaced by one or more of these: Fe^{3+}, Fe^{2+}, Mg^{2+}, or Zn^{2+}. These ions, in high concentrations in the solution, "substitute" because of similar size and + charge. Notice that most substitutions are by ions with lower charge (less positive) than the ones being replaced. Because the total negative charge from the anions (the oxygens) remains unchanged, the lower positive charge because of substitution results in an excess negative charge at that location in the structure.

The excess negative charge from isomorphous substitution attracts positive ions that are in the surrounding solution (see Fig. 5-4), but these ions do *not* become a part of the clay structure. These excess negative charges in the crystal lattice, caused by isomorphous substitution, are called **cation exchange sites.** These sites attract and hold cations somewhat loosely. Other cations in solution can compete with and often replace originally adsorbed cations. The total amount of the negative sites is referred to as the soil's **cation exchange capacity (CEC).** The cation exchange sites adsorb and hold mostly Ca^{2+}, Mg^{2+}, and K^+ in soils

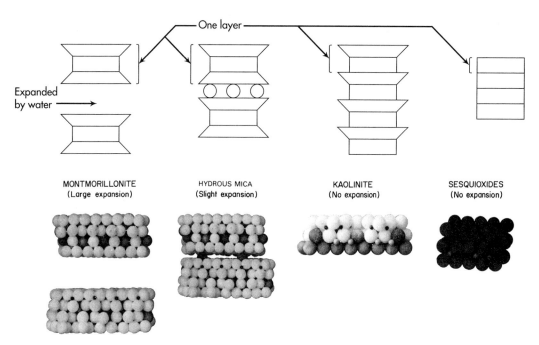

FIGURE 5-3 Schematic drawing of clay minerals. The shapes for sheets are identical to those given in Fig. 5-2. The lattice layers of montmorillonite within a single clay particle can be greatly expanded by water. In contrast, hydrous mica with strong bonding by potassium ions has limited swelling in some layers, probably where potassium has been removed by leaching. Kaolinite, the potters' clay, has no swelling. Sesquioxides also have low or no swelling, and they usually have irregularly stacked sheets, making them x-ray amorphous. The coordination is octahedral, as suggested in the photo, but the crystallinity is quite poor.   (Courtesy of Raymond W. Miller, Utah State University.)

of pH greater than about 6 or 7. In more-acidic soils, the sites are mostly occupied by $Al(OH)_2^+$ or Al^{3+} ions. Lesser amounts of other cations (Na^+, NH_4^+, Zn^{2+}) occur on some of the exchange sites.

Besides the charge deficit that causes cation exchange, the *amount* of cation exchange also depends on the surface of the colloid exposed to the soil solution. Ionic attraction for cations is effective only at distances of a few oxygens' thicknesses from the soil solution. Clays whose layers are spread apart to allow the soil solution to pass between clay layers have exchange sites also along this "internal" surface. These clays have weak bonding between layers so that water can spread the layers apart and cause swelling of the clay particle. Such clays are therefore unsuitable for supporting highways or buildings.

Other clays (kaolinite) have tightly bonded layers and do not swell when wet. Kaolinite is used to make pottery, tile, and other fired-clay items because it does not shrink, crack, or deform when kiln-baked. Kaolinite has a low cation exchange capacity with no internal surfaces exposed.

5:2.4 The Silicate Clays

If clay minerals are examined under a high-powered microscope, evidence of their crystalline structure can be seen (Fig. 5-5). Each crystalline clay is like a partial deck of magnetic cards—that is, the clay is thinner in one dimension than in the other two dimensions. Each card represents a layer, and the layer is nearly an exact replication of each other layer in that clay.

Table 5-3 Ionic Radii, Charge, and Coordination Numbers for Ions Common in the Crystal Lattice of Soil Clays[a]

Ionic Radii for Common Soil Ions (nm)

Nonhydrated				Hydrated	
Si^{4+}	0.042	Na^+	0.074	Na^+	0.790
Al^{3+}	0.051	Ca^{2+}	0.099	Ca^{2+}	0.960
Fe^{3+}	0.064	Mg^{2+}	0.066	Mg^{2+}	1.08
Fe^{2+}	0.074	K^+	0.133	K^+	0.530
O^{2-}	0.140				

Ionic Radii, Coordination Numbers, and Crystal Geometries

r_{Cation}/r_{Anion}	Coordination No.	Crystal Geometry
0.155–0.255	3	Trigonal planar
0.255–0.414	4	Tetrahedron
0.414–0.732	6	Octahedron
0.732–1.0	8	Close-fit lattice

[a] Based on nonhydrated sizes. Water molecules of hydration are stripped away when crystal lattices form. The "free" ions in water or soil solution will be "hydrated" and effectively be larger ions.

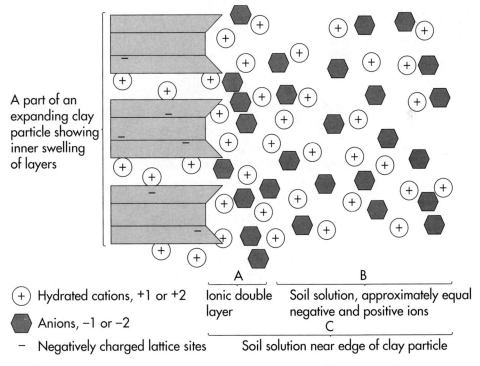

A part of an expanding clay particle showing inner swelling of layers

A — Ionic double layer
B — Soil solution, approximately equal negative and positive ions
C — Soil solution near edge of clay particle

(+) Hydrated cations, +1 or +2
⬢ Anions, −1 or −2
− Negatively charged lattice sites

FIGURE 5-4 Illustration of the movement, mostly of cations, to the outer surfaces and interlayer areas as might occur in montmorillonite, vermiculite, and some expanded interlayers of illite. Adding a new cation, such as ammonium fertilizer, puts ammonium ions on many of the adsorption sites. (Courtesy of Raymond W. Miller, Utah State University.)

(a)

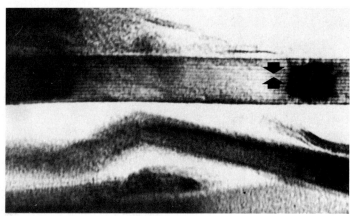

(b)

FIGURE 5-5 (a) This electron micrograph of Utah Dickite, a kaolinite-like clay mineral is magnified 2720 times. Each thin portion comprises many clay particles stacked. A soil sample would most often have the individual particles less oriented, like decks of cards thrown into a pile along with many large irregularly shaped silt and sand particles of all sizes. In (b), a magnification of over 700,000 times makes it possible to see "layers" of chlorite clay. The distance between arrows is about 1.4 nm (14 angstroms), the thickness of about 6 oxygen planes. Chlorite is a 2:2 or 2:1:1 sheet clay. (*Sources:* (a) B. F. Bohor and R. E. Hughes, "Scanning Electron Microscopy of Clays and Clay Minerals," *Clays and Clay Minerals,* **19** (1971), pp. 49–54; (b) J. L. Brown and M. L. Jackson, "Chlorite Examination by Ultramicrotomy and High Resolution Electron Microscopy," *Clays and Clay Minerals,* **21** (1973), pp. 1–7.)

The many cards adhere together to make up a clay particle called a **micelle.** Each layer is made up of two oxygen sheets in some clays, of three or four in others, and of six in still others (chlorites). A few hydroxyl ions (OH^-), which are almost the same size as oxygen ions, replace some of the oxygen ion sites. In some clays up to one-fourth of the silicon ion positions are substituted with aluminum. Similarly, other ions—such as magnesium, zinc, and iron—with atomic sizes similar to that of aluminum may fit into aluminum sites in place of aluminum.

The most common silicate clays in soils are categorized in the following paragraphs.

Amorphous silicate clays (allophane) are mixtures of silica and alumina that have not formed well-oriented crystals; they lack crystallinity. Even mixtures of other weathered oxides (iron oxide) may be a part of the mixture. Typically, these clays occur where large amounts of weathered products existed but have not had the conditions or time for good crystal growth. Amorphous clays are common in soils forming from volcanic ash. Amorphous clays are not well characterized but do exist in many soils in varying amounts. Their properties are often quite unusual, such as having high cation exchange capacity. Because almost all of their charge is from accessible hydroxyl ions (OH^-), which can attract a positive ion or lose the H^+ attached, these clays have a *variable charge* that depends on how much H^+ is in solution (the soil acidity).

Only the most intensively weathered soils have had extensive destruction of the kaolinite, washing away the soluble silica to leave the less soluble aluminum hydroxide clays and iron-oxide clays—the sesquioxide clays—as the dominant clays.

Kandites (kaolinite, nacrite, halloysite) are residues from extensive weathering in high-rainfall, acidic soils. Kaolinite, the most common clay of this group, has only one sheet of silica tetrahedra per sheet of alumina octahedra per layer. Thus, it is a 1:1 lattice clay. Almost no substitution of Al^{3+} for Si^{4+}, or Mg^{2+} for Al^{3+} has occurred in kaolinite, so the net negative charge (cation exchange capacity) is low. However, each layer has one plane of oxygens (O^{2-}) replaced by hydroxyls (OH^-), which results in strong hydrogen (—H—) bonds to the oxygen plane of the adjacent layer. Kaolinites have such strong hydrogen bonding that they do not allow water to penetrate between the layers and have almost no swelling. These are the types of clays used for pottery work because they do not shrink and swell.

Kaolinite is predominant in the clay fraction in acidic, humid, warm, well-drained soils, such as in the southeastern United States and in the humid soils in the tropics and subtropics, if they are well drained and have weathered for several hundred thousand years.

Halloysite is similar to kaolinite but has water layers between kaolinite layers. Thus, halloysite curls and forms tubelike particles.

Acidic soils occur where rainfall is relatively high. As minerals weather, much of the silica and basic cations are dissolved and slowly leached from the profile. Less silica present in the remaining solution means the proportion of alumina will be high enough to result in kaolinite formation rather than the higher silica-containing montmorillonite.

Smectites (montmorillonite, saponite) are the swelling, sticky clays.[3] They are often referred to as 2:1 lattice, or expanding lattice, clays. The 2:1 refers to the number of silica sheets per alumina sheet per clay layer. In montorillonites, water easily penetrates between planes of adjacent oxygens, causing the individual particles of clay to swell. If the solution has mostly sodium as the cations, the clay may swell 3 to 10 times its dry volume and become like a gelatinous mixture. **Bentonite,** an impure deposit of montmorillonite, is used to seal earthen ponds (see Fig. 5-6), to spread water over fires when the gelatinous slurry is dropped by planes, to act as solution stiffeners or gels in well-drilling muds, and to act as thickeners in paints and lipstick. Bentonite refers to any deposit of highly colloidal clay.

---

[3] The clay called *montmorillonite* was first described near Montmorillon, a town in France. Its name came from the town. *Illite,* which in this text is referred to as *hydrous mica,* was named by geologists from the Illinois Geological Survey. The term *kaolinite* comes from the Chinese *kauling,* meaning "high ridge," the name of the hill from which the first kaolinite was shipped to Europe for making fine "china" dishes.

(a)

(b)

FIGURE 5-6   (a) Pond near Winthrop, Washington, built on Owhi fine sandy loam, a porous soil. The pond is "sealed" with a high-sodium montmorillonite clay (bentonite) that has swelled and now seals all pores, inhibiting much water loss. The pond was 15 years old at the time the European swans and peafowl were photographed. (b). Nonswelling clays can be used for bricks, tiles, and ceramics (china dinnerware and vases). Kaolinitic clays are found in tropical climates where weathering has been long and intensive, as in this site in Guatemala. The clay is molded, dried, and sometimes fired to harden the bricks or tile.   (*Source:* (a) USDA—Soil Conservation Service, photo by William Brewster; (b) Raymond W. Miller, Utah State University.)

Montmorillonite is most common in soils that have had little or no leaching. Leaching removes relatively more silica (SiO_2- hydrated) than it does alumina [$Al(OH)_3$]. Thus, leaching (water moving down through the soil) reduces the relative amount of silica left that is needed for montmorillonite clay formation. Soils of the arid regions, poorly drained soils, and soil developed from alkaline parent rocks such as limestone have mostly montmorillonite clays.

Hydrous mica and **illite** are obsolete terms but without terms to replace them. *Mica-like* is sometimes used. For lack of replacements and because illite and hydrous mica are still extensively used, both names will be used in this text. **Hydrous micas** have a gross structure similar to that of montmorillonite—a 2:1 lattice clay of silica and alumina sheets. However, it also has large potassium ions holding adjacent layers together so tightly that water cannot penetrate between layers. Thus, hydrous mica has slight to moderate swelling, depending upon how many of the planes of potassium ions have been weathered out, allowing some clay layers to be separated and the clay to expand, somewhat like montmorillonite.

Because hydrous mica is quite similar to the structure of the primary micas, it is believed to form by limited alteration of primary mica. Hydrous mica is found in soils still high in primary minerals (not extensively weathered). Both hydrous mica and montmorillonite may occur in similar slightly leached environments.

Vermiculite clay minerals are similar in structure to hydrous mica but have lost the interlayer potassium ions. (The Latin *vermiculari* means "to breed worms," from the curled wormlike shapes formed upon heating and expanding vermiculite.) Vermiculite has the layers held weakly together by hydrated magnesium (6 water molecules in octahedral coordination with Mg^{2+}), less tightly together than occurs with potassium in hydrous mica. Thus, vermiculite has swelling but not as much as montmorillonite; it has a high cation exchange capacity.

The **chlorite** group of clays are common in some soils. They are often called 2:2 lattice (or 2:1:1 lattice) clays because they are similar to the unit lattice of vermiculite, except the hydrated Mg in chlorite is a firmly bonded, complete, magnesium hydroxide, octahedral sheet. Thus, a layer of chlorite has 2 silica tetrahedra, an alumina octahedra, and a magnesium octahedra sheet (2:2 or 2:1:1). Chlorites do not swell when wetted and have low cation exchange capacities.

5:2.5 Sesquioxide Clays (Metal Oxide and Hydrous Oxides)

Under conditions of extensive leaching by rainfall and long-time intensive weathering of minerals in humid, warm climates, most of the silica and much of the alumina in primary minerals are dissolved and slowly leached away. The remnant materials, which have lower solubilities, are sesquioxides. **Sesquioxides (metal oxides)** are mixtures of aluminum hydroxide, $Al(OH)_3$, iron oxide, Fe_2O_3, or iron hydroxide, $Fe(OH)_3$. (The Latin word *sesqui* means "one and one-half times.") Sesquioxides refer to the clays of iron and aluminum because their formulas can be written $Al_2O_3 \cdot xH_2O$ and $Fe_2O_3 \cdot xH_2O$, one and one-half times more moles of oxygen than of Al or Fe. However, TiO_2 and MnO_2 are also often included. These clays can grade from amorphous to crystalline.

Small amounts of these clays exist in many soils, even in soils with a relatively low extent of weathering. However, they will be the predominant clay and occur in large percentages only in soils formed in the humid, hot, well-drained soils of tropical areas where intense weathering has occurred for perhaps hundreds of thousands of years. Iron oxide and iron hydrate commonly color the soils various shades of red to yellow, respectively.

Sesquioxide clays do not swell, are not sticky, and have many differences from the silicate clays. The sesquioxides coat large particles and form stable aggregates. Soils with

30–40% sesquioxide clays may absorb water almost as if they were fine sands. During World War II in some tropical islands of the Pacific, vehicles moved through muds nearly hub deep, something quite impossible in sticky, montmorillonite clays. Mixtures of sesquioxides and kaolinite often have low stickiness. The high percentage of iron and aluminum hydrous oxides furnishes an enormous surface for adsorption of phosphorus. Insoluble phosphates form with soluble iron and aluminum ions. Such soils have a high phosphorus adsorption capacity. This often results in lower phosphorus efficiency of added fertilizers.

5:2.6 Relationship of Kinds of Clays to Climate

Mineral weathering sequences show the likelihood of finding certain clays in various environments. The most resistant products of mineral weathering are those that are least soluble, such as the iron oxides. From the *more soluble* to the *least soluble,* the clay-forming constituents are: (1) hydrated basic cations, (2) silica, (3) aluminum hydrous oxides, and (4) iron hydrous oxides.

The first clays to form from weathering products in solution would be the *amorphous* silicate clays, such as **allophane.** With increased wetting (solubilization) and drying cycles (reprecipitation and crystal growth), crystalline clays increase in amount. During formation, the kind of clay that forms depends on the relative amounts of silica, alumina, iron hydroxides, cations, solution pH, and other soluble materials. Thus, arid regions, because of minimal leaching and minimal weathering, form **smectites (montmorillonite clays)** and **vermiculites.** Some clay minerals are very similar to certain primary minerals and may be formed by partial solubilization and/or other slight structural modification. Micas (a primary mineral) seem to weather to produce **hydrous mica** and/or **vermiculite clays,** which have similar structures. If the area has more rainfall, relatively more silica will leach away, the pH will be more acidic, and **kaolinitic clays** will form.

The long yearround weathering periods in wet, warm climates in the tropics "decompose" most of the *original* or *primary* minerals. Most of the basic cations, silica, and some alumina are leached out of the upper soil profile. The residues left are high in alumina and particularly high in iron hydrous oxides. The clays that form are abundant in **metal oxides,** mostly **sesquioxides.** Some kaolinite forms, also. A summary of clays is given in Table 5-4.

5:3 Organic Colloids (Humus)

Humus is a temporary intermediate product left after considerable decomposition of plant and animal remains—temporary because the organic substances remaining continue to decompose slowly. The humus is often referred to as an organic colloid; it consists of various chains and loops of linked carbon atoms.

Humus is amorphous, dark brown to black, nearly insoluble in water, but mostly soluble in dilute alkali (NaOH or KOH) solutions (Fig. 5-7). It contains about 30% each of the nitrogen-rich proteins, the slow-to-decompose lignins, and complex sugars (polyuronides). The polyuronides comprise much of the varied organic substances that cement soil aggregates together. Humus has about 50% carbon, some oxygen, 5% nitrogen, and lesser amounts of sulfur, phosphorus, and other elements. It also has a cation exchange capacity on a dryweight basis many times greater than that of clay colloids. On a weight basis, a few percent humus exerts much greater influence than does several times more percent clay, especially for properties such as the cation exchange capacity.

Table 5-4 Summary of Some Common Soil Clays, Bonding between Layers, Range of Charge on the Accessible Clay Surfaces, and Major Conditions Necessary for Formation of Various Clays

Clay Group	Ratio Tetrahedral: Octahedral[a]	Bonding Strength between Layers and Swelling[b]	Platelet Charge (CEC in $cmol_c$ kg^{-1})	Origin of and Conditions Necessary for Clay Formation
Crystalline Silicate Clays				
Kaolinite	1:1	Strong bonding, structural–OH, no swelling	Low, 3–15	Crystallization in acidic soil where *some* silica and basic cations are leached
Hydrous mica (illite)	2:1	Strong bonding through K^+ ions, slight swelling	High but access blocked, 20–40	Alteration of micas or recrystallization from soluble products of weathered micas and vermiculites
Vermiculite	2:1	Moderate bonding, moderate expansion	High, 80–150	Alteration of micas by partial solubility and reformation
Smectites (montmorillonite)	2:1	Weak bonding, high swelling	High, 60–100	Crystallization from solutions; minimal leaching loss; arid, low permeability
Chlorite	2:1:1 or 2:2	Strong bonds, no swelling	Moderate to high but access blocked, 2–5	The conditions for kaolinite plus high Mg (sea-bottom muds, serpentine rock weathering products, other)
Amorphous Silicate Clays				
Allophane	Amorphous	Amorphous	High, 50–150	Rapid solubility (large surface area and soluble amorphous silica, as in volcanic ash), so precipitation is too fast to allow crystallization
Metal Oxide Clays: Amorphous and Crystalline				
Metal oxides (sesquioxides)	Octahedral coordination	No swelling, low negative and positive charges	Low, 0–3	Precipitation from soluble ions at the site where silica and cations are leached

(Courtesy of Raymond W. Miller, Utah State University.)

[a] Ratio of tetrahedral "sheets" per "layer" to octahedral sheets.
[b] Number of lower-charged ions in substituted sites and the nearness of the site to the layer surface. Charges originating from tetrahedral substitutions are stronger sites than are octahedral-substituted sites.

▬▬ 5:4 Cation Exchange

Plant growth is altered by modification of many chemical properties, such as soil acidity, availability of nutrients, and toxicities. These concerns involve cation exchange, soil acidification, and nutrient balance.

5:4.1 The Mechanisms of Cation Exchange

Any electrically charged colloidal surface area will attract mobile substances of opposite charge in the soil water. The dominant residual charge on most soil colloids is *negative*. These negatively charged sites attract positively charged ions in the soil water. Positively charged ions are called **cations.**

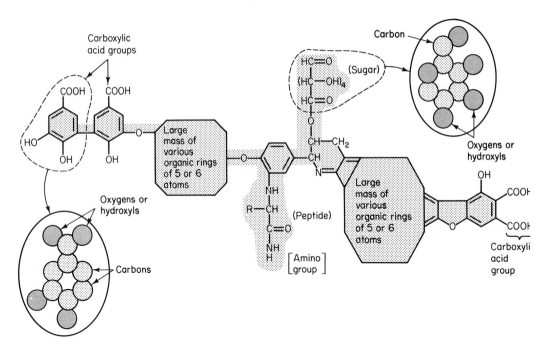

FIGURE 5-7 Schematic illustration of what a *portion* of a "molecule" of humus might look like. One molecule of humus would be from 10 to more than 100 times greater than the portion shown. Humus will be more porous and "stringy" than are clay particles. The humus can attach to clays through bonding, thus helping to hold small particles into larger masses (aggregates). All humus molecules will be different from each other and are constantly changing as they are attacked by microbes and further decomposed. Distinctive features of all humus are (1) the many "active" groups, mostly **carboxylic acid groups** and **hydroxyl (–OH) groups,** (2) the large portion of humus that consists of rings of carbon atoms, and (3) the appearance of humus as a very large, knobby cord folded around and over itself somewhat. (Courtesy of Raymond W. Miller, Utah State University.)

To illustrate cation exchange by an equation, represent the colloid (clay or humus) by any of several symbols, such as a rectangle, parallelogram, or a capital *X*. When the equation is completed, the total charges on the left side of the equation equal the charges on the right side.

$$2Na \begin{array}{|c|} \hline 12Ca \quad\quad 4Mg \\ \text{Colloid} \\ \hline \end{array} + 9KCl \rightleftharpoons \begin{array}{|c|} \hline 10Ca \quad\quad 3Mg \\ \text{Colloid} \\ \hline \end{array} + 2CaCl_2 + MgCl_2 + HCl$$

$$2K \quad 6[Al(OH)_2{}^+] \text{ (fertilizer)} \quad\quad 11K \quad 5[Al(OH)_2{}^+]$$

$$+ 2NaCl \quad \text{(in solution)}$$

$$+ Al(OH)_3^0 \quad \text{(precipitates)}$$

Adsorbed cations resist removal by leaching water but can be replaced (exchanged) by other cations in solution by **mass action** (competition for the negative site because of the large *number* of ions present). This exchange of one positive ion by another is called **cation exchange.** Thus, when a potassium fertilizer with its cation, K⁺, is added to soil, many of the numerous potassium ions replace other cations already adsorbed to the exchange sites. Cation exchange takes place on the surfaces of clay and humus colloids as well as on the surfaces of plant root cell walls.

The cations most numerous on exchange sites in soils are calcium (Ca^{2+}), magnesium (Mg^{2+}), hydrogen (H^+), sodium (Na^+), potassium (K^+), and aluminum (Al^{3+}).[4] The proportions of these cations on the colloid surfaces are constantly changing as ions are added from dissolving minerals or by additions of lime, gypsum, or fertilizers. Losses by plant absorption or by leaching also change cation proportions.

The strength of adsorption increases as (1) the valence of the cation increases, (2) the cation's hydrated size is smaller, and (3) the strength of the site's negative charge increases. Thus, adsorption strength of cations increases approximately in this order:

$$Na < K = NH_4 < Mg = Ca < Al(OH)_2 < H.$$

5:4.2 The Importance of Cation Exchange

Cation exchange is an important reaction in soil fertility, in causing and correcting soil acidity and basicity, in changes altering soil physical properties, and as a mechanism in purifying or altering percolating waters. *The plant nutrients calcium, magnesium, and potassium are supplied to plants in large measure from exchangeable forms.* In fact, the usual "soil test" to predict a soil's ability to furnish potassium to the plant is a measure of its exchangeable potassium content. Cation exchange is very important in soils because of the following relationships:

1. The exchangeable K is a major source of plant K.

2. The exchangeable Mg is often a major source of plant Mg.

3. The amount of lime required to raise the pH of an acidic soil is greater as the CEC is greater.

4. Cation exchange sites hold Ca^{2+}, Mg^{2+}, K^+, Na^+, and NH_4^+ ions and slow their losses by leaching.

5. Cation exchange sites hold fertilizer K^+ and NH_4^+ and greatly reduce their mobility in soils.

6. Cation exchange sites adsorb many metals (Cd^{2+}, Zn^{2+}, Ni^{2+}, Pb^{2+}) that might be present in wastewaters. Adsorption removes them from the percolating water, thereby cleansing the water that drains into groundwaters or surface waters.

The amounts of cations in the soil solution are intimately related to the exchangeable ions. Any change in concentration of a cation in the solution (fertilizer or lime) forces a change in proportions of all exchangeable ions.

---

[4]Aluminum forms octahedral coordination with six water molecules or hydroxyls, which have essentially the same size. As the soil water solution becomes less acid (has more OH^- ions), one or more aluminum-held water molecules ionizes H^+(s), which are less attracted to the oxygen of water molecules held to the aluminum; the ionization leaves a hydroxyl ion attached to the aluminum, which neutralizes some of its charge. The aluminum ion becomes successively less positively charged by such ionization. At different pHs, these are the approximate forms of aluminum in soil solution:

$Al(H_2O)_6^{3+}$	predominant form below pH 4.7
$Al(H_2O)_4(OH)_2^+$	predominant between pH 4.7 and 6.5
$Al(H_2O)_3(OH)_3^0$	predominant between pH 6.5 and 8.0
$Al(H_2O)_2(OH)_4^-$	predominant between pH 8.0 and 11

Notice the noncharged hydrated aluminum that exists at the near-neutral pH and that precipitates as the noncharged hydroxyl form; it is usually written as $Al(OH)_3$. The $Al(H_2O)_5(OH)^{2+}$ form does not exist in dominant amounts, so it has been omitted from this list. See Detail 8-1 for more detail.

In its broadest sense, **chromatography** refers to processes that permit the separation of components in a mixture. This is possible in soils as a consequence of differences in rates at which the individual components of that mixture migrate through the stationary soil under the influence of mobile water. As soluble ions in water percolate through soil, some of these ions will adsorb to the cation exchange sites (and anions to anion exchange sites), replacing ions already adsorbed. $Al(OH)_2^+$ and H^+ ions would be adsorbed strongly, Ca^{2+} and Mg^{2+} less strongly, and K^+ and Na^+ least strongly. Some H_3BO_3 would adsorb to anion exchange sites (few sites), but NO_3^- and Cl^- ions would adsorb weakly and move nearly at the rate of water flow.

Chromatographic movement of ions is the flow of different ions through the soil at different speeds, partly because of adsorption differences. What happens? Suppose that a solution of water containing calcium, sodium, boric acid, and chloride is added onto the soil surface. Immediately, some of the calcium and sodium ions replace ions already adsorbed on the cation exchange sites. The replaced ions join the remaining ions still in solution and move deeper into the soil. Continuously, the ions in solution encounter new exchange sites and compete with each nearby adsorbed ion to occupy its site. The downward rate of movement of each ion will be different and will depend on (1) the rate of water flow, (2) the strength with which the ion is adsorbed to the site, and (3) the number and nature of competing ions in the solution (which compete with it for the site). The most strongly adsorbed ion will tend to move the most slowly down through the soil (it may appear not to move). Eventually, some ions will drain out the lower part of the soil. The relative rates of movement should be, approximately (where H_3O^+ is hydrated H^+):

Cations: Slowest movement; $H_3O^+ = Al(OH)_2^+$
Faster movement; $Ca^{2+} = Mg^{2+} < K^+ < Na^+$

Anions: (Slow) $H_2PO_4^- < H_3BO_3 < SO_4^{2-} < NO_3^- = Cl^-$ (Fast)

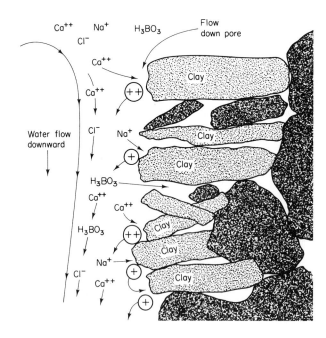

Soil acts as a cation exchanger. The cation exchange sites immobilize cations but also keep them exchangeable and thus available to plant roots. Any water moving through the soil (exchanger) will lose many of its soluble cations to the soil (exchanger) and pick up those cations that have been replaced from exchange sites. Cation nutrients—such as potassium

(K+), ammonium (NH$_4$+), and calcium (Ca^{2+})—will not move far in soil before large portions of them are adsorbed to the exchange sites.

The mechanism of cation exchange is illustrated in Detail 5-3. Different ions in solution move at different speeds through a soil. However, for every cation that adsorbs, the cation originally on that site moves out into solution as it is replaced. The percolating water constantly carries some cations with it, as well as many anions. Certain cations (calcium) and anions (phosphates, for example) may form *insoluble salts* and precipitate out of solution. This precipitation may reduce the amount of ions moving with the percolating water. Table 5-5 lists the leaching losses of some common soil ions from various soils. Notice that Udolls, Udalfs, and Aquepts are soils of humid climates or wet conditions. Observe that large losses of calcium occur, not because it is not strongly adsorbed, but because so much calcium exists in soils from the various weathering minerals. Plants absorb nitrate as it is made available, so plant-covered soil loses less nitrogen than similar barren soils. Less sulfur is needed by plants, so plant cover doesn't affect sulfate losses as dramatically. Nonorganic soils have little loss of phosphates. Chlorides would easily leach with water.

Water pollutants—such as heavy metals (lead [Pb^{2+}], cadmium [Cd^{2+}], mercury [Hg^{2+}], and many others—can be disposed of on soil. Instead of leaching easily to pollute ground and surface waters, the heavy metals are almost all adsorbed to the cation exchange sites. They may finally form insoluble hydroxides, carbonates, or other forms, but cation exchange helps initially to make them immobile and "cleanses" the percolating liquid.

Liming a soil to correct the adverse acidity is a cation exchange reaction. When lime is added to neutralize acidic soils, most of the exchangeable H+ and Al(OH)$_2$+ are neutralized to alter the soil pH. The amount of lime required for neutralization is determined directly or indirectly by the amounts of exchangeable H+ and Al(OH)$_2$+ that must be replaced by calcium or other cations.

A sodium saturation percentage of over about 10 to 20% on exchange sites can cause poor soil structure and reduced aeration when the salinity is low. If, however, the salinity is moderate to high, the soil remains strongly flocculated. The effect of sodium to cause breakdown of aggregates (soil dispersion) is discussed in a later chapter.

Table 5-5 Leaching Losses of Cations and Anions under Various Conditions

Condition	Loss per Year in kg ha^{-1}				
	Ca	Mg	K	N	S
Soil in Scotland (6-year average)	56	17	10	8	—
Uncropped well-drained (3.5-year average)					
Udoll in Illinois	101	52	1	86	12
Udalf in Illinois	42	21	3	—	3
New York Udalf, 10-year average					
Barren soil	446	71	81	77	59
Grass-covered soil	291	56	69	3	49
New York Aquept, 15-year average					
Barren soil	362	46	72	48	39
Rotation of crops	280	30	64	7	37
Canadian soil, 5-year average					
Grass under orchard	51	19	19	—	—
Clean-cultivated under orchard	374	104	45	—	—

When phosphorus was measured, only a "trace" was reported in leaching waters.

Source: Selected data from N. L. Brady, *The Nature and Properties of Soils,* 10th ed., Macmillan, New York, 1990, p. 621.

As the world's nations try to agree on a uniform system of weights and measures, the United States is gradually converting to a metric system. The use of customary U.S. units—such as pounds, feet, and gallons—is being eliminated, and metric units—such as kilograms, meters, and liters—are replacing them (see the conversion unit tables inside the book covers).

Cation exchange capacity is reported in chemical terms. The unit used prior to 1982 was *milliequivalents of exchangeable cations per hundred grams of soil* (meq/100 g). The metric system no longer uses the term *equivalents;* now **moles** are the accepted chemical unit. All the calculations and concepts of "equivalents" are still mentally used by those who learned them, but the notation must be written differently. The old "equivalent" is represented by *moles (+)* or *mole$_c$*, which indicates a monovalent ion portion; for example,

to write 6 meq/100 g in the newer metric system, do one of the following:

Old nomenclature	New accepted nomenclature
6 meq/100 g	= 60 mmol$_c$ kg^{-1} (where c = one charge)
	= 6 cmol$_c$ kg^{-1}
	= 6 cmol (+) kg^{-1} of soil (= centimoles)
	= 60 mmol (+) kg^{-1} of soil (= millimoles)
	= 60 mmol ($1/2$ Ca^{2+}) kg^{-1} (if Ca used)

Any of these five methods could be used to report the 6 meq/100 g. Also, the solidus (/) can be used in place of "$^{-1}$"—such as writing the second one above as 6 cmol$_c$/kg, or the first one as 60 mmol$_c$/kg.

5:4.3 Cation Exchange Capacity (CEC)

Cation exchange capacity (CEC) is the amount of exchangeable cations per unit weight of dry soil. It is measured in *centimoles$_c$ of cations per kilogram of soil* (cmol$_c$ kg^{-1}) (see Detail 5-4). The term *centimoles$_c$* is used because the *number* of negative sites in a given soil sample does not change, but the *weights* of the cations that may be adsorbed to those sites at one time do change. One centimole$_c$ of cation X occupies the same number of cation exchange sites as one centimole$_c$ of cation Y (a different cation). (If chemical *weight* units were used, one gram of cation X would *not* occupy the same number of cation exchange sites as would one gram of cation Y.) Measuring the CEC in centimoles$_c$ of cations per kilogram of material keeps the CEC value the same regardless of what cations occupy the sites.

Representative exchange capacities of soils in the United States are shown in Tables 5-6 and 5-7. Note the very low CECs of sands and loams compared to those of clays. CEC values for the various colloids are given in Table 5-8.

The amounts of exchangeable cations in most soils are surprisingly large. Usually to depths of normal rooting (60–90 cm or 2–3 ft), the amounts of exchangeable ions range from

Table 5-6 Generalized Relationship between Soil Texture and Cation Exchange Capacity

Soil Texture	Cation Exchange Capacity (centimoles$_c$ per kg of soil) (Normal Range)
Sands	1–5
Fine sandy loams	5–10
Loams and silt loams	5–15
Clay loams	15–30
Clays	>30

Table 5-7 Typical Amounts of Exchangeable Cations of a Variety of Soils and the Amounts in Soil in kg/ha–30 cm or lb/a–ft Depths

Typical Centimoles$_c$ of Cations per Kilogram of Soil

Soil (Names of Suborders)	CEC	Ca^{2+}	Mg^{2+}	K$^+$	Na$^+$	H$^+$, Al^{3+}
			cmol$_c$ of Cations per kg			
Psamment (sandy soil, pH 6.4, Kansas)	5.2	1.9	1.2	0.3	tr	1.8
Argid (sandy loam, pH 6.3, California)	4.2[a]	2.2	1.1	0.8	0.2	1.8
Udoll (silt loam, pH 6.7, Illinois)	25.4	17.1	3.1	0.4	0.1	4.7
Ustert (clay loam, pH 6.4, Texas)	28.6[a]	23.0	4.3	0.8	0.3	4.5
Humod (**Oe** layer, pH 3.6, Alaska)	105.7	5.8	6.5	0.5	1.3	91.6
Aquult (sandy loam, pH 3.5, North Carolina)	24.0[a]	2.7	0.6	0.06	0.02	20.6
Orthox (clay, pH 4.9, Puerto Rico)	26.5	8.1	2.1	0.6	0.1	15.6
Aqualf (Si.cl.l.,[b] pH 5.4, Ohio)	37.8	8.6	4.3	0.6	0.0	4.3
Andept (Si.cl.l.,[b] pH 5.3, Oregon)	103.6	6.7	1.2	0.4	0.4	94.9

Amount of Exchangeable Cations for 1 cmol$_c$/kg of Soil CEC

Ion on the Exchange Complex	Approximate kg/ha for 30-cm Depth	Approximate lb/a for 1-ft Depth
Ca^{++}	800	800
Mg^{++}	480	480
K$^+$	1560	1560
Na$^+$	920	920
H$^+$	40	40

Source: Profile data from Soil Survey Staff, *Soil Taxonomy,* Agriculture Handbook 436, USDA—Soil Conservation Service, 1975, p. 754.
[a] Often the sum of cations does not equal the measured CEC because of pH of determination, solubility of certain minerals, and the adsorption strength of some of the cations involved.
[b] Silty clay loam.

Table 5-8 Representative Cation Exchange Capacities of the Common Soil Colloids Responsible for Most Soil Cation Exchange Capacity

Soil Colloid	Cation Exchange Capacity (cmol$_c$/kg of Colloid)
Humus	100–300
Vermiculite (similar to hydrous mica)[a]	80–150
Smectites (montmorillonite)	60–100
Hydrous mica	15–25
Kaolinite	2–8
Sesquioxides	0–3

[a] Vermiculite has extensive tetrahedral substitution as does hydrous mica, but, without the potassium to keep the layers collapsed, it has moderate swelling and a high cation exchange capacity.

hundreds to thousands of kilograms per hectare (see Calculation 5-1 and Table 5-7). In Table 5-7 the high CEC of the Humod is due to a high humus content; that of the Andept is due to its volcanic-ash origin (amorphous clays) plus its high humus content.

Exchangeable K$^+$ plus soluble K$^+$ are the major immediate sources of potassium to plants. A CEC measurement of exchangeable K$^+$ allows an estimate of whether or not fertilizer potassium needs to be added. As an example, a CEC measurement of less than 250 kg/ha of K$^+$ to a depth of 30 cm (about 222 lb/a per 1-ft depth) indicates that a crop having a high potassium requirement would need added potassium fertilizers.

As an example of the large quantity of exchangeable ions in soils, the following problem demonstrates the method of calculating amounts of exchangeable cations (see Table 5-7).

Problem In laboratory analyses, the following values were obtained for exchangeable potassium. Calculate whether potassium fertilizer will be needed for a corn crop in each soil.

Soil Description	cmol of K/kg soil
Well-weathered Paleudult sandy loam, Johnston Co., North Carolina, pH 4.9	0.22
Highly weathered Orthox sandy clay loam, Belém, Brazil, pH 4.2	0.06
Arid Paleargid sandy loam, Cochise Co., Arizona, pH 6.6	0.78

Solution

1. Whether or not potassium fertilizer is needed depends on the crop grown, its expected yield, and the soil test correlation for the area. For this problem, assume fertilizer is needed for the corn if the test for K^+ in the top 30 cm (1 ft) depth is less than 240 kg of exchangeable K^+/ha–30 cm.

2. **Soil weight.** Usually an "average" soil weight per hectare–30 cm is used for all soils in routine analysis. A hectare area 30 cm deep has a volume of about 100 m $\times$ 0.3 m;

$$100 \text{ m} \times 100 \text{ m} \times 0.3 \text{ m} = 300 \text{ m}^3$$

If a bulk density of 1400 kg/m³ is used as an average value, the weight of that soil would be

$$(3000 \text{ m}^3)(1400 \text{ kg/m}^3 \text{ bulk density}) = 4,200,000 \text{ kg of soil/ha–30 cm}$$

The value of 4 million kg/ha–30 cm is often used. Remember it.

3. **Weight of K^+.** To convert centimoles of K^+ to weight units,

$$1 \text{ cmol}_c \text{ of } K^+ = \frac{1 \text{ cmol}_c \text{ K}}{\text{valence of K}} \left| \frac{\text{atomic wt of K}}{1 \text{ mol K}} \right| \frac{1 \text{ mol K}}{100 \text{ cmol K}}$$

$$= \frac{1}{1} \left| \frac{39 \text{ g}}{1} \right| \frac{1}{100} = 0.39 \text{g}$$

$$= 390 \text{ mg of } K^+ \text{ per cmol}_c \text{ of } K^+$$

4. **Kilograms of K^+ per hectare–30 cm** is calculated using the 4 million kg/ha–30 cm approximation of "average" soil weight.

$$1 \text{ cmol}_c \text{ K}^+/\text{kg}^{-1} \text{ soil} = 390 \text{ mg K per kg of soil (see step 3)}$$

$$\text{kg of K per ha–30 cm} = \frac{390 \text{ mg K}}{1 \text{ kg soil}} \left| \frac{1 \text{ kg}}{1,000,000 \text{ mg}} \right| \frac{4,000,000 \text{ kg soil}}{1 \text{ ha–30 cm of soil}}$$

$$= 1560 \text{ kg of } K^+ \text{ per hectare–30 cm depth}$$

Thus, 1 cmol of K^+/kg of soil is equivalent to 1560 kg of K^+ per hectare taken to a 30-cm depth. (This is the way the values in Table 5-5 were derived for the 30-cm depth of soil.)

Continued.

For the Paleudult soil from North Carolina, with 0.22 cmol$_c$ K$^+$/kg,

$$\frac{1560 \text{ kg K}}{\text{ha--30 cm}} \left| \frac{0.22 \text{ cmol}_c}{1 \text{ cmol}_c} \right. = 343 \text{ kg K}^+ \text{ per ha--30 cm}$$

For the Orthox soil from Brazil, with 0.06 cmol$_c$ K$^+$/kg,

$$\frac{1560 \text{ kg K}}{\text{ha--30 cm}} \left| \frac{0.06 \text{ cmol}_c}{1 \text{ cmol}_c} \right. = 94 \text{ kg K}^+ \text{ per ha--30 cm}$$

For the Paleargid from Arizona, with 0.78 cmol$_c$ K$^+$/kg,

$$\frac{1560 \text{ kg K}}{\text{ha--30 cm}} \left| \frac{0.78 \text{ cmol}_c}{1 \text{ cmol}_c} \right. = 1{,}217 \text{ kg K}^+ \text{ per ha--30 cm}$$

5. The soil from North Carolina is above the 240 kg K$^+$/ha--30 cm value previously set as the critical value. Additional K is not needed for this soil. Obviously, the Orthox from Brazil is quite low in potassium; corn would respond to the addition of appreciable potassium fertilizer. The Arizona soil, as typical of most arid-region soils, has adequate potassium available in it.

5:4.4 The pH Dependence of Cation Exchange Capacity

Is the cation exchange capacity (CEC) always the same in a given soil? The answer is yes, approximately, if the soil pH, humus, and clay contents remain the same. The soil's CEC does change as these properties change (see Detail 5-5).

5:4.5 Estimating CEC and Exchangeable Cations

The measurement of CEC is time-consuming and often not measured because of this cost. One method used to estimate CEC is simple and straightforward (see Calculation 5-2). First, estimate or measure the clay and humus percentages. Second, assign an "average" CEC value to each 1% of clay or humus. Divide the average CEC value for that material (=100%) material) by 100. This is the CEC for 1% of the material. Third, add up the CEC contributions for the clay and humus. In general, each percentage of humus contributes about 1.5–2.0 cmol$_c$/kg of CEC, and each percentage of clay contributes about these amounts in cmol$_c$/kg: montmorillonite = 0.6–1.0, kaolinite = 0.02–0.08, and sesquioxides = 0.03. Because soil contents of hydrous mica, chlorite, or vermiculite are usually quite small, it is difficult to know how to estimate these.

The kinds and proportions of the various cations are discussed later. In general, the calcium ions dominate the CEC except in quite acidic soils. The Al(OH)$_2$$^+$ ion dominates in soils of pH about 4–6. Below pH 4, more Al^{3+} and H$^+$ occur on the exchange sites.

5:5 Anion Exchange and Adsorption

Note that the discussion of cation exchange capacity has considered primarily the plant nutrients calcium, magnesium, ammonium, and potassium. **Anions**—the negatively charged nutrients such as sulfate, nitrate, phosphate, molybdate, borate, and chloride—are not held

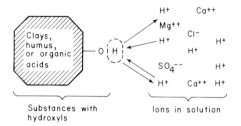

The cation exchange capacity of a soil changes with a change in pH (acidity or basicity). The majority of the negatively charged exchange sites are from isomorphous substitutions and are a permanent charge. However, sesquioxides (metal oxides) and kaolinite have only a few lattice sites with isomorphous substitution. The various hydroxyls of clays, humus, and organic acids ionize H^+ into the soil water solution, thereby producing negatively charged cation exchange sites on these soil particles as the pH rises.

In acidic solutions (high H^+ concentration), fewer H^+ ions ionize off the R—OH. In basic solutions, fewer H^+ are in solution and more R—OH sites are ionized to R—O$^\ominus$. Nonionized sites do not react as exchangeable sites.

In more strongly acidic solutions, the H^+ hinders ionization of the R—OH groups. At lower pH, there will be *fewer* pH–dependent CEC sites ionized, which allows them to react as exchange sites.

As much as 10–40% of the soils' CEC may be from pH-dependent sites. Most CEC from humus is pH dependent. As the pH rises, pH-dependent CEC also increases.

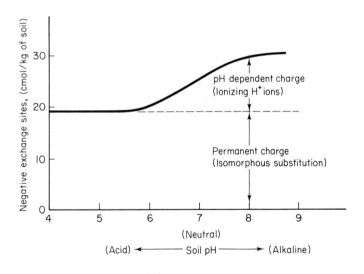

on cation exchange sites. Also, few of the other micronutrients are held there; they form hydroxides or carbonates of low solubility and are available to plants directly from the small amounts solubilized into the soil solution. However, most soils have a low anion exchange capacity.

Colloids that have an appreciable anion exchange are those that also have low cation exchange. Multicharged positive ions, such as iron and aluminum, have hydroxyls (OH$^-$) that can be exchanged with sulfate (SO_4^{2-}), phosphate ($H_2PO_4^-$ or HPO_4^{2-}), molydbate (MoO_4^{2-}), and some other anions. The highest anion exchange capacities (AEC) occur in amorphous silicate clays (volcanic-ash soils) and in iron and aluminum hydrous oxide clays; lesser exchange occurs in kaolinite. Phosphates particularly can be held firmly on anion exchange sites.

Calculation 5-2 Estimating Cation Exchange Capacities

Estimating cation exchange capacity is done using the data of Table 5-8 and assuming that all cation exchange is on clay and soil humus. First, from the range of CEC values given, select an *average* CEC value for the humus and for each clay type. Second, divide the value from Table 5-6 by 100 to get the CEC value contributed by 1% of that material.

Problem From data of soils selected from Table 5-5, estimate the CEC for each horizon described below:

Soil	Predominant Colloid	Clay (%)	Humus (%)	Measured CEC ($cmol_c/kg$)
Sandy Entisol Ap	Montmorillonite	2.6	0.5	5.2
Subhumid-area Vertisol Ap	Montmorillonite	36.0	1.7	28.6
Spodosol **Oe** layer	Organic	24.0	85.0	105.7
Oxisol Ap	Kaolinite and metal oxides	73.1	4.7	26.5

Solution Using average values from Table 5-7, assume that each 1% of humus (10 g of humus/kg of soil) contributes 2 $cmol_c$ of CEC per kilogram of soil (divide the 200 $cmol_c/kg$ for pure humus by 100%). In a similar manner, 1% of montmorillonite contributes about 0.8 $cmol_c$ (80 $cmol_c/kg$ average for 100% clay). Similarly, 1% of kaolinite contributes about 0.5 $cmol_c$ and 1% metal oxides contribute about 0.02 $cmol_c$ of CEC.

Sandy Entisol Ap: With 0.5% humus, it should have 2 $cmol_c$ x 0.5 = 1.0 $cmol_c/kg$ of CEC from humus. Assuming that the clay is montmorillonite, the 2.6% clay contributes 2.6 x 0.8 $cmol_c$ = 2.08 $cmol_c/kg$. Adding the CEC from clay and from humus, this soil has an estimated CEC of 1.0 + 2.1 = **3.1 $cmol_c/kg$ of soil.**

Subhumid-area Vertisol Ap: The Ustert, a swelling soil, has montmorillonite clay, so its 36% clay contributes 36% x 0.8 $cmol_c$ CEC/kg = 28.8 $cmol_c/kg$. Its 1.7% humus contributes 1.7% x 2.0 $cmol_c$ of CEC/kg = 3.4 $cmol_v/kg$, for a total of **32.2 $cmol_c$ of CEC/kg of soil.**

Spodosol Oe horizon: This organic layer with 85% humus has 2 $cmol_c/kg$ x 85% humus = 170 $cmol_c$ of CEC/kg contributed by humus. The 4.1% kaolinite clay provides 0.05 $cmol_c/kg$ x 4.1% = 0.21 cmolc of CEC per kg of soil.* The total is **170.2 $cmol_c$ per kg of** soil for its estimated CEC.

Oxisol: This soil of 73% clay probably has mixed kaolinite and sesquioxide clays (assume half of each). The soil would have (73%/2) x 0.05 $cmol_c$ CEC/kg for kaolinite and (73%/2) x 0.02 $cmol_c$ CEC/kg for metal oxides. This is a total of 2.5 $cmol_c/kg$ of CEC from its clay. It also has 4.7% x 2.0 $cmol_c/kg$ = 9.4 $cmol_c/kg$ from humus. The total estimated CEC is 2.5 + 9.4, for a total of **11.9 $cmol_c$ per kg of soil.**

When comparing estimated values to measured values, there will be considerable error in some soils. This may be due to a poor choice of the "average" CEC values, the extent of humus decomposition, an incorrect estimate of the kind of clay, or the presence of an unusual colloid in that soil. It seems that extremes in percentage clay or humus produce less accurate estimates. However, using an estimation for more-common soils that have 10–25% clay and 1–6% humus can provide a fair approximation of the soil's CEC. Consider that clays in humid climates are mostly kaolinite and hydrous mica, and soils in climates with less than 600 mm (24 in.) annual precipitation are montmorillonite plus hydrous mica or vermiculite.

Source: Courtesy of Raymond W. Miller, Utah State University.

* The 24% kaolinite is 24% of the mineral portion, which is only 15% of the whole soil. So on a whole-soil basis, the soil has only (24%) x (15%) or 4.1% kaolinite.

Soil pH is the *negative logarithm of the active hydrogen ion (H+) concentration in solution.* When water (HOH) ionizes to H+ and OH− (a neutral solution), both H+ and OH- are in concentrations of 10^{-7} mole* per liter. The [ ] = concentration.

$$HOH \rightleftharpoons H^+ + OH^-$$

$$\frac{[H^+][OH^-]}{[HOH]} = 1 \times 10^{-14}, \quad [H^+] = [OH^-] = 1 \times 10^{-7}$$

Thus, the negative logarithm of [H+] is 7, or pH 7. When the H+ concentration is greater (more acidic), such as 10^{-4} mole per liter, the pH is lower (e.g., pH 4). In basic solution the OH−concentration exceeds the H+ concentration. The product of the H+ and OH− concentrations equals 10^{-14} mole per liter. When H+ is 10^{-5}, OH− is 10^{-9}, for example. (The weight and volume units in chemistry are always expressed in the centimeter-gram-second system, and conversions to the U.S. units are not applicable here.)

*A mole is one molecular weight of ion or of the molecule.

Anion exchange capacities are generally low, usually only a *few tenths* of a $cmol_c/kg$ of soil, in contrast to much larger cation exchange capacities. In soils with high percentages of iron oxides, values of a few $cmol_c/kg$ have been measured. A few rare values as high as 40 $cmol_c$ of sulfate/kg of soil have been reported.[5]

5:6 The Soil Reaction (pH)

The term *pH* is from the French *pouvoir hydrogéne,* or "hydrogen power." **Soil reaction (pH)** is an indication of the acidity or basicity of the soil and is measured in pH units (Detail 5-6). The pH (logarithmic) scale ranges from 0 to 14 with pH 7 as the neutral point. At pH 7, hydrogen ion concentration (H+) equals the hydroxyl ion concentration (OH−). From pH 7 to 0 the soil is increasingly more acidic; from pH 7 to 14 the soil is increasingly more alkaline (basic). The H+ concentration, which is the substance measured when determining pH, has a tenfold change between each whole pH number. Thus, a soil of pH 5 has 100 times more H+ in solution than a soil solution with a pH of 7. See Fig. 5-8 for some pH relations among various materials and environments.

5:6.1 Importance of Soil pH

The soil pH is easily determined and provides various clues about other soil properties. The soil pH greatly affects the solubility of minerals. Strongly acidic soils (pH 4–5) usually dissolve high, even toxic, concentrations of soluble aluminum and manganese (Fig. 5-9). Azaleas, tea, rhododendrons, cranberries, pineapple, blueberries, and several conifer timber species tolerate, or may even require, a strong acidity and grow well. In contrast, alfalfa, beans, barley, and sugar beets do well only in slightly acidic to moderately basic soils because of a high calcium demand or inability to tolerate soluble aluminum. Most minerals are more soluble in acid soils than in neutral or slightly basic solutions. Soluble $Al(OH)_4^-$ may exist at these higher pH values. On mineral soils most agricultural crops do best in slightly acidic soils (pH 6.5); on organic soils, about pH 5.5.

---

[5] H. Gebhardt and N. T. Coleman, "Anion Adsorption by Allophanic Tropical Soils: II. Sulfate Adsorption," *Soil Science Society of America, Proceedings,* **38** (1974), pp. 259–262.

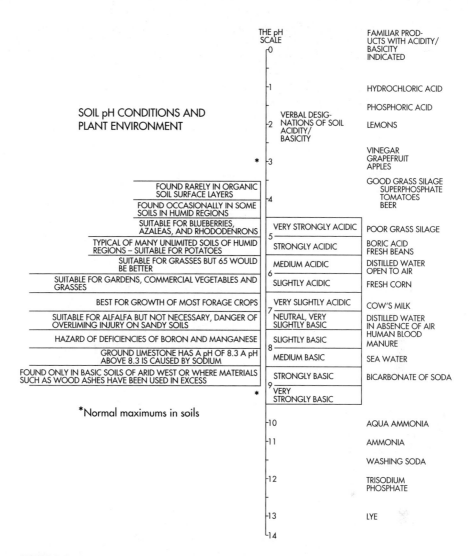

SOIL pH CONDITIONS AND
PLANT ENVIRONMENT

	THE pH SCALE	VERBAL DESIG-NATIONS OF SOIL ACIDITY/BASICITY	FAMILIAR PRODUCTS WITH ACIDITY/BASICITY INDICATED
	0		
	1		HYDROCHLORIC ACID
			PHOSPHORIC ACID
	2		LEMONS
*	3		VINEGAR GRAPEFRUIT APPLES
FOUND RARELY IN ORGANIC SOIL SURFACE LAYERS	4		GOOD GRASS SILAGE SUPERPHOSPHATE TOMATOES BEER
FOUND OCCASIONALLY IN SOME SOILS IN HUMID REGIONS			
SUITABLE FOR BLUEBERRIES, AZALEAS, AND RHODODENRONS	5	VERY STRONGLY ACIDIC	POOR GRASS SILAGE
TYPICAL OF MANY UNLIMITED SOILS OF HUMID REGIONS – SUITABLE FOR POTATOES		STRONGLY ACIDIC	BORIC ACID FRESH BEANS
SUITABLE FOR GRASSES BUT 65 WOULD BE BETTER	6	MEDIUM ACIDIC	DISTILLED WATER OPEN TO AIR
SUITABLE FOR GARDENS, COMMERCIAL VEGETABLES AND GRASSES		SLIGHTLY ACIDIC	FRESH CORN
BEST FOR GROWTH OF MOST FORAGE CROPS	7	VERY SLIGHTLY ACIDIC	COW'S MILK
SUITABLE FOR ALFALFA BUT NOT NECESSARY, DANGER OF OVERLIMING INJURY ON SANDY SOILS		NEUTRAL, VERY SLIGHTLY BASIC	DISTILLED WATER IN ABSENCE OF AIR HUMAN BLOOD
HAZARD OF DEFICIENCIES OF BORON AND MANGANESE		SLIGHTLY BASIC	MANURE
GROUND LIMESTONE HAS A pH OF 8.3 A pH ABOVE 8.3 IS CAUSED BY SODIUM	8	MEDIUM BASIC	SEA WATER
FOUND ONLY IN BASIC SOILS OF ARID WEST OR WHERE MATERIALS SUCH AS WOOD ASHES HAVE BEEN USED IN EXCESS	9	STRONGLY BASIC	BICARBONATE OF SODA
*		VERY STRONGLY BASIC	
*Normal maximums in soils	10		AQUA AMMONIA
	11		AMMONIA
			WASHING SODA
	12		TRISODIUM PHOSPHATE
	13		LYE
	14		

FIGURE 5-8 The entire pH scale ranges from 0 to 14, but soils under field conditions vary between pH 3.5 and 10.0. Few soils have pHs outside this range. In general, most plants are best suited to a pH of 5.5 on organic soils and a pH of 6.5 on mineral soils. (*Source:* Adapted from Winston A. Way, "The Whys and Hows of Liming," University of Vermont Brieflet 997, 1968.)

The soil pH can also influence plant growth by the pH effect on activity of beneficial microorganisms. Most nitrogen-fixing legume bacteria are not very active in strongly acidic soils. Bacteria that decompose soil organic matter, releasing nitrogen and other nutrients for plant use, are hindered by strong acidity. Fungi usually tolerate strong acidity better than do other microbes.

Soils become acidic as rainfall leaches away the basic cations (Ca^{2+}, Mg^{2+}, K^+, Na^+) and many of them are replaced by H^+ from carbonic acid (H_2CO_3) formed in water by dissolved carbon dioxide. Eventually, $Al(OH)_2^+$ replaces H^+.

Soil basicity, although more difficult to alter than soil acidity, may be just as undesirable for plants. Nonleached soils or those high in calcium (low rainfall areas) may have pH values to 8.5. With increased exchangeable sodium, soils may reach values of over pH 10.

FIGURE 5-9 As a soil becomes more acidic (pH is lower), more soluble aluminum is available for absorption by plants. The more aluminum absorbed, the greater is its toxic effect and the greater will be the reduction in the root growth of corn sorghum. (*Source:* Redrawn from Eduardo Brenes and R. W. Pearson, "Root Responses of Three Gramineae Species of Soil Acidity in an Oxisol and an Ultisol," *Soil Science,* **116** (1973), pp. 295–302. © The Williams & Wilkins Co.)

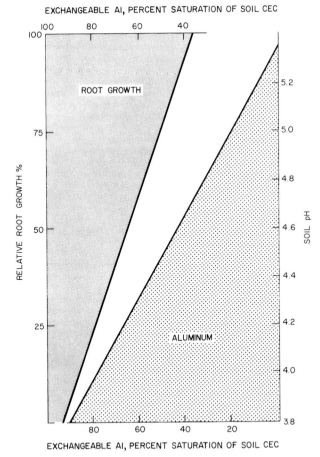

Plants on soils of pH greater than about 9 usually have reduced growth or even die. However, some plants (halophytes) are tolerant of high salt or pH.

The major effect of a basic pH is to reduce the solubility of iron, zinc, copper, and manganese. Also, phosphate is often not readily available to some plants because of its precipitation in the soil solution by calcium or precipitation on *solid* calcium carbonate. Iron deficiency, associated with wet, clayey soils high in carbonates, has long been recognized (although not well understood); it is referred to as *lime-induced iron chlorosis.* Solutions of high pH have low solubilities of iron, zinc, manganese, and copper; the addition of phosphorus often further decreases the availability of those metals at the root surface or just inside the root by precipitating them as insoluble phosphates.

Most micronutrient problems caused by high soil pH are solved by adding special fertilizers, such as water-soluble **chelates** (key'-lates). Chelate ligands are "stable," soluble complexers of the metal ions (although susceptible to microbial decomposition).

5:6.2 *Basic Cation Saturation Percentage (BCSP)*

The cations commonly adsorbed on exchange sites of soil colloids can be divided into acid-forming cations (aluminum and hydrogen) and base-forming cations (calcium, magnesium, potassium, sodium, and some others). The proportion of basic cations, in percentage, to the total cations on the cation exchange complex is the **basic cation saturation percentage (BCSP).** For example, if a soil has a cation exchange capacity of 16 cmol$_c$/kg and 4.2 cmol$_c$ of those cations are aluminum and hydrogen, the remaining cation exchange sites always contain the basic cations. The basic cations = 16.0 − 4.2 = 11.8 $cmol_c$/kg. The basic cation saturation percentage is (11.8/16) (100) = 73.7%. The more acidic a soil is, the lower its percentage of basic cation saturation (Fig. 5-10). At pH 7 or higher, most soils are essentially 100% basic cation saturated (very little exchangeable hydrogen and soluble aluminum). Compare proportions in Table 5-7.

Basic cation saturation does not provide much information that is not already available from pH values, but it does provide numerical values of the amount of exchangeable hydrogen and aluminum ion species, and this aids in predicting the amount of lime needed to neutralize the soil acidity.

5:7 Buffering in Soils

Most soils can resist appreciable pH changes when large amounts of a material either strongly acidic or basic are added, such as an acid-forming or base-forming fertilizer. This ability to resist a change in pH is the **buffering capacity** of the soil. The buffering capacity increases as the cation exchange capacity increases. To exhibit buffering, the soil must "remove" hydrogen ions (H^+) of added acids or neutralize the hydroxyls (OH^-) of added bases. This occurs by cation exchange and neutralization:

$$\text{colloid-H} + NH_4OH \text{ (aqueous ammonia)} \rightleftharpoons \text{colloid-}NH_4^+ + HOH \text{ (water)}$$

$$\text{colloid-2H} + CaCO_3 \text{ (lime)} \rightleftharpoons \text{colloid-Ca} + CO_2 \uparrow \text{ (gas)} + HOH$$

In these two examples, neither the aqueous ammonia fertilizer nor the lime greatly changes the soil pH because the *base* is "neutralized" to either exchangeable NH_4^+ or Ca^{2+} plus the neutral water. A small pH change does occur, and the pH will gradually change as more and more base is added.

A similar buffering reaction occurs with an added acid:

$$\text{colloid-Ca} + 2H_2CO_3 \text{ (carbonic acid)} \rightleftharpoons \text{colloid-2H} + Ca(HCO_3)_2$$

Carbon dioxide dissolves in water to form weak carbonic acid (H_2CO_3), which ionizes to produce free H^+ (Detail 5-7). The free H^+ goes on the exchange sites and the calcium replaced forms calcium bicarbonate that is slightly basic. Eventually, the bicarbonate will precipitate as lime ($CaCO_3$) and leave neutral water again or will be leached from the soil.

The concept of buffering can be broadened to include a *resistance to change in the concentration of any ion in the solution* that is also adsorbed to the colloid. If calcium in solution is precipitated, additional exchangeable calcium will be exchanged off exchange sites into solution so that only a slight change in soluble calcium will occur. Conversely, adding soluble calcium will force much of that calcium onto exchange sites, again lowering the soluble calcium to a low concentration. Exchangeable ions are always in dynamic equilibrium with soluble ions in the soil solution.

Buffering action should be effective in controlling soluble concentrations of H^+, Al^{3+}, Ca^{2+}, Mg^{2+}, K^+, and Na^+. In all instances, the larger the CEC of the soil, the greater should be the buffering capacity. Soils high in humus and/or clay, particularly montmorillonite or vermiculite clays, will have high buffering. Organic soils and clays have higher CEC values and are more strongly buffered than are sandy soils of the same geographic area.

The term **free H+ in solution** refers to hydrogen ions in the solution. This is also the form of H+ that is measured as pH. A much larger amount of acidity is the **bound** or **reserve acidity.** This refers to exchangeable cations that will, during reactions, supply more H+ into the solution. These bound acidic cations include H+ (in very acidic soils), and iron and aluminum ions (mostly as hydroxyl forms, e.g., $Al(OH)2^+$). As H+ in solution is neutralized, the pH rises. This causes another of the waters attached to the aluminum to ionize a hydrogen and form an insoluble hydroxide.

$$Al(H_2O)_4(OH)_2^+ + OH^- = Al(H_2O)_3(OH)_3 \downarrow + HOH$$

(exchange   (basic   (precipitates)   (water)
aluminum)   solution)

Neutralizing the solution H+ would cause the pH to rise and some nonionized R—OH groups on humus and clays to ionize their H+, leaving a R—O− charged site. Thus, the cation exchange capacity also increases.

The amount of bound acidity is usually many times (10–1000 times) greater than the amount of free acidity in solution at any one time.

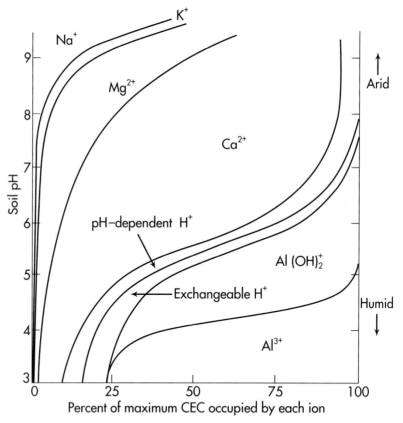

FIGURE 5-10 Generalization of the soil pH and the approximate proportions of various cations on the cation exchange sites for a clay loam soil and 3–5% organic matter. The amount of H+ on pH-dependent H+ sites (ionizable organic matter and metal hydroxides) varies with the organic matter and metal hydroxide contents. A major portion of the exchangeable ions in acidic soils is some form of hydrated aluminum. Only a small portion of aluminum occurs as $Al(OH)^{2+}$, but $Al(OH)_2^+$ still predominates at those pH values, so $Al(OH)^{2+}$ is not shown in the diagram.   (Courtesy of Raymond W. Miller, Utah State University.)

The role of metallic impurities is to increase the disorder of silica, as long as it is being organized upon itself. But when these cations become more abundant, they constrain the silica into another type of order; this is the genesis of clay minerals.

—G. Millot

Questions

1. (a) What are colloids? (b) How large are they? (c) What substances are the colloids in soils?
2. What are three different meanings of the word *clay*?
3. (a) What are clay minerals? (b) Are they inherited in some soils?
4. From what materials do clays form in (a) neoformation and (b) degradation?
5. (a) What elements comprise clays? (b) Which elements occupy the most space in minerals?
6. Do clays contain basic cations (a) in their structures? (b) adsorbed to their structures?
7. (a) What charge do most clays carry? (b) What causes the charge?
8. How does (a) montmorillonite differ from kaolinite? (b) allophane differ from sesquioxides?
9. What is (a) hydrous mica? (b) allophane? (c) vermiculite?
10. What are the metal oxides?
11. In what climatic conditions are each of the various clays expected to form? Explain for each clay.
12. (a) What is the nature of organic colloids? (b) To what extent are all organic colloids similar?
13. (a) Do all soils have organic colloids? Explain.
14. (a) Define cation exchange. (b) Illustrate cation exchange by an equation. (c) What causes the exchange? (d) Are all ions involved in cation exchange?
15. (a) What ions are involved in cation exchange? (b) Which ions are dominant in strongly acidic soils? (c) What is meant by *chromatographic flow*?
16. (a) Is cation exchange important? (b) For what items or reactions is it important?
17. List some typical CEC ranges for (a) pure clays, (b) soils, and (c) humus.
18. In what units is CEC expressed?
19. How does the soil pH alter CEC?
20. (a) What is the basis used for estimating a soil's CEC? (b) Explain why estimates might be used. (c) How accurate are such estimates?
21. Compare CEC and AEC as important soil processes.
22. (a) What kinds of soils might have appreciable AEC? (b) Which clays have largest AEC values?
23. (a) What is soil pH? (b) Give the extreme pH values found in most soils.
24. In what ways does soil pH adversely affect soil properties and plant growth?
25. (a) What can be done about soil pH? (b) Are the changes permanent?
26. (a) Define *BCSP*. (b) What is the BCSP (high or low) in arid-region soils?
27. (a) What is a buffer? (b) What kinds of soils are highly buffered? (c) Which are poorly buffered? (d) Does buffering apply only to pH control? Explain.

6

Soil as a Water Reservoir

Born in a water-rich environment, we have never really learned how important water is to us. . . . We have spent it with shameful and unbecoming haste. . . . Everywhere we have poured filth into it.

—**William Ashworth**

6:1 Preview and Important Facts

PREVIEW

Water is required for all life. Soils are water reservoirs that supply water to plant roots and perform many other roles. Water "lubricates" the soil, allowing root penetration; it is necessary for microbial mobility and action; and it allows nutrient mobility. In a dry portion of soil, water uptake stops, nutrient absorption essentially ceases, and the root growth practically stops. A typical wetted and drained loam soil will contain water in about 25% of its volume, with only about half of that water available to plants.

Silt loams usually hold the most available water. Water entry into soils (infiltration) is too fast in sands and too slow in clays for optimum crop production. Inadequate air exchange will occur in the small, water-filled pores in clayey soils. Sandy soils hold low "plant-available" amounts of water following wetting; thus, they will require frequent rains or irrigations to supply adequate water. Clayey soils will have slow drainage of excess water, but sandy soils will be well drained, if drainage is possible.

The soil is very important to the plant as a water supply, daily or even for weeks between rains or irrigations. Soil texture, humus, and compaction will greatly modify this water reservoir and its capacity to supply water to the plants.

IMPORTANT FACTS TO KNOW

1. The forces causing soils to retain water
2. The terminology for water: water potential, matric potential, plant-available water, field capacity, permanent wilting percentage, mass water ratio, mass water percentage, saturation percentage, volume water ratio, and volume water percentage

3. The approximate amounts of water held in soils of different texture at their field capacities and their permanent wilting percentages
4. The water available to plants from soils
5. How to measure water contents on a mass and a volume basis, and depths of water held in soil
6. How to estimate water contents using tensiometers, resistance blocks, and neutron meters
7. The relationship of evapotranspiration to evaporation of water from a pan of water
8. Understand evapotranspiration and what causes it to vary in amount
9. The rates of water flow in "saturated" and "unsaturated" soil conditions, in centimeters per hour
10. Some methods to retain more water on and in soils

6:2 How Soil Holds Water

Soil is a leaky water reservoir. When too much water is added, the excess runs off over the surface or into deeper layers. Why does the soil hold some of the water, yet allow part of it to drain deeper? Water is held in soils because of **hydrogen bonding** (Fig. 6-1).

Hydrogen bonding is the attraction of the positively charged hydrogens of water to both its oxygen and to nearby negatively charged ions, often another oxygen. Most soil minerals are composed of 70–85% by volume of oxygen. Hydrogens of water bond strongly to these surface oxygen atoms in minerals by **adhesive bonding** (the attraction of unlike molecules.) The hydrogens of water are also attracted (*bonded*) to oxygens of other water molecules, including those already adsorbed to the soil particle surfaces. The attraction of like molecules for each other is **cohesive bonding.** When fatty or oily substances, which are low in oxygens, coat the soil particles, water is not attracted to and held to the coated surface. Such soils are called **water-repellent soils.** They are formed in nature under many plant covers and after

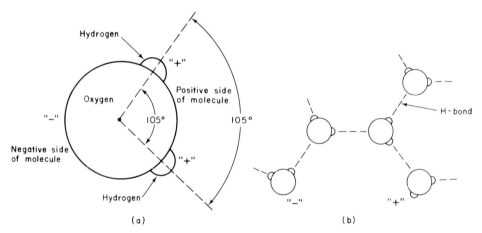

FIGURE 6-1 (a) Polar water molecule, H_2O. Because of the nonlinear positions of the H+s, water is polar. Water has one portion that is more negative than positive and an opposite side that has two hydrogens that are more positive than negative. Polar means there is no center of zero charge. In (b) the bonding of water to itself through H bonding is shown. The H of water in soils may bond to oxygen ions of soil mineral surfaces, thereby holding the water tightly to soil. The bonds in (b) become more rigid (less flexible) in colder temperature (ice or snow). A distance between molecules is shown for illustration purposes. In actual shape, molecules are adjacent to each other in close-fit orientation. (Courtesy of Raymond W. Miller, Utah State University.)

forest fires, which tend to drive vaporized oils and resins into the soil, where they coat the soil particles and cause them to resist wetting.

Strong combined adhesion and cohesion forces cause water films of considerable thickness to be held on the surface of soil particles. Because the forces holding water in soil are surface-attractive forces, the more surface (the more clay and organic matter) a soil has, the greater is the amount of adsorbed water.

To indicate the strength with which water is held, several concepts have been used. The concept of pressure—the pressure required to force the water off soil—was used in early studies and was measured in *atmospheres of pressure* needed to remove water. The opposite of pressure—moisture suction or tension—has also been used and was measured in *atmospheres of suction* or *tension*. Currently, **soil water potential** is used; it is defined as the work the water can do when it moves from its present state to a pool of water in the defined reference state (Fig. 6-2).

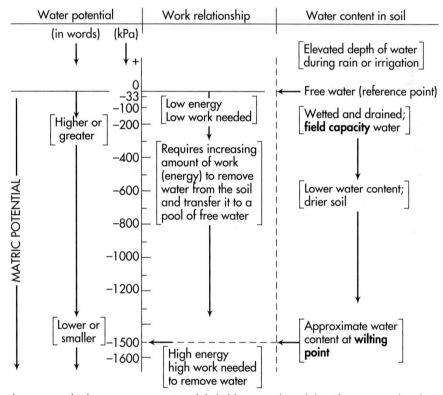

Dry soil = low potential = low water content = tightly held water = low ability of water in soil to do work
Wet soil = high potential = high water content = weakly held water = higher ability of water in soil to do work

FIGURE 6-2 Schematic representation of the terminology for matric potential and for water potential in general. Water with a high potential has more mobility in a soil than has water with lower potential. The *matric potential* is zero at soil saturation and is a positive number only when water sits on the soil surface. Notice that as the water potential becomes lower or smaller, the kilopascals (kPa) are more negative. (Courtesy of Raymond W. Miller, Utah State University.)

6:3 Terminology for Soil Water

The strength with which water is held in soils has been indicated in several ways in past decades as follows:

- **Pressures:** the pressure required to force the different portions of water off or out of the soil, given in *atmospheres* (until the mid-1960s)
- **Negative pressures:** moisture *suction* or *tension*, given in "negative" values of pressure (until the mid-1970s)
- **The pF scale:** the force needed to remove water, given in logarithmic units (similar to a pH scale) of the centimeters of height for a water column of equal force (developed in the United Kingdom)
- **Water potential (present system):** the ability of the water to do work, in energy units

These various systems are intermixed and include a change to use the International System (SI) of metric units. The result is a complex set of units and terms used in various textbooks. Table 6-1 tabulates a brief summary of units and their equivalents used to report numerical values for water potential.

Water potential is defined as the **work** the water can do when it moves from its *present state* to a *pool of water* (at a specified elevation, temperature, soluble salt content, etc.). Because it refers to work, it is an *energy* term, not a *pressure* term. Adsorbed water in soils is less free to move than is water in a pool of water (which has zero potential by definition). Thus, soil water has less **free energy** (less ability to do work than does water in a pool). The free-energy value of less than zero is indicated by a *negative sign. Negative free energy means that work must be done on the water to remove it from the soil to a pool of water.* The more tightly water is held by soil, the more negative is the number (Fig. 6-2).

Table 6-1 Terms and Relationships between Terms Used to Express Water Potential (the Driving Force for Water flow)

Currently used SI metric terms:	Basis for the term:
Joules per kilogram (J kg^{-1} or J/kg)	Energy per unit mass
Pascals (Pa)	Energy per unit volume

Relationships of the older and the new units:

1. At water density of 1 Mg/m^3, 1 J/kg = 1 kPa
   (water is within 0.2 to 0.5 percent of 1 Mg/m^3 regardless of temperature in the soil)
2. The older unit bar = 100 kPa = 100 J/kg
   (1 bar = 0.9869 atmosphere of pressure)
3. Relationships to older units at water density = 1 Mg/m^3:
   1 kPa = 1 J/kg
   = 0.01 bar
   = 0.009869 atmosphere
   = weight of a 10.2-cm column of water
   = 10,000 dynes/cm^2
   = weight of a 0.7501-cm column of mercury
   = 1000 newtons (N) per square meter

Source: Agronomy News, Mar.–Apr. 1982, p. 11.

This smorgasboard of terms, symbols, and units used during various time periods in the development of water nomenclature has not been blithely ignored by the national scientific organizations. The American Society of Agronomy indicated in 1972 when publishing the SI units that water potential was one of the four major areas in which adoption of the SI metric units creates difficulties. Nevertheless, there is always the need for good terminology; scientists continue to search for better terms and more accurate representations.

The **soil water potential** is a combination of the effects of (1) the surface area of soil particles and small soil pores that adsorb water (**matric potential**), (2) the effects of dissolved substances (**solute** or **osmotic potential**), and (3) the atmospheric or gas pressure effects (**pressure potential**). In nonsalty, well-drained soil, the *matric potential is almost equal to the water potential.* An additional effect of the position of the water (such as being elevated) compared to the *reference* state (the reference free-energy state = 0 and is at a specified elevation) is called the **gravitational potential.** Gravitational potential is not related to soil properties, only to the water's elevation in comparison to a reference position. These potentials are defined or interrelated by the following symbols.[1]

$$\psi_w \quad = \quad \psi_m \quad + \quad \psi_s \quad + \quad \psi_p$$

$$\text{water potential} = \text{matric potential} + \text{solute potential} + \text{pressure potential}$$

$$\psi_t \quad = \quad \psi_w \quad + \quad \psi_g$$

$$\text{total water potential} = \text{water potential} + \text{gravitational potential}$$

Because most productive soils do not have a depth of water standing on them for long time periods and have few salts,

$$\psi_{total} \quad \simeq \quad \psi_w \quad \simeq \quad \psi_m$$

for most soils. Matric potential ψ_m is the dominant portion (about 95 percent or more) of total water potential (ψ_{total}) in most situations.

Water from irrigation or rainfall moves into and through saturated soil by gravity flow, often very rapidly. Slower water movement, in all directions, occurs when the soil is not saturated and because of the forces (other than gravitational forces) holding the water. **Water flows from areas of high water potential (usually wetter soil) to low water potential (usually drier soil).**

The water potential of different portions of water in a soil varies. Water that is adjacent to particle surfaces has a lower (more negative) potential (e.g., −800 MPa) than has water on the outside portion of thick water films (which might have a potential of about −10 to −30 kPa, almost loosely enough for gravitational flow). Note that *water is held more tightly as the negative value increases in magnitude.* Fig. 6-3 represents this concept of water potential changing with distance from the particle surface.

The force with which water is held in a water film varies through the film. If a soil holds water with a potential of −1000 kPa, it means that *the most weakly held water* in that sample is at −1000 kPa; the other portions of water in the films are at smaller potentials, such as −1500 kPa, −3000 kPa, and others up to −800 MPa or −1000 MPa.

---

[1] R. J. Hanks and G. L. Ashcroft, *Applied Soil Physics,* Springer-Verlag, New York, 1980, pp. 21–31.

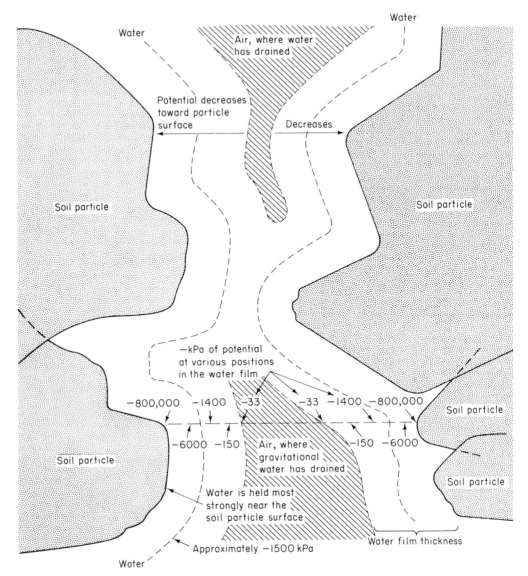

FIGURE 6-3 Cross section of a soil pore and the solid soil particles that make up its walls, showing the increase in force with which water is held with changing distances from the soil particle surface. At some distance from the particle surfaces, water is held so weakly that the pull of gravity causes some of it to drain. The water layers depicted are arbitrary thicknesses that have been selected for illustration purposes only.   (Courtesy of Raymond W. Miller, Utah State University.)

▨ 6:4 Soil Water Classification for Water Management

The most useful classifications of water are those relating to water that plants can and must use. Terms include *gravitational (drainage) water, plant-available water, field capacity,* and *permanent wilting point.*

   Gravitational water is water held at a potential greater than –33 kPa, that portion of the soil water that will drain freely from the soil by the force of gravity. In many soils, this reference value is closer to –10 to –20 kPa.

6:4.1 Ecological Classification of Soil Water

Plant available water is the portion of stored soil water than can be absorbed fast enough by plant roots to sustain life. It is defined as the weight percentage of total soil water held with a water potential between −33 and −1500 kPa (for sands, −10 to −1500 kPa). The water held within these potentials makes up most of the storage water used by plants. (Gravitational water also is equally available as it flows past the plant roots if adequate aeration is maintained. Because gravitational water is present only for short periods of time in permeable soils, it cannot be depended upon in calculating plant water requirements.) Some plant species can extract water held at water potentials as dry as −6000 kPa, but water held at potential drier than about −1500 kPa is absorbed only slowly and in small amounts by most plants, especially crop plant species. Crop plants wilt if only water near −1500 kPa is present because the water loss through transpiration (loss through the stomata of the leaves of the plant) is faster and greater than that amount absorbed by the roots at these low moisture potentials.

Drought-tolerant and arid-zone plants such as cacti have special transpiration-resistant leaf coatings or other adaptations that permit them to survive relatively long time periods when the soil is at or below the permanent wilting percentage (−1500 kPa). The permanent wilting percentage for such plants is at a much drier soil water potential than −1500 kPa.

Some soils hold appreciable water at water potentials of −10 to −30 kPa for a long enough time to have it considered part of plant-available water. This occurs particularly in sandy soils if they overlie coarser sands or gravels or slowly permeable (clayey or cemented) subsoils.

The **wilting point** (sometimes called **permanent wilting point percentage**) is the content of soil water in the soil at a water potential of −1500 kPa. Water at the permanent wilting point (PWP or WP) is held so strongly that plants are not able to absorb it fast enough for their needs. On a hot, dry day, a plant such as corn may transpire excessively and temporarily wilt even when the water is only at −100 or −200 kPa (the water is available but cannot be absorbed fast enough). In this instance, however, the plant easily recovers at night when transpiration losses are less. In contrast to such *temporary* wilting, the *wilting point* indicates low water availability; in such conditions wilting plants do not recover, even at night, except when additional water is added to the soil.

Field capacity is the percentage of soil water that is held when the water potential is −33 kPa. The field capacity is a measure of the greatest amount of water that a soil can hold, or store, under conditions of complete wetting followed by free drainage. Field capacity values are used to determine the amount of irrigation water needed and the amount of stored soil water available to plants. **Plant-available water is equal to the difference of water percentage at field capacity and at permanent wilting point** (Fig. 6-4).

Field capacity approximates the amount of water in a soil after it has been fully wetted and all gravitational water has drained away, usually in a day or two. Field soils reach this condition only momentarily because as excess water is still slowly draining from deeper soil layers, "plant-available" water from surface layers is being evaporated from the soil and absorbed and transpired by plants. Determining when the profile reaches an overall "average field capacity" is difficult to do precisely. The rate at which soils approach an average field capacity is slower with clayey soils than with sands. Figure 6-5 illustrates the rate at which two typical saturated soils drain to field capacity, assuming no evaporation or loss through plant uptake during that time.

Capillary water moves through the soil pores because of a water potential gradient. **Water potential gradient** is the difference in total water potential between two locations in the soil; for example, the amount at 10 cm (4 in.) deep and at 30 cm (12 in.) deep. The gradient difference usually occurs when one location in the soil is drier than another location because of

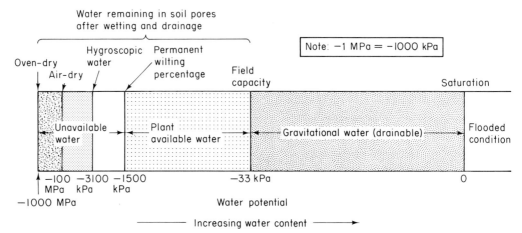

FIGURE 6-4 Soil water constants and their approximate equivalents in kPa of water potential as they affect the relative availability of water to plants. At water potential near –33 kPa, more water is held too loosely to overcome the effect of gravity and drains away. Capillary water (that held by capillary pressure) remains for plant use; this is held in water potentials ranging from –33 kPa (field capacity) to –3100 kPa or lower (hygroscopic), depending on the pore sizes of the soil. Plants can only use capillary water held by not more than –1500 kPa (the point of permanent wilting). Soil water at the air-dry state is held by water potentials that vary from –100 to –30 MPa, depending on humidity.   (Courtesy of Raymond W. Miller, Utah State University.)

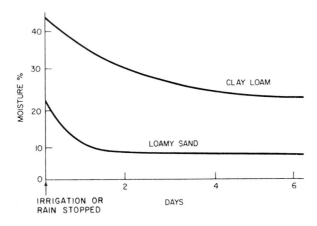

FIGURE 6-5 Speed with which two soils of different texture drain to field capacity. Field capacity is reached when the curve tends to level off to a constant moisture percentage. Drainage is faster in coarse-textured soils than in fine-textured ones. The difference in water content after wetting ceases and the content after drainage ceases is equal to the gravitational water that drains out of the wetted soil.   (Courtesy of Raymond W. Miller, Utah State University)

evaporation or transpiration losses or because of recent wetting of one of the soil locations. Water moves from larger water potential (wetter soil) toward areas with lower water potentials (usually drier soil, although salt content and temperature differences can affect it). For example, capillary movement of water at a 30-cm (12-in.) depth and –150 kPa water potential would slowly move to drier soil near the surface having a water potential of –1000 kPa. A uniform soil with the same water content throughout but more salt in part X than in part Y would have water seeping toward the portion with greater salt contents (lower total water potential).

The **saturation percentage** is the water content of soil when all pores are filled with water. It is approximately *double the field capacity value*. The saturation percentage is approximated by the water in a soil "paste" prepared for laboratory analysis of soluble salts.

6:5 Soils as Water Reservoirs

Which soils hold the most or the least water? Water in soil is held as films on particle surfaces and in small pores. Large pores (in sands and between large aggregates) allow water to drain by gravitational flow. Small pores (clays) retain water by capillary force. Generally, the more clayey the soil and the higher the humus content, the larger the amount of water the soil can store (Table 6-2 and Figs. 6-6 and 6-7). Not only does the quantity of water retained in clay soils remain large, but much of it is tightly held to the large clay surface area. This means that the clay soil will hold a large amount of water at both its field capacity and its permanent wilting point percentage (Fig. 6-6). Smectites (swelling montmorillonitic clays) hold more water than a soil with similar amounts of kaolinite and sesquioxide clays.

Medium-textured soils have the unique combination of pores small enough to hold large amounts of water at high water potentials (held loosely) and a relatively small amount of total surface, which holds low amounts of water at low water potentials (held tightly as in clays). Because of these conditions, the largest amount of *plant-available water* is held in silt loams and in other high-silt soils (Fig. 6-7). Soil humus, compaction (which alters pore sizes), and the kinds of clays will modify water contents.

6:6 Precise Measurement of Soil Water Content

The water content of soil is measured in several ways. The reference and classical method is *gravimetric*. Other techniques include tensiometers, electrical conductivity, and neutron and gamma ray attenuation. Most of these methods measure some property of water but often do not measure the pascals of water potential directly.

Table 6-2 Permanent Wilting Point, Field Capacity, and Plant-Available Water-Holding Capacity of Various Soil Textural Classes[a]

	Water per 30 cm[b] of Soil Depth					
	Permanent Wilting Point		Field Capacity		Plant-Available Water Capacity	
Soil Texture	%	cm	%	cm	%	cm
Medium sand	1.7	0.7	6.8	3.0	5.1	2.3
Fine sand	2.3	1.0	8.5	3.7	6.2	2.7
Sandy loam	3.4	1.5	11.3	5.0	7.9	3.5
Fine sandy loam	4.5	2.0	14.7	6.5	10.2	4.5
Loam	6.8	3.0	18.1	8.0	11.3	5.0
Silt loam	7.9	3.5	19.8	8.7	11.9	5.2
Clay loam	10.2	4.5	21.5	9.5	11.3	5.0
Clay	14.7	6.5	22.6	10.0	7.9	3.5

Source: "Water," *Yearbook of Agriculture*, USDA, Washington, D.C., 1955, p. 120.

[a] There is a variation in the amounts and kinds of sand, silt, and clay within any one textural group (such as within loam soils); therefore, there is also a variation in the water constants; for purposes of simplification, an average value is given in this table.

[b] Centimeters times 0.39 equals inches.

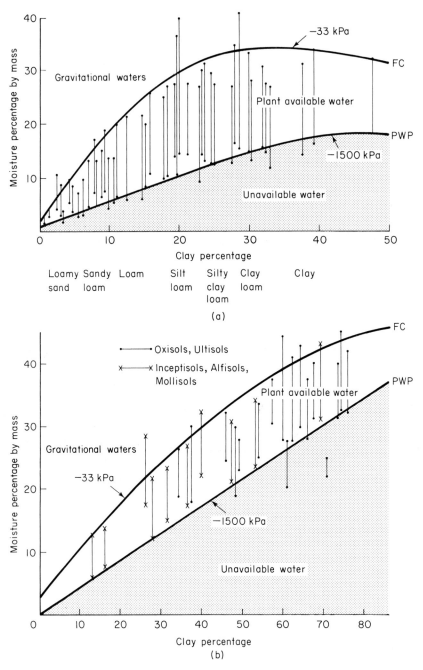

FIGURE 6-6 Actual soil moisture constants for selected soils. For each vertical line the upper end is the moisture content at −33 kPa (field capacity) and the lower end is the moisture content at −1500 kPa (permanent wilting percentage). The length of each line is the plant-available water percentage (from −33 kPa to −1500 kPa). The soils in (a) are from Utah, Nebraska, Iowa, California, and the northeastern United States (silicate clays). The soils in (b) are from Puerto Rico and the Virgin Islands and contain mostly sesquioxide and kaolinite clays. Notice the differences in clay percentages. Textural classes are only approximate. (Courtesy of Raymond W. Miller, Utah State University.)

FIGURE 6-7 Under the same amount of annual rainfall (76 cm, 30 in.) and approximately the same steepness of slope, the soil in (1) supports only xerophytic (drought-tolerant) plants such as Eriogonum, a shrub, and an herb, Phacelia; the soil in (2) supports western yellow pine. The difference is available water. In (1) the soil is a volcanic cinder cone, which has an available water capacity less than that of coarse sand, perhaps 1.25 cm/30 cm (0.5 in./ft) of soil depth. By contrast, the clay loam soil (2) has an available water capacity of perhaps 5 cm/30 cm (2 in./ft) of soil depth. Arizona.   (Courtesy of Michigan State University.)

6:6.1 Measuring Soil Water Percentage

The **gravimetric** method is the classical procedure used as the check for all other methods. A soil is sampled, put into a container, weighed in the sampled (moist) condition, oven dried, and weighed again after drying. Drying is done at 105–110°C (221-230°F) to constant weight (2 hours for small samples, but as much as days for bulky clayey samples and low airflow in the oven).

The **mass water ratio** (θ_m) is the decimal fraction that equals the weight of water divided by the weight of oven-dried soil. The **mass water percentage** (P_m) is the mass water ratio times 100:

$$\textbf{mass water ratio} \; = \; \theta_m \; = \; \frac{\text{mass of water}}{\text{mass of oven-dry soil}} \tag{1}$$

$$\textbf{mass water percentage} \; = \; P_m \; = \; \theta_m(100) \tag{2}$$

See Calculation 6–1 for an example calculation.

6:6.2 Measuring Water Volumes

Plants are dependent on water from *a volume of soil*. Thus, it is often useful to calculate water content based on a volume of the soil. The volume water contents are defined as follows:

$$\textbf{volume water ratio} \;=\; \boldsymbol{\theta_v} \;=\; \frac{\text{volume of water}}{\text{volume of soil}} \tag{3}$$

$$=\; \frac{\text{weight of water/density of water}}{\text{weight of oven-dry soil/soil bulk density}}$$

$$=\; \frac{\text{weight of water}}{\text{weight of dry soil}} \;\left|\; \frac{\text{bulk density of soil}}{\text{density of water}} \right. \tag{4}$$

$$\textbf{volume water percentage} \;=\; \boldsymbol{P_v} \;=\; \theta_v(100) \tag{5}$$

Water volume values are used to estimate the reservoir of water in a soil volume or the water needed to wet that soil by irrigation or rainfall. See Calculation 6-1 for an example calculation.

6:6.3 Water Contents, Gains and Losses of Water

Water volume is used to determine changes in the amount of water in a soil. For example, how much irrigation water needs to be added, or how much water has been evaporated, or how deeply will a rainfall or irrigation wet a soil? The soil's water is usually given as a *depth of water* in a depth of soil, as though the water measured were pulled out of the soil, placed in a container that has the same cross-sectional area as the soil containing the water, and had its depth recorded. Rainfall is measured in this manner; for example, "rainfall was 1.3 inches" means there was rainfall equivalent to cover the area of the rain gauge to a depth of 1.3 in.

The general equation for these kinds of calculations follows:

$$\frac{\text{depth of water}}{\text{depth of soil}} \;=\; \theta_v \;=\; \theta_m\left(\frac{\text{bulk density of soil}}{\text{density of water}}\right)$$

Thus,

$$\text{depth of water} \;=\; \theta_v(\text{depth of soil})$$

$$=\; \frac{\theta_m}{}\left|\; \frac{\text{bulk density of soil}}{\text{density of water}} \;\right|\; \text{depth of soil}$$

Written in symbols, this equation is

$$d_w \;=\; \theta_v(d_s) \;=\; \frac{\theta_m}{}\left|\; \frac{\rho_b}{\rho_w} \;\right|\; d_s \tag{6}$$

See Calculation 6-2 for sample calculations using this equation.

Problem A soil sample taken from a field was placed in a can, weighed, dried in an oven at 105°C (221°F), and reweighed. The measurements were:

Moist soil plus can weight	=	159 g
Oven-dried soil plus can weight	=	134 g
Empty can weight	=	41 g
Bulk density of the soil	=	1400 kg/m³

Calculate (a) the mass water ratio and percentage and (b) the volume water ratio and percentage of this soil when it was sampled.

Solution (a) The mass water ratio (θ_m) is the portion of the soil mass occupied by water. All calculations must involve only soil and water weights, so the weight of the can must first be subtracted.

$$\text{Moist soil only} = 159 \text{ g} - 41 \text{ g} = 118 \text{ g}$$

$$\text{Dried soil only} = 134 \text{ g} - 41 \text{ g} = 93 \text{ g}$$

Then,

$$\text{mass water ratio} = \theta_m = \frac{\text{moist soil} - \text{oven dry soil}}{\text{oven dry soil}}$$

$$= \frac{118 \text{ g} - 93 \text{ g}}{93 \text{ g}} = 0.269$$

$$\text{Mass water percentage} = P_m = \theta_m (100) = (0.269)(100) = 26.9 \text{ percent}$$

(b) The volume water ratio (θ_v) is the fraction of the soil volume occupied by water. Using the equation (4) from the text for θ_v yields

$$\theta_v = \frac{\text{weight of water}}{\text{weight of dry soil}} \left| \frac{\text{bulk density of soil}}{\text{density of water}} \right.$$

$$= \frac{25 \text{ g}}{93 \text{ g}} \left| \frac{1400 \text{ kg/m}^3}{1000 \text{ kg/m}^3} \right. = 0.376 \text{ volume water ratio}$$

$$\text{volume water percentage} = P_v(100) = (0.376)(100)$$

$$= 37.6 \text{ pecent water by volume}$$

Compare water percentages by mass and by volume.

Calculation 6-2 Water Volume Calculations

Problem Calculate these values: (a) the *total* water currently contained in the top 30 cm (12 in.), (b) the *depth* to which 27.5 mm (1.1 in.) of irrigation water would wet this uniform soil, and (c) the *available* water the soil contains in the top 30 cm when the soil is at field capacity. The soil's measurements follow:

Present mass water percentage	= 18%
Mass water percentage at field capacity	= 23%
Permanent wilting percentage	= 9%
Bulk density of the surface soil	= 1300 kg/m^3 (= 1.3 g/cm^3)

Solution Equation (6) is involved in all of the solutions. The major problem is to decide what value is to be used for θ_m in each instance.

$$d_w = \theta_v(d_s) = \frac{\theta_m \left| \rho_b \right| d_s}{\left| \rho_w \right|}$$

(a) To calculate the total water depth in the 30 cm (12 in.) of soil, use the present mass water content given, which is 0.18 (= $P_w/100$ = 0.18).

$$d_w = \frac{\theta_m \left| \text{bulk density of soil} \right| d_s}{\left| \text{density of water} \right|}$$

$$= \frac{0.18 \text{ cm} \left| 1300 \text{ kg} \right| m^3 \left| 30 \text{ cm deep} \right.}{\left| m^3 \right| 1000 \text{ kg} \left| \right.}$$

$$= \textbf{7.02 cm of total water} \text{ in the top 30 cm of soil}$$

(b) To calculate the depth of wetting by a 27.5 mm (1.1 in.) irrigation, d_w is 27.5 mm and θ_m is the difference between present moisture and field capacity (= 0.23 − 0.18 = 0.05). This is because when water is added, it wets the soil from its present condition of 18% to its field capacity before all additional water (gravitational water) drains deeper. The equation with the numbers substituted is:

$$27.5 \text{ mm} = \frac{\theta_m \left| \text{bulk density of soil} \right| d_s}{\left| \text{density of water} \right|}$$

$$= \frac{(0.23 - 0.18) \left| 1300 \text{ kg} \right| m^3 \left| d_s \text{ cm} \right.}{\left| m^3 \right| 1000 \text{ kg} \left| \right.}$$

$$d_s = 423 \text{ mm} = 42.3 \text{ cm, the depth of soil wetted}$$

The **soil will be wetted 42.3 cm** (about 16.5 in.) **deep.**

(c) To calculate the total possible plant-available water in the top 30 cm, when the soil is wetted, it equals field capacity minus the permanent wilting percentage, which is 0.23 minus 0.09 in this example. So the plant-available water is

$$d_w = \frac{\theta_m \left| \text{bulk density of soil} \right| d_s}{\left| \text{density of water} \right|}$$

$$= \frac{(0.23-0.09) \left| 1300 \text{ kg} \right| m^3 \left| 30 \text{ cm} \right.}{\left| m^3 \right| 1000 \text{ kg} \left| \right.}$$

$$= \textbf{5.46 cm of available water} \text{ in the top 30 cm of soil}$$

Source: Raymond W. Miller, Utah State University.

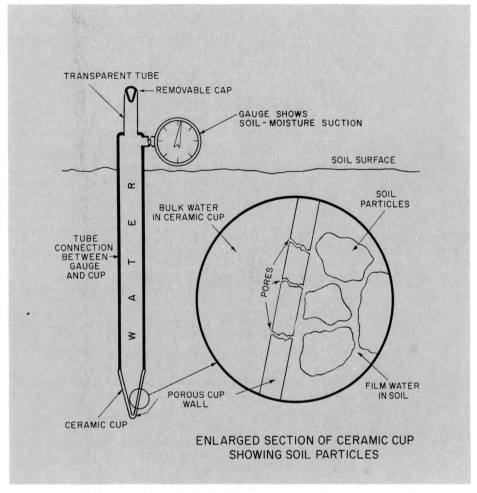

FIGURE 6-8 Sketch of a tensiometer inserted vertically into the surface of the soil, with an enlarged schematic diagram of a section of the porous ceramic cup in contact with soil water. Water in the soil should usually be kept sufficiently high in the dominant root depth to maintain a reading of 50 or below on the vacuum gauge. (*Source:* S. J. Richards and R. M. Hagan, "Soil Moisture Tensiometer," Extension Service Leaflet 100, California Agricultural Experiment Station, 1958.)

6:7 Estimating Water Contents

For rapid and frequent estimations of soil moisture contents at many depths, tensiometers, resistance blocks, and neutron probes are sometimes more convenient and quicker to use than the gravimetric method. **Tensiometers** measure the matric potential of soil moisture with the use of a porous-clay cup attached to a tube filled with water. The water in the cup and tube is attached to a vacuum gauge or a mercury manometer (Fig. 6-8). As the soil dries, water moves out through the porous cup, creating a suction or vacuum on the water column. These suction readings are then calibrated on the tensiometer gauge to interpret the percent of moisture in each soil.

6:7.1 Tensiometers Measure Matric Potential

Tensiometers can be used to schedule irrigation by placing one instrument at a depth of maximum root density (and activity); a second instrument may be placed near the bottom of the active root zone. A need for irrigation could be indicated by a reading or calculation of −500 cm of water (−50 kPa), for example, for the tensiometer in the active root zone and −400 cm or a little wetter (−40 kPa) for the deeper tensiometer. Automatic irrigation systems would have a vacuum-sensing valve attached to permanently installed tensiometers. Sugarcane and potatoes are two crops with which water control by tensiometer measurement has been successfully used. Orchards, ornamental nurseries, and turf farms have used tensiometers extensively, particularly in California. To relate vacuum to percentage water, calibration must be done for each soil used.

The principal limitation of tensiometers is that they do not measure soil matric potential values as low as the usual wilting values. The actual range of effective measurement is only from 0 to −85 kPa. Tensiometers are more useful for measuring moisture in sandy soils than in fine-textured ones because the matric potentials of most of the plant-available water are higher in sandy soils than in clayey soils.

6:7.2 Resistance Blocks Measure Electrical Conductivity

Resistance blocks are based upon the principle that electrical conductivity decreases with a decrease in soil moisture. In 1940 G. J. Bouyoucos of Michigan State University introduced a gypsum block, inside of which were two electrodes a fixed distance apart. The blocks were buried in soil, and the conductivity (or resistance) across the electrodes was measured with a modified Wheatstone bridge. With calibration, the moisture percentage from field capacity to wilting point was readily determined (Fig. 6-9). Newer, improved Bouyoucos blocks are made of nylon or fiberglass, which do not deteriorate in the soil as the gypsum blocks do. Bouyoucos blocks are not satisfactory in wet soils, but are useful in rangelands, dryland farming areas, or where irrigation is needed.

How is moisture content related to electrical conductivity in resistance blocks? Soluble salts (ions) in water carry electrons, which produce an electrical current; the more ions present, the more electrical current that flows. (Gypsum blocks furnish their own ions as they slowly dissolve.) The thicker the water films, the more ions that flow between the block's screen electrodes. (With nonsoluble blocks, electrical-current flow depends on the soluble salts in the soil and water film thickness.) For use, the conductivity (or resistance) reading for various moisture contents must be calibrated *for each soil* that the blocks would be used in. Each soil has different contents of soluble salts and of water at a given matric potential. Calibration of the Bouyoucos blocks should be redone periodically because the salt content of the soil may change or the gypsum blocks may deteriorate.

6:7.3 Neutron Attentuation Measures Slowed Neutrons

Neutron probes look like flashlight cylinders with long cords attached (Fig. 6-10). The probe contains radioactive material (radium–beryllium, americium–beryllium) that emits *fast-moving neutrons*. As the neutrons emitted from the probe collide with or deflect from equal-size hydrogen ions (of which water is a major source), they are slowed. Some of the *slowed* deflected neutrons are deflected back to the probe, where a counter measures only the slowed neutrons. The more slowed neutrons that return (indicating a large number of collisions), the greater the water content of the soil.

The neutron probe is used by lowering it into a hollow, vertically buried metal or plastic pipe, which is about 5 cm (2 in.) in diameter and as long as needed. The probe is suspended at the soil depth to be measured, the neutron count is taken for several seconds to a minute,

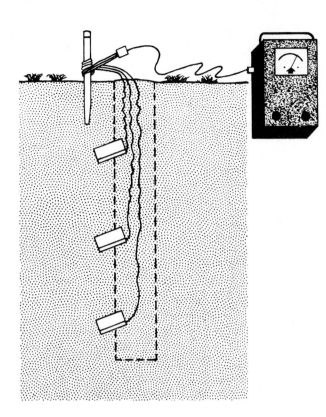

FIGURE 6-9 Electrical conductivity between two electrodes set a fixed distance apart inside a small block is an indirect measure of soil moisture from the field capacity to the wilting point. The blocks may be made of gypsum, nylon, fiberglass, or other porous material and can be buried in the soil at any desired depth, with wires from each block extending to the surface of the soil for easy reading of conductivity with the Bouyoucos bridge. (*Source:* USDA, Agricultural Handbook 107, 1957.)

FIGURE 6-10 The neutron probe (A) can be lowered into any open cylinder set vertically into the soil. The counter (B) can record a moisture reading in about 1 minute. When not in use, the probe is kept inserted in the shield (C). The neutron probe allows moisture readings to be taken as frequently as wanted without disturbing the soil. (Courtesy of Raymond W. Miller, Utah State University.)

and the moisture percentage is read. Although the neutron probe is easy to use, it is expensive to purchase, has legal restrictions because of the radioactive contents, must be calibrated for soil texture groupings by comparisons to samples measured by the gravimetric method, has delicate electronics that may be difficult to maintain, involves a fairly large soil volume (15-cm, or 6-in., radius) in its measured readings, and is not very practical in surface soil.

6:7.4 Gamma Ray Absorption in the Laboratory

Gamma rays are high-energy rays not easily absorbed by thin masses of matter. Gamma rays are produced by several sources; the two most used are cesium-137 (^{137}Cs, half-life of 30 years) and americium-241 (^{241}Am, half-life of 470 years). During calibration of the system, the soil solids in the volume tested remain about the same when dry and when wet. Some bulk density reduction may occur when the soil is wet, which would lower absorption of the gamma rays by the solids. The difference in gamma ray absorption by the dry soil and by the wet soil is due to absorption by water; thus, the "reading" is proportional to water content in the soil. Soil sample thickness used must be about 5–25 cm (2–10 in.). The optimum thickness depends on the gamma ray energy source used. The procedure is more adaptable for laboratory studies than for field work. It has been used on plant tissue, also.

6:8 Water Flow into and through Soils

Most rapid movement of water into and through soil is caused by gravity pull. **Saturated-flow** water moves because of water potentials larger (less negative) than −33 kPa. A much slower flow, **unsaturated flow,** occurs when the matric potential is low enough that the gravitational force is no longer strong enough to cause flow. Matric and osmotic potential forces predominate at water potentials lower (more negative) than −33 kPa. Unsaturated flow can be in any direction. Soils with layers of varied textures, such as alternating sands and low-permeability clays, may have water retained at potentials from −10 to −30 kPa; less is drainable by gravity flow than is expected in a uniformly textured soil.

6:8.1 Saturated Flow

Saturated flow is water flow caused by gravity's pull. It begins with **water infiltration,** which is water movement into soil when rain or irrigation water is on the soil surface. When the soil profile is wetted, the *movement of more water flowing through the wetted soil is termed* **percolation.** It is percolating water moving through the soil and substrata that carries away the nutrients and other salts dissolved from the soil (leaches the soil).

Water infiltration is rapid into large, continuous pores in the soil. It is reduced by anything that decreases either the size or amount of pore space or wettability, such as structure breakdown, pore clogging by lodged particles, and the slower movement of deeper water as it reaches denser subsoils. The factors that control the rate of water movement *into* the soil include these:

1. *Percentage of sand, silt, and clay.* Coarse sands permit rapid infiltration.

2. *Soil structure.* Fine-textured soil with large water-stable aggregates (granular structure) have higher infiltration rates than massive (structureless) soils.

3. *Amount of organic matter in the soil.* The greater the amount of organic matter and the coarser it is, the more water that enters the soil. Organic surface mulches are especially helpful in keeping infiltration high because they protect soil aggregates from breakdown by reducing the impact of raindrops and by continuing to supply the cementing agents for aggregates, such as gums, as the mulch decomposes.

FIGURE 6-11 This soil in Nevada has a hardpan cemented by carbonates and perhaps silica (*caliche* or *duripan*). The pan is so hard that this soil surveyor must use an air-driven jack hammer to dig through the pan to study the profile. Some pans are very shallow and hinder water infiltration. In all arid areas this is of limited concern until irrigated; in wetter climates a hardpan could cause overwetness. The hardpan begins at about 55 cm (22 in.) deep. (*Source:* USDA—Soil Conservation Service; photo by Vern Hugie.)

4. *Depth of the soil to hardpan, bedrock, or other impervious layers.* Shallow soils do not permit as much total water to enter as do deep soils, if they are similar in other respects, such as texture and structure (Fig. 6-11).

5. *Amount of water in the soil.* Wet soils do not have as high an infiltration rate as do moist or dry soils. This is partly because pores or cracks are fewer or smaller because clays have already wetted and swelled.

6. *Soil temperature.* Warm soils take in water faster than do cold soils. Frozen soils may or may not be capable of absorbing water, depending upon their porosity and water content when freezing took place.

7. *Compaction.* This usually reduces pore space and slows infiltration.

A typical water infiltration (intake) curve is shown in Fig. 6-12. Notice that with increased time and wetted depth, the infiltration *rate* decreases because (1) clays swell, which makes pores smaller and reduces water movement, (2) loose particles that flow with water become lodged in "bottleneck" pores and plug them, and (3) resistance to flow increases as water passes deeper through long pores (see Detail 6-1). A dry, cracked, clayey soil may take up a large amount of water during the first hour but very little during each succeeding hour, particularly if rainfall is dispersing the surface soil particles, and they move downward to plug pores.

Four infiltration rates have been classified by the National Cooperative Soil Survey.

1. *Very low:* soils with infiltration rates of less than 0.25 cm (0.1 in.) per hour; soils in this group are very high in percentage of clay.

2. *Low:* infiltration rates of 0.25–1.25 cm (0.1–0.5 in.) per hour; most of these soils are shallow, high in clay, or low in organic matter.

3. *Medium:* infiltration rates of 1.25–2.5 cm (0.5–1.0 in.) per hour; soils in this group are loams and silts.

4. *High:* rates of greater than 2.5 cm (1.0 in.) per hour; these are deep sands, deep well-aggregated silt loams, and some tropical soils with high porosity.

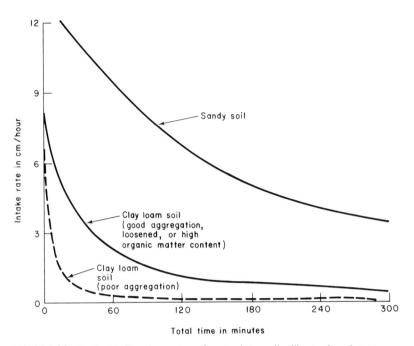

FIGURE 6-12 Typical infiltration rates of water into soils, illustrating the reduced intake rate with increased time of infiltration. The curves for clay loam soils may have various slopes (changes) in infiltration rates, as indicated by the dashed line, if particles begin to plug pores, or soil swelling reduces pore sizes. When no more loose particles are moving and swelling has reached a maximum, porosity remains constant and the curves of intake rate level off. (Courtesy of Raymond W. Miller, Utah State University.)

In the more humid regions percolation is common (Fig. 6-13). Excess water moves through the profile, dissolving soluble ions and carrying them into groundwater. The water also moves small soil particles (clays and organic colloids) downward until they lodge in pores or adsorb to ped surfaces. The leached soil layers lose most soluble salts and much of their adsorbed (exchangeable) calcium, potassium, magnesium, and sodium. Acidic (hydrogen) ions from carbonic acid (formed from carbon dioxide dissolved in water) and several other soil acids and soluble $Al(OH)_2^+$ ions replace these adsorbed basic cations. This percolation (leaching) causes humid-region soil gradually to become more acidic. Plant macronutrients lost by leaching (depending upon the solubility and total amounts of nutrients present in the soil in different forms) are generally:

- **Calcium (Ca):** the ion in largest amount in most leaching waters.

- **Magnesium, Sulfur, Potassium (Mg, S, K):** next largest amounts, but amounts depend upon soil composition.

- **Nitrogen (N):** low from cropped soils, unless there were recent nitrogen fertilizer additions. Large losses can be expected if the soil has a high natural fertility, was recently burned, or is barren of crops.

- **Phosphorus (P):** very little is leached because soil phosphorus forms are of low solubility.

The most common mathematical expression for the vertical water flow rate through soil is called **Darcy's Law**. Darcy stated that the rate of flow (Qw) was *increased* with an increased depth of water (dw) and the soil area through which it flowed (A). The flow decreased with an increased depth of soil (ds) through which the water flowed. Because each soil has a different combination of pore sizes and numbers of pores, each soil has a different flow rate constant (K). Because flow is downward and water loses potential (work must be done to put the water back where it started from), the downward values are negative. Darcy's equation may be written:

$$Qw = -K\,\frac{(dw)At}{ds}$$

or, to solve for K,

$$K = -\frac{Qw(ds)}{At(dw)}$$

Qw: water quantity, cm^3
K: rate constant, cm/s
dw: water height (head), cm

A: soil area, cm^2
t: time, (any time units may be used as long as units are given)
ds: soil depth used, cm

Some typical K values are given in the following examples. The 10 to the negative exponent means to move the decimal point that many places to the left. For example, 1.4×10^{-4} equals 0.00014. The K value indicates the soil's permeability to water flow. Small decimal values (e.g., 10^{-2}, 10^{-1}) indicate fast flow rates, while large negative exponent values (e.g., 10^{-9}) mean very slow rates.

Material	K (cm/s)
Gravel	1.5×10^{-1} to 2.0×10^{-2}
Sand	1.2×10^{-1} to 2.0×10^{-3}
Loam	1.7×10^{-4} to 1.7×10^{-7}
Clay	2.5×10^{-8} to 1.0×10^{-9}
Superstition sand (Arizona)	1.8×10^{-3}
Sarpy loam (Colorado)	1.4×10^{-3}
Millville silt loam (Utah)	4.7×10^{-4}

[a] Symbols used in this note are from R. J. Hanks and G. L. Ashcroft, *Applied Soil Physics*, Springer-Verlag, New York, 1980, pp. 62–65.

Water flow rate through tubes (uniform pores) increases with about the square of the diameter (d^2) of the tube. Thus, doubling the diameter of water-filled pores increases water flow rate about four times. However, soil pores are irregular shapes and constantly change size. Simple flow rate comparisons are less accurately known, although the general concepts should still be valid—larger pores have greatly increased flow rates. Unsaturated water movement is rapid through fine sand or well-aggregated loams (larger pores) and slower through very fine and poorly aggregated clayey soils (small pore sizes). Maximum flow rates may be as high as a couple of centimeters per hour but are usually much less.

Gravitational water partly drains from large pores, and the water is replaced by large air pockets that greatly reduce water flow rates. Water does not move appreciably from small water-filled pores into large air-filled pores, just as a wet sponge suspended in the air does not drip into

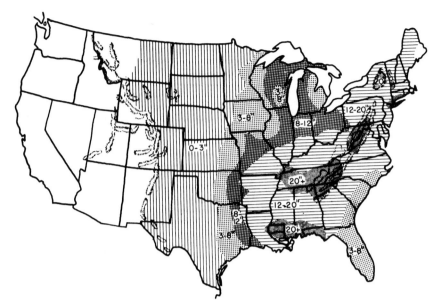

FIGURE 6-13 Average annual percolation. (*Source:* L. B. Nelson and R. E. Uhland, "Factors That Influence Loss of Fall Applied Fertilizers and Their Probable Importance in Different Sections of the United States," *Soil Science Society of America, Proceedings,* **19** [1955], no. 4.)

the air around it until gravitational water exists at the drip edge. Water flow in stratified soils is like this. Water flow from a fine-textured layer (such as clay loam) into a coarse-textured one (sand) is slow, except at water potentials near 0 (free-flowing water). Even portions of gravitational water may flow slowly from one layer to another of contrasting texture (stratified). Because of this phenomenon, nearly all *stratified* soils hold more water, as much as 50–60% more plant available water, than if the profile were of a single uniform texture. A stratified field soil usually holds more water available to plants than laboratory measurements of −33 kPa values would suggest. Laboratory measurements of −33 kPa water content for individual soil layers usually underestimate the amount of water the entire soil *profile* will hold when the field is wetted.

6:8.2 Unsaturated Flow

Unsaturated flow is the flow of water held with water potentials lower (more negative) than about −20 to −33 kPa. *Water will move toward the "drier" region (lower potential or toward the greater "pulling" force).* In a *uniform* textured and structured soil this means that water moves from wetter to drier areas. The water movement may be in *any* direction (Fig. 6-14). The *rate of flow is greater as the water potential gradient (the difference in potential between wet and dry) increases and as the size of water-filled pores also increases.* Unsaturated flow is much faster when the soil is wetted to near field capacity; it may be several centimeters per hour. When the soil is near permanent wilting percentage, the flow is slower, such as a few millimeters in several hours.

6:8.3 Water Flow in Small Containers

The importance of understanding water flow principles is illustrated by water retention in greenhouse pots or other small containers. Flow at water potentials less (more positive) than about −10 to −33 kPa (−0.1 to −0.3 bar) occurs only as long as water has contact with other water. The water

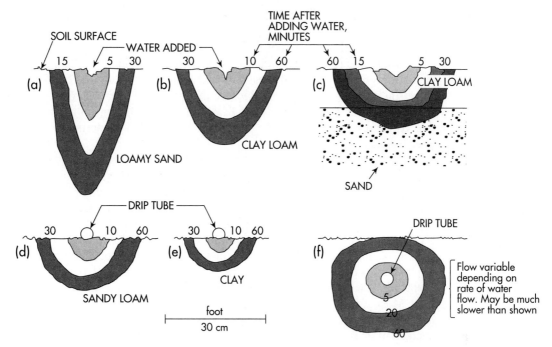

FIGURE 6-14 Typical wetting patterns expected for saturated and unsaturated flow as influenced by texture and location of the water source. *Saturated flow,* which always has unsaturated flow occurring at the same time, (a) is mostly vertical in coarse-textured soils, (b) has more lateral movement in fine-textured soils than occurred in (a), and (c) has a temporary inhibition to flow when sand is below a finer-textured soil. In (d) and (e), flow is still mostly saturated flow from drippers. In (f), water movement (unsaturated flow upward) is in all directions. Only the speed of flow, mostly because of pore size, varies with texture and water content. (Courtesy of Raymond W. Miller, Utah State University. See also Walter H. Gardner, "How Water Moves in the Soil," *Crops and Soil Magazine,* Nov. 1979, pp. 13–18.)

at the bottom of a flower pot with a hole in it is only in contact below it with air; the water does not easily drain from the bottom until the water potential approaches zero (saturated). Thus, drainage occurs only when the soil nears saturation. The lower part of the soil may often be too wet for good aeration unless its water is replaced daily by freshly aerated water (irrigation).

For well-drained pots, it is essential to avoid fine-textured soil in shallow pots. Coarse-textured materials—such as sand, sawdust, peat, bark, or perlite—mixed with the soil may permit better drainage because of large pore sizes. Approximately 3 volumes of fine material to 10 parts of coarse material (less of peat or bark) are needed for a suitable mixture with large pores that will drain adequately.

6:9 Water Uptake by Plants

The storage and release of water in the soil form only half the soil water story. Other vital questions are: (1) How do the plants absorb water? (2) From what soil layers is it obtained? and (3) When is water most needed?

6:9.1 Water Absorption Mechanisms

Although a number of mechanisms are known to affect root absorption of water, *over 90%* of the total water absorption is by passive absorption. **Passive absorption** is the pulling force on soil water by the continuous water column up through the plant cells as water is lost by transpiration. The plant can be thought of as a wick or sponge losing water at one end by transpiration while dipped in water at the other, the water moving slowly through the permeable membranes and through porous cell walls and xylem tubes.

Root extension is important to the absorption mechanism. Plant root systems are not static; some roots die as new ones grow and expand into new areas of the soil, encountering more soil water as roots extend. Because the water movement to roots by *unsaturated flow* occurs only over short distances of a few millimeters (fractions of an inch) per day, root extension is important in aiding the plant's absorption of water, particularly when the soil root zone holds no gravitational water.

Active absorption requires that the plant expend energy to absorb the water. The selective accumulation of soluble ions by the plant cells increases a plant's soluble salt content (osmotic potential). This osmotic potential aids plants in water uptake but requires plant energy to produce the osmotic potential. That is, the plant must use energy to bring ions into root cells. Water enters the plant roots because of net movement toward areas of high salt (the osmotic effect). Active absorption accounts for a considerable portion of absorbed water only during *low water needs*. Plants can also obtain some moisture from fog, rain, or dew by absorption through the leaf stomata.

A few plants called **epiphytes,** such as orchids and Spanish moss, absorb water and nutrients from the air (also some nutrients from decaying organic materials in tree forks, and the like). Most of these plants grow in rain forest–type moist regions. Yet many plants can absorb an appreciable amount of water through the leaves from dew. An interesting effect of using condensation, such as dew, is described in Detail 6-2.

6:9.2 Depths of Water Extraction

Most water for irrigated crops is extracted from shallow soil depths, where most roots reside. In the few days following wetting of the soil, 40% or more of a plant's water may come from the top 30 cm (12 in.) (Fig. 6-15). Golf greens may be irrigated with little water but irrigation must occur almost daily because of shallow-rooted grasses and because sand and gravel "soil" comprise the green's base. This coarse-textured base holds little water and few nutrients but resists deformation by people walking on it. In contrast to the grasses, mesquite trees, which grow in arid soils, can remove appreciable water from great distances laterally and from subsoils 2.5–3.0 m (8–10 ft) deep. Turf grasses on moist clayey soils take most of their water from a small area of the top 20–30 cm (8–12 in.) of soil most of the time. As the plants grow and the soil dries between wettings, an increasing portion of the water taken up by the plant comes from deeper soil layers until three-fourths of the water absorbed in a given day by deep-rooted plants could come from depths below 60 cm (2 ft). Keeping the top 30 cm (1 ft) of soil moist is important to plant growth because most roots (and nutrients) exist there.

Soil wetness and adequate aeration will always be competing conditions in some soils. Usually clayey and compacted soils are the most likely to have poor aeration as they become wetter. The larger and more continuous the pores, the more likely that layer is to have good air exchange and supply deep roots with adequate oxygen. Some examples are shown in Fig. 6-16.

6:9.3 When Plants Most Need Water

It is obvious that plants are in need of water when they begin to wilt, but by the time wilting is visible, plant growth has already been reduced significantly. Wilting usually is due to insufficient soil moisture. However, it can occur on a well-moistened soil on warm, dry days because the water loss by evapotranspiration is greater than the absorption rate of water by plant roots. Although plants usually recover from such temporary wilting during the cooler evening or night periods, some growth loss has usually happened. Many small-grain crops can exhibit temporary wilting during later growth without extensive loss in grain yield. However, drying to a condition of wilt will reduce *vegetative* growth of nearly all plants.

Some plants grow well in salty water; others do not. Some plants grow fairly well in salty water in coarse-textured soils but do not grow well with salty water in fine-textured soils. In the first example, the salt tolerance of plants is the dominant factor determining growth. In the second example, the phenomenon may be the processes of distillation (vaporization) and condensation (forming liquid) at work.

When coarse materials (sands and gravels) are wetted with salty water, the plant roots may not remain submerged in salt water. After normal drainage, the roots extending into large soil pores may be mostly in soil air, not submerged in salty water (see diagram a).* The soil air is near 100% relative humidity whether the soil is dry or moist. Slight temperature fluctuations allow water vapor to condense in cool periods on the exposed root surface (diagram b). This condensed water is distilled; thus, it contains no salts and can be absorbed easily by plants. Thin salt water films continue to supply a source of evaporating water and some nutrients.

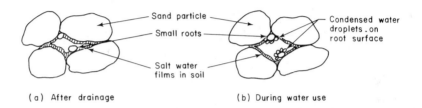

(a) After drainage   (b) During water use

Water use calculations by French scientists caused them to estimate that common plants such as peas, tomatoes, corn, and cabbage have the capacity to absorb in their hair roots more than 55 times the maximum water the plants actually need for transpiration. A single small crop plant may accumulate one to two liters of "dew" on its roots per morning.

* The diagram in this note was drawn by Raymond W. Miller (Utah State University), from a concept by Hugo Boyko (ed.), *Salinity and Aridity: New Approaches to Old Problems*, Dr. W. Junk, Publishers, The Hague, The Netherlands, 1966.

Sometimes damage to the plant can occur even with only a few dry days. Drought stress for 10 days (during the 30- to 40-day period after flowering) reduces cottonseed germination from 100% to 20%.[2] Drought stress before or after this period has almost no effect on germination. Drought stress seems to affect germination most during the critical stage of "protein filling" in the seed.

Most plants have critical periods of growth during which a lack of water is most damaging. For plants producing seed, the period from flowering to seed set (fertilization) is the most critical. If drought occurs during this period, grain or seed yield will be greatly reduced. Table 6-3 shows yield reduction of several crops caused by drought. Similar reductions also apply to fruits and fiber-producing plants such as cotton.

The effects of **water stress** (lack of adequate water) vary with the kind of plant. Small reddish spots 1–3 mm across (up to 0.1 in.) on lower leaf surfaces of the Indian rubber plant (*Ficus elastica* 'Robusta') are now thought to result from water stress and high light effects rather than from inadequate potassium.[3]

---

[2] E. L. Vigil, "Deadliest Days of Drought Stress," *Agricultural Research*, **37** (1989), no. 7, p. 19.
[3] Timothy K. Broschat and Henry M. Donselman, "Effects of Light Intensity, Air Layering, and Water Stress on Leaf Diffusive Resistance and Incidence of Leaf Spotting in *Ficus elastica*," *HortScience*, **16** (1981), no. 2, pp. 211–212.

For maximum vegetative growth (alfalfa, sugarcane, pastures), water is most important during rapid size increase; for fruits and tubers, water is needed during enlargement; for tillering grain crops (oats, wheat), water is critical during tillering and seed fertilization periods.

6:10 Consumptive Use and Water Efficiency

The amount of water needed for plant growth includes the water lost in two ways: (1) The **transpiration** loss and (2) the **evaporation** loss from the soil. The two losses are usually combined and called **evapotranspiration loss** (**ET** or E_t). **Consumptive use** (**CU**) is the

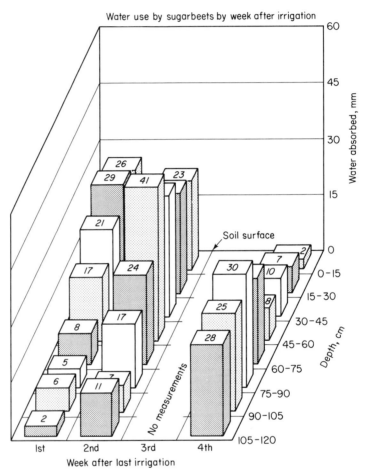

FIGURE 6-15 Amounts of water absorbed by sugar beet roots from different soil depths during 1-week periods following irrigation. Soil is a highly calcareous soil of pH 8.0. Notice that when the soil is wet, 76 mm (67%) of the water absorbed is taken from the top 45 cm (17.5 in.) of soil (first week). In the second week after irrigation, 80 mm (but only 46%) is absorbed from the top 45 cm. During the fourth week, the surface soil is quite dry. The deep portion of soil (75–120 cm; 29–47 in.) supplies 63% (83 mm), and the top 45 cm supplies only 14% (19 mm). The kind of crop, rooting characteristics, and other factors will alter this extraction pattern. (1 mm = 0.039 in.) (Source: Redrawn by Raymond W. Miller from data of S. A. Taylor, "Soil," *Yearbook of Agriculture*, USDA, 1957, p. 65.)

FIGURE 6-16 Typical curves for changes in oxygen content with changes in soil depth for four soil conditions. (A) a moist loamy sand with large, continuous air-filled pores; (B) a clay loam with smaller and less continuous pores; (C) a loam surface **A** horizon with a clayey **Bt** horizon and clay loam subsoil layers; and (D) a clay loam soil after being wetted in which water fills many pores, thus blocking oxygen exchange to deeper soil depths. A clayey or wet soil becomes, in essence, a shallow soil because of poor aeration in its deeper layers. (Courtesy of Raymond W. Miller, Utah State University.)

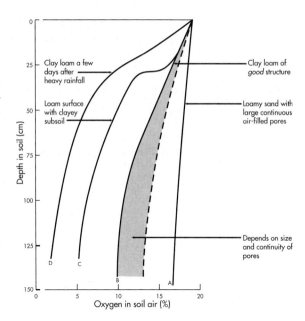

Table 6-3 Effects of Water Stress (Dryness) on Crop Yields

Condition or Time of Water Deficiency (Stress)	Yield Reduction (%)
Five weeks prior to ear emergence	
Wheat grain	70
Total wheat plant tops	52
Stress 20 days during male meiosis (chromosome numbers halved)	
Total corn plant tops	29
Corn grain	0
Stress 20 days during grain filling	
Total corn plant tops	30
Corn grain	47
Texas, sorghum grain, emerging grain head, reached −12.9 bars	17
Texas, sorghum grain, emerging grain through seed set, −13.0 bars	34
Texas, sorghum grain, milk through soft dough stage, reached −12.4 bars	10

Growth Stage When Water Stress (Dryness) Reduces Yields the Most[a]	Crops that Fit This Category
1. Need high water levels constantly	Castor bean, cauliflower
2. From flower initiation to early stages of developing fruit or seed	Apricots, barley, beans, citrus, corn, cotton, oats, olives, peanuts, peas, small grains, soybeans, sunflower, tomatoes, wheat
3. During rapid fruit or tuber growth to harvest	Cherries, lettuce, peaches, potatoes, strawberries, turnips, watermelon
4. During head formation and enlargement	Broccoli, cabbage, lettuce
5. During tillering,[b] prior to heading	Oats, sorghum
6. Period of maximum vegetative growth rate	Alfalfa, sugarcane, tobacco

[a] Food and Agricultural Organization of the United Nations, "Crop Water Requirements," Irrigation and Drainage Paper 24, FAO, Rome, 1975.
[b] Shoot development.

quantity of water lost by evapotranspiration (ET or E_t) plus that contained in plant tissues. The latter water is only about 0.1% of the total consumptive use for slow-growing plants and may be up to about 1% for plants efficient in water use. Measurements of water loss often require elaborate setups for accurate readings (Fig. 6-17).

6:10.1 Evapotranspiration

Evapotranspiration, the water lost by evaporation from soil and transpiration from plants, *increases* when the air is dry (low relative humidity), warm, and moving (windy), and if the surface soil is near its field capacity. Evaporation *decreases* when the opposite conditions occur: air has high relative humidity, temperatures are cool, there is not wind, and soil is quite dry (Fig. 6-18). ET_p is the potential ET loss that would occur if the soil was kept fully moistened for maximum ET.

Evapotranspiration involves a large amount of water, but the water loss is usually less than the amount of water that would evaporate from ponded water covering an equal area under similar atmospheric conditions. The amount of water lost is estimated using a weather station **Class A evaporation pan.** The water loss from mature plants is about 0.5 to 0.9 as much (50% to 90% as much) as is lost from a free water surface with the same area (Table 6-4). The amounts of these daily losses of water range from perhaps a tenth of a centimeter per day during early growth in cool weather to more than 1.3 cm (about 0.5 in.) daily during hot, dry weather (Table 6-5).

6:10.2 Water Use Efficiency (WUE)

The amount of water (transpiration, plant growth, evaporation from the soil, drainage loss) required to produce a unit of dry weight material (a kilogram of corn, for example) is a measure of **water use efficiency.**[4] How efficiently can water be used by plants? The **transpiration ratio (TR)** is the amount of water used per unit of dry matter produced. The transpiration ratio is the weight of water transpired divided by the weight of dry plant material produced (only aboveground plant portions are used, except for root crops, such as sugarbeets and potatoes). Transpiration ratios range from 200 to 1000 (200:1 to 1000:1), with 300 to 700 the most common values. Extremely hot, dry, irrigated regions typically have ratios of 600 to 1000.

Transpiration ratios may be used to compute water needed for irrigation of particular crops. For example, if a field of alfalfa produces 10 metric tons per hectare (4.46 t/a) and has a transpiration ratio of 500, the water used would be 500 times 10,000 kg, which equals 5,000,000 kg of water needed to produce the crop—about 50 hectare-cm (about 20 in.) of water. (An acre-inch of water weighs 226,500 lb; a ha-cm of water weighs 100,000 kg, or 220,500 lb.)

6:11 Reducing Water Losses

Reducing water losses from a local area (farm, small watershed) can only be accomplished by altering the conditions effecting the losses: evapotranspiration, runoff waste, and percolation through soil into groundwaters. Movement of water by runoff and percolation is not always a "loss," as the water is useful elsewhere.

---

[4] Engineers may use a different definition: the net amount of water added to the root zone as a fraction of the amount of water taken from some source. (Source: Daniel Hillel, ed., *Optimizing the Soil Physical Environment toward Greater Crop Yields*, Academic Press, New York, 1972, p. 90.)

(a)

(b)

FIGURE 6-17 Consumptive use of water and leaching losses are determined with precision by research studies using lysimeters, which are sealed containers of various sizes with access or drainage tubes. (a) The 20,430-kg (45,040-lb) soil core in this lysimeter in Pawnee, Colorado, was lowered into a metal-lined hole onto large balance beams to weigh water losses and gains. (b) The lysimeter in place. Lines shown are power and other electrical lines to an adjacent trailer laboratory for making various recordings and measurements. (*Source:* International Biological Programs Grassland Biome Study, Colorado State University; photos by Alan Brooks.)

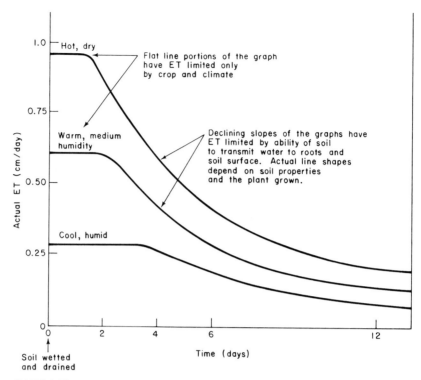

FIGURE 6-18 Illustration of the rates of evapotranspiration as affected by climatic conditions and by the ability of the soil to supply water to the plant. (Courtesy of Raymond W. Miller, Utah State University.)

6:11.1 Reducing Evapotranspiration

Evapotranspiration losses can be reduced by selecting more growth-efficient plants, by reducing total growth, by shading or cooling the area, or by using a moisture barrier.

Mulches   Mulches (straw, peat, gravel, formed aggregates, transparent and opaque plastic sheeting, asphalt) act as barriers to moisture movement out of the soil. Most mulches also keep the soil temperature lower by **shading** it. Except for nurseries or landscaping purposes, mulches have not been considered economically practical for the sole purpose of evaporation control. Most often they are used for the multiple purposes of weed control, soil erosion control, an aid to water infiltration, a barrier that prevents fruits and berries from rotting because of soil contact, a need to use crop residues, and a control of evaporation. For example, early-season irrigation in the San Joaquin Valley of California accentuates the drainage problem. Clear-plastic mulch placed over Californian soils planted in cotton in March saved enough water for germination and avoided a preseason irrigation.[5] For growing Pima cotton, this plastic mulch increased yields, and profits were calculated to be about $450 per acre (about $1112 per hectare).

A common misconception in the use of surface mulches is that they always result in a final moisture saving. A mulch will reduce evaporation loss in the first days or a week or so following

---

[5] E. Fereres and D. A. Goldhamer, "Plastic Mulch Increases Cotton Yield, Reduces Need for Preseason Irrigation," *California Agriculture*, **45** (No. 4): .1991.

Table 6-4 Water Required for Evapotranspiration of Selected Crops Given as a Percentage of the Amount of Water Evaporated from an Open Water Surface (Class A Weather Station Pan)[a]

	Percentage of Free Water Evaporation in:			
Crop	January	April	August	November
Alfalfa	55	85	90	70
Citrus	50	55	55	50
Deciduous orchard	15	50	65	15
Grapes	15	40	60	25
Pasture grass	40	65	70	50
Pangola grass[b]	115	100	95	100
Plantain (cooking banana)[b]	80	85	75	110
Sugarcane[c]	75	50	75	90

ET as a Percentage of Maximum ET Values[d]

	For Western U.S. Areas[d]			
Crop	Seedling Stage 20 Days after Planting	Has Developed 50% of Full Cover	When the Full Plant Cover Is Reached	50 Days after Reaching Full Cover (Maturing)
Alfalfa	41	79	100	100
Beans	21	51	107	59
Corn	21	49	49	68
Pasture (full cover)	87	87	87	87
Peas	22	51	105	20
Potatoes	12	41	91	38
Small grains	17	51	104	19
Sugar beets	12	41	91	90

Sources: Selected and modified data from (1) J. E. Christiansen and G. H. Hargreaves, "Irrigation Requirements from Evaporation," *International Commission on Irrigation and Drainage, Seventh Congress*, 1968, R. 36, Question 23, p. 23.593. (2) M. E. Jensen, ed., *Consumptive Use of Water and Irrigation Water Requirements*, Report of ASCE Committee on Irrigation Water Requirements, American Society of Civil Engineers, New York, 1973, 215 pp.

[a] The depth of water used by the plant is estimated by multiplying the percentage shown, in decimal form, times the depth of water evaporated from a Class A weather station pan.

[b] South America (Colombia), tropical area with rainy weather May to August, very dry January to March.

[c] Planted in March, harvested a year later in April, with maximum growth October to February.

[d] These values assume that pan evaporation equals potential (maximum) evapotranspiration. This may not always be true; pan values may need adjustment first. Pan values vary such that potential evapotranspiration is 0.7 to 1.1 times the pan evaporation data.

wetting, but over a *long* drying period, the water loss from a mulch-covered soil approaches the same amount as that from uncovered soil. Barrier-type mulches (plastic sheeting, asphalt covers) are much more effective than natural organic (porous) ones and will greatly reduce evaporative water loss for a long time (Fig. 6-19). Any shading cover (plant residues, seed hulls, gravel, rocks, wood chips) will reduce evaporative water losses for several days, and if these days are during the sprouting and new-growth periods, the moisture saving can be critical to the survival of the plants.

Petroleum mulches sprayed as narrow thin strips above the seed row increased both temperature and moisture retention in southern California. Strips 15–22.5 cm (6–9 in.) wide and 10–20 mm (0.4–0.8 in.) thick helped seeds planted in February and March to germinate and develop more-vigorous seedlings.

Table 6-5 Some Consumptive Use Rates of Various Plants in Various Locations and Situations

Plant	Use Condition and Location	Consumptive Use Rate (ET)	
		mm/Day	in./Day
Cotton	Peak use (blooming), July, Salt River Valley, Arizona	10.1	0.40
Oranges, navel	July, Orange County, California	11.7	0.46
Potatoes	Most active growth, warm, dry	6.1–8.1	0.24–0.32
Safflower	Peak growth, May, Mesa, Arizona	12.7	0.50
Soybeans	Peak use, Nebraska	11.7	0.46
Sugar beets	Peak use, Davis, California	7.6	0.30
Wheat, winter	Late fall, seedling stage	1.0–2.0	0.04–0.08
	Dormant winter period	0.2–1.0	0.008–0.04
	Peak growth, windy, arid	8.1–9.1	0.32–0.36

Source: J. S. Robins, J. T. Musick, D. C. Finfrock, and H. F. Rhoades, "Grain and Field Crops," Chapter 32; D. W. Henderson, "Sugar, Oil, and Fiber Crops, Part III—Oil Crops," Chapter 33; and J. R. Stockton, J. R. Carreker, and M. Hoover, "Sugar, Oil, and Fiber Crops, Part IV—Irrigation of Cotton and Other Fiber Crops," Chapter 33, all in *Irrigation of Agricultural Lands,* R. M. Hagan, H. R. Haise, and T. W. Edminster, eds., No. 11 in the Agronomy Series, American Society of Agronomy, Madison, Wis., 1967.

FIGURE 6-19 Plastic mulches can conserve water and warm the beds earlier. These plastic covers are having holes punched for transplanting peppers near Irvine, California. Buried drip system lines help keep the water costs down ($494/ha–30 cm or $200/a–ft). The plastic mulch allows early warming, reduces water evaporation, and prevents low-hanging fruit from touching a wet soil and rotting. The system is also used for strawberries and tomatoes. (*Source: Ag Consultant Magazine,* Aug. 1987, pp. 4–5; photo by Parry Klassen.)

Fallow One of the distinctive features of dryland grain production is the use of fallow. Fallow is the practice of leaving land unplanted, commonly in alternate years, to accumulate a little extra water in the fallow year to be used in the next growing year. During fallow, the fields are clean-cultivated to hinder weed and volunteer grain development and so save water that would have been lost by transpiration.

The water saved by fallowing is small, but the conservation of even a small part of the year's water can mean the difference between a good or poor yield of grain, particularly if soil water is low without fallow during germination and early growth periods.

Approximately 10 cm (4 in.) of *available water* are required to produce wheat plants from seed to maturity. Each additional inch of water available in the soil produces an additional 280–470 kg/ha (4–7 bu/a) of wheat. Yield increases also depend upon climate, soil characteristics, and adequate soil fertility.

Plant Selection   Altering the crop can reduce ET. Sorghum uses less water than corn on the same site. Short-season crops such as lettuce and beans can be used to reduce water needs. Even watershed areas can have vegetation altered to save water. Between 4 and 8 million hectares (about 10–20 million acres) of land in California lie in foothill areas, receive 380–510 mm (15–20 in.) of annual precipitation, and support mainly low-value trees and brush. There are similar areas in most of the Western states. The brush species are usually deep rooted and use all the available moisture in the soil. Furthermore, the brush, leaves, and stems intercept a large amount of precipitation that is evaporated directly back into the atmosphere. Killing sagebrush by spraying in Wyoming reduced moisture loss 24% between June and September.

Research on the conversion of such brush land to adapted grasses has demonstrated that:

1.   Grasses root less deeply than trees and brush.

2.   Grasses become dormant earlier in the fall and thus leave more stored water in the soil for early-spring growth, if winters are dry.

3.   Grasses intercept less precipitation, thus permitting more water to enter the soil.

4.   Grasses protect the soil from erosion better than does brush.

5.   Grasses permit more runoff water to be caught behind dams for use by municipalities and in irrigation. On some watersheds, replacing trees and shrubs with grasses has saved as much as 5 cm (2 in.) more water in runoff per year.

Forest trees, like all other plants, transpire large amounts of water as a vapor into the atmosphere through their leaves. Forests also intercept considerable quantities of rain or snow and permit it to be evaporated back into the atmosphere before the water ever reaches the soil. By removing part of the forest cover, more water reaches the soil and less is transpired; although evaporation from the soil might be increased by the cutting, the net result is less water loss. Clearcutting (total harvesting) of hardwoods in West Virginia resulted in only about one-third as much water loss by evapotranspiration as occurred in uncut forests.

6:11.2 Evaporation Loss from Water Reservoirs

Evaporation from reservoirs can exceed well over a meter (3.3 ft) per year (about 2 m, or 6.6 ft., at Lake Mead). Although no general recommendations can be made to fit all circumstances, surface evaporation can be reduced by both organic surface coatings and by floating solids.

Certain long-chain organic alcohol molecules (hexadecanol, octadecanol), which have high evaporation temperatures (so they do not readily evaporate), spread themselves out in a thin film on the water. The nonevaporative coating makes a barrier to the evaporation of the reservoir water. Winds or fish can disturb the water, continually breaking the film, but it reforms; constantly choppy water also reduces film effectiveness. The cost of coating reservoirs is high, and some films are so airtight that fish in the water beneath suffocate.

Floating bits of plastics, other floating granules, or bubbled concrete (which floats) reduce evaporation. These are costly and not available yet; also, they are not yet studied adequately.

6:11.3 Reducing Liquid Water Losses

Irrigation techniques can conserve water. Runoff losses in furrow irrigation can be reduced by cutting down the size of the stream for the duration of soaking after the water has reached the furrow end. Surge flow reduces deep percolation. Land leveling increases surface irrigation water efficiency by allowing better water control. Irrigating more frequently, but only wetting alternate furrows each time, encourages maximum water use and minimum evaporation and drainage loss. The greatest water conservation, particularly on sandy soils is with drip and sprinkler irrigation. Both drip and sprinkling irrigation are efficient watering systems adaptable to hilly or irregular slope areas, shallow soils, sandy soils, and slowly permeable clay soils.

In dry areas level terraces with the lower end closed maximize water retention. In the 510-mm (20-in.) rainfall belt of Texas, closed-end terraces impounded more water and resulted in a 6% increase in the yield of cotton compared with cotton grown on land not terraced, based on average yields over a 26-year period.

Conservation terraces are small one-field watersheds that collect runoff on sloped areas for use on an adjacent more level area. Conservation terraces have about two to four times as much area in watershed as in the planted level terrace area, or 2:1 to 4:1 watershed: terrace planting area ratios. The crop on the flat water-receiving area should be of sufficiently high value to justify the cost if much soil needs to be moved in preparation of the area. Conservation terraces with *collection:cultivated* land area ratios of 20:1 to 30:1 are used in the very dry areas of modern Israel, similar to those found in ancient agricultural relics of the Negev Desert. In a rainfall area of 150–200 mm (6–8 in.) per year, runoff into basins supplies enough water to grow peaches, apricots, grapes, figs, pomegranates, almonds, grain crops, and pastures.

Reuse Waste and Runoff Waters
Waste waters include liquids from various industries and food-processing plants. The most reused waste is irrigation runoff tailwaters. Pump the tailwater drainage back to the upper fields. Recent legislation enacted in California presents the irrigator with a dilemma: If the drainage water has accumulated considerable salts, it must be desalted on the farm. The new laws restrict the dumping of such high-salt waters back into rivers.

Water is more critical than energy. We have alternative sources of energy. But with water, there is no other choice.

—Eugene Odum

A river is more than an amenity—it is a treasure.

—Oliver Wendell Holmes

If there is magic on this planet, it is in water.

—Loren Eisley

Questions

1. How does hydrogen bonding hold water to soil mineral surfaces? Illustrate.
2. Define (a) *water potential* and (b) *matric potential.*
3. Relate to each other these three terms that have been used to indicate the water potential in soils: (a) atmospheres of suction, (b) bars of suction, and (c) kilopascals of pressure.
4. Which water in a film of water on a soil particle is held most strongly to the soil, the water near the particle or the water at the outer edge of the film? Explain.
5. Define (a) *plant-available water* and (b) *field capacity.*
6. Define (a) *saturation percentage* and (b) *permanent wilting percentage.*
7. Give some approximate mass water percentages for the field capacity and wilting percentage conditions for (a) a loamy sand, (b) a loam, and (c) a clay soil.
8. Calculate mass water ratio, if the field capacity of a wetted soil plus can weighs 248 g, the dry soil plus can weighs 233 g, and the can weight is 141 g.
9. If the loam soil's bulk density is 1.3 Mg/m^3, calculate the volume water ratio for the soil in question 8.
10. (a) How much water (in cm of water per 30 cm depth of soil) does the soil described in questions 8 and 9 contain at field capacity? (b) About how many centimeters of plant-available water will the soil hold in the top 90 cm? Assume that it is uniform in depth.
11. How deep would a 4-cm rain wet the soil described in questions 8, 9, and 10 if the soil at the time of the rain was at 10% mass water percentage?
12. Describe how each device is used to estimate water content and what is measured in each method: (a) resistance blocks, (b) tensiometers, and (c) neutron probes.
13. (a) Define *unsaturated flow.* (b) How fast and how far might unsaturated flow wet a dry loam from an adjacent wetted area of the loam?
14. (a) Briefly discuss the soil factors that influence water infiltration rates. (b) Discuss the factors that can be changed by the land user.
15. Explain why some potted plants may be "too wet" in much of the root zone even when the top soil seems well drained.
16. Define and discuss *passive water uptake* by plants.
17. From which soil depths do plants extract the most water (a) during the few days after irrigation and (b) during the few days prior to the next irrigation (assuming 15 days between irrigations)? Explain your answers.
18. Discuss the statement, "plants wilt even when they can still extract soil water." [Consider (a) hot, windy days and (b) rates of uptake in a nearly dry soil.]
19. (a) Does water shortage for 2 weeks have the same effect on all crop yields? (b) Does the stage of growth when water was short have any effect? Explain.
20. (a) Define *ET.* (b) Give some typical daily ET rates for several crops and locations.
21. (a) How do the loses by ET compare to water losses from a pan of water at the same location and time? (b) Why is this relationship important?
22. If an average transpiration ratio is about 450:1, why is the TR of some crops higher and that of other crops lower than this?
23. (a) How do mulches influence ET losses? (b) Explain how each kind of mulch accomplishes this: (1) plastic, (2) plant residue, and (3) dry soil = fallow.
24. How can water losses (ET and liquid losses) be reduced?

Organisms and Their Residues

Waste is a human concept. In nature nothing is wasted, for everything is part of a continuous cycle. Even the death of a creature provides nutrients that will eventually be reincorporated in the chain of life.

—Denis Hayes

7:1 Preview and Important Facts

PREVIEW

Soils appear to be inert masses of minerals, but, in fact, they teem with microorganisms and plant roots. Living organisms and dead organic matter in soils are constantly changing. Soil organisms mix and aerate soil, fix atmospheric N_2, decompose dead organic substances, oxidize many elements, and recycle nutrients. Some organisms cause horrendous damage by reducing or destroying plant yields and by spreading animal infections and disease. For example, in the United States, parasitic nematodes (small roundworms) in the soil cause an estimated plant loss of about $2 billion annually; soil-borne diseases, such as *Pythium* root rot of cereals and *Fusarium* wilts of fruits and vegetables, cause an estimated billion-dollar damage annually; and various large larvae (grubs, cutworms, wireworms, root maggots, and sod webworms) destroy several million dollars worth of crops annually. These losses are much greater in areas of the world where herbicides, fungicides, and insecticides are not commonly used.

Although the detrimental soil organisms are many, the beneficial ones exceed them in effect. Fungi and bacteria are the most important organic matter decomposers. Both fungi and bacteria form **symbiotic** relationships in which two different organisms live in association, with both benefiting. **Mycorrhizae,** the symbiotic root-fungi association is found most dramatically on roots of trees but now is known to be possible on most plants. Mycorrhiza aid nutrient and water uptake. **Rhizobia** (N_2-fixing bacteria) form nitrogen-fixing nodules most commonly on legumes; similar microbe–plant relationships exist on many other plants, even on corn and other grasses. Certain bacteria called **autotrophs** get their energy from oxidizing important soil elements, including nitrogen, sulfur, iron, manganese, and carbon monoxide. Some **algae** fix atmospheric nitrogen and become part of the organic matter. **Actinomycetes** fix N_2 and help decomposition of organic materials. **Protozoa** are also decomposers and act as predators on bacteria.

Although most cultivated soils contain only 1–5% organic matter, which is mostly in the top 25 cm (10 in.) of soil, that small amount can modify a soil's physical properties and strongly affect its chemical and biological properties. **Soil organic matter,** from living or dead plant and animal residue, is a very active and important portion of the soil. It is the nitrogen reservoir; it furnishes large portions of the soil phosphorus and sulfur; it protects soils against erosion; it supplies the cementing substances for desirable aggregate formation; and it loosens up the soil to provide better aeration and water movement. For maximum benefit, organic matter must be readily decomposable and continuously replenished with fresh residues—roots, tops, and manures.

Microorganisms decompose organic materials. **Enzymes** are produced by the decomposer and are directly responsible for the decomposition. Enzymes reduce the **activation energy** necessary to break the bonds of the organic materials. Many different enzymes are needed for decomposition of the complex variety of organic substances. The amount of available nitrogen is one control on the soil: It greatly affects the rate of decomposition. Residues with carbon/nitrogen ratios wider than about 30:1 will lack adequate nitrogen for optimum decomposition rates. Other factors that promote optimum organic-matter decomposition are near-neutral soil pH, moist soil, temperatures about 29–35°C (85–95°F), and adequate levels of all nutrients. After the active decomposition has taken place, organic-matter residues are collectively called **humus.**

Organic matter (humus) is constantly undergoing change and must be replenished continuously to maintain soil productivity. The sources already mentioned can be supplemented by bringing in organic amendments, such as animal manures, municipal sewage sludge and septage, logging and wood-manufacturing refuse, industrial organic residues, and food-processing residues. Of the total residues produced, crop residues comprise about 70%, animal

FIGURE 7-1 Animal manures are excellent soil amendments, but very little is applied to cultivated fields in the tropics and subtropics. Only about 20% of the collected animal manures in India is applied to fields; the other 80% is made into manure cakes for fuel used to cook meals. (Courtesy of Roy L. Donahue.)

manures about 23%, and logging and wood manufacturing wastes about 5%. In many countries some of the residues available are used for purposes other than to improve the soil, such as for building structures and for fuel (Figs. 7-1 and 7-2).

IMPORTANT FACTS TO KNOW

1. The value of earthworms to the growth of plants
2. The problems caused by nematodes and the practical solutions to those problems
3. The differences between the "bulk soil" of a field and the soil of the rhizosphere, particularfly in pH, microbial activity, and in nutrient availability
4. The importance of fungi and the many roles they play, particularly in decomposition, production of toxins, and as mycorrhizae associated with plant roots
5. The importance of bacteria in soils, particularly their role in decomposition, autotrophic roles, and as diseases. Compare their sizes to sizes of pores and clay particles, and know rates of population doubling

FIGURE 7-2 A villager in Afghanistan digs alfalfa roots for use as fuel. Woody residues have already been used so intensively that they are difficult to obtain and alternative materials are sought. The same tillage method is used to loosen the soil prior to planting a new crop. (Courtesy of University of Illinois.)

6. The particular value of symbiotic N_2 fixers, including how much nitrogen is fixed by several common crops per year, how it is fixed (what are "nodules"), and how it benefits plants
7. The nature and value of actinomycetes
8. The general conditions favoring the growth of microbes and the methods and limitations involved in controlling unwanted microbes
9. The unique nature of viruses and viroids, particularly whether or not they are living organisms, how they cause damage, and why they are so difficult to control
10. The nature of soil humus: its original sources, what changes have occurred, and its properties
11. The importance of the C:N ratio to soil fertility and to organic-matter decomposition rates
12. The composition of enzymes and their importance in soil organic-matter decomposition
13. The various products of organic-matter decomposition and their importance to crops
14. The benefits of soil organic matter, particularly (a) as a source of nitrogen, phosphorus, and sulfur, (b) in soil aggregation, and (c) as a contributor to soil cation exchange capacity
15. The detrimental effects of some soil organic matter, especially (a) as an energy source to disease organisms and (b) as a source of allelopathic substances
16. The typical humus contents in arid soils (1–4%) and in humid soils (3–8%) and how much nitrogen these soils would release per season
17. The values and properties of (a) animal manures, (b) crop residues, (c) sewage sludges, and (d) composts, particularly in a LISA program

▬▬▬ 7:2 General Classification of Soil Organisms

There is no single accepted classification for microorganisms. The most expansive system has five *kingdoms:* **Animalia, Plantae, Fungi, Protista,** and **Monera** (Detail 7-1). *Bacteria* and *fungi* are of particular interest in the study of soils because they do most of the organic-matter decomposition, fix nitrogen, oxidize or reduce elements, and include most of the disease organisms. Table 7-1 lists some details about the soil organisms in numbers and biomass.

▬▬▬ 7:3 Animalia: Rodents, Worms, and Insects

The large animal life—Animalia (formerly, macrofaunae)—that inhabit the soil range in size from large burrowing animals, such as badgers, down to mites (the tiny arachnids barely visible to the eye).

7:3.1 Burrowing Animals

Large burrowing animals (e.g., moles, prairie dogs, gophers, mice, shrews, rabbits, badgers, woodchucks, armadillos, chipmunks) aerate the soil and alter its fertility and structure, but they eat and destroy vegetation also, which makes them more detrimental than beneficial.

7:3.2 Earthworms

Earthworms are important, but not necessary, soil organisms. Earthworms feed on animal and plant residues and on most deciduous leaves, but not on the waxy and resinous conifer needles. The ingested organic matter and fine-textured soil are excreted as small granular

Detail 7-1 Partial Outline of the Classification of Living Organisms with Emphasis on the Organisms Most Active in Soils*

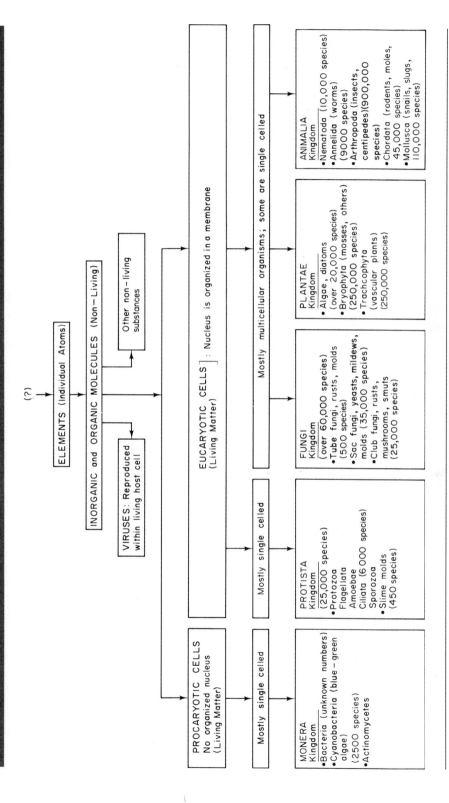

Source: Raymond W. Miller, Utah State University, from data by Edward W. Wilson, et al., *Life on Earth*, Sinauer Associates, Sunderland, Mass., 1973.
*Most subkingdoms listed are common names within the five kingdoms, but are not phyla names in most instances.

Table 7-1 Approximate Numbers, Biomass, and Carbon/Nitrogen Ratios of Microorganisms Common in Soils

Organism	Number per Gram of Soil[a]	Biomass per Hectare–15cm (kg)		Carbon/Nitrogen Ratio of the Organism
		Live Weight	Dry Weight[b]	
Bacteria	10^7–10^9	400–5000	80–1,000	5/1 to 8/1
Actinomycetes	10^6–10^8	300–4000	60–800	5/1 to 8/1
Fungi	10^5–10^6	1000–10,000	200–2000	8/1 to 20/1
Algae	10^4–10^5	50–600	10–120	5/1 to 9/1
Protozoa	10^4–10^5	15–200	3–40	5/1 to 10/1
Nematodes	10–50	10–100	2–20	5/1 to 10/1
Earthworms	—	100–1000	20–200	5/1 to 10/1
Other soil life	—	15–200	3–40	5/1 to 15/1

[a] One gram equals about 1/5 teaspoonful of soil.
[b] Dry weight estimated as 20% of live weight.

aggregates, which resist rupture by raindrop impact and provide abundant and readily available plant nutrients. Earthworms can spread plant diseases.

Earthworms aerate and stir the soil, which allows better water infiltration and easier root penetration. Some earthworm species burrow to 6 m deep (about 20 ft) in the soil, but most live in the common root zone, which averages 2 m (6.6 ft) in depth in the temperate regions. Earthworms prefer moist, well-aerated, warm (70°F, or 21°C) soils with soil pH between 5.0 and 8.4, with plenty of palatable organic matter, with low salt concentrations but high available calcium, with fairly deep soil of medium or fine texture, and undisturbed by tillage.

Earthworms are hindered by heavy farm machinery; sandy, salty, arid, acid, cold or hot, bare or barren soils; mice, moles, mites, millipedes; and "strong" insecticides.

7:3.3 Arthropods and Gastropods

Arthropods are joint-footed invertebrate organisms and include mites, millipedes, centipedes, and insects such as spring-tails, proturans, diplurans, and the larvae of beetles, flies, ants, and termites. They feed mostly on decaying vegetation and help to aerate the soil with their burrows; however, many species also can be pests because they are phytophagous (from the Greek *phyto*, "plant," and *phagos*, "to eat").

Ants and termites can radically change soil structure and "till" the soil; the net result can be beneficial or harmful (Fig. 7-3).

Slugs and snails are important members of the **gastropods** (belly-footed organisms); they feed on decaying vegetation but will eat and damage living plants. In infested areas counts as high as 243,000 slugs per hectare with a live weight of 450 kg/ha (401 lb/a) have been reported.

7:3.4 Nematodes

Nematodes are microscopic, unsegmented, threadlike (which is the meaning of *nema*) worms; they are classified according to their different feeding habits. *Omnivorous* nematodes live mainly on decaying organic matter and are the most common of the soil nematodes. *Predaceous* nematodes prey on soil bacteria, fungi, algae, and even other nematodes. *Parasitic* nematodes infest plant roots, causing the conspicuous knots that give visible proof of their presence. Parasitic nematodes are so prevalent that nearly all field and vegetable crops and trees are infected. The entry of nematodes into a plant allows easy entry for other pathogens

FIGURE 7-3 Termite activities are a common disturbance on large areas in tropical climates. (a) termite mounds in Guinea-Bissau on cultivated land and (b) a cross section of a termite mound in northern India. (Photos by permission of (a) Raymond W. Miller, Utah State University, and (b) Julian P. Donahue.)

(a)

(b)

(disease organisms), which may cause even more extensive damage than the nematodes themselves. Sugar beets and corn are particularly susceptible; an infested untreated field can result in crop losses of over 50%. Nematodes rank as the primary pest of soybean roots.

Nematodes are controlled by chemical fumigants (nematocides, which are expensive), hardwood bark, or by rotation with more-resistant plant species or the use of resistant plant varieties (Fig. 7-4). Control of nematodes by fumigation increased marketable yields of cantaloupes in California 50–100%.[1] Hardwood bark mixed with soil is a newer, lower-cost technique for the control of plant-parasitic nematodes. So far this control measure has been demonstrated with scientific accuracy on container-grown tomato and forsythia plants.

(a)

(b)

(c)

FIGURE 7-4 Many plants are susceptible to injury by parasitic nematodes, and crop losses often exceed 50%. Control may be by specific chemicals, hardwood bark, rotation with resistant plant species, or selection of more-resistant varieties. (a) A field of sugar beets in Idaho infested with the nematode Heterodera schachtii and controlled, on the left, by an organophosphate nematocide. (b) Nematodes on blue spruce in a Wisconsin greenhouse were controlled, on the left, by steam sterilization. (c) Ranger alfalfa variety is susceptible to nematodes, but Lahontan variety is resistant. Nevada. (*Source:* USDA—Forage and Range Research Station, Logan, Utah; photos by Gerald Griffin.)

---

[1] Erik Likums, "Cantaloupes vs. Nematodes," *Agricultural Research,* **28** (1980), no. 9, p. 15.

7:4 Plantae: Plants and Algae

The **Plantae** organisms obtain energy from the sun and can exist as stationary life. Algae and diatoms are examples of *micro*plantae; the *macro*plantae range from the 250,000 species of Bryophytes (mosses) to the 250,000 species of Tracheophytes (vascular plants).

7:4.1 Plants

Most familiar plants have root systems, some of which are often 30–50% of the total plant mass, comprising thousands of kilograms of biomass per hectare. Older roots develop thick protective coverings of a mucilage. In contrast, **root hairs** are single cells of the root surface, with thin walls that allow water and nutrient absorption.

Roots exude, or secrete, many substances, including at least 18 amino acids, 10 sugars, 10 organic acids, various proteins, growth substances, growth inhibitors, microbe attractants, and repellants.[2] As early as 1832 DeCandolle ascribed the problem of "soil sickness" to the toxic exudates produced by certain crop plants. Yet it is only recently that extensive research has been able to identify many of these chemical excretions. The quantities of exuded material vary from about 2% of the "organic matter flowing into the root system" from plant tops to values of 7% exuded from plant roots in sand cultures. Amounts approaching 15% of the total radioactive carbon added were recovered in the rooting media when sorghum was given radioactive $^{14}CO_2$ for 48 hours. [3,4]

The products around a root in the soil constitute a "gold mine" of microbial substrates: carbohydrates and other chemicals mentioned earlier plus dead and sloughed root cells. As a result, the environment surrounding the root (1 mm or so distance) will teem with microbes, each performing its function(s) of humus decomposition, N_2 fixation, or other activity. Some organisms will also be producing their own protective chemicals to help them survive. The area in the soil near the roots is called the **rhizosphere.** This portion of soil may be quite different in chemical properties from the bulk of the soil. The rhizosphere may also be one to two pH units more acidic than the neutral or alkaline soil as a whole.

7:4.2 Algae

Soil **algae** are microscopic organisms that carry on photosynthesis. The main groups are the green algae, yellow-green algae, and diatoms. What were once called blue-green algae are now being reclassified into the Monera kingdom as cyanobacteria (also called blue-green bacteria). Algae are not important as decomposers of organic matter but are producers of new photosynthetic growth. In soils kept moist and fertile, algal growth on the soil can produce considerable amounts of organic material—hundreds of kilograms per hectare annually.

7:5 Fungi: Molds, Mushrooms, Yeasts, Rusts

Fungi are organisms without the ability to use the sun for energy; they live on dead or living plant or animal tissue. Fungi are a curious assortment of one-celled organisms (*yeasts*), multicellular filamentous *molds, mildews, smuts* and *rusts,* and the well-known *mushrooms,* to

---

[2] M. G. Hale, L. D. More, and G. J. Griffin, "Root Exudates and Exudation," in *Interactions between Non-pathogenic Soil Microorganisms and Plants,* Y. R. Dommergues and S. V. Krupa, eds., Elsevier Science Publishing, New York, 1978, pp. 163–204.

[3] A. Haller and H. Stolp, "Quantitative Estimation of Root Exudation of Maize Plants," *Plant and Soils,* **86** (1985), pp. 207–216.

[4] Chung-Shih Tang, "Continuous Trapping Techniques for the Study of Allelochemicals from Higher Plants," Chapter 7 in *The Science of Allelopathy,* A. R. Putnam and C.-S. Tang, eds., Wiley, 1986, p. 114.

name a few. "As with other microorganisms, the classification of the fungi is difficult, sometimes ambiguous, vehemently debated, and subject to constant revision."[5]

7:5.1 Fungal Organic Matter Decomposers

One of the first visual evidences of decomposition of some materials is the appearance of fungal **mycelia,** a vegetative mass of threadlike branching filaments (**hyphae**). Molds on bread, on cheeses, on many rotting foods, and in forest litter exhibit mycelia. Fungi are vigorous decomposers of organic matter and readily attack cellulose (woody materials), lignins, gums, and other complex compounds, even in quite acidic conditions (unusual for most living organisms). Fungi also compete with economic plants for nutrients released from organic-residue decomposition, particularly nitrogen, phosphorus, and sulfur. Fungi also secrete substances that aid in the formation of water-stable soil aggregates.

7:5.2 Deleterious Fungi

Some fungi exist as predators on living cells. Hyphae of fungi can penetrate protozoa and, when they become immobile, slowly digest them. Even nematodes can be ensnared in mycelia and be devoured.

Although relatively few types of soil fungi are predatory, their importance is major. Various fungi cause *smuts* and *rusts* on grains and lawn grasses, many of the *wilts, powdery* and *downy mildews, leaf spots, cankers, scabs, clubfoot, blight, black wart, takeall, leaf curl,* and some *root rots.* The *amanita* fungi look like mushrooms but are very poisonous.

Cucumber fruit rot (belly rot) is a hazard in warm, humid climates, such as in the Southern states. All management control measures attempt to encourage drying (drainage, defoliants), incorporating surface plant debris (plowing), and allowing more open space (lower population density). The use of composted hardwood bark in place of sphagnum peat in pot culture has suppressed poinsettia crown and root rots. *Potato blight* caused the Irish potato famine of 1845-1849. The famine caused a mass migration to the United States in 1847 to 1854.

Even the fungal decomposers are not all innocuous. A recent flurry of interest has centered on **aflatoxins,** toxins produced by molds growing on grains (mostly *Aspergillus flavus*); some of these are carcinogens (cancer-causing). One of them, Aflatoxin B1, is the most potent liver carcinogen known. In Hong Kong and Thailand a research team sampled grains and found 80% of the peanut samples and 50% of the rice, corn, beans, and other cereals were infected with toxin-producing molds; one-third of those foods with mold were toxic to rats. In 1974, 106 persons died and 297 became ill after eating contaminated corn. North Carolina alone estimated losses from contamination of corn to be nearly $32 million. Mycotoxins may produce birth defects, abortions, tumors, cancers, and other effects.

7:5.3 Mycorrhizae: The Root-Fungi Association

Mycorrhizae (my-koe-rye'-zee) means "fungus root." Mycorrhizae are mutually beneficial (**symbiotic**) relationships between fungi and plant roots. Some fungi form a kind of sheath around the root, sometimes giving it a hairy, cottony appearance. The plant roots transmit substances (some produced by exudation) to the fungi, and the fungi aid in transmitting nutrients and water to the plant roots. The fungal hyphae may extend the roots' lengths 100 times. These hyphae reach into additional and wetter soil areas and help absorb many nutrients for transmission to the plant, particularly the less mobile nutrients, such as phosphates,

---

[5] R. M. Atlas, *Microbiology: Fundamentals and Applications,* 2nd ed., Macmillan, New York, 1988, p. 314.

zinc, copper, and molybdenum. Because they provide a protective cover, mycorrhizae increase the plant seedling's tolerance to drought, to high temperatures, to infection by disease fungi, and even to extreme soil acidity.

Two kinds of mycorrhizae are described by the fungal growth habits. **Ectomycorrhizal** fungi (*ecto* = outside) "sheath" the host root but penetrate only the outer cell layers of the root cells walls (Fig. 7-5); the hyphae of **endomycorrhiza** fungi (*endo* = inside) actually penetrate into some host cells. Some are called vescicular--arbuscular hyphae (VAM). VAM are particularly helpful in phosphate absorption and they are the most common form found on plants, although the least visually obvious mycorrhizae.

When nursery soils are sterilized, or nonnative plants are grown in them, it is usually beneficial to inoculate the soil with the appropriate mycorrhiza. For years, pines transplanted from the United States to Puerto Rico would grow only a few inches, turn yellow, show severe phosphorus deficiency, then usually die. When pine-growing soil from the United States

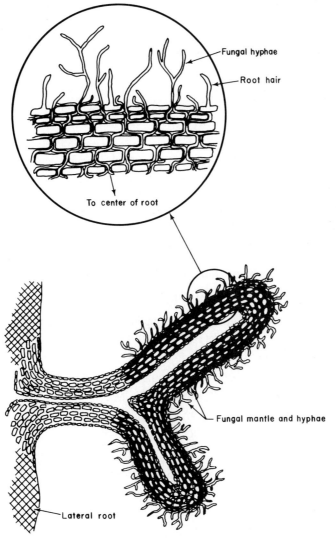

FIGURE 7-5 Ectomycorrhiza surrounding a root and penetrating around cells (replacing the middle lamella between cells). Root exudates stimulate the fungi. The fungus mat helps in nutrient absorption and water uptake; it also buffers the root against harsh soil environment. (Courtesy of Raymond W. Miller, Utah State University.)

Fungal hyphae

Root hair

To center of root

Fungal mantle and hyphae

Lateral root

was used to inoculate the Puerto Rican soil, the treated pines grew to 2.4 m (8 ft) in 3 years, whereas the noninoculated pines were only 30 cm (1 ft) tall.

The practice of sterilizing greenhouse-nursery soil with methyl bromide gas led to plant zinc deficiences, partially due to the unwitting destruction of zinc-absorbing mycorrhizae. Inoculated peach tree seedlings had higher levels of zinc than those that had only been fertilized with zinc chelate. Fumigated nursery soil produced severely stunted citrus trees due to low phosphorus absorption resulting from reduced mycorrhizae. The addition of several hundred kilograms of phosphorus per hectare corrected that problem.

Studies on (*Glomus*) vesicular-arbuscular endomycorrhizal fungi (VAM) have used the fungi as aids to increase phosphorus uptake from the soil.[6] A stimulant to VAM formation—*Rhizotropin*—will soon be marketed.

The greatest growth responses to mycorrhizal fungi are probably to plants in highly weathered tropical soils because the leached Oxisols and Ultisols are low in basic cations, are acidic, are low in phosphorus, and may have toxic levels of aluminum. Plants that have coarse or limited root systems should benefit the most. The mycorrhizal association improves on these soil conditions and helps protect such root systems from the hostile environment. Benefits from mycorrhizae have been shown for many crops, including pasture and forage legumes, corn, wheat, barley, tomatoes, many vegetables, herbaceous and tree fruits, onions, citrus, grapevines, sweetgum, pineapple, coffee, tea, cocoa, oil palm, papaya, cassava, and rubber trees.

The fungi–plant root symbiotic association, called mycorrhizae, is summarized in these statements.

1. Mycorrhizae, or fungus–root associations, are expected on most vascular plants and are probably present on most plants to some extent.

2. The major aid from the fungi seems to be in increasing phosphorus uptake from low-phosphorus soils. This effect may be caused by several actions: (a) exploration of a larger volume of soil is increased by fungal hyphae, (b) organic acids exuded by fungi increase phosphorus solubility, and (c) the fungi alter the concentrations of cations in solution and thereby increase the phosphate solubility.

3. An increase of phosphorus to adequate levels by mycorrhizae may be necessary before N_2-fixing bacteria in low-phosphorus soils can nodulate the roots.

4. Mycorrhizal associations aid zinc and copper absorption.

5. Mycorrhizal fungi reduce stress due to drought.

6. Protection of roots against plant pathogens may be a common benefit. Colonization of roots by mycorrhizae has reduced root pathogen damage by *Fusarium* on tomato and the "stunting" of cotton.

7. Extracts from mycorrhizal tomato roots caused 50% mortality in nematode larvae in 4 days. This observation indicates that it is likely that many new factors affecting plant growth will be discovered in the future.

8. Mycorrhizal formation is reduced by fertilization or poor aeration.

9. Mycorrhizal fungi probably are protective to roots against many stress conditions (drought, adverse pH, pathogens, low nutrition, and adverse temperatures).

---

[6] J. M. Barea, "Vesicular-Arbuscular Mycorrhiza as Modifiers of Soil Fertility," *Advances in Soil Science,* **15** (1991), pp. 1–40.

7:6 Protista

Protozoa and slime molds are the main phyla of the **Protista** kingdom. A **protozoan** is a unicellular organism without a true cell wall. Protozoa ingest bacteria, fungi, other microbes, nematode larvae, and eggs, and even smaller protozoa.[7] The protozoan classes are named for their particular methods of movement in the soil water: *amoeboid* moving by pseudopodia, *flagellar* moving by whiplike units, and *ciliate* moving by hairlike cilia that "wave." The protozoans are numerous in soil; they help to control other microbes but also cause critical diseases (e.g., malaria). Protozoan digestion of bacteria and fungi influences microbial populations and hastens the recycling of plant nutrients.

Although protozoa cause few serious plant diseases, they are the cause of many animal problems. Some of these are *sleeping sickness, Clogas's disease, severe diarrhea, amoebic dysentery, Texas cattle fever,* and *malaria.*

7:7 Monera: Soil Bacteria and Actinomycetes

The soil microorganisms **bacteria** and **actinomycetes** belong to the **Monera** kingdom. Both decompose organic matter, although actinomycetes are not as effective as bacteria or fungi. Actinomycetes are the source of numerous beneficial antibiotics, and some produce musty tastes and odors to waters.

7:7.1 Soil Bacteria

Bacteria are unicellular microorganisms; the name comes from the Greek word for "rod," designating the usual bacterial shape. The numbers of bacteria in the soil usually exceed all other microorganisms, although fungi may exceed bacteria in weight.

One gram of soil (about 1/5 teaspoon) can contain 100 million bacterial cells, 1 million actinomycetes, and 5 m (16.4 ft) of fungal mycelia. All of this is less than 0.05% of the dry-soil weight. The larger bacteria are nearly one micrometer long; one centimeter (0.4 in.) of distance could have 10,000 to 100,000 bacteria, end to end.

As small as bacteria are, most clay particles will be *much smaller* than the bacteria. Some bacteria double numbers in as little as 30 minutes, especially when the soil has an abundance of organic residues. However, most bacteria require several hours or days to double in the natural environment.

Bacteria are classified by nutritional patterns, oxygen needs, and symbiotic relationships (Table 7-2). **Autotrophic** (self-nutritive) bacteria manufacture their food by the synthesis of inorganic materials, such as plants do in photosynthesis. **Heterotrophic** (meaning other, or different, nutrition) bacteria derive their food (carbon) and energy directly from organic substances. Fungi, protozoa, animals, and *most* bacteria are heterotrophs. Bacterial groups are further defined as **photo-** or **chemo-** to designate their energy source.

7:7.2 Autotrophic Bacteria

Autotrophic bacteria obtain their nutritive carbon from carbon dioxide, and specific groups can oxidize ammonium, nitrites, sulfides, sulfur, ferrous ion, manganous ions, hydrogen gas, and carbon monoxide. The oxidation transforms nitrites, sulfides, and carbon monoxide to useful nitrates, sulfates, and carbon dioxide. Other oxidations eliminate toxic forms of carbon and manganese.

---

[7] K. M. Old and J. F. Derbyshire, "Soil Fungi as Food for Giant Amoebae," *Soil Biology and Biochemistry,* **10** (1978), pp. 93–100.

Table 7-2 Subdivisions of Bacteria on the Basis of Their Reactions, Energy Requirements, and Other Properties

Divisions based on method of obtaining nutrition and energy:

Photoautotrophs   Energy from sunlight; nutritive carbon from carbon dioxide
Photoheterotrophs   Energy from sunlight; carbon from organic matter
Chemoautotrophs   Energy from oxidation of inorganic substances such as nitrogen, iron, or sulfur; carbon from carbon dioxide
Chemoheterotrophs   Energy and nutritive carbon from organic matter

Divisions based on oxygen requirement:

Aerobic   Require free gaseous oxygen source; includes most active decomposers; most populous bacteria
Anaerobic   Can use electron acceptors other than oxygen such as NO_3^- or SO_4^{2-}; do not require free oxygen
Facultative anaerobes   Can be either aerobic or anaerobic in presence or absence of O_2

Dinitrogen fixers divided on presence or absence of symbiotic relationships:

Symbiotic N_2 fixers   Associated with a host plant; both host and bacteria benefit; fixes N_2 from atmosphere
Nonsymbiotic N_2 fixers   Exist as free bacteria without a host but fix N_2

Probably the most important groups of autotrophic soil bacteria are those that oxidize ammonium to nitrites (a toxic, transitory form of nitrogen) and then to nitrates. These *nitrifying* organisms achieve maximum growth under the following conditions:

1. The presence of proteins to release ammonium as they decompose, or the presence of ammonium salts, such as from ammonium sulfate or hydrolysis of urea
2. Adequate aeration
3. A moist but not overly wet soil that will hinder aeration
4. A large amount of calcium (not strongly acidic)
5. Optimum temperature between 20 and 40°C (68–104°F), depending on the soil and the bacterial adaptation

The bacterial nitrification process is shown diagrammatically as follows:

$$\underset{\substack{\text{ammonium}\\\text{(utilized by}\\\text{plants)}}}{NH_4^+} \xrightarrow[\text{bacteria}]{\textit{(Nitrosomonas)}} \underset{\substack{\text{nitrite}\\\text{(transitory and}\\\text{apparently toxic)}}}{NO_2^-} \xrightarrow[\text{bacteria}]{\textit{(Nitrobacter)}} \underset{\substack{\text{nitrate}\\\text{(utilized}\\\text{by plants)}}}{NO_3^-}$$

Nitrification is of great concern because NO_3^- flows into groundwaters. Although high levels of nitrate in waters may be beneficial to plants being irrigated, large concentrations of nitrate in drinking water or foods can cause severe health problems to babies of mammals. **Eutrophication,** an increase in nutrient concentrations in water, is evident by the increased algal growths that, when dead and decomposing, require oxygen that otherwise could be used by aquatic life.

Several groups of autotrophic bacteria can either oxidize or reduce carbon monoxide to carbon dioxide or methane, all of which are gases. These bacteria are literally life giving. The world's population adds 220 million tons (200 million Mg) of carbon monoxide to the atmosphere each year. With no conversion to carbon dioxide or methane, the atmosphere would become lethal to animal life within a few years. This bacterial conversion is anaerobic (without free oxygen).

7:7.3 Heterotrophic Bacteria

Heterotrophic bacteria are those that depend upon organic matter for their nutrition; most soil bacteria are in this group. Heterotrophic bacteria include both nitrogen-fixing and non-nitrogen-fixing groups. The nitrogen-fixing bacteria are further subdivided into symbiotic and nonsymbiotic and are commonly associated with leguminous plants. Heterotrophic bacteria that do not fix nitrogen are the most prevalent soil bacteria and account for much of the decomposition of organic materials.

7:7.4 Symbiotic Bacteria [8,9]

The heterotrophic bacteria that fix atmospheric dinitrogen gas in plant root nodules (**symbiotic bacteria**) have a mutually helpful relationship with their host plants. The ancient Greeks knew that **legumes** (pod-bearing plants such as peas, beans, alfalfa, and clovers) have a beneficial effect upon both the companion crop and whatever crop was planted next in the same soil. It was not until 1838 that Boussingault, a French chemist, demonstrated that the beneficial effect was due to the fixation of atmospheric nitrogen in the legume **root nodules.** In 1879 Frank proved that artifical inoculation with specific bacteria resulted in legume root nodule formation and that the bacteria in these nodules fixed atmospheric nitrogen.

Symbiotic bacteria begin by infecting root hairs, causing an invagination (enclosing-like sheaths) inward through several cells. Surrounding plant cells proliferate rapidly, perhaps because of auxin hormone produced by the infecting bacteria. As the bacteria enter the nodule cells, they form enclosing membranes and produce methemoglobin, an oxygen-carrying pigment (the nodule may be pink in cross section). The hemoglobinlike material may be an oxygen sink or trap to keep the bacteria in an anaerobic environment, which is necessary for N_2 fixation. Other nonsymbiotic bacteria include both aerobic and anaerobic organisms, which are capable of fixing N_2. The plant roots supply essential minerals and newly synthesized substances to the bacteria and eventually benefit from the atmospheric nitrogen that the bacteria fix and either release for plant use or use to build bacteria protein.

The fixation of the inert dinitrogen is accomplished by the enzyme **nitrogenase.** This enzyme lowers the **activation energy** (the energy needed to cause the reaction to occur). The fixation progresses in reduction stages from dinitrogen ($N \equiv N$) through uncertain intermediates $HN=NH$ and $H_2N—NH_2$ to produce $2NH_3$. Finally, the amide group ($—NH_2$) is transformed into some organic compound such as amino acids. All of this takes place while the nitrogen is bonded to the enzyme(s).

The life-span of a single bacterium may be only a few hours, and the bodies of a portion of the bacterial population are continuously dying, decomposing, and releasing ammonium and nitrate ions for use by the host plant. In legumes a few entire nodules are sloughed during the season. Most of the nitrogen fixed is excreted by the bacteria and made available to the host plant and to other plants growing nearby. The amount of available nitrogen that leguminous crops add to the soil is phenomenal (Table 7-3). Figure 7-6 shows nodulation on

[8] David R. Benson, Daniel J. Arp, and R. H. Burns, "Cell-Free Nitrogenase and Hydrogenase from Actinorhizal Root Nodules," *Science,* **205** (Aug. 1979), pp. 688–689.
[9] John D. Tjepkema and Lawrence J. Winship, "Energy Requirement for Nitrogen Fixation in Actinorhizal and Legume Root Nodules," *Science,* **209** (July 11, 1980), pp. 279–280.

Table 7-3 Approximate Quantities of Nitrogen (N_2) Fixed by Selected Legumes, Non-legumes, and Soil Systems under Ideal Environmental Conditions

Plant/Condition	Amount of N_2 Fixed Each Growing Season		
	kg/ha	Range	lb/a
Legumes (symbiotic)			
Sesbania[a]	540	505–581	
Popinac[a]	450	110–548	
Alfalfa	224	128–600	200
Red clover	129	117–154	115
Kudzu	123		110
Soybean	112	157–200	100
Cowpea	101		90
Peanut	45		40
Bean	45		40
Lupines	—	150–169	—
Nonlegumes (symbiotic)			
Alder, red	168	40–300	150
Buckthorn	67		60
Casuarina	62		55
Sweet gale	9		8
Lichens	—	39–84	—

[a] Sesbania (*sesbania rostrata*); Popinac (*Leucaena leucocephala*).

field peas; nodulation is more complete when all plant nutrients except nitrogen are available in sufficient quantities to the leguminous host plants.

The best known symbiotic bacteria belong to the genus ***Rhizobium***. Symbiotic heterotrophic bacteria specific to the crop to be grown are frequently added, or inoculated, in a dried, powdered form to the crop *seed* to ensure that nitrogen-fixing organisms are present (Fig. 7-7). Applied rates are often about 5 g of inoculum per kilogram of seed (1 part inoculum to 200 parts of seed). Heavier applications of inoculum mixed into peat granules trickled into soil as the seed is planted is an alternative technique to encourage nodulation. The same bacterial specie will not inoculate all legumes. A specie that inoculates lupines will not inoculate lespedezas or trefoil. But some bacterial species can react symbiotically with several similar legumes.

Some plants that have recently been found to have symbiotic relationships with various N_2-fixing bacteria, including blue-green bacteria (**cyanobacteria**), are *Digitaria* (grass species), corn, sorghum, pearl millet, water fern (with blue-green bacteria), the tropical herb *Gunnera macrophylla* (with blue-green bacteria), Douglas fir (the bacteria are found on the needles), and white fir.

The blue-green cyanobacteria also fix N_2 without association with plants. The optimum pH range for N_2 fixation is between 7.0 and 8.5. In flooded Philippine rice fields, *Anabaena spiroides* and *Aulosira fertilissima* help to maintain the nitrogen level of the soil by utilizing atmospheric nitrogen. In desert soils blue-green bacteria are prominent microorganisms and may increase the nitrogen content in some of them. Special species of blue-green bacteria of the genera *Nostoc, Scytonema,* and *Anabaena* are capable of fixing atmospheric nitrogen in solution cultures.

In the early 1970s two Brazilian scientists reported a symbiotic bacterium associated in N_2 fixation with several tropical grasses and corn. Since then, bacteria of the genus *Klebsiella* have been shown to be associated in N_2 fixation with numerous grasses, winter wheat, and

FIGURE 7-6 Nodules on root of a field pea plant grown at the Matanuska Research Farm, Alaska. The pea seed was inoculated with a commercial Rhizobium bacterial culture when planted in early June. Photo taken September 13, near end of the growing season. (*Source:* L. J. Klebesadel, "Biological Nitrogen Fixation in Natural and Agricultural Situations in Alaska," Agroborealis [Jan. 1978], p. 11. Photo courtesy of researchers at the University of Alaska Agricultural Experiment Station.)

FIGURE 7-7 All legume seed should be inoculated before planting to ensure sufficient bacteria for maximum nodule formation. Left: soybeans that were not inoculated. Right: soybeans that were properly inoculated. (Courtesy of Nitragin Co.)

Kentucky bluegrass.[10] However, the hope that common crops might be able to be developed to fix N_2 or use less nitrogen and thus to reduce the need for extensive nitrogen fertilizers seems still to be many decades away.[11]

7:7.5 Nonsymbiotic N₂-Fixing Heterotrophic Bacteria

Unlike symbiotic bacteria, the **nonsymbiotic** nitrogen-fixing organisms do not need a host plant; they live free from any association with another organism.

The anaerobic *Clostridium* are usually more abundant in soils than the aerobic *Azotobacter*. In aerobic conditions in tropical soils the acid-tolerant N_2-fixer *A. beijerinckia* is most abundant. *Clostridium* develop best in poorly drained, acid soils; *Azotobacter* are more abundant in well-drained, neutral soils. The amounts of atmospheric nitrogen fixed are variable, but under ideal conditions the total nitrogen fixed varies from 1 to 15 kg/ha (about 1–14 lb/a) per year.

7:7.6 Bacterial diseases

Bacteria cause numerous plant diseases, although not as many as do fungi. Some of these are *wildfire* of tobacco; *blight* of soybeans, rice, and peas; *moko* of bananas; *wilt* of carnations, corn, and tomatoes; *slippery skin* of onions; *gumming* of sugar cane; *soft rot* of fruit; *leafy gall* of ornamentals; *pox* of sweet potatoes; *cane gall* of raspberries; aster *yellows;* and corn *stunt.* Human diseases are also of serious concern and include bacterial food poisoning (such as *salmonella*), *typhoid fever, Rocky Mountain spotted fever* from ticks, *parrot fever, rheumatic fever, cholera,* and *dental caries.*

7:7.7 Actinomycetes

Actinomycetes are taxonomically and morphologically related to both fungi and bacteria but are usually classified with bacteria. They are characterized by branched mycelia, similar to fungi, and resemble bacteria when the mycelia break into short fragments. Actinomycetes aid in decomposition of organic matter, especially cellulose and other resistant organic molecules. Like fungi, actinomycetes aid in the development of water-stable soil structure by secreting non-water-soluble gummy substances. In recent years actinomycetes have attracted worldwide attention after it was discovered that they produce many useful antibiotics. Nearly 500 antibiotics have been isolated from actinomycetes, the most common of which are streptomycin, Aureomycin, Terramycin, and Neomycin.

In the late 1970s actinomycetes were found to form symbiotic N-fixing relationships with a diverse selection of plants. Definite actinomycete–plant symbiosis has been established for many plants,[12] including boysenberry, soapberry, alders, bayberry, sweet fern, sweet gale, New Jersey tea, coffeeberry, buffaloberry, bitterbrush, mountain mahoganies, Australian pine, European autumn, Russian olive, and flooded rice. The great interest in symbiotic actinomycetes is partly due to the fact that actinomycetes infect at least seven different botanical families, whereas *Rhizobia* bacteria infect only the family Leguminosae (with a few exceptions).[13] Fixation in red alder (a tree) has been measured as high as 168 kg of nitrogen per hectare (150 lb/a) per year.

---

[10] L. V. Wood, R. V. Klucas, and R. C. Shearman, "Nitrogen Fixation (Acetylene Reduction) by *Klebsiella pneumoniae* in Association with 'Park' Kentucky Bluegrass (*Poa pratensis* L.)," *Canadian Journal of Microbiology,* **27** (1981), pp. 52–56.

[11] A. Quispel, "A Critical Evaluation of the Prospectus for Nitrogen Fixation with Non-legumes," *Plant Soil,* **137** (1991), pp. 1–11.

[12] D. L. Hensley and P. L. Carpenter, "The Effect of Temperature on N_2 Fixation (C_2H_2 Reduction) by Nodules of Legume and Actinomycete-Nodulated Woody Species," Botanical Gazette, **140** (supplement) (1979), pp. S58–S64.

7:8 Soil Viruses and Viroids

Viruses are unique substances, many of which cause serious plant and animal diseases (*potato leaf roll, tobacco mosaic, raspberry ringspot, cacao mosaic, foot-and-mouth disease,* and *bovine leukemia*). The word *virus* was used by ancient Romans to mean "poison venom and/or secretion." In early medical use it meant "any microscopic etiologic agent of disease." **Viruses** are nonliving nucleic acids often surrounded by a protein coat. Most viruses are ribonucleic acid (RNA) material, genetically reproducible, nucleoprotein substances, and have a protective coat of a different protein that must be shed before the virus can be made "active." Currently, some scientists list viruses as *acellular microorganisms,* meaning they are without cells and cannot carry out physiological functions on their own. Yet, they store genetic information that, inside a living organism cell, can be replicated. Viruses are capable of only a few "life functions."

Virallike substances have been classified into three kinds of materials that carry genetic material that can be replicated[14]:

Prions Various proteins without a protective coat
Viroids No protective coat around RNA protein (naked RNA)
Virus Has a protective protein coat around its RNA or DNA protein; sometimes even a lipid-containing envelope surrounds the nucleic acid

Viruses and viroids cause many plant diseases in addition to those mentioned earlier. Some of these are *carnation latent virus, beet yellows virus, citrus tristerza virus, mosaics of cucumber, turnip yellow, tomato bushy stunt, rice dwarf virus, tomato spotted wilt, lettuce necrotic yellows,* and *wound tumors.*

Enzymes in the soil attack viruses and decompose them, so they are not viable in soils for long periods of time. The spread of viruses can be controlled to some extent by removing the host carriers, some of which are nematodes, fungi, and roots of some plants. Few viruses overwinter in soil but are commonly hosted and disseminated by nematodes or fungi. Otherwise, common plant viruses in soils may survive only about 1 to 4 weeks. There is almost no chemical control in the field for virus infections. Denaturation (decomposition, high salt, high temperature, etc.) inactivates the virus (viroid).

7:9 Optimum Conditions for Microbial Activity

Microbes are in constant competition for organically combined carbon and other nutrients. Their ability to procure these growth materials depends on temperature, moisture, soil acidity, soil nutrient levels, suitable energy source, and the competition by other microbes, among other factors. There is intense competition. Generally, microbes are most numerous in soils with moisture very near field capacity, with near neutral pH, with a high nutrient content, and at temperatures near 30°C (86°F).

7:9.1 Optimum Soil Water and pH for Soil Microorganisms

The water content near or just greater than the field capacity (wet but adequately aerated) is near optimum for most microorganisms. Dryness kills many microbes, and many others

---

[13] Peter Del Tredici, "Legumes Aren't the Only Nitrogen-Fixers," *Horticulture* (Mar. 1980), pp. 30–33.
[14] R. M. Atlas, *Microbiology: Fundamentals and Applications,* 2nd ed., Macmillan, New York, 1988, 807 pp.

tolerate dryness by developing resistant strains or entering into a dormant stage. A few anaerobes, because they are hindered by free-oxygen gas, grow best at saturation conditions.

The optimum soil pH is near pH 7. This is the pH of microbial cytoplasm (the cell material). Bacteria and actinomycetes are usually less tolerant of acidic soil conditions than are fungi, and few grow well at soil pHs below pH 5. One exception is the sulfur-oxidizing genus *Thiobacillus,* which produces sulfuric acid; it tolerates soil pHs even down to pH 0.6. Many fungi survive in forested and organic soils with pHs as low as 3.0 . Localized microenvironments near roots or decomposing residues can produce locales of lower pH than that of the soil as a whole, differences of as much as 1 or 2 pH units.

7:9.2 Optimum Temperatures and Other Conditions for Microorganisms

Microbial activity accelerates rapidly as temperature rises. Biological (enzymatic) reaction rates nearly double as temperature increases from 10° to 20° C (50° to 68°F). Plants and most microbes are nearly dormant at freezing (Fig. 7-8). There are exceptions: Some microbes are able to tolerate very cold temperatures (psychrophiles, cold lovers), and others tolerate relatively high temperatures (thermophiles, heat lovers). The majority of soil bacteria and actinomycetes, however, have optimum activity temperatures similar to those of the mesophiles (middle group). The temperature tolerances of these three general groups are:

- *Psychrophiles:* can grow at temperatures below 5°C (41°F), but have optimum temperatures near 15–20°C.
- *Mesophiles:* grow slightly near 0°C (32°F) and show little if any growth above 40°C (104°F); many die at this higher temperature; optimum temperature usually between 25 and 37°C (77–99°F)
- *Thermophiles:* can tolerate 45–75°C (113–167°F), with optima between 55 and 65°C (131–149°F) (e.g., some composting microbes)

Microorganisms have high nutritive needs, especially for nitrogen, phosphorus, sulfur, and calcium. The carbon source (from organic substances) is most easily attacked when it is plentiful and from succulent young plants.

Organisms compete for nutrients with other species and even with each other. Protozoa consume other microbes, particularly bacteria; one amoebic cell may devour several thousand bacteria per cell division (once or twice a day). Some microbes produce **antibodies,** substances that retard or kill other nearby organisms and acquire more nutrients by eliminating the competition. Even viruses can be detrimental to microbes. **Bacteriophages** are viruses that parasitize bacteria and cause their death and partial breakdown. Optimum growth conditions for microbes imply not only adequate moisture, temperature, pH, and nutrients, but also an absence or low level of microbial and viral enemies.

7:10 Encouraging Beneficial Microorganisms

The best approach to encouraging beneficial organisms is to maintain optimum conditions for them. Fortunately, most conditions good for plant growth also favor growth of beneficial microorganisms. Unfortunately, it does *not* follow that good conditions for plants and beneficial microorganisms are poor conditions for harmful ones. It can be as important to *make conditions unfavorable for harmful microorganisms* as it is to promote favorable conditions for beneficial ones; the beneficial ones may become expendable in the more desperate need to eradicate the harmful. Ways to encourage an active and usually good microbe population include these actions:

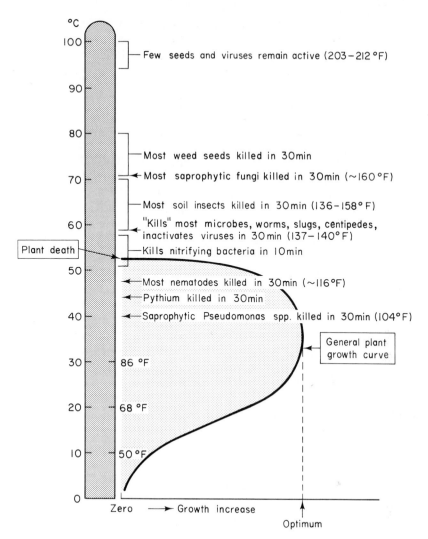

FIGURE 7-8 Approximate temperatures at which many living organisms and viruses in the soil are killed or inactivated. The plant growth curve is only approximate for many common crops, typically corn. (*Source:* Drawn by Raymond W. Miller from data of K. F. Baker and R. J. Cook, Biological Control of Plant Pathogens, W. H. Freeman, San Francisco, 1974.)

1. Inoculate the soil with the desired symbiotic organism (e.g., *Rhizobia,* fungi for mycorrhizae) in a soil where the host plant has not grown for many years or has never grown. Some *Rhizobia* can survive 10 years in soil without a host plant, but inoculation is so simple and inexpensive that taking a chance on the microbe's presence is not worth the risk.

2. Lime the soil to values above pH 6. Do not overlime.

3. Minimize soil fumigation or sterilization, which kills both harmful and beneficial biota. Many greenhouse mixtures must be sterilized because the harmful microbes are too hazardous in the optimum conditions in greenhouses.

4. Maintain as large a soil organic-matter level as is practical, taking into account cost, time, and other management problems.

5. Try to avoid all contamination. Do not carry contaminated soil from infected fields to clean ones on uncleaned equipment. Burn or remove infested plants; do not discard by burial in the field or on compost piles.

6. Avoid causing stress conditions, such as drought, salt accumulation, water-logging, or excess fertilizer additions.

7:11 Controlling Harmful Microorganisms

Controlling harmful soil microorganisms is more difficult than controlling aboveground pests because of the problem of distributing biocides (killing chemicals) within soils. Gases heavier than air (e.g., methyl bromide) can be injected into preplant soil with some success but aren't usable on turf grasses because of phytotoxicity to the grass. Soluble solid biocide materials can be applied either to the soil and moved by irrigation or applied in solution to most soil areas to depths of 15 cm (6 in.) or more, the location of most microbial life.

Unfortunately, few chemical treatments are specific. Fungicides kill most fungi and nematocides kill most nematodes, the good along with the harmful, so the case should be desperate that calls for their heavy use. Chemical pesticides are about the only extensive control available, but caution and restraint in their use are essential.

Other methods of control are based on introducing natural enemies (viruses, insects, others) or in changing the soil conditions that favor the harmful growth. Keeping the pH below about 5.2 controls potato scab actinomycetes. Rots (fungi) in sweet potatoes can be ameliorated by careful maintenance of a moderately to strongly acid soil. Many wilts and rots are favored by hot, wet, limed soils. Reducing surface litter (mulch) or allowing adequate interirrigation drying may be enough to reduce the damage to acceptable levels. To be effective in control, knowledge of the conditions favoring each problem is needed.

Because different problems are caused by different organisms, it is difficult to generalize. Some sample recommendations for general management are itemized in the following list:

1. Always start with clean, disease-free plants and use varieties resistant to diseases known to be a problem in your area.

2. Maintain careful sanitation practices. Diseases from fungi, bacteria, and viruses are easily spread on digging or pruning tools, clothing that contacts plants, mud on shoes, chewing and sucking insects, and wind. Soil sterilization between pot or greenhouse bench crops may be essential. Control of insect vectors is important.

3. Minimize mechanical damage to plants on stems or leaf tissue. These damaged areas become entry points for the omnipresent disease spores and reproductive bodies of disease organisms.

4. Carefully control water. Most fungi and bacteria require high moisture conditions. Excessively frequent sprinkling that leaves the foliage wet for long time periods favors disease. Often, allowing surface drying between irrigations and exposure to bright sunlight where plants tolerate it will decrease some bacterial and fungal diseases. Good ventilation (for drying) may also help.

5. Control soil acidity. Some wilts are particularly prevalent if soil is basic or has too much lime (calcium carbonate) in it. Usually, pH values near pH 6 are preferred.

6. Where antibiotic controls are known, use them. They are specific to certain bacteria or fungi and will not kill other beneficial organisms.

7. Control infestations immediately.

The residue left after extensive decomposition of organic materials in soils is called **humus.** It is extremely variable in composition and is quite resistant to further microbial decomposition. In most formulas written to try to depict "humus," the major "ingredients" are (1) many active chemical functional groups exposed to the surrounding solution for reaction with other substances in the solution and (2) a very large cross-linked and "folded" molecule with molecular weights in the hundreds of thousands of grams per gram molecular weight. An example of one structure was given in Fig. 5-7.

As examples of the resistance of humus to decay, some of the carbon atoms stay for hundreds of years in the soil. Radioactive carbon dating values are the "average" age of all carbon atoms in the material (fresh material = zero age). Because some of the soil carbon is plant residues recently deposited, some of the humus carbon must be much older than the average age. Some reported values* are:

Material	Average Age of Carbon in the Organic Matter
Frozen peat, Barrow, Alaska, 2.1 m (6.9 ft) deep	25,300 yr ± 9%
Arctic Brown soil humus	3000 yr ± 4.3%
Surface soils, Iowa and North Dakota	210–440 yr ± 120 yr
Cheyenne grassland soil, 0–15 cm (0–6 in.) deep	1175 yr ± 100 yr
Manured plum orchard, Cheyenne soil	880 yr ± 75 yr
Bridgeport soil, Wyoming cultivated land	3280 yr
Carbon of substances adsorbed to clay, virgin soil	6690 yr
Humus in continuous wheat plots	1895 yr

The mechanisms hindering decomposition of soil carbon to CO_2 are many. Some of them are (1) adsorption of humus to clays, which hinders access of enzyme sites to the molecule's bonds, (2) lignin content, which reacts with proteins forming complex, hard-to-attack substances, (3) tannin–protein complexes, (4) insolubility of phenolic–amino acid substances, and (5) carbohydrate–amino acid derivatives of low accessibility to microorganisms and enzymes.

Source: Raymond W. Miller, Utah State University.
* F. E. Allison, *Soil Organic Matter and Its Role in Crop Production,* Elsevier Science Publishing, New York, 1973, pp. 157–158.

7:12 Composition of Organic Matter

The number of organic substances is immense, and they are as variable in composition as they are numerous. Organic matter is composed of about 45–50% carbon with lesser amounts of oxygen and hydrogen plus small quantities of nitrogen, phosphorus, sulfur, and many other elements. Carbon atoms joined together into carbon chains of many lengths and linkages with **H**s attached are the basic "skeleton" of organic compounds. The remaining elements fill out the skeletons to make different groups of organic-matter substances called proteins, lignins, carbohydrates, oils, fats, waxes, and many other materials. Humic substances are the colloidal, amorphous, polymeric, dark-brown components of soil organic matter (Detail 7-2).

Soil **humus,** the complex array of substances left after extensive chemical and biological breakdown of fresh plant and animal residues, makes up 60–70% of the total organic carbon in soils. Because of its complexity, humus is often divided by solubility separations into fulvic acid, humic acid, and humin. Both **fulvic acid** and **humic acid** are soluble in dilute sodium hydroxide solutions, but humic acid is larger and will precipitate out (be insoluble) when the solution is made acidic. **Humin** is the portion of humus that is insoluble in dilute sodium hydroxide.

The nature of soil humus is extremely complex. In addition to humin, humic acid, and fulvic acid, some of the other specific substances comprising soil humus are sugar amines, nucleic acids, phospholipids, vitamins, sulfolipids, and **polysaccharides**—the chains of sugar molecules that help to cement soil aggregates together. All of these substances are of complex nature. They are residual materials from plant tissues, substances synthesized by microbes, or residues of microbial degradation.

7:13 Decomposition of Organic Matter

When organic substances are manufactured by plants, the process of photosynthesis stores energy from the sun in the plant's organic substances. When the substances are decomposed, the stored energy is again released. However, the decomposition process has an energy barrier, called the **activation energy,** which must be overcome. When wood is burned, the activation energy is the elevated heating provided by the match or flame. In nature, only a few reactions, such as lightning, are available to provide this heat energy. Most biological processes require a means to reduce this activation energy so that the process can occur in natural conditions; enzymes, which lower the activation energy between chemical bonds, provide this means.

7:13.1 Enzymes and the Biological Reaction

An **enzyme** is a substance that is able to lower the activation energy of selected other compounds enough to allow the breaking or formation of a particular bond in a given natural environment (Detail 7-3). Such enzyme-influenced reactions are called **biological reactions.** The enzyme makes splitting the bond easier, but the process does not consume or destroy the enzyme. When one reaction is completed, the changed molecule diffuses into the solution and the enzyme is the same as it began and can split another similar bond. An activator, or any process that is not consumed or changed by the process, is termed a **catalyst.** Enzymes are catalysts to aid decomposition of organic materials.

There is a different enzyme for breaking each kind of bond. Each enzyme is given a name descriptive of the particular reaction it does, plus the ending **-ase.** Table 7-4 lists a number of common enzymes and their particular reactions.

Enzymes are produced by plants, animals, and microorganisms, and some are functional even when outside the living cell. Many different enzymes are produced by a single organism, and many organisms produce the same enzymes.

Free enzymes in the soil have several fates. First, they may function awhile before they themselves are decompsed by other enzymes and undergo further chemical breakdown. Second, they may become **denatured**—inactivated permanently. Denaturation can occur because of substances or reactions that change the protein shape; this may result from a high concentration of soluble salts or high temperature in the environment. Third, enzymes may become inactive because some reaction blocks access of the active parts of the enzyme to the substrate—such as the adsorption of enzymes onto humus colloids or clay particles, by the bonding of enzymes to each other, or by bonding to other chemicals that do not split off as does the usual compound.

When enzymes have been experimentally added to soils, they have existed free and active only for a short time (hours or days); yet periods of temporary inactivity (e.g., adsorption to humus colloids) allow enzymes to remain active for years. Measured activity of a selected enzyme (*phosphatase*) in peats that have existed in frozen conditions for at least 9000 years was found to be greater than such activity measured in many "fresh soils."[15] Active phosphatase enzyme has also been found in buried soils and in lake sediments covered over about 13,000 years ago.

---

[15] T. W. Speir and D. J. Ross, "Soil Phosphatase and Sulphatase," Chapter 6 in *Soil Enzymes,* R. G. Burns, ed., Academic Press, New York, 1978, pp. 209–210.

The two substrates (materials involved in the process) are (1) a chain of amino acids (polypeptides) to be broken into shorter units and (2) water molecules. The *carboxypeptidase* enzyme in I has a configuration that allows electron bonding (weakens nearby bonds) at sites **A, B, C,** and **D.** In II, as the polypeptide chain attaches to the enzyme, the **H** from site **B** goes to the **N** in the polypeptide chain, as shown by the arrow. Other bonds are altered as shown by dotted lines in several places. Water supplies the **H** to regenerate the enzyme site at **B** and supplies **OH** to form the carboxyl in the polypeptide residue near site **A.** The two portions of the polypeptide chain are now split apart and leave the enzyme; the enzyme is rejuvenated and ready to split another similar bond.

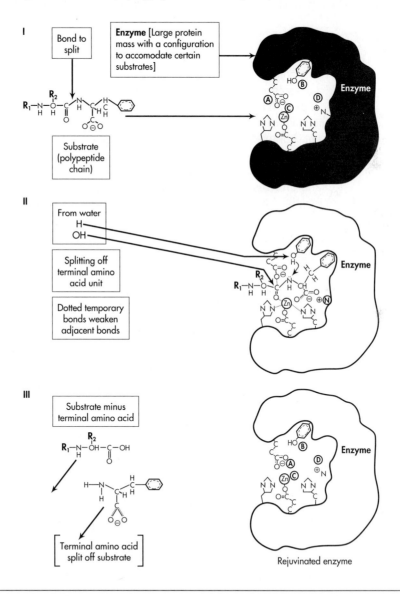

Source: Raymond W. Miller, Utah State University; drawn from details in E. O. Wilson, et al., *Life on Earth,* Sinauer Associates, Stamford, Conn., 1973, pp. 122–130.

7:13.2 Products of Decomposition

In well-aerated soils, the end products of decomposition are CO_2, NH_4^+, NO_3^-, $H_2PO_4^-$, SO_4^{2-}, H_2O, resistant residues, and numerous other essential plant nutrient elements in smaller quantities. If soil is not well aerated, less desirable products result. For example, in anaerobic conditions significant amounts of methane (CH_4), also called "swamp gas," are produced. Also, some organic acids (R—COOH), ammonium (NH_4^+), various amine residues (R—NH_2), the toxic gases hydrogen sulfide (H_2S), dimethyl sulfide, and ethylene (H_2C=CH_2), plus the resistant humus residues are produced.

One example of nutrient release by decomposition is the release of nitrogen. Table 7-5 illustrates the amount of nitrogen released annually from soils of different textures and different organic-matter contents. Usually decomposition is more rapid in the sandy soils.

7:13.3 Decomposition Action

The death of organisms eventually results in the rupture of cell membranes and the expulsion of many soluble cell substances into the soil and water solution. The organic solids (most of the nonwater portion of a dead organism) are altered only very slowly until microorganisms attack them. The decomposing organisms excrete a variety of enzymes to begin breakdown of the materials. The decomposition rate is directly proportional to the numbers of microbes present. Microbes absorb the nutrients released during decomposition—particularly nitrogen and carbon—and use them for growth and reproduction.

Nitrogen most often controls the rate of organic matter decomposition; it is needed to build proteins in new bacterial and fungal populations. The nitrogen content in the microorganisms and in organic materials is given as the **carbon: nitrogen ratio (C:N ratio).** A

Table 7-4 Some Common Enzymes and the Reactions They Influence

Enzyme	What It Does
Cellulase	Breaks celluloses (cell-wall fibers, "wood"), which are chains hundreds of sugar units long, into those component sugars. Important in organic-matter decay.
Urease	Breaks down urea (H_4N_2CO) to water, carbon dioxide, and ammonia. Makes animal urine and urea fertilizer nitrogen more available to plants. It requires the presence of Ni^{2+} ion.
Phosphatase	By involving water, it breaks the "humus—O—P(=O)—$(OH)_2$" bond to produce "humus—OH" and H_3PO_4, which helps to decompose humus, making phosphorus available to plants.
Sulfatase	By involving water, it breaks the "humus—O—S(=O)(=O)—OH" bond to produce "humus—OH" and H_2SO_4.
Protease	By involving water, it breaks the bond linking two amino acids $\left[R_1—(NH)—\overset{\displaystyle O}{\overset{\displaystyle \|}{C}}—R_2 \right]$ to form separate amino acids $\left(R_1—NH_2 \text{ and } HO—\overset{\displaystyle O}{\overset{\displaystyle \|}{C}}—R_2 \right)$ or parts of proteins. This is a digestive process of living tissues. R_1 and R_2 are distinct portions of organic material.

Table 7-5 Nitrogen Released from Soil Organic Matter from Three Soil Textural Classes during the Growing Season

Soil Organic Matter (% of total soil weight)	Nitrogen Released (kg/ha) [a]		
	Sandy Loam	Silt Loam	Clay Loam
1	50	20	15
2	100	45	40
3	—	68	45
4	—	90	75
5	—	110	90

[a] Soils in Southern regions release more nitrogen and soils in Northern regions release less nitrogen than that shown. When a good legume stand has been turned under, the crop that follows may have an additional 45–56 kg (88–123 lb) of available nitrogen per hectare.

wide organic-carbon:total-nitrogen ratio indicates a material relatively low in nitrogen content. Table 7-6 lists some common organic materials and their carbon and nitrogen contents.

Bacteria, requiring 1 pound of nitrogen for each 5–6 pounds of carbon (C:N ratio of 5:1 or 6:1), are heavy users of nitrogen. If straw with its low nitrogen content (C:N ratio of 80:1) is incorporated into a soil low in nitrogen, bacteria will multiply slowly because the straw is a low nutrient "food" for the decomposing microorganisms. The process of decay can be speeded up by adding more nitrogen (usually from fertilizers) to supply microbial needs. Bacteria (or fungi) will use any available nitrogen in the soil. Plants growing in a nitrogen-deficient soil are deficient in nitrogen because the soil microorganisms, which are more abundant and in more intimate contact, are able to use most available nitrogen before it can become accessible to plant root surfaces. The same is true for phosphorus and other nutrients.

Table 7-6 Approximate Percentages of Organic Carbon and Total Nitrogen and the C:N Ratio of Common Organic Materials Applied to or Growing on Arable Soils

Organic Material	Organic Carbon (C) (%)	Total Nitrogen (N) (%)	C:N Ratio
Crop residues			
Alfalfa (very young)	40	3	13 : 1
Clovers (mature)	40	2	20 : 1
Bluegrass	40	1.3	30 : 1
Cornstalks	40	1	40 : 1
Straw, small grain	40	0.5	80 : 1
Sawdust	50	0.1	500 : 1 [a]
Soil microbes			
Bacteria	50	10	6 : 1
Actinomycetes	50	8.5	6 : 1
Fungi	50	5	12 : 1
Soil humus	50	4.5	12 : 1

[a] Some sawdust may reach C:N ratios of nearly 800:1.

All plant residues contain materials easy to decompose and difficult to decompose. A mixture of residues from Austrian and Scots pines and English oak *(Pinus nigra, p. sylvestris,* and *Quercus robur)* had the following composition and rates of decomposition.

Original Litter	Portion of Whole (%)	Percentage lost by decomposition by:			
		1st year	*2nd year*	*5th year*	*10th year*
Sugars	15	99	100	—	—
Cellulose	20	90	100	—	—
Hemicellulose	15	75	92	100	—
Lignins	40	50	74	97	100
Waxes	5	25	43	77	95
Phenols	5	10	20	43	70
Whole litter materials		55.1	79.6	87.1	98.2

Within the first year just over half of the material is decomposed. By the end of the second year 80% is gone, and at the end of five years 13% of the original matter is still not lost. When organic residues are mixed in soils, some adsorption to clays slows decomposition rates of some materials (often proteins) even more than is shown above.

As decay of organic materials progresses, much of the carbon released escapes into the atmosphere as carbon dioxide (CO_2). This narrows the C:N ratio in the organic matter (which includes microbe bodies) because only a little of the nitrogen is lost while large quantities of carbon are expelled into the atmosphere.

Eventually the easily decomposable residues are gone and only materials that are decomposed more slowly remain (Detail 7-4). The food and energy source is now in short supply and more of the bacteria and fungi die. Their bodies, having a high nitrogen content, are decomposed by other living microorganisms, evolving carbon as carbon dioxide and releasing some nitrogen to the soil solution. Most of this "released" nitrogen is available to growing plants.

Dense populations of microorganisms inhabit the upper soil surface and have ready access to the soil nitrogen sources. Plant residues with C:N ratios of 20:1 or narrower have sufficient nitrogen to supply the decomposing microorganisms and also to release nitrogen for plant use. Residues with C:N ratios of 20:1 to 30:1 supply sufficient nitrogen for decomposition, but not enough to result in much release of nitrogen for plant use. The first few weeks after incorporation, residues with C:N ratios wider than 30:1 decompose slowly because they lack sufficient nitrogen for the microorganisms to use for increasing their numbers; this causes microbes to use nitrogen already available in the soil. If environmental conditions are favorable, the rate of decomposition of plant residues is most rapid during the first two weeks after incorporation into the soil. This pattern is given diagrammatically in Fig. 7-9 and shows the radical differences in the amounts of nitrogen released during decomposition of a narrow C:N ratio material (alfalfa) and a wide ratio material (oat straw). Sometimes residues with a wide C:N ratio may not have nitrogen immobilization, when incorporated into soils. If rates of decomposition are slow (large particle size, cool weather, quite dry), the need for nitrogen by smaller populations of decomposing microorganisms is low.

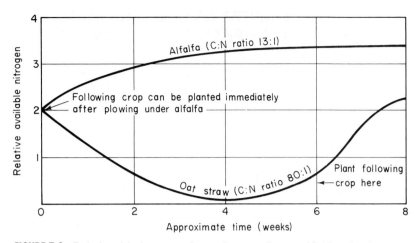

FIGURE 7-9   Relationship between the carbon-to-nitrogen (C:N) ratio of young alfalfa and mature oat straw and the time of planting a following crop after their incorporation in the soil. Notice that if the available soil nitrogen increases, there will not be as much nitrogen deficiency as with the oat straw.

7:13.4 Factors Affecting Decomposition

The most important conditions that alter the accumulation of soil humus are:

1. **Temperature.** Cold periods retard plant growth and organic-matter decomposition. Microbial activity generally follows the plant growth curve (Fig. 7-8). If temperatures are warm enough to produce considerable vegetation during the growing season, but are cold for long periods at other times of the year, organic-matter accumulation in and on the soil will be high. Continuous *cold* temperatures lower soil humus because little plant material is grown. Continuous warm temperatures aid high plant production, but also promote faster decomposition.

2. **Soil water.** Both plant growth and organic-matter decomposition require water; water contents near or slightly wetter than field capacity are most favorable for both processes. Extremes of both arid (dry) and anaerobic (waterlogged) conditions reduce growth of most plants and microbial decomposition. Poorly drained soils with growing vegetation have high humus contents, as indicated by formation of organic peats and mucks (Histosols).

3. **Nutrients.** Lack of nutrients, particularly nitrogen, reduces plant growth more than it slows decomposition because microorganisms use the nutrients in the dead organic material before plant roots can absorb them.

4. **Soil pH.** Most microorganisms grow best at pH 6–8 but are severely inhibited below pH 4.5 and above pH 8.5. Strongly acid soils are even more inhibiting than are strongly alkaline soils to microbe growth.

5. **Soil texture.** Soils higher in clays tend to retain larger amounts of humus, other conditions being equal. Most organic substances adsorb to clay surfaces by many kinds of bonds. The many active bonding sites of clays and humus include the $=0$, —OH, —Al—OH, —Fe—OH, and cation exchange sites of minerals, and the —NH_3^+,

—SH, —OH, and —COOH portions of organic materials.[16] When oily or waxy molecules (such as come from conifer needles) are adsorbed, the soil may become **water repellent** as these materials coat soil solids.

6. **Other factors.** Other decomposition inhibitors include toxic levels of elements (aluminum, manganese, boron, selenium, chloride), excessive soluble salts, shade, and organic phytotoxins (toxic to plants) in plant materials. The type of plant is also important, as succulent plants are more readily decomposed than are woody plants.

The accumulated organic matter in a soil is the net result of all the growth minus decomposition. The loss by decomposition is increased by mixing organic matter into the soil. Mixing the plant residue into soil (tillage) speeds decomposition and lessens accumulation. Cultivation and mixing by rodents or other organisms reduces soil humus content. Cultivation of grasslands has lowered soil humus 40–50% in 30 or 40 years.

7:14 Benefits of Soil Organic Matter

The list of benefits from having organic matter in soil is so varied and extensive that it makes one think of extravagant claims printed on old-time patent medicine labels; fortunately, these for organic matter are all true:

1. Organic matter is the source of 90–95% of the nitrogen in unfertilized soils (Detail 7-5).

2. Organic matter can be the major source of both available phosphorus and available sulfur when soil humus is present in appreciable amounts (about 2% or more) (Detail 7-5).

3. Organic matter supplies directly, or indirectly through microbial action, the major soil aggregate-forming cements, particularly the long sugar chains called **polysaccharides.**

4. Organic matter contributes to the cation exchange capacity, often furnishing 30–70% of the total amount. The large available surfaces of humus have many cation exchange sites that adsorb nutrients for eventual plant use and temporarily adsorb heavy-metal pollutants (lead, cadium, and the like), which are usually derived from applied wastewaters. Adsorption of pollutants helps clean contaminated water.

5. Organic matter commonly increases water content at field capacity, increases available-water content in sandy soils, and increases both air and water flow rates through fine-textured soil. The latter effect is probably due mainly to soil aggregation, which produces larger soil pores.

6. Organic matter acts as a chelate. A ligand is any organic compound that can bond to a metal (usually iron, zinc, copper, or manganese) by more than one bond and form a ring or cyclic structure by that bonding, called a **chelate** (key-late). The *soluble* chelates probably help mobilize these micronutrient metal ions, increasing their availability to plants and general mobility in soils. The chelate mechanisms are not fully known at present.

7. Organic matter is a carbon supply for many microbes that perform other beneficial functions in soil (e.g., free dinitrogen fixers, denitrifiers).

---

[16] M. Schnitzer, "Humic Substances: Chemistry and Reactions," in *Soil Organic Matter*, M. Schnitzer and S. U. Khan, eds., Elsevier Science Publishing Co., New York, 1978, pp. 53–54.

Whether or not a fertilizer is good depends on its furnishing (1) the needed nutrients (2) in adequate amounts (3) at the right time (4) in an efficient (low-loss) manner. The major lack from organic matter is the inadequate amount of nutrients supplied during stages of peak growth.

The table below illustrates the amount of fresh organic residue left following certain crops. Obviously, each harvested crop leaves different amounts of residues to supply nutrients to succeeding crops. Supplemental residues or careful crop rotation is necessary to supply the amounts of residues wanted. The rates of nutrient release from the residues depend on the microbial decomposition rate and the nutrient content of the residues. Decomposition rates increase as the weather warms and furnishes maximum plant growth conditions. To some extent, nutrients are released in larger amounts as crop needs increase. However, this is generally correct only in temperate areas where crops are planted in the cool spring and reach high nutrient demands in the early warm periods of summer. In the tropics or subtropics, where less distinct cold–warm cycles regulate planting and plant-growth dates, nutrient release rates do not correspond as well with plant nutrient needs. Late-planted crops also lack a good need–supply relationship.

Whether or not adequate nutrient amounts are released depends on the amount and kind of organic matter, soil texture, and crop needs. Because these factors vary, no one soil condition can supply optimum nutrients for the many different crop situations.

Perhaps the major advantages of soil organic matter as a nutrient source are its conservation of nutrients against leaching losses and its continual release of nutrients. Because only small portions of organic matter decompose in a few days, heavy rainfall or irrigation is able to leach only small amounts of solubilized nutrients. Immediately after a rainfall further organic-matter decomposition begins to supply new reservoirs of the many nutrients in the humus.

Organic materials should not be added indiscriminately to soil in large amounts in an effort to supply adequate nitrogen and/or phosphorus. There may be attendant problems that are worse than the nutrient deficiencies. The production of toxins (Allelopathy, the presence of soluble salts in manures, and the introduction of potentially toxic heavy metals, e.g., cadmium, mercury, zinc, and lead) in sewage sludge are all accumulative as amounts added to soils are increased.

Estimated Addition of Organic Matter to Soils by Crops Grown and Residues Left on the Field

Crop and Harvested Yield[a]	Organic Matter Additions (lb/a)		
	Root Portion	Plant Tops	Total
Corn, grain harvested (100 bu/a)	2200	4100	6300
Soybeans, grain harvested (32 bu/a)	1000	2200	3200
Wheat, grain harvested (45 bu/a)	1700	3100	4800
Field beans, seed harvested (1500 lb/a)	500	1400	1900
Alfalfa, hay removed (3 t/a)	3400	800	4200

[a] Corn, bu/a times 62.78 = kg/ha; soybeans and wheat, bu/a times 67.26 = kg/ha; beans and organic matter, lb/a times 1.121 = kg/ha; alfalfa, t/a times 2242 = kg/ha.

8. When left on top of soil as a mulch, organic matter reduces erosion, shades the soil (which prevents rapid moisture loss), and keeps the soil cooler in very hot weather and warmer in winter.

9. Most soils above the Arctic Circle and below the Antarctic Circle depend on a thick layer of organic matter to stabilize them. When this layer is destroyed by fire or construction activities, soils may become warmer by 9°C (20°F). In summer this causes melting of permafrost and results in very severe surface and pothole (vertical) soil erosion.

FIGURE 7-10 A chisel plow only partially incorporates grain stubble, leaving part of the straw exposed to protect the soil against erosive action of rainfall, flowing water, or wind. The soil also absorbs water faster in this condition. (Courtesy of J. I. Case Co.)

10. Humus buffers the soil against a rapid change in acidity, alkalinity, and salinity and damage by pesticides and toxic heavy metals.

11. Humus reduces the crystallization and hardening of the plinthite (laterite) layer of soils in the humid tropics that are rich in soluble iron and aluminum. Humate complexes with iron and aluminum reduce crystallization. Organic matter also reduces hardening by maintaining more uniform soil temperature and soil moisture.

Mulch tillage (Fig. 7-10) involves organic matter as a mulch. Mulch tillage is used on grain lands where wind or water is likely to cause extensive erosion of exposed soil. A coarse surface mulch increases the percentage of water that seeps into the soil. On one bare Texas soil, a 7-cm (2.7-in.) rain wetted the soil to a depth of 39 cm (15.2 in.); the addition of a 36-Mg/ha (16-t/a) straw mulch increased the water penetration to 75 cm (29 in.), even though the straw itself absorbed some water.

Surface organic matter can insulate soil, retarding heat flow between the atmosphere and the soil. In hot summers this benefits some plant roots, but in cool areas it slows soil warming in the spring. In cold areas one should leave small amounts of plant residues on the soil surface to allow the maximum rate of warming in the spring, but leave only enough residue to control soil erosion.

Some rather curious effects may emanate from soil humus. The release of various vitamins, hormones, amino acids, and other substances exhibit unexpected effects on plants.

A newly discovered extract from rape plant pollen (called *brassinolide*) in concentrations of one-billionth of a gram causes cells to elongate, and when applied on the eyes of seed potatoes has increased potato yields 24%.[17]

---

[17] Judy McBride, "Pushing Plants to Full Potential," *Agricultural Research,* **28** (1979), no. 2, pp. 14–15.

Ries at Michigan State University reported in 1985 that triacontanol, an alcohol occurring in alfalfa, serves as a plant hormone. As small an amount as 1 mg (portion of one drop) per hectare stimulates the growth of corn, tomatoes, lettuce, and rice.[18]

The list of "growth substances" produced by microorganisms in soils is very long; it includes indole acetic acid (IAA), vitamins, gibberellic acid, and numerous unidentified auxins (growth promotors). In large amounts some of these substances can produce abnormal growths in plants. In one study two of these growth substances were effective in helping roots penetrate a compacted subsoil nearly as well as where the subsoil was not compacted.

▪ 7:15 Detrimental Effects of Soil Organic Matter

The many benefits of organic matter in the soil are counterbalanced in certain situations by detrimental influences. Organic matter is an energy and carbon source for many disease organisms, ensuring their longer survival in soils. Excessive amounts of organic matter are physically difficult to incorporate into the soil and hinder easy planting. Residues (or virgin vegetation on land being cleared) are often burned to reduce bulk. Although it is common practice, burning off crop residues is not desirable except in a few selected situations (e.g., sugarcane harvests). Burning is harmful because (1) it removes organic material that protects the soil against erosion, (2) some ash that contains the plant nutrients can be lost easily by wind or water erosion, (3) most nutrients in the ash are soluble and some are easily leached through the soil, and (4) the organic matter whose decomposition results in the cements for soil aggregates is lost in burning.

If planting is delayed for weeks or months (or over the winter) after plant residues are placed on the soil, some decomposition takes place and reduces the problem of residues clogging equipment.

Numerous plants contain or produce phytotoxins (plant poisons; for example, juglone from black walnuts) (Detail 7-6). The phytotoxins make such residues undesirable organic matter. Any and all plant materials should not be added indiscriminately to soils. Unfortunately, the problem of phytotoxin production cannot always be avoided; the decomposition of many plant residues produces such toxins. This production of toxins is a form of allelopathy. **Allelopathy** is any direct or indirect harmful effect of one plant on another through the production and liberation of toxic or inhibiting chemical compounds into the environment. For example, drilling cereal seed into crop residues can cause poor stands of seedlings. Acetic acid, which is produced in the cool, wet, anaerobic decomposition of wheat straw, is toxic to barley seedlings.[19,20] Most of the toxins identified so far have been short-chain organic acids. A list follows of a few of the observed allelopathic effects of plants or their decomposing residues.[21] One research problem is to separate the effect of competition from that of allelopathy (Detail 7-7).

[18] Stanley K. Ries, "Regulation of Plant Growth with Tricontanol," *Critical Reviews in Plant Science,* **2** (1985), no. 3, pp. 239–285.

[19] Joan M. Wallace and L. F. Elliott, "Phytotoxins from Anaerobically Decomposing Wheat Straw," *Soil Biology and Biochemistry,* **11** (1979), pp. 325–330.

[20] M. J. Krogmeier and J. M. Bremner, "Effects of Water-Soluble Constituents of Plant Residues on Water Uptake by Seeds," *Agronomy Abstracts* (1986), p. 181.

[21] Dirk C. Drost and Jerry D. Doll, "Allelopathy: Some Weeds Use It against Crops and Could We Use It to Fight Weeds?" *Crops and Soils Magazine,* **32** (Mar. 1980), pp. 5–6, and (Apr.–May 1980), pp. 5–6.

Detail 7-6 Which Chemicals Are Hazardous?

The view that "natural" chemicals are safe whereas "synthetic" or "manufactured" chemicals are hazardous is a fallacy. Many toxic natural substances (poison hemlock, toadstools, cyanic acid in almonds and lima beans, hemagglutinins in castor beans) are as poisonous as, or more so, than manufactured ones. Botulism toxin occurs naturally on many foods when they are not preserved properly; it is 100 times more toxic than anything humans have fabricated. **Aflatoxins**—potent carcinogens—are found as fungal growths on many grains and nuts. Aflatoxin B1 is the most potent, naturally occurring, cancer-producing substance known. Antibiotics, toxins produced by microorganisms, are well known as biocides in the practice of medicine.

These facts do not lessen the hazard of pesticides, synthesized by people. The great hazard of the chlorinated hydrocarbons (DDT, chlordane, BHC) is not their high toxicity to humans, but their toxicity to lower forms of animal life and because these chemicals remain in the environment for decades. More-powerful toxic chemicals, such as the organophosphates used now, can be decomposed by soil microbes to inactive substances within a few days or weeks after application. According to a 1975 report of the National Research Council, the U.S. population was consuming at that time about 40 mg (0.0014 oz) of pesticide residue per person per year in food. This total consumption was estimated to have the acute toxicity equivalent to the chemicals in one aspirin tablet or to the caffeine in one cup of coffee.

Some of nature's toxins are stranger than fiction. The numbers and examples known increase rapidly. More than 1500 plant species are toxic to one insect or another. In fact, this relationship is used partially for insect control by organic gardeners. Many plant substances are also toxic to people. The hemagglutinins in two castor beans can cause clumping of red blood cells and be nearly lethal for some humans. A single Monarch butterfly larva may concentrate enough cardiac glycosides from milkweed, their primary food, to kill several human beings if it were ingested. People in Wales are said to have the world's highest rate of bladder and intestinal cancer; researchers believe the cause is the wild bracken fern that cows eat, which transfers a carcinogen to the milk.

A substance called acetylandromedol, isolated from rhododendron plants, is used to lower blood pressure. It is known to cause severe hives and even to paralyze skeletal muscles. Rhododendron tea has been used for centuries with various claims. Some tree varieties secrete a virulently toxic nectar. Ancient reports of two different armies near the Black Sea coast of Turkey—Greek soldiers in 401 B.C. and Roman troops in 66 B.C.—describe the severe effects of this toxin. After the soldiers ate honey in large quantities that they found there, they behaved like drunk men, others like madmen, and some like dying persons for 1 to 3 days. While the Romans were affected, they were attacked and massacred by the Pontic army.

Natural toxins are widespread, and some less obvious in their effects—as evidenced by birth defects months later—are still discovered regularly. Many of these natural toxins in common foods are only toxic under selected diets or in certain eating combinations.

Sources: (1) Robert P. McIntosh, "The Ecology Program," *Environmental Biology,* **2,** no. 4, July 4, 1978, (2) "Toxicants Occurring Naturally in Foods," National Academy of Sciences, National Research Council Publication 1354, NAS, NRC, Washington, D.C., 1966, 301 pp., (3) I. Antice Evans and J. Mason, "Carcinogenic Activity in Bracken," *Nature,* **208** (1965), pp. 913–914, (4) David G. Leach, "The Ancient Curse of the Rhododendron," *Garden,* **2,** no. 4, July–Aug. 1978, pp. 4–9.

1.   Corn and giant foxtail were grown in greenhouse pots connected by recirculating nutrient solution. Root exudates of living foxtail and leachates of dead foxtail plants both inhibited the growth of corn (Illinois).

2.   Soil samples taken from heavy growth of the weed leafy spurge reduced growth of other plants. Tiny amounts of leafy spurge litter (only 0.1% of leaves and stems mixture) mixed into the soil reduced the growth of tomatoes (Colorado).

A weed called *dyer's woad (Isatis tinctoria)* is a spreading nuisance and serious weed problem in several Western states.[a] Used in Europe for thousands of years to make blue dye, yellow-flowered dyer's woad uses allelopathy to aid its spread. When seed pods fall and rot, they exude a toxin into the soil that kills the roots of nearby grass plants. The only known resolution is for the toxin to be washed away by rainfall. Optimistically, the toxin, when identified, could be synthesized and used as an environmentally safe herbicide. In the meantime, the meter-tall dyer's woad spreads to new areas, competitive over weeds and crops alike because of its allelopathic toxin and its big, leafy rosettes that shade nearby plants.

[a] J. A. Young, "Dyer's Woad Wages Chemical War in the West," *Agricultural Research,* **36** (1988), no. 7, p. 4.

3. Dried and pulverized foliage or tubers of the weed yellow nutsedge reduced the dry weight of corn and soybeans. Quackgrass also reduced corn growth (Wisconsin).

4. Grain sorghum residues grown in sterile laboratory media (agar) exuded toxins, which later reduced the growth of test plants grown on the agar (Michigan).

7:16 Economical Levels of Organic Matter

How much soil humus should be maintained in soils?

Plants can grow well in mineral soils without any organic matter at all and also in "soils" that are 100% organic matter, if fertilizers are added. Commercial greenhouses and nurseries use mixed potting media having as much as 30–60% peat moss or other organic materials. The optimum level of organic matter that should be maintained in soil has never been determined, nor is it any single value for all soils.

Wide ranges in organic-matter contents in soils (Table 7-7) indicate effects of climate, plant cover, and management variations that occur in a single soil order, soil texture, or in an area. Some cropping systems tend to cause greater soil organic-matter changes than others. Tillage also affects organic-matter concentrations. The greater the tillage, the less the amount of humus that accumulates. Uncultivated soils are higher in total soil organic matter (*on* and *in* soil) than they are after cultivation. The obvious exception is arid soil that supported almost no plants before cultivation but due to irrigation has an increase in humus content—resulting from moisture and plant growth rather than tillage.

Generally, organic matter in the soil improves plant growth. To a farmer, however, the bottom line for deciding a good level of organic matter to maintain depends on *cost* and *convenience.* It is usually too costly and inconvenient to purposely build up organic matter in soils by means other than incorporating plant residues and disposing of available manures, plant residues, or sewage sludge onto nearby soil. A practical rule of thumb is to *use all available crop residues* by incorporating them into the soil (rather than burning them), apply the minimum fertilizer economically feasible to produce maximum plant size for larger harvests, and add other suitable residues that might be available.

7:17 Animal Manures and Sewage

How important are animal manures as additives to land areas? Thirty-nine percent of all livestock in the United States are fed in confinement, and safe environmental disposal of their manure is an important public concern. An estimated 159 million metric tons (175 million tons, dry-weight basis) of manures are excreted by all domestic livestock and poultry in the United States each year. (As voided by the animal, the excreta average 85% water and 15% solids.)

Table 7-7 Organic-Matter Contents of Surface Soils from a Variety of Locations, under Various Plant Cover, and in Various Climates

Soil and Site Description	Organic-Matter Content (%)
Arizona–Nevada soils, 280 mm (11 in.) rainfall yearly	
Aridisol (loamy coarse sand)	0.60
Aridisol (fine sandy loam)	1.64
Aridisol (silty clay), pH 8.4	1.19
India, 1000–1500 mm (39–58.5 in.) rainfall, cultivated	
Acid sulfate paddy soil, pH 6.2	0.75
Alluvial paddy soil, pH 6.3	2.88
Oxisol (clay), pH 5.2	5.52
Hawaii, cultivated soil	
Oxisol (clay) from basalt, pH 5.9, 1100 mm (43 in.) rainfall	0.7
Oxisol (clay) from basalt, pH 5.6, 2400 mm (94 in.) rainfall	6.0
Inceptisol (loam) from volcanic ash, pH 5.1, 2500 mm (98 in.) rainfall	12.0
Iowa, U.S., 800 mm (31 in.) precipitation, cultivated	
Entisol (silt loam)	2.59
Mollisol (loam)	2.83
Mollisol (silty clay)	4.9
Nebraska, U.S., 450–800 mm (18–31 in.) precipitation, cultivated	
Mollisol (silty clay loam) (33% clay), pH 5.2	3.8
Entisol (loamy fine sand) (6% clay), pH 5.9	1.4
Liberia, Africa, 2000+ mm (78+ in.) rainfall, virgin soils	
Ultisol (sandy clay loam), pH 4.9	3.45
Ultisol (sandy loam), pH 4.3	2.28
Turkey, 900 mm (35 in.) rainfall, cultivated	
Inceptisol (clay) (61% clay), calcareous	1.1
Vertisol (clay) (54% clay), calcareous	0.9
Mollisol (loam) (20% clay), 65% lime	4.3
Entisol (sandy loam) (6% clay), from volcanic ash	0.6
Costa Rica, low elevation, 1500–2000 mm (58.5–78 in.) rainfall	
Ultisol (clay) forested, pH 5.5, low elevation	3.74
Inceptisol (clay loam) forested, pH 5.2, from volcanic ash	3.97
Inceptisol (silty clay) forested, pH 4.1, from volcanic ash	18.94
Inceptisol (silty clay) forested, pH 6.2, from volcanic ash	10.25
Ultisol (clay) forested, pH 5.1, low elevation	4.34
Santa Barbara, California, 480 mm (19 in.) rainfall	
Mollisol (loam), pH 7.3	7.85
Mollisol (very fine sandy loam), pH 7.8	11.3
Michigan–Indiana soils, cultivated, 750–1000 mm (29–39 in.) rainfall	
Mollisol (sand) 2.9% clay, pH 6.6, imperfectly drained	6.15
Alfisol (sand) 3.4% clay, pH 7.5, well drained	1.81
Mollisol (loam) 19.8% clay, pH 6.3, imperfectly drained	8.24

The most pragmatic means of disposing of this massive quantity of manure is by additions to the soil, where its nutritive value can be utilized by growing plants. Cattle manures (dry basis) average 3% N, 0.8% P (1.8% P_2O_5), 2% K (2.4% K_2O), 25% organic carbon, plus varying amounts of other elements essential for plant growth. Barnyard manure, because of water content, has a fertilizer grade often less than 1–0.3–1 (Table 7-8).

Table 7-8 Typical Composition of Selected Animal Manures (Dry-Weight Basis)[a]

Constituent	Beef/Dairy (%)	Poultry (%)	Swine (%)	Sheep (%)
Nitrogen (N)	2–8	5–8	3–5	3–5
Phosphorus (P)	0.2–1.0	1–2	0.5–1.0	0.4–0.8
Potassium (K)	1–3	1–2	1–2	2–3
Magnesium (Mg)	1.0–1.5	2–3	0.08	0.2
Sodium (Na)	1–3	1–2	0.05	0.05
Total soluble salts	6–15	2–5	1–2	1–2

[a] Data were obtained from many sources.

7:17.1 Manure Composition and Use

Manures differ in composition, partly as a result of differences among the kinds of feeds that animals consume (Table 7-8). In comparison with chemical fertilizers, all manures supply relatively small quantities of plant nutrients per unit of dry weight. However, the *micronutrient* content in manures is usually higher than in chemical fertilizers to which micronutrient fertilizers have not been intentionally added. Manures have high soluble salt contents (6–15% of the total dry weight in beef/dairy cattle manure). This is partly because salt (sodium chloride) often is fed to livestock to increase appetite and to reduce kidney stones; much of the salt is voided in the manure. If wisely used, manures are good fertilizers, although they are low in phosphorus.

Today manure is often added to soil to be rid of it. But how much manure can be applied to various soils without causing damage to plants or to animals eating the plants? Leaching may move some contaminants from manure to the soil surface or to surface waters in runoff. There are complications when manures are applied at rates exceeding about 200 metric tons of dry weight per hectare (over 90 tons per acre) annually. Optimum rates of addition seldom exceed 40–50 dry metric tons per hectare (18–22 t/a). Heavy applications may add too much soluble salt, heavy metals, or available nutrients. For example, copper and arsenic fed to poultry for disease control or "growth-promoting" purposes are defecated as part of the manure and accumulate when such manures are applied to soils. Nitrates can move slowly through the earth's substratum to groundwater or be eroded into surface waters. Some nitrates in California have been predicted to reach groundwater 30 m (98 ft) deep in 10 to 50 years when 18–25 cm (7–10 in.) of water each year leaches through the soil. Excess nitrates also accumulate in the crops grown on soil with excess manure added. High nitrate forage (reports suggest from 0.21% to 0.48% nitrate-nitrogen) can be harmful to animals and high nitrate waters can cause "blue-baby" disease (methemoglobinemia) in humans. These levels can be reached with application rates of over 179–224 dry metric tons per hectare (80–100 t/a) annually. In spite of these potential hazards, relatively few serious cases of pollution have occurred.

Recommended manure application rates solely for plant growth enhancement range from about 33.6–56 dry metric tons per hectare (15–25 t/a) on irrigated lands in the Western states for a single yearly high-value crop. Only 20% of this amount can be used on nonirrigated dryland grains. The manure releases from 15% to 45% of its nitrogen the first year and 5% to 10% the second year. In warm Southern regions, such as Arizona and southern California, the percentages released could be somewhat higher in irrigated soils.

For greatest environmental integrity, manure should be spread daily on unfrozen and nearly flat lands. Because this is frequently impossible, storage in lagoons (ponds) is common in humid areas. The ponds may become anaerobic when not artificially stirred. Aeration reduces odors, hastens decomposition, and reduces pathogens.

FIGURE 7-11 This beef cattle feedlot in Nebraska has 6000 cattle, which produce more than 300,000 pounds (136,079 kg), wet weight, of manure each day, excluding bedding, that must be disposed of without polluting the water and air environments. Who can supply an acceptable answer to this problem? (*Source:* USDA.)

7:17.2 Environmentally Safe Management of Manure

To reduce pollution hazards, feedlots should be established with the following guidelines:

1. All outdoor runoff and retention facilities should have the capacity to control any maximum 10-year, 24-hour storm water.

2. All lagoons, ponds, and other animal waste storage facilities should be watertight to eliminate runoff and deep seepage to surface water or groundwater.

3. Animal wastes should be incorporated into the soil or, if surface-applied, be retained on sloping fields by terraces.

4. The maximum safe application rate of animal wastes on fields should be established by research (Fig. 7-11).

7:18 Sewage Sludge and Wastewaters

Sewage sludges (the solids settled out in sewage treatment plants) are much like farm manures (Table 7-9). They are especially well adapted for application to turf grasses on golf courses, in cemeteries, and around public buildings. Commercial nurseries are also potential users of sludge. Larger quantities of sludge are expected to be used for the revegetation of soils

Table 7-9 Human Effluents in the United States: Total Produced and Their Major Plant Nutrient Content (in Thousands of Tons, Dry-Weight Basis)[a] (Estimated for 1975 and Predicted for 1990)

| Human Effluent | Total Produced | | Major Plant Nutrients[b] | | | | | |
| | | | Nitrogen (N) | | Phosphorus (P) | | Potassium (K) | |
	1975	1990	1975	1990	1975	1990	1975	1990
Sewage sludge	4300	5418	172	217	86	108	17	36
Septage	694	972	18	25	11	15	3	4

Source: Adapted from U.S. Environmental Protection Agency, *Composting of Municipal Solid Wastes in the United States,* EPA, Washington, D.C., 1971.

[a] Tons × 0.9072 = metric tons.

[b] Based on typical sewage sludge composition of 4.0% N, 2.0% P (4.6% P_2O_5), and 0.4% K (0.5% K_2O) and on septage composition of 4.4% N, 1.6% P (3.7% P_2O_5), and 0.4% K (0.5% K_2O) (dry-weight basis). Organic-carbon percentage of both sludge and septage is estimated at 25% (dry-weight basis) because of the amounts of nonorganic materials (soil, plastics, precipitated salts) in sewage.

drastically disturbed by surface mining and construction. Forests, pastures, and rangelands are also suitable sites for applying large amounts of sewage sludges and their waste waters.[22]

Although variable in chemical composition, sludges contain about 4.0% nitrogen (N), 2.0% phosphorus (P) (4.6% P_2O_5), and 0.4% K (0.5% K_2O) on a dry-weight basis. They also contain toxic and heavy metals, such as boron, cadmium, copper, mercury, nickel, lead, selenium, and zinc. There is real concern that toxic quantities of these metals will be detrimental to plants, animals, and people. Cadmium is of special concern because it is relatively soluble, mobile in plants, and toxic.

Pathogenic (disease-producing) organisms present in some sewage wastewaters and sludges may cause cholera, diarrhea (amoebic and bacterial), hepatitis, pinworms, poliomyelitis, and tapeworms. This is another example where natural organic systems of agriculture may not be totally advantageous.

Four techniques are used to reduce to near zero the hazard of pathogens from soil-applied sludges.

1. Compost the sludge outside for at least 21 days. During composting, the heat of microbial decomposition normally reaches 55°C (131°F).

2. Store the semiliquid, anaerobically digested sludge for at least 60 days at 20°C (68°F), for 120 days at 4°C (39°F), or for some combination in between.

3. Treat the sludge, when moist, with lime [CaO or Ca(OH)$_2$] for at least 3 hours. This pH adjustment also helps reduce plant uptake of heavy metals.

4. Pasteurize for 30 minutes at 70°C (158°F).[23]

Environmentally safe disposal of large tonnages of sewage and septage (human and industrial wastes) and their much larger volumes of effluent wastewaters is one of the most

---

[22] Alex Hershaft and J. Bruce Truett, *Long-Term Effects of Slow-Rate Land Application of Municipal Wastewaters,* EPA-600/57-81-152, Environmental Protection Agency, Washington, D.C., 1981.

[23] Norman E. Kowal, *Health Effects of Land Application of Municipal Sludge,* EPA/600/1-85/015, Environmental Protection Agency, Washington, D.C., 1985.

urgent technical and economic problems of modern society (Table 7-9). Cities near oceans have been dumping human and other solid wastes into the ocean, but this polluting practice legally stopped at the end of 1981. Sewage sludge can be burned, but the fuel costs to do so are now prohibitive and air is polluted thereby. (Sludge contains too much water to be burned without additional fuel.) Sludge has been buried in landfills along with other solid wastes; but again, costs of hauling are increasing and suitable soils nearby are scarce. The principal remaining alternative is to spread sludge on soils, either before or after composting. Before composting, sewage sludge has inherent hazards of spreading pathogens, polluting surface and underground waters, and being toxic to plants, animals, and people.

Treated sewage wastewater can be recycled back onto the land. Sewage wastewater can substitute for expensive irrigation water, and sewage solids and dissolved nutrients can replace part or all of the application of chemical fertilizers. If water is dispersed by sprinkler irrigation on forests and croplands and applied at a rate no faster than the soil's infiltration capacity, the use can be environmentally safe. One recycling program in Pennsylvania disposed of wastewater and produced these positive results (Fig. 7-12):

1. Field crop yields were increased. A 2.5 cm (1 in.) depth of effluent per week increased grain yields 24% over land with 672–1120 kg/ha (600–1000 lb/a) of 10-10-10 commercial fertilizer added annually for 5 years.

2. Forest tree growth of species such as white pine and white spruce were increased (but some species such as red pine were injured by heavier applications) (Fig. 7-12).

3. Wastewaters replaced both chemical fertilizers and usual irrigation waters.

4. Groundwaters were adequately recharged.

7:19 Composts and Composting[24,25]

Composting is the microbial decomposition of piled organic materials into partially decomposed residues, which are called *compost* or *humus*. Composting is not a common process in nature because usually there are no piles of organic materials to decompose. Where thick accumulations of organic matter do occur naturally (e.g., in stagnant water or under cold acid forests), conditions are far from optimum for the decomposition of the residues, unlike those of well-maintained compost piles. The process of enzymatic decomposition of organic matter by microbes in composting is similar to that which occurs as plant remains in and on the soil decompose; various nutrients (nitrogen, phosphorus, sulfur, calcium, and others) are released as the less complex compounds decompose, while more-complex residues continue to decompose at slower rates.

Composting requires conditions that are favorable for microbial growth. The process can be either anaerobic or aerobic, but it is much faster and less odoriferous if done aerobically. Anaerobic composting may also produce plant-toxic organic acids, ethylene, and others. For optimum decomposition rates, the materials must be kept moist and warm, and with sufficient nutrients available. To destroy pathogens effectively, exact temperatures for determined lengths of time must be reached inside the composting mixture. The complete destruction of plant pathogens by composting is nearly impossible because the outside of the compost heap is cooler than the required killing temperature.

---

[24] Jerry Minnich, Majorie Hunt, and editors of *Organic Gardening Magazine, The Rodale Guide to Composting,* Rodale Press, Emmaus, Pa., 1979.

[25] Staff of *Organic Gardening Magazine, The Encyclopedia of Organic Gardening,* Rodale Press, Emmaus, Pa., 1978, pp. 235–248.

(a)

(b)

FIGURE 7-12 Sewage wastewater being applied through a sprinkler irrigation system to a forest in summer (a) and in winter (b). A forest seems to be an ideal ecosystem for discharge of sewage effluent. To avoid damage, a scientific assessment must be made of each site as to the water capacity of the soil and the tolerance of the vegetation to chemicals in the effluent. Selection of spray sites must be done carefully to avoid runoff into surface water if freezing seals the soil, causing ponding or runoff. (Courtesy of William E. Sopper, Pennsylvania State University.)

If composted materials are known to be deficient in particular plant nutrients, the missing ingredients can be added to the compost. Nitrogen, phosphates, and lime are the most frequently added. Fertilizers, legumes, seed meals, manures, and animal slaughter wastes (dried blood, fish scraps, tankage) are commonly recommended materials.

Water is essential for all enzymatic processes, so composts must be kept moist to ensure rapid decomposition. For small compost piles, plastic or other moisture barriers can be used to cover composts during dry, hot periods.

Composted materials often are not sources of large amounts of available nutrients. Fresh organic residues that are high in nutrients (young legume plants, most manures, vegetable wastes) can be spread directly on the soil without composting because they rapidly decompose and release nutrients that plants can absorb. Low-nutrient materials (sawdust, mature woody plants, straw) should be composted first, where it is practical to do so.

Public health problems encountered in compost materials are viruses, pathogens, and parasite eggs in the wastes. Composting would seem to favor the growth of these harmful organisms, but during decomposition the compost can heat up enough that only thermophilic (heat-tolerant) microbes are active in the interior portions. At temperatures near 65–75°C (149–167°F), most hazardous pathogens are destroyed in a few days but may not be killed at the edges of the compost. If temperatures are as low as 55°C (131°F), the time necessary to destroy pathogens lengthens to weeks or even months. Pathogens are likely to be in sewage sludges and some municipal garbage.

Composting does not remove nonorganic materials such as lead, mercury, cadmium, arsenic, cyanide, strong acids, and other inorganics, nor do these prevent the potential formation of **mycotoxins**—substances that are toxic to plants or animals and formed by fungal action of decomposing organic materials. Also, *teratogenic* (deformity-causing), *carcinogenic* (cancer-causing), or other harmful organic substances may be produced naturally by the composting process.

Phytotoxins (substances toxic to plants) and bad odors are produced during early stages of decomposition of some organic materials, particularly in anaerobic conditions. These can often be dissipated by evaporation of the substances at the elevated temperatures while in the composting bed.

Practically any organic substance can be composted, but some are less suitable than others. Large pieces of shells, wood, cornstalks, greases, oils, pesticides, large masses of wet materials, weeds with seeds, diseased plants, newspaper, and pine and other conifer needles are not generally good for composting. Some of these materials can be used if they are first shredded and well mixed with other substances.

Large amounts of low-cellulose-content materials, such as food wastes and animal tankage, are difficult to compost because they form large masses of gelatinous, anaerobic material in piles. The addition of fibrous materials (sawdust, straw, shredded paper, leaves, wood chips, mature plant materials) to gelatinous or dense masses improves composting action by increasing aeration.

Good composting usually requires several weeks of aerobic decomposition at elevated temperatures, high moisture, and high nutrient levels. Commercial operations can speed the natural process by continually mixing the compost and adding heat from an external source when necessary. Some facilities composting municipal garbage turn out a good compost in a month. Where mixing is infrequent and little is done to assist decomposition, the compost may require several months to reach "finished" conditions. Composting is such an inexact process that there are no certain criteria for determining when the compost materials are ready for use. Although the principles are the same, the methods of composting vary from the small, backyard, enclosed compost bin to those of large-scale operations (Fig. 7-13).

Composts should be added to soils in manners similarly to using humus or animal manures. They are generally low-analysis materials, commonly with less than 2% of nitrogen, phosphorus, or potassium. Perhaps 40–60% of the nutrient content is mineralized during the first growing season.

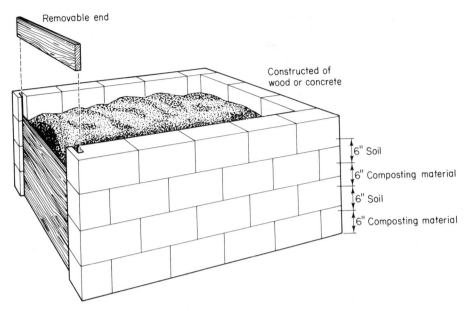

Removable end

Constructed of
wood or concrete

6" Soil
6" Composting material
6" Soil
6" Composting material

FIGURE 7-13 Well-constructed "backyard" type of compost bin made from concrete blocks. (Courtesy of Vocational Instructional Services, Texas A&M University.)

7:20 Low-Input Sustainable Agriculture

The high cost of energy and limited future supplies of petroleum in the world have caused new interest in an agricultural program with lower energy inputs. Fixation of N_2, crop rotations, and fewer passes with equipment over the land will reduce energy inputs. In fact, **low-input sustainable agriculture (LISA)** is very much like farming with low chemical and machinery uses. LISA is very much like organic farming in many respects, but LISA has important differences. LISA will lessen chemical and fertilizer use to minimize costs. Organic farmers similarly restrict the use of most chemicals and fertilizers, but they do it because the materials are not "natural" or because they believe the materials are health hazards.

Agriculture in the 1960s and 1970s was expensive in its energy use. Fields were traversed many times a season—to plow, to disc, to harrow, to smooth, to furrow, to plant, to apply fertilizers and pesticides, and to harvest. Energy was cheap and yields were high. Suddenly, energy costs increased, and they will continue to rise. The overworked, barren soils eroded easily and extensively. That soil erosion was severe, and it needs to be controlled. We must take a new look at the desirable levels of tillage and the economics of water, fertilizer, pesticide, and tillage inputs that we use to produce crops.

Which compromises allow us to evolve the best balance of production costs with the need to minimize our use of limited natural resources, particularly energy reserves? Any low-input sustainable agricultural system usually involves (1) better management of organic materials as nutrient reservoirs and (2) less chemical pesticide and fertilizer use. It is essential to schedule crop rotations involving N_2 fixers to build nitrogen soil supplies. Mixtures of crops or selected rotations will help reduce weeds and other pests with minimal pesticide use. Crop varieties resistant to certain diseases will minimize pesticide needs. Many of these principles have been studied for decades by organic farmers and numerous Amish peoples in the Midwestern United States. Agriculture before the machinery age was one of low input (there were few fertilizers and pesticides, and the equipment used was small and pulled by horses).

Table 7-10 Comparison of Yields between Conventional Farming and Organic Farming Methods

Comparison	Yield Organic/Conventional [a]
Aggregate of crops and livestock production	
Illinois Amish vs. conventional	0.62
Pennsylvania Amish vs. conventional	
Old-order Amish	1.03
Nebraska sect	0.56
Wisconsin Amish vs. conventional	0.70
Wheat yields, 20 New York and Pennsylvania farms	0.78
Midwest, 14 organic vs. 14 conventional farms	0.87
Corn yields, 5 organic vs. 5 conventional farms	0.85
Soybean yields, 3 organic vs. 3 conventional farms	1.07
Wheat yields, 4 organic vs. 4 conventional farms	0.91
Miscellaneous crop yields, 18 farms	0.97
Virginia, four gardens of 7 vegetables	0.20
Estimated national average if exports were such as to require use of most cropland	
Wheat	0.47
Corn	0.50
Other feed grains	0.30
Soybeans	0.49
Cotton	0.44

Sources: (1) Council for Agricultural Science and Technology, "Organic and Conventional Farming Compared," CAST Report 84, CAST, Ames, Iowa, 1980, p. 32. (2) R. C. Lambe and J. G. Petty, "'Chemical' Garden Out-Yields 'Organic' Garden," *Agri-News Newspaper,* **4,** no. 2, Feb. 1973, pp. 1, 3.

[a] Values of 1.0 mean yields of both methods are equal; less than 1 means organic farms produce less than conventional. Most yields were farmers' estimates, which can be good but may be inaccurate as well.

Agriculturalist in the new age of LISA will try to marry the best practices of the organic farmer with the inexpensive but helpful inputs of better crop varieties, the use of smaller quantities but more effective pesticides and fertilizers, reduced tillage, effective crop rotations, and integrated pest management. These changes will probably result in lower crop yields. But if input costs are less, profits may still be higher. Many areas where reduced tillage is used have maintained yields near those with conventional tillage. It will require greater attention to many details of production to keep yields high with the techniques of LISA.

Some comparisons of crops grown under organic and conventional farming are shown in Table 7-10. Amish farms are considered here because they are similar in operation to organic farms (few chemicals are used) and have been established for a long time. It is expected that well-managed organic farms could produce as effectively as many conventional farms, but this is not often observed. Organic farms more often yield only 60–90% as much as conventional practices. Modeled projections, as given in the last five items of Table 7-10 suggest that conversion to organic methods on a wide scale is likely to produce less than half as much of several important world crops.

Organic farms usually produce nearly as much or more energy than is used in their crop production. In contrast, conventional farms overall tend to produce only 30–60% as much energy as needed to produce the crops. Conventional farming currently requires a net energy input; in its present form it is not a self-sustaining agriculture.

It is perhaps best to consider viruses not as organisms, but as interesting and frequently dangerous biochemicals. They do not live or have life cycles by themselves. They can even be crystallized like common table salt.

—Anonymous

Questions

1. (a) In what ways can earthworms be beneficial to the growth of plants and yet be nonessential? (b) What are optimum conditions for earthworms?
2. (a) What problems can be caused by nematodes? (b) What are the practical solutions to the nematode problem?
3. (a) Describe the nature of the rhizosphere. (b) Does the rhizosphere have a different pH and/or microbial activity than is in the bulk soil? Explain.
4. How important are fungi as (a) organic matter decomposers? (b) diseases? (c) toxins to plants? and (d) toxins to animals and people?
5. (a) Define *mycorrhizae.* (b) How extensive are mycorrhizae in nature?
6. (a) How many *large* bacteria may be on one *large* clay particle (or vice versa)? (b) In good conditions, how much could the bacterial population grow in a day?
7. Of what particular value are autotrophic bacteria?
8. What is the special value to soil fertility of symbiotic N_2 fixers?
9. How much nitrogen is fixed by several common plants per year?
10. (a) How are nitrogenase, *Rhizobium,* and activation energy interrelated? (b) What are legume nodules?
11. (a) What are actinomycetes? (b) What beneficial actions in soils do they perform?
12. (a) Are viruses living substances? (b) How do viruses cause damage?
13. If viruses are not living, how are they "inactivated" so that they do not cause diseases or can be "controlled"?
14. For the majority of soil microorganisms, what are the optimum conditions of water, temperature, pH, and nutrients?
15. What management practices (a) encourage growth of beneficial microbes? (b) help control harmful organisms?
16. What are the origins and the general composition of soil humus?
17. (a) What is an enzyme? (b) Are enzymes "living"?
18. Enzymes lower activation energy. What does this mean?
19. (a) Define a *biological reaction.* (b) How are biological reactions related to organic-matter decomposition?
20. (a) How stable are enzymes in soils? (b) What can "inactivate" them?
21. Answer the following questions about decomposition of soil organic matter. (a) What initiates the breakdown? (b) What are the products produced? (c) What is the solid residue that is left called?
22. The C:N ratio of crop residues left in soils is important. Explain why.
23. Discuss the relationship between organic-matter decomposition rates, waterlogging (swamps), and formation of peats and mucks (organic soils).

24. State how these affect organic-matter decomposition rates: (a) temperature, (b) pH, and (c) nutrients.
25. Although soil organic matter will supply some of all nutrients, what is the nutrient for which it is most important? Explain why this is so.
26. What is the relationship between soil aggregation, organic-matter decomposition, and polysaccharides?
27. In general, state how soil humus affects (a) water retention of soils and (b) the quantity of "ligands" and chelates.
28. Discuss briefly the soil humus as sources of hormones and other growth substances.
29. Briefly discuss some detrimental effects of some soil organic materials.
30. Is there a "best-content value" for soil humus? Explain.
31. Evaluate animal manures as a soil amendment (a) to supply nutrients and (b) that may contain materials detrimental to crops or soil.
32. What are some suggested practices for handling manures to reduce contamination of the environment?
33. (a) Define *sewage sludge* and list some of its contents. (b) Should sewage sludge be used on food crops? Discuss.
34. (a) List the conditions and materials for making good compost. (b) Does old compost release more nutrients than plowed-under green-manure crops?
35. Although the text says that plants do not need organic matter to grow, most soils are much more productive if humus is high in the soil. Explain.
36. (a) Define *LISA*. (b) Why is the concept of *LISA* important?

Acidic Soils and their Modification

I cannot say whether things will get better if we change; what I can say is they must change if they are to get better.

—G. C. Lichtenberg

8:1 Preview and Important Facts

PREVIEW

Even today the two elements H^+ (pH) and Al ions are still considered to be the most important causes of toxicity to freshwater biota and in soil solutions. Soils in humid regions become acidic as waters containing weak acids flow through them, replacing basic cations (Ca^{2+}, Mg^{2+}, Na^+, K^+) from cation exchange sites with hydrogen and aluminum ions. Soils high in humus and clay have the potential of causing strongly acidic soils. These colloids have large surface areas with proportionally larger amounts of cation exchange sites than do silts and sands.

Lime is material added to acidic soils to raise their pH (lessen the acidity). Most agricultural lime used is impure, crushed calcium carbonate (powdered limestone). Wood ashes, burned limestone (forming CaO), and marl (soft calcium carbonate with clays) also are sometimes used on acidic soils to increase plant growth. Columella, a Roman philosopher, recorded the use of lime in A.D. 45 to enhance plant growth. However, Edmund Ruffin, a Virginia farmer-scientist from 1825 to 1845, may have been the first person to apply lime on the soil specifically to correct a condition that he called soil acidity. Today lime is one of the most common agricultural soil amendments in humid regions.

Crop response to added lime does not occur immediately, perhaps not for several weeks or months, and large amounts are usually needed to be effective. Crops respond to liming because it improves microbial activity; it may correct calcium, magnesium, and molybdenum deficiencies; it lessens aluminum, manganese, and iron toxicities; and it makes phosphorus more available and potassium more efficient in plant nutrition. Ammonium fertilizers and urea make soil more acidic; the continued use of these fertilizers on acidic soils without also adding lime may decrease soil productivity. Plants differ in their ability to grow in acidic soils and in their response to added lime.

IMPORTANT FACTS TO KNOW

1. Why some soils become acidic but others remain alkaline
2. The reasons most plants grow poorly in strongly acidic soils
3. The substances that can be used as "limes" and the composition of the most commonly used agricultural lime
4. The neutralizing index of lime materials
5. The changes caused by lime addition to acidic soils
6. The final pH value wanted after liming is finished
7. The increases expected from crops growing on limed soils
8. The "typical" amounts of lime needed in some soils to correct acidity
9. The effects of overliming and why those effects occur
10. The preferred ways to add lime to soils
11. The methods to acidify soils and the soils on which they might be practical

8:2 Why Some Soils Are Acidic

Most soils become acidic because of leaching. Carbon dioxide (CO_2) dissolved in water, plus excreted H^+ (= H_3O^+) from roots and some organic acids from humus decomposition, furnish H^+ in percolating water. As soil solution pH becomes acidic, aluminum hydroxides interact and some soluble hydrated $Al(OH)_2^+$ ions form (Detail 8-1). As percolating water moves these $Al(OH)_2^+$ and H_3O^+ ions through the soil, many adsorbed basic cations (Ca^{2+}, Mg^{2+}, Na^+, K^+) are replaced by these acidic cations. Such a leached soil becomes more acidic after decades or centuries of leaching. At pH values below about 4.7, appreciable amounts of Al^{3+} ions will exist in solution and on cation exchange sites.

Most soils in rainfall areas exceeding about 500 mm (20 in.) yearly will develop some acidic soils. In areas with higher rainfall, increased acidity is expected, with values of pH 4.5–5.5 common. Extreme values are near pH 3.0–3.5. The sources of H_3O^+ ions are as follows (Fig. 8–1):

1. Carbon dioxide from humus decomposition and root respiration
2. Oxidation of NH_4^+ from fertilizers (Fig. 8–2)
3. Oxidation of added elemental sulfur
4. Excreted H^+ ions by plant roots
5. Acid rain (sulfur and nitrogen oxide pollutants)
6. Crop removal of the basic cations (Ca, Mg, K, Na) and excretion of H^+ by roots

Details of the mechanisms of soil acidification by natural processes are given in Detail 8–2. The actual equations representing soil acidification are shown in the two following equations. The (–65) is an arbitrary charge that indicates the total negative charge for that colloid. The dots at each cation represent the valences for that cation attracted to the negative sites of the colloid.

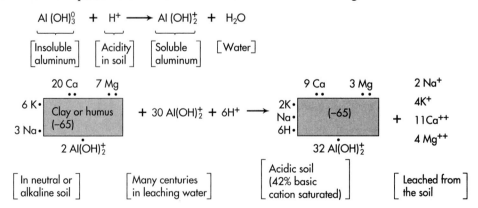

Detail 8-1 Aluminum Ions in Soil Solution

In water solutions aluminum always forms octahedral coordination with some combination of water molecules and hydroxyl ions, much as it does with six oxygens or hydroxyls in clay minerals. Aluminum's natural coordination is with six of these size molecules (octahedral coordination); the size of hydroxyls, oxygen, and water are similar. If the soil is not too strongly acidic, one or more of the water molecules ionize, releasing hydrogen, H^+, to the solution, increasing the solution acidity. The possible chemical forms simply indicated are as follows:

$$[Al(H_2O)_6]^{3+} \longrightarrow [Al(H_2O)_5(OH)]^{2+} + H^+ \qquad \text{(pH about 4.5–5) (limited)}$$

$$Al(H_2O)_5(OH)^{2+} \longrightarrow [Al(H_2O)_4(OH)_2]^+ + H^+ \qquad \text{(pH about 5–6.5)}$$

$$Al(H_2O)_4(OH)_2^+ \longrightarrow [Al(H_2O)_3(OH)_3]^0 + H^+ \qquad \text{(pH about 6.5–8.5)}$$

Most often the hydration is not indicated, and the ions are simply written Al^{3+}, $Al(OH)^{2+}$, $Al(OH)_2^+$, or $Al(OH)_3$. The water molecules are ignored.

Iron has similar reactions and products in soil solutions, but pH values differ.

Aluminum and iron tend to form larger units than described above, mostly of hydroxides.[a] These may be "rings" or "clumps" of Al^{3+} and OH^-, such as

$$[Al_6(OH)_{12}]^{6+} \qquad \text{and} \qquad [Al_{10}(OH)_{22}]^{8+}$$

At pH near 8 and reaching a maximum concentration near pH 10, two new soluble aluminum species, described simply as $[Al(H_2O)_2(OH)_4]^-$ and $[Al(H_2O)(OH)_5]^{2-}$, occur.[b] These may cause toxicity in excessively basic soils. As soils are leached, some basic cations are exchanged off the CEC and the exchange sites are increasingly occupied by aluminum species.

The dominant aluminum species at the various solution pHs are shown in the diagram below.[c] Notice that $Al(OH)_2^+$ will be the dominant *soluble* aluminum form from about pH 4.7 to pH 7.5. Thus, most exchangeable acidity is the $Al(OH)_2^+$ ion. At pH less than about 4.5, the Al^{3+} ion and increasing amounts of H^+ ions will be in solution and on the cation exchange sites as the exchangeable acidity.

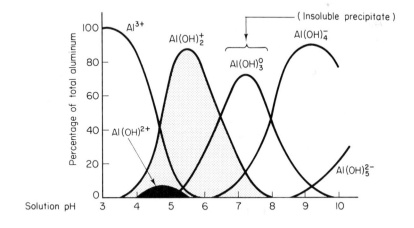

[a] (1) P. H. Hsu and T. F. Bates, "Fixation of Hydroxyl-Aluminum Polymers by Vermiculite," *Soil Science Society of America, Proceedings,* **28** (1964), pp. 763–769. (2) H. L. Bohn, B. L. McNeal, and G. A. O'Conner, *Soil Chemistry,* John Wiley, New York, 1979, p. 200.

[b] P. R. Hesse, *A Textbook of Soil Chemical Analysis,* Chemical Publishing Co., New York, 1971, p. 341.

[c] Modified by Raymond W. Miller after G. M. Marion, D. M. Hendricks, G. R. Dutt, and W. H. Fuller, "Aluminum and Silica Solubility in Soils," *Soil Science,* **121** (1976), p. 76.

FIGURE 8-1 (a) Soils are acid partly because ammonium nitrogen fertilizers are used on them. In (a) the center plot received 200 lb/a (224 kg/ha) of ammonium sulfate per year for 21 years. In (b) the center plot received 200 lb/a (224 kg/ha) per year of ammonium sulfate for 21 years, the same as in (a) but in addition, it received 230 lb of limestone per acre (258 kg/ha) per year for the same period. Alabama. (*Source:* Alabama Agricultural Experiment Station.)

(a)

(b)

Acidification in nature requires (1) removal of some cations yearly for many centuries of time and (2) high rainfall for leaching water to wash out basic cations. Acidic soils in the United States are found primarily in the high-rainfall areas of (1) the states east of the Mississippi River, (2) the Pacific coastal soils of the Northwest, and (3) some of the mountain areas. For example, Logan in northern Utah averages about 400 mm (15.6 in.) of rainfall annually and seldom has soils more acidic than pH 6.5. Mountains within 50 km (31 mi) will often have 750–900 mm (29–35 in.) of precipitation, and soils are acidic (pH 5–6).

8:3 Composition of Lime

Liming materials are usually the carbonates, oxides, hydroxides, and silicates of calcium and magnesium. More than 90% of the agricultural lime used is impure calcium carbonate; next used are carbonates of calcium plus magnesium (dolomitic lime); a much smaller quantity is composed of calcium oxide or calcium hydroxide. In the building trades lime refers to calcium oxide, a caustic powder that is slaked in water for use in various brick-laying mortars. Although this material is usable as agricultural lime, it is caustic and dusty to handle.

The common liming materials used include:

1. **Calcic limestone** ($CaCO_3$), which is ground fine for use.

2. **Dolomitic limestone** [$CaMg(CO_3)_2$], a ground limestone high in magnesium.
   Although state laws vary, the average composition of dolomitic limestone sold in the United States is about 51% $CaCO_3$, 34% $MgCO_3$, and 15% soil and other impurities.

The hydrogen ions in soil water, which over centuries work to produce acidic soils, are from these six sources:

1. *Carbon dioxide* from decomposing organic matters and root respiration dissolves in water to form weak carbonic acid.

$$CO_2 + HOH \longrightarrow H_2CO_3 \longrightarrow HCO_3^- + H^+$$

    These acidified waters percolate through the soil to gradually cause soil acidity. Percolating waters continuously move small amounts of H^+, which replace solubilized basic cations of calcium, magnesium, potassium, sodium, and other elements. The replaced basic elements are leached from the root zone.

2. *Ammonium-containing fertilizers* are oxidized by bacteria to form nitrate and hydrogen ions (Fig. 8-3). For each NH_4^+ cation oxidized, two H^+ result.

$$NH_4^+ + 2O_2 \xrightarrow{\text{(nitrifying bacteria)}} NO_3^- + H_2O + 2H^+$$

    This reaction applies to any source of NH_4^+, including urea after hydrolysis and the mineralization of NH_4^+ from organic materials.

3. *Sulfur,* an ingredient in some fungicides and fertilizers, oxidizes to sulfate and hydrogen ions. Note that the sulfate ion itself is not acid-forming.

$$2S + 3O_2 + 2HOH \xrightarrow{\text{\textit{Thiobacillus} bacteria}} 2SO_4^{2-} + 4H^+$$

4. Some hydrogen ions are *excreted by plant roots,* which are exchanged for other nutritive cations. Scientific evidence of the amounts of hydrogen ion released has been measured in water cultures. In soils, studies have shown pH values as much as 1.2 pH units lower in soil near roots than in the general mass of soil.[a]

    Acidity increases as soils are progressively leached and the soils lose their exchangeable basic cations, Ca^{2+}, Mg^{2+}, Na^+, and K^+. Hydrated aluminum ions in solution at a pH below 5.5 become part of the exchangeable cations and increase acidity. However, soils high in carbonates (lime) are slow to become acidic because the supply of calcium is continuous as the lime reserves and calcium minerals in the soil dissolve.

5. *Acidic rain* in the United States results from an annual discharge of millions of metric tons of sulfur oxide and nitrogen oxide pollutants into the atmosphere. Acidic rain is also a major pollutant in Canada, Norway, and Sweden. The original sources of these polluting oxides are the burning of fossil fuels—such as wood, coal, and petroleum products—and from forest and range fires. Acidic rain falls when airborne sulfur oxides (mostly sulfur dioxide, SO_2) and nitrogen oxides (mostly nitric oxide, NO) are converted to sulfuric acid (H_2SO_4) and nitric acid (HNO_3) through oxidation and dissolved in raindrops. Rainwater made acidic by these strong acids may have a pH as low as 2.[b] The amounts of sulfur dissolved in rainfall yearly range from 10 kg/ha (8.9 lb/a) for light

Continued.

---

[a] R. W. Smiley, "Rhizosphere pH as Influenced by Plants, Soils, and Nitrogen Fertilizers," *Soil Science Society of America, Proceedings,* 38 (1974), pp. 795–799.

[b] (1) Environmental Protection Agency, *Research Summary: Acid Rain,* EPA-600/79-028, EPA, Washington, D.C., 1979. (2) William M. Lewis, Jr., and Michael C. Grant, "Acid Precipitation in the Western United States," *Science,* 207 (Jan. 11, 1980), pp. 176–177. (3) C. R. Frink and G. K. Voigt "Potential Effects of Acid Precipitation on Soils in the Humid Temperate Zone," in *Proceedings of the First International Symposium on Acid Precipitation and the Forest Ecosystem,* USDA—Forest Service; General Technical Report NE-23, Upper Darby, Pa., 1976, pp. 685–709.

industrial areas to 30 kg/ha (27 lb/a) and as much as 100 kg/ha (89 lb/a) for soils in more industrialized states and near large plants, respectively. Acidity in rain has been enough to eradicate fish from many lakes; soils are less affected than surface waters because most soils are buffered.

6.   *Crop removal* helps make soils more acidic by depleting the reserves of calcium, magnesium, and potassium. For example, a yield of 13 metric tons per hectare (6 t/a) of alfalfa removes 45 kg (100 lb) of calcium (Ca) and 9 kg (20 lb) of magnesium (Mg). Normal yields of tobacco and soybeans each remove about half as much calcium and magnesium. Corn and the small grains are less lime-depleting.[c]

---

[c] "Liming Soils: An Aid to Better Farming," *USDA, Farmers' Bulletin 2124*, 1966, p. 28.

3.   **Quicklime** (CaO), which is burned limestone.

4.   **Hydrated** (slaked) **lime** [$Ca(OH)_2$], from quicklime that has changed to the hydroxide form as a result of reactions with water.

5.   **Marl** ($CaCO_3$), from the bottom of small freshwater ponds in areas where the soils are high in lime. The lime has accumulated by precipitation from drainage waters high in lime. Some marls contain many shell remains from ancient marine animals, which are major sources of the carbonate.

6.   **Chalk** ($CaCO_3$), resulting from soft limestone deposited long ago in oceans.

7.   **Blast furnace slag** ($CaSiO_3$ and $CaSiO_4$), a by-product of the iron industry. Some slags contain phosphorus and a mixture of CaO and $Ca(OH)_2$. This product is called *basic slag* and is used primarily for its phosphorus content.

8.   Miscellaneous sources, such as ground **oystershell, wood ashes,** and **by-product lime** resulting from paper mills, sugar beet plants, tanneries, water-softening plants, fly ash from coal-burning plants, and cement-plant flue dust.

9.   **Fluid lime,** a relatively new product that is becoming popular because it can be spread with the same equipment that is used to apply fluid fertilizers. Typically, fluid lime is the suspension in water of *any* suitable liming material that has a fineness of <60-mesh (250 um). A sieve of 60-mesh means a screen with 60 divisions per linear inch in each of two directions. When applied with urea–ammonium nitrate fluids, only the *carbonate* form of lime should be used so ammonia is not volatilized.[1,2,3]

Gypsum ($CaSO_4$) is not a lime, but is sometimes added to the soil to supply calcium, which might alleviate somewhat the toxicity from soluble aluminum. In Georgia, gypsum increased alfalfa yield 25%, but calcium carbonate (which precipitates more Al) increased it 50%.[3] Calcium silicate ($CaSiO_3$) slag has been used successfully as a source of silicon for rice and sugarcane. Its value as a liming material has not been demonstrated.

---

[1] Gary W. Colliver, "Liquid Lime," *Crops and Soils Magazine* (Aug.–Sept. 1979), pp. 14–16.

[2] K. T. Winter, et al., "Liming Has No Miracles," *Solutions* (Mar.–Apr. 1980), pp. 12, 14, 18, 24, 28, 30, 32, 34.

[3] D. L. Anderson, D. B. Jones, and G. H. Snyder, "Response of a Rice-Sugarcane Rotation to Calcium Silicate Slag on Everglade Histosols," *Agronomy Journal*, **79** (1987), pp. 531–535.

All the liming materials mentioned supply either calcium or both calcium and magnesium and make aluminum, manganese, and iron less toxic. The choice of a particular liming material involves a combination of the cost, the material's purity, its ease in handling, and the speed with which the lime reacts in the soil.

8:3.1 Chemical Guarantees of Lime

There are several methods of expressing the relative chemical value of lime. The most common ones are the following two:

Calcium Carbonate Equivalent   This is sometimes known as the **total neutralizing power.** If a lime is chemically pure calcium carbonate (calcite), the calcium carbonate equivalent would be 100. If all the lime were in the calcium carbonate form, but it was only 85% pure, the calcium carbonate equivalent would be 85. Limestone is seldom pure; it formed in ocean bottoms and collected clay and silt sediments. The content of other forms of lime can be calculated to the calcium carbonate equivalent by the use of atomic and molecular weights.[4] The question is: How effective is 100 g (3.5 oz) of the liming material used compared to 100 g of pure $CaCO_3$?

$$\% \ CaCO_3 = \left(\frac{\text{grams of pure } CaCO_3 \text{ equal to } 100 \text{ g of lime used}}{100 \text{ g of pure } CaCO_3} \right)(100) \qquad \text{(EQ 1)}$$

If we are comparing CaO material, add gram atomic weights,

$$CaO \text{ gram molecular weight} = 40 + 16 = 56 \text{ g}$$
$$CaCO_3 \text{ gram molecular weight} = 40 + 12 + 48 = 100 \text{ g}$$

Therefore, 56 g of CaO neutralizes the same amount of acid as does 100 g of $CaCO_3$. If we represent soil acidity by H^+,

$$CaO + HOH + 2H^+ \rightarrow Ca_2^+ + 2HOH$$
$$CaCO_3 + CO_2 + HOH + 2H^+ \rightarrow Ca_2^+ + H_2CO_3$$

and the H_2CO_3 breaks down into CO_2 gas and water. We can now write the relationship between CaO and $CaCO_3$:

$$\frac{100 \text{ g } CaCO_3}{56 \text{ g } CaO} = \frac{? \text{ g of } CaCO_3}{100 \text{ g } CaO}$$

$$100 \text{ g } CaO = 179 \text{ g } CaCO_3$$

To calculate the "% $CaCO_3$ equivalent" question asked earlier (EQ 1 above):

$$\% \ CaCO_3 = \frac{179 \text{ g of } CaCO_3 \text{ is equal to } 100 \text{ g } CaO}{100 \text{ g of pure } CaCO_3} \ \bigg| \ 100\%$$

$$= 179\% \text{ for } CaO$$

(see Table 8–1).

---

[4] The "Model Agricultural Liming Material Bill" proposes these minimum calcium carbonate equivalents: quicklime 140, hydrated lime 110, ground limestone 80, blast furnace slag 80, and ground oystershell 80. Reference: *Abstract of State Laws and ACP Specifications for Agricultural Liming Materials,* 3rd ed., National Limestone Institute, Fairfax, Va. 1977, p. 75.

Table 8-1 Lime Conversion Factors

To Convert from This Material[a] (Column A)	To Each of These:		
	Ca	CaO	CaCO₃
	Multiply column A by:		
Calcium (Ca)	1.00	1.40	2.50
Calcium oxide (CaO)	0.71	1.00	1.78
Calcium hydroxide [Ca(OH)₂]	0.54	0.78	1.35
Calcium carbonate (CaCO₃)	0.40	0.56	1.00
Magnesium (Mg)	1.65	2.31	4.12
Magnesium oxide (MgO)	0.99	1.39	2.48
Magnesium hydroxide [Mg(OH)₂]	0.69	1.00	1.72
Magnesium carbonate (MgCO₃)	0.48	0.67	1.19
Dolomite, pure (CaCO₃ · MgCO₃)	0.43	0.63	1.09

[a] Calculated using the following atomic weights: Calcium—40.08, oxygen—16.00, carbon—12.01, and hydrogen—1.00. For example: In line 1 above, to convert from Ca to CaO equivalent, divide the molecular weight of CaO by the atomic weight of Ca = 56.08/40.08 = 1.40. Therefore, to convert Ca to CaO), multiply the kilograms (pounds) of Ca by 1.40 to obtain equivalent kilograms (pounds) of CaO.

Elemental Percentage of Calcium and/or Magnesium   The second method of expressing the lime guarantee is determined in a similar way. If the material quality were to be reported as elemental calcium, the calculations would be:

$$\% \text{ Ca} = \frac{\text{atomic weight of Ca}}{\text{molecular weight of CaCO}_3} \left| \frac{179\% \text{ of CaCO}_3}{100\% \text{ CaCO lime}} \right| 1 \text{ unit}$$

$$= \frac{40 \text{ g}}{100 \text{ g}} \left| \frac{1.79}{1.0} \right| 100$$

$$= 71.6\% \text{ Ca equivalent}$$

A material with 100 g of CaO would have 179% CaCO₃ equivalent and 71.6% Ca equivalent, all for the same material.

8:3.2 Physical Guarantees of Lime

The chemical activity of liming material is determined by the solubility of the chemical compounds in the lime. For example, calcium oxide is more soluble than calcium carbonate, whereas calcic limestone is more soluble than dolomitic limestone; calcium silicate is the least soluble of all these liming materials. It is obvious that the finer lime particles react faster in the soil (Fig. 8–2).

There are no U.S. laws governing a commercial supplier's physical guarantee of lime, this regulation being left to the states. Thirty-nine states have lime laws, enforced by each state department of agriculture. In general, the physical guarantees of lime in the respective states may be averaged roughly in this way: *85% of the lime particles must pass through a 16-mesh (1.18 mm) sieve and 30% must pass through a 100-mesh (150 μm) sieve.*[5]

---

[5] Sieves are designated as number of openings per linear inch. An 8-mesh (2.36-mm) sieve has 8 openings per linear inch, and a 60-mesh (25-n or 250-μm) sieve has 60 openings per linear inch.

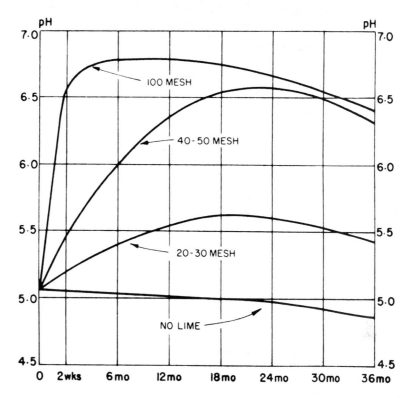

FIGURE 8-2 The finer the limestone, the more quickly it reacts with the soil to raise the pH and the more quickly some calcium becomes available to the plant. A satisfactory fineness is 85% through a 16-mesh (1.18-mm) sieve and 30% through a 100-mesh (150-μm) sieve. *Note:* 20 mesh = 850 μm, 30 mesh = 600 μm, 40 mesh = 425 μm, 50 mesh = 300 μm. (*Source:* The Fertilizer Institute and Purdue University.)

A unique technique for evaluating the *effectiveness* of liming materials is to calculate the calcium carbonate equivalent and multiply this by an arbitrary fineness factor. This **neutralizing index** gives a better evaluation of the usefulness of the material than would either factor by itself (Calculation 8–1). Finer material and larger calcium carbonate equivalents are more effective than coarse, less pure limes and have a neutralizing index near 70 or higher.

8:4 Reactions of Lime Added to Acidic Soils

Strongly acidic soils greatly restrict the growth of most plants. Notable exceptions are blueberries, cranberries, watermelons, white potatoes, tea, and pineapples; these crops do well on strongly acidic soils. By contrast, alfalfa and sweet clover yield their maximum harvest only when the soils are nearly neutral to slightly basic.

On strongly acidic soils the majority of crop plants produce yields less than their potential when grown on soils less acidic for one or more of the following reasons:

1. **Aluminum toxicity** is perhaps the most important cause of reduced plant growth (Fig. 8–3; Details 8–1 and 8–2).
2. **Reduced microorganism activity**
3. **Manganese toxicity**
4. **Iron toxicity** in a few soils

An effective technique for assessing a liming material is to consider both chemical and physical factors together. First, determine the calcium carbonate equivalent as explained in Sec. 8:3.1, then use the sieve analysis (which expresses fineness) to calculate the **effective calcium carbonate,** also called **neutralizing index.**

Problem Calculate the neutralizing index of this lime:

Percent calcium carbonate equivalent: 90%

Sieve analysis: Retained on 8-mesh sieve =	10%
Retained on 60-mesh sieve =	20%
Passing 60-mesh sieve =	70%
Total	100%

Solution Calculating **fineness factor:**

1. Greater than 8-mesh sieve size lime is presumed too coarse to neutralize soil acidity within 3 years after application and so has zero effectiveness. Therefore, 10% of sample > 8 mesh $\times$ 0 effectiveness = 0.

2. Lime that passes between 8- and 60-mesh sieve is presumed 50% as effective as finer lime: 20% sample $\times$ 0.50 effectiveness = 10.

3. Presuming < 60-mesh lime is 100% effective: 70% $\times$ 1.00 effectiveness = 70.

4. Total of fineness factors = 80. Effective calcium carbonate (neutralizing index) = percent calcium carbonate equivalent $\times$ fineness factor = 0.90 $\times$ 80 = 72.

Sources: Adapted from (1) L. S. Murphy and Hunter Follett, "Liming—Take Another Look at the Basics," *Solutions* (Jan.–Feb. 1978), pp. 53, 54, 56, 58, 60, 62, 64–67. (2) K. T. Winter, et al., "Liming Has No Miracles," *Solutions* (Mar.–Apr. 1980), pp. 12, 14, 16, 24, 28, 30, 32, 34. (3) K. A. Kelling and E. E. Schulte, "Liming Materials—Which Will Work Better?" *Solutions* (Mar.–Apr. 1980), pp. 52–54, 56, 60.

5. **Calcium deficiency**
6. **Magnesium deficiency**
7. **Molybdenum deficiency,** especially for legumes and the cabbage family
8. **Nitrogen, phosphorus, and/or sulfur deficiency** because of very slow organic-matter decomposition

The addition of lime raises the soil pH, thereby eliminating most major problems of acid soils, including excess (toxic) soluble aluminum [6,7] and very slow microbial activity. The process of changing pH by the addition of lime is illustrated in Fig. 8–4 and is a reversal of acidification (Fig. 8-1): The adsorbed acidic aluminum ions are replaced with calcium ions from the lime; the released H^+ are neutralized by the carbonates or hydroxides added as lime.

---

[6] Although most studies have shown marked growth increases when soils are limed to pH 6.0 to 6.5, hydroxy aluminum, $Al(OH)_2^+$, at pH 6.8 is still toxic to plants if it is in soluble form. (R. J. Bartlett and D. C. Riego. "Toxicity of Hydroxy Aluminum in Relation to pH and Phosphorus," *Social Science,* **114** [1972], pp. 194–200.)

[7] F. Ahmad and K. H. Tan, "Effect of Lime and Organic Matter on Soybean Seedlings Grown in Aluminum-Toxic Soil," *Soil Science Society of America Journal,* **50** (1986) pp. 656–661.

(a)

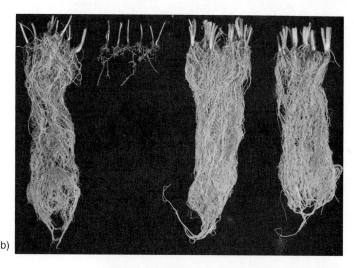

(b)

FIGURE 8-3 Excess soluble aluminum, which occurs in soils of pH about 5 or lower, is toxic to most plants. In (a) and (b), the two wheat cultivars on the left, Thorne and Redcoat, respectively, at pH 4.3 show different tolerances to acidity. The same cultivars on the right at pH 5.8 (after adding 3000 ppm calcium carbonate). (c) shows tall fescue grass grown in nutrient solutions containing, from left to right, 0, 2, and 4 ppm soluble aluminum. (*Source: Agronomy Journal,* vol. 66, by permission of the American Society of Agronomy; photos by A. L. Fleming, J. W. Schwartz, and C. D. Foy.)

(c)

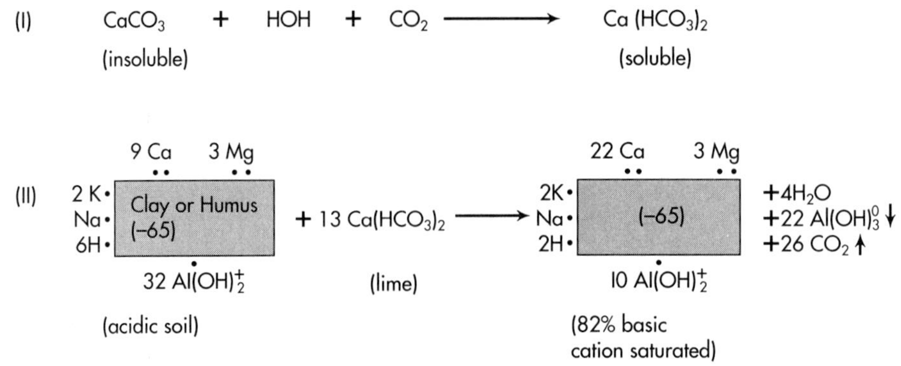

(I) $CaCO_3$ + HOH + CO_2 $\longrightarrow$ $Ca(HCO_3)_2$

(insoluble)   (soluble)

(II)

9 Ca   3 Mg

2 K·
Na· | Clay or Humus (-65) | + 13 $Ca(HCO_3)_2$ $\longrightarrow$
6H·

32 $Al(OH)_2^+$ (lime)

(acidic soil)

22 Ca   3 Mg

2K·
Na· | (-65) | $+4H_2O$
2H· $+22\ Al(OH)_3^0 \downarrow$
                       $+26\ CO_2 \uparrow$

10 $Al(OH)_2^+$

(82% basic
cation saturated)

FIGURE 8-4 Lime (calcium carbonate) added to soil dissolves in the solution as calcium bicarbonate. The calcium in the lime exchanges for the exchangeable acidity [$Al(OH)_2^+$ and H^+ ions]. The $Al(OH)_3$ formed in the higher pH solution is insoluble and precipitates. Any H^+ exchanged is neutralized by forming dilute carbonic acid with bicarbonate, which is unstable, giving off gaseous carbon dioxide and water. (Courtesy of Raymond W. Miller, Utah State University.)

Liming also has benefits other than relieving toxic Al levels and speeding microbial processes:

1. The raised pH reduces *excess* soluble manganese and iron by causing them to form insoluble hydroxides.

2. The calcium and magnesium that are deficient in many acidic soils are added if the lime is *dolomitic* (it contains both calcium and magnesium carbonates) rather than calcic lime (only calcium carbonate).

3. Lime makes phosphorus in acidic soils more available. In strongly acidic soils iron and aluminum combine with the fertilizer phosphates to make insoluble compounds. *Liming* reduces the solubility of iron and aluminum, and, therefore, less of the added phosphorus will combine with them to form insoluble iron and aluminum phosphates.[8-10]

4. Lime makes potassium more efficient in plant nutrition. When potassium is plentiful, all plants absorb more of it than they need. Lime reduces the excessive uptake of potassium. Economically, the practice of liming is desirable because the plant absorbs more of the cheaper calcium and less of the costlier potassium. Because calcium is often deficient in animal rations and potassium is in excess, it is desirable to increase the percentage of calcium in the plant.

5. Lime increases the availability of nitrogen by creating a more favorable environment for microbes, which hastens the decomposition of organic matter (the soil bacteria are more active at pHs higher than strongly acidic).

6. Lime on acidic soils increases plant-available molybdenum.

[8] Larry Unruh and David Whitney, "Soil Acidity and Aluminum Toxicity: An Important Factor in Winter Wheat Yields," *Better Crops with Plant Food* (Summer 1986).

[9] E. Kamrath, "Soil Acidity in Well-Drained Soils of the Tropics as a Constraint to Food Production," in *Priorities for Alleviating Soil-Related Constraints to Food Production in the Tropics,* Proceedings of a Conference at IRRI, Los Baños, The Philippines, 1980.

[10] M. E. Sumner, et al., "Amelioration of an Acid Soil Profile through Deep Liming and Surface Application of Gypsum," *Soil Science Society of America Journal,* **50** (1986), pp. 1254–1258.

7. Liming an acidic soil above pH 6.5 reduces the solubility and plant uptake of potentially toxic heavy metals, such as cadmium, copper, lead, nickel, and zinc, from added sources, such as sewage sludge. On high-lime soils growing plants and grazing cattle absorb *less* radioactive strontium-90 (from atmospheric fallout) because calcium is absorbed in preference to strontium.

8:5 Crop Response to Lime

On mineral soils below pH 5 most crops respond to the judicious use of lime. Both corn and oats increase in yield, but not with equal increases when a soil with a pH of 5 is limed to about pH 6.5. Above 6.5 both crops rarely respond to additional lime. Red clover shows only a slight response to lime at any pH between 5 and 7.4, whereas alfalfa yields increase greatly from pH 5 to 7.

On adequately fertilized soils in Maryland the proper amount of lime applied on soil with a pH of 5.6 increased corn yields valued at $3.95 for each dollar spent on lime. In Puerto Rico, sugarcane yielded 50 metric tons per hectare (22.3 tons per acre) when exchangeable aluminum provided more than 70% of the exchangeable ions; adding lime until the exchangeable aluminum was less than 30% increased the yield more than four times. Soybean yields in Brazil went from 1753 kg/ha (1565 lb/a) with no liming to 3960 kg/ha (3536 lb/a) with application of 28 metric ton/ha (12.5 ton/a) of lime. Raising the pH to 5.2 was sufficient for these high yields because the soils are Oxisols. Although price changes in lime, fertilizers, and the crop produced determine the net benefit derived, lime is a profitable soil additive on most strongly acidic soils. Although liming to pH 6.5 is commonly recommended, there are specific soils, crops, or cropping systems that are better adjusted to lower pH values.

Three facts about liming soil are particularly important. First, phosphorus additions *with* lime additions frequently give much larger increases in yields than with lime alone.

Second, toxic levels of soluble and exchangeable aluminum can be almost eliminated by raising the pH to 5.2–5.5; further liming—from pH 6.0 to 6.5—usually, although not always, still increases yields. The beneficial effects of raising the pH from 5.3 to 6.5 may be due to an increase in biological activity, which increases the available nitrogen, molybdenum, and other nutrients.

Third, excess lime (raising pH higher than 6.0 or 6.5) is often detrimental to plant growth, perhaps due to reduced solubility of boron, zinc, and other micronutrients. Some crops (small grains, sugar beets, alfalfa) are not injured by overliming, especially on fine-textured humid-region soils.

8:6 Lime Requirements of Crops

To arrive at a satisfactory solution to the problem of how much lime to apply, the lime requirements of the proposed crop as well as the actual pH of untreated soil should be considered. Soil acidity and lime level have a fairly good correlation in temperate humid regions, so the pH of the soil can be used as an index of the differing lime needs of various crops. However, the *soluble and exchangeable acidity,* particularly the exchangeable acidity, must be partly neutralized. Buffered solutions are used in soil tests to estimate the lime needed to neutralize portions of the exchangeable hydrogen ions. In tropical and subtropical humid lowlands (Oxisols and Ultisols) the percentage of exchangeable aluminum is sometimes used to determine lime requirement.[11]

---

[11] R. Hunter Follett and Ronald F. Follett, "Soil and Lime Requirement Tests for the 50 States and Puerto Rico," *Journal of Agronomic Education,* **12** (1983), pp. 9–17.

The relative lime requirements of selected crops are listed in Table 8–2; alfalfa, barley, cotton, sugar beets, and sweetclover have the highest lime requirement; corn, tobacco, and wheat have a medium requirement; buckwheat, potatoes, rice, and rye have a low requirement; and blueberries, cranberries, and pineapples have the lowest lime requirement. If liming costs for a particular soil are high, a crop with a lower lime requirement sometimes can be grown as an alternative economic possibility.

8:7 Lime-Soil-Nutrient Relationships

To evaluate the lime requirement, both the pH requirement of the crop to be grown and the pH and buffer capacity (cation exchange capacity, CEC) of the cultivated soil should be determined. The lime requirement can be interpreted more accurately when the soil series is known because the series defines soil texture, structure, mineralogy, and other root-zone characteristics, such as humus content and permeability, which may affect the lime response.

The relationships of texture, cation exchange capacity, and buffer capacity (resistance to a change in ion concentration) are shown in Fig. 8-5. The more clay and organic matter there is in a soil, the more lime that is needed to change the pH because the soil colloids contain large quantities of exchangeable aluminum and hydrogen ions due to their high cation exchange capacities. It is true that the greater the amount of organic matter in a soil the lower the pH required for greatest plant nutrient availability. For example, Ohio State University recommends (Fig. 8-6):

$$< 10\% \text{ organic matter, ideal pH } = 6.5$$
$$10\% \text{ organic matter, ideal pH } = 6.0$$
$$20\% \text{ organic matter, ideal pH } = 5.5 \text{[12]}$$

The amount of pH change desired and the type of clay present also cause variation in the amount of lime needed to change the pH. The relative lime requirement in soil of the same initial pH with the principal clay minerals are in this order: vermiculite > montmorillonite > illite > kaolinite > sesquioxides (metal oxides). Oxisols and Ultisols respond differently to pH change when they are limed because the predominant clay minerals are sesquioxides in Oxisols and kaolinite in Ultisols. The pH of Oxisols rises faster (becomes less acidic) than it does in Ultisols at the same rate of lime up to about 45 Mg/ha (20 t/a); above this rate the pH of Ultisols rises faster than it does in Oxisols.[13] As the pH to be achieved increases in value closer to 7, the amount of lime required to effect the same numerical amount of pH change becomes greater.

The general relationship between soil pH and plant nutrient availability is provided in Fig. 8-6, which shows that the primary nutrients—nitrogen, phosphorus, and potassium—as well as the secondary nutrients—sulfur, calcium, and magnesium—are as available or more available at a pH of 5.5 and 6.5 for *organic* and *mineral* soils than at any other pH; molybdenum, copper, and boron availabilities are also relatively high at pH 5.5 and 6.5, respectively. The micronutrients iron, manganese, and zinc are less available at a pH of 5.5 and 6.5 than at more acidic reactions. At these soil pH ranges, nutrients in commercial fertilizers are also most readily available.

8:8 Methods of Applying Lime

The most *efficient* way to use lime is to apply small amounts every year or two. However, frequent applications generally increase the cost. The *usual* liming practice consists of a compromise between what is most effective and what is cheapest per ton of lime applied. Add lime less often, but add larger amounts.

---

[12] "Agronomy Guide, 1976–1977," *Bulletin 472*, Cooperative Extension Service, Ohio State University, Columbus, 1977.
[13] R. L. Fox, R. S. Yost, N. A. Saidy, and B. T. Kang, "Nutritional Complexities Associated with pH Variables in Humid Tropical Soils," *Soil Science Society of America Journal*, **49** (1985), pp. 1475–1480.

Table 8-2 Relative Lime Requirement of Selected Plants

High Lime Requirement	Medium Lime Requirement	Low Lime Requirement	Very Low Lime Requirement
Alfalfa	Blackberry	Alsike clover	Azalea
Asparagus	Cabbage	Buckwheat	Blueberries
Barley	Cantaloupe	Oats	Coffee, arabica
Beans, field	Corn	Peanut	Cranberries
Cotton	Crown vetch	Potatoes	Kudzu
Kentucky bluegrass	Fescue grass	Raspberries	Lespedeza
Peas	Grain sorghum	Rice	common
Red clover	Grasses (most)	Rye	Korean
Soybeans	Lettuce	Strawberries	sericea
Spinach	Peanut	Vetch	Napier grass
Sugar beets	Sweet potato	Watermelon	Pineapples
Sunflower	Tobacco		Rhododendron
Sweet clover	Trefoil		Stylosanthes
	Wheat		(tropical legume)
	White clover		Tea

Sources: (1) R. G. Hanson, "Corrective Liming of Missouri Soils," Science and Technology Guide 9102, *Agronomy,* **11** (1977). (2) D. R. Christenson and E. C. Doll, "Lime for Michigan Soils," Michigan State University Extension Bulletin 471, 1979.

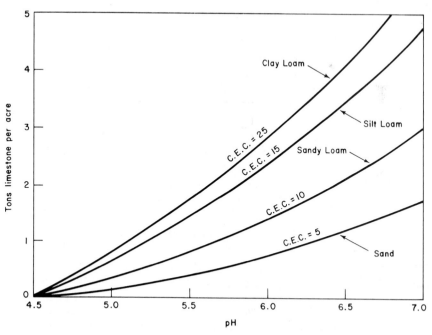

FIGURE 8-5 Approximate tons of limestone required to raise the pH of a 17-cm (7-in.) layer of soil of four textural classes with typical cation exchange capacities (CEC) in milliequivalents per 100 g of soil. (*Sources:* modified from D. R. Christenson and E. C. Doll, "Lime for Michigan Soils," Michigan State University Extension Bulletin 471, 1979; and R. G. Hanson, "Corrective Liming of Missouri Soils," Science and Technology Guide 9102, *Agronomy,* **11** [1977].)

ORGANIC SOILS

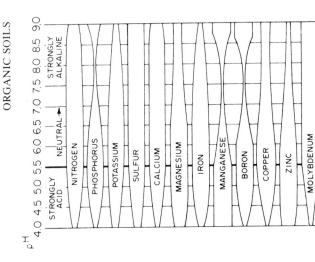

MINERAL SOILS

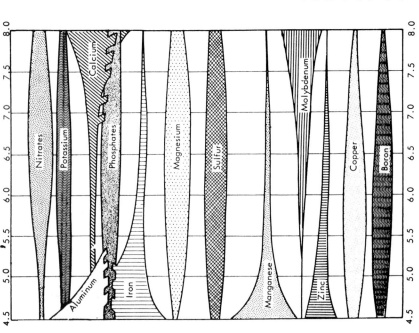

FIGURE 8-6 Theoretical relationship between soil pH and relative plant nutrient availability (the wider the bar, the greater the plant availability). (a) Organic soils. Note that the pH is about 5.5 for greatest plant availability of most nutrients in organic soils. (b) Mineral soils. Note that the pH is about 6.5 for greatest availability for the most nutrients. The lime requirement is one pH unit higher than for organic soils. Where elements are shown interlocking, the two elements at that pH combine to form insoluble compounds, which reduces phosphate solubility. (Sources: (a) Department of Crop and Soil Sciences, Michigan State University; (b) Soils Handbook, Kentucky Agricultural Experiment Station, Miscellaneous Publication 383, 1970, p. 28.)

Lime can be applied to advantage at any stage in the cropping system, but it is best applied 6–12 months in advance of seeding a legume or a few months before planting a high-value crop that responds well to lime. The application rate should be determined by soil testing. Applying 4.5 Mg/ha on a field that needs 9 Mg/ha is shortsighted economy because there may be little or no return on a significant cash outlay for legume seed, fertilizer, and lime. Similarly, liming a field that needs no lime provides no benefit at all, may be injurious, and is a waste of money.

Newly spread lime should be well mixed within the whole plow layer. On strongly acidic soils, where 6.7–13 Mg/ha or more of lime are required, one-half the amount may be applied before plowing and the other half applied and disked in after plowing. When not more than 4.5 Mg/ha are needed, the entire amount may be applied and disked in before seeding the legume or legume-grass mixture.

The usual method of applying lime is to spread it on the soil surface by a truck with a specially built, V-shaped bed and a spreading mechanism in the rear.

When both surface soils and subsoils are strongly acidic, as are Ultisols, it sometimes pays to incorporate lime to a depth of about 30 cm (12 in.).[14]

8:9 Liming No-Till Fields

Lime reaction is faster and greater when mixed into the soil rather than simply spread on the surface, so the recent increase in **no-till cropping** does slow the incorporation process. Adequate rainfall ensures the downward movement of some lime on no-till fields. The mulch of crop residues on no-till soil and lack of tillage mean that soil microbial activity takes place at a much shallower depth; there is more root activity, pH change, and fertilizer buildup in the surface (10–20 cm; 4–8 in.) than in deeper levels.

Although lime applications to no-till fields may not be as effective as application to an equivalent cultivated soil, liming no-till acid soils is well worth the cost. On a no-till field, surface applications of lime in late winter for eight consecutive years in Virginia (in a medium-rainfall area) produced greater corn yields on a no-till area than on a tilled and limed area (Fig. 8-7).

Nitrogen fertilizers applied on the soil surface in no-till farming acidify the *surface* 2.5 cm (1 in.) of soil. This reduces the effectiveness of certain herbicides, such as atrazine and simazine, and limits plant growth and yield. However, field research in Kentucky on two contrasting soils showed that surface applications of lime are effective enough to increase crop yields.[15]

Surface-applied lime induces alfalfa roots to grow more deeply, to 84 cm (33 in.), even though the soil pH at this depth was 4.1 and exchangeable Al was 225 mg/kg.[16] This emphasizes the strong need for Ca^{2+} ions in order for alfalfa to develop properly and grow into soils with limited Ca^{2+}.

8:10 Lime Balance Sheet

When a soil has had its acidity corrected by lime, how often must lime be added and how much is needed to keep the soil pH suitable? The answers depend upon the rate of lime loss. Lime is neutralized or lost from the soil by six activities.

1.   *Neutralization by acid-forming fertilizers (ammonium):* a rapid change

---

[14] B. D. Doss, W. T. Dumas, and Z. F. Lund, "Depth of Lime Incorporation for Correction of Subsoil Acidity," *Agronomy Journal,* **71** (July–Aug. 1979), pp. 541–544.

[15] R. L. Blevins and L. W. Murdock, "Effect of Lime on No-Tillage Corn Yields," *University of Kentucky Agronomy Notes,* **12,** no. 1 (Feb. 1979).

[16] J. E. Rechcigl and R. D. Reneau, Jr., "Effect of Subsurface Acidity on Alfalfa in a Tatum Clay Loam," *Communications in Soil Science and Plant Analysis,* **15** (1984), pp. 811–818.

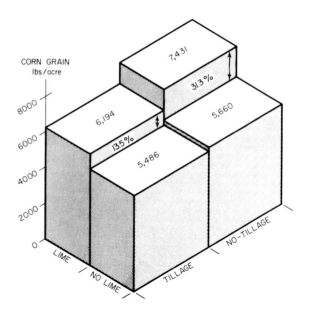

FIGURE 8-7 Surface-applied lime, which is a necessary method of addition in no-till operations, resulted in greater growth during 8 years on Frederick silt loam in Virginia than did lime incorporated by conventional operations. These results indicate that in at least some instances the surface application of lime will be a suitable method of application. (*Source:* Drawn from data by W. W. Moschler, D. C. Martens, C. I. Rich, and G. M. Shear, "Comparative Lime Effects on Continuous No-Tillage and Conventionally Tilled Corn," *Agronomy Journal,* **65** [1973], pp. 781–783.)

2. *Neutralization by the acid formed by carbon dioxide in water (from air, respiration, and organic matter decomposition):* one of the most important causes of lime neutralizations; it is a slow but continual process
3. *Leaching:* a relatively slow change
4. *Removal in harvested or grazed crops:* relatively slow loss
5. *Erosion:* as topsoil is lost with its higher base saturation, erosion often leaves more acidic subsoil to be limed
6. *Neutralization by acids dissolved in precipitation (rainfall):* results from oxides of sulfur fumes from manufacturing plants; a slow process

If compensating lime is not included, ammonium fertilizers may neutralize an average of 45.5 kg (100 lb) of field-applied lime per year. Each kilogram of nitrogen (N) from ammonium fertilizers generates soil acidity equal to 1.8 kg of pure calcium carbonate. Thus, to maintain soil pH, for each 100 kg of ammonium nitrogen (N) fertilizer applied, 180 kg of calcium carbonate equivalent must be applied. Each year in the United States more soil acidity is generated by nitrogen fertilizers than is neutralized by lime, by a ratio of 4:3. Fluid lime is often used with fluid ammonium-nitrogen fertilizer, never with any fluid phosphorus fertilizer.[17,18]

The carbonic acid formed by carbon dioxide dissolved in water helps to solubilize, and thus leach, limestone:

$$H_2CO_3 \quad + \quad CaCO_3 \quad \longrightarrow \quad Ca(HCO_3)_2$$

carbonic acid	calcium carbonate	calcium bicarbonate
	(limestone)	(more soluble and
	(less soluble)	more leachable)

---

[17] K. A. Kelling and E. E. Schulte, "Liming Materials—Which Will Work Better?" *Solutions* (Mar.–Apr. 1980), pp. 52–54, 56, 60.
[18] Gary W. Colliver, "Lime as a Fluid," Science and Technology Guide 9103, University of Missouri, Columbia, 1979.

258 *Chapter 8 Acidic Soils and their Modification*

In high-rainfall areas leaching losses may average 112 kg/ha (100 lb/a) per year. Harvested and grazed crops may remove the calcium equivalent of another 56–224 kg/ha (50–200 lb) of lime. Erosion, if active, may remove another 45–112 kg/ha (40–100 lb/a) of unreacted lime plus high-pH soil in the lost topsoil. These losses approximate 336–560 kg/ha (330–500 lb/a) yearly. To maintain the desired pH in these conditions would require the application of an additional 2.2 Mg/ha (1 t/a) of lime every fifth year.

■ 8:11 Acidifying Soils

Some plants actually grow better or have fewer problems if the soil is moderately acidic rather than neutral or with a basic pH. It is possible to increase soil acidity, but, as with liming, the pH change requires months, years, or decades. To be effective the soil must also be leached to remove some of the basic cations (mostly calcium) that have been exchanged. Elemental sulfur, iron, and aluminum compounds, and sulfuric acid are the most common materials used to acidify soils. Waste sulfuric acid, collected from industrial scrubbing smoke stacks (which prevent sulfur oxides from entering the atmosphere), is frequently utilized because it allows beneficial use of a waste by-product. The most effective materials are listed in Table 8–3. The reactions of soil acidification by additions of these materials follow.

$$2S + 3O_2 + 2H_2O \xrightarrow[\text{bacteria}]{\textit{Thiobacillus}} 4H^+ + 2SO_4^{2-} \quad \text{(sulfuric acid)}$$

$$2Fe^{3+} + 6H_2O \longrightarrow \underset{\text{(insoluble)}}{2Fe(OH)_3} + 6H^+$$

$$2Al^{3+} + 6H_2O \longrightarrow \underset{\text{(insoluble)}}{2Al(OH)_3} + 6H^+$$

Table 8-3 Materials Added to Soils to Make Them More Acidic (Lower pH) and the Approximate Quantity Needed for Various Soils to Alter the pH

Amendment Added	Kg Equivalent to 1 kg of Sulfur
Sulfur	1.0
Lime-sulfur solution (about 24% S)	4.2
Sulfuric acid (98%)	3.1
Iron Sulfate [$Fe_2(SO_4)_3 \cdot 7H_2O$]	8.7
Aluminum sulfate [$Al_2(SO_4)_3 \cdot 18\,H_2O$]	6.9

The pH Change Wanted	kg of Sulfur Needed per Hectare[a]		
	Sand	Loam	Clay
8.5 to 6.5	2200	2800	3300
7.5 to 6.5	550	900	1100
7.0 to 6.5	100	170	350

Source: Selected and recalculated data from *Western Fertilizer Handbook,* Interstate Printers and Publishers, Danville, Ill., 1975, pp. 231–232.

[a] Assumes noncalcareous soils and average organic matter for temperate-region soils. Soils high in organic matter and in swelling (montmorillonite clays) may have slightly higher requirements than shown.

The H⁺ produced in each reaction above can be consumed, as they are produced, by neutralization in the alkaline soils with bases such as $Ca(OH)_2$, $Mg(OH)_2$, $NaOH$, and the bicarbonates of these cations. Examples are

$$2H^+ + Ca(OH)_2 + SO_4^{2-} \rightarrow HOH + CaSO_4$$

and $\quad Ca^{2+} + 2HCO_3^- + SO_4^{2-} + 2H^+ \rightarrow CaSO_4 + 2H_2CO_3$

$$(H_2CO_3 \text{ is unstable and evolves } CO_2)$$

Making soil more acidic or less basic may be necessary for such reasons as lessening infection from pathogens (rots, scabs) or making micronutrient metals more soluble for better plant absorption. Sweet potatoes grown on Mississippi River terrace soils in Louisiana can be devastated by soil rot if the soil pH exceeds 6. These particular soils have toxic amounts of manganese which produce "crinkle-leaf" symptoms if the pH falls below 4.9.[19] The management of such soils is to lime carefully when needed to grow soybeans and various other crops, but to keep the pH between pH 5 and 6 when used for sweet potatoes. Soils with free lime usually require too much amendment to neutralize the lime to justify adding acidifying amendments.

Questions

1. Most soils, if all substances remained, would weather to produce neutral or alkaline soils. Explain, then, why so many soils are acidic and why some soils remain alkaline.
2. Show how (a) humus decomposition, (b) root respiration, and (c) ammonium nitrification produce acidity.
3. What are the dominant two ions on the cation exchange sites in soils of about pH 5.5?
4. Why are strongly acidic soils poor growing media for most plants?
5. (a) Define *lime*. (b) What actual material is most used as agricultural lime?
6. (a) Define the *neutralizing index*. (b) What are the important considerations in selecting a lime?
7. (a) In general terms, what does lime do when added to acidic soils? (b) How does liming improve the soil for growth of most plants?
8. Compare the importance of lime and fertilizers.
9. (a) Which crops respond to lime? (b) To what extent do all crops require lime on acidic soils?
10. Can too much lime be added (other than cost)?
11. What is the preferred method of adding lime to soils? Why?
12. (a) Is lime effective on no-till lands? (b) Explain how it would neutralize soil at depths of 10–15 cm (4–6 in.).
13. As a general guide, how much and how often might lime be applied?
14. Which soils (clays, sands, low CEC, high CEC) would have the highest lime requirements if the initial and final pHs were the same? Explain.
15. To what extent might different soils (mineral vs. organic, for example) be limed to different pH values (for different crops, as one example)?
16. (a) How can soils be deliberately acidified? (b) What materials are used?
17. How economical would it be to try to acidify (a) alkaline soils having no lime? and (b) alkaline soils containing lime? Explain.

---

[19] L. G. Jones, R. L. Constantin, J. M. Cannon, W. J. Martin, and T. P. Hernandez, "Effects of Soil Amendment and Fertilizer Applications on Sweet Potato Growth, Production, and Quality," Louisiana Agricultural Experiment Station Bulletin 704, 1977.

9

Nitrogen and Phosphorus

Nitrogen is the key to successful organic matter management and thereby to successful soil management.

—R. L. Cook and B. G. Ellis

9:1 Preview and Important Facts

PREVIEW

Plants are known to need at least 16 essential elements to grow, although more than 90 elements can be absorbed by plants. The 16 essential elements are carbon, oxygen, hydrogen, nitrogen, calcium, potassium, magnesium, phosphorus, sulfur, chlorine, iron, boron, manganese, zinc, copper, and molybdenum. The elements cobalt, nickel, silicon, sodium, and vanadium are also needed by some plants. Humans and other animals require 15 of the 16 essential plant nutrients (boron is not required by animals); in addition, animals require sodium, iodine, selenium, and cobalt. There is some evidence that most mammals may also need fluorine, chromium, nickel, vanadium, silicon, tin, arsenic, and cadmium.

From the air and water, plants utilize hydrogen, oxygen, and carbon. The other **macronutrients,** those absorbed in large amounts from soil and fertilizers, are nitrogen, phosphorus, and potassium (the three primary fertilizer nutrients) plus calcium, magnesium, and sulfur. The **micronutrients,** those absorbed in lesser quantities (formerly called *trace elements*), are chlorine, copper, boron, iron, manganese, molybdenum, and zinc. Table 9-1 lists these elements and their ionic forms that are available for plant use.

The principal soil storehouse for large amounts of the nutrient anions is soil organic matter. Decomposition of organic matter releases nutrient anions. Organic matter holds more than 95% of the soil nitrogen, often half or more of the total soil phosphorus, and as much as 80% of the soil sulfur. Boron and molybdenum reserves are stored both in organic matter and adsorbed to iron and aluminum oxides and other solids through hydroxyl (OH) groups.

As nutrients are absorbed from the soil solution, they are replenished from several sources, such as exchangeable (adsorbed) ions on clay minerals and humus, the slow decomposition of soil minerals, and the decomposition of soil organic matter. Seldom is the rate of renewal for all essential elements from untreated soils fast enough to achieve maximum crop production; to augment these insufficient supplies, fertilizers are added.

Table 9-1 The 16 Plant Nutrients and 5 Additional Elements, Their Chemical Symbols, Content in Plants, and the Form(s) Common in Air, Water, and Soil and Available for Plant Uptake

Element and Symbol	Portion of Plant [a] (%)	Ion or Molecule
Carbon (C)	41.2	CO_2 (mostly through leaves)
Oxygen (O)	46.3	CO_2 (mostly through leaves), H_2O, O_2
Hydrogen (H)	5.4	HOH (hydrogen from water), H^+
Nitrogen (N)	3.3	NH_4^+ (ammonium), NO_3^- (nitrate)
Calcium (Ca)	2.1	Ca^{2+}
Potassium (K)	0.80	K^+
Magnesium (Mg)	0.42	Mg^{2+}
Phosphorus (P)	0.30	$H_2PO_4^-$, HPO_4^{2-} (phosphates)
Sulfur (S)	0.085	SO_4^{2-}
Chlorine (Cl)	0.011	Cl^- (chloride)
Iron (Fe)	0.0066	Fe^{2+}, Fe^{3+} (ferrous, ferric) = Fe(II), Fe(III)
Boron (B)	0.0045	H_3BO_3, (boric acid)
Manganese (Mn)	0.0036	Mn^{2+} = Mn(II)
Zinc (Zn)	0.0009	Zn^{2+} = Zn(II)
Copper (Cu)	0.0007	Cu^{2+} = Cu(II)
Molybdenum (Mo)	0.000005	MoO_4^{2-} (molybdate)
Cobalt (Co)	—	Co^{2+} = Co(II)
Nickel (Ni)	—	Ni^{2+} = Ni(II)
Silicon (Si)	—	$Si(OH)_4$ (nonionized)
Sodium (Na)	—	Na^+
Vanadium (V)	—	VO_3^- (vanadate)

Source: Selected data and additions by Raymond W. Miller from B. G. Ellis and B. D. Knezek, "Adsorption Reaction of Micronutrients in Soils," p. 68; and L. O. Tiffin, "Translocation of Micronutrients in Plants," p. 204, both in *Micronutrients in Agriculture*, R. C. Dinauer, ed., Soil Science Society of America, Madison, Wis., 1972.

[a] Estimates of element contents taken from oven-dried alfalfa. Various plants will have different values, particularly in content of potassium, nitrogen, calcium, phosphorus, and sulfur.

Nitrogen *is most often the limiting nutrient in plant growth;* it is a constituent of chlorophyll, plant proteins, and nucleic acids. Nitrogen can be utilized by plants as the ammonium cation or as the nitrate anion. Atmospheric dinitrogen (N_2) is made available by **nitrogen fixation,** which requires the action of specific microorganisms. Other soil nitrogen is made available by **mineralization,** which is the microbial decomposition of organic matter that releases nitrogen as ammonium ions. Ammonium ions are also adsorbed on cation exchange sites.

Nitrification is the bacterial oxidation of ammonium cations to nitrate anions, which in turn are used by plants, lost by **leaching,** or **denitrified** by bacteria to volatile N_2 and N_2O. Ammonium can be volatilized as ammonia gas. Significant amounts of soil nitrogen are used by microorganisms and become part of the soil's organic substances (are **immobilized**). Nitrogen in microbial bodies is temporarily unavailable for plant use.

Phosphorus *is the second most often limiting nutrient.* It is contained in plant cell nuclei and is part of energy storage and transfer chemicals in the plant. Soils have low total and low plant-available phosphate supplies because mineral phosphate forms are not readily soluble. Phosphorus used by the plant is taken up as the HPO_4^{2-} and $H_2PO_4^-$ anions. Unfortunately, most soluble phosphates become fixed (precipitated or adsorbed to form insoluble compounds) before plants can absorb them. Organic phosphates are important—even major—phosphate sources in most soils.

IMPORTANT FACTS TO KNOW

1. The general mechanism of nutrient movement to root surfaces and the process of nutrient absorption into root cells
2. The importance of nitrogen to plants and its availability in soils
3. The form(s) of nitrogen used by plants
4. The general mechanism of N_2 fixation and its importance
5. The relative importance of soil organic matter as a nitrogen reservoir and how soil nitrogen is made available to plants
6. The process of nitrification and why it is "acidifying"
7. The mechanisms by which nitrogen is lost from soils (a) in water and (b) as gaseous forms
8. The purposes for using nitrification inhibitors and their proven benefits
9. The conditions necessary to cause denitrification losses
10. The conditions necessary to initiate ammonia volatilization losses
11. Some approximate "average soil" nitrogen values, using different crop covers, for (a) leaching losses, (b) nitrogen made available from mineralized nitrogen, and (c) denitrification losses
12. Some important characteristics of these fertilizers: anhydrous ammonia, urea, ammonium nitrate, and ammonium phosphate
13. The nature of "controlled-release" nitrogen fertilizers, their value, and relative costs
14. The conditions that decrease phosphate absorption by roots
15. The properties of phosphorus that cause it to be so often deficient in soils
16. The dependence of phosphorus availability on (a) soil organic matter and (b) anaerobic (waterlogged) growing conditions, as in rice paddies
17. The amounts of phosphorus (a) lost in leaching, (b) removed in crops, and (c) mineralized from soil humus
18. Some properties of these phosphorus sources: rock phosphate, concentrated superphosphate, ammoniated phosphates, and ammonium polyphosphate

9:2 Mechanisms of Nutrient Uptake

Prior to their absorption into root cells, nutrients reach the surface of roots by three mechanisms: mass flow, diffusion, and root interception. **Mass flow,** the most important of these mechanisms quantitywise, is the movement of plant nutrients in flowing soil solution. **Diffusion** is movement by normal dispersion of the nutrient from a higher concentration (such as near its dissolving mineral source) through soil water by its kinetic motion to areas of lower concentration of that nutrient. **Root interception** is the extension (growth) of plant roots into new soil areas where there are untapped supplies of nutrients in the soil solution. All three processes are in constant operation during growth. The importance of each mechanism in supplying nutrients to the root surface varies with the chemical properties of each nutrient. Nevertheless, mass flow involves large amounts of water flowing to roots as the plant transpires water. Mass flow is the dominant mechanism carrying nutrients to roots and supplies about 80% of nitrogen, calcium, and sulfur to root surfaces. Diffusion is the dominant transport mechanism for phosphorus and potassium.

The mechanisms of absorption into the root cells are not well understood. The cell walls are porous, and the soil solution can move through some or all of the cell walls, causing intimate contact of the soil solution with the outer membranes of the cells. For a nutrient to cross a cell membrane into the cell, it is believed that each nutrient ion must be attached to some *carrier*. The carrier-nutrient complex can pass through the membrane or in some other unknown manner move the ion into the cell. The necessary carriers are different for many of the nutrients. This means of nutrient absorption allows the root to have some selectivity in the kinds of elements absorbed. Some elements can be partially but not entirely excluded from absorption; others can be preferentially absorbed, even against a *concentration gradient* (can be absorbed from a low-concentration soil solution and transferred to a higher concentration in the plant cell) (Fig. 9-1).

A very poorly understood mechanism of electrical balance also seems to be involved in ion absorption and accumulation. As nutrient cations are absorbed, H^+ ions are excreted into the soil solution or more organic acid anions are produced inside the cell to balance the

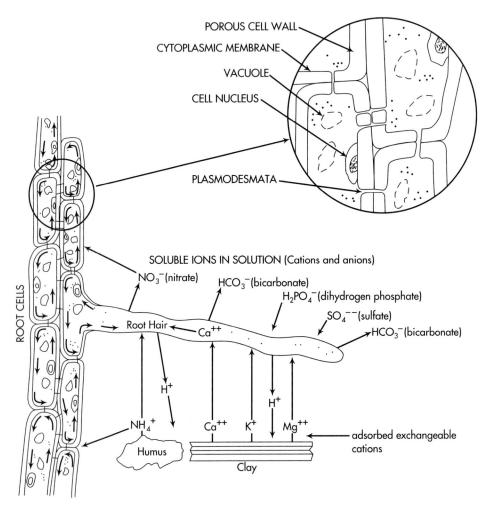

FIGURE 9-1 Diagrammatic scheme showing root structure and illustrating that a root hair absorbs nutrients from the soil solution and from adsorbed (exchangeable) ions on a clay crystal or humus colloid. A root hair is an extension of one of the epidermal (surface) cells of the plant root and is thought by some scientists to absorb nearly all the plant's water and nutrients. However, much evidence indicates that older and larger roots are also active in water absorption. Water can move within and through the cell walls and pore spaces between cells and thus furnish the cells with large amounts of contact between soil solution and the cell membranes enclosing the active cell protoplasm. Plasmodesmata are fine strand connections of cytoplasm between cells through which absorbed water and nutrients move. In the insert they are shown exaggerated in comparative size. Note that to maintain electrical balance within the plant, approximately equivalent amounts of HCO_3^- must be exuded to soil to balance the total *anion* uptake. Similarly, sufficient H^+ ions must be exuded to balance the total *cation* uptake.

absorbed cations. Likewise, as nutrient anions are absorbed by the plant, more compensating cations are absorbed and/or HCO_3^- ions are excreted into the soil solution in order to maintain an electron balance in the cell. Perhaps the H^+ ions and HCO_3^- ions are excreted into the soil solution first in order to aid solubility of soil nutrients. The processes involved are slowly being discovered and the mechanisms clarified.

Plants also absorb nutrients through small openings in leaves, the **stomata.** Carbon enters almost entirely through the stomata as carbon dioxide, and the plant releases the oxygen (O_2) produced during photosynthesis out through the stomata. Hydrogen, as a part of water molecules, is absorbed through stomata, but this intake is usually small compared to the amount entering through the roots. Other nutrients are also absorbed through the stomata; soluble ions from fertilizer-enriched water from overhead sprinkler irrigation or other sprays are absorbed to some extent. It is believed that direct absorption through the leaves seldom exceeds a few kilograms per hectare per application. However, the small concentrations of micronutrients needed can be satisfactorily added by a foliar spray. Such sprays are widely used to supply fruits and berries with iron, zinc, manganese, molybdenum, boron, or copper needs.

■■■■ 9:3 Soil Nitrogen Gains and Transformations

Nitrogen is the key nutrient in plant growth. It is the most often deficient nutrient, and it thereby is the controlling factor in most growth. Nitrogen is a constituent of plant proteins, chlorophyll (the green plant pigment important to photosynthesis), nucleic acids (the regenerative portions of the living cell), and other plant substances. Adequate nitrogen often produces thinner cell walls, which results in more tender, more succulent plants; it also means larger plants (greater crop yields).

A deficiency of nitrogen causes poor plant yields. There is enough total nitrogen in most soils, but not enough of the nitrogen is in a chemical form that can be utilized by plants. Large quantities of nitrogen, in the form of N_2 gas, reside in the atmosphere above the surface of the earth, but this form of nitrogen cannot be utilized by the majority of plants; N_2 must first be changed by microorganisms into other forms (Fig. 9-2).

Nitrogen is a unique plant nutrient. Unlike the other essential nutrient elements, plants can absorb nitrogen in either the cationic form (ammonium ion, NH_4^+) or the anionic form (nitrate, NO_3^-). Only a small part of soil nitrogen occurs in these forms at any one time. Nitrate nitrogen is soluble and mobile in soils and is easily leached. Both nitrate and ammonium forms may be consumed by microorganisms or converted to gaseous nitrogen forms (N_2 or NH_3, respectively) and lost to the atmosphere. These relationships are shown in the nitrogen cycle in Fig. 9-2. *Learn this cycle well.*

Are both nitrate and ammonium ion forms equally good and available to the plant? A definitive answer is still sought. Many studies have shown that selected plants seem to grow better with one form or the other. If the NH_4^+ form is used, it does not have to be reduced again inside the plant to the amino form (—NH_2) as does NO_3^-. This would save the plant energy. Perhaps the flavors or tastes of foods to people and other animals are influenced by the form of the nitrogen plants use. For example, French beans fed ammonium ion were smaller, were 3 to 10 times higher in amino acid contents, and expended only half as much metabolic energy (ATP) per gram of material produced as did plants given nitrate ions. In contrast, beans supplied with nitrate ion rather than ammonium ion were 50% larger and had 10–30 times more organic acids.[1] Numerous studies have described increased yields when both NH_4^+ and NO_3^- forms are available, as itemized on page 267.

---

[1] S. Chaillou, J.-F. Morot-Gaudry, C. Lesaint, L. Salsac, and E. Jolivet, "Nitrate or Ammonium Nutrition in French Bean," *Plant and Soil,* **91** (1986), pp. 363–365.

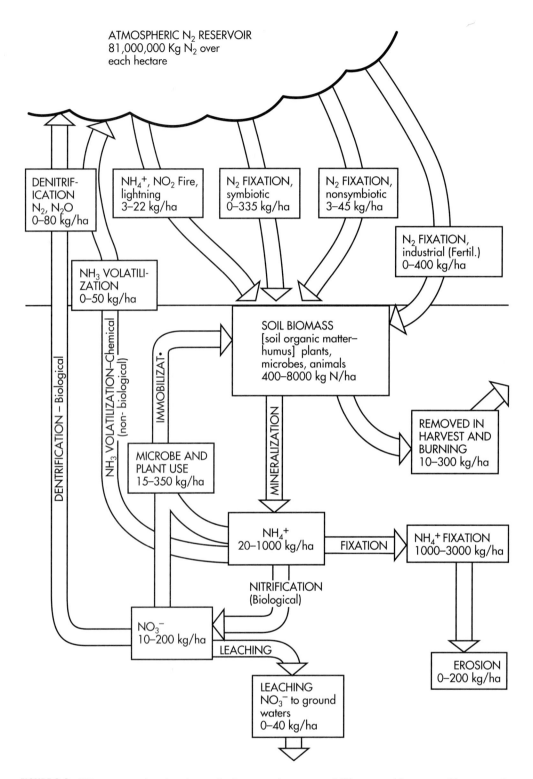

FIGURE 9-2 Nitrogen cycle, showing soil nitrogen changes, additions, and losses. (Courtesy of Raymond W. Miller, Utah State University.)

- NH_4^+ plus NO_3^- increased wheat yields 7–47% in 14 studies. Tillers also were increased by having both nitrogen forms, compared to having NO_3^- only.
- In hydroponic studies, corn fertilized with 50-50 NH_4^+ and NO_3^- averaged about 12.6 Mg/ha (200 bu/a). With only nitrate available, the corn averaged 11.3 Mg/ha (179 bu/a).

The question of which nitrogen form is best, nitrate or ammonium, is not yet answered. It seems evident that the form used by the plant will alter somewhat the plant's chemical composition.

9:3.1 Fixation of Dinitrogen Gas (N₂)

A major source of soil nitrogen comes from **nitrogen fixation,** a microbial action in which the relatively *inert* dinitrogen (N_2) is taken from the soil air and changed into forms used by the plants. The nitrogen fixation by microorganisms is either *symbiotic* or *nonsymbiotic*. In **symbiotic fixation** bacteria and actinomycetes cause the formation of the root nodules (abnormal root growth) in certain host plants and then inhabit those growths, where they fix nitrogen.

The amounts of N_2 fixed by soil bacteria and actinomycetes vary enormously and are usually less in soils that have high nitrogen levels or have had nitrogen fertilizers added. In Denmark three-year averages of two pea cultivars fixed 165 and 136 kg of N/ha and one field bean cultivar fixed 186 kg/Ha even though 50 kg of N/ha was added as fertilizer at planting.[2] Blue-green bacteria (also called blue-green algae) fixed only 25 kg/ha, alfalfas ranged from 128 to 600 kg/ha, and nonlegume fixers fixed from about 40 to 300 kg/ha. Only a few crops, such as alfalfa, can usually fix enough nitrogen to produce optimum growth, without fertilizer N additions. In one study good-growth legumes with no added nitrogen fertilizer, measured "fixed N_2" was only 80% of the crop's needs; at least 20% was supplied from decomposing soil humus. Some nonlegumes, such as alder (*Alnus*), also are associated with *Frankia* as N_2 fixers.

In free or **nonsymbiotic N₂** fixation, specific types of microorganisms exist independently in soil and in water, convert nitrogen (N_2) into body tissue nitrogen forms, and then release it for plant use when they die and are decomposed. Nitrogen fixed by nonsymbiosis yearly varies from a few kilograms per hectare to over 45 (40 lb/a) with 5–8 kg/ha (5–7 lb/a) as an average.

9:3.2 Mineralization of Nitrogen

The major source of nitrogen in nonfertilized soils is released from decomposition of organic material. The conversion of organic nitrogen to the ammonium form is termed **mineralization.** Soil organic matter contains an average of approximately 5% by weight of nitrogen. Only about 1–3% of the total organic matter is decomposed yearly (as much as half of *fresh* residues may be decomposed). The decomposition rate is fastest in warm, well-aerated, moist soils, such as in sands in summer, and is slower in clays in the cool spring. As an example, mineralizing 2% of the organic matter (humus) in a soil having an organic matter content of 4% would release about 93 kg/ha (83 lb/a) of nitrogen as ammonium. This is only about one-third to one-fourth of the nitrogen needed by a good corn crop. More soil humus content and a greater percentage decomposed (in warmer climates) would provide more released nitrogen. Some rates of mineralization are faster than the 2% percent per year "average" just described. Four examples given by Ross[3] all have higher decomposition rates (Table 9-2). It is

[2] E. S. Jensen, "Symbiotic N₂ Fixation in Pea and Field Bean Estimated by ¹⁵N Fertilizer Dilution in Field Experiments with Barley as a Reference Crop," *Plant and Soil,* **92** (1986), pp. 3–13.
[3] S. Ross, *Soil Processes: A Systematic Approach,* Routledge Publishers, New York, 1989, p. 59.

Table 9-2 Turnover Rates of Organic Carbon in Four Plant Cover Systems

Land Cover	Soil Depth (cm)	Production Mg C/ha-yr	Total C in Soil Mg C/ha-yr	Decomposed %/yr
Continuous wheat	0-23	2.6	26	4
Continuous meadow	0-23	2.7–3.2	77	3
Tropical rain forest	0-30	9–10	44	11
Beech forest (cold)	0-30	7.1	72	3

Source: Modified from S. Ross, *Soil Processes: A Systematic Approach,* Routledge Publishers, New York, 1989, p. 59.

obvious that rates of humus decomposition, amounts of humus in soils, and the nitrogen content in the humus all influence the amount of nitrogen released each year through decomposition. Notice that the year-round warm tropical forest has a high rate of 11% decomposed yearly, even of a high yearly biomass production.

9:3.3 Nitrification of Ammonium

Nitrification is the oxidation of ammonium cations to nitrate anions by bacteria or other organisms. Nitrification is a microbial transformation, and the process is rapid. Most small amounts of mineralized ammonium ions are nitrified within 1 or 2 days, unless the soil is strongly acid, cold, or waterlogged; these conditions slow nitrification markedly. Thus, the mineralized ammonium ions usually exist for a short time. Some ammonium ions are adsorbed temporarily to the negatively charged cation exchange sites of clay or organic particles; some ammonium ions are fixed in clay lattices (**ammonium fixation**); and other ammonium ions are used directly by plants. Eventually, most of the ammonium ions in the soil are oxidized by selective bacteria to NO_2^- (*Nitrosomonas*) and then to NO_3^- (*Nitrobacter*), as shown here:

$$2NH_4^+ + 3O_2 \xrightarrow[\text{bacteria}]{\text{Nitrosomonas}} 2NO_2^- + 4H^+ + 2H_2O + \text{Energy}$$

$$2NO_2^- + O_2 \xrightarrow[\text{bacteria}]{\text{Nitrobacter}} 2NO_3^- + \text{Energy}$$

Seldom do large amounts of nitrite (NO_2^-) accumulate, which is fortunate because it is toxic to living organisms, including plants. Notice that the oxidation of each NH_4^+ to NO_3^- also releases $2H^+$, which is important in acidification of soils. About 1.8 kg of pure lime is required to neutralize the acidity of 1 kg of urea-nitrogen or ammonium-nitrogen.

Nitrification is slowed by conditions unfavorable to the bacteria: dryness, cold, or toxic chemicals. Figure 9-3 illustrates two locations and their reported nitrification rates. In Fig. 9-3, the desirable ammonium thiosulfate fertilizer, temporarily producing NH_3 plus SO_2, inhibits the second step involving *Nitrobacter*. This could produce an accumulation of toxic nitrate ion (NO_2^-) in the soil. Notice the amount of nitrite produced. In many of the soils, nitrification was complete within 1 or 2 weeks after large applications of ammonium had been made.

9:3.4 Other Fixation Reactions Involving Soil Nitrogen

Soil nitrogen is involved in changes additional to those previously mentioned. Nitrogen can be **immobilized,** which is the use of soluble nitrogen (mostly NH_4^+ and NO_3^-) by plants or

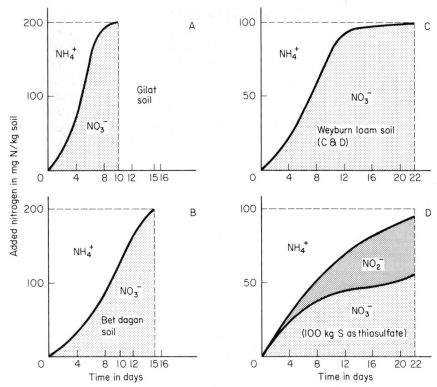

FIGURE 9-3 Rates of nitrification for three soils (a, b, c) and retardation of nitrification by ammonium thiosulfate (d). Soils (a) and (b) are in Israel; (c) and (d) (Weyburn loam) are in Alberta, Canada. (*Sources:* Aviva Hadas, Sala Feigenbaum, A. Feigin, and Rita Portnoy, "Nitrification Rates in Profiles of Differently Managed Soil Types," *Soil Science Society of America Journal,* **50** [1986], pp. 633–639; H. H. Janzen and J. R. Bettany, "Influence of Thiosulfate on Nitrification of Ammonium in Soil," *Soil Science Society of America Journal,* **50** [1986], pp. 803–806. Redrawn by Raymond W. Miller, Utah State University.)

microbes. The nitrogen again becomes complex organic compounds in new organism bodies. Much of the soil nitrogen added as fertilizer or mineralized is immobilized into proteins in higher plants and into new microbe bodies.

Ammonium fixation is the process by which certain clays bond NH_4^+ tightly between mineral lattices. Most of this "fixed" ammonium is neither exchangeable nor available to plants. Fixation is most common in vermiculite, hydrous mica, and montmorillonite clays. The ammonium ions fit into locations between clay layers within the clay particle similar to where potassium ions fit in natural hydrous mica (illite). Ammonium fixation occurs in small amounts compared to the quantities adsorbed to cation exchange sites, which is still available for release to plant roots.

▮▮▮ 9:4 Nitrogen Losses from the System

Nitrogen losses occur as leaching losses in percolating water and two gaseous losses. Although fixation and immobilization are losses from *available* forms, they still exist in the soil for possible conversion later to available forms again and are not considered as losses here.

9:4.1 Leaching of Soil Nitrogen

Nitrate (NO_3^-) is the most readily leached form of nitrogen. Both ammonium (NH_4^+) and nitrate ions are very soluble in water, but the positively charged ammonium ion is held to cation exchange sites and resists leaching. Leaching losses of nitrates are increased as the quantities of percolating water increase and when there is little or no growing crop cover to absorb the nitrates as rapidly as they are produced.

Losses of nitrogen from the soil covered with an actively growing crop are usually only a few kilograms per year unless large amounts of a fertilizer was recently added. In contrast, it is not unusual to find losses that exceed 20 kg/ha of nitrogen on crops heavily fertilized. On fields where fertilizer application is poorly timed for plant use or coincides with heavy rainfall and leaching, leaching losses can be 50–80 kg/ha of nitrogen yearly.

9:4.2 Nitrification Inhibitors[4,5,6]

If oxidation of ammonium ions to the nitrate form is slowed or hindered, less nitrate is produced and less nitrogen is lost by leaching or denitrification. Several dozen chemicals have been tested for their ability to inhibit nitrification. The best known **nitrification inhibitor** (**NI**) is **Nitrapyrin** [2-chloro-6-(trichloromethyl)pyridine], also called **N-Serve®**. Dicyandiamide (**DCD**) and **ATC** (**4-amino-1,2,4-triazole**) are two other materials. Potassium azide (KN_3) has been used extensively in Japan.

Nitrapyrin (as do many of the inhibitors) inhibits the *Nitrosomonas* species of bacteria that convert ammonium to nitrite, the first step of nitrification. Nitrapyrin has the disadvantage of being a volatile liquid, which can be lost rapidly from sandy soils. DCD and ATC, in comparison, are easier to handle, are water-soluble, do not volatilize, and can be applied as a coating to granular fertilizers.

A recent fertilizer—ammonium thiosulfate—has the thiosulfate act as a "nitrification inhibitor." The conversion of ammonium ion to the easily leached nitrate is slowed down (Fig. 9-3d). The retardation action occurs on the *Nitrobacter* organisms (the second step, nitrite to nitrate) rather than on the *Nitrosomonas,* which accomplish the first step (ammonium ion to nitrite ion).

Nitrification inhibitors should be placed with the nitrogen to be most efficient. Rates of about 0.2–2.0 kg/ha of active NI material or 0.5–1.0% of added fertilizer are satisfactory to retard nitrification, but results vary excessively with conditions of application and the soil. Some studies also show poor results with additions of Nitrapyrin of less than 2–5 kg/ha (1.8–4.5 lb/a). Conservation of added urea or ammonium nitrogen after a month may be 50–90% of the amounts added; after three months, retention as NH_4^+ may be from nearly zero up to only 20–30%.

Although NIs are used commercially, the results are not consistent, even in duplicate circumstances. Even when oxidation of ammonium ions is inhibited, as shown by higher ammonium-to-nitrate ratios, there may not be consequent yield increases, indicating that other unidentified causes are more influential than just the conserved ammonium fertilizer. It should also be noted that NIs have a gradually reduced effectiveness over time because of their volatilization, adsorption, leaching, or microbial breakdown. The effective life of NIs is also shortened by warmer temperatures and use on sandy soils, which increase volatilization and leaching losses above those occuring in finer textured soils.

---

[4] T. F. Guthrie and A. A. Bomke, "Nitrification Inhibition by N-Serve and ATC in Soils of Varying Texture," *Soil Science Society of America Journal,* **44** (1980), pp. 314–320.

[5] George W. Bengtson, "Nutrient Losses from a Forest Soil as Affected by Nitrapyrin Applied with Granular Urea," *Soil Science Society of America Journal,* **43** (1979), pp. 1029–1033.

[6] R. J. Goos, B. E. Johnson, and W. H. Ahrens, "New Uses for Ammonium Thiosulfate," *Solutions,* **34** (1990), no. 3, pp. 32–33.

Some studies even suggest that use of NIs can be detrimental to plant growth. Some of the observed effects are a toxic level of ammonium ion on plant roots (radishes) and a claimed reduction in cation uptake, because NH_4^+, a cation, is absorbed in considerable amounts. Research work presents conflicting results on these and other claims. For example, one application of 120 lb of N with nitrapyrin on cauliflower was more effective in increasing yields than split applications without nitrapyrin.[7] Nitrification inhibitors currently have some valid uses, such as retaining fall-applied ammonium fertilizers through the winter with minimum leaching losses. Few general-use recommendations are validated by confirmed research.

There is increased interest in using NI materials to minimize the NO_3^- leaching into surface and groundwaters, thus avoiding pollution as well as avoiding nitrogen losses from the soil. Nitrates are a health problem where food supplies are adequate. The health problem is discussed in a later chapter.

Searches are continual for new and better NIs. A recent one, **CMP** [1-carbomoyl-3 (5)methylpyrazol] worked well in a calcareous soil.[8] Without CMP, ammonium sulfate was nitrified in 3 weeks. With CMP, only 10% was nitrified after 3 weeks and 42% was nitrified after 8 weeks during warm growing conditions.

9:4.3 Gaseous Losses of Soil Nitrogen

In addition to nitrogen lost by leaching, immobilization, and ammonium fixation, soil nitrogen can be lost through two mechanisms producing gaseous forms that escape into the atmosphere. **Denitrification** is the change by bacteria of nitrate to a nitrogen gas (*mostly N_2, some N_2O, and less of other oxides*). Usually denitrification is *the most extensive gaseous nitrogen loss.* When poor aeration limits the amount of free oxygen in the soil, a few specifically adapted bacteria are forced to use the nitrogen in NO_3^- as an electron acceptor (normally, O_2 is used). The end products of the process are dinitrogen gas (N_2) and/or nitrous oxide (N_2O); these gases volatilize from the soil into the atmosphere.

Denitrification is rapid. Even when conditions favorable for denitrification exist for only a day or less, appreciable losses of nitrogen as N_2 can occur. Estimates of total losses by denitrification on cropped lands average 10–20% of all nitrates formed from added fertilizers and, in extreme conditions, can be as much as 40–60% of added nitrate-nitrogen. A loss of 10–15% is common.

Denitrification requires, for highest losses by denitrification, (1) a lack of adequate free gaseous oxygen in the soil or solution, (2) an energy source of oxidizable organic matter—"food"—for the bacteria, and (3) warm, slightly acidic soils. Waterlogging for even a few hours in warm soil that contains decomposable humus may develop anaerobic conditions.[9] Even interiors of soil aggregates after wetting by rain or irrigation may develop anaerobic interiors. Spherical aggregates (granules) of a silty clay loam and of a silt loam from Iowa often had anaerobic interiors in aggregates larger than 20 mm (0.8 in.) diameter. One granule of only 8 mm (0.31 in.) diameter had an anaerobic zone (Fig. 9-4).[10] All aggregates involved in denitrification also had a measured zone that was anaerobic. Poorly structured clayey soils, especially subsoils, may have large peds (50–100 mm, or 2–4 in., diameter) that have poor porosity and are temporarily anaerobic after they are wetted.

---

[7] N. C. Welch, K. B. Tyler, D. Ririe, and F. E. Broadbent, "Nitrogen Uptake by Cauliflower," *California Agriculture,* **39,** nos. 5/6, (1985) pp. 12–13.

[8] P. I. Orphenos, "Inhibition of Ammonium Sulphate Nitrification by Methylpyrazol in a Highly Calcareous Soil," *Plant and Soils* **144** (1992), pp. 145–147.

[9] N. C. Welch, K. B. Tyler, D. Ririe, and F. E. Broadbent, "Nitrogen Uptake by Cauliflower," *California Agriculture,* **39** (1985), nos. 5/6, pp. 12–13.

[10] Alan J. Sexstone, Niels Peter Revsbech, Timothy B. Parkin, and James M. Tiedje, "Direct Measurement of Oxygen Profiles and Denitrification Rates in Soil Aggregates," *Soil Science Society of America Journal,* **49** (1985), pp. 645–651.

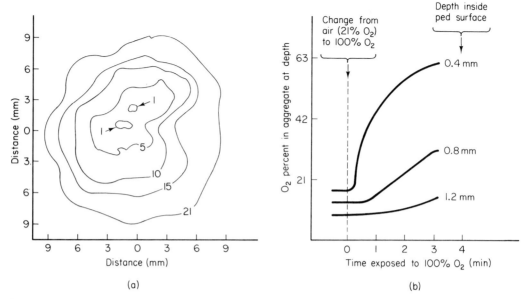

FIGURE 9-4 Typical O_2 characteristics in wetted spherical soil peds: (a) cross section of a ped showing O_2 levels from the outside (21%) inward to centers having only 1%; (b) rate of O_2 concentration changes at three distances from the ped surface as air O_2 is increased. Pores filled with water would transmit almost no O_2 that was not already dissolved in the water. (*Source:* A. J. Sexstone, N. P. Revsbech, T. B. Parkin, and J. M. Tiedje, *Soil Science Society of America Journal,* **49** [1985], pp. 645–651. Redrawn by Raymond W. Miller, Utah State University.)

Denitrification losses increase as temperature increases to 25°C (77°F), then speeds up above 25°C to even above 60°C (140°F). Large amounts of easily oxidized organic matter enhances the rate at which the soil develops anaerobic conditions and the rate of N_2 gas volatilized. The pH maximum favoring denitrification varies, but the maximum loss seems to be from slightly acidic to near neutral pH values. In acidic soils more nitrogen may evolve as N_2O even though the major form lost is still N_2.

Ammonia volatilization losses occur when ammonium is in a *basic* solution. The greatest losses occur from surface applications of any ammonium or urea fertilizer on calcareous (high carbonate content) soils. Small losses of ammonia also occur on nonfertilized soils. Ammonia volatilization losses from applied ammonium or urea fertilizers can be as much as 30% but usually are less than 10%. To minimize volatilization, cover the fertilizer with soil or leach it in with irrigation or rainfall. Broadcasting on the surface may cause large losses, especially on a turf. Urea, plus the enzyme *urease,* in soil forms ammonium carbonate. In basic soils calcium hydroxide reacts on the ammonium carbonate to form ammonium hydroxide:

$$Ca(OH)_2 + (NH_4)_2CO_3 \longrightarrow 2NH_4OH + CaCO_3 \downarrow \text{ (precipitated)}$$

NH_4OH easily decomposes to $NH_3 \uparrow$ (gas) $+ H_2O$

The loss of NH_3 gas from NH_4^+ and NH_4^+-forming fertilizers depends on the final pH of the NH_4^+ solution at the soil surface. Up to about pH 9 the ratio of NH_3 partial pressures in air versus that in water increases about 10-fold for each pH unit increase. Thus, the NH_3 lost from a solution of pH 8.2 will be much less than if the pH were 9 or higher. As an NH_4^+ ion is added to calcareous soil, two reactions occur:

$$(NH_4)2Y + CaCO_3 = (NH_4)_2CO_3 + CaY \qquad (Y = \text{various anions})$$

$$(NH_4)_2CO_3 + HOH = 2NH_3 \uparrow (gas) + 2HOH + CO_2 \uparrow (gas)$$

It has been determined that if the CaY salt is very *insoluble,* the NH_3 loss is greatest. For example, adding ammonium fluoride to the soil forms calcium fluoride (CaF_2), which is quite low in solubiity. This system can reach a pH of 9 (from the large amount of ammonium carbonate produced) and have 30–40% ammonia loss in 1 hour.[11] Salts such as ammonium chloride and ammonium nitrate do not form insoluble calcium salts; thus, little ammonium carbonate is produced to raise the pH (see the second equation above). The amount of ammonia lost depends on the soil solution pH.

Ammonia is unstable in water and is evolved in increasing amounts as the soil solution pH nears and exceeds pH 9. Actually, small ammonia losses have been reported from soils with a pH of 4. If ammonium carbonate can form, the chance for loss is great. Losses of ammonia gas from soils can be summarized as follows:

1. Losses are greatest on high-pH calcareous soils
2. Losses are greatest when fertilizer is left on the soil *surface*
3. Losses increase with higher temperatures, especially as surface soil dries out after being wetted (drying concentrates ammonia)
4. Losses are greatest in soils of low cation exchange capacities
5. Appreciable losses of urea applied on grass or pastures are possible because fertilizer on pastures is applied on the surface

The increased use of urea fertilizer can easily increase losses of nitrogen as ammonia gas. Urea is hydrolyzed by the enzyme *urease* to ammonium carbonate. If urea is spread on a calcareous soil surface or on another alkaline soil and left several days to dissolve in the morning dew or from soil water, great losses of ammonia are likely. To slow this enzymatic action until the soluble urea can be leached or tilled into the soil where the NH_4^+ can be adsorbed onto soil particles, inhibitors of urease action are needed. The inhibitors are usually substances that react with the urease, thereby blocking the enzyme from reacting with urea. Typical inhibitors of urease include phenols, quinones, benzoquinones, various insecticides, and substituted ureas. Applied inhibitors are adsorbed to soil materials gradually, decomposed by microbial activity, or react with metals, such as iron.

The use of urease inhibitors on sandy soil significantly reduces urea conversion to ammonium carbonate. Addition of 2.3 kg of *p*–benzoquinone per 100 kg of urea (a costly addition rate) decreased urea loss as NH_3 gas (after 14 days, at 20°C) from 63% with no inhibitor to 0.1% with inhibitor added.[12] Use of urease inhibitors is limited presently to exploratory greenhouse work or small-plot studies.

9:5 The Nitrogen Balance

More than any other major plant nutrient, nitrogen in the soil is subject to a complex system of gains, losses, and interrelated reactions. Intelligent management demands a working knowledge of these relationships and their comparative magnitudes. Figure 9-5 depicts

---

[11] L. B. Fenn and L. R. Hossner, "Ammonia Volatilization from Ammonium or Ammonium-Forming Nitrogen Fertilizers," *Advances in Soil Science,* **1** (1985), pp. 123–169.

[12] K. L. Sahrawat, "Control of Urea Hydrolysis and Nitrification in Soil by Chemicals—Properties and Problems," *Plant and Soil,* **57** (1980), pp. 335–352.

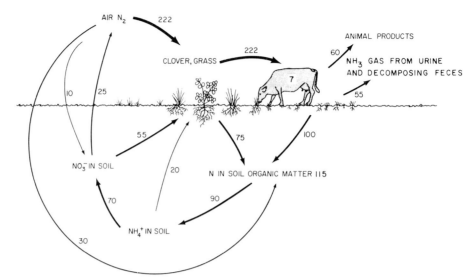

FIGURE 9-5 Example of nitrogen changes, in pounds per acre, in a clover-grass system used as pasture (grazed). Pounds per acre times 1.12 equals kg/ha. (*Source:* Redrawn from data by D. C. Whitehead, 1970, quoted by R. A. Date, "Nitrogen, a Major Limitation in the Productivity of Natural Communities, Crops, and Pastures in the Pacific Area," *Soil Biology and Biochemistry,* **5** [1973], p. 10. Courtesy of Raymond W. Miller, Utah State University.)

some values for nitrogen changes of a grazed clover/grass pasture; Tables 9-3 and 9-4 list amounts of nitrogen (and the relative percentage of change) from various soil sources. For example, if solid nitrogen fertilizer materials are added before planting, enough fertilizer must be added to supply plant needs plus some for the expected losses by leaching, immobilization, denitrification, and volatilization. Knowing the expected losses helps to resolve the question of whether or not the increased fertilizer cost outweighs the convenience of an early application. In Table 9-3 eight particular crop systems are shown. Notice the appreciable nitrogen in the irrigation water regularly added to cotton, and N_2-fixation by soybeans, which eliminates the need for nitrogen fertilizer. Also, notice the variable loss of nitrogen by leaching, which is increased by high fertilizer additions and extensive leaching (when it rains frequently). Mineralization values are variable, and nitrogen turnover time in the two forested areas is much longer; in cultivated crops nitrogen is released faster. Most of the nitrogen in dense forests is recycled. Ammonia volatilization in acidic soils is slight.

9:6 Materials Supplying Nitrogen

The production of nitrogenous fertilizers has increased faster than that of any other type of chemical fertilizer. The principal nitrogenous fertilizer materials and their percentages of nitrogen are given in Table 9-5; the list does not include many specialty and experimental materials.

9:6.1 Ammonia and Aqueous Nitrogen

Anhydrous ammonia, NH₃, is the principal nitrogenous fertilizer, with more than 90% of all nitrogenous fertilizers consisting of ammonia or compounds made from ammonia.

Table 9-3 Examples of Nitrogen Gains, Losses, and Transformations (in kg/ha/yr) for Eight Different Cropping Systems[a]

Description of Change	Grazed Bluegrass (N.C.)	Corn Grain (Ind.)	Soybean Seeds (Ark.)	Wheat (Kansas)	Irish Potatoes (Maine)	Cotton (Calif.)	Loblolly Pine (Miss.)	Douglas Fir (Wash.)
Additions								
Added fertilizer	168	112	0	34	168	179	—	—
Irrigation, floodwater	—	10	—	—	—	50	—	—
Sediments added	10	10	10	6	6	3	11	10
N_2-fixation	—	—	123	—	—	—	8	—
Removals								
Harvested product	38	85	90	36	80	79	12	10
Denitrification	5	15	15	5	15	20	1	1
Volatilization of ammonia	98[b]	—	—	—	—	—	—	—
Leaching loss	—	15	10	4	64	83	1	1
Erosion and runoff	14	16	16	5	15	50	3	2
Recycling process								
Uptake from soil	151	126	120	56	145	127	20	35
Manure from grazing	60	—	—	—	—	—	—	—
Plant residues left	113	41	30	20	65	48	9	25
Mineralization from humus	48	50	15	28	65	48	6	—

Source: Data from M. J. Frissel, ed., *Cycling of Mineral Nutrients in Agricultural Ecosystems,* Elsevier Science Publishing, New York, 1978, pp. 202–243.

[a] A dash means that no measurement was made or the item does not apply to the system.

[b] Losses from voided animal urine and feces as ammonia gas.

Ammonia is a colorless gas containing one atom of nitrogen (atomic weight 14) and three atoms of hydrogen (atomic weight 1), so the percent N in pure NH_3 is:

$$\% \ N = \left(\frac{14}{14 + 3}\right)(100) = 82.35$$

Commercial grade ammonia is 99.5% pure and, therefore, contains 82% nitrogen.

In recent years, the use of anhydrous ammonia has increased tremendously. It is applied by injecting it into the soil behind chisels set at about 12 cm (5 in.) deep and using tractor- or truck-mounted pressure tanks. Anhydrous ammonia has physical properties somewhat like butane or propane gas; it is liquid when under tank pressure but gaseous at atmospheric pressure. Anhydrous ammonia is the *least expensive* nitrogen source fertilizer.

Many safety precautions must be observed when handling anhydrous ammonia. The most serious threat from anhydrous ammonia is *blindness.* Most accidents occur when the anhydrous ammonia is being transferred from one tank to another. Some other precautions are these:

1. Use and wear proper equipment. Use goggles. Anhydrous ammonia will corrode copper, brass, and galvanized parts.[13]

2. Keep flame away from mixtures of 16–25% ammonia because they will burn.

---

[13] W. Mueller, "The Nasty Fertilizer," *Agrichemical Age,* **33** (1989), no. 8, pp. 8, 9, 12.

Table 9-4 Nitrogen Balance for Productive Cropping Systems[a]

Nitrogen Resources	Percentage of Original Source Changed	Amount of Nitrogen	
		lb/a	kg/ha
Initially in soil		3564	3992
Nitrogen available			
Mineralized N (from decomposition of organic matter to yield nitrates)	2% of initial N	71	80
Additional sources			
Fixation by bacteria and/or algae (free + symbiotic)		18	19
Addition from fertilizers		89	100
Precipitation (rain, dew)		7	8
Total available:		185	207
Nitrogen losses			
Denitrification (bacterial change of nitrates to volatile N_2) and ammonia volatilization			
From fertilizer	15% of original fert. N	13	15
From mineralized nitrates	5% of total mineralized	4	4
Removed in crop plants			
Absorbed from fertilizer	55%	49	55
Absorbed from mineralized N	45%	32	36
Absorbed from fixation	50%	9	10
Absorbed from precipitation	100%	7	8
Leaching and runoff		4	4
Total losses:		118	132
Nitrogen immobilized by microbial activity			
From fertilizer N not denitrified	45%	27	30
From mineralized N not denitrified	55%	36	41
From fixation of N	50%	9	10
Total immobilized N:		72	81

Source: Modified and calculated from data by R. D. Hauck, "Quantitative Estimates of Nitrogen-Cycle Processes—Concepts and Review," in *Nitrogen-15 in Soil-Plant Studies,* Panel Proceedings Series of International Atomic Energy Agency, Vienna, 1971, IAEA-PL-341/6, p. 77.

[a] A hypothetical system, averaged from numerous studies using tracer ^{15}N; these values are an approximate guide for "average" soils.

Table 9-5 Principal Nitrogenous Materials

Material	Nitrogen Content (N) (%)
Anhydrous ammonia	82
Urea	45–46
Ammonium nitrate	33.5
Aqua ammonia	20–24
Ammonium sulfate	20–21
Diammonium phosphate (plus 46–53% available P_2O_5)	18–21
Ammonium phosphate sulfate (plus 20% available P_2O_5)	16
Ammonium polyphosphate (having 34–37% P_2O_5)	10–11
Sodium nitrate	16
Potassium nitrate (also containing 44% K_2O)	13
Urea-formaldehyde (slow-release materials)	30–40
Sulfur-coated urea	15–45
Organic products (animal manures, sewage sludge, meat meal, cottonseed meal, fish meal)	1–12

3. Have water available, even in a squeeze bottle; it is the best first-aid material.

4. Keep away from ammonia if it escapes into the atmosphere. Ammonia is very soluble in water; living tissues, especially the eyes, are high in water content. Ammonia causes severe irritation of the eyes, nose, throat, and lungs. The skin can be easily burned; rubber goggles and gloves give some protection. Ammonia that contacts plants kills them because of the high concentration of NH_4OH produced in plant tissues.

5. Store only in pressure tanks that are designed to withstand pressures of at least 17.5 kg/cm^2 (250 lb/in^2) and do not overfill tanks, leaving no room for vapor.

6. Paint all ammonia tanks white to help reflect heat, and store tanks in a cool, shady place.

7. Arrange for an inspection of all tanks at least once a year, and check equipment for worn parts; see that fittings are tight.

Because anhydrous ammonia is difficult to handle, water solutions of ammonia, urea, ammonium phosphate, or other soluble solid-nitrogen materials are being more widely used each year. Urea ammonium nitrate (UAN) solutions are widely used. These solutions can be spread on the soil surface and cultivated or injected into the soil. Anhydrous ammonia is also used to prepare protein feeds for cattle and sheep and as a defoliant to hasten the shedding of cotton leaves to facilitate mechanical harvest.

Anhydrous ammonia is manufactured from atmospheric nitrogen and natural gas to supply hydrogen. Energy shortages have increased the proportion of ammonia used compared to other nitrogenous forms because it is the *first industrial product* of the conversion of atmospheric nitrogen to usable form. Other nitrogen fertilizers require further processing and utilization of more energy. Urea, for similar reasons, has become increasingly the most popular *solid* nitrogenous material on the market.

9:6.2 Urea, Ammonium Nitrate, and Ammonium Sulfate

Urea, $CO(NH_2)_2$, is a synthetic organic fertilizer, now cheaper per pound of N than any other solid nitrogenous material; it contains 45–46% N. Unlike anhydrous ammonia or ammonium salts, urea cannot be absorbed by plants until the nitrogen it contains is converted by the enzyme urease to ammonium.

Urea is readily soluble and leachable when it is first applied to the soil, but when it is changed to ammonium by the action of urease, it is held as exchangeable ammonium cations by clay and humus in a form readily available to plants. Under favorable warm temperatures and moist soil conditions, urea can be hydrolyzed to ammonium carbonate and then by bacterial action to nitrate within less than a week.

Biuret, a manufacturing contaminant of urea, is one possible hazard because it is toxic to sensitive plants, such as tobacco, in concentrations of more than 1%.

Urea is a popular nitrogen fertilizer because it is usually the cheapest solid nitrogenous fertilizer and is readily soluble in water, making it a convenient material for application in sprinkler water, as sprays, or as aqueous nitrogen solutions.

Ammonium nitrate, NH_4NO_3, is a good, relatively cheap source of solid nitrogen fertilizer, analyzing 33.5% N. Half of the nitrogen content is in the ammonium form, and the other half is in the nitrate form. When added to a cool soil, the nitrate ions are immediately available for plant use and are mobile in the soil. The ammonium cations are adsorbed on the exchange sites in the soil and are available to plant roots although they are not mobile. The ammonium ions nitrify rapidly to nitrate.

Ammonium sulfate, $(NH_4)_2SO_4$, is manufactured mostly from recovered coke-oven gases; it contains 21% N. Because of its relatively high cost, it is less popular than ammonium nitrate and urea. For use on rice, however, ammonium sulfate is one of the best forms of nitrogenous fertilizers because the nitrogen, as the ammonium ion, is all potentially available to the plant. If the nitrate instead of the sulfate form were used, under the anaerobic conditions of a flooded rice field, the nitrate would be denitrified. This means only about half of the N (the ammonium ions only) of ammonium nitrate can be utilized, compared to all of the nitrogen in the ammonium sulfate form. Recent Environmental Protection Agency regulations on air pollution control have resulted in reclamation of considerable ammonia and sulfuric acid. These waste products may result in more price-competitive ammonium sulfate coming into the fertilizer market.

9:6.3 Controlled-Release Nitrogen Fertilizers

Crop recoveries of the usual fertilizer nitrogen are commonly 40–70% of that added to soil. In porous (sandy) soils and high rainfall areas, recovery can be even less because of more leaching. A less soluble, slow-release fertilizer would reduce some of the nitrogen losses responsible for low nitrogen recoveries. Unfortunately, slow-release materials, which do reduce leaching losses, also dissolve too slowly at the time of high nitrogen need (in rapid plant growth periods, such as just before the formation of the grains for cereals). Slow-release fertilizers are expensive, perhaps two or three times the cost of regular nitrogen fertilizers. A satisfactory solution for the controlled-release nitrogen fertilizer problem is not yet in sight, but some useful "compromise" materials are available. These materials have proved useful on many slower-growing grasses, permanent pastures, and greenhouse crops (Table 9-6).

Sulfur-coated urea is one of the more successful of the controlled-release nitrogenous fertilizers. Granular urea is spray-coated with molten sulfur; then a wax coating and later a clay film coating are applied to improve handling characteristics. Water solubility of *untreated urea* is 100% within a few minutes; in contrast, from one formula of *sulfur-coated urea,* only 1% of the coated urea dissolved about every 5 days.

Sulfur-coated urea pellets are *not* effective on flooded rice fields because an insoluble coating of iron sulfide (FeS_2) forms around each pellet, making them so slowly soluble that they are almost valueless.

Various nitrogen materials are made into polymers using urea and formaldehyde (or isobutylaldehyde) or are coated by polymer material. These produce materials of high percentage nitrogen and various rates of release. Materials such as Nutricote,® described in Table 9-6, can have almost any nutrient composition and yet the controlled release rates can range from relatively short to long time periods. Such materials are well adapted to nursery and greenhouse plants as well as turf and pastures. Release rates of several slow-release nitrogen fertilizer materials are shown in Fig. 9-6. A study using Nutricote™ and Osmocote™ materials grew better chrysanthemums than the treatments fertilized with liquid fertilizer.[14] (The liquid fertilizer treatment had not had early phosphorus added when it was apparently needed.) Slow-release fertilizers should be selected carefully to fit the nutrient release rate needed by each crop. Slow-release fertilizers are also more expensive and cost must be considered.

---

[14] James W. Boodley, "New Japanese Fertilizer Shows Dramatic Results," *Florists' Review,* **168,** no. 4354, May 14, 1981, pp. 10–11.

Table 9-6 Some Slow-Release Nitrogen Sources and Mixed Fertilizers and Their Properties

Material	Origin	Properties	Comments
1. Natural organic sources	Manures, crop residues, sewage, sludge, composts	Low N (1–3%); released by microbial decomposition; released more rapidly in warm, moist, fertile soil	Expensive to store and add; may have toxic salts, weed seeds, and heavy metals; relative cost is 5–6 × [a]
2. Polymers a. Isobutylidene diurea (IBDU)	Polymerization of urea with isobutylaldehyde	Hydrolyzes in water to urea; smaller particles hydrolyze fastest; slow in dry soil	Most is made available within 80–100 days; 8–20 mesh sizes suggested for turf; R.C. = 3–4 × [a]
b. Polyform UF	Polymerization of urea with formaldehyde	Two-thirds water soluble, 1/3 water insoluble; requires microbes for dissolution	A fast-release nitroform material; fastest in warm, moist conditions; R.C. = 3–4 × [a]
c. Ureaform ™ and Uramite ™	Polymerization of urea with formaldehyde	One-third water soluble, 2/3 water insoluble; requires microbes for dissolution	Slower release nitroform materials; some residues even after one year sometimes; R.C. = 3–4 × [a]
d. Nutricote ™ type 100, 140, 180, 270, 360	High-quality mixed fertilizer coated with various polyolefin resins and release-controlling additive	Release rate is increased by higher temperature; type 100 and other numbers indicate the days during which controlled nutrient release occurs	Is used for supplying all three nutrients: N, P, K; not affected by pH, water content, or microbes
e. Osmocote ™	Contains mixed fertilizer in an organic resin coating	Nutrients released as water vapor enters coating; nutrients diffuse out or burst coating	Coming in varying mixtures with N only or with various proportions of N, P, and K
3. Sulfur-coated urea (SCU)	Urea granules coated by molten sulfur, then wax, then clay conditioner	Releases urea as microbes decompose sulfur coating or water penetrates cracks in coating; about 20–30% is released in water at 100°F during 7 days; acidifies soil	Least uniform release rate of slow-release materials; depends on sulfur thickness; can be crushed by equipment

Sources: (1) Richard L. Duble, "Nitrogen, Fertilizers for Turfgrass," Parts One and Two, *Turf-Grass Times,* **14** (1978), no. 1, pp. 14–23, and **14** (1978), no. 2, pp. 18–19. (2) James W. Boodley, "New Japanese Fertilizer Shows Dramatic Results," *Florists' Review,* **168**, no. 4354, May 14, 1981, pp. 10–11.
[a] Relative cost (R.C.) compared to cost for urea. ("4 ×" means 4 times more expensive than urea alone.)

9:7 Soil Phosphorus

Phosphorus is the second key plant nutrient; it is the second most often deficient nutrient. Phosphorus is an essential part of nucleoproteins in the cell nuclei; these molecules carry the inheritance characteristics of living organisms. Phosphate to phosphate bonds are the major energy storage and energy transfer bonds as ATP (adenosine triphosphate) and ADP (adenosine diphosphate). In its many compounds phosphorus has roles in cell division, in stimulation of early root growth, in hastening plant maturity, in energy transformations within the cells, and in fruiting and seed production. For people and other animals eating the plants,

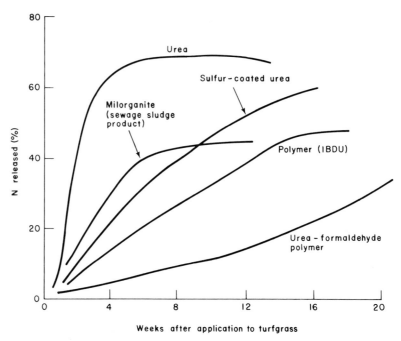

FIGURE 9-6 Examples of rates of nitrogen release from various slow-release nitrogen fertilizers added to turfgrass and compared to urea.
(*Source:* Redrawn from Richard L. Duble, ed., "Nitrogen Fertilizers for Turfgrass," *Turf-Grass Times,* **14** [1978], no. 1, pp. 14, 15, 21, 23.)

phosphorus is critical for growth of bones and teeth, which are mostly calcium phosphates (Fig. 9-7).

Plants differ in their ability to compete for soil phosphorus. Young plants absorb phosphorus rapidly, if it is available. Winter wheat absorbs about 70% of its phosphorus between tillering and flowering. A cold, wet spring usually results in a retardation of plant growth, often because of inadequate phosphorus absorption. The peak phosphorus demand for corn is just 3 weeks into the growing season, and the root system is still relatively small (Fig. 9-8).[15]

9:8 The Phosphorus Problem

Plants use perhaps one-tenth as much phosphorus as they do nitrogen, yet getting adequate phosphorus into plants is a widespread problem. Why? The answer is complex; in a short generalization, *soil phosphates are insoluble.* Nothing is *insoluble,* so this statement really means that the soil forms are of very low solubility. Added soluble phosphates will readily combine with cations in soil solution to form low-solubility substances. Some examples of soil solution concentrations of some nutrients are given in Table 9-7. The data show that the plant needs about one-fifth as much phosphorus as nitrogen and potassium, but the concentration of phosphates in solution is only about one-twentieth as high—or even less—as the concentrations of nitrogen and potassium.

Data of this type caused scientists to wonder if all or most of the nutrients used by the root would "flow" to the root in the water absorbed by the roots. This carrying of nutrients to the root in the absorbed water is called **mass flow.** Scientists soon calculated that the

---

[15] Bob Coffman, "Roots: Feed Them Where They Are When They Want It," *Farm Journal,* (Apr. 1978), pp. J-1, J-2.

FIGURE 9-7 The forage available to these cows in Minnesota does not contain sufficient phosphorus because the soil is deficient in this mineral. The result is weakened animals that chew on bones or on the bark of trees. (*Source:* Extension Service, University of Minnesota.)

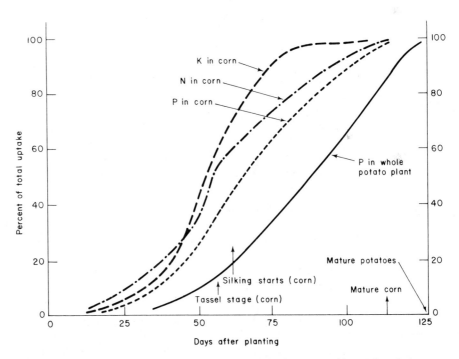

FIGURE 9-8 Uptake rates of phosphorus by corn and potato, with uptake of nitrogen and potassium by corn shown for comparison. A straight line diagonally from 0 to 100% would indicate a uniform rate of uptake with time. The S-shaped curve is typical of plant growth: slow at both the beginning and end and rapid in midgrowth. Variable rates for different species of plants and different environments will alter the curve shapes.

Table 9-7 Examples of the Soil Solution Concentrations of Macronutrients[a]

| Nutrient | Fertile Indiana Alfisol | | Range in Average Soils (mg/L) | Needed for 9500 kg of Corn/ha (kg/ha) |
	Total Available (kg/ha–20 cm)	Amount in Solution (mg/L)		
NO_3^-	200	60	6–1240	190
NH_4^+			1.8–36	
$H_2PO_4^- + HPO_4^{2-}$	100	0.8	0.1–1.9	40
K^+	400	14	3.9–39	195
Ca^{2+}	6000	60	2.0–100	40
Mg^{2+}	1500	40	1.2–60	45
SO_4^{2-}	100	26	9.6–960	22

Source: Combined and calculated data from Stanley A. Barber, *Soil Nutrient Bioavailability,* Wiley-Interscience, New York, 1984, pp. 94–95.
[a] Values in the table are amounts of the element, not of the ion.

amount of phosphorus dissolved in the soil solution and carried to the root in the absorbed water was not enough phosphorus to supply even low needs by plants. It is now believed that most phosphorus and potassium are supplied to root surfaces by **diffusion** rather than by mass flow (Table 9-8). Supplying adequate phosphorus requires a mechanism other than simple mass flow. However, movement by diffusion is very slow. Movement by diffusion is only approximately 0.02 to 0.1 mm per hour for HPO_4^{2-}. Perhaps soil water "turbulence" by temperature gradients and flowing water that isn't adsorbed by all the roots that it passes may help explain the mechanism. Whatever the facts, *low solubilities of soil phosphates are the major problem in getting and keeping soil phosphates available to plants.* A phosphorus cycle is shown in Fig. 9-9.

9:8.1 Inorganic Phosphorus

Making phosphorus available to plants is critical because (1) the supply of phosphorus in most soils is low and (2) the phosphates in soils are not readily available for plant use. The

Table 9-8 Percentage of Each Macronutrient Supplied to Corn from a Fertile Alfisol Soil by the Three Supply Mechanisms

| Nutrient | Approximate Percentage of Nutrient Supplied by: | | |
	Root Interception	Mass Flow	Diffusion
Nitrogen	1	79	20
Phosphorus	3	5	92
Potassium	2	18	80
Calcium	150	375	0
Magnesium	33	222	0
Sulfur	5	295	0

Source: Calculated percentages from data by Stanley A. Barber, *Soil Nutrient Bioavailability,* Wiley-Interscience, New York, 1984, p. 96.

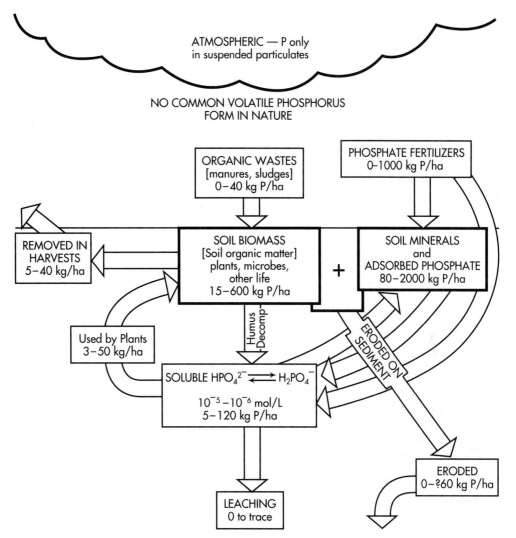

FIGURE 9-9 The phosphorus cycle showing gains, losses, and transformations in the soil. Depositions to form phosphate deposits in ocean bottoms not shown. (Courtesy of Raymond W. Miller, Utah State University.)

total phosphorus in an average arable soil is approximately 0.05% by weight (400–2000 kg/ha), of which only an infinitesimal part is available to the plant at any one time.

The original natural source of phosphorus is the mineral apatite, a calcium phosphate of low solubility and typical formula $Ca_5(PO_4)_3F$. Apatite (**rock phosphate**) and small amounts of other phosphates of iron, aluminum, manganese, and zinc comprise the inorganic phosphate minerals. These rock phosphates, when powdered, can be used directly as fertilizers; unfortunately, all mineral forms in nature have low solubilities.

The soluble ion $H_2PO_4^-$ (ortho phosphate) rapidly reacts in soil to form insoluble phosphates, a process loosely termed **phosphate fixation** (precipitation and adsorption). In acid soils, the phosphate ions react with soluble iron and aluminum ions to form insoluble phosphates, also phosphates adsorb to surfaces of insoluble iron, aluminum, and manganese hydrous oxides. In alkaline soils low-solubility calcium triphosphate is formed; soluble phosphate ions also adsorb on solid calcium carbonate surfaces. Phosphorus is most available near pH 6.5 for mineral soils and about pH 5.5 for organic ones (Fig. 9-10). There is no

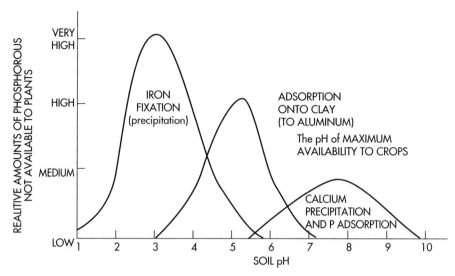

FIGURE 9-10 Soil phosphorus is precipitated and adsorbed to soil minerals, or made less available, by the formation of less-soluble phosphates of iron, aluminum (from clays), and calcium. At the soil pH below 5.5 (acidic), both iron and aluminum precipitate phosphorus. At a pH above 7.0 (basic), calcium precipitates phosphorus. Maximum phosphorus availability is at a pH of 6.5 for mineral soils and about 5.5 for organic soils and Oxisols. Addition of lime (a basic substance) to soils of pH 5.5 or less improves phosphorus availability to crops. (*Source:* George D. Scarseth, *Man and His Earth,* Iowa State University Press, Ames, 1962, p. 143.)

efficient mechanism in the soil to retain $H_2PO_4^-$ or HPO_4^{2-} ions in large quantities as exchangeable anions. Thus, much of the phosphorus used by plants, other than that from applied phosphate fertilizers, is believed to come from organic phosphates released by decomposition of organic matter.

The description of phosphate reactions from an added superphosphate fertilizer pellet is informative. The monocalcium phosphate (MCP) in the pellet is water soluble and dissolves readily in soil water. The soluble $H_2PO_4^-$ ion moves by diffusion outward from the concentration area of the dissolving pellet. The solution of phosphate may have a pH near 0.6–1.4, with high concentrations of both P (3.4–5 molar) and Ca (1 molar). If the soil is *calcareous* with a high pH, various low-solubility calcium phosphates form as below:

Phosphate Name	Formula	Solubility in Water
Monocalcium phosphate (MCP)	$Ca(H_2PO_4)_2 \cdot H_2O$	1.8 g/100 mL
Dicalcium phosphate (DCPD)	$CaHPO_4 \cdot 2H_2O$	0.0316 g/100 mL
Tricalcium phosphate (TCP)	$Ca_3(PO_4)_2$	0.002 g/100 mL
(There are many additional phosphates of calcium)		

As the phosphate diffuses outward from the pellet, some of it precipitates, mostly as DCPD. Some of the $H_2PO_4^-$ ions diffuse farther and are adsorbed to solid lime and to iron or aluminum hydrous oxides.

If the soil is *noncalcareous,* the soluble phosphate forms various insoluble aluminum phosphates, such as taranakites [$H_6K_3Al_5(PO_4)_8 \cdot 18H_2O$ and $H_6(NH_4)_3Al_5(PO_4)_8 \cdot 18H_2O$] and Al-Fe phosphates [$H_8K(Al,Fe)_3(PO_4)_6 \cdot 6H_2O$], as well as some calcium phosphates.

The area of the fertilizer pellet has three zones: (1) the dissolving pellet, (2) the second zone of concentrated phosphate-causing precipitation, and (3) the outer zone, where phosphate adsorption is dominant.

A Gilat soil in Israel had applications of organic materials and/or mineral fertilizers for 24 years (1961–1984). The additions caused large differences in phosphate reserves (Fig. 9-11). The control (= soil with no phosphorus additions) had enough phosphorus for about 320 cropping years if each crop needed 40 kg of phosphorus and half of that was left in the roots in a 120-cm (47-in.) soil depth. This assumes that all depths to 120 cm could supply phosphorus equally well. It is evident that the 24 years of additions have nearly *doubled* the phosphorus content in the surface soils. If most of the phosphorus was supplied by the top 30 cm (12 in.) of soil, phosphorus in the nonfertilized soil would be completely exhausted in 80 years. Supplies of phosphorus in soils are not usually enormous, and the low solubilities of phosphate compounds aggravate the problem. Phosphate is a mined resource that has finite limits. Some persons already say that phosphorus reserves may be unable to supply cheap and plentiful amounts even for the next century.

In anaerobic conditions (e.g., paddy rice), phosphates are more soluble than in aerated soils. When paddy rice is flooded, much of the iron becomes soluble (as ferrous iron) in anaerobic conditions and some iron phosphates may be solubilized.[16] Interestingly, the solubilized iron reprecipitates when the soil is drained (and reaerated). This fresh iron precipitate is somewhat disorganized (amorphous) and in one study did not adsorb added phosphate as strongly or as much as did the iron compounds before they solubilized in the anaerobic soil. A drained soil may, within 2 or 3 years, change back to the unwanted high phosphorus adsorption that it had before it was initially flooded.[17]

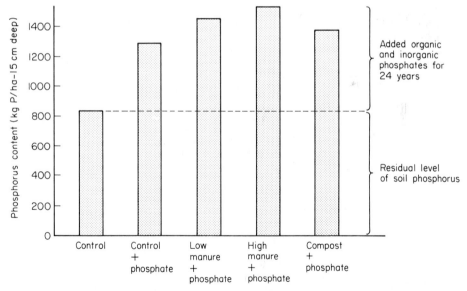

FIGURE 9-11 Levels of soil phosphorus before and after 24 years of application of manures, compost, and inorganic phosphate fertilizers. (*Source:* Unpublished data by Raymond W. Miller and Amos Feigin, 1985.)

[16] R. A. Khalid, W. H. Patrick, Jr., and F. J. Peterson, "Relationship between Rice Yield and Soil Phosphorus Evaluated under Aerobic and Anaerobic Conditions," *Soil Science and Plant Nutrition,* **25** (1979), no. 2, pp. 155–164.
[17] R. N. Sah and D. S. Mikkelsen, "Sorption and Bioavailability of Phosphorus during the Drainage Period of Flooded–Drained Soils," *Plant and Soil,* **92** (1986), pp. 265–278.

9:8.2 Organic Phosphorus

Phosphorus is a part of complex plant proteins (DNA, RNA) and has a unique role in the adenosine triphosphate (ATP) energy storage bond. *Phosphatase* (an enzyme splitting off phosphates from complex organic molecules) in the plant can split off $H_2PO_4^-$ from organic phosphates. Some of the $H_2PO_4^-$ is exuded by the plant roots (along with some phosphatases). The phosphatases (also produced by bacteria and other organisms) continue to hydrolyze organic phosphorus from dead organic residues, making it available for absorption into living plants and microorganisms. Other enzymes active in the decomposition of organic matter release various organic molecules containing phosphates; these phosphate compounds are more soluble than either the complex organic phosphates or most phosphate minerals. Phosphates moved from soils, but which are not attached to erosion sediment, are mostly these soluble organic phosphates, which may comprise as much as half or more of the soluble soil phosphorus.

It is not known conclusively if plants absorb the organic phosphates directly or whether the phosphatases found in normal soils split off the phosphates before the plants absorb them. Soils high in organic matter (4–5%) and with conditions favorable to microorganisms (which speed decomposition) have better supplies of organic phosphates for plant uptake than do soils low in organic matter.

The rate of turnover of organic phosphorus is rapid in conditions favorable for microorganisms. Although critical concentrations are not well established, it is believed that as the carbon : organic-P ratio is about 200 : 1 or narrower, phosphorus is readily released (mineralized) into the soil solution. If ratios are 300 : 1 or wider (about 0.2% P in organic matter), the microorganisms use most of the phosphorus, immobilizing it into their cells instead of releasing it for plant use.

9:8.3 Managing Soil Phosphorus

Soil phosphorus availability is low in many sandy low-humus soils. Oxisols and other soils high in sesquioxide clays, iron oxides, and aluminum oxides often have high phosphorus adsorption capacities, rapidly adsorbing added soluble phosphates.

Soil pH, because of its influence on the presence and solubility of calcium, iron, and aluminum, and because of its effect on bacterial growth, greatly influences available phosphorus. Optimum phosphorus availability in mineral soils is believed to be near pH 6.5, although some scientists question this conclusion. For example, on cotton at soil pH between pH 6.0 and 7.0, 11.2 kg/ha (10 lb/a) of fertilizer phosphorus was as efficient as double that amount at a soil pH of 5.0.

Phosphate deficiencies are usually remedied by application of phosphate fertilizers. Because phosphorus precipitates and adsorbs, phosphates are often placed in a band about 5 cm (2 in.) to one or both sides of the seed and 5 cm below it to minimize contact with the soil but to be close to the first young roots. Even with these precautions, only about 20% of the phosphorus added is used by a crop during the season the phosphorus was applied.

Although excess phosphorus is mostly retained in soil, excess additions can cause problems, such as zinc deficiency. In susceptible plants, such as corn, beans, and flax, too much soluble phosphate causes zinc deficiency. Because zinc phosphate should still be soluble enough to supply adequate zinc, the problem may be *inside* the plant or in interactions not clearly understood.

For maximum phosphorus efficiency:

1. Maintain the soil between pH 6.0 and 7.0, if practical.

2. Promote as much relatively fresh organic matter in the soil as is economically practical, to release phosphorus as it decomposes.

3. Band phosphorus fertilizers for row crops. Broadcast some or all phosphate fertilizer and incorporate it if the crop is not planted in rows (e.g., pastures) or if the soil phosphorus level is very low (roots need some phosphorus throughout the soil for good growth).

4. Expect to need more phosphorus on aerobic soils than on anaerobic ones.

5. Anticipate less root absorption of phosphorus in cold weather.

Some examples of phosphorus needs, gains, and losses are shown in Table 9-9. Notice that much less phosphorus is needed than was needed of nitrogen, that almost no phosphorus is lost by leaching, and that relatively little data are given on the mineralization of phosphate (See Fig. 9-9, the P cycle). The mineralization data shown suggest that the phosphorus available from mineralization is usually inadequate to meet crop needs.

9:9 Materials Supplying Phosphorus

The phosphorus ores used for fertilizer manufacture in the United States come mostly from Florida, where ancient oceans have left millions of years' accumulation of marine shell organisms. Extensive deposits of apatite, a phosphorus-bearing mineral, exist in the western United States and are used to some extent as a fertilizer. Rock phosphate ore is mined and ground to make the commercial **rock phosphate;** when mixed with sulfuric acid, it is made into **superphosphate,** 8–9% P, (Table 9-10). If rock phosphate is mixed with phosphoric acid, it makes **triple superphosphate,** 20–22% P (45–50% P_2O_5).

Table 9-9 Examples of Phosphorus Gains, Losses, and Transformations (kg/ha/yr) for Eight Different Cropping Systems[a]

Description of Change	Grazed Bluegrass (N.C.)	Corn Grain (Ind.)	Soybean Seeds (Ark.)	Wheat (Kansas)	Irish Potatoes (Maine)	Cotton (Calif.)	Loblolly Pine (Miss.)	Douglas Fir (Wash.)
Additions								
Added fertilizer	24	30	19	13	101	14	—	—
Irrigation, floodwater	—	tr	—	—	—	tr	—	—
Sediments added	tr	—	tr	tr	tr	tr	tr	tr
Removals								
Harvested product	—	15	10	7	10	13	1	2
Leaching loss	—	tr	tr	tr	5	tr	tr	—
Erosion, runoff	tr	3	3	3	tr	1	1	—
Recycling processes								
Uptake from soil	20	22	13	10	16	19	2	6
Manure from grazing	15	—	—	—	—	—	—	—
Plant residues left	—	7	3	3	6	6	1	5
Mineralization from humus	5	7	—	3	—	6	—	—

[a] A dash means that no measurement was made or does not apply to the system; *tr* means that a trace amount was measured.

Table 9-10 Phosphate Fertilizers in Common Use in the United States

Material	Total P[a] (%)	Available P[a] (%)
Rock phosphate (mined phosphate minerals)	13–17	0.8–2.2
Superphosphates		
Ordinary (sulfuric acid treated)	9	9
(85% of P is water soluble)		
Triple (phosphoric acid treated)	20–22	20–22
(most of P is water soluble)		
Concentrated (about same as triple)	24	24
Superphosphoric acid (polyphosphoric acid)	30–33	30–33
Wet-process phosphoric acid, crude	13	13
(used in many solid mixed fertilizers)		
Basic slag (also acts as a lime)	3.5–5.2	(?) 2–3
Bone phosphate (steamed bonemeal)	10–13	(?) 2–3

[a] To obtain P_2O_5 equivalent, multiply P content by 2.29. "Available phosphorus" is that which is soluble in a dilute ammonium citrate solution.

The average composition of granulated ordinary superphosphate is:

	Percent
Gypsum ($CaSO_4$)	48
Monocalcuim phosphate, $Ca(H_2PO_4)_2$, water soluble	30
Dicalcium phosphate, $Ca_2(HPO_4)_2$, not water soluble	9
Iron oxides, aluminum oxides, and silica	9
Tricalcium phosphate, $Ca_3(PO_4)_2$, slowly soluble	2
Moisture	2
Total:	100

Other phosphorus fertilizer materials are **diammonium phosphate** (46–53% P_2O_5 and 18–21% N), **monoammonium phosphate** (48% P_2O_5 and 11% N), **ammonium phosphate sulfate** (20% P_2O_5 and 16% N), and **basic slag** (approximately 10% citrate-soluble P_2O_5). A fertilizer that is fast finding favor is ammonium polyphosphate, which in liquid forms analyzes 10-34-0 and 11-37-0 (Table 9-11).

When the price of sulfur, and therefore of sulfuric acid, increases, there is renewed interest in using nitric acid to make several kinds of **nitric phosphates,** the most common of which contains 20% P_2O_5 and 20% N.

9:10 Mixed Nitrogen–Phosphorus Fertilizers

Many of the solid fertilizers used are mixtures (formulations) containing two or three of the elements nitrogen, phosphorus, and potassium. Many also contain sulfur. These are usually made from high-analysis liquids such as superphosphoric acid and gaseous ammonia, resulting in fertilizers with high concentrations of several nutrients. Most mixtures are nitrogen-phosphorus ones that can be mixed with naturally occurring mined potassium compounds (KCl or K_2SO_4) to obtain a variety of grades of fertilizers (Table 9-11).

Ammonium phosphates include monoammonium phosphate (MAP) and diammonium phosphate (DAP). Although MAP in water has a pH of about pH 6.5, the DAP will have a

Table 9-11 Multinutrient Fertilizers in Common Use in the United States

Material	Comments	Grade
Ammonium polyphosphate	Used in liquid fertilizers	11-37-0 and 10-34-0
Ammonium phosphates	Water-soluble, good for fast-growing crops	11-48-0 to 18-46-0 to 21-53-0
Ammonium phosphate nitrate	Ammonium nitrate, anhydrous ammonia, and phosphoric acid; water soluble	30-10-0 27-12-0 22-22-0
Urea-ammonium phosphate	In developmental stage, damaging to germinating seeds	25-35-0 to 34-17-0
Ammoniated superphosphate	Well-known material	4-16-0 to 8-32-0
Nitric phosphates	No established advantage over other materials	12-35-0 to 17-22-0
Ammonium thiosulfate	Used in liquid fertilizers	12-0-0-26S

Source: Tabulated from details in U. S. Jones, *Fertilizers and Soil Fertility,* Reston Publishing, Reston, Va., 1979, pp. 151–166.

pH between 9 and 10. In high concentrations and/or poorly buffered soils, pH above 9 can cause evolution of appreciable amounts of NH_3 gas.

There are two ways of avoiding extreme fixation of the phosphorus in soluble solid fertilizers. Granulation provides the first method and banding the second.

—R. L. Cook and B. G. Ellis

Questions

1. Is an element of low solubility likely to be supplied to roots mostly by diffusion or mostly by mass flow? Explain.
2. Draw a nitrogen cycle including these "boxes" or "stages" in it: N_2 fixation, mineralization, nitrification, denitrification, and leaching.
3. How much N_2 is fixed by plants and by "free-fixers"?
4. About how much nitrogen is released during decomposition of soil humus each season?
5. Write the equation illustrating acidification produced by nitrification.
6. About how many kilograms of lime are neutralized by the acidity from nitrification of 200 kg of anhydrous ammonia?
7. (a) How fast is nitrification: completed in hours, or days, or weeks? (b) Explain why nitrification rates differ in different systems.
8. (a) Is leaching loss from soil and fertilizer nitrogen a large loss, typically? (b) In what conditions will the losses be large, and when will the losses be small?
9. (a) Why are nitrification inhibitors used? (b) How do they work? (c) Are they effective and profitable?

10. Tabulate the necessary conditions for each of these: (a) ammonia volatilization, and (b) denitrification losses of nitrogen.

11. How might so-called aerobic clayey soils still have some denitrification at various times during growth?

12. Briefly tabulate some typical "N additions" and "N losses" to show how a soil's nitrogen balance changes. Remember, nitrogen losses include removal in crops.

13. Why is anhydrous ammonia the most used nitrogen fertilizer?

14. Urea has several important properties related to its use in soils. Discuss: (a) urea's solubility and mobility in soils, (b) the requirements for conversion of urea to plant-usable forms, and (c) the products of initial urea breakdown.

15. (a) Why are nitrogen fertilizers only 40–70% efficient? (b) What happens to that part not used?

16. (a) What are the purposes for using controlled-release fertilizers? (b) Indicate how several of these fertilizers work. (c) Why aren't these materials used more?

17. Briefly discuss the problem of adequate phosphate uptake by plants as influenced by (a) solubility, (b) mass flow, (c) mobility in soil, and (d) cold temperatures.

18. Explain why plant efficiency of phosphorus fertilizers is only about 10–30%, much lower than efficiency of nitrogen fertilizers.

19. (a) What happens to phosphorus fertilizer not used by the plant? (b) About how much of that phosphorus left might be used by the next year's crop?

20. (a) How much phosphorus is lost by leaching? (b) How much phosphorus is lost by erosion?

21. How important is soil humus in furnishing plants with (a) nitrogen, and (b) phosphorus?

22. Explain why phosphorus is more available at pH 6–7 than at pH values of 4 or 8.

23. Discuss briefly the effect of anaerobic conditions on phosphate solubility in soils.

24. Tabulate four of the most important management practices that will help keep the phosphorus in soils most available.

25. (a) How do superphosphate and monoammonium phosphate (MAP) differ? (b) Are both of these materials good phosphorus fertilizers? (c) What is concentrated superphosphate?

26. Draw the P cycle and emphasize some of the major differences of the P cycle and the N cycle.

10

Potassium, Sulfur, and Micronutrients

The trace element content of a soil is dependent almost entirely on that of the rocks from which the soil parent material was derived and on the processes of weathering . . .

—Fireman E. Bear

10:1 **Preview and Important Facts**

PREVIEW

Potassium is the third most commonly added fertilizer nutrient. In soils potassium is released from weathering minerals and from cation exchange sites. Primary minerals containing potassium have very low solubility, so most potassium available to plants during a growing season is supplied from the soil's *exchangeable* potassium reservoir. Potassium deficiencies are most common in leached soils (humid climatic areas) where soluble and much of the exchangeable potassium has been leached out. Sulfur, in contrast to potassium, is supplied from decomposing organic matter and from several moderately soluble minerals, but not generally from the anion exchange capacity. Sulfur is less often deficient than is potassium, but soils having a deficiency are increasing.

Seven micronutrients are essential to plants. Boron and molybdenum are least available in strongly acidic soils because their soluble forms are leached out. In contrast, copper, iron, manganese, and zinc are *least available in alkaline (basic) soils,* mostly because of the low solubility of the hydroxides and carbonates they form. Chlorine, as a nutrient, is almost never deficient.

Generally, fertilizer potassium and sulfur are easily added to correct deficiencies. Iron and, to a lesser extent, zinc are the most difficult element deficiencies to correct by added fertilizers. Iron substances precipitate in soil solution and have very low solubilities.

IMPORTANT FACTS TO KNOW

1. The climatic areas most likely to have either adequate or deficient amounts of each of these: potassium, sulfur, boron, iron, and zinc
2. The high relative solubilities of K^+, $SO_4{}^{2-}$, and H_3BO_3 forms but the *low mobility* in soils of K^+, because K^+ is a *cation*
3. Sulfur release during decomposition of humus is similar to the reactions involving nitrogen release from humus
4. The amounts of soil iron and zinc available to plants are more a function of the soil's pH and organic substances than of the kinds of iron and zinc minerals in the soil. Correcting the problem of inadequate available iron is more difficult than correcting deficiencies of zinc, copper, or manganese
5. The choice of fertilizers purchased to supply plant nutrients is based mostly on costs rather than other factors for potassium and sulfur. For micronutrients, the method of application and its effectiveness are often more important than cost in selecting the fertilizer material to use
6. The most used potassium and sulfur fertilizer materials
7. The soil chemistry of and value of metal chelates as fertilizers
8. The kinds of fertilizer materials used to supply micronutrients for plant use

▮▮▮▮ *10:2* Soil Potassium

Potassium and sulfur are a contrasting pair, even more so than are nitrogen and phosphorus. Soil humus supplies little potassium but may be a major sulfur source. **Potassium, K^+,** is a very soluble cation in solution, yet it moves only slowly in soils. In contrast, sulfur forms the soluble and quite mobile sulfate ion ($SO_4{}^{2-}$).

Potassium is an enigma. First, it is one of the elements whose usual chemical compounds are highly soluble, yet its soil mineral forms—micas and orthoclase feldspar ($KAlSi_3O_8$)—are only very slowly soluble. Second, it is the most abundant metal cation (often up to 2 or 3% of dry weight) in plant cells, but soil humus furnishes very little –potassium during decomposition. Third, decomposition of *fresh* plant residues supplies whatever potassium the plant absorbed for growth. Because the potassium occurs in plants only as a mobile soluble ion, K^+, rather than as an integral part of any specific compound; yet it is known to affect cell division, the formation of carbohydrates, translocation of sugars, various enzyme actions, the resistance of some plants to certain diseases, cell permeability, and several other functions. Over 60 enzymes are known to require potassium for activation. It is particularly important in plant control of water (regulation of osmosis in the plant).

10:2.1 Forms of Soil Potassium

The *total* amount of potassium found in most soils is sufficient to last several decades, even centuries, yet the low solubility of soil micas and feldspars (the soil minerals that contain potassium) supply only very small amounts of it during a growing season. Soil humus also supplies very little potassium for plant use because the soluble K^+ in decaying organic matter is rapidly leached into soil solution and then onto cation exchange sites on clay particles and humus.

Most K^+ used by plants in a given season comes from exchangeable K^+ and soluble K^+. In neutral and basic soils soluble K^+ alone may be adequate to supply modest plant needs. In most soils, particularly acid ones, **exchangeable K^+ is the major source of potassium to plants**. The exchangeable K^+ accumulates as the mica and feldspars weather and as potassium in plant residues is released into the soil solution. Although some soils may contain as much as 2% total potassium (90,000 kg/ha–30 cm or about 81,000 lb/a-ft), high-yielding crop plants depend on the exchangeable supply, which often is a *small* reservoir of readily available potassium. Exchangeable plus soluble potassium in the top 15 cm (6 in.) may be less than 100–200 kg/ha (89–178 lb/a) in many acidic soils, a level that is inadequate or marginal for plant growth. About 170–200 kg/ha (150–180 lb/a) is about the minimum amount considered necessary for a good cultivated crop.

Soluble or exchangeable potassium may be taken up in excess amounts by plants; this is called **luxury consumption.** Excess uptake may reduce Mg absorption in the plant; if the uptake results from overapplication of fertilizers, it is an expensive waste. Other than being absorbed by plants, soluble K^+ can be (1) immobilized into microbe bodies, (2) lost in leaching waters, and (3) entrapped between layers of hydrous mica (illite) and similar clays during drying and is similar to ammonium entrapment between clay layers **(potassium fixation).**

In Table 10-1 potassium losses, gains, and changes are shown for several crop systems. Notice the relatively high erosion and leaching losses. Although no potassium release is shown for mineralization of humus, the addition of undecomposed organic materials to soil does replace large amounts of potassium (dry, fresh plant material contains 1–2% K). This potassium in the plant material can be used almost as fast as water flows through the material and then to roots. If the potassium is not used by plants, some of it attaches to cation exchange sites in the humus or on clays.

The potassium cycle is shown as Fig. 10-1. It is a simple cycle. Microorganisms are involved directly only in decomposition of fresh plant and animal residues. There is only one ion form of potassium, K^+, and there is no volatile form at temperatures found in nature. Exchangeable K has a dominant role in the soils' potassium fertility. Potassium moves through soil *chromatographically,* as described in Chapter 5, Detail 5-3.

Table 10-1 Examples of Potassium Gains, Losses, and Transformations (in kg/ha/yr) for Eight Different Cropping Systems[a]

Description of Change	Grazed Bluegrass (N. C.)	Corn Grain (Ind.)	Soybean Seeds (Ark.)	Wheat (Kansas)	Irish Potatoes (Maine)	Cotton (Calif.)	Loblolly Pine (Miss.)	Douglas Fir (Wash.)
Additions								
Added fertilizer	46	65	37	0	207	0	—	—
Irrigation, floodwater	—	4	—	—	—	50	—	—
Sediments added	4	—	4	2	3	1	4	4
Removals								
Harvested product	—	20	22	6	117	30	7	11
Leaching, loss	—	15	15	5	15	10	2	1
Erosion, runoff	6	10	10	5	tr	10	3	1
Recycling processes								
Uptake from soil	150	111	37	50	177	67	9	14
Manure from grazing	127	—	—	—	—	—	—	—
Plant residues left	—	91	15	44	60	37	2	4
Mineralization from humus	—	—	—	—	—	—	—	—

Source: Data from M. J. Frissel, ed., *Cycling of Mineral Nutrients in Agricultural Ecosystems,* Elsevier Science Publishing, New York, 1978, pp. 202–243.

[a] A dash means that no measurement was made or the item does not apply to the system.

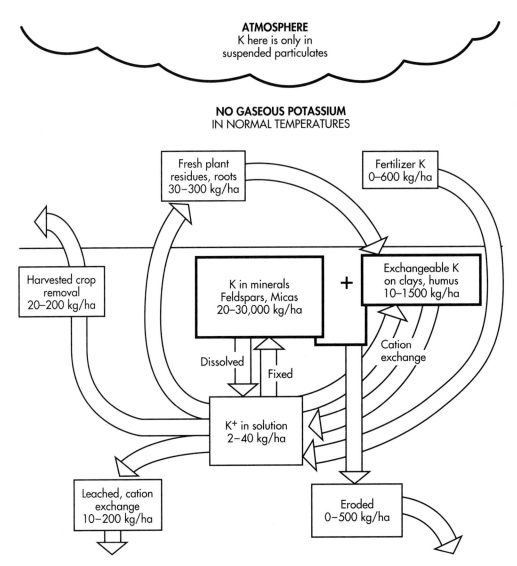

ATMOSPHERE
K here is only in
suspended particulates

NO GASEOUS POTASSIUM
IN NORMAL TEMPERATURES

Fresh plant
residues, roots
30–300 kg/ha

Fertilizer K
0–600 kg/ha

Harvested crop
removal
20–200 kg/ha

K in minerals
Feldspars, Micas
20–30,000 kg/ha

+

Exchangeable K
on clays, humus
10–1500 kg/ha

Dissolved

Fixed

Cation
exchange

K⁺ in solution
2–40 kg/ha

Leached, cation
exchange
10–200 kg/ha

Eroded
0–500 kg/ha

FIGURE 10-1 The potassium cycle, showing transformations, forms, and amounts of potassium in various soils. (Courtesy of Raymond W. Miller, Utah State University.)

10:2.2 Some Characteristics of Soil Potassium

Potassium chemicals are usually very soluble substances. Surprisingly, the two major mineral sources of potassium in soils (micas and feldspars) have very low solubility. However, even the soluble potassium in soil solution is not very mobile in soils. As the water containing potassium flows through soil, the K^+ adsorbs to cation exchange sites, replacing other cations; this cation exchange slows the movement of K^+ through soil. Some other important characteristics of potassium are the following:

1. Potassium is not supplied from decomposing soil humus, except as an exchangeable ion on humus exchange sites. *Fresh plant residues* do contain the 1–3% potassium common in plants, but that potassium is soluble and is immediately leachable from the dead plant tissue. Manures do have some potassium, but most of it is excreted in the urine.

2. Potassium, because it is a positive ion, moves slowly through the soil. Potassium fertilizers should be placed in the soil where roots have good access to the potassium.

3. The immediate crop obtains *practically all of its potassium from soluble and exchangeable forms.* In acidic soils these forms are often in deficient amounts. In more arid regions potassium deficiencies in the soils are likely only in sands and a few other soils where heavy crop demands over the years have depleted exchangeable potassium.

4. There is some competition for uptake among potassium, calcium, and magnesium ions. The extent or exact reasons are not yet clear. Because all of these ions are cations, the plant's internal "charge balance" is expected as one cause for competition among these cations.

Even with the guides given above, few scientists would have expected potassium deficiency in cotton when soil tests suggest there is adequate potassium. Yet potassium deficiency symptoms—bronzing of thick leathery leaves, with no necrosis, that snap when bent—are found in cotton in the San Joaquin Valley of California. Also, the symptoms appear on *younger, not older, leaves.* The explanation for this shows the uniqueness of this condition. The carpel wall of the cotton bolls contains about 4% K and accounts for 60% of all the potassium accumulated by the plant. As the bolls develop, they become a *sink* for potassium during peak boll growth. The cotton bolls probably use translocated potassium at the expense of some potassium that the younger leaves would otherwise get.[1] Prune fruits also seem to be a potassium sink, which can cause potassium deficiency in a similar manner.

This example emphasizes several factors that should be kept in mind about nutrient requirements. First, plants may differ greatly in the *amounts* and in the *proportions* needed of various nutrients. Second, each plant usually has a growth stage during which the need for one or more nutrients is much higher than during the rest of the plant's growth cycle. If nutrients are limiting during this critical peak-use period, yields will be reduced. Third, seeming abnormalities in symptoms or response to a nutrient (the deficiency of potassium on younger leaves, for example) often have been logically explained when enough is known. One should expect to have some differences in nutrient needs (timing, amounts, proportions) for each plant species.

Additional examples of gains, losses, and transformations are shown in Table 10-1. Notice the heavy potassium needs for potatoes (also for sugar cane and bananas). Leaching losses occur; potassium does move chromatographically through soil as large amounts of water percolate. Notice the lack of any volatilization product and *no* potassium mineralized from humus.

Cost per kilogram of potassium is the dominant criterion for choosing a potassium fertilizer. The fertilizer KCl is usually the cheapest. Yet the more expensive sulfate or nitrate forms may sometimes be chosen in preference to KCl. The sulfate, nitrate, or phosphate supplies that nutrient plus the potassium. Also, some plants, such as avocados and potatoes, may be injured by large amounts of chloride. Excess chloride in potatoes lowers starch and makes them poor for French fries. Damage can also occur if chlorides in foliage spray or irrigation water are left on the leaves. Tobacco "burns" better when the high potassium need is partly supplied as the sulfate rather than all as the chloride.

---

[1] Bill L. Weir, Thomas A. Kerby, Bruce A. Roberts, Duane S. Mikkelsen, and Richard H. Garber, "Potassium Deficiency Syndrome of Cotton," *California Agriculture* (Sept.–Oct. 1986), pp. 13–14.

The cost of allowing potassium deficiency to develop in prunes in California was given in dollar values as follows[2]:

Extent of K Deficiency	Value per Hectare
Trees with no K deficiency	$2470–$6175
Trees with severe K deficiency	Less than $617
Slight K deficiency symptoms	$2643 average
No K deficiency symptoms	$4384 average
Net value if fertilized with K[a]	$1541

[a] Cost for K fertilization/year about $200.

Generally, even "hidden hunger" (= no visual symptoms) will reduce production and cost the producer in lost yields.

10:2.3 Managing Soil Potassium

Sugar-producing plants, potatoes, and tobacco are heavy users of potassium (Figs. 10-2 and 10-3). It is likely that one of the problems with "worn-out lands" in the southeastern United States in pre–Civil War days was a result of exhausting the available potassium, along with nitrogen and/or phosphorus. Leaving the soil to "rest" a few years allows some buildup of available potassium as the low-solubility mineral forms weather and the K^+ is adsorbed onto active exchange sites.

Plant potassium requirements are high during early growth. As with phosphorus, it is important to have adequate potassium available near young seedlings, but not in amounts large enough to cause salt damage.

Little can be done to alter the available potassium in deficient soils other than adding potassium fertilizers. Deficiencies should be expected in soils that are low in micas (the more soluble mineral source), soils that are low in clay (few exchange sites), and soils of pH 4–6 (leached by high rainfall, which is common to areas of acidic soils). The soil test for potassium is simple and quite good; it should be used to predict deficiencies.

Management of soil potassium emphasizes (1) maximizing efficient use of added potassium, (2) minimizing luxury consumption, and (3) trying to make maximum use of natural potassium sources. The following practices are most useful:

1. Avoid heavy applications of potassium fertilizers; use smaller split applications. Heavy additions allow unnecessary and expensive luxury consumption. High chloride levels from KC1, if used, may be toxic to some plants. Split applications are especially important on sandy soils where leaching losses are likely.

2. Maintain soil pH near 6–6.5 with lime reduces potassium losses from leaching (more calcium rather than potassium is leached).

3. Returning crop residues and manures adds large amounts of potassium, some of which remains in the soil as exchangeable forms or, in less leached soils, as soluble potassium. Because of the large amounts of potassium in grasses, the removal of lawn clippings in areas of marginal potassium soils can finally cause potassium deficiency unless fertilizer-K is added.

[2] William H. Olson, Kiyoto Uriu, Robert M. Carlson, William H. Krueger, and James Pearson, "Correcting Potassium Deficiency in Prune Trees Is Profitable," *California Agriculture* **41** (May–June 1987), pp. 20–21.

FIGURE 10-2 Tobacco in the southeastern United States has fairly high potassium requirements because the whole plant is harvested and high rainfall often leaches the soil extensively. In some instances such large amounts of potassium chloride are added to correct the potassium deficiency because high chloride concentrations in the tobacco cause slow curing and poor burning qualities. Other potassium salts (K_2SO_4 and KNO_3) should be used for part of the need to avoid this problem.   (*Source:* USDA--Soil Conservation Service.)

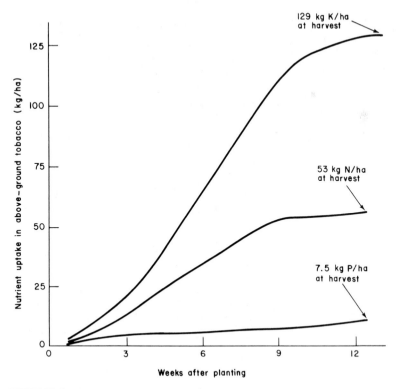

FIGURE 10-3 Comparison of the potassium, nitrogen, and phosphorus requirements of tobacco.   (*Source:* Recalculated and redrawn from C. D. Raper, Jr., and C. B. McCants, *Tobacco Science,* **10** [1966], p. 109.)

10:3 Materials Supplying Potassium

Muriate of potash (potassium chloride, KC1) is the principal fertilizer material supplying potassium; second in importance is **sulfate of potash** (potassium sulfate, K_2SO_4). **Sulfate of potash-magnesia** (potassium-magnesium sulfate) is increasing in significance because it supplies the additional nutrient magnesium. **Potassium nitrate** (KNO_3) is also an excellent fertilizer. Muriate of potash is usually 95% KC1, equivalent to 60% K_2O. In Saskatchewan, Canada, the largest known potash reserves in the world are now being mined; it is also mined in New Mexico, Utah, and California.

Other materials less popular, but *not less suitable,* are **potassium nitrate** and **potassium polyphosphates.** (Table 10-2 lists common potassium fertilizer materials.)

10:4 Soil Sulfur

Sulfur, as a constituent in three of the nearly two dozen amino acids, is an essential part of proteins. It is found in several vitamins and in oils of plants in the mustard and onion families. In protein materials sulfur content should be about 6–8% as high as nitrogen content, although it is 10–15% of nitrogen values in most soil humus.

Sulfur has been considerably neglected by researchers for various reasons. With the common additions of sulfates as ammonium, potassium, and regular superphosphate fertilizers, it was unintentionally added. Few crops exhibited obvious sulfur deficiency symptoms. Also, the lack of a good simple analytical method to measure sulfur made it more attractive to study nitrogen, phosphorus, potassium, or lime needs.

At least three factors have increased the need for sulfur fertilizers: (1) the lower amounts of sulfate added incidentally with other nutrients, (2) the lower pollution from sulfur oxides into air, later brought down in precipitation, and (3) higher plant yields, which put greater nutrient demands on the soils. The availability of sulfur to plants is hard to predict because, like nitrogen, major portions may come from soil humus. Sulfur from soil humus depends on microbial action and its dependency on climate.

10:4.1 Sources of Sulfur

Soils contain many sources of sulfur. The mineral pyrite (FeS_2, fool's gold) is common to most soils and oxidizes to sulfuric acid and ferric oxide. Moderately soluble gypsum is found in many arid soils; even quite soluble sodium sulfates are found in some arid soils.

Rainfall dissolves the sulfur oxides evolved during the burning of plant-derived fuels (wood, coal, and oil) from range and forest fires and from the roasting of ores in smelters. Rainwater combined with sulfur oxides produces sulfuric acid. This acidic rainfall is

Table 10-2 Common Potassium Fertilizers Given in Order of Percentage Use in the United States (Beginning with the Most Used)

Material	K_2O (%)	Other Nutrients (%)
Muriate of potash (KCl)	50–60	Unneeded Cl—may be toxic
Sulfate of potash (K_2SO_4)	45–50	18% sulfur (S), as sulfate
Potassium-magnesium sulfate	18–22	18–23% sulfur (S), 11% magnesium (Mg)
Potassium nitrate	37–44	13% nitrogen (N), as nitrate
Potassium polyphosphate	20–40	25–50% phosphorus (P_2O_5)

corrosive enough to damage metals and other building materials. It can, through a sequence of reactions, also cause the death of fish in bodies of water where large amounts of the acid fall.

Sulfur-bearing pesticides and fertilizers—such as ammonium sulfate, superphosphate, and potassium sulfate—are agricultural chemicals that add sulfur to soils. The incidental addition of sulfur to soil in agricultural chemicals, particularly in fertilizers, is the major reason why sulfur deficiencies are not more common.

The sulfate ion is relatively soluble and generally is leached downward by water. In arid soils sulfate often precipitates as moderately soluble gypsum ($CaSO_4 \cdot 2H_2O$). In acidic soils most moderately soluble sulfur minerals are leached out, and much of the sulfur for plants comes from mineralization of organic matter, particularly in highly weathered soils, such as Ultisols and Oxisols. The sulfur cycle is shown in Fig. 10-4.

10:4.2 Some Characteristics of Soil Sulfur

Sulfur and nitrogen share many features in common. First, in leached soils decomposition of organic matter can release major portions of sulfur. Mineralization supplies a large fraction of the plants' sulfur needs. Second, both nitrogen and sulfur have several chemical forms under different soil conditions. The following table itemizes some of these comparisons.

Item	Sulfur	Nitrogen
Oxidized negative ion	SO_4^{2-}	NO_3^-
Intermediate oxidations	S^0, SO_3^{2-}	N_2, NO_2^-, others
Reduced ion forms	S^{2-}	NH_4^+
Amino acid form	—C—SH	—C—NH_2
Most leachable form	SO_4^{2-}	NO_3^-
Several gaseous forms	CH_3SCH_3, H_2S	NH_3, N_2O, NO
	SO_2, SO_3	N_2, NO_2

Although sulfur can be supplied from decomposing soil humus, the various inorganic sulfur minerals may also supply considerable sulfur to plants. Because of the inorganic forms, available sulfur is less dependent than is nitrogen upon microbial action.

The C/S ratio operates much as the C/N ratio does. If residues low in sulfur are added, some soil sulfate will be immobilized into microbes and humus residues. The estimated ratio thresholds are indicated below:

C/S ratio in the residue added to soil

< 200	200 – 400	> 400
(sulfate released to soil solution)	(neither gain nor loss of sulfate to solution)	(immobilization of soil sulfate into humus)

Sulfur-oxidizing bacteria, mostly the genera *Thiobacillus,* oxidize elemental sulfur (S^0) or the sulfide (S^{2-}) in reduced and iron pyrite forms (FeS and FeS_2) to sulfuric acid (H_2SO_4). From pyrites the reactions can occur both by purely chemical and by several biological pathways to form sulfuric acid. In anaerobic conditions microbes can reduce sulfate to sulfide, where it precipitates or causes toxicity to plants as H_2S gas (*Akioki* disease in some paddy rice). When aeration occurs or when elemental sulfur is added, oxidization of the sulfides or sulfur to sulfate results in acidification of the soil. (Detail 10-1).

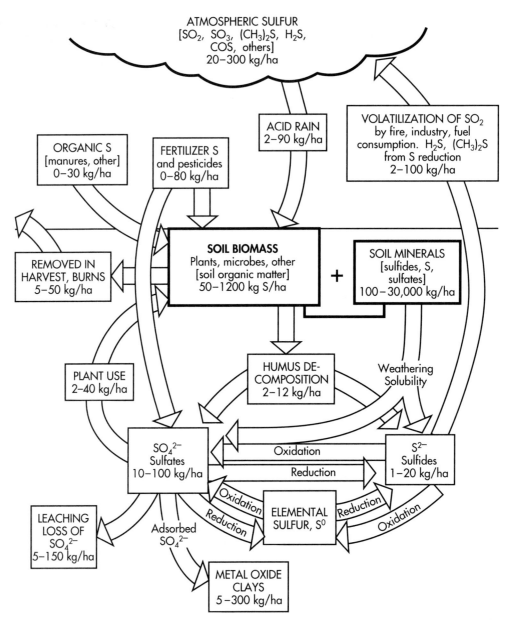

FIGURE 10-4 The sulfur cycle, showing transformations, forms, and amounts of sulfur in various soils. (Courtesy of Raymond W. Miller, Utah State University.)

10:4.3 Sulfur Fertilizers or Amendments

Common ways to add sulfur are to select *ammonium sulfate* or *potassium sulfate* deliberately if you expect sulfur to be low. *Ammonium thiosulfate* [$(NH_4)_2S_2O_3$] is a popular liquid fertilizer. It is noncorrosive and has good storage features. In soil it forms ammonium sulfate and elemental sulfur. *Ammonium polysulfide* (NH_4S_x, where x may take on various integers) is used as both a fertilizer and to reclaim high-pH soils in arid regions. **Elemental sulfur**

Soil with enough sulfides (FeS_2, others) to become strongly acidic when drained and aerated enough for cultivation are termed **acidic sulfate soils** or, as the Dutch refer to them, *Katteklei* (**cat clays**). When allowed to develop acidity, these soils are usually more acidic than pH 4. Before drainage, such soils may have normal soil pHs. Lands inundated with waters that contain sulf*ates,* particularly salt (ocean) waters, accumulate sulfur compounds, which in poorly aerated soil are bacterially reduced to sulf*ides.* Such soild are not usually very acidic when first drained of water.

When the soil is drained and then aerated, the sulfide is oxidized to sulfate by a combination of chemical and bacterial actions, forming sulfuric acid. The extent of acid development depends on the amount of sulfide in the soil and the conditions and time of oxidation. If pyrite (iron sulfide, FeS_2) is present, the oxidized iron accentuates the acidity, but not as much as aluminum in normal acid soils because the iron oxides are less soluble than aluminum oxides and so hydrolyze less.

The slow oxidation of mineral sulfides in soils is nonbiological until soil pH reaches an acidity of pH 4. Below pH 4, the bacteria *Thiobacillus ferrooxidans* are the most active oxidizers and the acidity builds up rapidly.[a] The chemical reactions are as follows.[b]

Nonbiological

$$2FeS_2 + 2H_2O + 7O_2 \longrightarrow 2FeSO_4 + 2H_2SO_4$$
(pyrite)   (ferrous sulfate)   (sulfuric acid)

Accelerated by bacteria (Thiobacillus ferrooxidans)

$$4FeSO_4 + O_2 + 2H_2SO_4 \longrightarrow 2Fe_2(SO_4)_3 + 2H_2O$$
(ferrous sulfate)   (ferric sulfate)

Rapid in acid pH (nonbiological)

$$FeS_2 + 7Fe_2(SO_4)_3 + 8H_2O \longrightarrow 15FeSO_4 + 8H_2SO_4$$
(ferrous sulfate)

Acidic sulfate soils contain a **sulfuric horizon,** which has a pH of the 1 : 1 soil : water mixture of less than 3.5 plus some other evidences of *sulfide* content (yellow color, mineralogy). *Sulfaquepts, Sulfihemists, Sulfohemists,* and *Sulfaquents* great groups include all these acid sulfate soils. (See Chapter 3 for general soil nomenclature.)

Strong acidity can result in possible toxicities of aluminum and iron (if the solution is acid enough), soluble salts (unless leached), manganese, and hydrogen sulfide (H_2S) gas. Hydrogen sulfide, often formed in paddy soils, causes the rice disease known by its Japanese name, *akiochi,* which prevents rice plant roots from absorbing nutrients.

Management techniques are extremely variable and depend on many specific facts—that is, the extent of acid formation, the thickness of the sulfide layer, leaching possibilities, and the value of the land area. The general approaches to reclamation are as follows.

1. *Keep the area flooded.* Flooded (anaerobic) soil inhibits acid development, which requires oxidation. This solution almost limits the use of the area to paddy rice growing.

Continued.

---

[a] G. J. M. W. Arkesteyn, "Pyrite Oxidation in Acid Sulfate Soils: The Role of Microorganisms," *Plant and Soil,* **54** (1980), pp. 119–134.

[b] Darwin L. Sorenson, Walter A. Kneib, Donald B. Porcella, and Bland Z. Richardson, "Determining the Lime Requirement for the Blackbird Mine Spoil," *Journal of Environmental Quality,* **9** (1980) no. 1, pp. 162–166.

2. *Control the water table.* If a nonacidifying layer covers the sulfuric horizon, drainage to keep only the sulfuric layer under water (anaerobic) is possible.

3. *Lime and leach.* The primary way to reclaim these soils, as for any acid soil, is to lime them. This solution is possible but not always practical. Normal soils may require from 11 to 45 Mg/ha (5–20 t/a) of lime in a 20-year period, whereas acid sulfate soils may need from several metric tons per hectare per year up to even 224 Mg/ha (100 t/a) within a 10-year period or less.

If these soils are leached during early years of acidification, lime requirements are lowered. Leaching, however, is difficult because of the high water table common to these soils and low permeability of the clay. Because acid sulfate soils are often in reclaimed swamps and salt marshes, seawater is sometimes available for *preliminary* leaching.[c]

Source: C. Ckharoenchamratcheep, C. J. Smith, S. Satawathananont, and W. H. Patrick, "Reduction and Oxidation of Acid Sulfate Soils of Thailand," *Soil Science Society of America Journal,* **51**(1987), pp. 630–634.
[c]Charles R. Lee, et al., "Restoration of Problem Soil Materials at Corps of Engineers Construction Sites," Instruction Report EL-85-2, U.S. Army Corps of Engineers, May 1985, pp. H-1 to H-14.

suspensions (S^0) and gypsum ($CaSO_4 \cdot 2H_2O$) are also used as sulfur amendments. One source[3] lists 62 sulfur carriers used as amendments.

10:4.4 Managing Soil Sulfur

Increasing restrictions to reduce air pollution and greater use of higher-analysis (purer) fertilizers will reduce incidental sulfur additions. The result will be an increased incidence of sulfur deficiency.

The factors that affect nitrogen release from soil humus also affect sulfur release. Sulfate, as an anion, is easily leached. In arid climates where drainage water flows through high-sulfur soils, the irrigation water often carries adequate sulfate for plant needs.

Soils most likely to have sulfur deficiency are those that are sandy, low in organic matter, nonirrigated (no added sulfate in water), well leached, and far removed from highly industrialized areas (less airborne sulfur). Crops with high sulfur requirements (corn, sorghum, peanuts, tobacco, and cotton) and fertilized with nonsulfur-containing materials might also have sulfur deficiency. Sulfur deficiencies are observed in several rice-producing areas where a shift from sulfur-containing fertilizers (e.g., ammonium sulfate) to high-analysis nitrogen (urea) has caused sulfur deficiency. Without well established soil tests for sulfur deficiency, it would seem a good precaution to use some sulfate-containing fertilizers periodically on acid, sandy soils low in humus.

10:5 Soil Micronutrients

The micronutrients are boron, iron, manganese, zinc, copper, chlorine, and molybdenum. These elements are essential to plant growth, but they are utilized only in minute quantities, in contrast to the macronutrients, which comprise a proportionally larger percentage of plant

[3] S. L. Tisdale, W. L. Nelson, and J. D. Beaton, *Soil Fertility and Fertilizers,* 4th ed., Macmillan, New York, 1985, pp. 324–325.

weight. Except for chlorine, the dominant role of micronutrients is as activators in numerous enzyme systems. **Chlorine** affects root growth, but little more is known of its use by plants. **Boron** is another micronutrient whose function is not clearly understood. It is supposed that boron acts as an electron scavenger, collecting free electrons from around positive charges created in cell membranes, allowing a positive charge to last "long enough to get a growth process started."

The origin of micronutrients in soils is slowly weathering minerals; the micronutrients come from contaminant minerals mixed with the common primary minerals. Many chlorine and some boron salts are soluble; the metals zinc, copper, manganese, and iron are more soluble in acidic solution (and hence more available in acidic soils), becoming less soluble as pH increases. Molybdenum is more soluble in basic soil (it reacts much as does phosphate). In strongly acid soils, manganese, zinc, and copper may dissolve to form toxic concentrations that actually hinder plant growth.

Boron is probably one of the most added micronutrients to correct a deficiency. This is a common need in humid areas because available boron can be leached. In contrast to the need for boron in humid areas, zinc and iron are the nutrients most often deficient in soils of arid regions, especially on calcareous soils. The deficiencies of manganese, copper, and molybdenum are less common. Although chloride is probably adequate for nutrition, higher addition levels of chloride do seem to benefit growth by reducing certain plant diseases.

10:6 Soil Boron

Boron is essential for the growth of new cells. It is not readily mobile in the plant, and a boron deficiency causes the terminal bud to cease growth, followed by death of young leaves. Without adequate boron, the number and retention of flowers is reduced and pollen germination and pollen tube growth are less. The result is that less fruit develops.

Boron (B) forms a weak acid. In the soil solution at most soil pH values it will occur as *nonionized* H_3BO_3 and at pH values greater than about 8.5, it will occur as $B(OH)_4^-$. Boron is a nonmetal. Deficiencies are most common in high-rainfall areas, particularly (1) the Atlantic coastal plain and the Southeastern states, (2) northern Michigan, Wisconsin, and Minnesota, (3) the Pacific coastal area, and (4) the Pacific Northwest. The most prominent boron mineral in leached soils is *tourmaline,* a very slightly soluble mineral. In less leached soils, various soluble borates may exist.

In soils boron has four major forms: (1) in its primary rocks and minerals, (2) combined in soil organic matter, (3) adsorbed on colloidal clay and hydrous oxide surfaces, much as is phosphorus, and (4) as the boric acid (H_3BO_3) or $B(OH)_4^-$ ion in solution. Freshly precipitated aluminum hydroxides adsorb large amounts of boron, so that liming acidic soils frequently causes a boron deficiency as soluble boron adsorbs on the new metal oxide precipitates. Boron deficiency in grape vines in the San Joaquin Valley of California drastically reduces fruit set. The cost of adding enough boron is relatively inexpensive, and it would seem logical to add a little *insurance* boron. *Too much boron must be carefully avoided.* The concentrations between adequate and toxic boron is not very large.

Some disorders attributed to boron deficiency are canker of beets, hollow stem of cauliflower, cracked stem of celery, water core of rutabagas, and stem-end russet of tomatoes.

Boron can accumulate to toxic concentrations. In some arid regions the accumulation of boron to toxic levels already exists. Careless application of extra boron to be sure of having enough is a hazardous practice. Tables 10-3 and 10-4 list minimum desired boron contents and toxicity thresholds for selected crops.

The most common boron amendment is borax (**sodium tetraborate,** $Na_2B_4O_7 \cdot 5H_2O$), which is 14% boron. It is sold by various names and is quite soluble. **Solubor** (20% boron)

Table 10-3 Minimum Needed Boron Concentrations for Optimum Yields

0.2 mg B/ha–15 cm[a]	0.2 to 1.0 mg B/ha–15 cm	1.0 to 2.0 mg B/ha–15 cm
Small grains	Tobacco	Apple
Corn	Tomato	Alfalfa
Soybean	Peach, pear, cherry	Clover
Pea, bean	Peanut	Mustard
Potato	Carrot	Celery

[a] Boron extracted from soils by the "hot water" test.

Table 10-4 Threshold Concentrations for Boron Concentrations in Field Capacity Water

Sensitive Crops		Semitolerant Crops		Tolerant Crops	
0.6 mg B[a] ↓	Citrus fruits	↓	Sesame Pea	↓	Sorghum Alfalfa
1.0 mg B	Grape Pecan Onion Wheat	2.0 mg B	Carrot Potato Leafy vegetables Barley	8.0 mg B	Oat Tomato Sugar beet Cotton
1.6 mg B	Strawberry Beans	4.1 mg B	Corn Tobacco	20.1 mg B ↓ 30 mg B	Asparagus

Source: Selected data from R. Karen and F. T. Bingham, "Boron in Water, Soils, and Plants," *Advances in Soil Science,* **1** (1985), pp. 229–265.

[a] Toxicity above these levels begins to decrease yields. Units are in milligrams of boron per hectare–15 cm deep, assuming that a ha– cm weighs 2 million kilograms.

is a modified borate completely water soluble and used in many liquids and mixes. Low-solubility **frits** are made by mixing glass with boron and other nutrients, melting this mixture, and then cooling and shattering it. These frits have a low rate of nutrient release but are fast enough. They protect against excessive solubility (toxic) and limit leaching losses in sands and in high-rainfall areas.

10:7 Soil Iron

Iron is an important part of the plants' oxidation-reduction reactions. As much as 75% of the cell iron is associated with chloroplasts. Iron is a structural component of cytochromes, hemes, and numerous other electron-transfer systems, including *nitrogenase* enzymes necessary for the fixation of dinitrogen gas.

In aerated soils iron oxides are one of the *least soluble* of soil minerals. When soil is limed, any soluble ferric iron (Fe^{3+}) readily forms one of many hydrous oxides, all of which have low solubility. In intensively weathered soils (e.g., Oxisols and some Ultisols) most of the primary minerals have been weathered and the more soluble materials leached away. The hydrous iron oxides remain because they are the *most resistant residues.* Most of the basic cations, silica, and even considerable alumina have been leached away. Some Oxisols have 50–80% clay consisting primarily of iron hydrous oxides. *The problem with soil iron is that iron has very low solubility in soil solutions and waters.* Even added soluble-iron chemicals readily precipitate as low-solubility minerals. *The major problem with iron availability is how to keep iron sufficiently soluble for plants to absorb enough of it.* In strongly acidic solutions, below pH 5, iron becomes increasingly soluble and is rarely deficient.

The reduced ferrous iron (Fe^{2+}), formed in anaerobic conditions, is considered more soluble than the ferric ion. Yet, the ferrous iron is easily and rapidly oxidized in aerated soils and its solubility becomes that of hydrous iron oxides. Alternating aerobic-anaerobic conditions cause some solubilization and precipitation cycles. Iron solubilization occurs in anaerobic periods (water-logged), then precipitation occurs when the soil becomes aerated (dried or drained) and allows iron oxide cementation to develop soft *plinthite* forms *ironstone*.

With these solubility problems, some major questions about iron are (1) How is iron made available? (2) Why isn't iron deficient in practically all soils? and (3) What materials and methods are used for supplying needed iron?

10:7.1 Iron in Soil Solution

The solution pH has a dominant influence on iron solubility. At about pH 3, iron is soluble enough to supply plant needs, but plants are usually "poisoned by toxic levels of aluminum." The solubility of Fe^{3+} decreases about 1000-fold per pH unit rise. In soils of pH suitable for most crops, simple solubility of iron will not supply enough iron for plants (Fig. 10-5).

10:7.2 Chelates and Nutrient Availability

Because the total soluble mineral iron is inadequate for plants, and yet most plants are not deficient in iron, *what makes iron available?* Many soluble organic substances can react with iron to bond it into soluble forms that are quite mobile in the soil solution. These many substances come from root exudates, humus decomposition, microbial cell exudates, animal manures, and even polyphenols excreted from leaf surfaces and dissolving in rainfall. When

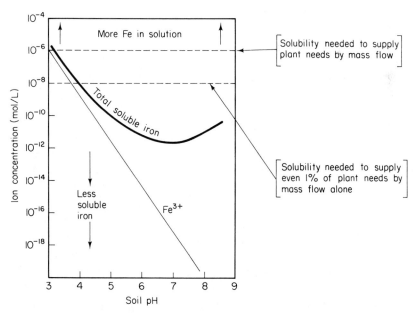

FIGURE 10-5 Illustration of iron solubility in aerobic soil solution. Only near pH 3–4 will most soils supply enough soluble iron from mineral sources to supply plant needs. (*Source:* Modified after Willard Lindsay, Chapter 17 in *The Plant Root and Its Environment,* E. W. Carson, ed., University Press of Virginia, Charlottesville, 1974; redrawn by Raymond W. Miller, Utah State University.)

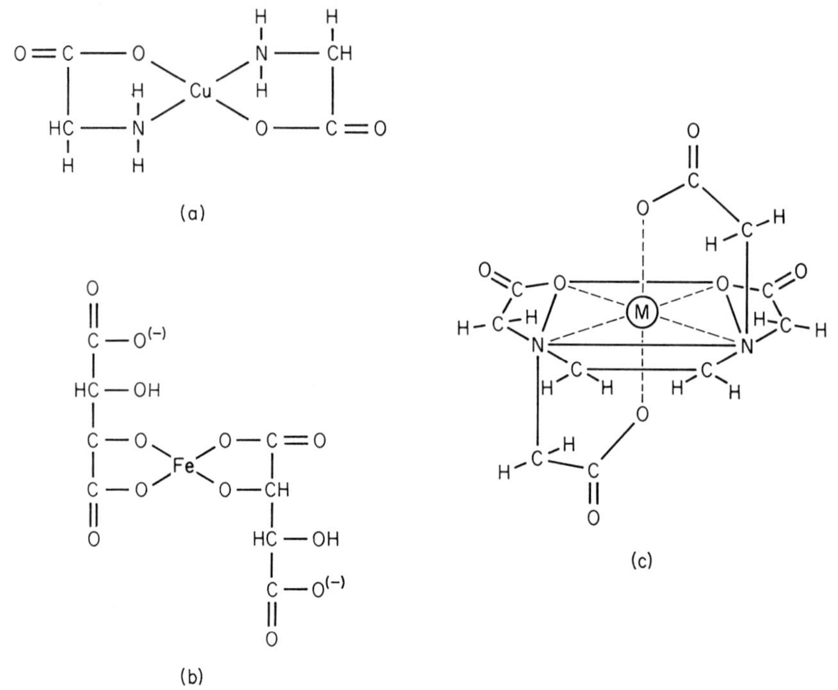

FIGURE 10-6 Example of ligands bonding to a metal. The resulting substances are called chelates. (a) Copper chelate involving the common amino acid glycine; (b) iron with tartaric acid; (c) the ligand ethylenediaminetetraacetic acid (EDTA) is shown bonding to any of several metals (M) forming chelates. Common metals bonding with ligands are Fe, Ca, Mn, Mg, Cu, Zn, and heavy metals. Other well-known natural ligands are citric acid, proteins, amino acids, polyphenols, and fulvic and humic acids. (Courtesy of Raymond W. Miller, Utah State University.)

the organic substance bonds to the metal by two or more "contacts," the organic substance is called a chelating **ligand** (Fig. 10-6). The ligand plus metal is called a **chelate**[4] (pronounced "key-late"). Metals in these chelates are carried in soil water. Many natural and synthetic chelates are available; some of the chelates are insoluble in water. Some plants when under iron stress (iron deficiency) can produce substances that solubilize or chelate soil iron. These substances have been called by various names, such as **sideraphores** and **iron chelate reductases.**[5] Production of these substances by plants make the plant more competitive to obtain iron; deficiencies of iron will be less frequent for siderphore-producing plants.

One ligand may supply four "arms" from one molecule that form "rings" by bonding to the metal. Multiple bonds make the chelate very stable. The stability of a chelate is affected by the number of rings formed, the metal ion, and the solution pH. Usually, bonds are stronger with metals in approximately this order:

$$Fe^{3+} > Al^{3+} > Cu^{2+} > Co^{2+} > Zn^{2+} > Fe^{2+} > Mn^{2+} > Ca^{2+} = Mg^{2+}$$

---

[4] Chelate is derived from the Greek word *chele,* which means a "crab's claw." The intent was to refer to the pincherlike manner in which the ligand bonds and holds the metal.

[5] D. A. Konstant, "Root Enzymes Are in Control," *Agricultural Research,* **39** (1991), no. 9, p. 21; M. Treeby, H. Marschner, and R. Romheld, "Mobilization of Iron and Other Micronutrient Cations from a Calcareous Soil by Plant-Borne, Microbial, and Synthetic Metal Chelators," *Plant and Soil,* **114** (1989), pp. 217–226.

Table 10-5 Crops Sensitive to Low Iron Availability in Soils[a]

Sensitive Crops (Deficiency Is Most Likely)	Moderately Tolerant Crops	Tolerant Crops (Infrequently Have Deficiency)
Berries	Alfalfa	Alfalfa
Citrus	Small grains	Barley
Several fruits	Some fruit trees	Corn
Sorghum	Many grasses	Cotton
Grape	Rice	Potatoes
Many ornamentals	Soybeans	Rice
Soybeans	Vegetables	Vegetables

Source: Selected data from J. J. Mortvedt, "Do You Really Know Your Fertilizers? Part 4: Iron, Manganese, and Molybdenum," *Farm Chemicals,* **143** (1980), no. 12, p. 42.

[a] Some crops are listed in more than one category because of inadequate testing, varying conditions, or different plant varieties.

Chelates keep large amounts of metals in a mobile form in which the metal can move to the roots (mass flow or diffusion). Roots are able to use the metal ion held in the chelate, although the mechanism(s) is (are) not yet clear. Soil humus is, therefore, usually important, but not necessarily sufficient, for increasing iron availability to plants. Synthetic chelates are often used in foliage sprays or as soil additions to plants to provide iron, zinc, manganese, and copper.

10:7.3 Problem Soils and Susceptible Plants

Iron deficiencies are most common in calcareous soils, in arid soils, and in soils cropped to high-iron-demand plants (nursery ornamental trees and shrubs, fruits, corn, soybeans, sorghum, and beans) (Table 10-5). High levels of bicarbonate and phosphates lower iron availability to plants because of the formation of relatively insoluble iron salts. All iron salts are less soluble in basic media, which is the reason plants grown in arid soils are the most often deficient in iron.

10:7.4 Iron Amendments and Their Use

Broadleaved plants that are iron deficient exhibit the typical interveinal chlorosis-light-colored areas between the darker veins on young leaves (Fig. 10-7). Additions of soluble

Iron	11	18	27	32	43	ppm
Chlorophyll	.3	.7	1.3	1.6	1.8	mg/g

FIGURE 10-7 A deficiency of iron causes chlorosis (whitening) of citrus leaves in California. The amount of iron (parts per million of dry weight) is associated with the amount of chlorophyll (in milligrams per gram of fresh weight)—the more iron, the more chlorophyll. (Courtesy of Ellis F. Wallihan, University of California.)

Table 10-6 Partial List of Common Carriers of Iron

Fertilizer Material	Formula	Percent Iron
Ferrous sulfate	$FeSO_4 \cdot 7H_2O$	19
Ferrous ammonium phosphate	$Fe(NH_4)PO_4 \cdot H_2O$	29
Ferrous ammonium sulfate	$(NH_4)_2SO_4 \cdot FeSO_4 \cdot 6H_2O$	14
Iron frits (crushed glass)	—	Varies
Iron chelates	Many kinds	5–14
Iron-sul	Mixture of $FeO(OH)$, $KFe_3(OH)_6(SO_4)_2$, FeS_2, and $CuFeS_2$	20

chelates usually correct iron deficiencies. Fast-acting, foliar, chelate sprays are usually not effective very long. They may need to be added two to four times a year for year-round growth. Applications to the soil last longer but act slower and require 5 to 15 times more of the expensive chelates ($15–$100 per kilogram). Some organic amendments (manures and sludges) may supply soluble chelates that help nutrition.

Several of the most common iron carriers are given in Table 10-6. The added *soluble inorganic iron sources* revert rapidly to insoluble forms, unless they are sprayed on foliage or injected into tree trunks or limbs.

10:8 Soil Zinc

Zinc is essential for numerous enzyme systems and is capable of forming many stable bonds with nitrogen and sulfur ligands. Unique among the quartet of metals—copper, iron, manganese, and zinc—only zinc does not exhibit multiple valences. Zinc is not subject to oxidation-reduction reactions in the soil–plant system.

10:8.1 Zinc in the Soil Solution

Zinc occurs in solution as Zn^{2+}. As a positive ion, it is quite immobile in soil. Above pH 7.7 it becomes $Zn(OH)^+$, and at pH 9.1 it precipitates as $Zn(OH)_2$. Zinc will form soluble carbonates. Generally, zinc does not form particularly insoluble inorganic forms, but it does bond strongly with sulfide, forming ZnS (sphalerite). In rice paddy soils zinc can become deficient because of its tendency to combine with the sulfide produced from decomposing humus under anaerobic conditions. Zinc, in contrast to iron or manganese, *is less soluble in anaerobic conditions than in aerobic soil.*

10:8.2 Problem Soils and Susceptible Plants

Zinc deficiencies occur mostly in basic soils, in limed soils, and in soils cropped to high-zinc-demand plants, such as corn, onions, pecans, sorghum, and deciduous fruits (Table 10-7). Calcareous soils have both a high pH and carbonates to which zinc adsorbs. Where topsoil (with its humus) has been eroded or cut off in land grading, zinc deficiency is more likely than if cut-and-fill had not occurred. Sands and anaerobic soils may also be likely to have lower soluble zinc than clays or aerobic soils.

Visual symptoms of zinc deficiency are sometimes distinctive. The low zinc mobility in plants causes some interveinal chlorosis, *but in both younger and some older leaves.* A zinc enzyme is involved in auxin production. Low auxin production reduces stem elongation, which results in shortened internodes and "bunched" leaves on the ends of branches

Table 10-7 Crops Sensitive to Low Levels of Available Zinc

Sensitive Crops (Often Deficient)	Intermediate Sensitivity	Insensitive (Seldom Deficient)
Beans	Alfalfa	Carrots
Citrus	Barley	Forage Grasses
Corn	Clovers	Mustards
Deciduous fruits	Cotton	Oats
Grapes	Potatoes	Peas
Onions	Tomatoes	Rye
Pecans	Wheat	Safflower
Rice		
Soybeans		

Sources: Selected data from (1) *Zinc in Crop Nutrition,* International Lead-Zinc Research Organization and the Zinc Institute, New York, 1974; (2) J. J. Mortvedt, "Do You Really Know You Fertilizers? Part 3: Zinc and Copper," *Farm Chemicals,* **143** (1980), no. 11, p. 56.

(rosette). Leaves are often smaller and thicker than normal leaves and early leaf fall may occur. Names of many zinc deficiencies indicate some of the visual symptoms: "white bud" of corn, "little leaf" of cotton, "mottle leaf" in citrus, and "fern leaf" in Russet Burbank potato (Fig. 10-8).

Solubility of soil zinc is increased by stronger acidity. Solubility increases about 100-fold for each unit that pH is *lowered,* although one study measured only changes of 30-fold per pH unit between pH 5 and 7. Zinc deficiencies are most expected at high pH, particularly in calcareous soils.

10:8.3 Zinc Amendments and Their Use

Various zinc sources are available to correct zinc shortages (Table 10-8). Most zinc salts are soluble enough to supply needed zinc if their total amount is increased in deficient soils. Zinc sulfate is widely used.

The small amount of zinc needed permits use of foliar sprays of zinc chelates for rapid correction of deficiencies. These sprays are used mostly on trees and ornamentals. Rates are about 0.5–2 kg Zn/ha for foliar chelate sprays but may be 10–20 kg Zn/ha for inorganic amendments added to soil.

10:9 Soil Manganese

Manganese is involved in many enzyme systems and in electron transport. In solution it occurs as the Mn^{2+} ion. When in oxidized soils, most of the manganese precipitates as insoluble MnO_2 (pyrolusite). As with zinc, manganese solubility increases about 100-fold per pH unit more acidic. Organic-matter decomposition aids manganese solubility by furnishing electrons to reduce manganese as the decomposition proceeds. Toxic concentrations of manganese are more likely than are toxic levels of zinc, iron, or copper. This is partly because some soils have high total Mn contents and partly because strongly acidic soils can dissolve toxic concentrations of manganese.

10:9.1 Problem Soils and Susceptible Plants

Manganese deficiencies are most common in sands, organic soils, high-pH calcareous soils, and in soils growing fruits, small grains, and leafy vegetables. Manganese-deficient plants have interveinal chlorosis of younger leaves. The deficiency has been given such descriptive

(a)

(b)

FIGURE 10-8 Zinc deficiency in snap beans in Idaho. In (a), the rows in the foreground had no applied zinc and grew poorly. The taller plants to the back had been given 10 lb of Zn per acre as an application to the soil. In (b), note the chlorosis in the younger leaves. (Courtesy of Dale T. Westerman, USDA—ARS Soil and Water Management Research Station, Kimberly, Idaho.)

names as "marsh spot" of peas, "gray speck" of oats, and "speckled yellows" of sugar beets. Toxic levels usually occur only in strongly acidic soils. Lime addition will control the toxicity. Table 10-9 lists plants tolerant of low levels of available manganese.

10:9.2 Manganese Amendments and Their Use

The most used manganese inorganic material is manganese sulfate (in fact, iron, zinc, manganese, and copper sulfates are all the most used *inorganic salts* of these nutrients). Several

Table 10-8 Sources of Zinc

Source	Formula	Percent Zinc
Zinc sulfates	$ZnSO_4 \cdot xH_2O$	23–35
Zinc oxide	ZnO	78
Zinc carbonate	$ZnCO_3$	52
Zinc sulfide	ZnS	67
Zinc in glass frits	—	Varies
Zinc chelates	Various kinds	9–14
Zinc phosphate	$Zn_3(PO_4)_2$	51

Table 10-9 Crops Sensitive to Low Levels of Available Manganese[a]

Sensitive Crops (Often Deficient)	Moderately Sensitive	Tolerant Crops (Seldom Deficient)
Alfalfa	Corn	Corn
Citrus	Cotton	Cotton
Fruit trees	Potatoes	Field beans
Oats	Rice	Fruit trees
Onions	Vegetables	Rice
Potatoes		Vegetables
Sugar beets		

Source: Selected data from J. J. Mortvedt, "Do You Really Know Your Fertilizers? Part 3: Iron, Manganese, and Molybdenum," *Farm Chemicals,* **143** (1980), no. 12, p. 42.

[a] Some crops are listed in more than one column because results show different conclusions, perhaps because of the plant variety and/or growing conditions.

materials used are shown in Table 10-10. Rates of addition may range from 0.5 kg Mn/ha in some foliar sprays to 20–25 kg Mn/ha for additions to some soils.

10:10 Soil Copper

Copper exists in soils mostly as **cupric(Cu^{2+})** and less as **cuprous (Cu^+)** ions. Copper is essential in many plant enzymes (*oxidases,* for example) and is involved in many electron transfers. Plants absorb copper as the cupric ion, but the solution forms are Cu^{2+} in strongly acidic soil, $Cu(OH)^+$ in mildly acidic soil, and $Cu(OH)_2$ at pH values near neutral and more alkaline. The most common copper mineral in soils is chalcopyrite ($CuFeS_2$), with copper in the Cu^+ form. Other copper sulfides also exist. Most copper minerals are of very low solubility. It is

Table 10-10 Sources of Manganese Amendments

Fertilizer Material	Formula	Percent Manganese
Manganese sulfate	$MnSO_4 \cdot 4H_2O$	26–28
Manganous oxide	MnO	41–68
Manganese dioxide	MnO_2	63
Glass frits	—	Variable
Manganese chelates	Numerous kinds	5–12

believed that some available copper comes from exchangeable forms, some from *less exchangeable* forms, and some from soluble organic complexes or chelates. Copper is strongly adsorbed to many solids (clays, aluminum and iron hydrous oxides, and manganese oxides). *Copper generally forms stronger Cu-organic bonds than do the other metal ions.* Copper solubility is also pH dependent, increasing about 100-fold for each pH unit lowering (more acidic).

10:10.1 Problem Soils and Susceptible Plants

Copper deficiency exists because of the following:

1. Copper bonds strongly to organic substances. Excessive straw additions may cause copper immobilization. Organic soils are often Cu deficient. Newly cultivated organic soils have had copper shortages frequently enough to have the problem given the name *reclamation disease.*

2. Sandy soils often have low total copper contents.

3. Calcareous soils, with the high pH of 8.0–8.4, have low copper solubility. Seldom is the availability reduced enough to cause deficient levels.

4. There is competition of copper with other metals (mostly with aluminium, zinc, iron, and phosphate) for uptake by the plant.

Copper deficiencies are fewer than for most other micronutrients, except perhaps of molybdenum and chloride. *Toxicities* occur mostly near copper ore deposits or where copper is smelted and volatile copper or solid wastes accumulate.

Visual deficiency symptoms vary. Yellowing of younger leaves, some off-colors (bluish greens), some small dead spots, and leaf curl are common symptoms. Plants sensitive to low available copper include alfalfa, barley, rice, carrots, citrus, onions, wheat, and oats. Plants that seldom show deficiencies include beans, asparagus, peas, potatoes, rye, and soybeans. Rye and triticale seem to be very tolerant of low available copper levels; in contrast, wheat may exhibit deficiency at relatively average available copper levels.

10:10.2 Copper Amendments and Their Use

Copper applications, like those of zinc, have been quite successful. Often only a few kilograms per hectare have been adequate to correct copper shortages for many years, and in a few instances for several decades. The most used carrier is copper sulfate, commonly known as *blue vitriol,* which has been used to control algal growth in water (even in swimming pools). This material and other copper sources are shown in Table 10-11. Of those materials in the table only the sulfates and chelates are considered as "soluble."

Typical application rates are from 0.2 kg Cu/ha in foliar chelate sprays up to as high as 20 kg Cu/ha for the mineral forms applied to the soil. In Western Australia, one location with only 1.2–2.5 kg Cu/ha applied to the soil aided growth and plant content for up to 35 years.

10:11 Soil Molybdenum

Molybdenum occurs in the soil solution as MoO_4^{2-} (molybdate) ion. It exists in very low amounts in soil but is needed by plants in very small quantities. Most plant molybdenum exists as part of the enzyme *nitrate reductase.* In plants fixing N_2 molybdenum is also needed in the nitrogen-fixing enzyme *nitrogenase.*

Table 10-11 Sources of Copper Used for Fertilizers

Fertilizer Material	Formula	Percent Copper
Tenorite, copper oxide	CuO	75
Copper sulfate	$CuSO_4 \cdot 5H_2O$	25–35
Basic copper sulfates	$CuSO_4 \cdot xCu(OH)_2$	12–50
Copper ammonium phosphate	$Cu(NH_4)PO_4 \cdot H_2O$	32
Glass frits	—	Varies
Copper chelates	Various kinds	8–14

Molybdate ion has many reactions similar to those of phosphate. It is strongly adsorbed to iron and aluminum hydrous oxides. Molybdenum is more soluble (available) as the pH rises to values of 7 or 8 as the hydroxyl ion competes with the molybdate ion for adsorption. Solubility of molybdenum increases about 10-fold per unit rise in pH above about pH 7. In more acidic soils the solubility may be nearer a 100-fold increase with each unit increase in pH. Lime addition increases available molybdenum. High concentrations of soluble manganese and/or copper reduce molybdate absorption by plants.

10:11.1 Problem Soils and Susceptible Plants

Molybdenum deficiencies will be most common in acidic sandy soils, where leaching losses, strong molybdate adsorption, and few molybdenum minerals exist. Soils high in metal oxides (sesquioxides) have low molybdenum availabilities. In Australia and New Zealand, large soil areas are molybdenum deficient; acidic sandy soils of the U.S. Atlantic and Gulf coasts are likewise low in molybdenum.

Crops sensitive to low molybdate levels include legumes, crucifers (cauliflower, brussels sprouts, broccoli), and citrus. Moderate sensitivity to low molybdate levels include cotton, leafy vegetables, corn, tomatoes, and sweet potatoes. "Whiptail" of cauliflower and "yellow leaf spot" of cashew are molybdate deficiencies. When legumes are grown, with their high molybdate demands, chances for deficiency increase.

Even molybdenum, with normally low concentrations in soils, can occasionally exist in toxic concentrations. The toxicities are usually to *grazing animals,* not to plants. Such soils are typically high in organic matter (Histosols) and have a neutral to alkaline pH. The toxicity is really an imbalance between copper and molybdenum. The low copper causes stunted animal growth and bone deformation called *molybdenosis.* Feeding or injecting copper or adding copper fertilizer to the grazing area usually corrects the problem.

10:11.2 Molybdenum Amendments and Their Use

The low amounts of molybdenum needed in plants and the adequate solubility of most molybdate sources make correcting molybdenum deficiency relatively simple (Table 10-12). Only 40–400 *grams* (0.04–0.4 kg) per hectare are needed. The fertilizer may be applied as a foliar spray, or even "dusted or adsorbed to seed" before planting. Spraying cashews with 0.03% foliar spray (weight/volume) corrected symptoms, but not quickly. Up to three months was required before the symptoms were gone.[6] Liming the soil to about pH 5.3 also corrected the deficiency but took even longer than did the molybdenum spray.

---

[6] C. C. Subbaich, P. Manikandan, and Y. Joshi, "Yellow Leaf Spot of Cashew: A Case of Molybdenum Deficiency," *Plant and Soil,* **94** (1986), pp. 35–42.

Table 10-12 Sources of Molybdenum

Source	Formula	Percent Molybdenum
Ammonium molybdate	$(NH_4)_6Mo_7O_{24} \cdot 2H_2O$	54
Molybdenum trioxide	MoO_3	66
Glass frits	—	1–30
Sodium molybdate	$Na_2MoO_4 \cdot 2H_2O$	39

10:12 Soil Chloride

Chlorine exists in soils almost entirely as **chloride ion** (Cl^-), a very soluble and mobile ion. Chloride has little tendency to react with anything in soil. Its role in plants is believed to be osmotic and in balancing cell cationic charges. Amounts of chloride in plants range from values similar to sulfur (0.2%) to that of nitrogen (2%). In some salt-tolerant plants, up to 10% chloride has been measured. If plants sensitive to chlorides have more than 1–2% chlorides, yields are often reduced. Some of these sensitive crops include fruit trees, berry and vine crops, many woody ornamental plants, tobacco, and avocados.

10:12.1 Some Unique Features of Chloride

Chloride cycles easily in the environment. It is supplied to the air by volcanoes and sea spray, to water by water softener wastes, industrial effluents, road de-icing salt, and food wastes (sewage). It is also added to the soil in animal manures, KC1 fertilizers, and rainfall or irrigation waters.

Chloride may accumulate in toxic amounts. Soluble salts, which hinder plant growth, usually have chloride as the most numerous anion. Irrigation water containing high chloride contents, when sprayed and left to dry on the foliage, may cause salt burn.

Some diseases, particularly "take-all root rot," have been decreased by using chloride-containing fertilizers.[7] Some others ("stripe-rusts," "leaf rust," and "tan spot of wheat") seem to be reduced by adequate chloride (above nutritional needs). For example, banding about 40 kg/ha of chloride has been recommended on winter wheat to reduce "take-all root rot." Total additions of 100–130 kg/ha of chloride have been used on winter wheat, part at planting time in the fall, but most of it in February or March.

10:12.2 Chloride Amendments and Their Use

Very little has been done with chloride amendments because no field plots exhibit deficiency and most fertilizers contain some chloride as a contaminant. With the additional benefits attributed to concentrations of chloride higher than needed as a nutrient, more study of large additions (30–50 kg/ha) is needed. Potassium chloride is the most used fertilizer containing large amounts of chloride. Other soluble chlorides are available, such as those of ammonium, calcium, magnesium, and sodium. Each one has disadvantages (cost, physical condition, etc.) or side effects.

---

[7] R. J. Goos, "Chloride Fertilization," *Crops and Soils Magazine,* **39** (1987), no. 6, pp. 12–13.

Availability of Cu, Fe, Mn and Zn ... decreases as soil pH increases, so most deficiencies may occur in neutral and calcareous soils. Conversely, toxicities of these micronutrients, especially Mn, may occur in very acid soils.

—John J. Mortvedt

Questions

1. (a) Give the ionic form of potassium in solution. (b) How does being a cation affect mobility of the potassium in the soil?
2. (a) What sources of potassium do plants use during a growing season? (b) What plants have high potassium requirements? Low requirements?
3. Explain why humid areas are likely to have inadequate available potassium, whereas arid regions are likely to have adequate available potassium.
4. Discuss briefly some "management suggestions" to maintain adequate potassium available to plants.
5. Both potassium chloride and potassium sulfate are extensively used fertilizers. (a) When might chloride be used rather than the sulfate? (b) When might the sulfate be preferred?
6. (a) Is there a volatile potassium form? (b) Is potassium released during humus decomposition? Explain. (c) What does *chromatographic* movement of K^+ during leaching mean?
7. Tabulate four of the many similarities between soil sulfur and soil nitrogen.
8. To what extent is sulfur deficiency likely to be more common in the future? Explain.
9. (a) What ion forms of sulfur are most common in the soil solution? (b) What natural soil sources supply sulfur for plants?
10. Generally, where do acidic sulfate soils develop? Explain.
11. How do these soils differ from each other: acidic sulfate soils, cat clays, sulfuric horizons, and *potential* acidic sulfate soils?
12. In words, explain the nonbiological oxidation of sulfides to produce acidity.
13. How does each of these management systems or techniques allow use of acidic sulfate soils: (a) growing paddy rice and (b) controlling the depth of the water table?
14. If neither system listed in question 13 is practical, how feasible is it to drain and lime these soils?
15. Briefly discuss a combination management scheme that would involve crop selection, controlled water table depth, and liming. Rely on information derived from this chapter and from previous chapters.
16. (a) What is the form of boron in soil solution? (b) In which climatic region is deficiency most likely? Explain.
17. If soluble boron amendments are easily applied to correct a boron deficiency, why is it still hazardous to apply enough for several years in one addition?
18. (a) Although iron occurs in large amounts in soils, iron is often deficient. Explain why. (b) In what pH range are iron compounds least soluble?
19. (a) If iron compounds have such low solubility, why are there not more crops exhibiting iron deficiency? (b) How is iron deficiency corrected?
20. (a) Name a few plants that are (i) sensitive to and (ii) tolerant of low available iron levels? (b) What is the ionic form of iron in normal soil solution?
21. Give the visual deficiency symptoms for iron.
22. (a) Give the ionic form of zinc in the soil solution. (b) How mobile is zinc in soils? Explain.
23. (a) In what soils and (b) in which soil pH range is zinc deficiency most likely?

24. (a) Give the visual zinc deficiency symptoms. (b) List some plants susceptible to zinc shortage. (c) What materials are used to correct a zinc deficiency?

25. (a) In what soils are deficiencies of manganese and/or copper most expected? (b) List the common ion forms of manganese and copper in soil solution.

26. (a) Give the ionic form of molybdate. (b) What other plant nutrient has somewhat similar ion form and chemical reactions?

27. What crops are most likely to exhibit a molybdenum deficiency? Explain why.

28. Even though chloride does not exist in deficient amounts, soil amendments containing chloride have benefited the growth of some crops. Explain how this could be.

11

Salt-affected Soils and Their Reclamation

Natural systems can take a lot of stress and abuse, but there are limits.

—Anonymous

11:1 Preview and Important Facts

PREVIEW

Salt is the savor of foods but the scourge of agriculture: In excess it kills growing plants. As early as 3500 B.C., the people of Mesopotamia farmed some of the richest land in the world—the Fertile Crescent of the Tigris and Euphrates Rivers (in Turkey and Iraq). Over nearly 5000 years they grew wheat and barley, then only salt-tolerant barley; then salt took over and nothing grew. The land was abandoned. About 2100 years ago the Romans plowed the fields of conquered semiarid Carthage and applied salt to ensure that the Carthaginians could not reestablish their powerful metropolis; efforts to recolonize the area 24 years later failed because the salted fields were still unproductive.[1]

Soluble-salt contamination of soil has caused problems for all of recorded history, primarily in arid regions of the world where inadequate rainfall leaches few or no salts from the soil. Salt devastates many areas and is increasingly a serious limitation to plant growth as water supplies become more limited and are increasingly polluted with soluble salts. Losses due to salt damage just in the Colorado River basin (eastern Utah, western Colorado, and Arizona, but also supplying water to southern California) cost agriculture more than $100 million per year in the last decade.[2] The area of salt-affected soils will increase as irrigation increases. Managing and reclaiming salt-affected soils is, indeed, an increasing worldwide concern.

Soluble salts are those inorganic chemicals that are more soluble than gypsum ($CaSO_4 \cdot 2H_2O$), which has a solubility of 0.241 g per 100 mL of water at 0°C (0.032 oz/per gal). Common table salt (NaCl) has a solubility nearly 150 times greater than gypsum (35.7 g per 100 mL or 47.7 oz/gal). Most soluble salts in soils are composed of the cations sodium (Na^+), calcium (Ca^{2+}), and magnesium (Mg^{2+}) and the anions chloride (Cl^-), sulfate (SO_4^{2-}), and

---

[1] Moses Hadas, *Imperial Rome*, Time, New York, 1965, pp. 38–39.
[2] Anonymous "California's Problem Passing Water," *Agrichemical Age,* **33** (1989), no. 11, pp. 24, 25, 29C.

bicarbonate (HCO_3^-). Relatively smaller quantities of potassium (K^+), ammonium (NH_4^+), nitrate (NO_3^-), and carbonate (CO_3^{2-}) also occur, as do many other ions. In some soil solutions soluble-salt concentrations are higher than in seawaters, which are 3–4% total salts.

The cations and anions that form soluble salts come from dissolved minerals as they weather. If precipitation is too low to provide leaching water, usually less than about 38 cm (15 in.) annually, most or all of the soluble salts remain in the soil. As water evaporates from the soil surface, the soil salts move toward the surface but remain within or on the soil.

Large contents of soluble salts act osmotically to make it harder for plants in **saline soils** to absorb water from the soil solution. Soils with a high exchangeable sodium percentage (over about 15% = **sodic soil**) cause the soil to be dispersed, making it slowly permeable or impermeable to water. Irrigation can cause salt accumulation because all surface waters and groundwaters contain soluble salts.

Reclaiming saline soils requires leaching out of the soluble salts. In theory, removing salt is easy: The salts are dissolved in irrigation water and leached out of the soil profile. Reclamation of salt-affected soils requires (1) adequate internal drainage, (2) replacement of excess exchangeable sodium (and some potassium) in sodic soils, and (3) leaching out of the soluble salts. Establishing internal soil drainage may be the most difficult requirement.

If the salt content is not too high, soils sometimes can be used for plant growth by careful management. In managing salty soils, salt-tolerant plants suitable for that soil, climate, and farm operation are selected; frequent irrigations are used to keep salts diluted; nongrowing-season irrigation is used to leach salts partially downward; and seeds are planted in the low-salt areas of seedbeds.

The area of salt-affected soils will increase as irrigation increases. Managing and reclaiming salt-affected soils is, indeed, an increasing worldwide concern.

IMPORTANT FACTS TO KNOW

1. The origins and composition of soluble salts
2. The taxonomy of salt-affected soils
3. How soluble salt contents are measured and reported
4. The approximate salt contents that cause damage to plants
5. How damage to plants caused by soluble salts and excess exchangeable sodium is lessened
6. The principles used in the reclamation and management of salt-affected soils

11:2 Soluble Salts and Plant Growth

Soluble salts in water cannot be seen. Salts are solids when dry and sometimes can be seen on the surface of soils during drying conditions (Figs. 11-1 and 11-2). Descriptive names have been given for various salty soils, such as *white alkali, black alkali, slick spots,* and *summer snow.* These names come from the surface appearance of some salty soils. The white appearance of salts explains the names *white alkali* and *summer snow*. If the soil has high exchangeable sodium, its pH will be around pH 9; soil humus colloids disperse, coloring puddles of surface water black, like puddles of oil. After drying, the soil has black crusts over its surface and ped faces (black alkali).

11:2.1 Measuring Soluble Salts

Soluble salts are measured by electrical conductivity, and the units of conductance (International System, SI) are **siemens per meter.** Some sciences and recent literature still report in the older units of **mhos per centimeter.** The relationships of the units of conductance are given in the following:

Basic Unit	Units Most Used in Soils	Old Units Still in Literature
Siemens meter^{-1} **S m^{-1}**	Decisiemens meter^{-1} **dS m^{-1}**	Millimhos centimeter^{-1} **mmhos cm^{-1}**
	1 S m^{-1} = 10 dS m^{-1} = 10 mmhos cm^{-1}	

The range of plant tolerance to salt is approximately as given below for the conductivity of the soil's **saturation paste extract**:

Extract conductivity (dS m^{-1})	Growth Reduction by Salt in Soil
0–2	Few plants are affected
2–4	Some sensitive plants affected (strawberries)
4–8	Many plants affected
8–16	Most crop plants affected
16 +	Few plants grow well

Some approximate conversions from electrical conductivity to other relationships used to measure soluble salts are given below:

$$(\text{dS m}^{-1})\ (640) = \text{total dissolved solids (TDS) in mg L}^{-1}$$
$$(\text{dS m}^{-1})\ (0.36) = \text{osmotic pressure of solution in -bars}$$
$$(\text{dS m}^{-1})\ (10) = \text{mmol}_c\ \text{L}^{-1} \text{ of total cations or of total anions}$$

FIGURE 11-1 Toxic salt accumulation (white layer on ridge at arrow No. 1) has prevented the growth of all plants on this irrigated cotton field in southwestern Texas. The accumulation of excess salt could have been caused by a slight surface depression and/or a concentration of clay in the soil profile that reduced infiltration of irrigation water, as seen in the insert (arrow No. 2). (*Source:* Texas Agricultural Experiment Station, El Paso.)

FIGURE 11-2 Nonproductive area caused by salt accumulation (white areas) in northeastern Montana. The saline-seep develops when water infiltrates into soil below root depth to an almost impermeable dense clay, which causes water to move laterally downslope. Eventually, at lower elevation on slopes, the water seeps laterally at a shallow depth below the soil surface and then to the surface by capillarity. As water evaporates on the surface, it leaves calcium, magnesium, and sodium sulfate salts. The use of deep-rooted crops in rotation and intensive cropping with minimum summer fallow so that soil water gets used up in areas above the low seep area have been recommended for reducing seepage movement. (*Source:* A. D. Halvorson and A. L. Black, "Saline-Seep Development in Dryland Soils of Northeastern Montana," *Journal of Soil and Water Conservation,* **29** [1974], no. 2, pp. 77–81; also *North Dakota Farm Research,* **33** [1976], no. 4, pp. 3–9.)

11:2.2 Effects of Salt Concentration

Although specific toxicities due to high concentrations of sodium, chloride, or other ions can occur, salts usually reduce plant growth by an **osmotic effect**. High salt concentration increases the forces that hold water in the soil and require plant roots to expend more energy to extract the water. During a drying period salt in soil solutions may become concentrated enough to kill plant by "pulling" water from them (**exosmosis**).

Salt in the soil solution forces a plant to exert more energy to absorb water and to exclude the excess ions of salt from metabolically active sites. The saltier a soil is, the wetter it must be kept to "dilute" the salt if it is to cause the least salt hinderance to the growing plant.

Salts are usually most damaging to *young* plants, but not necessarily at the time of germination, although high salt concentrations can slow seed germination by several days or completely inhibit it. Because soluble salts move readily with water, evaporation moves salts to the soil surface, where they accumulate, sometimes becoming visible as powdery white salt crusts. Plant species have variable tolerances to the salt in soils, and the specific effects on different parts of a plant also vary (Figs. 11-3 and 11-4). The minor disagreements in salt content at which a particular plant experiences growth reduction are often due to differences in the test conditions, plant age, or cultivar used. Usually, plant tolerances increase with maturity. Thus, the data of Figs. 11-3 and 11-4 should be used with some reservations about their exactness.

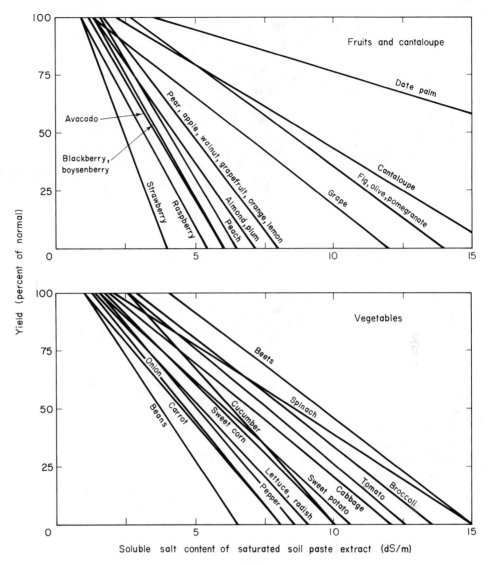

FIGURE 11-3 Yield reduction of fruits and vegetable crops by soluble salts in soils. Notice that this salt scale is expanded more than that of Fig. 11-4. (*Source:* Drawn from data in *Soil Survey Investigations for Irrigation,* FAO Soils Bulletin 42, Food and Agricultural Organization of the United Nations, Rome, 1979, pp. 72–74. Courtesy of Raymond Miller, Utah State University.)

Most plants are least affected by soil salts when in their *mature* stages, but plants in the germination and seedling stages may be quite sensitive to salt damage. The data of Figs. 11-3 and 11-4 are for salt damage to plants *already germinated and in the later seedling stage.* Crop yields may be greatly reduced for some crops at lower salt levels than shown in Figs. 11-3 and 11-4 *if* that salt level reduces the number of germinated seeds or number of seedlings that survive to older stages of growth.

The effect of salt on crop yield must also consider the productive part, which is important for yield. Fig. 11-2 shows that salt levels must be kept lower when producing corn for grain (a 50% yield loss at 6 dS m^{-1}) than when producing corn for forage (a 50% yield loss at 9 dS m^{-1}).

Both barley and cotton have considerable salt tolerance, but high concentrations of salt affect the *vegetative* growth (stems, leaves) more than the seed heads of barley or the bolls

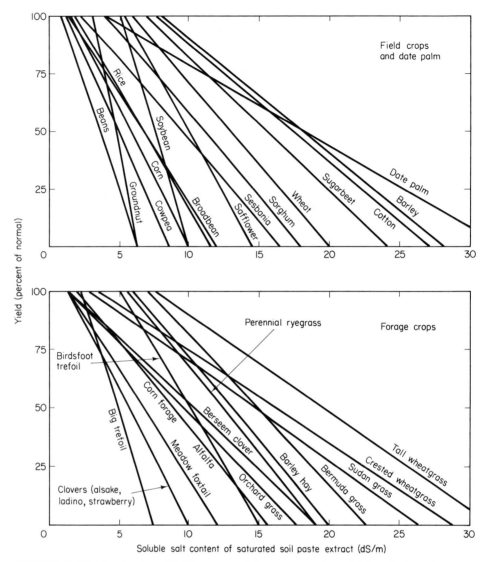

FIGURE 11-4 Yield reduction of field crops and forage crops by soluble salts in soils. Notice that this salt scale is double the salt range of Fig. 11-3. (*Source:* Drawn from data in *Soil Survey Investigations for Irrigation,* FAO Soils Bulletin 42, Food and Agricultural Organization of the United Nations, Rome, 1979, pp. 72–74. Courtesy of Raymond Miller, Utah State University.)

of cotton. This is fortunate because these are the products desired. Rice is an exception to this generality; in rice grain, yields are reduced before vegetative growth is affected. However, because rice can be grown in *ponded water* (the most dilute condition possible for soil salts), the crop is often grown on high-salt soil in the early stages of reclamation (salt removal).

Commercial tomatoes—which are moderately tolerant to salinity—grown in saline soils in Israel had yields reduced 10% for each 1.5 dS m^{-1} that the saturation extract was above 2 dS m^{-1} (at which no yield reduction occurred). This is about 50% yield reduction at a conductivity of about 9.5 dS m^{-1} (9.5 mmhos/cm).

Valencia oranges in California, considered to be salt sensitive, have important yield reductions even at conductivities of 2.5 dS m^{-1}. Measurements over an 8-year period showed that conductivities above 3 dS m^{-1} indicated excessive salt effects.

These two examples suggest that even though a soil has a conductivity lower than 4 dS m^{-1} and is not *termed* saline, it may have enough salt to lower the yields of salt-sensitive plants.

11:2.3 Effects of Specific Ions

Exchangeable sodium in concentrations above about 15% (or SAR = 13) exerts its greatest effect on plant growth by dispersing the soil (Detail 11-1). As low as 10% exchangeable sodium in fine-textured (clayey) soils and 20% in sandier soils have caused dispersion damage. Colloid dispersal makes the soil less permeable, or even impermeable, and causes it to form hard surface crusts when dry (Fig. 11-5).

Not all soils have dispersion problems at the same exchangeable sodium percentage. Montmorillonite clays are the most easily dispersed. Some clayey soils disperse with only 9–10% exchangeable sodium. Kaolinitic soils, however, may be quite stable to even 25–35% exchangeable sodium. In many Oxisols and Ultisols, the kaolinite–metal oxide clay mixtures

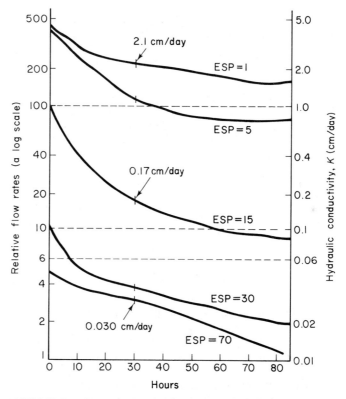

FIGURE 11-5 Example of the change in water flow (hydraulic conductivity) through a soil as influenced by the exchangeable sodium percentage. The curves are for the same soil. Each curve is for the soil with a different exchangeable sodium percentage. Note that *water flow in all soils decreases with time as clays swell*. The swelling reduces pore sizes. Also, flowing water carries soil particles that lodge in pores, further reducing flow through the soil. Percolation of water through soil with an ESP of only 5 is 47 times faster at the end of 30 hours flow than flow in that soil with ESP of 70 and is over 8 times faster than at an ESP of 15.

The cause of soil dispersion with increasing exchangeable sodium percentage is the *net sum of attraction and repulsion forces* between the soil particles. The *attraction forces* are called **van der Waals** and **Londonary** forces. Even though they are not well understood, they exist—much like gravity exists and can be measured, but is not easily explained. The *repulsive forces* are the charged surfaces of the clays, caused by the density of positive ions near clay surfaces. When particles are very close together (within a few nanometers), the attractive forces exceed repulsion forces (see the figure below). But if the thickness of

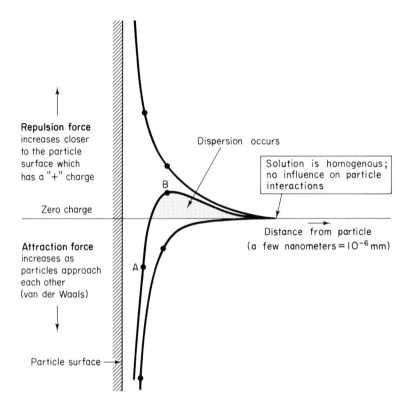

Repulsion force increases closer to the particle surface which has a "+" charge

Zero charge

Attraction force increases as particles approach each other (van der Waals)

Particle surface →

Dispersion occurs

Solution is homogenous; no influence on particle interactions

Distance from particle (a few nanometers = 10^{-6} mm)

hydrated adsorbed cations, such as Na^+, can keep the particles farther apart, the attraction forces are reduced by a factor of the distance squared. The net result is a lower attraction and a higher repulsion resulting in a net repulsion of particles from each other. This net repulsion causes **dispersion** or **deflocculation.** Dispersed particles tend to stay separated and suspended in water longer than do nondispersed ones.

If the particles have a net *attraction,* the particles will **flocculate,** which is the "clumping together" of particles. Flocculation can be caused by (1) a high salt concentration and (2) by adsorption of most cations other than sodium or potassium. The high salt content "pushes" the **ionic double layer** of adsorbed cations closer to the particle surface by its high concentration. In more dilute solutions and with adsorption of *large-diameter, low-charge hydrated ions,* such as that of sodium, the ionic double layer expands out from the colloid surface. Particles approaching each other are repelled by the weak net positive charge on the surface of each. The particles cannot get close enough to each other for strong attractive forces to become effective.

may be fairly stable even to values of nearly 45–50% exchangeable sodium. Soils of sandy texture (and low clay content) can also tolerate higher exchangeable sodium percentages because of their good permeability and large pores. Dispersion is destruction of soil structure. The upper soil pores become filled with lodged dispersed particles, and both air and water exchange into and out of the soil are reduced. The hardened crusts can physically inhibit seedling emergence.

Many deciduous fruits can be injured by as little as 5% exchangeable sodium; citrus, stone-fruits, and blackberries are among the sensitive ones; grapes are quite sodium tolerant. Rapid absorption of sodium or chloride by the leaves of stone fruit and citrus trees makes some water that might be satisfactorily added as surface irrigation *unsatisfactory* to use if sprinkler-applied. Strawberry leaves, which absorb sodium and chloride more slowly, and avocado leaves, which absorb little or none, can be sprinkle-irrigated safely with those same waters.

11:3 Saline and Sodic Soils

Salted soils are classified on the basis of two criteria: (1) the *total soluble salt content,* and (2) the *exchangeable sodium percentage* or, more recently, *sodium adsorption ratio.* Because ions in water conduct electrical current, electrical conductivity (EC) is a fast, simple method of estimating the amount of total soluble salts in a soil sample. To measure a soil's conductivity, a weighed soil sample is mixed with water to form a saturated paste; the liquid is then removed by pressure or suction filtration and the conductivity of this extract is measured.

11:3.1 Sodium Adsorption Ratio

The **sodium adsorption ratio (SAR)** is used to estimate what the exchangeable sodium percentage of a soil is, or what it is likely to become if the water of known SAR is used for years on that soil. The SAR has a good correlation to the exchangeable sodium percentage (ESP) and is much easier to calculate exactly or to estimate from a few simple analyses than is the ESP. The SAR defined, in mmoles/L, is

$$SAR = \frac{Na^+}{\sqrt{Ca^{2+} + Mg^{2+}}}$$

If values are in $mmoles_c/L (= meq/L)$:

$$SAR = \frac{Na^+}{\sqrt{\dfrac{Ca^{2+} + Mg^{2+}}{2}}}$$

The units left are $(mmoles_c\ L^{-1})^{1/2}$ and are commonly ignored in typical usage (Calculation 11-1).

When the SAR is 13, the soil probably will lose permeability as salts are removed. An SAR of 13 replaces an ESP of 15% in the criteria in the United States. This approximate relation is shown below:

Sodium – Absorption – Ratio (SAR)

10   20   30 40 50   100 150 250

0   10   20   30   40   50   60   70

Estimated Equilibrium
Exchangeable – Sodium – Percentage (ESP)

To calculate the value of the SAR of a soil solution or water sample, simple insert the values in the following equation as $mmol_c/L$:

$$SAR = \frac{\text{millimoles}_c \text{ of } Na^+}{\sqrt{\text{millimoles}_c \text{ of } (Ca^{2+} + Mg^{2+})/2}}$$

Problem   Calculate the SAR for soils 1, 2, and 4 from Table 11–2.

Solution   The $mmol_c/L$ is calculated by dividing the mg/L of the cation by the ion's atomic weight and multiplying by the cation's valence (oxidation number). For example, calculate the $mmol_c/L$ for Ca^{2+} in soil 1 of Table 11-2. From the table, there are 1301 mg of Ca^{2+} and the gram atomic weight of Ca^{2+} (from chemical periodic table) is 40.08 g:

$$mmol_c \text{ of } Ca^{2+} \ L^{-1} = \frac{1301 \text{ mg}}{L} \left| \frac{1g}{1000 \text{ mg}} \right| \frac{1 \text{ mol } Ca^{2+}}{40.08 \text{ g}} \left| \frac{2 \text{ mol}_c \ Ca^{2+}}{1 \text{ mol } Ca^{2+}} \right| \frac{1000 \text{ mmol}_c}{1 \text{ mol}_c}$$

$$= \frac{2602 \text{ mmol}_c}{40.08} = 64.9 \text{ mmol}_c \ L^{-1} \text{ of } Ca^{2+}$$

for Soil 1:

$$SAR = \frac{15.26 \text{ mmol}_c/L \text{ of } Na}{\sqrt{\dfrac{65.0 + 34.2 \text{ mmol}_c/L}{2}}} = \frac{15.26}{\sqrt{49.6}} = \frac{15.26}{7.04} = 2.24$$

for Soil 2:

$$SAR = \frac{79.5 \text{ mmol}_c/L \text{ of } Na}{\sqrt{\dfrac{6.7 + 9.9 \text{ mmol}_c/L}{2}}} = \frac{79.5}{\sqrt{8.3}} = \frac{79.5}{2.88} = 27.6$$

for Soil 4:

$$SAR = \frac{29.2 \text{ mmol}_c/L \text{ of } Na}{\sqrt{\dfrac{1.1 + 0.33 \text{ mmol}_c/L}{2}}} = \frac{29.2}{\sqrt{0.71}} = \frac{29.2}{0.84} = 37.8$$

Source: Raymond W. Miller, Utah State University.

Table 11-1 Classification of Salt-Affected Soils[a]

Name for Soil	Electrical Conductivity of Saturation Extract EC_e (decisiemens meter $^{-1}$)	Sodium Adsorption Ratio (SAR)
Normal soils	Less than 4	Less than 13
Saline soils	More than 4	Less than 13
Sodic soils	Less than 4	More than 13
Saline-sodic soils	More than 4	More than 13

Source: *Glossary of Soil Science Terms,* Soil Science Society of America, Madison, Wis., 1979, pp.14–15.

[a] Although the salt content division concentration is left at 4.0 decisiemens meter $^{-1}$ (= 4dS m^{-1}), plants sensitive to salts may be affected by contents as low as 2.0 dS m^{-1}; salt-tolerant plants may not be affected below 8.0 dS m^{-1} salt content.

[b] Formerly called *white alkali* soils and *solonchak.*

[c] Formerly called *black alkali* because of dispersed black organic-matter coatings on peds and the soil surface.

[d] Formerly called *white alkali* or *black alkali,* depending on the visual appearance of the individual soil.

11:3.2 Salty Soil Classification

Salt-affected soils may have one of three names: saline, sodic, or saline-sodic (Table 11-1).

- **Saline** soil has a saturation extract conductivity of 4.0 decisiemens per meter or greater and has a low SAR. Formerly, these soils were called *white alkali* and *solonchak.*
- **Sodic** soil has an SAR of the saturation extract of 13 or more but has low salt content. Formerly, these soils were called *black alkali.*
- **Saline-sodic** soil has *both* the salt concentration to qualify as saline and SAR of 13 or more needed to qualify as sodic. Formerly, these were called either *white alkali* or *black alkali.*

Table 11-2 lists some typical soluble-salt data for four salted soils: Soil 1 is saline, 2 and 3 are saline-sodic, and 4 is sodic soil. However, Soil 1 of Table 11-2, the highest in salt of the four soils listed, has only about 0.7% soluble salts; many soils are much saltier than this.

11:4 The Salt Problem and Salt Balance

Salt buildup is an existing or potential danger on *almost all* of the 17 million hectares (42 million acres) of irrigated land in the United States, and it is an increasing problem on non-irrigated semiarid and arid cropland and rangeland. Much of the world's unused land that is expected to supply future increased food is in arid and semiarid regions where irrigation would be necessary. Continual application of water, all of which contains salts, will continually increase the soluble salts in all soils unless leaching of the soil occurs.

The demand for water and water quality legislation, such as the Porter-Cologne Water Quality Act in California, promote the reuse of drainage water; eventually, the water is no longer suitable for agriculture. This salty wastewater must eventually be recycled to groundwaters or streams. This is permitted only after the water's salt content is lowered to an acceptable level, a value of 500 ppm in California.

Table 11-2 Soluble-Salt Constituents in the Saturation Extracts from Three Selected Salty Soils and One Sodic Soil, Showing the Relative Amounts of the Various Ions[a]

Soil 1 (pH 7.8; EC 7.6) Saline				Soil 2 (pH 7.3; EC 9.2) Saline-Sodic			
Cations		Anions		Cations		Anions	
Na^+	351	Cl^-	4329	Na^+	1828	Cl^-	2556
Ca^{2+}	1301	SO_4^{2-}[b]	452	Ca^{2+}	134	SO_4^{2-}[b]	965
Mg^{2+}	411	HCO_3^-	117	Mg^{2+}	119	HCO_3^-	146
K^+	41	CO_3^{2-}	0	K^+	20	CO_3^{2-}	0

Soil 3 (pH 8.1; EC 7.9) Saline-Sodic				Soil 4 (pH 9.6; EC 3.2) Sodic			
Cations		Anions		Cations		Anions	
Na^+	1661	Cl^-	75	Na^+	672	Cl^-	266
Ca^{2+}	278	SO_4^{2-}	4325	Ca^{2+}	22	SO_4^{2-}	221
Mg^{2+}	71	HCO_3^{2-}	183	Mg^{2+}	4	HCO_3^-	1141
K^+	23	CO_3^{2-}	12	K^+	160	CO_3^{2-}	0

Source: Soils 2, 3, and 4 selected and modified from J. D. Rhoades and Leon Bernstein, "Chemical, Physical, and Biological Characteristics of Irrigation and Soil Water," in *Water and Water Pollution Handbook,* vol. 1, L. L. Ciaccio, ed., Marcel Dekker, New York, 1971, p. 160.
[a] Values are milligrams of the ion per liter. To get millimoles$_c$ divide each value by the ion's atomic or polyatomic weight and multiply by the ion's valence. Soil 1 is saline, soils 2 and 3 are saline-sodic, and soil 4 is sodic. EC is electrical conductivity in dS m^{-1} (decisiemens per meter), a measure of salt content. Ion concentrations are given in milligrams per liter; multiply by 0.000134 for ounces per gallon. As an example: to get mmol$_c$ for sulfate in soil 1, divide 452 by weight of sulfate (= 96) and multiply by it valence of 2 = 9.4 mmol$_c$ of sulfate per liter.
[b] Actually, sulfate plus nitrate, calculated as only sulfate.

The **salt balance** is making *outgoing* salt equal to *incoming* salt. Because salts are continually added in applied waters, especially irrigation waters, some leaching must be caused by the addition of more water than needed to wet the plant root zone. This can maintain the salt balance. This *additional* water needed for leaching, over that needed to wet the profile, is called the **leaching requirement (LR)**, and is defined as

$$LR = \frac{EC_{iw}}{EC_{dw}}$$

where EC_{iw} is the electrical conductivity of the irrigation water and EC_{dw} is the electrical conductivity of the soil saturation extract at which a 50% decrease in yield is obtained in uniformly saline soil. Usually, the effort to achieve adequate deep wetting from irrigation at the lower end of the field means that enough excess water for leaching is already being added. With increased use of saltier water and with less water available, more attention will need to be given to the leaching requirement than is now given rather inadvertently by overirrigation. Notice that as irrigation water becomes saltier, the fraction (*LR*) becomes larger, meaning that more water must be added for leaching to avoid salt buildup (see Calculation 11-2). The use of brackish water to grow salt-tolerant crops is proposed as one way to find more water in some water-tight areas.[3]

---

[3] Don Gardner, "Irrigated Land May Include Widespread Use of Saline Waters," *Irrigation Age,* **18** (1984), no. 5, pp. 6–7.

Calculation 11-2 Leaching Requirement Calculation

Problem Assume that an irrigation water has a conductivity of 108 mS m⁻¹ (= 1.08 dS m⁻¹ or 1.08 mmhos/cm). The field corn planted has a 50% yield reduction at a soil saturation extract conductivity of 6 dS m⁻¹ (from Fig. 11–4). Calculate the additional amount of water to add if the water needed to wet the profile is 6.35 cm (2.5 in.).

Solution Substituting in the leaching requirement equation yields

$$LR = \frac{EC_{iw}}{EC_{dw}} = \frac{1.08 \text{ dS m}^{-1}}{6 \text{ dS m}^{-1}} = 0.18$$

This decimal (or fraction) is that fraction of the amount of water needed to wet the soil that must be added additionally. The total water needed is

$$6.35 \text{ cm} + (0.18)(6.35) = 7.49 \text{ cm } (2.95 \text{ in.})$$

Some scientists claim that the leaching requirement may be satisfactorily reduced to only 25–40% of the amount given in the previous equation.[4] However, the limit of the leaching requirement is not simple to determine. The salt sensitivity of the plant must be considered. For example, in California the more-salt-tolerant wheat and sorghum can have a leaching requirement as low as 0.08 before growth reduction occurs with water containing 1350 ppm of salts. Lettuce, a more salt-sensitive plant, must have a leaching requirement of 0.20, two and a half times more.[5]

In the Wellton-Mohawk project in southwestern Arizona, a leaching requirement (*LR*) of 0.42 on soil using Colorado River water (150 metric tons of salt per thousand cubic meters of water) caused salt removal in drainage water of 22.3 metric tons per hectare (9.95 t/a); when the *LR* was reduced to 0.10 in an effort to comply with the international agreement to reduce the Colorado River water salt load, only 9.96 metric tons per hectare (4.44 t/a) of salt were removed. Obviously, the low leaching requirement results in more salt accumulation in time. The more uneven the water application is, the greater the *LR* must be to keep all of the soil area low in salts.

11:5 Reclaiming Salty Soils

Irrigation of arid lands is increasing salt problems in many soils. Three general rules to reclaim soils affected by salt are given in the following:

1. **Establish internal drainage**. For some soils, drainage is already adequate. In other soils, drainage might require the installation of drainage systems (tile lines, open ditches, etc.). Drainage may be impractical (too costly) or impossible (too flat, no near outlet, cost prohibitive, or illegal according to Environmental Protection Agency [EPA] regulations).

---

[4] J. van Schilfgaarde, L. Bernstein, D. Rhoades, and S. L. Rawlins, "Irrigation Management for Salt Control," *Journal of Irrigation and Drainage Division,* **100** (1974), no. IR3, p. 321.
[5] G. J. Hoffman, S. L. Rawlins, J. D. Oster, J. A. Jobes, and S. D. Merrill, "Leaching Requirement for Salinity Control: [Part I.] Wheat, Sorghum, and Lettuce," *Agricultural Water Management,* **2** (1979), pp. 177–192.

2. **Replace excess exchangeable sodium**, if needed. This is necessary for some sodic and saline-sodic soils. The extent of this need varies with soil texture, kind of clay, water quality of available water, extent of present damage, and other things.

3. **Leach out most of the soluble salts**. Saline soils must have the salt content lowered, at least in part of the root zone. In sodic soils, most of the *replaced sodium* must also be leached from the root zone. *Without leaching there is no reclamation for long.* It is desirable to use good-quality irrigation water for leaching out salts.

Although these three general guides seem simple, often the physical problems and costs are great. Special situations require different approaches. When economics is considered, as the land owner will certainly do, additional compromises and different techniques may be tried in each instance.

11:5.1 Reclaiming Saline Soils

Saline soils are *relatively* easy to reclaim for crop production if adequate amounts of low-salt irrigation water are available, internal and surface drainage are present, and salt disposal dump areas (sinks) are available. The main problem is to leach most of the salts downward and out of contact with subsequent irrigation water.

Frequently, saline soils have a high water table or a dense gypsum layer or are fine textured. These conditions reduce the movement of irrigation water downward and make it difficult to leach the salts to the desired depth below the plant root zone. In salty soils that have a high water table, artificial drainage is necessary before excess salts can be removed. Deep chiseling or deep plowing may be used on soils with impervious layers to open the soil for the desired downward movement of percolating salty water. This process is expensive and may be repeated several times.

Reclamation of saline soils, particularly when only rainfall or limited irrigation is used, can be hastened by the application of a surface organic mulch, as reported from the Rio Grande Valley of Texas. Because mulch slows surface evaporation, salt movement to the soil surface in evaporative water is decreased and the *net downward movement* of salt is increased. Cotton gin trash and chopped woody plants are equally effective when applied at the rate of 67.2 metric tons per hectare (30 t/a). With mulch, the surface soil salt content becomes less, whether the area receives only natural rainfall or supplemental sprinkler irrigation. Crop residues have also helped in North Dakota.

The quantity of water required to remove salts from the soil depends on many things: how deep the salts are to be washed, what percentage of the salts are to be removed, and how the leaching is done (ponding constantly or by intermittent sprinkling). A general guide is that with ponded water about 30 cm (12 in.) of water are required to remove 70–80% of the salt for each 30 cm depth of soil to be leached of salt (Fig. 11-6). Intermittent water additions are more efficient and reduce the water applied to about 70% of that needed with ponding (continuous) leaching methods.

Where soil boron is sufficiently high to be a special problem as a specific toxic element, its removal by leaching is slower. Boron is weakly adsorbed by some soil constituents, and leaching may require three times more water than is needed to remove other soluble salts to the same extent.

11:5.2 Reclaiming Sodic and Saline-Sodic Soils

The reclamation of sodic soils may require a technique modified from that used for reclamation of saline soils. In sodic soils the exchangeable sodium is so great that the resulting dispersed soil is almost impervious to water (Fig. 11-7). But even if water could move downward freely in

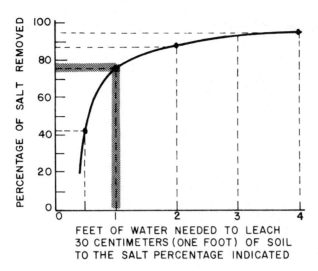

FIGURE 11-6 Estimated amount of normal (low-salt) water required to remove salts from saline soils. About 30 cm (1 ft) of water through 30 cm (1 ft) of soil removes nearly 80% of the salt. (*Source:* Data of R. C. Reeve; redrawn and modified from J. D. Rhoades, "Drainage for Salinity Control," in *Drainage for Agriculture*, J. V. Schilfgaarde, ed., No. 17 in the Agronomy Series, American Society of Agronomy, Madison, Wis., 1974, pp. 433–461.)

(a)

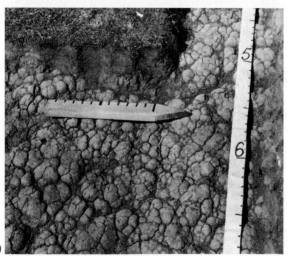

(b)

FIGURE 11-7 This sodic soil in North Dakota occurs in spots where high exchangeable sodium (16–20% at a 25-cm [10-in.] depth) on the clayey soil makes it impermeable to water. (a) These "gumbo" or "scab spots" hold ponded water for days. (b) A close-up shows the shallow surface soil (an **A**) and the rounded tops of *columns* making up the **Bt** clay horizon. The scale on the stake is one foot (30 cm) divided into inches. (*Source:* Fred M. Sandoval and G. A. Reichman, "Some Properties of Solonetzic (Sodic) Soils in Western North Dakota," *Canadian Journal of Soil Science,* **51** [1971], pp. 143–155; photos by Fred Sandoval.)

sodic soils, water alone will not leach out the excess *exchangeable* sodium. The sodium must first be replaced by another cation and then leached downward and out of the root zone.

By cationic exchange reactions, calcium is used to replace sodium in sodic soils. Of all calcium compounds, calcium sulfate (gypsum, $CaSO_4 \cdot 2H_2O$) is considered the most convenient and cheapest for this purpose. Calcium solubilized from gypsum replaces sodium, leaving soluble sodium sulfate in the water, which is then leached out. Rarely, the leaching waters themselves may contain enough cations to replace the sodium. In a calcareous saline-sodic soil in India, leaching with a good low-salt irrigation water (SAR = 0.6, pH 7.5) was effective in removing exchangeable sodium without added gypsum. Without gypsum, about 50% more water was required to remove the same amount of sodium as removed by the 35 cm (13.7 in.) of water used with gypsum. The limiting factor in reclamation of sodic soils is usually permeability; *if the soil is permeable, gypsum addition is less critical to reclamation.*

When sulfur is added to the soil, *Thiobacillus* bacteria slowly oxidize the sulfur to sulfuric acid (H_2SO_4). The hydrogen ions of sulfuric acid can replace sodium ions on the soil cation exchange sites; or if the soil contains lime ($CaCO_3$), the sulfuric acid may react to form gypsum, which then has the same effect as applied gypsum.

$$CaCO_3 + H_2SO_4 + H_2O \longrightarrow CaSO_4 \cdot 2H_2O + CO_2 \uparrow (gas)$$
$$\text{(lime)} \qquad\qquad\qquad \text{(gypsum)}$$

Applications of about 40.3 Mg/ha (18 tons/a) of gypsum in Nevada increased water infiltration and the depth of water penetration.[6] Three years after applying the gypsum, the water penetrated to a depth of 48 cm (19 in.) in the soil receiving the gypsum and only 25 cm (10 in.) into the soil that did not receive gypsum. This method reduced the exchangeable sodium percentage from 42% to 18% during a 3-year period on the gypsum-treated soil. At the same time, the plot without gypsum increased in exchangeable sodium from 50% to 53%. Yields of hay increased by amounts of 0.1–2.3 Mg/ha (0.05–1.02 t/a) per year as a result of the application of gypsum.

The **gypsum requirement (GR)** is the calculated amount of gypsum necessary to add to reclaim the soil (Calculation 11-3). Its numerical definition depends upon the weight used for the volume of soil reclaimed and the claimed gypsum efficiency. Using a soil bulk density of 1340 kg/m for an average soil, the gypsum requirement is described by these two equations:

$$\textbf{GR} = \frac{\text{metric tons of gypsum needed}}{\text{hectare of soil to some fixed depth}}$$

$$\textbf{GR} = (Na_x)\ 4.50 \text{ metric tons of gypsum per hectare–30 cm}$$

where Na_x is the centimoles/kg of exchangeable sodium *to be replaced by calcium* from the added gypsum. When calculated in short tons per acre,

$$\textbf{GR} = (Na_x)\ 1.80 \text{ tons of gypsum per acre-foot)}$$

The conversion factors (4.50 or 1.80) change as soil bulk density, soil depth reclaimed, or gypsum efficiency change. These values assume about 75–80% efficiency (about 25% more gypsum is added than is calculated by chemical formulas) (see Calculation 11-3).

---

[6] The confusion in using "tons" in English, U.S., and metric systems has led to the currently used suggestion to indicate "metric ton" (= 1000 kg) by the symbol *Mg* (megagrams = 10^6 grams = 1000 kg).

Problem   A sodic soil has an average exchangeable sodium percentage (ESP) of 24% in the top 45 cm (17.5 in.) and a cation exchange capacity (CEC) of 18 centimoles$_c$ per kg of soil. The average exchangeable sodium to be left in the top 30 cm (12 in.) of soil is selected as 6%. Calculate the amount of gypsum and the amount of sulfur required to reclaim the soil in its top 30 cm.

Solution   The gypsum requirement factor is calculated as follows, using a soil bulk density of 1340 kg/m³:

$$GR = \frac{1 \text{ cmol}_c \text{ of Na to exchange}}{1 \text{ kg of soil}} \; \Bigg| \; \frac{\text{mol}_c \text{ weight of gypsum}}{\text{mol}_c \text{ weight of sodium}} \; \Bigg| \; \frac{4 \times 10^6 \text{ kg soil}}{1 \text{ ha–30 cm}} \; \Bigg|$$

$$\frac{\text{kg of Na}}{1 \text{cmol}_c \text{ of Na}} \; \Bigg| \; \frac{1 \text{ metric ton}}{1000 \text{ kg}}$$

$$= \frac{\text{Na}_x}{} \; \Bigg| \; \frac{171/2 \text{ g gypsum}}{23 \text{ g Na}} \; \Bigg| \; \frac{4 \times 10^6 \text{ kg}}{1 \text{ ha–30 cm}} \; \Bigg| \; \frac{0.00023 \text{ kg Na}}{1 \text{ cmol}_c \text{ of Na}} \; \Bigg| \; \frac{1 \text{ Mg}}{1000 \text{ kg}}$$

$$= \text{Na}_x \, (3.42) \text{ Mg of gypsum per ha–30 cm depth}$$

where the (Na$_x$) is the cmol$_c$ of exchangeable sodium per kilogram of soil to be replaced by calcium from gypsum.[a] If the gypsum is not pure, or for other reasons is not 100% efficient, more than the amount calculated must be added. Experience suggests that the gypsum is only 75–80% efficient, so the amount added must be increased accordingly. Adding 30% more gives these approximate equations for the gypsum requirement (GR):

$$GR = (\text{Na}_x) \, 4.50 \text{ metric tons of gypsum per hectare–30 cm}$$

or

$$GR = (\text{Na}_x) \, 1.80 \text{ tons of gypsum per acre-foot (tons are short tons)}$$

To solve the given problem, do these steps:

1. The cmoles$_c$ of Na needing replacement are calculated as the total exchangeable Na minus the exchangeable Na to be left.

(CEC)(ESP/kg) = cmol$_c$ of exchangeable Na per kilogram of soil

$$= \frac{18 \text{ cmol}_c}{1 \text{ kg}} \; \Bigg| \; \frac{24\% \text{ of sites have Na}}{100\% \text{ of exchange sites}} = \left(\frac{432 \text{ cmol}_c}{100 \text{ kg}} \right)$$

$$= 4.32 \text{ cmol}_c \text{ per kilogram of soil is exchangeable Na}$$

The cmol$_c$ of exchangeable Na to leave in soil is

$$= \frac{18 \text{ cmol}_c}{1 \text{ kg}} \; \Bigg| \; \frac{6\% \text{ exchangeable Na to be left}}{100\% \text{ total exchange sites}} = \left(\frac{108 \text{ cmol}_c}{100 \text{ kg}} \right)$$

$$= 1.08 \text{ cmol per kg of soil to be left}$$

Continued.

---

[a] In the SI metric system, millimoles (+) or mmol$_c$ [or centimoles (+) = cmol$_c$] is used in place of milliequivalents (meq). For sodium, 1 meq = 1 mmol because the valence of sodium is 1. One correct way to write the units is as follows, for millimoles per kilogram of soil:

$$\text{mmol}(+) \text{ kg}^{-1} \text{ or mmol}(+)/\text{kg or mmol}_c/\text{kg}$$

The "+" or subscript "c" is used to indicate an amount (an equivalent) that will *react with* 1 mole of a salt, base, or acid of a monovalent element.

The difference in total exchangeable Na and exchangeable Na to be left is the Na to be replaced = 4.32 − 1.08 = 3.24 cmol.

2. Putting the values from step 1 into the equation shown above for gypsum requirement,

$$GR = 4.50(Na_x) = 4.50(3.24 \text{ cmol}_c)$$
$$= 14.6 \text{ metric tons per hectare of gypsum needed}$$

3. For the amount of sulfur needed, refer to Table 11–3. It shows that only 0.18 times as much sulfur by weight is needed compared to gypsum, so the needed sulfur is:

$$14.58 \text{ tons gypsum } (0.18) = 2.62 \text{ metric tons of sulfur/hectare}$$
$$(= 1.05 \text{ short tons of sulfur/acre})$$

Gypsum dissolves slowly, and use efficiencies, such as uneven spreading, are seldom 100%. The preceding equations assume that about 30% extra gypsum or sulfur is added. The equations finally used have already incorporated those corrections. Added elemental sulfur is even slower in reaction, often needing many months or longer to be oxidized in moist soil.

Source: Raymond W. Miller; see inside book covers.

Sulfur, which oxidizes to sulfuric acid during several months, is used also in the reclamation of sodic soils and for lowering the pH of the soil. The relative required amounts of several amendments for reclaiming sodic salts are shown in Table 11-3.

A significant innovation in reclaiming sodic and saline-sodic soils is the initial use of *salty* water for leaching. Such a use appears paradoxical but is scientifically sound. A high salt content in water keeps sodic soils flocculated. The floccules have large pores between them, as do aggregates, and allow penetration of the leaching waters. Thus, the first water used for leaching may be moderately salty water. After most of the exchangeable sodium is removed by the calcium in the salts in the water or from gypsum additions, water of lower salt content can be used for final leaching.

11:6 Salt Precipitation Theory

The elimination of salts and exchangeable sodium from soils by leaching is one satisfactory method of reclaiming salted soils, but the leached salts are washed into groundwaters or streams, making those waters more salty, thus polluting them. More costly water, legal re-

Table 11-3 Estimated Efficiencies for Various Materials Used to Reclaim Sodic Soils Compared to Gypsum

Material	Tons of Material Equivalent Material to 1 Ton of Gypsum
Gypsum	1.00
Sulfuric acid	0.57
Sulfur	0.18
Lime-sulfur	0.75
Iron sulfate	1.62

strictions on disposal of wastewaters, legal restrictions on additions of salts to groundwaters, and excess salts in downstream river waters have all focused attention on a relatively new concept in managing salty soils: **precipitation of salts**. Briefly, this idea suggests that instead of leaching salts completely away into groundwaters, they can be leached to only 0.9–1.8 m deep (3–6 ft) or a little deeper, where much of the salt would form slightly soluble gypsum ($CaSO_4 \cdot 2H_2O$) or carbonates ($CaCO_3$, $MgCO_3$) during dry cycles and not react any longer as soluble salts. Any water eventually reaching groundwater or surface water would be carrying less total salts than it contained initially.

A number of questions immediately arise about using the salt precipitation approach. First, how much salt will precipitate? Second, how much will the remaining salt hinder plant growth? Third, what are the techniques and hazards in using this method? And fourth, what are the costs?

The amount of salt precipitating out will vary with the cation and anion composition of those salts. The ions precipitating will be mostly those of calcium, magnesium, carbonate, bicarbonate, and some sulfate. Estimates are that an average of about 30% of the total salts may eventually precipitate.

The Huntington Power Plant in Utah was faced in the 1970s with high costs to clean salty waste cooling waters. The plant managers considered adding it to soil. Although salt accumulation was expected, the accumulation was not as rapid as expected. One explanation is that a large portion of the added salt precipitates to relatively low solubility substances during drying cycles during the year. A desalting plant built to clean this waste water, rather than dispose of it onto soil, was estimated to cost about $12 million. The project is still being monitored and is now over 14 years old. Boron accumulation is now being monitored to see the effect of some toxic levels.

If only one-third of the salt is actually "removed" from activity by precipitation, how will the remaining salt affect plant growth? The answer is "just as much as that amount of soluble salt ever did," *but* the position of the salt in the root zone has been altered and the salt will be higher in sodium and chloride. The salt has been moved to the lower root zone, some of it even below the root zone. Most plant roots proliferate in the upper 30–60 cm (1–2 ft) from which they absorb water. Only for short times during hot summer days at maximum evapotranspiration losses will plants need much water from the deeper root zone. Thus, if the salts affect only water uptake rather than exerting a toxicity, normal yields should be obtainable as long as the upper root zone is moist and sufficiently low in salts. Little effect on yield of corn and tomatoes occurred even when *two-thirds of the root zone* (lower part) was too salty. Alfalfa, normally tolerant of only 6–8 mmhos/cm conductivity, yielded *normally* with its *lower* root system in solutions of 30–35 mmhos/cm.

The third question, concerning techniques and hazards, appears to have obvious answers, but limited research has been done. The management technique is simply to *add less water, but to do it more carefully to ensure uniform depth of wetting*. The leaching requirement can be reduced by 60–75% of that calculated by the equation in Calculation 11-2 by more careful watering. Drip and sprinkler methods of applying water will work best for precise movement of salts to predetermined depths.

The hazards in this technique include (1) careless water application causing inadequate downward salt movement and (2) possible toxic effects on some plants of sodium, boron, and chloride. Soils with shallow water tables and seep areas are not suitable for this technique.

Salt controls will be expensive and "best management practices" are still in developmental stages for many situations. Eventually, some salt must be leached beyond the root zones and will accumulate in groundwaters or surface waters. Until now the problem of removing excess salts from wastewaters has not been placed upon the individual farmer. Various states now have regulations on releasing water having high salt contents back into water resources. Groundwaters or surface waters in California having more than about 500 ppm salts must have salts removed or be ponded on the farm. Salt removal may require desalinization, ponding to allow some salts to precipitate, or mixing with less salty water before disposal. Treatment of sewage wastewaters from large cities can cost millions of dollars annually for treatment.

11:7 Managing Salty Soils

It is not always possible or practical to eliminate all salts from soil, but managing the soil to *minimize salt damage* is a necessary part of using salted soils. Although in sodic soils some of the exchangeable sodium must be removed, with slightly saline soils, the control of water, the proper techniques of planting, and the choice of tolerant crops are essential for their successful use in crop production.

11:7.1 Water Control

Maintaining a high water content in the soil, near field capacity, dilutes salts and lessens their toxic and osmotic effects. If soil is irrigated lightly but frequently to keep it at a high moisture content during the salt-sensitive germination and seedling stage, plants are able to survive to the more-tolerant mature stage of growth.

Limited leaching done before planting (when water is usually more plentiful than later) or a light irrigation by sprinklers after planting will move salts below the planting and early-rooting zone. Later, when the salt gradually moves upward with water, the plant will be more mature and more salt tolerant. Sprinkler-applied water after planting in the Imperial Valley of California increased lettuce germination 20%. Soils that develop hard crusts (1) must not be irrigated after planting until the seedling emerges or (2) the crust must be sprinkled at the time of emergence to soften it so seedlings can "break" through (Fig. 11-8).

The Imperial Valley of California has a severe salt problem. It has naturally salty areas, inadequate natural drainage, and irrigation being done with salty Colorado River waters.

FIGURE 11-8 Where salt is a problem, lettuce may avoid some salt injury by being planted on the slope of the ridge, because salt concentrates on the ridge crest. The white areas are salt crusts. Notice the vacant spots in many of the rows in this Imperial Valley, California, field where plants have been killed by the salt. (Courtesy of Alvin R. Southard, Utah State University.)

Correct water management practices in combination with drainage for salt control are imperative. To keep the average salt content in the root zone as low as 8.0 dS m^{-1} (moderately high) requires 17% excess irrigation water to ensure leaching. A single initial sprinkling of winter vegetable cropland moves salt down below the seed germination zone; then conventional surface irrigation is used. Double rows and sloping beds are also used to reduce salt damage.

11:7.2 Planting Position

Salt moves with water, and some will accumulate in the surface soil or furrow ridge tops as water with its salt moves upward and evaporates. Figure 11-9 illustrates furrow cross sections and planting positions used to avoid damage from salt accumulation zones. Notice that these methods avoid the centers of wide ridges and the tops of narrow ridges where salt will be most concentrated from furrow irrigation. Sprinkler methods of irrigation will eliminate some of these problems by washing salt deeper into the soil profile, even from ridge tops.

11:7.3 Choice of Crops

The choice of crops is based on the (1) tolerance to salt, (2) adaptability to the climate or soil characteristics, and (3) value of the crop in the individual farm activity. The chances of a crop failure or loss is less if an adequately salt-tolerant crop is selected. Obviously, the soil must

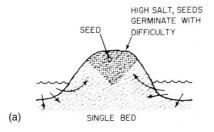
(a)

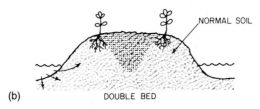
(b)

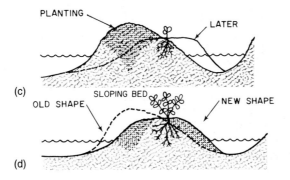
(c)

(d)

FIGURE 11-9 Methods of preparing seedbeds and of planting to reduce the effects of soluble salts on plants. Shaded areas illustrate locations where salt would normally concentrate because of water flow and evaporation. Darkest shaded areas are highest in salt. Diagrams (c) and (d) are bed shapes of the same planting at planting time and after the plant is growing well, respectively   (Courtesy of Raymond W. Miller, Utah State University.)

have a suitable texture, adequate depth, and exist in a climate suitable to the crop. The purpose of farming is to make a profit, so a high-value crop is preferred. Yet if a crop entirely different than is grown on the rest of the farm is selected, new seeders, harvesters, herbicides, and other equipment and farm programming will be needed for that crop.

11:7.4 Examples of Salt Management

Israel, a water-short nation, has pioneered the use of *brackish water*[7] (less salty than seawater). In 1975 brackish water comprised 10% of Israel's freshwater use. Waters of electrical conductivity of 3.7 dS m^{-1} and SAR of 8.5 have been used on cotton and other salt-tolerant crops.[8] With brackish water, cotton produced the same plant size and more bolls than with lower-salt-content water. The soil salt buildup, which could result rapidly from using brackish water, can only be controlled by regular leaching. Water is applied at a rate that always exceeds evapotranspiration. Extensive drying of the soil between irrigations must be avoided (which would move salt upward to the surface). Mixing brackish and fresh water to get water of suitably low salt content to use for irrigation is being practiced extensively in Israel. Whether the practice is a gain is still debated.

A second unusual situation in the management of salt deposition on soil is the treatment of saline seeps in Montana and the Dakotas (see Figs. 11-2 and 11-7). **Saline seeps** develop on several kinds of layered substrata, but all have water moving in a downslope direction within the soil (recharge area), usually above a low-permeability geologic layer, and probably dissolving more salts as it flows. The water seeps eventually to a soil surface at some lower elevation (seep area), where it is evaporated or transpired, leaving salt accumulations. Preventing salinization requires preventing water seepage. This has been done by using the water on the recharge area before it can seep away, usually by growing a plant cover to increase water use by transpiration.[9] Deep-rooted alfalfa on 80% of the recharge areas has hindered seep formation because there is less water to seep; wheat fallow areas had inadequate retardation of water flow and hastened development of new seeps.

11:8 What Makes Plants Salt Tolerant?

Scientists have tried for decades to understand why different plants tolerate different salt concentrations. The overriding hope is that a higher salt tolerance could some day be bred into plants or salt tolerances of plants increased. An Asian wild rice is reported that is said to survive in soils with saturated soil paste extract values of 30–40 dS/m.[10] Mangrove forest trees tolerate high salt concentrations along many ocean coastal waters.

Limited information is available on the plant salt tolerance mechanisms. Some of the mechanisms so far identified would seem to be very plant specific and difficult to transfer genetically. Perhaps genetic-engineering techniques can be utilized. Some of the observed

---

[7] Brackish water is water of salinity high enough to significantly restrict its direct use, yet not prevent its use completely. In Israel so much citrus is grown that chloride content (toxic to citrus) is used to define brackish water. Brackish water has 400–4000 mg/L (about 1000–10,000 ppm salts) of chloride.

[8] Marvin Twersky, Dov Pasternak, and Ilan Borovic, "Effects of Brackish Water Irrigation on Yield and Development of Cotton," in *International Symposium on Brackish Water as a Factor in Development,* Ben Gurion University of the Negev, and others, Beer-Sheva, Israel, 1975, pp. 135–140.

[9] A. D. Halvorson and C. A. Reule, "Alfalfa for Hydrologic Control of Saline Seeps," *Soil Science Society of America Journal,* **44** (1980), no. 2, pp. 370–374.

[10] A. R. Bal and S. K. Dutt, "Mechanisms of Salt Tolerance in Wild Rice (*Oryza coarctata Roxb*)," *Plant and Soil,* **92** (1986), pp. 399–404.

plant techniques to tolerate high-salt environments are briefly described in the following observed facts[11,12]:

1. Accumulation of high levels of sodium and chloride in shoots are associated with concentrations (usually increases) in certain plant organic compounds in cells.

2. Exclusion of salt ions by the root cells. Many important grains (wheat, barley, rye, triticale) seem to employ this ability.

3. Excretion of absorbed salts from the plant by means of "salt glands" that either burst, dripping out the high-salt solution, or finally drop off the plant. Many **halophytes** ("salt-loving" plants) have this mechanism.

11:9 Effect of Soluble Salts on Crop Quality

For perhaps centuries, growers and users have made claims that a crop product, grown in one location is better, sweeter, or more tart than the same food product grown in a different location. The major problem is that the specific substance (the sugar, acids, protein, or combinations of substances) needed to "improve" the food taste is not usually known. Measured differences in composition of many products have been attempted and found. Yet the importance of the measured items or how to increase them has usually been uncertain.

Israel, with its recognized perennial problems of soluble salt, has looked at salt effects on quality of tomatoes, melons, lettuce, and peanuts.[13] Tomatoes, irrigated with saline water, were rated as being of higher quality in solids and acidity. Melons irrigated with saline water had higher taste test ratings. In both crops salinity lowered yields, but it increased quality. With lettuce and peanuts, no observable differences in quality were measured.

11:10 Monitoring Salts in the Field[14]

There is a need to monitor salt in the field. Although taking soil samples for laboratory analysis is relatively simple, there are limitations:

1. Salt patterns in a field are not homogeneous and regular. Salt distribution and amount can change with each irrigation. Also, the number of samples needed and the cost to prepare the saturated paste extracts are enormous.

2. The laboratory values are not true field values. The laboratory sample has been dried and then rewetted with an amount of water about double its field capacity and then extensively stirred.

3. The process of using laboratory analyses is relatively slow. Lag times from sampling to the final data are usually several days, at best. A quick method is needed.

---

[11] Ibid.

[12] L. Gorham, R. G. Wyn Jones, and E. McDonnell, "Some Mechanisms of Salt Tolerance in Crop Plants," *Plant and Soil,* **89** (1985), pp. 15–40.

[13] Yosef Mizrahi and Dov Pasternak, "Effect of Salinity on Quality of Various Agricultural Crops," *Plant and Soil,* **89** (1985), pp. 301–307.

[14] J. D. Rhoades and D. L. Corwin, "Monitoring Soil Salinity," *Journal of Soil and Water Conservation,* **39** (1984), pp. 172–175.

At the present time the four most commonly used methods for salt measurements in situ are (1) vacuum extractors, (2) in-place sensors, (3) bulk soil electrical conductivity, and (4) the four-electrode salinity probe. The first two methods have limited usefulness. The **vacuum extractors** remove soil water with mild suction, but only when the soil is at nearly field capacity. Also, water from only a small volume of soil around each extractor is removed. The **in-place sensors** have the same limitation of measuring only a small localized portion of soil where the probe is located. Neither of these two methods is easily adapted to measuring large soil areas.

The two other techniques measure larger soil volumes and obtain "average" salt values for each soil volume analyzed. The **bulk soil** and **the four-electrode probe** measure the electrical conductivity. The bulk soil methods uses four metal stakes (called probes) that are pushed into the soil in a line. As electrical current from a battery is passed through two of the probes, the resistance to current flow *through the soil* to the other two probes is measured. For a uniform soil, the current penetrates a depth that is about one-third the distance between the two outer electrodes. The measurement averages the soil salinity to this depth. Thus, the volume and depth of soil measured can be varied by the distance between probes; this distance can be selected by appropriate probe placement.

This bulk soil electrical resistance method is fast and averages the salt content in a larger volume of soil than do the other methods. This method also measures salinity in the soil in its normal physical state and water contents. However, the *readings are specific for each soil and require calibration for each soil* in order to interpret the readings obtained to some particular salt condition.

For more localized measurement of salinity in field conditions, the **four-electrode salinity probe** can be used. This probe has four annular rings molded into a plastic probe. The rings are a fixed distance apart. The probe (about 15 cm long, 3 cm diameter, with a 1.5 m-long tube handle) is tapered slightly at the end. This allows the probe to be inserted to any depth into a hole of the probe's diameter (usually dug with a soil-coring tube). The electrical current is passed from a battery, and the measured salinity is an average inside a sphere of soil about 30 cm (1 ft) in diameter around the probe.

The major advantage of the four-electrode-probe method is the ability, once calibration is done, to make many very rapid readings. Rapid monitoring of salt changes and variation over large areas could be quick and easy.

> *Old alfalfa is more tolerant of salt-affected soils than young alfalfa, and . . . deep-rooted legumes show a greater resistance to such soils than the shallow-rooted ones.*
>
> **—N. C. Brady**

Questions

1. How long have salt damages to plants been known?
2. (a) Can soluble salts be seen? Explain. (b) Why was the term *summer snow* probably used for some salty soils?
3. (a) Explain what each of these tells about salt in soil: ESP, SAR, and EC (electrical conductivity). (b) Why is SAR replacing ESP? (c) Give the units in which electrical conductivity is reported.
4. (a) Define *soluble salts*. (b) What are the six most numerous ions of soluble salts?
5. (a) What is the source of soluble salts? (b) Why will all flowing surface waters have soluble salts?

6. (a) What is the effect of a high ESP? (b) How is ESP related to "slick spots"? (c) Is it desirable to have a high ESP?
7. (a) Is it desirable to have a high SAR? (b) What is the effect of a high SAR on soil?
8. Most calcium precipitates as insoluble carbonates. How will this change the soil solution SAR?
9. (a) Give definitions (give SAR, ESP, EC) for sodic, saline, and saline-sodic soils. (b) Could a "nonsaline" soil, by definition, have enough salt to damage some crops?
10. What is the major damage of high exchangeable sodium percentages on soil clays?
11. (a) What does a high salt content affect that reduces plant growth? (b) Are all plants equally affected by the same salt concentrations? Explain.
12. (a) Name two plants that are very sensitive to and two plants that are very tolerant of soluble salts. (b) At what growth stage are plants most sensitive to salts?
13. Define a salt "balance" for a soil.
14. (a) What are the three key requirements needed to accomplish reclamation of salt-affected soils? (b) How does reclamation of sodic soils differ from reclamation of saline soils?
15. (a) How important is leaching the profile in reclamation of salt-affected soils? (b) What happens without leaching?
16. (a) Why is gypsum sometimes used? (b) What will happen if gypsum-treated soils are not adequately leached?
17. (a) Which soluble salt ions precipitate? (b) Which ions are left in solution? (c) What factors limit the use of the salt precipitation management scheme?
18. (a) Why is "waste salt" disposal (reclamation waters, irrigation runoff water, others) a serious problem? (b) Will the problem become more serious? Explain.
19. (a) In what ways can water control without profile leaching be used to reduce damage from soil salts? (b) Are any of these techniques practical?
20. Diagram the pattern of salt accumulation in furrow-ridge surface flow irrigation and explain the importance of seed position to plant survival in saline soils.
21. (a) Do plant varieties and cultivars differ in sensitivity to soluble salts? Discuss. (b) What are some known mechanisms used by plants to help them tolerate high salt contents?
22. What are some management techniques for reducing salt buildup in irrigated soils?
23. (a) Do soluble salts affect crop quality? (b) How extensive is the evidence for this?
24. How is salt measured in situ (in the field soil at normal water content)?
25. Is salt in the field (question 24) still measured by conductivity? Explain.

12

Diagnosis of Soils and Plants

[W]hoever could make two ears of corn, or two blades of grass, to grow upon a spot of ground where only one grew before, would deserve better of mankind, and do more essential service to his country, than the whole race of politicians put together.

—King of Brobdingnag,
Gulliver's Travels, Jonathan Swift

12:1 Preview and Important Facts

PREVIEW

Soil and plant diagnoses are the basis to predict crop needs and to explain past crop yields. The plant itself is the final proof of what nutrients were or are available to the plant, but even plant analyses have limitations in their use. First, we only measure the total amounts of nutrients in the plant. Factors other than soil availability do affect the amounts of nutrients absorbed: weed competition, diseases, or poor management (dryness, drainage, nutrient balance, pH control). These specific effects are not identified. A second, and more important, limitation is that by the time a plant can be tested, it is usually too late to add fertilizer. Chemical analyses of the plants are helpful information in the overall evaluation of the soil conditions for plant growth over years of study. However, plant analyses are less satisfactory than are soil tests for making decisions about corrections for the current crop. Some important uses are made of plant analyses for long-growing crops (orchards, turf, sugar cane, and bananas). Only a few crops of high value (such as potatoes) have much baseline data permitting use of plant analysis for current crop decisions.

Soil tests are suitable to predict needed modifications to allow optimum plant growth. Soils are tested for pH, soluble salts, available nutrients, element toxicities, and various other properties. It is possible to predict with some degree of confidence the amount of phosphorus or potassium fertilizer needed. Recommendations of fertilizers and of lime to add are done with considerable, but not complete, accuracy.

Soil analysis, however, is no panacea. It will *not* supply answers to unsatisfactory plant growth when the cause is dry weather, compacted soils, critically low or high temperatures, inadequate soil drainage and low oxygen in the root zone, improper placement of fertilizer, salt accumulation, plant diseases, toxic elements, insect damage, competition from weeds or tree roots, or untimely operations. These limiting factors can alter interpretations of soil tests for a fertilizer recommendation.

IMPORTANT FACTS TO KNOW

1. The approximate accuracy of the recommendations made from the various soil tests done
2. The general guidelines for collecting a good-quality soil sample to be used for analyses and recommendations
3. The general tests used to measure lime requirement and the salt hazard in soils
4. Many laboratories do not make soil tests for "available" nitrogen, yet they make fertilizer nitrogen recommendations. Understand the basis they use for recommendations
5. The reason that dilute acids are used to extract phosphates in acidic soils but bicarbonates are used with alkaline soils. Neither solution works satisfactorily with the other group of soils
6. The general status of soil tests to evaluate micronutrient needs
7. The expected problems in greenhouse soil testing
8. The values of and disadvantages of plant total analyses for making fertilizer recommendations
9. The special sampling needs and plant parts to sample when doing plant total analyses
10. The definition of *critical nutrient range* and the reason for the continual research to get these values
11. The general visual nutrient deficiency symptoms for nitrogen, phosphorus, potassium, and iron
12. The concept of "hidden hunger" and its hazard
13. The techniques and approaches for making fertilizer recommendations
14. The benefits and limitations of using computers in making recommendations
15. Factors affecting yields or crop responses to fertilizers that soil tests do not evaluate

12:2 Soil Testing: How Good Is It?

How good are soil tests? The headlines of some magazines would suggest that the tests are not very accurate: (1) "How to save $42 an acre. It's easy. Just don't automatically follow soil test recommendations . . . " ;[1]; and (2) "Soil Test Sleuth Saves $28/A . . . now that he's making *his own* fertilizer recommendations."[2] Are such criticisms valid? For particular situations, perhaps yes. But they are an overall negative attitude when the majority of examples of using soil test recommendations are beneficial and encouraging.

12:2.1 Making Good Recommendations

Making a success of using a fertilizer recommendation requires accuracy and details from several sources. First, the farmer must send in an accurate cropping history, an accurate projection of his expected yield, and an accurately taken soil sample to be used for analysis. Second, the laboratory must analyze the soil sample correctly and have good field correlation data on which to evaluate the meaning of that soil's test data. Third, the test evaluator must be trained in the job and know the area for which he makes predictions. Fourth, the

---

[1] William C. Liebhardt and Martin Culik, "How to Save $42 an Acre," *New Farm,* (Feb. 1985).
[2] Mike Brusko, "Soil Test Sleuth Saves $28/A," *New Farm,* (Feb. 1985).

farmer must correctly select good seed, plant on time and at the correct density, manage the crop adequately, and follow the fertilizer recommendation. Finally, the recommendation cannot account for adverse weather (heavy rains, hail, drought, cold periods), pest damage, or poor management. It is easy to imagine, from this partial list, that some farmers will not be satisfied with the results they obtain for a given field in a given year. But what is the cause of the problem that brings dissatisfaction? Was the problem caused by an error in fertilizer addition, by the soil sampler, by the farmer's incorrect yield prediction, by the farmer's management of his crop, or by the laboratory's recommendation?

Soil tests, as now used, are generally quite good for phosphorus, potassium, soluble salts, pH, and lime requirements. Only a few extensively acceptable tests are in use for nitrogen and the other nutrients. Varying degrees of success are obtained using tests for nitrogen, magnesium, sulfur, boron, zinc, and iron. *Most nitrogen recommendations are made from data of many years of field trials.*

12:2.2 Why Recommendations May Be in Error

Why aren't the predictive tests more accurate, particularly for some elements? The *laboratory tests are easy to do well,* so the chances for most errors are in other parts of the process. The following are processes in the soil testing–recommendation procedure where errors may happen that affect the final recommendations:

1. **The soil sample taken for analysis.** All of the analyses and interpretations are based on a soil sample. Each laboratory lists directions for taking the sample in a certain way (depth, time, composite of subsamples). This sample must accurately represent the whole field or interpretations will be in error.

2. **Field history.** The history of each field's use and fertilization, particularly for the past year or so, is requested. There will be some fertilizer carryover from past crops; legumes may leave high nitrogen reserves; and other crop residues or added manures affect available nutrients. This history must be accurate. It is another place for error.

3. **Correlation of laboratory tests to field responses.** This is the most critical step and is the responsibility of the testing laboratory. *Laboratory data have no meaning until the test values are carefully correlated to crop response to different fertilizer levels in the field.* If a laboratory is depending on old correlation data, it is probably in error. Also, data being used may have been collected using obsolete plant varieties, have been based on smaller yield goals, and have the crop grown under different management (have different soil preparation, rotations, or plant densities or use sprinkler rather than surface irrigation, etc.). Up-to-date field correlations with current varieties, current fertilizers, and current management systems is essential.

4. **Good computer integration.** Many growth factors are involved in making predictions. Soil textures, soil depths, season lengths, average temperatures, projected yields, crop varieties, soil compaction, presence of hardpans, and soluble salts are some of the many factors needing to be considered. A person's mind cannot continually and consistently do the integration job that a computer can do. The vast computer memory plus a wise, trained scientist make a formidable combination. However, do not forget that all data and interrelationships in a computer were measured, interpreted, and installed in the computer by people. The output will only be as good as the quality of input.

5. **A wise analyst and predictor.** Even with extreme care in all work with natural systems, exceptions and "special cases" will continually be encountered. There is even a

lack of data for evaluating some factors of growth. Nevertheless, the predictor still must make a recommendation. The wise, experienced person has his own "accumulated" knowledge and "feel" of when and how an adjustment should be made for any sampled area. Many of these uniquely qualified persons have extensive farming experience themselves.

Unfortunately, not all persons in today's science have the necessary experience, abilities, and knowledge. Recommendations from inexperienced persons will more likely "conform to available data" but be less apt to have modifications made because of personal observation of "strange" data. Abnormalities in a sample's data are not seen by the computer unless clues are put into it for those abnormalities. The present age has computers, a great asset, but it also has fewer scientists working those computers who have farm experience.

12:2.3 Recommendations Are Good But Not Perfect

Obviously, there are many places for error in fertilizer recommendations, especially in the field correlation work. If most errors were to be eliminated, the correlations would require *thousands of field plots, for every crop and variety, in all soils and climates, and in all management levels.* Costs and time force us to reduce the numbers and kinds of field trials that are done. For example, accurate field plot testing may cost $20,000 per field test and consequently are difficult to fund today. The extent to which correlation of soil tests to field response are updated determines the quality of the recommendations. All these limitations are part of the "experimental error" that exists in making fertilizer predictions.

Fertilizer predictions are not exact, and they are done better by some persons and laboratories than by others. Certain fertilizer and lime predictions are valuable and quite accurate. Laboratories and their scientists usually suggest that the *recommendations should be modified by the individual users (1) according to their own knowledge of their fields, (2) according to their management level, (3) on the basis of responses they have observed in past years, and (4) by their prediction of the weather pattern for this crop season.* Being accurate is not easy, but soil testing helps.

12:3 Soil Sampling

Take care in sampling a soil. Good judgment is often better than any single set of rules. The overriding guide should be to *take your sample so that it represents what you want it to represent.* If the sample is to represent the top 30 cm (12 in.), take many subsamples that have equal portions of soil from all of the 30-cm depth. Some soils of uneven relief or that have been graded or leveled may have different depths of topsoil over much different subsoils, such as dense clayey **B** horizons. Some scientists recommend not mixing layers of soil of *obviously different color;* the lower 5–10 cm (2–4 in.) of a 30-cm-deep core may be in quite different soil than the top part of the core. Mix the subsamples (cores) from all parts of a "uniform" field that the sample is to represent. Avoid abnormal areas. Sample these abnormal areas separately. Send the samples of abnormal areas along with the soil samples from normal areas for comparisons. Use one composite sample to represent only a field that had similar crops, fertilizer, and so on, the past year or even longer. Some greater details are in the following sections.

12:3.1 Depth and Number of Samples

What soil depth should be sampled? Most early workers sampled to "plow depth," which was interpreted to be 15–20 cm (6–8 in.) deep. At present, many laboratories request samples to be from the top 30 cm (1 ft). It is important to *sample the depth that is recommended by the laboratory that will analyze your sample.* Its recommendation will be based on correlation studies that used the soil depth they have recommended. It is common to sample different depths for different analyses. For salt problems, the recommended depths may be 0–10 cm (0–4 in.) and other depths. For no-till fields and established sod crops, sampling depths recommended might be shallow. For nitrate measurements, at least two depths are needed, 0–30 cm and 30–60 cm (0–1 ft and 1–2 ft). Some laboratories even request a third depth of 60–90 cm (2–3 ft). Sometimes, these deep samples are poorly taken because of difficulty in getting them.

How many subsamples should make up the single composite sample sent to the laboratory? Unfortunately, cost, rather than sample quality, usually determines the number of subsamples used. One composite sample comprised of many subsamples mixed together may represent one field or only a portion of the total field. This one field may be as large as 5 ha (12 acres) in a uniform soil area. If the soil area is not very uniform, a composite sample should represent a much smaller area. Statistical studies have led to recommending that 15–20 cores or subsamples be collected to make one good composite sample. In practice, people often take fewer subsamples, sometimes as few as two to five. If only a few subsamples are used, there is risk that the sample is not representative. For the deeper soil depths, perhaps only one-third as many subsamples are necessary for the same level of confidence. A nonrepresentative sample cannot result in an accurate soil test assessment of that field.

The number of subsamples and depths taken may need to be tied to the element being tested.[3] Although the outline in Table 12-1 *has not been standardized by laboratories,* the suggestions indicate several items: (1) You should take more subsamples for good phosphorus data than are needed for potassium data, (2) sample for nitrates and sulfates to deeper depths, and (3) sample reduced-tillage areas differently from conventional tillage fields. The best recommendation is to *follow suggestions given by the laboratory that will do your analyses.*

12:3.2 Frequency and Time of Sampling

When should sampling be done? Many recommendations simply say to sample "anytime at least one month prior to planting." On new land or land new to you, test it yearly for a few years. As you get a picture of the land's fertilizer needs for each area and for the plants grown,

---

Table 12-1 Suggested Guide for Sampling Soil

Number of subsamples to make each composite:

Phosphorus (P)	Take about 25 subsamples/field
Potassium (K)	Take about 5 subsamples/field
Nitrate	Take about 15 subsamples/field

When to sample:

pH, P, and K	Every 3–5 years, at least 15–20 cm (6–8 in.) deep
Nitrate, sulfate	Annually at least to 60 cm (2 ft) deep in 30-cm increments

Changes for reduced tillage fields:

pH	Sample shallow, 5–10 cm (2–4 in.) deep
Phosphorus, potassium	Sample at least 15–20 cm (6–8 in.) deep

Source: Wayne E. Sabbe, "Let's Rebuild Confidence in Soil and Plant Tissue Testing," *Ag Consultant* (Jan. 1987), pp. 10–11.

---

[3] Wayne E. Sabbe, "Let's Rebuild Confidence in Soil and Plant Tissue Testing," *Ag Consultant* (Jan. 1987), pp. 10–11.

you can settle into the more "standard" recommendation of sampling once every 2 or 3 years. Sample prior to the crop most needing fertilizer.

When to sample is less well defined. You wish to see what reserves the soil can make available. This is best measured after the soil has "rested" from the previous crop. Usually, soils will test lowest at about the peak of a crop's growth. Although sampling at this time will indicate some "supplying capacity" for that nutrient, it will not tell what reservoir the soil will provide at the beginning of the next crop.

Most recommendations are to take the samples as soon before planting as convenient. If the sample must be sent to a distant laboratory, a month or 6 weeks may be needed, plus any lead time needed for fertilizer purchase, and so on. Probably the poorest time to sample is during early fall, when the crop is still maturing or near harvest. Nutrient tests are likely to be the lowest at this time.

12:3.3 Uniform Sampling Areas[4]

Before sampling a field, it should be examined for differences in soil characteristics and past treatment. Consider soil productivity, topography, texture, structure (such as surface crusts or tillage pans), drainage, color of topsoil, and past management. If these features are uniform throughout the field, each composite sample can represent up to about 5 hectares (12.5 acres). If there is a great variation in any of these features, the field should be divided and a composite sample taken from each delineated area. A composite of samplings from two distinctly different areas is not representative of either area (Fig. 12-1).

The county, parish, or area Soil Survey Report is an excellent source of information about the kinds of soils in each field on a farm and can be used to determine soil boundaries for sampling purposes.

For homeowners, areas near the house are likely to be different than those a few feet away because the backfilling near foundations may be deep subsoil, inferior "fill" soil, or contaminated with building debris or cement. For these reasons, a separate soil sample should be taken from each area that seems to have visual differences. More intensely used soils, such as in greenhouses and nurseries, may need each bench or batch of soil and soil mixture sampled.

12:3.4 Soil and Crop Background

The more complete the information provided, the better will be the fertilizer recommendation interpreted from the soil analysis. The following information should accompany the sample when it is sent to the soil testing laboratory: (1) previous crop grown, (2) crop or crops to be grown, (3) realistic yield goal, (4) when the field was last limed and fertilized and rates of application, (5) whether the field will be manured for the crop being grown (kind and rate of application), (6) depth of plowing, (7) soil series or management group, (8) whether drainage is good, intermediate, or poor, (9) whether irrigation is to be used, and (10) other special problems or conditions that may affect plant growth (such as temperature, geographic location, elevation, hardpans).

12:3.5 Summary on Sampling Soil[5]

The following general recommendations are a summary of suggestions for taking soil samples for fertilizer and lime recommendations:

---

[4] J. C. Shickluna, "Sampling Soils for Fertilizer and Lime Recommendations," MSU Ag Facts, Extension Bulletin E-498, Cooperative Extension Service, Michigan State University, East Lansing, 1981.
[5] J. T. Cope and C. E. Evans, "Soil Testing," *Advances in Soil Science,* **1** (1985), pp. 210–228.

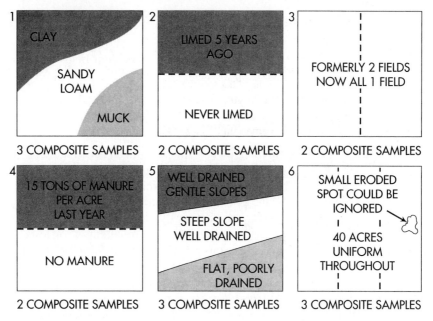

FIGURE 12-1 Conditions should be studied before sampling a field. Variations in soils, productivity, drainage, and past management will determine the number of samples that should be taken to obtain a representative soil sample. The two textural classes and muck in Field 1 require three composite samples. Due to differences in management of Fields 2, 3, and 4, two composite samples should be taken in each. Differences in topography and drainage represented in Field 5 necessitate three composite samples. Field 6 is a uniform 16-ha (40-acre) field; but because of its size, it should have three composite samples to ensure a representative soil analysis of the entire field. The examples of sample numbers to be considered are only approximate; the actual sampling area is determined by soil uniformity at the test site.

1. The sample should represent one uniform area or soil condition within a field.

2. Avoid mixing in subsamples from unusual areas. Sample and treat unusual areas separately, if they are large enough.

3. Sample row crop fields in the middles away from rows to avoid sampling old fertilizer bands.

4. Collect samples in the late summer or fall following summer crops, but not immediately at harvest time. For multiple cropping and year-round crops in warm climates, try to sample as near prior to seeding time as is practical. Unless the soil dries during a long period, continual plant growth (even weeds) constantly changes the nutrient status of such soils. *The best sample is one taken as soon before planting as will permit its analysis and your preparation for fertilizer purchase and addition.*

5. Subdivide fields into sampling units based on their differences in recent cropping or fertilization histories.

6. Subdivide large uniform fields into smaller units, probably not to exceed about 5 ha (12.5a), and collect a composite sample from each unit.

7. Adjust sampling depth to fit (a) laboratory suggestions, (b) the crops to be grown, and (c) the cultivation practices.

8. Sample annually until you understand your fields' needs and crop responses. Then sample every 2 or 3 years.

12:4 Soil Tests and Recommendations

To obtain maximum growth response from an added nutrient element, all other essential elements must be in adequate, but not injurious, amounts. Justus von Liebig (1803–1873) postulated this concept in the **law of the minimum,** which states that the growth of any plant is limited most by the essential plant nutrient present in the least relative amount.

No quick laboratory test can duplicate the actual uptake action by roots, and laboratory tests are arbitrary and empirical. Good correlations of soil tests with crop response in the field require hundreds of tests on each crop of interest, each different soil, each climatic variation, and even each different management system. Some of the better established tests throughout the United States, which can be expected to provide correct and useful data in most instances, include soil **pH, soluble salts, lime requirement, available phosphorus,** and **available potassium.** Other soil tests, including those for nitrogen, are less well established and may be consistent with field correlations only in certain states or regions where good correlations have been established.

12:4.1 Soil Acidity

The **pH** is usually measured with a pH meter on one of these suspensions: on a 1 : 1 or 1 : 2 soil-water or a 1 : 2 soil–0.01 molar$_c$ $CaCl_2$ suspension. In all instances, the laboratories should know how to evaluate the pH data, which varies slightly with the method. Usually, there will be a need for lime additions if pH is below 6. Above about pH 8 some nutrient deficiencies (iron, manganese, zinc) may be expected. The crop to be grown and other factors will determine the recommendations.

The **lime requirement** of a soil is the amount of lime required to bring about the desired change in soil pH. It is determined by measuring the total soluble and exchangeable hydrogen and aluminum or the change in pH as a buffer solution is added to a soil sample. One of the several lime requirement methods (SMP Buffer Method) first measures the pH of the soil in water. If the soil is acidic enough to need lime (below pH 6 for most soils), some buffer solution is added, the suspension well mixed, and the subsequent pH of the buffer + soil is measured. Table 12-2 provides an example of a table used to estimate the lime needed from the buffer-soil pH measured.

12:4.2 Soluble Salts

The reference method used for evaluating soluble salt hazard is the conductivity of the saturated soil paste extract. However, the 1 : 1 and 1 : 2 soil–water suspensions are frequently used instead of the saturated paste extract. The reason is simple: The saturation paste extract is more work to prepare and collect from the vacuum extractor. Thus, the more-dilute suspensions are preferred because of convenience and speed. These 1 : 1 and 1 : 2 suspensions are more dilute than the saturation extracts. Consequently, the diluted samples need to have their measured salt value multiplied by some factor in order to estimate what the salt value would be if measured on the saturated paste extract. An example of estimating the relationship between the paste extract values and the more dilute sample values is shown in Table 12-3.

The multiplication factors are not exact because the amounts and kinds of clays alter the saturation water content on which the factors are based. Also, in diluted 1 : 1 or 1 : 2

Table 12-2 Amounts of Lime Required to Bring Mineral and Organic Soils to the Indicated pH According to Soil Buffer pH of the SMP Buffer

| | Agricultural Ground Limestone [a] (t/a) [b] | | | |
| | Mineral Soils | | | Organic Soils[c] |
Soil Buffer pH	7.0 [d]	6.5 [d]	6.0 [d]	5.2 [d]
6.8	1.4	1.2	1.0	0.7
6.7	2.4	2.1	1.7	1.3
6.6	3.4	2.9	2.4	1.8
6.5	4.5	3.8	3.1	2.4
6.4	5.5	4.7	3.8	2.9
6.3	6.5	5.5	4.5	3.5
6.2	7.5	6.4	5.2	4.0
6.1	8.6	7.2	5.9	4.6
6.0	9.6	8.1	6.6	5.1
5.9	10.6	9.0	7.3	5.7
5.8	11.7	9.8	8.0	6.2
5.7	12.7	10.7	8.7	6.7
5.6	13.7	11.6	9.4	7.3
5.5	14.8	12.5	10.2	7.8
5.4	15.8	13.4	10.9	8.4
5.3	16.9	14.2	11.6	8.9
5.2	17.9	15.1	12.3	9.4
5.1	19.0	16.0	13.0	10.0
5.0	20.0	16.9	13.7	10.5
4.9	21.1	17.8	14.4	11.0
4.8	22.1	18.6	15.1	11.6

Source: "Recommended Chemical Soil Test Procedures for the North Central Region," Bulletin 499, North Dakota Agricultural Experiment Station, North Dakota State University, Fargo, 1975.
[a] Agricultural ground lime of 90% total neutralizing power (TNP) or $CaCO_3$ equivalent, and fineness of 40% < 100 mesh, 50% < 60 mesh, 70% < 20 mesh, and 95% < 8 mesh.
[b] To convert tons per acre to metric tons per hectare, multiply by 2.24.
[c] Because of lower mineral contents, organic soils are often suitable when limed only to pH 5.0 to 5.5.
[d] Desired pH level for the soil.

Table 12-3 Estimation of Saturation-Soil-Paste Values from Measured Values Using 1 : 1 or 1 : 2 Soil–Water Suspensions

| | Water Content of Sample | | |
Suspension Whose Solution Conductivity Is Measured in Laboratory Method	Paste (2X FC) (%)	Suspension (%)	Multiplication Factor to Estimate Paste Extract Conductivity
1:1 soil–water suspension			
Loamy sand texture	24	100	3.5–4
Loam texture	34	100	2.5–3
Clay loam texture	44	100	2–2.5
1:2 soil–water suspension			
Loamy sand texture	24	200	6–8
Loam texture	34	200	5–6
Clay loam texture	44	200	4–5

suspensions some salts, such as gypsum, dissolve more than they would have in the saturated paste.

For an example calculation of the factors, assume that a loam soil has a field capacity of 20% water. (Its saturation percentage is estimated [a general rule of thumb] to be about double the field capacity value, or 40%.) The 1 : 1 soil-water suspension is soil at 100% water content (= 20 g water, 20 g soil), so the salt in this loam's 1 : 1 suspension is dissolved in 2.5 times more water than would be in the saturated paste. Thus, conductivity of the 1 : 1 extract will be 2.5 times more dilute. To approximate the conductivity that would be in the saturation paste extract, multiply the 1 : 1 conductivity value by 2.5.

12:4.3 Nitrogen

Good predictive soil tests for nitrogen are not available, although some tests for residual nitrate in the top 60 or 90 cm (2 or 3 ft) of soil is extensively used for some estimates. The complex nitrogen transformations (biological denitrification, humus decomposition, immobilization, leaching, and ammonia volatilization) are rapid and sensitive to temperature and water; these make finding a satisfactory "quick soil test" for nitrogen difficult. The rapid immobilization of ammonium and nitrate ions and the high mobility of nitrate has also added to the problem of developing a suitable, convenient test. Where nitrogen tests are used, measurement of total soil nitrate to 60 cm (2 ft) is the most commonly used. This test is, essentially, a test of nitrate released by "field incubation of soil." An example of the recommendation table for nitrate measurements in North Dakota soils is shown in Table 12-4.

Most states give nitrogen recommendations based, at least partly, on data from many years of field trials with various crops, different soils, various management practices, and different fertilizers and fertilizer rates. A nitrogen fertilizer recommendation considers previous crops, estimates "carryover" nitrogen from the past-season additions, considers the crop, includes the farmer's projected yield, and evaluates the effect of the farm location (climate) (Table 12-5).

Table 12-4 Nitrate-Nitrogen (NO$_3$–N) Soil Fertility Values (lb/a–2 ft) and N Recommendations for Different Yield Goals for Oil and Confectionary Sunflowers (North Dakota)

Yield Goal: Oil or Confectionary (lb/a)[a]	Nitrogen Fertility Ratings					Soil Plus Fertilizer N Needed
	Very Low (VL)	Low (L)	Medium (M)	High (H)	Very High (VH)	
1000	0–12	13–24	25–36	37–49	50+	50
1200	0–15	16–30	31–45	46–59	60+	60
1400	0–17	18–34	35–52	53–69	70+	70
1600	0–20	21–40	41–60	61–79	80+	80
1800	0–22	23–45	46–70	71–89	90+	90
2000	0–25	26–50	51–80	81–99	100+	100
2200	0–27	28–55	56–85	86–109	110+	110
2400	0–30	31–60	61–90	91–119	120+	120
2600	0–33	34–66	67–101	102–134	135+	135
2800	0–36	37–72	73–112	113–149	150+	150
3000	0–41	42–81	82–123	124–169	170+	170

Source: *Fertilizing Sunflowers*, Cooperative Extension Service, North Dakota State University, Fargo.
[a] To convert pounds per acre to kilograms per hectare, multiply by 1.12.

Table 12-5 Guide for Estimating Nitrogen Needs (Pounds per Acre)[a] for Corn and Grain Sorghum Following Certain Crops

	Yield Levels (bu/a)[b]				
Previous Crop	100–110	111–115	126–150	151–175	176–200
	Lb of N to Add per Acre				
Good legume stand (alfalfa, red clover, sweet clover, 5 plants per square foot or more)	40	70	100	120	150
Average legume stand (alfalfa, red clover, sweet clover, 2–4 plants per square foot)	60	100	140	160	180
Soybeans, legume seeding of alfalfa, red clover, sweet clover	100	120	160	190	220
Corn, small grain, or grass crops	120	140	170	200	230

Source: R. K. Stivers, et al., mimeographed report, Department of Agronomy, Purdue University, Lafayette, Ind., 1979.

[a] To convert pounds of N per acre to kilograms per hectare, multiply by 1.12.

[b] To convert bushels per acre of corn and sorghum to kilograms per hectare, multiply by 56 X 1.12.

12:4.4 Phosphorus

Phosphorus soil tests have been quite reliable despite the fact that they extract little phosphorus from *organic* sources. Generally, at least two different kinds of extractants are needed. Dilute acids have been best for extracting phosphorus from the less soluble phosphate minerals in acidic soils. Sodium bicarbonate has been most used on alkaline pH soils. This is because acids will dissolve soil carbonates and thereby become neutralized and ineffective; carbonates (lime) are common in many alkaline (acid region) soils. The three solutions *most in use* in the United States are Bray-1, Mehlich-1, and Olsen's bicarbonate.

- **Bray No. 1**: 0.025 molar HCl + 0.03 molar NH_4F (for acidic soils)
- **Mehlich No. 1**: 0.05 molar HCl + 0.025 molar(+) H_2SO_4 (for acidic soils)
- **Olsen's:** 0.5 molar $NaHCO_3$ at pH 8.5 (for neutral and alkaline soils)

The Bray-1 method has been used to measure both available phosphorus and potassium in acidic soils. Because most available potassium is from soluble + exchangeable forms, the H^+ and NH_4^+ ions are adequate extractors of K^+. Table 12-6 illustrates some soil test extraction values and the recommendations for both phosphorus and potassium for corn and grain sorghum in Indiana. The NH_4F in the Bray-1 extract is a good extractant of phosphorus from aluminum phosphate sources.

A newer procedure, using a 1 molar ammonium bicarbonate extraction and containing dilute DTPA chelate, is being more widely used in alkaline soils as a *single extractant for K, P, Zn, Cu, Fe, and Mn.*[6]

---

[6] P. N. Soltanpour and A. P. Schwab, "A New Soil Test for Simultaneous Extraction of Macro- and Micro-nutrients in Alkaline Soils," *Communications in Soil Science and Plant Analysis,* **8** (1977), pp. 195–207.

12:4.5 Potassium

Except in the few soils with high mica contents, which can release appreciable potassium in a season, most soils supply available potassium for a season from the *soluble* (low amount) and the *exchangeable* reservoirs. Many tests, most of them fairly successful, have been tried: Strong boiling 1 molar nitric acid, cold 1.2 molar sulfuric acid, 1 molar ammonium acetate, the Bray-1 extract, and 1 molar ammonium bicarbonate, to name only a few. Table 12-6 illustrates the recommendations based on the potassium extracted by neutral 1 molar ammonium acetate. The critical level of soluble + exchangeable potassium varies with the test and crop but is around 180–280 kg K/ha. Above these levels, plant response is not usually expected, except for high-K-requiring crops such as tobacco, bananas, and potatoes.

More-recent work on potassium soil tests has encouraged some laboratories to include a factor that involves the cation exchange capacity (CEC) of a soil. The concept is that as the CEC is higher (soil is more clayey and has more humus), the potassium is less available. Thus, the critical level of potassium needed is higher. The optimum value for corn in Ohio is determined by this formula:

$$\text{optimum K soil test, lb K/acre} = 220 + (5 \times \text{CEC})$$

On a sandy soil of CEC = 9 cmoles_c/kg, the needed K value is 265 lb/a. On a clayey soil of CEC = 23 cmoles_c/kg, the needed K value is 335 lb/a.

The 1 molar ammonium bicarbonate + DTPA chelate extract is receiving a lot of interest for estimating the soil fertility of P, K, Zn, Fe, Mn, and Cu (and even NO_3^-) all in one simple extract. The values correlate well to values with sodium bicarbonate for available phosphorus (although only half as much phosphorus is extracted). Correlation for potassium also seems to be good. An example of recommendations is given in Table 12-7.

12:4.6 Calcium and Magnesium

Most calcium and magnesium tests are related to the need for lime. Well-limed neutral and alkaline soils contain sufficient calcium for normal plant growth. Even acidic soils in need of

Table 12-6 Phosphate and Potash Fertilizer Recommendations (lb/a)[a] for Corn and Grain Sorghum (Indiana)

Soil Test Level[b] (lb/a)			For Yield Goals (bu/a)[a]									
			100–110		111–125		126–150		151–175		176–200	
P		K	P_2O_5	K_2O	P_2O_5	K_2O	P_2O_5	K_2O	P_2O_5	K_2O	P_2O_5	K_2O
						Add These lb per Acre of P_2O_5 or K_2O						
0–10	Very low	0–80	100	100	110	120	120	150	130	180	150	200
11–20	Low	81–150	70	70	80	90	90	120	100	140	120	160
21–30	Medium	151–210	50	50	60	60	60	70	70	90	80	120
31–70	High	211–300	30	30	30	30	40	40	50	60	50	80
71+	Very high	301+	0	0	0	0	10	0	10	0	10	0

Source: R. K. Stivers, et al., mimeographed report, Department of Agronomy, Purdue University, Lafayette, Ind., 1979.

[a] To convert bushels per acre of corn and sorghum to kilograms per hectare, multiply by 56 X 1.12. To convert pounds per acre to kilograms per hectare, multiply by 1.12.

[b] Soil phosphorus and potassium are extracted with Bray P_1 solution and neutral normal ammonium acetate, respectively.

Table 12-7 Approximate Nutrient Levels in Soils as Estimated by Extraction of Soil with 1 Molar Ammonium Bicarbonate + 0.05 M DTPA Chelate

Concentration of Extracted Element (mg/kg soil)						Estimated Level of Availability in Soil [a]	Response to Added Fertilizer
P	K	Fe	Zn	Cu	Mn		
0–3	0–60	0–2.0	0–0.5	0–0.2	0–1.8	Low	Expected
4–7	61–120	2.1–4.0	0.6–1.0	(0.2)[b]	(1.8)[b]	Medium	Possible
8–11	120+	4.0+	1.0+	0.2+	1.8+	High	Unlikely
>11	—	—	—	—	—	Very high	Rarely

Source: P. N. Soltanpour and P. Schwab, "A New Soil Test for Simultaneous Extraction of Macro- and Micro-nutrients in Alkaline Soils," *Communications in Soil Science and Plant Analysis,* **8** (1977), pp. 195–207.

[a] The plant is expected to respond to fertilizer if the test shows "low." Response at higher concentrations is less sure.

[b] These estimates are for the critical level. Too few data are available for further separation into groups.

lime to change the pH commonly have adequate calcium for plant nutrition. There is very little danger of soils becoming calcium deficient, even the coarse sandy soils of the Coastal Plains region, as long as liming is done. If liming is not done (to minimize diseases), gypsum may be used to supply needed calcium.

Magnesium deficiency is common in plants growing on coarse-textured acidic soils having a sandy loam, loamy sand, or sand surface texture with subsoil as coarse as or coarser than the plow layer. Magnesium deficiency also occurs in similar soils limed with calcic limestone ($CaCO_3$) or marl because the added lime contains little or no magnesium.[7]

Three soil test criteria are used to interpret magnesium levels in the soil and to make magnesium recommendations: exchangeable soil magnesium, percentage magnesium saturation of soil colloids, and the ratio of potassium to magnesium. Interpretive values for magnesium levels in Indiana soils are[8]:

Humus and Texture	Inadequate Levels
Light colored (low humus), coarse textured	0–75 lb of Mg per acre (0–84 kg/ha)
Light colored (low humus), medium to fine textured	0–100 lb of Mg per acre (0–112 kg/ha)
All other Indiana soils	0–200 lb of Mg per acre (0–224 kg/ha)

Magnesium values of less than 75 lb per acre do not necessarily mean that yield or crop quality will always be reduced under such conditions, but the chances of getting lower yields and inferior-quality crops are greater on such soils than they would be on soils containing adequate levels of plant-available magnesium.

Suggested magnesium saturation percentage (of the CEC) range from the suggested minimum value of 3% in Michigan to 10% in Missouri and Pennsylvania. Researchers in Ohio suggest that soil used for the growth of animal forage have a minimum of 15% basic cation saturation of magnesium.

[7] D. D. Warncke and D. R. Christensen, "Fertilizer Recommendations for Vegetables and Field Crops in Michigan," Extension Bulletin E-550, Michigan State University, East Lansing, 1980.

[8] R. K. Stivers, et al., mimeographed report, Department of Agronomy, Purdue University, Lafayette, Ind., 1979.

Adequate levels of available soil magnesium are important not only for normal plant growth and maximum yields but for inhibiting nutritional/metabolic disorders referred to as *grass tetany* ("staggers" or hypomagnesemia) in grazing cattle. Potassium-induced magnesium deficiency may occur in plants if potassium exceeds magnesium as a percentage of the total available basic cations. Other factors related to grass tetany include levels of calcium, crude protein, carbohydrate, and free amino acids in the feed ration.[9]

Acidic soils that are deficient in available magnesium can be corrected by the use of dolomitic limestone ($MgCO_3 \cdot CaCO_3$). Soluble sources of magnesium, such as magnesium sulfate and sulfate of potash magnesia, are recommended for soils that are not acidic but are deficient in available magnesium.

12:4.7 Sulfur and Micronutrients

Sulfur Sulfur tests are less well developed than those for phosphorus, potassium, and lime. Fewer soils exhibit sulfur deficiency, yet the same needs exist in developing a good test: a large number of experimental sites and plots for field comparison. Developing a test for sulfur meets with some of the same problems as developing a test for nitrogen; much of the sulfur is released from decomposing humus, but how can this release be measured ahead of time? The tests used to evaluate sulfur measure mostly the extractable sulfate, which is quite soluble.

Boron *Hot-water-extractable boron* is generally considered the best measure of the available form of boron in soils, and the following levels have been established: less than 1.0 ppm, *too low* for normal plant growth; 1.0–5.0 ppm, *adequate* for normal plant growth; greater than 5 ppm, *may be excessive or toxic* to certain plants.

Soils derived from marine sediments generally contain more boron than soils formed from igneous rocks. The total boron content of soils may range from as little as 1 ppm to as much as 270 ppm, and averages 20–50 ppm.

Crops exhibit considerable variability in their response to boron. Under Michigan conditions, application rates of 1.5–3 lb of elemental boron per acre (1.7–3.4 kg/ha) are recommended for highly responsive crops and 0.5–1 lb/a (0.6–1.1 kg/ha) for low to medium-responsive crops.[10]

Zinc, Iron, Manganese, and Copper Considerable effort has been made to develop good predictive tests for the micronutrient metals, but with limited success. The relatively fewer locations of deficient soils limits the ease and extent of field correlation study. Because the metals have increased solubility in acids, it was logical to use an acidic extractant (0.1 or 1 molar HCl is common). Some laboratories make predictions of these nutrient needs from data using such extractants. For example, the Mehlich-1 extract has been used for zinc and manganese evaluation.

A more recent use of a chelate extraction (several have been tried) has been useful in several areas. The ligand DTPA has been extensively studied in Colorado and in a few other states.[11] Its use has spread to many laboratories (Table 12-8). Unfortunately, many laboratories depend on the correlation data developed by other areas for different soils, climates, and crop varieties. Part of this problem is caused by the few locations of deficiency existing on

---

[9] C. B. Elkins, C. S. Hoveland, R. L. Haaland, and W. A. Griffey, "Grass Tetany," Crops and Soils Magazine (1977).

[10] M. L. Vitosh, D. D. Warncke, B. D. Knezek, and R. E. Lucas, "Secondary and Micronutrients for Vegetables and Field Crops," Extension Bulletin E-486, Cooperative Extension Service, Michigan State University, East Lansing, 1981.

[11] W. L. Lindsay and W. A. Norvell, "Development of a DTPA Soil Test for Zinc, Iron, Manganese and Copper," *Soil Science Society of America Journal,* **42** (1978), pp. 421–428.

Table 12-8 Zinc Application (Recommended lb/a) for Corn, Sorghum, Soybeans, and Pinto Beans (Kansas)[a]

Management and Crop	Area of State	Zinc Soil Test (ppm Zn) [b]		
		Low (0–0.5)	Medium (0.51–1.0)	High (Above 1.0)
		Lb/a of Zn to Add		
Irrigated corn, sorghum, soybeans, and pinto beans	Entire	8–10	2–5	None
Nonirrigated corn	Eastern (humid)	8–10	2–5	None
Sorghum, soybeans, and pinto beans	Entire	2–5	None	None

Source: D. A. Whitney, R. Ellis, Jr., L. Murphy, and G. Herron, *Identifying and Correcting Zinc and Iron Deficiency in Field Crops,* Cooperative Extension Service, Kansas State University, Manhattan, 1975.
[a] Based on the use of zinc sulfate as source of zinc.
[b] DTPA-extractable zinc.

which research studies can be made to develop the correlations. The fact remains, however, that the accuracy of DTPA-extractable metals to predict nutrient adequacy or deficiency is not proven for many areas and crops.

The tests to predict micronutrient deficiencies should, in most instances, be regarded as indications of need or adequacy and not as detailed predictions. Many years of seeing field deficiency symptoms and response to additions of the micronutrient is a very practical aid to evaluating the laboratory tests.

Molybdenum Soil tests for molybdenum are not generally done routinely because such deficiencies are not common and testing for molybdenum is a sensitive procedure.

The total molybdenum content of soils is low, ranging from traces to 24 ppm and averaging 1–2 ppm (mg/ha). Similar levels have been reported for soils of humid tropical regions.

Oxalates, water, and anion exchange resins have been used to extract available molybdenum from soils. Soils containing 0.1–0.2 ppm of oxalate-extractable molybdenum generally are not deficient in this element.

12:5 Testing Greenhouse Soils

The production of greenhouse crops requires an intensive fertility program for maximizing growth and quality in a short time period. A monitored fertilizer program using chemical soil tests is basic to ensure that essential nutrients are not limiting plant growth and, conversely, are not in excess, creating an imbalanced or a toxic soil condition. The majority of all problems encountered in greenhouse production are related to soil fertility.

12:5.1 Sampling Soils for Greenhouse Crops

To ensure accurate soil tests, it is imperative that samples to be tested are representative of the entire area or conditions in question and are made at the best time.

When to Sample Sample prior to adding fertilizers and planting. Bench soils may be sampled every 3 months or when a problem is evident.

Table 12-9   General Guidelines for Greenhouse Soil Nutrient Levels and Their Interpretation[a]

Nutrient	Deficient	Optimum	Excessive
	(ppm in Media Extract)		
Nitrate-nitrogen	Below 39	100–279	Above 280
Phosphorus	Below 3	8–13	Above 20
Potassium	Below 59	150–249	Above 350
Calcium	Below 79	200–349	Above 500
Magnesium	Below 29	60–99	Above 150

Source: D. D. Warncke, "Testing Greenhouse Growing Media: Update and Research," Crop and Soil Sciences Department, Michigan State University, presented at the Seventh-Soil-Plant Analysts' Workshop, Bridgeton, Mo., 1979.
[a] Based on a saturated media extraction procedure.

How to Sample   Sample to the depth of the pot or to 30 cm (1 ft), whichever is reached first.

General Information   Use a soil probe or similar tool for taking samples. Wait at least 6 hours after watering or at least 5 days after applying dry fertilizer. Delay sampling 24 hours when fertilizer has been applied through a liquid injector system.

12:5.2 Soil Nutrient Levels for Greenhouse Crops

Table 12-9 shows nutrient levels in soils to be used for greenhouse floricultural crops. Levels considered to be deficient, optimum, or excessive (toxic) are shown for nitrate-nitrogen, phosphorus, potassium, calcium, and magnesium. The nutrient values are expressed in parts per million (ppm or mg/kg) in the media extract based on a saturated media extract procedure.

12:5.3 Soluble Salts in Greenhouse Soils

Any soluble fertilizer is also a soluble salt. Where heavy fertilizer applications are used, their action as soluble salts must be considered. The soluble salt may be a combination of fertilizers and of other salts added in irrigation waters. The evaluation of the salt hazard in greenhouse soils is approximately as shown in Table 12-10. Plants differ in their sensitivity to salt, so simple salt tolerance tables, such as Table 12-10, will always have some plants more tolerant and a few less tolerant than those listed in the table. Unfortunately, overwatering to ensure regular leaching of any accumulated salt also removes most of the soluble fertilizers from the soils for a period of time.

12:6 Analyses of the Plant

The only way to be sure that nutrients are adequately available to plants is to measure what the plants take up. Many conditions, however, may alter nutrient uptake, even though the nutrient is adequately soluble. Some of these conditions are (1) low soil temperature, (2) rapid plant growth, (3) low soil water contents, (4) poor drainage (poor aeration), (5) antagonistic or synergistic interactions among nutrients, and (6) root damage. Thus, soil tests that indicate that adequate nutrients should be available will not ensure that the plant will not be deficient in the nutrient or that the addition of the nutrient will not be helpful to the plant.

Table 12-10 Soluble Salt Levels (Electrical Conductivity) in Soils for Greenhouse and Bedding Plant Production in Relation to Plant Growth

Conductivity Readings (dS/m)			
Saturated Media Extract	2:1 (Weight Basis) Water : Soil	5:1 (Volume Basis) Water : Solid	
All Soils	Mineral Soils	Organic Soils	Relation to Plant Growth
0–0.74	0–0.25	0–0.12	Very low salt levels; indicates very low nutrient status
0.74–1.99	0.25–0.50	0.12–0.35	Suitable range for seedlings and salt-sensitive plants
2.00–3.49	0.50–1.00	0.35–0.65	Desirable range for most established plants; upper range may reduce growth of some sensitive plants
3.50–4.99	1.00–1.50	0.65–0.90	Slightly higher than desirable; loss of vigor in upper range; suitable for high nutrient-requiring plants
5.00–5.99	1.50–2.00	0.90–1.10	Reduced growth and vigor; wilting and marginal leaf burn
6.00+	2.00+	1.10+	Severe salt symptoms—wilting; crop failure

Source: "Chemical Controls for Michigan Commercial Greenhouse and Bedding Plant Production," Extension Bulletin E-1275, Cooperative Extension Service, Michigan State University, East Lansing, 1978.

12:6.1 Plant Analyses vs. Soil Tests

Should we change from using soil tests and use plant analyses for predicting fertilizer needs? Because the plant is the final arbiter of whether a nutrient is truly available, this might at first seem to be a logical change. However, it is not good, in most instances. The plant cannot be analyzed until it is growing; using the plant analyses means there is no information for pre-plant or during-planting fertilization. Further, you will not know that the plant is deficient in an element until you see it or measure it. By that time plant growth is reduced and it may be too late to adequately correct the damage done to yields. Only with a few long-growing crops—such as sugar cane, turf, citrus, forests, and bananas—might plant analyses alone be suitable to predict fertilizer needs.

12:6.2 Total Plant vs. Quick Tissue Analyses

Total plant analyses are accurate laboratory measurements of plant contents of each element analyzed. In contrast, **quick tissue tests** are usually approximate measurements done on plant tissues (crushed cells—plant sap) in the field. Unfortunately, quick tissue tests have had limited success because they are subject to operator error and climatic changes and are not quantitative. Usually, the concentrations of nutrients in quick tissue tests are reported only as *very low, low, medium, high,* and *very high,* based on the intensity of the color of the test paper or solution used (Fig. 12-2).

The increasing ability to make total plant analyses quickly and easily has increased the use of plant analyses. Quantitative values of any of the plant nutrients is possible within a day, even hours, but more commonly within a few days. This fast turnaround time makes it possible to make fertility adjustments within a few days by applying fertilizers in irrigation water, as foliar sprays, in drip systems, or even as top dressings. Even though plant tests cannot

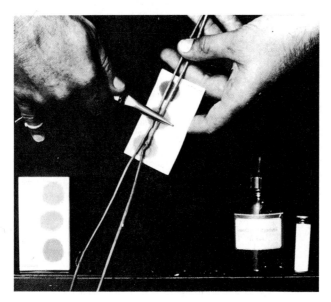

FIGURE 12-2 Potassium is determined in the plant by squeezing the green tissue on each of the orange dots imbedded in the paper (the dots contain dipicrylamine) and adding reagent PK-1. The disappearance of orange color from all the dots indicates a very low test for potassium; a high level of potassium for most crops is indicated when the orange color persists in all three dots. (Courtesy of Potash and Phosphate Institute.)

replace soil tests, the knowledge of the plant's nutrient contents during growth under different conditions or treatments will help modify future management and fertilizer applications.

12:6.3 Sampling for Total Plant Analyses

The tissue sampled for analysis should be selected on the basis of *physiological age* (development stage) rather than chronological (calendar) age. Ideally, a sample from a normal healthy plant should be obtained at the same time the problem plants are sampled so that the diagnostician can base conclusions on comparative tests.

Factors other than those associated with the soil can have pronounced influence on the chemical composition of leaf tissue. Orange leaves from *nonfruiting* terminal branches contained more than 50% more nitrogen and phosphorus, 220% more potassium, and 35% less magnesium than leaves behind young fruits. Micronutrient contents varied within individual corn leaves. Boron, iron, and manganese accumulate in leaf margins, so the portion of a leaf used for analysis influences test results.

Usually, the element composition in the leaf varies during growth. Nitrogen, phosphorus, potassium, and zinc levels in the leaves usually decrease as the growing season progresses; contents of calcium, magnesium, boron, manganese, and iron increase.

Sampling instructions for selected vegetable crops, fruit and nut crops, field crops, and ornamentals and flowers are presented in Tables 12-11, 12-12, 12-13, and 12-14. Even concentrations of nutrients in bark may be useful in determining nutrient needs for some crops, such as cassava.[12]

Although some nutrients are quite mobile in plants (nitrogen, potassium, and phosphorus), others are not (calcium, iron, and zinc). It would seem necessary to sample different parts of the plant for different nutrients, but special sampling for each nutrient is awkward. It is found generally that a particular plant part can be used for most of the nutrients, at least until more diagnostic nutrient levels in plants are known.

---

[12] W. Goldfrey, Sam Aggrey, and M. J. Garber, "Bark Analysis as a Guide to Cassava Nutrition in Sierra Leone," *Communications in Soil Science and Plant Analysis, 10* (1979), no. 8, pp. 1079–1097.

Table 12-11 Suggested Sampling Instructions for Selected Vegetable Crops

Stage of Growth	Plant Part to Sample	Number of Plants to Sample
Potato, prior to or in bloom	3rd to 6th leaf from growing tip	20–30
Head crops (cabbage, etc.), prior to heading	1st mature leaves from center whorl	10–20
Tomato, prior to or during bloom	3rd or 4th leaf from growing tip	20–25
Root crops (carrots, onions, beets), prior to root or bulb enlargement	Center mature leaves	20–30
Celery, midgrowth (30–70 cm tall)	Petiole of youngest mature leaf	15–30
Leaf crops (lettuce, spinach, etc.), midgrowth	Youngest mature leaf	35–55
Sweet corn, at tasseling	Entire leaf at the ear node	20–30
Melons (watermelon, cucumber, etc.), early stages of growth prior to fruit set	Mature leaves near the base portion of plant on main stem	20–30

Source: Modified from John E. Bowen, "Plant Parts to Sample," *Crops and Soils Magazine,* **31,** no. 3, Dec. 1978.

Table 12-12 Suggested Sampling Instructions for Selected Fruit and Nut Crops

Stage of Growth	Plant Part to Sample	Number of Plants to Sample
Apple, apricot, almond, prune, peach, pear, cherry, midseason	Leaves near base of current year's growth	50–100
Strawberry, midseason	Youngest mature leaves	50–75
Pecan, 6–8 weeks after bloom	Leaves from terminal shoots, pairs from middle of compound leaf	30–45
Walnut, 6–8 weeks after bloom	Middle leaflet pairs of mature shoots	30–35
Lemon, lime, midseason	Latest mature leaves on nonfruiting terminals	20–30
Orange, midseason	Spring cycle leaves, 4–7 months old	20–30
Grapes, end of bloom period	Petioles adjacent to fruit clusters	60–100
Raspberry, midseason	Youngest mature leaves on 1st canes	20–40

Source: John E. Bowen, "Plant Parts to Sample," *Crops and Soils Magazine,* **31,** no. 3, Dec. 1978.

The most useful plant parts to sample vary with the species and the plant development stage. In young seedlings of annual crops (corn, grains, beans), the *whole plant* is sampled. As the plant becomes more mature, the *youngest mature leaf* seems to be the preferred plant part. When this leaf is near the fruit body (corn ear, fruit terminal, or grape cluster), nutrients it might have contained are often "robbed" to help feed the growing fruit body. For ornamental trees, only the developed leaves on the current year's growth are recommended for sampling.

As crops and their fruits (seed, fruit, nuts) mature, the nutrient level in the tissues becomes less stressed and is less diagnostic. Many plants should not be sampled for analyses later than the early stages of fruiting.

Table 12-13 Suggested Sampling Instructions for Selected Field Crops

Stage of Growth	Plant Part to Sample	Number of Plants to Sample
Corn, tasseling to silking	Entire leaf at ear node	20–30
Beans, soybeans, initial flower	2 or 3 fully developed leaves	20–30
Small grains, rice, prior to heading	4 uppermost leaves	50–100
Hay, pasture, forage grasses; optimum forage stage	4 uppermost leaf blades	40–50
Alfalfa, clovers, prior to 1/10th bloom	Mature leaf blades, top 1/3 of plant	40–50
Sugarbeets, midseason	Fully mature leaves midway between younger center leaves and oldest leaf whorl on the outside	30–40
Tobacco, before bloom	Uppermost fully developed leaf	8–12
Grain sorghum, heading or prior	2nd leaf from top of plant	15–25
Sugar cane, up to 4 months old	3rd or 4th mature leaf from top	15–25
Peanuts, bloom stage or prior	Mature leaves	40–50
Cotton, prior to or when first squares appear	Youngest mature leaves on main stem	30–40

Source: Modified from John E. Bowen, "Plant Parts to Sample," *Crops and Soils Magazine,* **31,** no. 3, Dec. 1978.

Table 12-14 Suggested Sampling Instructions for Selected Ornamentals and Flowers

Stage of Growth	Plant Part to Sample	Number of Plants to Sample
Ornamental trees, current-year's growth	Fully developed leaves	30–100
Ornamental shrubs, current year's growth	Fully developed leaves	30–100
Turf, during normal growing season	Leaf blades (hand harvest)	1/2 pint
Roses, during flower production	Upper leaves on flower stems	20–30
Chrysanthemums, flowering or prior	Upper leaves on flower stems	20–30
Carnations, unpinched plants	4th or 5th leaf pair from base	20–30
Carnations, pinched plants	5th and 6th leaf pairs from top of primary laterals	20–30
Poinsettias, flowering or prior	Newest fully mature leaves	15–20

Source: Modified from John E. Bowen, "Plant Parts to Sample," *Crops and Soils Magazine,* **31,** no. 3, Dec. 1978.

12:6.4 Preparation of Plant Samples

It is important that the plant material be air dried for at least 24 hours to prevent spoilage prior to mailing in the container supplied by the laboratory. Samples should not be mailed in air-tight containers (plastic) because the sample will spoil if the material contains any water.

Where a plant sample has been contaminated from soils, sprays, or residues of other materials, it is imperative that it be cleaned before drying, especially when micronutrient analyses are to be done. The cleaning is usually a quick washing on the fresh green tissue (one minute or less) because nutrients such as potassium, sodium, nitrate, and chloride may be readily leached from the plant material. Wash the plant tissues in 0.1 – 0.3% detergent solution using a *phosphate-free* material to avoid phosphorus contamination. The tissue should be rinsed thoroughly but quickly in distilled water as soon after sampling as is reasonably possible. Tissues should be dried as rapidly as possible at 65–70°C (150–160°F) to minimize chemical and biological transformations and to destroy respiratory enzymes without burning or charring the plant material.

12:6.5 Interpreting Plant Analyses

It is essential that the diagnostician have a complete history of the management factors in order to make an accurate and practical interpretation of test results. This necessary information includes items such as plant variety, planting density, geographic location (for climate), date planted, and many other descriptive items.

When an essential element is classified as deficient on the basis of plant analyses, the diagnostician must determine the primary cause(s) before a treatment can be prescribed. The first difficult problem is to know *threshold (critical)* nutrient levels. These levels change with plant variety, plant part, time of season, climate, other nutrient levels, soil pH, root health, disease, and almost any other plant growth factor. This interpretation, not the laboratory analyses, is the most difficult problem in using plant analyses.

Nutrient sufficiency levels for various crops are listed in Tables 12-15 and 12-16. The assumption is that plant contents below the lower values indicate a deficiency of that nutrient in that plant. There is no general agreement on these threshold values for any crops because so many factors are involved. Even the same scientist may arrive at different values on data collected over different years. Climate, management, and soil variations cause differences.

12:6.6 Critical Nutrient Range or Threshold Values

The **critical nutrient range (CNR)** has been referred to by various names. In the preceding section, the *threshold* or *critical* nutrient level was used to refer to that CNR range. Figure 12-3 illustrates the relationship among several terms, including CNR.

The difficulty in pinning down a single critical value for each nutrient in each plant variety has made it necessary to use a range of values for CNR and nutrient *ratios.* An approach to using ratios is the **DRIS** (diagnosis and recommendation integrated system) approach. The concept is that the *balance of nutrients* is more critical than the *total amounts.* Certain interactions and relationships occur that can be indicated by various ratios. For example, N/P, N/K, and K/P are used. Both a *deficient* element and a *toxic* level of an element are determined. The approach sounds simple, but it is complex, varies in values, and will require extensive testing to prove its accuracy and value.[13,14] It does show recognition of the complex interrelationships of the soil–plant system.

12:7 Visual Nutrient Deficiency Symptoms

Plants exhibit external symptoms of starvation as a result of nutrient deficiency or imbalance. Certain nutrient deficiencies cause a reduction in the formation of the green pigment, chlorophyll, giving deficient plants a distinctive yellowish to whitish appearance (**chlorosis**). Unfortunately, chlorosis can be caused by diseases, insect damage, salt accumulation, and other growth stresses, as well as by nutrient deficiencies. Some visual deficiency symptoms in plants are easily diagnosed; other visual changes are symptomatic of many deficiencies. Abnormal growth (color and/or size) is a clue that something is wrong in growth, but it isn't

---

[13] C. A. Jones and J. E. Bowen, "Comparative DRIS and Crop Log Diagnosis of Sugarcane Tissue Analyses," *Agronomy Journal,* **73** (1981), pp. 941–945.

[14] Milton B. Jones, D. Michael Center, Charles E. Vaughn, and Fremont L. Bell, "Using DRIS to Assay Nutrients in Subclover," *California Agriculture,* Sept.–Oct. 1986, pp. 19–21.

Table 12-15 Plant Nutrient Sufficiency Levels (Threshold or CNR Values) for Selected Crops

Nutrients	Units	Alfalfa: Top 6 in. (15 cm) Sampled Prior to Initial Flowering	Corn: Ear Leaf Sampled at Initial Silk	Potatoes: Petioles from Most Recently Matured Leaf Sampled in Midseason	Soybeans: Upper Fully Developed Leaf Sampled Prior to Initial Flowering	Sugarbeets: Center Fully Developed Leaf Sampled in Midseason	Vegetables: Top Fully Developed Leaves	Wheat: Upper Leaves Sampled Prior to Initial Bloom
Nitrogen	%	3.76–5.50	2.76–3.50	2.50–4.00	4.26–5.50	3.01–4.50	2.50–4.00	2.59–3.00
Phosphorus	%	0.26–0.70	0.25–0.40	0.18–0.22	0.26–0.50	0.26–0.50	0.25–0.80	0.21–0.50
Potassium	%	2.01–3.50	1.17–2.50	6.0–9.0	1.71–2.50	2.01–6.00	2.00–9.00	1.51–3.00
Calcium	%	1.76–3.00	0.21–1.00	0.36–0.50	0.36–2.00	0.36–1.20	0.35–2.00	0.21–1.00
Magnesium	%	0.31–1.00	0.16–0.60	0.17–0.22	0.26–1.00	0.36–1.00	0.25–1.00	0.16–1.00
Sulfur	%	0.31–0.50	0.16–0.50	0.21–0.50	0.21–0.40	0.21–0.50	0.16–0.50	0.20–0.40
Manganese	ppm[a]	30–100	20–150	30–200	21–100	21–150	30–200	16–200
Iron	ppm	30–250	21–250	30–300	51–350	51–200	50–250	11–300
Boron	ppm	31–80	4–25	15–40	21–55	26–80	30–60	6–40
Copper	ppm	11–30	6–20	7–30	10–30	11–40	8–20	6–50
Zinc	ppm	21–70	20–70	30–100	21–50	19–60	30–100	21–70
Molybdenum	ppm	1.0–5.0	0.1–2.0	0.5–4.0	1.0–5.0	0.15–5.0	0.5–5.0	0.03–5.0

Source: M. L. Vitosh, D. D. Warncke, B. D. Knezek, and R. E. Lucas, "Secondary and Micronutrients for Vegetables and Field Crops," Extension Bulletin E-486, Cooperative Extension Service, Michigan State University, East Lansing, 1981.
[a] ppm means parts per million, or the micrograms of element per gram of dry plant material.

Table 12-16 Plant Tissue Nutrient Levels for Various Floricultural Crops (Critical Levels below Which the Plant Should Respond to Added Nutrient)

Plant	Dry Weight (%)					Micrograms per Gram Dry				
	N	P	K	Ca	Mg	Mn	Fe	Cu	B	Zn
Roses	3.0	0.2	1.8	1.0	0.25	30	50	5	30	15
Carnations	3.0	0.05	2.0	0.6	0.15	30	30	5	25	15
Chrysanthemums	4.5	0.18	2.2	0.5	0.14	200	125	5	25	7
Poinsettias	3.5	0.2	1.0	0.5	0.2	—	—	—	—	—
Geraniums	2.4	0.28	0.6	0.8	0.14	9	60	5.5	18	6

Source: Selected data from R. A. Criley and W. H. Carlson, "Tissue Analysis Standards for Various Floricultural Crops," *Florists' Review,* **146** (1970), no. 3771, pp. 19–20.

always easy to determine the cause. A general pattern for some nutrient deficiencies are shown in Table 12-17.

12:7.1 Nutrient Deficiencies in Specific Crops

Nitrogen in Corn   The corn plant may become light green or yellow (chlorotic), and the oldest leaf by this time will have died (**necrotic**) and turned light to dark brown.

Phosphorus in Corn   Young phosphorus-deficient corn plants are stunted and *dark green* in color because of their high nitrogen content. Sugars accumulate and increase anthocyanin

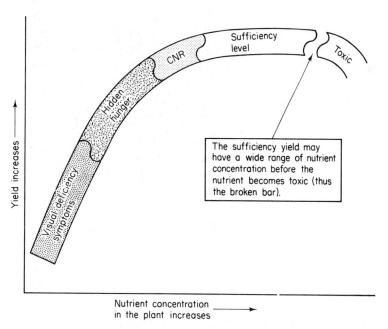

FIGURE 12-3   Relationships of plant growth and visual deficiency symptoms, hidden hunger, critical nutrient range (CNR), sufficiency levels, and toxic concentrations.   (Courtesy of Raymond W. Miller, Utah State University.)

Table 12-17 Abbreviated Outline of Visual Symptoms of Nutrient Deficiencies

	Nutrient
A. *Older leaves* mostly affected, stunted growth, slender stems	
1. Chlorosis of tips and then whole leaf, stunted	Nitrogen
2. Dark-green overall, reddish-purple colors, stunted	Phosphorus
3. Spots of dead tissue (dead brown tip and margins), stunted growth, tall grains may lodge easily	Potassium
4. Spotty or slender streaks of interveinal chlorosis, tips cupped, may redden	Magnesium
5. Interveinal spots, thick waxy leaves, short internodes (rosette of leaves)	Zinc
B. *Younger leaves* or buds mostly affected	
6. Terminal bud dies or is deformed	Calcium, Boron
7. Overall chlorosis of leaves	Sulfur
8. Dead spots scattered in leaves, veins green, a checkered effect on leaves	Manganese
9. Veins green, intricate interveinal chloroses, even to white in extreme deficiency	Iron
10. Leaves permanently wilted, maybe some chlorosis	Copper

(reddish-purple) pigment. With a decrease in chlorophyll (green pigment) content, the anthocyanin pigment predominates and the leaf has purple coloration. The purpling starts at the tip of the leaf and proceeds along the edges. This coloring may disappear when the plant develops a more extensive root system, and the only symptom remaining is the stunted condition of the plant. In some instances a bronze discoloration replaces the purpling effect. In some plants factors other than phosphorus deficiency can cause the purpling effect. Some varieties have natural purple coloring on the stems, particularly.

Potassium in Corn The plant is stunted and internodes are shortened when potassium is inadequate. Yellowing of the leaves starts at the tip of *older* leaves and proceeds along the edges to the base of the leaf. Eventually, the edges become brown and die (are necrotic). It has often been termed *leaf scorch*, but the term is not preferred. An excess of soluble salt produces a somewhat similar yellowing at the tips and may be confused with potassium deficiency.

Nitrogen in Cotton A light-green or yellow chlorosis becomes general over the entire leaf. The nitrogen-deficient leaf is smaller than leaves from normal, healthy plants. Yellowing starts with the older, basal leaves and progresses up the plant. In advanced stages of nitrogen deficiency the leaves turn brown and eventually fall from the plant.

Potassium in Cotton The leaf is chlorotic with yellow spots between veins and along tips and margins of the leaf. The leaf tends to curl downward at the margins. At the time of boll formation, reddish spots may appear on leaves near developing bolls.

Manganese Deficiency in Beans On navy beans, soybeans, and garden beans, a manganese deficiency appears first as a mottled (spotty) effect on new leaves (Fig. 12-4) and is generally uniform over the leaf with the veins remaining green. If symptoms develop early in the season, when the bean plants are small, a side-dress application of manganese sulfate or a spray application of a soluble manganese salt can generally correct the deficiency.

12:8 Making Fertilizer Recommendations

The objective in making fertilizer recommendations is to predict accurately the amount of each nutrient needed for growth of a particular plant, both in quality and quantity of plant material wanted. It is important to understand the techniques used in making recommendations.

FIGURE 12-4 On soil with pH of 7.5 and deficient in available manganese, a normal green color of navy bean leaves was exhibited only after 134 kg/ha (120 lb/a) of manganese (as manganese sulfate) was applied (right); 45 kg/ha (40 lb/a) of manganese did not supply sufficient manganese for normal color (center); and where no manganese was applied to the soil (left), the leaves were nearly white, except for the veins.   (Courtesy of J. Rumpel, B. G. Ellis, and J. F. Davis, Michigan State University.)

12:8.1 Basis for Making a Recommendation

Recommendations for nutrients can be based on different plant production schemes. Depending on the scheme used for soil management, different laboratories analyzing the same soil might suggest that a different amount of fertilizer be applied. The three following schemes are commonly in use.

Prescription Method   Recommend only the nutrients needed for the year's optimum crop yield. This would supply just the crop's need. It takes into account other nutrient additions (manure, residues, etc.) that you have said you will be adding to the crop.

Prescription Plus Buildup   In this scheme the nutrients recommended for the crop are given *plus* nutrients that are considered important to help build up the soil's nutrient level to improve growth or plant quality. This is often used to build up levels of phosphorus, micronutrients, and perhaps potassium. It would not be common to use it for nitrogen or sulfur because too much of these may be lost by leaching or gaseous loss.

Rotation Need   A recommendation may be to apply the total fertilizer needed during a full rotation. As an example, in a corn-soybean rotation, you may add enough nitrogen, phosphorus, and potassium to the corn to carry over to the soybeans that follow. This reduces two fertilizer application passes to just one. The producer must decide if the cost of two applications is cheaper than the lowered efficiency of fertilizer added so far ahead of the soybean crop. This program sometimes is used for double-cropping to reduce the time needed between planting.

From the discussion of these three schemes for making fertilizer recommendations, it is clear that the concept used by the laboratory to which you send your samples must be understood. Usually the information given is clear in its intent.

12:8.2 Developing a Fertilizer Recommendation

To develop a correlation between soil test values on a soil and the amount of nutrient needed to be added, a large number of field plot trials are essential. For example, different rates of the nutrient (N, P, K, etc.) are applied to small plots in the field and a crop is grown for yield data. The untreated soil is tested with a "quick soil test," a rapid laboratory procedure. The values of the soils and application rates are done for many locations (various field soils). The many test results are sorted according to the soil test values obtained and are compared to yields from the various plots having different fertilizer rates added. If a "phosphorus test" was being established, the large number of soil sites tested and the soil test values and crop yields might be as given in Table 12-18. From this data it would be obvious that if your soil tested 2.2 mg/kg of P, you would get increased yields by adding at least 90 kg/ha of P_2O_5; there may be a yield increase if you added even more P. For soils that had tests of 11–16, there would be little yield increase with more than a 60 kg/ha addition of P_2O_5. Further, soils with tests in the 17–24 mg/kg phosphorus range, only 30 or 40 kg/ha of P_2O_5 additions would increase yields very much.

Development of tables, such as Table 12-18, are needed for each crop (or even variety), for a selection of soils, for different climatic areas, and for each nutrient. To develop good correlation tables, an extraordinary number of field plots are needed and the data compiled and grouped for average values of yield increases.

Once the general amount of "additional fertilizer needed" is known, the person preparing the recommendation has various details to be considered. Some of these are listed below:

1. Subtract a certain amount of recommended fertilizer for each ton of manure to be added (as 10 kg P_2O_5 per ton).

2. Subtract a certain amount of recommended fertilizer for the residual P or N left over from certain previous crops. Tables of these additions or subtractions are prepared from previous experimental data and are probably programmed into the computer.

3. Add or subtract, from the recommended fertilizer base, the amount of fertilizer needed to meet yield goals that are larger or smaller than the reference value. There may be a built-in sliding scale.

4. Add or subtract fertilizer depending on your area's climate (short season, extra cool early season, expect heavy leaching).

5. Add or subtract recommended nitrogen, phosphorus, or sulfur because of unusually high or low humus or crop residue levels in the soil. These nutrients will be supplied from decomposing organic matter, but soil tests do not measure it.

Table 12-18 Example of a Correlation Tabulation of Soil-Test P and Alfalfa Yield Response to Fertilizer P in a Variety of Soils

Soil-test P (mg/kg)	Alfalfa Yields at Different P_2O_5 Rates (kg/ha)			
	0	30 kg/ha	60 kg/ha	90 kg/ha
0–3	1880	2800	3660	4500
4–6	2480	3410	4240	4750
7–10	3140	4050	4610	4900
11–16	3860	4530	4800	4880
17–24	4210	4820	4910	4950

Numerous other considerations are involved, such as the amount and kind of residue left that will need additional nitrogen added for early decomposition so that the crop will not suffer nitrogen deficiency.

12:8.3 Computers in Recommendations

Computers are now widely used in recommendations. They allow for integration of many interactions and cautions. For example, the computer could be coded to print out a warning of possible sodic soil conditions if the pH entered is over about 8.5. Several factors could be tied together so that a warning occurs whenever the *combination of factors* all occur. In the past, people also did this integration of data, but mentally and more slowly, and they often forgot to do so. The computer does not forget any of the combinations put into it. But a word of caution: *The computer readout is only as good as the data put into it.* Setting up the computer with poor or incorrect data cannot result in correct interpretations. An example of a computer readout and part of an evaluation table is given in Fig. 12-5 and Table 12-19.

12:8.4 Some Factors That Soil Tests Do Not Measure

Some assumptions are made in interpreting soil tests. The fact that some of these assumptions are not valid for the soil tested can be serious enough to make the recommendations inaccurate for the situation. Some of the assumptions used are the following:

Assumption 1   The soil is as deep as the plant normally roots. This is from about 1 m (39 in.) for shallow-rooted crops to nearly 2 m (79 in.) for corn, alfalfa, and other deep-rooted crops. If your soil is only 50–80 cm (20–31 in.) deep, there is only a partial root zone available to supply the nutrients and water.

Assumption 2   There are no conditions inhibiting root growth or root penetration. If inhibiting layers exist, such as plowpans, genetic hardpans, or compacted and poorly aerated soils, they may effectively limit the root zone as in the first assumption above.

Assumption 3   Other growth factors are assumed to be adequate. It is assumed that periods of drought, excess wetness, unusually cold or hot periods, and toxic levels of salts or other elements do not occur. It is usual in recommending nitrogen fertilizer for dryland crops to give a range of rates for fertilizer addition—high rates for high-rainfall years, medium rates for average years, and low rates for drought years. This allows the grower to make his own prediction of what he thinks the year's rainfall will be.

Assumption 4   Good management is assumed. Suitable control of pH is assumed. A grower may plant too late, not control weeds well, select a poor variety or poor-quality seed, have poor timing in adding split fertilizer applications and pesticides, irrigate too infrequently or too little, and get too thin or too dense a stand. All of these actions can reduce yields and make the farmer feel that the fertilizer recommendation was not good.

With all these problems in getting a good representative soil sample and considering all factors necessary to make accurate fertilizer predictions, it is amazing how often and well the recommendations work. Soil tests predict many situations well; their use can be and will get better.

Table 12-19 Typical Evaluation Tables Used to Make Predictions About the Fertilizer Requirements for a Soil[a]

Table 1 Basic Nitrogen Recommendations for Utah (lb N per acre)[a]

	Length of Growing Season[b]			
	Very Long	Long	Medium	Short
	Dryland			
Small grains	Irrigated 35–50	35–50	35–50	35–50
Alfalfa	0	0	0	0
Corn	—	200	150	100
Pasture				
All grass	—	150	100	75
Grass + legume	—	0–75	0–75	0–75
Pinto beans	0–40	0–30	0–30	—
Potatoes	—	300	200	150
Small grains				
Standard varieties	—	100	75	75
Dwarf varieties	—	150	100	75
Sorghum	150	—	—	—
Sudan grass	200	150	—	—
Sugar beets	—	150	100	—
Melons	35 lb preplant + 30 lb N after set of blossoms			
Tomatoes	50–75 lb N preplant + 25–50 lb N after first fruit set			

[a] The basic nitrogen recommendation is the application rate assuming no nitrogen carryover. To get the net (or adjusted) nitrogen recommendation, subtract the effective carryover obtained from Table 4.
[b] Growing season: Very long: Bluff, Green River, Kanab, Moab, St. George; long: Box Elder, Davis, Salt Lake, Tooele, Utah, Weber; medium: Beaver, Cache, Carbon, Duchesne, Emery, Iron, Juab, Millard, San Juan, Sanpete, Sevier, Uintah; short: Daggett, Garfield, No. Kane, Morgan, Piute, Rich, Summit, Wasatch, Wayne.

Table 2 Phosphorus Recommendations for Utah (lb P_2O_5 per acre)[a]

Soil tests P (ppm)	Potatoes	Alfalfa, beets, legume pastures	Corn, small grains	Grasses, lawns	Wet meadows	Dryland grains
1	250	200	150	100	50	60
2	235	160	125	80	50	60
3	220	130	100	60	50	50
4	200	100	80	60	40	50
5	180	80	70	40	40	40
6	160	70	60	40	40	40
7	150	60	40	0	25	40
8	140	0–50	0–40	0	0	0
9	130	0–50	0–40	0	0	0
10	120	0–50	0–40	0	0	0
11	100	0	0	0	0	0
12	90	0	0	0	0	0
13	80	0	0	0	0	0
14	70	0	0	0	0	0
15	60	0	0	0	0	0
16–20	50	0	0	0	0	0
20–40	0–50	0	0	0	0	0
40+	0	0	0	0	0	0

[a] Manure provides approximately 5 lb P_2O_5 per ton.

Table 3 Potassium Recommendations for Utah (lb K_2O per acre)[a]

Soil Test K (ppm)	Potatoes	Alfalfa, Corn, Grain and Other Intensively Managed Crops	Sugar Beets, Grass
0–50	200–250	150–200	0
50–75	200–250	100–150	0
75–100	200–250	100–150	0
100–150	100–200	0	0
150–200	50–200	0	0

[a] Manure applied since last crop provides approximately 5 lb K_2O per ton.

Table 4 Effective Carryover from Previous Crop[a]

| | N applied | Soil Texture | | |
		Coarse	Medium	Fine
Alfalfa last year				
4–5 t/a	0	120–180	120–180	120–180
2–3 t/a	0	50–100	50–100	50–100
1–2 t/a	0	25–50	25–50	25–50
Corn or grain				
With stubble removed	150–200	0–25	25–75	50–100
	100–150	0	0–50	25–75
	0–100	0	0	0
With stubble plowed down	150–200	0	0–50	25–75
	100–150	0	0–25	25–50
	0–100	(−25)	(−25)	(−25)
Potatoes or beets	200–250	75–125	100–150	125–175
	150–200	0–50	25–75	50–100
	0–150	0	0–50	0–50
Pasture				
All Grass	0–100	0	0	0
Grass + legume	0	25–50	50–70	50–70

[a] Manure applied since last crop provides approximately 5 lb N per ton.

Source: Raymond W. Miller, Utah State University.
[a] Tables 1, 2, and 3 are for N, P, and K, respectively. Table 4 is used to estimate amounts of nitrogen carried over from previous applications in the crop residues. These tables are incomplete. For example, there is no table shown with which to calculate adjustment of fertilizer recommended based on the projected yield.

NAME JOHN SMITH

STREET 55120 BIRCHWAY

SAMPLE IDENT.	CROP TO BE GROWN	SOIL TEXTURE	LAB. NO.
1) 1	CORN SIL.	SI LOAM	500
2) 2			501
3) 3	WHEAT STD.	LOAM	502
4) 4	ALFALFA	SI CL LO	503

Copy sent to Extension office in CACHE County.

SOIL TEST REPORT and FERTILIZER RECOMMENDATIONS

SOIL TEST RESULTS		VERY LOW	LOW	ADEQUATE/NORMAL	HIGH	VERY HIGH	RECOMMENDATIONS	NOTES
NITRATE-NITROGEN N ppm	1) ___	••••••	•				190 − 210	4
	2) 10						N lbs/A	
	3) ___						130 − 150	4
	4) ___						0	1,2,4
PHOSPHORUS P ppm	1) 6	••••••	••				50 − 70	6
	2) ___						P₂O₅* lbs/A	
	3) 16	••••••	••••••	•••••			0	
	4) 4	••••••	•				90 − 110	6
POTASSIUM K ppm	1) 70	••••••	•••				45 − 95	6
	2) ___						K₂O* lbs/A	
	3) 180	••••••	••••••	••••••••••			0	
	4) 200	••••••	••••••	••••••••••••			0	
SALINITY EC$_e$ mmhos/cm	1) .8	••••••	••••••	•••			___	
	2) ___						___	
	3) 4.5	••••••	••••••	••••••••••••••••••••••••	••		___	10a
	4) .6	••••••	••••••	•••			___	
pH	1) 8.0	••••••	••••••	••••••••••••••••••			___	
	2) ___						___	
	3) 8.3	••••••	••••••	••••••••••••••••••••••••	••••		___	10a
	4) 8.0	••••••	••••••	••••••••••••••••••			___	
LIME	1) ++	••••••	••••••	••••••••••			___	
	2) ___						___	
	3) ++	••••••	••••••	••••••••••			___	
	4) ++	••••••	••••••	••••••••••			___	
	) ___						___	
	) ___						___	

Notes

1. There is no indication that N fertilizer will increase yield or quality of alfalfa. If grain is to be seeded with new alfalfa, do not apply more than 50 lbs N/acre.

2. **Pasture and meadows.** Split N applications help to maintain yield and protein content throughout the season. Half of the year's application can be done in the fall if it is watered in immediately or injected directly into the sod (early spring application is also effective). The second half can be broadcast after the first cutting in the spring just before irrigating. Do not apply more than 75 lbs. of N at one time. See also Note 4 below.

   Mixed legume-grass pastures containing more than 1/3 legume may not benefit from added N.

3. A valid N test requires sampling at least 0-1 and 1-2 feet, and quick drying of the sample (see sampling instructions).

4. N recommendations are based on your crop and fertilizer history. A valid test for N requires special sampling procedures. Fertilizer recommended is based on your stated yield goal.

5. **Potatoes.** For potatoes, apply 1/3 of N preplant, the rest during the growing season. Follow petiole N. Avoid high N late in the season. See also Note 4.

6. **Phosphorus (P) and potassium (K).** Plowdown or band applications are preferred for all new seedings. For established perennial crops such as alfalfa and pasture, broadcast recommended fertilizer at earliest possible date.

   Subsoil P and K levels can affect crop responses to fertilizer P or K.

8. **Dryland production.** Response to fertilizer on drylands is highly dependent on available moisture. Fall applications are usually most effective.

   Phosphate must be incorporated into the soil by tillage or drilled with the seed.

10a. This sample shows a slight to moderate accumulation of salt, sufficient to affect growth of sensitive crops. If subsoil drainage is adequate, applying an excess of good quality water can reduce salts to an acceptable level. If pH is also HIGH, special treatment may be needed to reduce sodium.

FIGURE 12-5 Portion of a computer readout chart for soil test results and fertilizer recommendations. The test results are shown both in laboratory numbers and by an "asterisk line." The line shows what the number means in terms of low, medium, or high contents. Explanatory information is given as "Notes." As the examples show, the notes may be cautions (note 10a), information about the test interpretations (notes 3 and 4), or general guides to fertilizer use (notes 2, 5, 6, and 8).   (Courtesy of Raymond W. Miller, Utah State University. Printout supplied by Utah State Soil Testing Laboratory.)

A soil chemical test must be correlated with field studies to be of value.

—R. L. Cook and B. G. Ellis

Questions

1. (a) List six of the analyses made on soil samples sent to routine soil-testing laboratories. (b) What is each test to indicate?
2. Good, simple "quick soil tests" for nitrogen are not yet available. Why is such a test so eagerly sought by soil scientists?
3. If different extractants for phosphate remove different amounts of phosphate from the *same soil,* how can any of these tests be used to make fertilizer recommendations?
4. In a brief sentence, summarize the guide to taking a soil sample for analysis.
5. In details, tell what to do regarding each of these items concerning taking a soil sample: (1) depth to sample, (2) number of samples, (3) areas to avoid, and (4) when to take the sample.
6. Taking the soil sample correctly is very critical in fertilizer recommendations. Why? (*Hint:* Upon what basis is the recommendation made?)
7. Changes in laboratory procedures, particularly soil–water ratios used, are very critical in interpreting soluble-salt data. Explain.
8. (a) Despite hundreds of studies, a widely used and universally accepted "quick soil test" is not available for nitrogen. Why not? (b) On what basis are fertilizer N recommendations made where soil tests are not done?
9. Explain why (a) acidic solutions are used to extract phosphorus in acidic soils, (b) bicarbonate solutions are used in alkaline soils, but (c) bicarbonates are ineffective in extracting phosphates from acidic soils.
10. (a) What important source of phosphorus in soils is little measured by the "quick tests"? (b) How might this fact help you to modify phosphate recommendations for a soil you know to be abnormally high in humus?
11. (a) How do the soil sources of available potassium make it easy to understand why many procedures have been satisfactorily used as potassium soil tests? (b) What are some recent modifications added into the potassium evaluation?
12. As soils are limed, any calcium deficiency will probably be solved, but any magnesium deficiency may not be solved. Explain.
13. What is the general status of soil tests for micronutrient metal deficiencies?
14. What does the nature of the extractant in the soil test for boron indicate as a reason why boron may be the most deficient micronutrient in acidic soil regions?
15. How do the problems and techniques in testing greenhouse soils compare to normal field soils?
16. Why is a "salt problem" a common concern in many greenhouse soils?
17. Plants are the final evaluator of whether or not nutrients are available. (a) Then why aren't plant tests used instead of soil tests? (b) When are plant analyses useful for predictions of fertilizer needs? (c) What factors other than nutrient availability may reduce plant nutrient uptake in any one given example?
18. In sampling plants for analyses, what is recommended for (a) age of plant to sample, (b) part of plant to sample, and (c) preparation of the plants following sampling?
19. (a) Briefly discuss the critical nutrient range (CNR). (b) How does the CNR relate to "hidden hunger"?
20. Describe briefly the general visual deficiency symptoms for inadequate (a) nitrogen, (b) phosphorus, (c) potassium, and (d) iron.

21. Why might three different laboratories measure the same soil test value for a soil, yet each "correctly" give a different fertilizer recommendation?
22. What is meant by "prescription plus buildup"?
23. Explain why each of these items, sent in by the farmer with his soil sample, alters a fertilizer recommendation: (a) manure is to be added, (b) the previous crop and fertilizer on that field, (c) the farmer's projected yield goal, and (d) the location (weather) of the field that year?
24. Discuss the value of using microcomputers in recommendations.
25. What are some of the assumptions made by those making fertilizer recommendations that may be untrue about an actual field soil? (Consequently, the recommendations given will be inaccurate, although not a fault of the testing laboratory.)

Fertilizer Management

The first word in war is spoken by guns. But the last word has always been spoken by bread.

—Herbert Hoover

13:1 Preview and Important Facts

PREVIEW

Fertilizers and water control are the easiest growth factors to manipulate for increased crop growth. Thus, fertilizers are extensively used. With low prices for some agricultural products and concern about pollution, fertilizers must be used carefully. The costs of fertilizers may be 20–50% (about 30% on average) of the variable cash inputs into a crop. It is wise to add correctly the kinds and amounts of fertilizer needed.

There are various benefits, many options, and some essential requirements when fertilizing soils. Most soils need some added nutrients to produce optimum yields and high plant quality. Nevertheless, many important facts must be sought. What nutrients are deficient and in what amounts? What materials (chemical fertilizers, manures, compost) should be used, and how much should be added? Should the material be spread on the surface (broadcast), placed in bands into the soil, or applied in some other manner? Are there nonnutrient factors hindering growth that limit the expected benefits from added fertilizers?

Additions of small amounts of nutrients and some other chemicals sometimes cause spectacular increases in growth. Because of this, there will always be a number of "magical" growth additives offered for sale that are accompanied by claims that the substances are unusually effective. Such products can be identified by a few key indicators. To lure buyers, persuasive salespersons use claims about these products such as "enzymatic," "hormones," "super," "wonder," "secret," "miracle," "biological," "organic," "bacteriological," "conditioner," and "stimulant." Ingredients, names, and claims are changed too frequently for the promoters to be prosecuted effectively by the various state departments of agriculture. Evidence from scientific tests about any growth-promoting substance should be sought from each state agricultural university or its local extension representative.

Although caution should be used when considering testimonial-promoted substances, there are legitimate microbial and hormone additives. Urease inhibitors are used to slow urea hydrolysis and nitrification inhibitors are used to hinder nitrification of added or produced ammonium ions. These sometimes help to increase nitrogen efficiency. *Rhizobia* inocula to ensure legume nodulation are essential additives in use for decades. Mycorrhizal associations are known to increase phosphorus available to plants; inocula of effective fungi are being sought, and some are already in use.[1] One patented fungus has been shown to be very effective in controlling sicklepod, a hard-to-control weed in soybeans.[2] A plant growth regulator, ethephon (Cerone®), used in amounts of about 2 liters per hectare (less than a quart per acre), reduces heights of cereal grain plants and increases stem diameter, reducing lodging of the plant.[3] Before buying products such as those just described, ask for some nonbiased test results, often done by federal or state government agencies and by universities.

IMPORTANT FACTS TO KNOW

1. The trends and changes in fertilizer use
2. Some conditions when fertilizer use will not be profitable
3. Plants differ greatly in nutrient needs and in the amounts of added fertilizers needed
4. The fertilizer grade and the exact meaning of each number
5. The important characteristics of fertilizers, such as their (a) salt affect, (b) present and potential acidity, and (c) solubility and mobility in soil
6. The terms used to describe various ways to add fertilizers: *starter, broadcast, deep banding, split applications, side* or *top dressing, fertigation,* and *foliar*
7. (a) The value of starter banding; (b) the hazards of chemigation
8. The value of split applications and the nutrients for which it is most suitable
9. The place of fertigation in fertilization and its problems
10. The values and uses of foliar applications
11. The unique fertilization problems in paddy rice
12. Factors that affect fertilizer efficiency
13. The calculation of elemental contents and the fertilizer grade
14. Calculation of weights of nutrient carriers to prepare fertilizer mixtures

13:2 Goals and Concerns in Using Fertilizers

Crop producers have several goals in using fertilizers: (1) to increase yields, (2) to reduce costs per unit of production, (3) to increase plant quality, and (4) to reduce certain diseases. The first two of these goals are the most common reasons for adding fertilizers.

Efficient land managers spend perhaps 20% of all production costs on chemical fertilizers and expect increases in yield of up to 50%, a usual return of $2–$3 for each dollar spent on fertilizers. Fertilization is *not profitable* when (1) water is the first limiting factor, (2) other growth hindrances—such as insects, diseases, strongly acidic soils, saline soils, or cold temperatures—regulate growth, and (3) the increased yield has less market value than the cost of buying and applying the fertilizer.

Agricultural practices often change. Some of the changes that influence fertilization practices are the following:

---

[1] M. V. Shantaram and N. Saraswathy, "Occurrence and Activity of Phosphate-solubilizing Fungi from Coconut Plantation Soils," *Plant and Soil,* **87** (1985), pp. 357–364.
[2] Anonymous, "Fungus Saves Soybeans from Weed," *Ag Consultant,* (Jan. 1987), p. 9.
[3] Ann J. Rippy and Frank J. Wooding, "Use of a Plant Growth Regulator on Barley to Prevent Lodging," *Agroborealis,* **18** (1986), pp. 9–12.

1. **Hectareage involving reduced tillage and no-till are going to increase.** What is the best way to apply fertilizer to many of these systems? Fertilizer may be added deeper into the soil or as liquids.

2. **Certain fertilizers dominate the market.** This alters the materials easily available, costs, and methods of adding them. For example, in Canada, anhydrous ammonia is about 45% of all nitrogen used, urea is about 40%, and ammonium nitrate is only about 15% (down from over 75% two decades ago). Although ammonium phosphates are much more important in the United States than in Canada, anhydrous ammonia and urea still dominate as the most widely used materials. The expanding liquid-fertilizer industry is a heavy user of urea and ammonium nitrate. Increased fertilizer additions in water makes the use of *water-soluble phosphates,* such as ammonium phosphates and phosphoric acid, increasingly in demand. Ammonium thiosulfate, a relative newcomer, is a popular sulfur source for many fluid formulations.

3. **Some fertilizer application programs are more efficient than others.** Old concepts of fertilizer application are challenged, and better techniques are learned. The use of slow-release materials and better timing of applications may help. No practice seems sacred. Any situation may have a unique solution.

4. **Computers are one of our new tools.** Computers are more than record keepers. They can be programmed to do our fertilizer-mix calculations, regulate automated sprinklers and fertilizer injections, give us long-time records of fertilizer recommendations in its memory, and tell us what our new soil test recommendations mean. All of these data will eventually be matched to the soil and crop history in each field, our management system, and our preferences. These potential aids are not all here yet, but they will come. This phase of change will be difficult because the computer programmer is seldom a person with years of practical field training and with data. More than ever, good grower input to get "field sense" into these management models will be increasingly important to the development of accurate programs.

5. **Environmental controls and penalties become management concerns.** Careless and polluting fertilizer applications can become costly errors, as environmental issues, regulations, and pressures increase. The pressure to be sure to control nonpoint sources of pollutants will increase. Research to justify fertilizer needs and to verify the extent of pollution by various practices may become a requirement rather than a voluntary action of the agricultural business community. In all of our innovations we must consider the possible impact on the environment.

13:3 Balancing the Soil's Nutrients

Fertilizing crops usually has as its major purpose to produce better and/or larger yields, often at decreasing cost per unit of production. To accomplish these goals, it is essential to know how much of the critical nutrients the crop will need. The amount of nutrients the soil will supply subtracted from the total plant needs and losses is the amount of fertilizer to add. The "excess" fertilizer to add to accommodate the losses is difficult to calculate. Although the losses may not be known, the total amount to be added for a certain response in a given field is measured in the field trials used for correlation to yields.

13:3.1 Nutrients Removed by Plants

Nutrients removed from soil by plant growth vary with the variety of plant, its stage of growth, and its yield (Table 13-1). Crop plants usually contain more nitrogen than any other

Table 13-1 Major and Secondary Essential Nutrients Contained in the Entire Plant, with Yield Indicated

Crop	Crop Yield per Acre[a]	Nutrients Contained in Crop (lb/a per Crop)[a]					
		Nitrogen (N)	Phosphorus (P_2O_5)	Potassium (K_2O)	Calcium (Ca)	Magnesium (Mg)	Sulfur (S)
Alfalfa	5 tons	250	60	225	160	25	23
Corn	150 bu	220	80	195	58	50	33
Cotton	1.5 bales	95	50	60	28	8	4
Coastal Bermuda grass	6 tons	150	60	180	33	22	40
Soybeans	40 bu	145	40	75	7	9	7
Rice	2500 lb	185	51	18	20	15	18
Tobacco	2800 lb	95	25	190	105	24	21
Wheat	60 bu	125	50	110	16	18	16
Oats	100 bu	100	40	120	14	20	20
Potatoes	400 bu	200	55	310	50	15	18
Peanuts	3000 lb	220	45	120	105	28	25
Sorghum grain	8000 lb	260	110	220	445	36	38
Banana	1200 plants	400	400	1500	300	156	b
Coffee	1784 lb	27	4	43	46	61	16
Oil palm	13,382 lb	80	18	120	64	18	b
Pineapple	15,000 plants	134	107	535	102	53	b

Source: Magnesium-Sulfur, *Essential Plant Nutrients*, International Minerals and Chemical Corp., Libertyville, Ill., undated.

[a] To convert lb/a to kg/ha, multiply lb/a by 1.12. Weight of a bushel of wheat, 60 lb; corn, 56; soybeans, 60; oats, 32; sorghum, 56; potatoes, 60; rice, 45 lb.

[b] No information.

fertilizer nutrient. The second highest concentration of a fertilizer nutrient is potassium, followed by calcium, phosphorus, magnesium, and sulfur.

The values in Table 13-1 emphasize the different nutrient requirements of various crops. Bananas, corn, potatoes, peanuts, and sorghum have high nitrogen needs (alfalfa fixes its own). Bananas, sorghum, pineapple, and corn have high phosphorus requirements. Bananas, pineapple, potatoes, and alfalfa have high potassium needs. These nutrient demands will modify the composition of fertilizer needed to supply plant needs and perhaps replenish depleted nutrients from the soil.

13:3.2 Nutrients Supplied by Soils

Soils vary in the nutrients they can supply to various plants. The soil test is intended to estimate this amount. Unfortunately, a widely satisfactory soil test for nitrogen is not available. Most of the nitrogen provided by soil comes from decomposing organic matter. Most quick soil tests for nitrogen or phosphorus do not measure the potential release from the organic source because this measurement usually requires an "incubation" time of a couple of weeks. The nitrate test sometimes used for the nitrogen test is an attempt to measure nitrogen released by soil organic matter in the days or weeks prior to sampling. But soil nitrate may be from different times of "incubation" in the soil and may not measure losses from leaching or denitrification.

Soil tests do not tell us exactly how many kilograms of a nutrient are actually available from an area of each soil. The test values allow us to group the soils into categories of soils that have *deficient, marginal, adequate,* or *excess* amounts of "available" nutrients. As an ex-

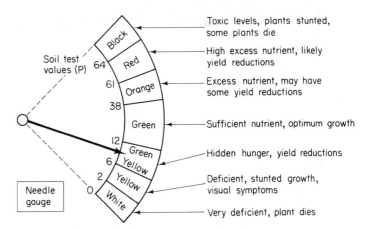

FIGURE 13-1 Representation of a soil test and its meaning. The phosphorus soil test values are hypothetical and are shown to illustrate that the numerical values are only used for purposes of grouping the soils into categories, such as deficient, marginal, or toxic levels of "available" nutrients. (Courtesy of Raymond W. Miller, Utah State University.)

ample, consider the "soil test numbers" obtained by a laboratory, as are shown in Fig. 13-1. The values are illustrated as a gauge. Soils with high soil test values may not require any additional fertilizer; some may even have toxic levels of nutrients, although this is somewhat rare. The lower the soil test values, the more fertilizer that will need to be added to optimize yields.

13:3.3 Nutrient Losses

It is difficult to estimate nutrient losses. Nevertheless, some fairly good estimates have been developed for certain areas. Although the rates of fertilizer added are still mostly determined by correlation of field trials to soil test values, as discussed in the previous chapter, improved predictions of fertilizer needs are made by estimating losses. If changes in management will reduce losses that were common in previous management (better efficiency), the amount of required fertilizer can be reduced. Many studies of "models" are attempts to quantify the "losses" under various crops, management changes, water conditions, temperature fluctuations, and other items. When these models are workable, adjustments in needed fertilizer can be predicted on many management factors, as well as on the soil test values.

13:3.4 Calculating the Fertilizer Needed

The amount of fertilizer to add is calculated by using field plot data and soil test groupings. The soil test value of a soil puts it into a grouping—say, the group with soil test P values between 4 and 7 mg P/kg of soil. Look at the accumulated data gathered over many years' field plot studies and tabulated. Suppose that plots whose soil test values were between 4 and 7 and were *in that location and climate* had shown a significant yield increase with increasing amounts of P up to 120 kg P_2O_5/ha. The value of 120 kg of P_2O_5/ha would be what we should add. But we must use additional fertilizer if it is needed to decompose straw residues or to build up soil levels of the nutrient. We must also subtract any fertilizer added in other forms, such as in manures or composts. We also need to subtract any residual benefit from previously added fertilizer that is not measured in the soil test value. In a general form we can write this equation, using Nt to represent the nutrient of interest, as

$$
\begin{aligned}
\text{Needed fertilizer Nt} = {} & \text{Field plot Nt value} \\
& \textit{plus} \text{ amount for soil build up} \\
& \textit{plus} \text{ amount for residue decomposition} \\
& \textit{minus} \text{ amount "available" from manure or} \\
& \qquad \text{compost} \\
& \textit{minus} \text{ amount residual from previous} \\
& \qquad \text{fertilizer}
\end{aligned}
$$

As an example, consider the following simple situation:

1. The soil test value for P is 6.2. We'll assume that correlation tests previously done suggest that this soil needs about 120 kg of P_2O_5 equivalent per hectare added for a 11,300-kg/ha (180-bushel) corn crop.

2. Subtract 10 kg P_2O_5 for each metric ton of manure added.

3. Add or subtract any other phosphorus credit or loss to the 120 kg value, such as residual P from last year's fertilizer.

4. The net phosphorus (120 kg minus or plus other additions or losses) is what is needed. The soil may be so low that the recommender also suggests additional **P** be added to "build up" the soil P level.

13:4 Fertilizer Guarantee (Grade)

All states require that fertilizers offered for sale be accurately labeled with the **grade** (minimum guaranteed percentage), the weight of the material, the manufacturer, and the manufacturer's address. Some states require additional information, such as the composition of the filler (inert materials sometimes added to make the fertilizer to convenient bulk weights) and the amount of acidity the fertilizer forms when added to soil.

The most useful information is the grade. The grade gives in order: (1) the **percentage total nitrogen,** measured as elemental N, (2) the **percentage available phosphorus,** listed as that phosphorus soluble in ammonium citrate solution and **calculated as phosphorus pentoxide,** P_2O_5, and (3) the **percentage potassium, calculated as water-soluble potassium oxide,** K_2O. These quantities are often listed on fertilizer labels as 10-16-12—meaning that this fertilizer contains 10% by weight total nitrogen, 16% available phosphorus (calculated as P_2O_5), and 12% water-soluble potassium (calculated as K_2O). There is a slow but definite effort being made to change these conventions and to report only N, P, and K *percentages.* One reason for the resistance to change is that P or K reported as percentages gives lower numbers and makes the fertilizer appear to be a lower grade. For example, a 0-45-0 triple superphosphate is only a 0-19-0 material if reported as percentage of element rather than as oxide. This is an unfortunate resistance, especially because fertilizers do not contain P_2O_5 or K_2O. It is convenient to remember that the order of listing is alphabetical by element: nitrogen, phosphorus, potassium. Figure 13-2 shows several examples of fertilizer labels. State chemists periodically sample and analyze commercial fertilizers to verify the grades listed. Fertilizer companies can be prosecuted if the grade displayed on the bags is higher than the laboratory analysis by more than that specified in the respective state laws.

Notice that the fertilizer grade only gives information about nitrogen, phosphorus, and potassium. Often now, sulfur is given, calculated as elemental sulfur. A value such as 10-20-5-4S has the last number —4S—meaning 4% sulfur. Obviously, agricultural minerals such as

AMMONIUM NITRATE PHOSPHATE
23-23-0

GUARANTEED MINIMUM ANALYSIS

TOTAL NITROGEN, N (All Water Soluble) **23%**
- Ammonia Form — — — — — 14%
- Nitrate Form — — — — — 9%

AVAILABLE PHOSPHORIC ACID (P_2O_5) **23%**

(a)

FERTILIZER COMPOUND
IDEAL FOR LAWNS, GARDENS, HOUSE PLANTS, SHRUBS, SHADE TREES, EVERGREENS

GUARANTEED ANALYSIS

NITROGEN 38%

derived from urea formaldehyde-38-0-0

One application as recommended releases a complete and uniform supply of nitrogen throughout the entire growing season.

- **SLOW RELEASE**
- **WON'T CAKE**
- **WON'T BURN**

when used in recommended amounts

(b)

SEQUESTRENE*
Na₂Cu
COPPER CHELATE

(CONTAINS 13% COPPER AS METALLIC)

FOR CORRECTION OF COPPER DEFICIENCY

DIRECTIONS FOR USE

(c)

(CONTAINS INORGANIC AND ORGANIC PLANT FOODS)

10 - 6 - 4
GUARANTEED ANALYSIS

Available Nitrogen	10.00%
Available Phosphoric Acid	6.00%
Potash Soluble in water	4.00%
Iron Sulphate	2.00%
Copper Sulphate	.70%
Zinc Sulphate	.60%
Manganese Sulphate	.50%

SEE DIRECTIONS INSIDE

50 Lbs. Net

(d)

FIGURE 13-2 Examples of fertilizer labels showing grades. In (a), the grade is 23-23-0, and additional information, the form of the nitrogen, is given in the box. In (b), a slow-release nitrogen fertilizer has 38% total nitrogen (grade is 38-0-0), but a small unlisted part of it is readily soluble. A chelate material (c) is not added to supply N, P, or K, and its grade is often 0-0-0. Grade information is not descriptive of such a material. A mixed (formulated) fertilizer containing many nutrients (d) is characterized by both grade and additional tabulated materials. (Courtesy of Raymond W. Miller, Utah State University.)

chelates and micronutrients usually have no "grade." They might be 0-0-0 but be a zinc chelate. Thus, *fertilizer grade gives information only about the nitrogen, phosphorus, and potassium contents,* unless the additional number with the element's chemical symbol is given.

13:5 Some Characteristics of Fertilizers

Fertilizers are soluble salts; some are "saltier" than others. Some fertilizers are acid-forming or initially basic. Not all fertilizers are easily soluble. Some nutrients are more mobile than others in soil, and not all are suitable for liquid fertilizers. All of these properties modify how and/or what we buy for use as fertilizer.

13:5.1 Fertilizers as Soluble Salts

Soluble salts are all soluble ions. Thus, most soluble fertilizers are soluble salts. Because of strong adsorption or precipitation of some ions (phosphates, calcium) and absorption or loss of others (nitrate), some materials cause greater short-time and long-time effects than others. Phosphates are usually low salt; chlorides, sulfates, and nitrates are high in their immediate salt effects. Use of chlorides, sulfates, and nitrates of ammonium and potassium should be managed as if soluble salts were being added—*they are.*

13:5.2 Acidity or Basicity

Most fertilizers are used on acidic soils, so the possibility that fertilizers might aggravate soil acidity is of great concern. Soils too highly acidic can have toxic levels of aluminum and manganese and greatly reduce microbial activity. All potassium fertilizers are neutral except potassium nitrate, which is residually slightly base forming. Superphosphate and triple superphosphate are neutral, but both monoammonium phosphate and diammonium phosphate are acid forming. Of the nitrogenous fertilizers, most are acid forming, except calcium nitrate, sodium nitrate, and potassium nitrate, which are base forming. Acidity is developed by microbial oxidation of ammonium cations to nitrate anions:

$$2NH_4^+ \; + \; 4O_2 \; \xrightarrow{\text{bacteria}} \; 2NO_3^- \; + \; 4H^+ \; + \; 2H_2O$$

(ammonium) (oxygen)   (nitrate)   (acid)   (water)

All ammonium materials and many organic nitrogen fertilizers are acid forming.

There is a tremendous difference in the acidifying properties of the nitrogenous fertilizer materials (Table 13-2). Continued addition of ammonium materials increases soil acidity. In some states, law requires that the potential acid-forming values be given in pounds of lime needed to neutralize the acidifying effect of the fertilizer. The amount of lime needed depends on the acidity of the fertilizer anion (Cl^-, SO_4^{2-}) and how much ammonium ion is oxidized to nitrate.

Although many fertilizers acidify soil over time, many are alkaline (basic) initially. Urea, when hydrolyzed by urease, may cause a pH of 9.5–10 in the area of the pellet or band. This is why NH_3 volatilization can occur rapidly from surface applications of urea. Diammonium phosphate initially also has a high pH, as does anhydrous and some liquid nitrogen.

13:5.3 Solubility and Nutrient Mobility

Most phosphates are of low solubility and can quickly form insoluble calcium, aluminum, or iron salts in mixtures. Mostly ammonium phosphates and phosphoric acids are used to sup-

Table 13-2 Relative Acid-Forming Characteristics of Various Nitrogen Fertilizers

Nitrogen Fertilizer Source	Theoretical Lime Requirement[a] (kg CaCO$_3$/kg N)	5-Year Michigan Study	
		Final Soil Surface pH	Lime Needed[b] (kg CaCO$_3$/kg N)
No fertilizer, soil only	—	5.7	0
Ammonium sulfate	5.35	4.0	12.7
Anhydrous ammonia	1.80	5.1	11.3
Ammophos (11% N)	5.00	—	—
Ammonium chloride	5.30	4.4	7.9
Urea	1.80	4.9	6.0
Ammonium nitrate	1.80	4.8	6.1
Ureaform (slow-release N)	1.80	5.0	3.8
Organic wastes (manure, tankage, seed meals)	0.5–1.5	—	—

Source: A. R. Wolcott, H. D. Foth, J. F. Davis, and J. C. Shickluna, "Nitrogen Carriers: I. Soil Effects," *Soil Science Society of America, Proceedings,* **29** (1965), pp. 405–410.

[a] Official method for neutralizing fertilizers.

[b] Measured lime requirement, of this sandy loam soil, to adjust the soil pH. Includes losses or chemical changes in the soil during the 5-year period.

ply phosphorus in "liquid" fertilizers. Urea and all ammonium, nitrate, and potassium salts are soluble.

Nutrient mobility is a function of ion charge, the ion's tendency to form insoluble compounds (precipitate), soil texture, water movement, and concentration of other ions. Phosphate ions precipitate or are adsorbed to soil solids; phosphates are the least mobile anions, usually moving only a centimeter or two from where they were placed in the soil. The K^+ and NH_4^+ ions also move very little because of their attraction to the cation exchange sites. In soils with low cation exchange capacity and in mixtures with other competing cations, K^+ and NH_4^+ ions can move many centimeters during a few weeks. Because of their attraction to the cation exchange sites, these cations are considered *immobile or slightly mobile* in soils. Nitrate and sulfate will be quite mobile, moving mostly wherever water moves. Thus, soluble nitrogen in soil eventually is oxidized to nitrate, a *mobile* ion. Because of this eventual mobility, nitrogen can be added to soil surfaces and be leached into the root zone as nitrate. Any surface-applied NH_4^+ will be oxidized in the soil surface where it was added, causing soil acidification before the nitrate moves into the soil.

In contrast to nitrate movement, neither phosphate nor potassium move very far nor very fast in soils. These nutrients should be placed in the root zone where the grower wants them to be; they stay where they are placed.

Soils with low cation exchange capacities—such as sands, Oxisols, and Ultisols—will have less ability to retain NH_4^+ or K^+. These cations are more mobile in coarse sandy soils.

13:5.4 Solid vs. Liquid Fertilizers

A limited amount of study has been made comparing reactions of liquid and solid fertilizers. The liquid fertilizer will often spread more, running through large cracks and contacting more soil than do *pellets* of dry fertilizers. When applied at the surface, liquid fertilizers may partly move into the soil. Where crop yield differences have been observed, usually the liquid fertilizer has produced higher yields than have solid materials or the liquids have been more *efficient.* However, equipment costs and handling techniques are different between liquid and solid fertilizers. Serious evaluation of the benefits should be made before making changes in the kind of fertilizer used.

13:6 Techniques of Fertilizer Applications

In this section we discuss various ways fertilizers are added, with some examples and evaluations given of each technique and its value. The various schemes discussed are the following:

Starter (Pop-Up)   Fertilizer is added with or near the seed (0–5 cm) for immediate use by the young seedling.

Broadcast   Fertilizer is uniformly spread on the soil surface. It may or may not be incorporated into the soil.

Deep Banding   Fertilizer is placed in a strip at a depth of 14–18 cm (or deeper); usually the strips are 30–70 cm apart, placed at planting or in preplant without regard to rows.

Split Application   The fertilizer is added in two or more portions at different times during the season.

Side Dressing, Strip Placement, and Top Dressing   Often these terms are used to mean fertilizer added after the crop is growing. Side dressing can involve insertion of the fertilizer into the soil beside the growing crop. Strip placement and top dressing may be placed as "strips" on the soil surface near plants rather than being broadcast over the entire soil surface.

Point Injector   A narrow cylindrical tube (injector) penetrates the soil and allows fertilizer to be inserted at the depth of penetration. Usually liquid fertilizer is used. The injectors may be on wheels (rolling-point injectors) at any frequency and spacing wanted (Fig. 13-3). This technique allows fertilizer to be inserted into the soil with minimal damage to roots.

Fertigation   Fertigation is the application of the fertilizer in water (as in irrigation water).

Foliar Sprays   This is the application of liquid fertilizer to the foliage of plants, just enough to wet the leaves.

13:6.1 Starter (Pop-Up) Fertilizers

The definition of **starter fertilizer** may vary. In this book we use the term to include any addition of fertilizer with the seed, dribbled in a strip near the seed or placed in a band anywhere within about 5 cm (2 in.) of the seed (Fig. 13-4). The key nutrients added are phosphates and potassium because of their low mobility in soils. Nitrogen will almost always be added, too.

Three conditions favor obtaining a good response from adding starter fertilizer: (1) a soil that is too cold for good nutrient absorption by roots, (2) a soil test value that is low for the added nutrients, and (3) a fast-growing plant. Crops on soils having "adequate" soil test values may still respond to starter fertilizer in cool periods.[4] Figures 13-5 and 13-6 and Table 13-3 illustrate examples of growth response to applied starter fertilizer.

Some applicators warn against adding fertilizer in contact with the seed because "salt damage" is possible, unless care is used. *Not more than about 10–12 kg/ha of nitrogen plus potassium should be put in contact with the seed.* Phosphates have much less salt affect and

---

[4] P. A. Costigan, "The Effect of Soil Temperature on the Response of Lettuce Seedlings to Starter Fertilizer," *Plant and Soil,* **93** (1986), pp. 183–193.

FIGURE 13-3 Rolling-point injector fertilizer application. The equipment can inject either liquid or gas into the soil, even during early stages of crop growth. It does not prune the roots nor require a very high tractive energy, as is needed by a knifing-in application. Also, surface residues are little affected. The injector shown is one of the early models using 1-cm-diameter spokes that inject the liquid about 10 cm (4 in.) deep. The liquid must be kept at sufficient pressure to prevent plugging of the outlet holes (about 3 mm in the side of the spoke near the solid end of each spoke). Some problems are getting uniform applications, keeping the wheel seals from leaking (because of the corrosive fertilizer solutions), and volatilization of some nitrogen from the holes, which do not always close. (Photo and information courtesy of James L. Baker, Department of Agricultural Engineering, Iowa State University, Ames.)

FIGURE 13-4 Side placement of fertilizer in two bands placed 5 cm (2 in.) below the level of lima bean seed. This is called banded "starter" fertilizer. (*Source:* USDA.)

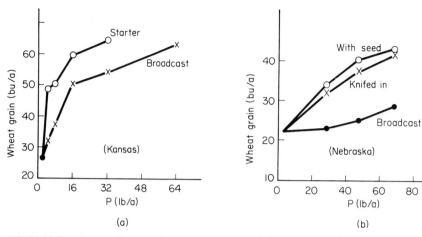

FIGURE 13-5 Effects of starter fertilizer on wheat yields compared to broadcast and to knifed-in applications: (a) 9-18-0 fertilizer, (b) 10-34-0 fertilizer. (bu/a times 67.4 = kg/ha; lb/a times 1.12 = kg/ha.) (*Source:* Modified from John L. Haolin, "Optimizing Wheat Inputs for Profit," *Solutions,* **30** [1986], pp. 50–60.)

FIGURE 13-6 Grapes, like many other crops, have trouble absorbing phosphorus when the soil is cold. Phosphorus deficiency—(a) the lower leaves yellow and (b) increasing red dot masses—occurs in grapes as growers expand plantings onto higher-elevation soils that are also shallow and acidic. The symptoms shown above are typical of those found in grapes in the hill areas of the Napa Valley and Sierra Nevada in California. Petiole phosphorus in deficient plants was about 0.04–0.14% compared to normal values of 0.3–0.6%. As little as 0.4 lb of phosphorus per vine caused much improved growth, but the best practices are yet to be determined. (*Source:* By permission of *California Agriculture;* James A. Cook, William R. Ward, and Alan S. Wicks, "Phosphorus Deficiency in California Vineyards," *California Agriculture,* [May–June 1983], pp. 16–18.)

Table 13-3 Yield Increases (kg/ha) of Corn Grain from Use of Starter Fertilizer on High-Soil-Test Land in Iowa

Starter Used[a]	1983	1984	1985	1986	Average
Check (no fertilizer)	155	152	165	207	170
5-17-0 fertilizer	157	157	173	213	175
6-18-6 fertilizer	163	158	168	213	177
7-18-6-3S-0.5Zn	165	166	182	220	183

[a] Starter was put on as a liquid at 7 gal/a (= 65 L/ha). Note that the last material in the table has 3% sulfur and 0.5% zinc.

have less restrictions for seed contact (see rates in Fig. 13-5 and Table 13-3). As the application rate of material is increased, it is increasingly important to have some soil between the seed and the fertilizer. With low fertilizer rates, about 2 cm of soil will usually protect the seed from salt effects; with high fertilizer rates, the thickness of soil between the seed and the fertilizer should be a minimum of 4–5 cm.

Starter fertilizer (1) is usually more efficient than other placements, (2) is only one part of many fertilizer programs, (3) is effective for helping early growth, (4) is particularly helpful on cold soils, and (5) is probably more important for phosphate and potassium than for nitrogen.

13:6.2 Broadcast Application

Broadcast applications are uniform surface applications of fertilizers. The fertilizer may be left on the surface or mixed into the soil to different degrees by plowing, disking, or having minimal covering as seeders "splash soil" over the fertilizer (as with cereal grain drills). Irrigation water or rainfall may move soluble materials, such as urea, sulfates, and nitrates, down into the soil. Usually, *the lowest fertilizer efficiency is obtained by broadcasting.*

If broadcasting is often so inefficient, why does it continue to be used? The many reasons for continued use include at least these four:

1. Often broadcast is the only practical method to use. How else could you put fertilizer on established pastures, on forests, on turf, on paddy rice during growth, or on partly grown crops? (See *Fertigation,* discussed later.) Some injection techniques on established crops are now being studied.

2. On low-fertility soils, it is often necessary to add large amounts of fertilizer or to build up available levels of the nutrient throughout the soil. This is best done by broadcast followed by incorporation tillage.

3. Broadcast is an easy and cheap method. Often this easy way out is chosen because of a need to hurry, a lack of injection equipment, or a preference. A grower may know that the added phosphate will be less efficient but may decide that it will not be lost and will, over many years, gradually build up soil reserves. In some developing countries limited equipment is available. Surface applications, with or without incorporation by hand work, is the only choice available.

4. Broadcast or a modification of it (top dressing, strip dressing, split applications) is the most feasible method of adding fertilizer after the crop has begun growth (except for fertigation). Any attempt to insert the fertilizer into the soil close to the plant will tear some roots. However, some shallow knifing of fertilizers is done carefully. On much of the early no-till acreage, fertilizers were applied broadcast. The problem of how to fertilize in no-till crops and in reduced-till farming is being studied.

Incorporation with narrow knives or fluted coulters or dribbling liquid fertilizers on the soil has helped improve the fertilizer efficiency by getting some phosphorus and potassium fertilizer down into the soil.

The efficiency of broadcast fertilizer usually increases under several conditions: (1) If the soil is shaded by vegetation, the soil surface will be shaded and stay moist longer. Thus, some roots will explore the soil at the surface, as happens in pastures, and use more of the immobile nutrients. (2) When the broadcast fertilizer is incorporated into the soil, more fertilizer is deep in the soil, where roots stay active (because the soil stays moist) most of the time. (3) If cool spring air temperatures are warming, the surface soil may be warmer and allow better nutrient uptake in the shallow surface soil. (4) If sprinkler irrigation or rainfall comes at the desirable times, the water will move nitrates down into the soil where roots have better access to it.

13:6.3 Deep Banding

Deep banding is the application of strips of fertilizer into the soil either (1) without regard to where the seed is planted exactly or (2) to the side and below the seed *at the same time when planting*. Most often, application at the time of planting puts the fertilizer 10–25 cm deep and about 5–8 cm to the side of the seeds. Anhydrous ammonia fertilizer may be applied in either of these two methods (Fig. 13-7).

The ammonia is transported under moderate pressure in tanks, and it changes to gas at normal atmospheric pressure. Normal knifed-in injections are placed about 8–20 cm deep, thus allowing the gas to be absorbed into the soil water (very effective) before much escapes into the atmosphere. Applications into quite dry soil, especially if it is sandy soil, should not be done; gaseous losses would be excessive. The ammonia gas dissolves readily in water to form ammonium hydroxide. This may cause a high pH, above 9–9.5, immediately, and some ammonia can volatilize. The gaseous ammonia is toxic to both plant roots and to crop foliage. Gaseous ammonia is seldom added except in preplant or during planting.

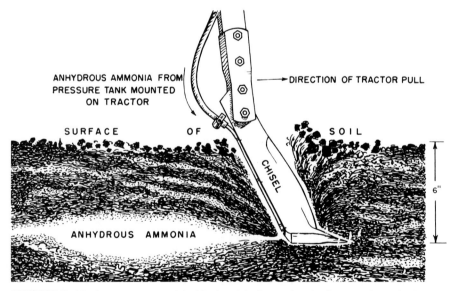

FIGURE 13-7 Under pressure and behind a chisel at a depth of approximately 15 cm (6 in.), anhydrous ammonia is released as a gas and is adsorbed on the surface of clay and humus particles for use by plants.   (*Source: Crops and Soils* magazine.)

Deep banding has the disadvantages of any deep tillage: the need for stronger equipment and higher energy costs to pull the equipment through the soil. Before deep banding is used, be sure that it will increase yields appreciably above those obtained with simpler and less costly methods.

It has been suggested that deep banding in *dryland* crops might be beneficial. The concept is that the nonirrigated soil dries in the surface soil layers for long time periods and deeply placed fertilizer might be in more-moist soil longer, and thus available to roots longer. The few studies done so far are inconsistent. The benefit certainly depends on the drying pattern of the surface layer of soil *during the critical nutrient uptake period.* In some dryland conditions it is conceivable that deep banding could be detrimental if the deeply placed fertilizer keeps the plant in a vegetative stage too long. If the plant runs out of profile water before it sets its seed, cereal grains may have low yields. This is one reason that nitrogen fertilization is reduced or omitted on dryland grains if yearly rainfall is quite low or expected to be lower than usual.

Banding of fertilizer is putting the most fertilizer where the most roots have access to it.[5] This is a most important concept for phosphorus. This guide suggests that applications close to and below the seed (starter banding) should be best. If the fertilizer is mixed into a little larger volume of soil, more of it would be available to more roots. Deep placement of fertilizer would be justified in cost only if the upper soil layer (to 10 or 15 cm deep) dried early in the growing season, restricting nutrient uptake.

Many persons have recommended keeping fertilizer phosphorus in bands to lessen contact of the phosphate with soil (Fig. 13-8). This minimum contact with soil is thought to reduce the rate and amount of soluble phosphate precipitated by calcium, iron, and aluminum and to reduce adsorption to mineral surfaces. The true benefit of this application pattern is still questioned by some. For example, the "ideal tight band" with minimal mixing of fertilizer with soil would have low soil contact but also have a low volume of phosphate accessible to roots and may have toxic salt levels in the band. Which factor is most important, having more accessible volume to roots or less contact with soil? Most banded fertilizers are not "tight" bands at all and perhaps already have extensive contact in the soil for precipitation and adsorption to soil particles. Generally, *banded phosphates have been shown to be more efficient than broadcast applications.*

13:6.4 Split Application

When the total applied fertilizer is split into more than one portion applied at two or more different times, the system is called a **split application.** The method of application is not specified and can include any combination of methods.

Split applications are usually done for several reasons, not always the same ones. First, if large applications are needed, it may be more efficient if some of the fertilizer is held back, to be added after the crop is growing well and has high demands. Second, in some conditions, as in sands, the soil may have low retention or high losses of nitrogen. Multiple but small additions are necessary to reduce losses and keep adequate amounts available to the plant. Third, control of vegetative growth in early stages may be desirable, as with cotton and tobacco. Higher fertilizer demands during later stages of growth may require supplemental fertilizer in midgrowth stages. Fourth, the most efficient fertilizer use (to reduce costs or minimize pollution of runoff waters) may require several well-timed but small applications. The choice the grower makes is a compromise in ease of addition versus efficiency. These choices become more varied as the crop season becomes longer (sugar cane, turf, rice). Fifth, the

---

[5] D. M. Sleight, D. H. Sander, and G. A. Peterson, "Effect of Fertilizer Phosphorus Placement on Availability of Phosphorus," *Soil Science Society of America Journal,* **48** (1984), pp. 336–340.

FIGURE 13-8 No-till system of tillage in Ohio, where corn is being planted into 3000 pounds of soybean residues per acre and fertilizer is banded to one side of the corn row at a depth of about 4–5 inches (10–13 cm).   (Courtesy USDA—Soil Conservation Service.)

number of applications may be related to the method of application or damages. If fertilizer is added in irrigation water, several smaller applications may be the safest and most efficient method.

Usually for nitrogen, efficiency is increased by split applications. For phosphorus and potassium, split applications may give lower efficiency because they are surface applied, unless added in irrigation water.

13:6.5 Side or Top Dressing

Side dressing is usually a surface or shallow-banded application put on after the crop is growing—another split application. These applications may be broadcast, stripped on the soil surface, inserted into the soil at shallow depths, and put on in liquid fertilizers (dribbled on the soil). Keep these things in mind with side and top dressing.

1. Losses of nitrogen from the soil surface by ammonia volatilization will be a major concern when urea or ammonium fertilizers are applied on calcareous and other alkaline soils. Irrigate the fertilizer into the soil as soon as possible after applying it.

2. Apply any fertilizer added to surface-irrigated crops in the furrows rather than to ridges so that irrigation water will move it into the root zone. If sprinkler irrigation or rainwater is used to move the fertilizer, location of the fertilizer applied is less critical because practically all of the soil surface will be leached under young crops.

3. Side or top dressing will not be very effective for phosphorus and potassium unless the crop *shades* the soil and keeps the soil surface moist so roots can explore it.

13:6.6 Point Injector Application

A point injection allows putting potassium and phosphate fertilizers down into soil where roots grow, yet avoiding great damage to roots. The concept has been used in small plots and gardens for a long time. A pointed stick or rod is used to poke a hole in the soil and fertilizer is poured into the hole before covering it over. This is done around fruit trees, vines, shrubs, large garden plants (tomatoes), and isolated berry, melon, and squash vines. For field use, mechanized equipment is needed. Rolling point injectors (Fig. 13-3) work fairly well, but they are still somewhat experimental and costly. Large units with 1.5-m-diameter wheels, 10 to 20 injectors per wheel, and 8 to 10 wheels are available.

13:6.7 Fertigation

Fertigation is the application of fertilizers by injecting them into irrigation water. The fertilizer is applied in large quantities of water, *not as a foliar spray.* (A foliar spray barely wets the leaves with the solution.) Fertigation has been very successful for nitrogen applied through center pivots, other large sprinkler systems, in drip irrigation, and as metered applications of liquid fertilizers to greenhouse plants.

If the irrigation water already contains appreciable salts, ammonium polyphosphates and anhydrous ammonia can cause salt precipitations, which plug lines and nozzles. The circular center-pivot sprinkler systems are easily adapted to fertilization in irrigation water. Fertilizer efficiencies by various application methods are listed below.[6]

Nutrient and Method of Application	Efficiency
Nitrogen by fertigation	95% possible
Nitrogen, in surface flow irrigation water	50–70%
Solid nitrogen, applied to soil surface	30–50%
Potassium, by fertigation	80% possible
Phosphorus, by fertigation	45% possible
Nitrogen, foliar spray (on citrus)	11–75%

Reports on the potential of applying fertilizers in irrigation water are enthusiastic. Nebraska has reported that adding nitrogen in irrigation water is 30–50% more efficient than any preplant application. Some Texas growers are putting 50% less nitrogen in various irrigation systems and are getting equal or better yields compared to yields from conventional methods of adding nitrogen fertilizers to soil.

The use of irrigation water to apply nutrients allows the addition of fertilizers when the crop can most benefit from it. If the need and conditions indicate more fertilizer can be beneficial, it can be applied immediately and conveniently.

Fertigation is most effective on crops where nutrient retention is low (sands and sandy soils with low humus) and for mobile nutrients, such as nitrate and sulfate. In sands, potassium, magnesium, and boron are also quite mobile.

---

[6] Tom Milligan, "Tooling Up for Fertilization," *Irrigation Age,* **6** (Nov. 1972), no. 3, pp. 6–8.

A new term—**chemigation**—is used to indicate the application of *any chemical* in irrigation water. Pesticides of various kinds can be applied. Some also call these additions **herbigation, insectigation, fungigation,** and **nemagation.**[7]

Even with all the interest in fertigation and chemigation, these are not simple substitutes for old techniques. They are good tools to use *for certain situations.* Some of the precautions and limits to chemigation are tabulated below:

1. Chemigation (includes fertigation) is a surface application. For phosphorus and potassium the surface applications may usually have less effectiveness than if placed *in the soil* before planting.

2. The technique is only "cheap" or economical if *you must or will irrigate anyway,* or at least if you already have the system set up.

3. Application of fertilizer is only as uniform as the water application. If the system is poorly designed (nozzles too small, too large, or too far apart), application of water will not be uniform. If it is windy, applications will not be uniform. Much of the chemical may even end up in adjacent fields. If the applications include pesticides, problems could be severe.

4. If pesticides are used, a number of concerns need to be considered.[8] Pesticides are dangerous and need to be applied (a) at carefully controlled rates, (b) only on the area specified, and (c) only during calm or low-wind periods. Currently, large sprinkler systems do not generally have the precision, nor is the care taken, to avoid problems in windy periods. Often application is continuous all day and/or all night without adequate attention to *changes* in wind. For fertilizer application this hazard is less serious, but the nonuniform application is still there. For certain pesticide applications the lack of supervision, stoppages, and the wind hazard make chemigation with pesticides a serious hazard.

Fertilization by sprinkler irrigation has the hazard of salt burn from residual salts left on the leaves. Overhead irrigation of alfalfa during the day near Salt Lake City, Utah, using only water without fertilizer, left enough salts on leaves to cause marked leaf damage.[9] Such damage is expected where salts in water are high and rapid evaporation between wetting cycles allows salt to accumulate and dry on leaves.

In addition, legal restrictions may limit the application of some pesticides and fertilizers by fertigation or chemigation. Also, the irrigation system must be designed to ensure proper safety precautions, such as "backflow" prevention devices. These would prevent possible contamination of groundwater with chemicals and fertilizers injected into the water.

The best time to put fertilizer in the sprinkler system is usually about halfway through irrigation, ceasing before irrigation is complete. With automatic moving systems, this is not possible.

13:6.8 Foliar Application

When only the foliage is wetted to allow maximum absorption through the leaf, the spray is referred to a **foliar application.** Nutrients from foliar sprays move into the plant both through the

[7] A. W. Johnson, J. R. Young, E. D. Threadgill, C. C. Dowler, and D. R. Sumner, "Chemigation's Strong Future," *Agrichemical Age,* (Feb. 1987), pp. 8–9.
[8] D. M. Clark, "Letters: Costing Out Chemigation," *Agrichemical Age,* (June 1987), p. 4.
[9] Rex Nielson and Orson S. Cannon, "Sprinkling with Salty Well Water Can Cause Problems," *Utah Science,* **36** (1975), no. 2, pp. 61–63.

leaf stomata (openings in leaves for gas exchange) and through parts of the epidermis (outer layer of cells). Sometimes translocation of absorbed nutrients within the plant may be slow. The relative mobility of nutrients in bean plants is an example of the variability of translocation; the order in each column of the following table is of decreasing mobility in the plant downward:

Mobile	Partially Mobile	Immobile
Potassium	Zinc	Magnesium
Phosphorus	Copper	Calcium
Chlorine	Manganese	
Sulfur	Molybdenum	

The nutrients in sprays may not be intended to be absorbed solely by the foliage; much of it falls on the soil and later is absorbed by roots. Florida citrus trees are sprayed with 3 or 4 applications per year of about 1.1 kg (2.4 lb) each of nitrogen and of potassium (K_2O) per tree per application; other nutrients are added as needed.

In California foliar-applied nitrogen is as effective as soil-applied nitrogen, but foliar application requires 3 to 6 applications yearly compared to one soil application. All the nitrogen needed for a growing crop is seldom applied only in spray form, but supplemental amounts are often included in sprays, supplying pesticides, micronutrients, or other materials.

In summary, the following are some facts and suggestions in the use of foliar sprays for adding plant nutrients:

1. Foliar sprays are best suited for applications of small amounts of nutrients (1–10 kg/ha) for quick plant uptake and response. This method is mostly used for micronutrients but can give a boost in plant growth by adding a few kilograms of nitrogen, potassium, or phosphorus per hectare. Pineapples have been "fed" up to 75% of their nitrogen requirement as urea and 50% of their phosphorus and potassium in foliar spray (Fig. 13-9). This requires very frequent applications.

2. Iron chelates, because of immobilization in soil, are frequently applied as foliar sprays to solve iron deficiencies (Fig. 13-10). Soil applications of zinc, manganese, copper, and iron require many times more material than when added in sprays.

3. Most sprays on foliage require a *wetting agent* to aid the liquid in spreading over the leaf and entering into the plant openings. Commercial wetting agents for this purpose are available. A *sticking agent* may help retain the liquid on the plant leaves.

4. Foliar sprays have been helpful when other conditions reduce root uptake (cold or diseases such as root rot or nematode damage).

5. Foliar sprays can give a quick response—in a few days, in many situations. The effects of foliar spray are also often short-lived because of the small amounts of nutrients absorbed. Several applications may be needed.

6. The spray left on the leaf can cause salt burn if it is too concentrated. Concentrations recommended seldom exceed 1–2%. Many phosphates are damaging, and the maximum orthophosphate ($H_2PO_4^-$) concentration tolerated in spray without plant damage was 0.5% for corn and 0.4% for soybeans.

13:6.9 Fertilizing in Paddy and Other Waterlogged (Anaerobic) Soils

Paddy rice is the growing of rice in anaerobic soil, kept anaerobic by maintaining a layer of water on the soil (Fig. 13-11). The depth of the water layer may vary, but it is usually between

FIGURE 13-9 Boron deficiency in Thompson seedless grapes (right) drastically reduces fruit set. The limited areas of deficiency in California occur on soils of granite origin in the San Joaquin Valley. Treatment cost is relatively low, so whole "blocks" of vineyards are fertilized where boron is known to be low. About 3 lb of B each 2–3 years is sprayed onto the *soil;* lesser amounts are sprayed on the foliage, applied by aircraft, or put on with a sulfur dusting machine. (*Source:* By permission of *California Agriculture;* Peter Christensen, "Boron Application in Vineyards," *California Agriculture* [March–Apr. 1986], pp. 17–18.)

(a)

(b)

FIGURE 13-10 Iron deficiency on peaches in a calcareous Utah soil. (a) Close-up of leaves showing no deficiency (right) to increasing deficiency (left). Note the darker veins. (b) View of peach trees; light foliage is chlorotic leaves from iron deficiency. (Courtesy Raymond W. Miller, Utah State University.)

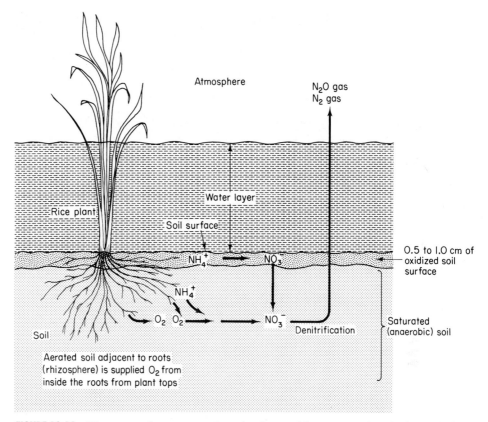

FIGURE 13-11 Diagrammatic cross section of a rice paddy, emphasizing the hazard of losing nitrogen by denitrification. Ammonium ion forms are oxidized to nitrate in the thin oxidized soil surface or near roots by oxygen brought down inside the plant to the roots. (Courtesy of Raymond W. Miller, Utah State University.)

about 7 and 15 cm (3–6 in.) deep. Although certain varieties of rice can be grown in normal, aerated soils, most rice is grown in water-covered, anaerobic soils. In areas of heavy rains it is often one of only a few crops that tolerate the excessive wetness, sometimes growing in water 30–120 cm (1–4 ft) deep.

The flooded condition favors a number of growth factors: (1) The pH becomes less acidic or less basic when flooded, (2) phosphorus and iron are more soluble and available, (3) fewer weeds can grow in anaerobic soils to compete with the rice, (4) N_2-fixation is increased, including more free-living algae that fix nitrogen, (5) fewer soil-borne diseases occur, (6) the water supplies some nutrients, (7) the terracing to hold a water layer greatly reduces soil erosion, and (8) yields of rice grain are higher. Soils that may have a pH between 4 and 5 when drained will often develop a pH higher than 6 when flooded for paddy. Soils whose tests for available phosphorus suggest a low phosphorus availability are often adequate in phosphorus when used for paddy.

Anaerobic soils do have some disadvantages. Soil organic materials can produce toxic decomposition products, especially in the first few weeks of inundation. Nitrogen is lost by denitrification. Optimum conditions for denitrification are mild acidity, high amounts of easily decomposable soil organic matter, a lack of free gaseous oxygen in the soil solution, and warm temperatures. Nitrogen added to paddy as nitrates may have over 50% loss by denitrification. In the presence of rapidly decomposing organic material, anaerobic conditions *can be developed* in warm soils in less than a day. In paddy with slow water replacement and

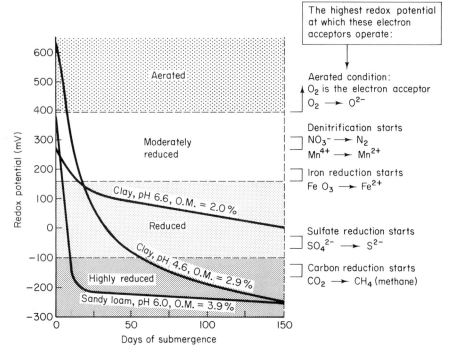

FIGURE 13-12 Combined graph illustrating (1) the rate at which three soils, following flooding, decreased in redox potential and became increasingly anaerobic, (2) the approximate extent of reduction of the different redox potentials, and (3) the approximate redox values at which NO_3^-, Mn^{4+}, Fe^{3+}, SO_4^{2-}, and carbon become electron acceptors in place of O_2, which is absent.   (Courtesy of Raymond W. Miller; data selected and combined from F. N. Ponnamperuma, "Some Aspects of the Physical Chemistry of Paddy Soils," and W. H. Patrick, Jr., "The Role of Inorganic Redox Systems in Controlling Reduction in Paddy Soils," both in *Proceedings of Symposium on Paddy Soil,* Institute of Soil Science, Academia Sinica, Science Press, Beijing, and Springer-Verlag, New York, 1981.)

stagnant conditions, most of the root zone and deeper water is anaerobic within a few days of flooding (Fig. 13-12). It may require 3 or 4 weeks to reach its most anaerobic level.

The extent to which soil around roots is made aerobic by oxygen brought down internally in the plant is not well documented. It is believed that the major oxidation of NH_4^+ to NO_3^- occurs in the thin, partially oxidized 1/2 cm of surface soil in the rhizosphere or in the water layer itself. In one laboratory study using tracer nitrogen ($^{15}NH_4Cl$) and a heavy nitrogen application (over 400 kg N/ha), the root zone caused a loss of 18% of the nitrogen during 40 days.[10]

The practical management of nitrogen in paddy is a problem. There is increasing use of urea and less use of ammonium sulfate (because it is more costly). Much of the urea is lost by ammonia volatilization from high-pH water caused by algae growth. The best efficiency of added nitrogen was with ammonium sulfate. When urea is used, best efficiency occurred when preplant urea was incorporated into soil *and* when later nitrogen additions were made

---

[10] K. R. Reddy and W. H. Patrick, Jr., "Fate of Fertilizer Nitrogen in the Rice Root Zone." *Soil Science Society of America Journal,* **50** (1986), pp. 649–651.

after the rice canopy was mostly closed (the closed canopy shaded the water and reduced algal growth).[11] The use of large urea granules about 10 mm (3/8 in.) in diameter reduced the nitrogen loss. Less than half as much urea-nitrogen was lost from large granules as when normal prilled urea was used.[12]

Soil phosphorus is more soluble in anaerobic conditions than in aerobic soil. Soil tests that suggest a phosphorus deficiency in normal soil is often adequate for paddy rice.

Flooded rice paddies may also require sulfur or zinc. Correcting zinc deficiency in rice in California is a good example. Zinc deficiency (alkali disease) is common on basic soils, sodic soils, and soils with calcium carbonate, especially in the Sacramento and San Joaquin valleys of California. Soils extracted with a chelate (DTPA) and testing lower than 0.50 ppm zinc usually respond to zinc additions. Because zinc moves very little in soils, it must be placed where it is needed. In flooded rice, zinc is most available in the top 2.5 cm (1 in.) of soil. Satisfactory correction of zinc deficiency is made with 9.0 kg/ha (8 lb/a) of zinc sulfate or zinc lignosulfonate (chelate) applied broadcast on the soil before flooding. Coating the water-sown seed with 0.9 kg (2 lb) calculated as actual zinc (as zinc sulfate or zinc oxide) per 45.5 kg (100 lb) of seed has also solved the problem.

13:7 Fertilizer Efficiency

Fertilizer efficiency is defined as *the percentage of added fertilizer that is actually used by the plants.* It is assumed that excess nutrient is not absorbed and that all absorbed nutrient was necessary for increased yields. These assumptions are not always true. In general fertilizer usage, *the expected efficiencies are approximately 30–70% of added nitrogen, 5–30% of added phosphorus, and 50–80% of added potassium.* These values can be improved by using special care; the values can also be even lower because of bad weather or carelessness.

Looking at the fertility problem from the growers' view, the addition of any amount of fertilizer is of interest only if it *profitably* enhances yields, either larger yields or better quality. The benefits from fertilizer additions must exceed the costs of purchasing and applying it. Sometimes, even the inconvenience or time needed at a busy period can be major concerns in evaluating the proposed fertilization.

In Fig. 13-13, an illustration of yields and profits from added fertilizer is shown. *Maximum profits are rarely at maximum yields* because the last increments of fertilizer to produce a little more yield costs more than the yield increase is worth. The diagram also illustrates that a good producer can afford to use higher amounts of fertilizer (and still make a profit) than could a producer not managing a farm well. Too often research work is reported with "significantly increased yields" from some experimental practice without any concern given as to whether the increase is a "profitable increase." The downward curve of Fig. 13-13, as excessive fertilization increases, is due to reduced crop yields. These may be caused by salt problems (too much fertilizer), imbalanced plant nutrition, or increased susceptibility to disease. When attempting to increase yields on farms, the economics of such increases must be assessed carefully.

A number of factors affect fertilizer efficiency and crop responses to added fertilizer. The plant itself will vary because of the nutrients it needs and its kind of root system. Water availability, temperatures, pests (weeds, insects, other organisms), and management are all

---

[11] A. C. B. M. van der Kruijs, J. C. P. M. Jacobs, P. D. J. van der Vorm, and A. van Diest, "Recovery of Fertilizer-Nitrogen by Rice Grown in a Greenhouse under Varying Soil and Climatic Conditions," in Institute of Soil Science and Academia Sinica, eds., *Proceedings of Symposium on Paddy Soil,* Science Press, Beijing, and Springer-Verlag, New York, 1981, pp. 678–688.

[12] K. Sudhakara and R. Prasad, "Ammonia Volatilization Losses from Prilled Urea, Urea Supergranules (USG) and Coated USG in Rice Fields," *Plant and Soil,* **94** (1986), pp. 293–295.

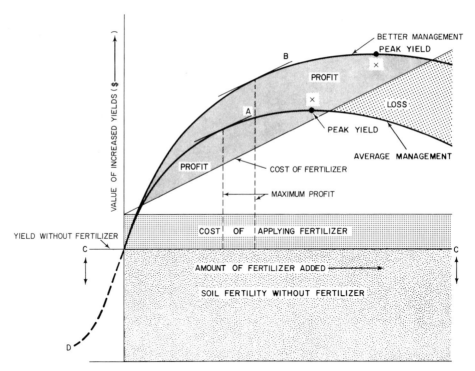

FIGURE 13-13 Representation of the increased profit resulting from added fertilizers. Notice that profit per dollar cost of additional fertilizer decreases as high fertility levels are reached. The initial soil fertility level (line C–C) varies with the soil and may be higher or lower than shown along the growth curve D–B or D–A. The maximum profits from fertilizer (where the line tangent to the curve is parallel to the fertilizer cost line) is higher and at a higher fertilizer addition for well-managed crops than for poorer management. Maximum yields (x) are seldom maximum profits because of slight growth increments per unit of fertilizer (or water or other treatment) added near maximum growth. (Courtesy of Raymond W. Miller, Utah State University.)

important. To evaluate the many factors, detailed crop-budget-analysis software is available. One of these is the **MEY** (**M**aximum **E**conomic **Y**ield) **analysis** package. This program attempts to consider the economic effects of each input in a crop's production[13] (Detail 13-1).

13:7.1 Plant Root Systems

Since early studies of why certain plants grew well on a given soil while other varieties or plant species did not, it has been known that some plants are better nutrient scavengers than others; for example, dicotyledons, especially legumes, feed strongly on divalent cations like Ca^{2+}, whereas grasses feed better on monovalent cations like K^+.

How well roots absorb fertilizers depends upon their distribution in the soil. The smaller the root system (shallow and/or few in number) and the shorter the growing season, the more dependent the plant is on fertilization. Plant growth rates and size must also be considered because small, slow-growing plants (short grasses) have low rates and low total demands.

---

[13] Harold F. Rutz, Jr., "MEY Analysis Software Looks at Cost Per Unit," *Solutions*, **32** (1988), no. 3, pp. 38–40.

As people attempt to estimate crop needs or a crop's response to various environmental conditions and additions, they develop simple "models" in their minds. For example, how much will the addition of 50 kg of nitrogen increase growth? Or how much will 2 cm of rainfall at corn tassel-time alter growth? Numerous persons and groups have been working for years to develop models. Most of these models are complex and allow some prediction of yield changes as the magnitudes of various growth factors are changed. All require computers.

For the convenience of readers, a few of these models and their sources are named below. (No slight to unmentioned models nor endorsement of those models listed is intended by the authors or publishers.)

1. Models from IBSNAT (International Benchmark Sites Network for Agrotechnology Transfer, 2500 Dole St., Krauss 22, Honolulu, HI 96822. These models cost about $500 each.)

CERES—Maize, ver. 2.1	CERES—Millet, ver. 2.1
CERES—Wheat, ver. 2.1	SUBSTOR—Potato, ver. 2.0
CERES—Rice, ver. 2.0	SOYGRO, ver. 5.42 (grain-legume)
CERES—Sorghum, ver. 2.1	PNUTGRO, ver. 1.02 (grain-legume)
CERES—Barley, ver. 2.1	BEANGRO, ver. 1.01 (grain-legume)

These models are designed to enable a user to match land characteristics and biological requirements of the crop. They include (1) soil, crop, weather, and management data, (2) validated crop models, and (3) application programs.

2. PLANTGRO: A. Retta and R. J. Hanks, Utah Agricultural Experiment Station Research Report 48, 1989. A computer model to predict dry-matter (and grain) production of field crops. Contact Department of Plants, Soils, and Biometeorology, Utah State University, Logan, UT 84322-4820.

When not restricted by compacted soils, rocks, or toxic salts, the root systems of some common plants are grouped as follows:

Deep and Numerous Roots, About 1.8 m (6 ft) Deep	Intermediate Root System, About 1.2 m (4 ft) Deep	Shallow and/or Few in Number, About 0.6 m (2 ft) Deep
Grapes	Corn	Potatoes
Sorghum	Peas	Beans
Alfalfa	Soybeans	Turf, short grasses
Tomatoes	Tall grasses	Lettuce
Orchards	Cotton	Peppers
Sweet clover	Small grains	Cabbage
		Onions

13:7.2 Fertilizer–Water–Application Interactions

Nutrients in soils move in water to root surfaces, where they are absorbed. The *availability of nutrients to plants is directly proportional to soil water content.* As surface soil—which is usually highest in nutrients—dries, those nutrients in the dry soil become less available. A wet surface soil may increase the hazard for diseases to crops, such as cucumbers and strawberries, but dry surface soil reduces available nutrients in the soil layer or maximum nutrients and root proliferation. Obviously, we must often compromise in the management we elect to follow.

Surface flow irrigation can alter nitrogen and sulfate levels in soils because of the excessive leaching that occurs at the tops of most fields. In contrast, sprinkler irrigation could be adjusted to avoid all leaching of nitrates and sulfates; the wetting depth can be controlled and is somewhat uniform. In the sandhills of Nebraska, the most efficient system involved fertigation that wetted the soil to about 45 cm (18 in.) deep in 12 hours, when irrigation was needed.

Some of the intriguing new management systems involve fertilizer added in *drip* irrigation systems. Efficiencies of both water and fertilizers are high. If the drip is surface applied and is adding phosphate, the constant high water content (near to field capacity) helps movement of some phosphate further into the soil. In one study on acidic sandy loam, placing phosphate fertilizer near a drip outlet increased available phosphate to at least 22 cm (8.6 in.) deep. Drip lines buried 20 to 25 cm (8–10 in.), with water filtered and adjusted to pH 6.0–6.5 to minimize plugging emitters, allowed fertilization with N, P, and K for years. Tillage was shallow to avoid damage to lines. Some lines are buried as deep as 45 cm (almost 18 in.), and irrigations can be as frequent as several times *daily* (each irrigation with 1 mm of water containing fertilizer).

Frequent irrigation with a drip system may cause periods of partially *anaerobic* conditions that could result in mobilization of phosphate and iron during the anaerobic conditions. Also, filtration of the water must be carefully done for buried emitters to minimize plugging problems.

Another concern, when using drip systems, is the higher costs of water-soluble fertilizers, particularly the phosphates. Additionally, a *fumigant* must usually be purchased to kill microbes that could produce emitter-plugging slime.

13:7.3 Variable-Rate Applications[14,15]

One utopian goal of fertilizer researchers is to be able to add variable amounts of fertilizer as the equipment moves across a single field. Fields vary in available nutrients from one position in the field to another. The computer age has made it possible to change, almost instantaneously, the application rate many times as an applicator moves across a field. Unfortunately, there is still a major limitation: *How do we rapidly test the nutrient's level in the soil while moving?*

The equipment designed to be a variable-rate applicator has a boom 8 to 15 feet (2.5–4.6 m) ahead of the fertilizer injector that has a boot or chisel with an analytical sensor moving along in the soil. The sensor "measures" the nutrient value, averaging the value in the soil over a few seconds as it moves, and relays the information to a computer. The computer changes the fertilizer flow rates. To allow rapid and accurate flow rates, most of these applicators have used liquid or gaseous fertilizers (suspensions will also work).

The major limitations, and they are major, are (1) the cost and delicacy of the equipment, and (2) the difficulty of measuring "available nutrient." What can be measured quickly that is an evaluation of the soil's nutrient status? Most soil tests are *wet chemical* measurements that are unsuitable for "instantaneous field measurements" by the probe as it moves through the soil. Chemists in recent years have greatly increased the elements that can be measured by infrared, ultraviolet, and other "radiation" waves of "light." These techniques require only a few

[14] Alissa Peitscher, "Variable Rate Fluid Technology Arrives for Dealers," *Solutions,* **36** (1992), no. 5, pp. 31, 33.
[15] Willie Vogt, "Soil Sensors Boost Management Precision," *Solutions,* **36** (1992), no. 3, pp. 26, 27.

seconds, so can "average" the measurement as the probe moves through the soil. This still does not solve the problem. How do you measure "available" nutrient? *Total nutrient* has never been very useful in correlation to growth response to added fertilizers. In studies sensing for nitrogen, the sensor measures *total organic matter,* not nitrate or ammonium ion, and the organic matter is measured by *soil color,* not organic carbon. In previous laboratory studies, total organic matter has not correlated well with soil nitrogen production.

One use made of this general kind of equipment is to measure soil water content on-the-go for the purpose of adjusting planting depth for best emergence conditions (as in planting dryland wheat).

13:7.4 Timing and Climate

The effectiveness of fertilizer depends on having it where the plant can get it when it needs it. To avoid excess losses by leaching or volatilization and reduce adsorption and precipitation "losses" of phosphorus, fertilizers are often applied in split applications. Much is known about when to apply fertilizer, but weather or irrigation timing can alter the benefits. Heavy rainfall following a top dress of urea may leach much of it away. A lack of rainfall may allow volatilization losses. A cold or extremely hot period following fertilization may reduce the plant growth more than expected. Applications too early or too late will reduce their effectiveness, even if you were late because another crop needed attention.

The plant cannot wait. It can absorb some nutrients in excess and translocate some from older parts of the plant. Usually a deficiency, however temporary, will affect some growth. Even though timing and climate are very important, many of the answers needed to make recommendations are not known. Further, the preferred dates to do irrigations or fertilizations cannot always be met. There is very limited control on climate.

13:7.5 Fertilizer Management Techniques

Some lower efficiency of fertilizer is often accepted to allow a saving in labor or to permit application in a more convenient manner or time. Maximum fertilizer efficiency is often not worth the changes required to get it; the value of the "lost" fertilizer caused by management changes may be traded for gains in time or savings in equipment. Sometimes the value of a crop justifies the addition of "insurance" fertilizer, small excesses that guarantee there will be enough for maximum production. Environmental-pollution concerns will require new evaluations of these concepts. In some instances there will not be a choice given, which could cause large polluting losses.

The goal in fertilization is to achieve *optimum fertilizer efficiency,* and the result is usually minimal pollution. Some guides to approach *optimum fertilization* are:

1. Avoid single large fertilizer additions of N or K, over 50 lb of element per acre (56 kg/ha), on sandy soils when rainfall or overirrigation can cause leaching. Use split applications when the crop has grown enough to get maximum uptake quickly.

2. Reduce ammonia volatilization losses by avoiding surface-broadcast applications of urea and ammonia solutions on moist calcareous soil during the warm growing period, unless the fertilizers are immediately incorporated by tillage or watered into the soil.

3. Reduce denitrification losses by avoiding heavy nitrate fertilizer additions on poorly drained soils, such as rice paddies and poorly drained clayey soils in wet climates. Use the ammonium forms and attempt to prevent cycles of aerated-nonaerated soil, which cause oxidation to nitrate, then loss by denitrification.

4. Band water-soluble phosphorus fertilizers; broadcast and incorporate low-solubility materials. On soils of very low available phosphorus, banding may be less productive than broadcast mixing. On high phosphorus-fixing soils (those high in iron and aluminum [acidic soils] or high in calcium [calcareous soils]), banding is most efficient.

5. Use a small amount (up to 11–22 kg of element/ha, 10–20 lb/a) of "starter" fertilizer with or near the seed, if fertilizer is not band-applied at planting. Avoid larger salt-damaging amounts of nitrogen and potassium fertilizers near seed or roots.

6. Nitrogen and potassium fertilizers are soluble salts. Avoid large additions of them nearer than 3 cm (about 1 in.), and preferably at least 5–7 cm (2–3 in.), from the seed, to avoid "salt burn."

7. Urea and nitrate are very soluble in water. They can be moved, even leached from the root zone, by rain or excess irrigation.

8. Combining ammonium and phosphates together in bands helps increase the efficient use of phosphates.

9. Foliar applications of nutrients (in concentrations of less than 1% solutions) provide rapid uptake and use of small quantities of nutrients (a few kilograms per hectare). Foliar applications are very effective for micronutrients, particularly for iron, zinc, and manganese.

10. Know the nutrient demands of the crop. Large amounts of nitrogen are needed for corn, sorghum, bananas, peanuts, and potatoes. Many legumes require large amounts of phosphorus but fix some or all of their nitrogen needs. Bananas, corn, and potatoes have high phosphorus needs. Crops producing sugars and starches need high levels of potassium (bananas, potatoes, sugarbeets, sugar cane, pineapples, corn). Tobacco is also a high potassium user.

11. Be willing to try "test strips" on your fields, which are modifications made by you of the recommended fertilizer to add. For example, add more fertilizer, add a micronutrient, reduce rates, add another nutrient, etc. Good fertilizer use requires integrating knowledge of your field continually over many years of different crops and their responses.

12. Fertilize and irrigate at the correct times. Timing may be as important as the variation in amount of fertilizer.

13. Remember *Liebig's law of the minimum:* The growth factor in the least relative amount will limit the growth. Fertilizer is only effective when other growth factors are adequate. These other factors include (a) using responsive varieties, (b) having correct plant densities, (c) employing correct water control, (d) timing properly the management operations, (e) being free of disease, (f) having no physical soil problems, (g) having adjusted the soil pH to an appropriate level for that crop, and (h) having optimum weather.

14. Large amounts of fertilizers in concentrated bands may hinder quick plant use because of (a) too high a salt content for nitrification and (b) hindrance to root entry into the band because of salt or ammonia toxicities in and near the bands.

15. Test soils frequently, some fields every year, and follow the recommendations. During some years, phosphorus, potassium, and/or other nutrients may be adequate in some fields for a year or more, especially if the crop is a low-demanding type; but soil testing annually or every second year provides better verification.

There are other considerations when planning fertilization. For instance, high soil levels of ammonium (NH_4^+) and nitrate (NO_3^-) minimize atmospheric nitrogen (N_2) fixation by microorganisms. The cultivar (cultivated variety) selected may have particular specific nutrient requirements that should be taken into account. For example, wheat cultivar UC 44-111 requires less nitrogen than the "Anza" cultivar.[16] Some rice cultivars produce satisfactory yields under *low* levels of soil nitrogen, phosphorus, potassium, sulfur, and zinc. At least three rice cultivars produce *high yields on low-phosphorus soils,* whereas others have low yields at low soil phosphorus levels.[17] **Grass tetany** (hypomagnesemia) is a serious disease of domestic animals that is caused by low levels of magnesium in their blood. Heavy applications of potassium (K^+) and/or ammonium (NH_4^+) fertilizers on pastures reduce plant uptake of magnesium (Mg^{2+}) and can cause grass tetany. The high concentration of potassium and ammonium in poultry litter can also cause grass tetany if large applications are made to pastures.[18] Lower rates or split applications of these fertilizers are recommended in such situations. Many methods of applying fertilizers for greater efficiency are yet to be learned. Other improvements in development include better fertilizers, optimum timing, and improved nutrient *balance.*

13:8 Fertilizer Calculations

A number of calculations on fertilizer materials may be necessary. The usual ones needed are (1) nutrient percentage, (2) weight of bulk fertilizer to use, and (3) weights of materials to use in making mixed bulk fertilizer for adding several nutrients. Such a mixture might be based on a fertilizer recommendation from a tested soil sample. The more complex these calculations are, the more suitably they can be adapted to computer programs by programming.

13:8.1 Calculating Nutrient Percentage

All nutrients should be expressed as the percentage of the element (percentage of zinc, nitrogen, phosphorus). Unfortunately, early chemists reported elements as "burned solids" that produced oxides. Thus, we are still using the convention of reporting phosphorus as P_2O_5 equivalent and potassium as K_2O equivalent. Efforts are being made to change these conventions to report only N, P, and K percentages. One reason for the resistance to change is that P or K reported as percentages gives lower numbers and makes the fertilizer appear to be a lower grade. For example, a 0-45-0 triple superphosphate is only a 0-19-0 material if reported as having a percentage of element rather than oxide. This is an unfortunate resistance, especially because fertilizers *do not contain P_2O_5 or K_2O.*

A close approximation of the percentage of a nutrient can be obtained using the fertilizer formula and chemical atomic weights. Urea has a formula of $(NH_2)_2CO$. The molecular weight (sum of all atomic weights) is 60.056. The *percentage nitrogen* is calculated as follows:

$$\frac{2N}{H_4N_2CO}(100) = \left(\frac{28.014 \text{ g}}{60.056 \text{ g}}\right)(100) = 46.6\%$$

[16] Calvin O. Qualset, John D. Prato, and Herbert E. Vogt, "Breeding Success with Spring Wheat Germplasm," *California Agriculture,* **31** (Summer 1977), no. 9, pp. 25, 27.

[17] M. Mahadevappa, H. Ikehashi, and F. N. Ponnamperuma, "Research on Varietal Tolerance for Phosphorus-Deficient Rice Soils," *International Rice Research Institute Newsletter,* **4** (Feb. 1979), no. 1, pp. 9–10.

[18] S. R. Wilkinson and J. A. Stuedemann, "Tetany Hazard of Grass as Affected by Fertilization with Nitrogen, Potassium, or Poultry Litter and Methods of Grass Tetany Prevention," in *Grass Tetany,* American Society of Agronomy, Crop Science Society of America, and Soil Science Society of America, Madison, Wis., 1979, pp. 93–121.

Problem Calculate the percentage of N, P, and P_2O_5 in diammonium phosphate if it were pure. Formula is $(NH_4)_2HPO_4$.

Solution

1. To calculate percentage N, put the atomic weight of two nitrogens over the molecular weight of the $(NH_4)2HPO_4$ and multiply by 100 to get percentage. Use values to the closest tenth.

$$\% \ N = \frac{2N}{(NH_4)_2HPO_4} \left| \frac{100}{} \right. = \frac{28 \text{ g}}{132 \text{ g}} \left| \frac{100}{} \right. = 21\%$$

Most commercial $(NH_4)2HPO_4$ fertilizer is less pure and less fully ammoniated, so it will have percentages of N about 16–18%.

2. To calculate percentage P, use the same procedure as for N, but remember that the formula has only one P atom, whereas there are two N in step 1 above.

$$\% \ P = \frac{P}{(NH_4)_2HPO_4} \left| \frac{100}{} \right. = \frac{31 \text{ g}}{132 \text{ g}} \left| \frac{100}{} \right. = 23.5\%$$

3. To calculate the percentage of P_2O_5 (fertilizer grades are calculated to this formula), use the relation of P_2O_5 (fertilizer grades are calculated to this formula), use the relation of P_2O_5 to two P. Using the atomic and molecular weights and the calculated 23.5% P from step 2 gives

$$\frac{23.5\%}{} \left| \frac{P_2O_5}{2P} \right. = \frac{23.5\%}{} \left| \frac{142 \text{ g}}{62 \text{ g}} \right. = (23.5)(2.29) = 53.8\% \ P_2O_5$$

Impure diammonium phosphate fertilizers usually have about 46–48% P_2O_5. Some may reach nearly 50–53%.

Because the fertilizers are seldom pure, the calculations should be rounded downward to the nearest whole integer: $(28 \text{ g}/60 \text{ g})(100) = 46\%$. Rounded-off values may still be high. The actual formulas and "contaminants" of superphosphates are seldom given, so one must depend on the percentage values given on the container. For another example of calculations, see Calculation 13-1.

13:8.2 Calculating Simple Fertilizer Mixtures

To combine various single-nutrient sources into a mixed fertilizer, do it one of two ways: (1) Mix the two or three materials to get a mixture with the proportions wanted, e.g., high N, low P, and low K or (2) mix the materials to an exact grade, adding a "filler" material to get a final weight previously selected.

For example, calculate the amount of ammonium nitrate (34-0-0) and treble superphosphate (TSP, 0-45-0) to make 1000 kg of a 15-10-0 mixture. The final mixture must have 15% of 1000 kg, which equals 150 kg of N. It also must have 10% of 1000 kg, which equals 100 kg of P_2O_5. To get 150 kg of N, each 100 kg of ammonium nitrate has 34 kg of N. Thus,

$$\frac{150 \text{ kg of N}}{1000 \text{ kg of mix}} \left| \frac{100\% \text{ of } 34\text{-}0\text{-}0}{34\% \text{ N}} \right. = \frac{15,000 \text{ kg}}{34}$$

$$= 441 \text{ kg of ammonium nitrate/1000 kg}$$

To get 100 kg of P_2O_5, each 100 kg of TSP has 45 kg of P_2O_5. So,

$$\frac{100 \text{ kg of } P_2O_5}{1000 \text{ kg of mix}} \mid \frac{100\% \text{ of } 0\text{-}45\text{-}0}{45\% \ P_2O_5} = \frac{10{,}000 \text{ kg}}{45}$$

$$= 222 \text{ kg of TSP needed}/1000 \text{ kg}$$

Adding the ammonium nitrate (441 kg) and the treble superphosphate (222 kg) is 663 kg of material needed. The total mixture was to be 1000 kg, so the additional material is 337 kg (1000 − 663 = 337) and is made up by filler or lime or whatever additive is wanted.

To mix a fertilizer with N, P, and K all in it, use the same pattern. However, if the sum of weights of the three carriers (the N source, the P source, and the K source) add up to more weight than you calculated it for (say, 1000 kg), it cannot be done. The sources used do not have a high enough percentage of nutrient to be mixed as you wished. If this happens, you have two choices: (1) lower the mixture's grade (and thus add more mix per hectare) or (2) select one or more different carriers with a higher nutrient content. For example, you could select urea with 46% N rather than ammonium sulfate with its 21% N.

13:8.3 Weights of Fertilizer to Add

If it is desired to add 80 kg of N and 40 kg of P_2O_5 per hectare, how much ammonium nitrate (34% N) and treble superphosphate (45% P_2O_5) should be added? For 80 kg of N,

$$\frac{80 \text{ kg of N}}{1 \text{ ha}} \mid \frac{100\% \text{ of } 34\text{-}0\text{-}0}{34\% \ N} = \frac{8000 \text{ kg of } 34\text{-}0\text{-}0}{34 \text{ ha}}$$

$$= \frac{235 \text{ kg of } 34\text{-}0\text{-}0}{1 \text{ ha}}$$

For 40 kg of P_2O_5, use the same procedure:

$$\frac{40 \text{ kg of } P_2O_5}{1 \text{ ha}} \mid \frac{100\% \text{ of } 0\text{-}45\text{-}0}{45\% \ P_2O_5} = \frac{4000 \text{ kg of } 0\text{-}45\text{-}0}{45 \text{ ha}}$$

$$= \frac{88.9 \text{ kg of } 0\text{-}45\text{-}0}{1 \text{ ha}}$$

Mix 235 kg of ammonium nitrate plus 89 kg of treble superphosphate times the number of hectares to be fertilized and set the drill to spread the 324 kg of mix per hectare. Keep in mind that kg/ha times 0.89 equals lb/a. Field applicators may not be adjustable to closer than about 10 kg/ha.

It is also easy to calculate the amount of fertilizer to apply to a small area. Consider adding the rate above to a garden plot that is 100 ft × 100 ft = 10,000 ft². The 10,000 ft² is

$$\frac{10{,}000 \text{ ft}^2}{1 \text{ plot}} \mid \frac{1 \text{ acre}}{43{,}560 \text{ ft}^2} = \frac{0.23 \text{ acre}}{1 \text{ plot}}$$

Adding the mixture calculated earlier, 324 kg of mix per hectare, the amount to add is

$$\frac{324 \text{ kg}}{1 \text{ ha}} \mid \frac{1 \text{ ha}}{2.47 \text{ acres}} \mid \frac{2.2 \text{ lb}}{1 \text{ kg}} \mid \frac{0.23 \text{ acre}}{\text{plot}} = \frac{66 \text{ lb}}{\text{plot}}$$

13:8.4 Calculating Mixed Fertilizers to Predetermined Weights

Often you want to make a particular amount of a mixture. Suppose you want to make a material that contains N and P in the ratio of 2 to 1 (2 kg of N per 1 kg of P_2O_5). Calculate for 100 kg of the mixture and then make as many 100-kg multiples as you wish to have. To do this, use **simultaneous equations.** Make as many equations that explain the information as you have variables. For only **N** and **P** (2 variables), do as follows:

$$X + Y = 100 \text{ kg of mix} \tag{1a}$$

where X = kg of N carrier and Y = kg of P carrier to use per 100 kg of mixture. If we use ammonium nitrate (34% N) and treble superphosphate (TSP, 45% P_2O_5),

$$X \text{ kg of AN} + Y \text{ kg of TSP} = 100 \text{ kg of mixture} \tag{1b}$$

We need to make a mixture with 2 kg of N per 1 kg of P_2O_5, so

$$\frac{X \text{ kg of AN}}{} \left| \frac{34\% \text{ N}}{100\% \text{ AN}} = \frac{X}{} \right| \frac{34}{100} \text{ kg of N} \tag{2}$$

$$\frac{Y \text{ kg of TSP}}{} \left| \frac{45\% \text{ } P_2O_5}{100\% \text{ TSP}} = \frac{Y}{} \right| \frac{45}{100} \text{ kg of } P_2O_5 \tag{3}$$

Since we want 2 kg of N for each 1 kg of P_2O_5, we must have twice as much of P_2O_5 to equal the amount of N wanted. Thus,

$$X \text{ kg of N} = 2(Y \text{ kg of } P_2O_5) \tag{4}$$

This is the key step in this problem. Substitute into equation (4) the values from equations (2) and (3) to give the following:

$$\frac{X}{} \left| \frac{34}{100} = \frac{2Y}{} \right| \frac{45}{100} \tag{5}$$

Multiplying each side by 100 and simplifying yields

$$34X = 90Y \qquad \text{or} \qquad 34X - 90Y = 0 \tag{6}$$

Recall equation (1); $X + Y = 100$. So equations (1) and (6) will be the two **simultaneous equations** used to solve the problem. Eliminate one of the unknowns; we will choose to eliminate X. To do so, multiply equation (1) by 34 (so that X will cancel out), rearrange equation (6), and subtract the second equation from the first (multiply the second equation by –1 and add the two equations):

$$
\begin{array}{rl}
34\,X - & 90\,Y = 0 \\
-34\,X - & 34\,Y = -3400 \\
\hline
-0\,X - & 124\,Y = -3400
\end{array}
\qquad Y = 27.42 \text{ kg of TSP needed/100 kg mix}
$$

Then

$$\text{kg of P}_2\text{O}_5 = \frac{27.42 \text{ kg of TSP}}{100 \text{ kg of mix}} \left| \frac{45\% \text{ P}_2\text{O}_5}{100\% \text{ TSP}} \right. = 12.34 \text{ kg of P}_2\text{O}_5/100 \text{ kg mix}$$

Substitute the value of Y into equation (1):

$$X + 27.42 \text{ kg} = 100 \text{ kg of mixture}$$
$$X = 100 \text{ kg} - 27.42 = 72.58 \text{ kg of AN needed.}$$

Then

$$\text{kg of N} = \frac{72.58 \text{ kg of AN}}{100 \text{ kg of mix}} \left| \frac{34\% \text{ N}}{100\% \text{ AN}} \right. = 24.68 \text{ kg of N}/100 \text{ kg mix}$$

To check the work, see if the ratios of N to P_2O_5 are 2 to 1. Remember, we rounded numbers off to the closest hundreth.

$$24.68 \text{ kg of N} = 2(12.34) \text{ kg of P}_2\text{O}_5$$

The ratios are correct.

Calculation 13-2 gives an example of the more complicated calculation for a mixture of all three nutrients. This requires three equations (always as many equations as there are unknowns). Use pairs of equations and solve for one unknown at a time.

Tomorrow's growth depends on the use we make of today's materials and experiences.

—Elmer Wheeler

We used to use one fertilizer rate for the county, then the farm, then the field, and now ... [we can] vary application rates within the field.

—Dick Stiltz (manager, fertilizer plant)

▢ Questions

1. How will increasing the reduced-tillage acreage affect the patterns of fertilizer use?
2. How do environmental concerns alter our picture of the future in fertilizer use?
3. What are some of the fertilizer use changes with respect to (a) use of dry mixed fertilizers, (b) use of fluid fertilizers, and (c) use of other nutrients, such as zinc and sulfur?
4. What are several conditions in which fertilizer use will *not* be profitable?
5. What are several high-nutrient-requiring crops?

Problem Calculate the weight of urea (46% N), treble superphosphate (TSP = 45% P_2O_5), and KCl (60% K_2O) to make up 100 kg of a mixture having a ratio of N-P_2O_5-K_2O of 4-2-1. When the mixture is made, calculate its grade.

Solution 1A Two methods for solving this will be shown. The first technique seeks to make a 4-2-1 ratio using simple mathematics and logic. Calculate the mixture needed to provide 4 kg of N, 2 kg of P_2O_5, and 1 kg of K_2O (a 4-2-1 ratio).

$$\frac{4 \text{ kg of N}}{} \left| \frac{100\% \text{ urea}}{46\% \text{ N}} \right. = 8.696 \text{ kg of urea} \tag{1}$$

$$\frac{2 \text{ kg of } P_2O_5}{} \left| \frac{100\% \text{ TSP}}{45\% \text{ } P_2O_5} \right. = 4.444 \text{ kg of TSP} \tag{2}$$

$$\frac{1 \text{ kg of } K_2O}{} \left| \frac{100\% \text{ KCl}}{60\% \text{ } K_2O} \right. = 1.667 \text{ kg of KC} \tag{3}$$

Add the three materials together:

$$8.696 \text{ kg urea} + 4.444 \text{ kg TSP} + 1.667 \text{ kg KCl} = 14.807 \text{ kg}$$

Now, the mixture of 14.81 kg has the 4-2-1 ratio wanted. To make 100 kg of a mixture with the same ratio,

$$\frac{100 \text{ kg}}{14.81 \text{ kg}} = 6.752 \text{ multiples}$$

The required 100 kg of mix is 6.752 times more than the 14.81 kg. So, multiply the amount of each ingredient used by 6.752. Thus,

$$8.696 \text{ kg of urea} \times 6.752 = 58.715 \text{ kg of urea needed}$$

$$4.444 \text{ kg of TSP} \times 6.752 = 30.006 \text{ kg of TSP needed}$$

$$1.668 \text{ kg of KCl} \times 6.752 = \underline{11.256 \text{ kg of KCl}}$$

$$\text{Total weight} = 99.977 \text{ kg}$$

Solution 1B This problem can also be calculated, as was done in the text, using simultaneous equations. With three unknowns (N, P, and K), three equations are needed. These three equations can be used (there is at least one other equation possible):

$$X \text{ kg} + Y \text{ kg} + Z \text{ kg} = 100 \text{ kg} \tag{1}$$

$$\frac{46\%}{100\%} \left| X \right. = \frac{2}{} \left| \frac{45\%}{100\%} \right| Y \qquad [1 \text{ kg N} = 2 \text{ (kg } P_2O_5)] \tag{2}$$

Clearing fractions and transposing X and Y to the same side, we get $4600X - 9000Y = 0$.

$$\frac{45\%}{100\%} \left| Y \right. = \frac{2}{} \left| \frac{60\%}{100\%} \right| Z \tag{3}$$

Continued.

Clear and transpose: $4500Y - 12,000Z = 0$.

Equations (2) and (3) could be simplified by dividing each by 100. The three equations, then, are

$$X \text{ kg} + Y \text{ kg} + Z \text{ kg} = 100 \text{ kg} \qquad (1)$$

$$46X - 90Y = 0 \qquad (2)$$

$$45Y - 120Z = 0 \qquad (3)$$

Eliminate, successively, the X (equations 1 and 2), then Y, and obtain the first value for Z (= kg KCl). Your answers will be the same as calculated by Solution 1A.

Solution 2   The grade of the material is obtained by multiplying the quantity of each material by its nutrient content:

$$\text{kg of N} = (58.715 \text{ kg of urea})(46\%/100\%) = 27.0 \text{ kg}$$

$$\text{kg of P}_2\text{O}_5 = (30.006 \text{ kg of TSP})(45\%/100\%) = 13.5 \text{ kg}$$

$$\text{kg of K}_2\text{O} = (11.256 \text{ kg of KCl})(60\%/100\%) = 6.75 \text{ kg}$$

The grade usually must be in *whole numbers, never rounded upward,* and is 27-13-6 (grade).

6. If a fertilizer grade is given as 16-20-10-5S-1Zn, indicate exactly what each number means.
7. (a) To what extent are fertilizers hazards as soluble salts? (b) Are all fertilizers the same pH in water? (c) What fertilizers have high *potential* acidity? (d) Which nutrients tend to be least soluble and mobile?
8. (a) Because many fertilizers are broadcast but not incorporated and are of low efficiency, why is broadcasting used? (b) Is broadcast plus incorporation better? Discuss.
9. State one major reason (two, if you can) for using each of these fertilizing techniques in preference to using a different technique: (a) deep banding, (b) split applications, (c) fertigation, (d) foliar application, and (e) starter.
10. State at least one disadvantage for each fertilizing technique given in question 9.
11. (a) Discuss the extensive use of starter-banding applications. (b) Is it the best method for most crops? Explain.
12. Evaluate the desirability of adding each of these nutrients in split applications: (a) phosphorus, (b) potassium, and (c) nitrogen.
13. Evaluate fertigation and chemigation as (a) economic and (b) "safe" techniques.
14. How does fertigation differ from foliar application?
15. What are major advantages and disadvantages of foliar applications?
16. (a) Describe the aeration conditions in paddy rice. (b) How do these conditions influence fertilization techniques?
17. Discuss some of the factors, other than fertilizer application techniques, which affect fertilizer efficiency.
18. How does Liebig's law of the minimum enter into the expected response from added fertilizer?

19. Why are individual farmers encouraged to set up some of their own simple and small fertilizer test strips? (Consider uniqueness of your soil, weather, and management.)
20. Calculate the amount of urea (46% N) and of triple superphosphate (0-45-0) to make 100 kg of a 10-20-0 mixed fertilizer.
21. Calculate the percentage P (element, not P_2O_5) in pure phosphoric acid (H_3PO_4).
22. Calculate the amount of urea (46% N), ammonium phosphate (11-48-0), and potassium sulfate (0-0-50) to make 1000 kg of mixed 20-10-5 fertilizer.
23. If 600 kg/ha of the mixture in question 22 is added, what is the equivalent in kg of N, P_2O_5, and K_2O added per hectare?

14

Tillage Systems

Nature did not provide means for stirring, or tilling, soil except through action of micro- and macro-organisms. Major operations are innovations introduced by farmers. We surely have evidence that, on the whole, these innovations resulted in detriment rather than good.

—R. L. Cook and B. G. Ellis

14:1 Preview and Important Facts

PREVIEW

For over 4000 years people have labored to till the soil to plant and grow crops. Tillage methods in much of the world are virtually unchanged from what they were centuries ago (Figs. 14-1 and 14-2). Sometimes ancient tillage techniques are used because of custom. But more often the tillage systems are used because of limitations in equipment, money, knowledge, and environment. In developed nations tillage has become a highly mechanized and sophisticated process.

The "zenith" in *extensive* tillage was reached in about the 1960s, when heavy equipment prepared large, clean fields, devoid of even fence row vegetation. All residues were mixed into the soil (or previously burned). Many operations could be done in one pass over the land (would plow, disk, and smooth in one pass; fertilize, plant, and spray preemergent herbicides in one pass). However, this pinnacle of success was expensive in energy, costly because of compacted and eroded soil, and polluting to the environment. Technological abilities often exceeded good reasoning. Increased energy costs, concern about soil erosion losses, and environmental concerns have focused new attention on our tillage practices. The concepts of *reduced tillage* quickly drew the attention of many scientists and farmers.

Reduced-tillage and no-tillage practices are increasingly popular (Figs 14-3 and 14-4). Less tillage reduces fuel costs and operation time. However, less tillage also increases problems of weed control, insect control, and seedbed preparation. Heavier, more rugged equipment is needed to plant in nontilled soil. Fields may appear to the farmer and others to be "trashy" and poorly managed. The truth is that the reduced-tillage farmer must, in fact, be a *better farmer* than the conventional-till farmer to produce the same yields.

FIGURE 14-1 Farmer in southern Senegal using a *caillande,* a special hoe to scalp the sod and turn it over to form a ridge from two sides. The weeds in the sod will gradually rot, and the bare strip will be used to plant peanuts, yams, and other crops. (Courtesy of C. K. Kline, Michigan State University.)

FIGURE 14-2 *Yuntas* (oxen plows) incorporating wheat and fertilizer in Carrasco Province, near Cochabamba, Bolivia. These parents of students are donating their time to plant school land to wheat to raise funds for the school. Annual precipitation is about 61 cm (24 in.), and the mean annual temperature is 61°F (16°C). (Courtesy of Walter Carrera M., Ing. Agron., Bolivia.)

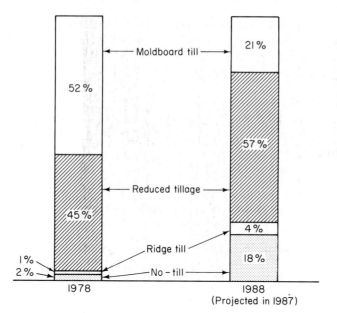

FIGURE 14-3 Changes in tillage systems in the United States during a recent decade. Notice that reduced tillage was nearly 60% in 1988 and no-till had reached almost one-fifth of all cultivated area. (Courtesy of Raymond W. Miller, Utah State University. Data from Fletcher and Lovejoy, *Crops and Soils Magazine,* **39** [1987], no. 2, p. 17.)

FIGURE 14-4 Changes in tillage practices between 1989 and 1991 on Midwestern lands managed by Capital Agricultural Property Services (CAPS). These are farms of absentee land owners managed by CAPS. (*Source:* Ray Brownfield, "Farm Managing," *Ag Consultant,* **48** [1992], no. 3, p. 13.

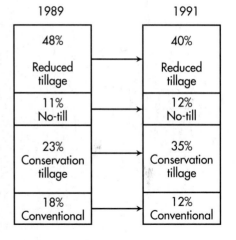

Reduced tillage is not for all crops or all soils. To decide whether to pursue reduced tillage, the land user must understand the tillage options, their benefits, and their disadvantages.

IMPORTANT FACTS TO KNOW

1. How conventional tillage differs from reduced tillage
2. The benefits of tillage
3. The damages from tillage, in some conditions
4. The reasons for wishing to use reduced tillage
5. The definition of *conservation tillage*
6. How "no-till" differs from reduced tillage
7. The problems in no-till, including (a) seedbed preparation, (b) planting, (c) fertilizer placement, (d) weed control, (e) insect control, and (f) liming

8. The recommended practices in reduced tillage related to (a) seedbed preparation, (b) planting, (c) weed control, (d) insect control, (e) erosion control, (f) fertilizer and lime application, and (g) irrigated soils

9. The effects of the various tillage practices on environmental contamination or pollution

10. Advantages and disadvantages of changing to reduced-tillage or even to no-till

▌ *14:2* Tillage Terminology

With the change from simple moldboard plow-harrow sequence to disk plows, rotary plows, chisel plows, disks, cultivators, subsoilers, and other equipment, it is necessary first to understand the terms used. Some examples of equipment are shown in Fig. 14-5.

Bedder A sweep that looks somewhat like a small moldboard plow with the curved sides on both sides (Fig 14-6). It is used to build ridges in ridge-furrow cultivation.

Blading *See* Rod weeding.

Chisel A narrow shank, usually with a narrow sweep (with no wings or only small wings) fastened to its tip. The chisel is pulled through the soil to "rip" it.

Chisel Plow Large chisels (see Fig. 7-10), usually about 0.6–1.0 m (2–3.3 ft) apart, pulled through the soil to rip it to 15–30 cm (6–12 in.) deep.

Conservation tillage (*See also* Reduced tillage.) The U.S. Soil Conservation Services defines *conservation tillage* as any tillage system that leaves at least 30% of the surface covered by plant residues for control of erosion by water; for controlling erosion by wind, it means leaving about 1000–2000kg/ha of flattened small-grain straw equivalent.[1] The amount of residue needed is based on its ability to control erosion; the amount depends on the kind of residue and whether it is standing or flat.

Conventional tillage Generally, the objective in conventional tillage was to burn or bury all residues and prepare a fine pulverized seedbed. For corn this would be a typical conventional tillage: Plow, disked once or twice, further break the clods by a spike harrow, incorporate lime and fertilizer during or prior to planting, and cultivate two to four times to control weeds before the corn was too tall to drive in the field.

Coulter A sharpened disk, with a straight fluted edge, often running in front of a chisel or small sweep to cut the surface vegetation and to open a slit in the soil.

Cultivator An implement fitted with several small sweeps that are pulled 5–10 cm (2–4 in.) deep between crop rows to cut and kill weeds. Wings 10–25 cm (4–10 in.) long.

Disk A combination of concave-shaped disks, usually a front set and a following set, both offset in directions opposite from each other and from being perpendicular to the direction of movement.

---

[1] Frank M. D'Itri, ed., *A Systems Approach to Conservation Tillage,* Lewis Publishers, Chelsea, Mich., 1985, p. 4.

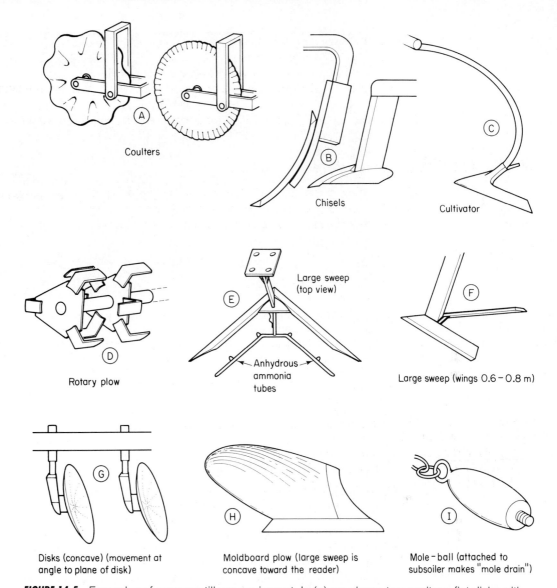

FIGURE 14-5 Examples of common tillage equipment. In (a) are shown two coulters, flat disks with wavy cutting edges to open a slit for a planter chisel or to cut surface litter to avoid plugging the planter. Chisels in (b) are used to loosen the soil without removing surface litter. Cultivators (c) are small sweeps to cut weeds 3–6 cm below the soil surface. Rotary plows (d) are "rototillers," which allow vigorous shallow mixing 10–25 cm deep. Large sweeps (e and f) are used to do shallow minimal tillage. Preplant anhydrous nitrogen could be added as shown in (e). Disks (g) are concave and pulled at slight angles to the plane of the disk, thus forcing the disk to dig into the soil. The moldboard plow (h) was the implement of the "tillage revolution" in the 1920s and 1930s. It inverts soil and buries all residues, producing a "clean-field" look. Various modifications, such as pulling the mole-ball (i), allow additional beneficial actions all in one pass over the field.   (Courtesy of Raymond W. Miller, Utah State University.)

FIGURE 14-6 Establishing the ridge tillage system using bedders. This system allows some drainage in soils with drainage problems. It also allows faster soil warming where seed will be planted and is used for some furrow irrigation, although ridges are narrow. The method involves extra tillage over nonridge systems. (Courtesy of the Buffalo Line, Fleischer Manufacturers.)

Disk harrow A combination set of closely spaced disks to break the large clods to smaller units. Sometimes a vertical tine harrow or unit is attached.

Disk plow A combination of large concave-shaped discs capable of turning soil to depths of 15–40 cm (6–15.6 in.).

Cross-slot drill This planter opens a slot (furrow), drops the seed, and then covers the slot and seed with soil to retain soil moisture and to firm the soil.

Fallow Land left without a crop for a period of several months to a year, with all weeds or other plants killed. Fallow is used in marginal rainfall areas (25 to 50 cm [10–20 in.], usually) to accumulate a few extra centimeters of water for the next crop. To kill weeds, the V-sweeps are used. They have wings 45–60 cm (17.5–23.5 in.) long, but some up to 100 cm (39 in.) are used.

Harrow A set of vertical tines that, when pulled over the soil, break the larger clods into smaller pieces. Often the harrow is the final equipment pulled over soil to prepare the seedbed.

Minimum tillage *See* Reduced tillage.

Moldboard plow A deep tillage implement (15–40 cm [6–15.5 in.] deep) with one side having a curved surface to turn the plowed strip almost upside down. This *inverts* the plowed layer and buries most of the surface plant residues.

No-till, no-tillage A farm management scheme using no tillage except to insert seed, and sometimes fertilizer, in a slit. Often a leading coulter is followed by a chisel. The seed, and sometimes the fertilizer is dropped through a small tube immediately behind the chisel.

Paraplow A series of chisels on a frame. Each chisel leg goes straight down and then bends to the side at a 45° angle. As the plow is pulled, soil flows over each bent wing (leg) and falls back, causing the soil to shatter but not to be mixed or inverted on itself.

Reduced tillage Any combination of tillage operations that do less tillage than all the operations used in "conventional" tillage. Reducing the number of operations may be accomplished by eliminating plowing or one pass with a disk, or it may be almost a no-till operation with only a single disking or pass with wide sweeps. Various terms are used for different degrees of reduced tillage: *minimum tillage, conservation tillage, zero tillage,* and *no-tillage.*

Ridge furrow The formation of alternate ridges and furrows, often with ridge tops 15–25 cm (6–10 in.) above the furrow bottom. This system is needed for most surface-flow irrigation of row crops. It is also useful for a better (larger or aerated) root zone (1) in shallow soils, (2) in poorly drained soils, (3) in cold soils where ridges warm faster and are drier than nonridged soil, and (4) on low-fertility soils. In the latter, surface ash from burning is scraped together in a pile or ridge where the plant roots grow most. Ridges may be used several years and are rebuilt by a special cultivation sweep (the bedder).

Rod weeding Pulling a sweep or blade through the subsoil at a depth of 6–10 cm (2.5–4 in.) to cut and kill weeds. Sometimes called *blading.*

Roller harrow A large set of "toothed" disks close together (10 cm, or 4 in., apart) on a wide-diameter drum. As the drum disk rolls over the soil, it pulverizes and smooths the seedbed.

Rotary plow Blades rotating into soil to mix it (rototill). Often, only *narrow strips* of soil are rotary-tilled slots in front of a seeder (*rotary strip tillage*).

Shovels A term often used for narrow-winged sweeps.

Slot mulcher Cuts a slot into the soil about 8 cm (3 in.) wide and 25 cm (10 in.) deep along the contour each 4–6 m (13–20 ft) apart. The mulcher fills the slot with grain straw mulch left in the fields. Flowing water is readily absorbed.

Strip tillage Tillage of only strips of soil, perhaps 15–25 cm (6–10 in.) wide, with nontilled strips between. The tilled strip is prepared into a loose seedbed, which allows incorporation of fertilizer and seed and leaves protective untilled strips between planted areas.

Stubble mulch Stubble mulching is one of the early erosion control practices in which only part of the crop residue is incorprated somewhat into the soil. Most of the residue was left anchored in the soil but exposed at the soil surface. Frequently, this was accomplished by pulling wide sweeps through the soil (see Fig. 15-15 in Chapter 15).

Subsoiler bedder *See* Bedder. When the bedder is built onto a subsoil chisel, it is called a *subsoiler bedder.*

Sweep plow A sweep, up to 2 m (6.5 ft) wide, fitted on its underside with four (or more) shanks about 15 cm (6 in.) long. The shanks incline toward the center, which breaks up the soil to a depth of 20–25 cm (8–10 in.).

Sweeps Any chisel tip that has "wings," curved surfaces to throw or disrupt the soil, or flat surfaces simply to cut through the soil to kill weeds without excessive soil disturbance. Small sweeps are the tips of cultivators; large sweeps are used to stubble-mulch or rod-weed fallow fields.

14:3 Tillage: Past and Future

Tillage is hard work and energy expensive. When people first began to cultivate crops, they did only what tillage was necessary to plant and to control weeds. As cultivation became more sophisticated, tillage operations and equipment were altered and specialized. Powered machinery and cheap fossil fuels brought in the age of "maximum" tillage—loose, fine seedbeds; weedless, trashless fields; and extensive mixing of soils, lime, plant residues, and sometimes fertilizers.

Now it is recognized that this extensive tillage often is not good. Compaction of soils, the need for soil conservation, and the increased costs of energy and labor have led to a reappraisal of our concepts about tillage.

14:3.1 The Purposes of Tillage

The major purposes for tillage are to prepare an adequate seedbed and to control weeds, as well as to improve aeration, increase water infiltration, make furrows for irrigation, and bury crop residues. The importance of the various reasons for tillage varies with geographic location, soil differences, crops grown, and the climate. A detailed list of reasons for tillage are given in the following paragraphs:

Seedbed Preparation Tillage to prepare the seedbed is most needed when planting small and/or high-cost seed (lettuce, tomatoes, sugar beets, alfalfa, clovers). It is less critical for vigorous large-seeded plants (corn, small grains, soybeans, dry beans, and sorghum). The soil-covering depth for a seed is about 2–5 times the seed diameter for large seeds and 1–3 times its length for a small seed. A planted seed needs to be firmly set into moist soil, where it can get adequate water to germinate. The soil must be loose or aggregated enough to allow good root penetration and to not form a thick, hard crust that hinders seedling emergence.

Weed Control Once crop seeds germinate, weed control is essential. Weeds compete with the crop for water, nutrients, and light. Until herbicides became available in the 1940s, tillage for weed control was an integral part of crop production. Weeds were plowed under in land preparation and killed by cultivation once or several times during the early stages of the crop's growth. But cultivation was only partially effective; many weeds were in the plant row, where they were not removed except by hand; also, the cultivator always cuts some shallow crop roots. Herbicides are more effective than tillage and eliminate extra trips over the field. Tillage for weed control is still effective and sometimes is used. Certain crop systems need to control weeds other than with herbicides. Crop rotation is one solution. Another solution recommends a rotation in tillage (deep-plow every few years; less tillage other years) as an improvement in management.[2] Throughout the world, particularly in developing countries, weed control by tillage and by hand are still common; often the area cultivated is limited by the ability to control weeds.

Incorporation of Surface Trash The "clean" field has been the "ideal" for many land users. Burial of plant and weed residues and seeds has reduced plugging of planters and has reduced weed and insect infestations.

---

[2] James V. Parochetti, "Rotate Tillage Instead of Crops," *Crops and Soils Magazine* (Oct. 1980), pp. 8–9.

Loosening the Soil   Equipment compacts soil, often developing pans, which may reduce or inhibit both root and water penetration. The heavier the machinery and the finer and more moist the soil, the greater will be the compaction problem. In one soil in the southern Great Plains, 35 years of cropping to grain sorghums reduced infiltration rates from 23 cm/h above the compaction zone to 0.5 cm/h in the compacted area; this was a 46-fold greater infiltration in the noncompacted soil.

Corrective tillage may be useful on several types of *natural* soil pans. These **soil pans** (hard, compacted, or impermeable layers) often develop from natural soil processes or parent material stratification. These include **claypans, siltpans, fragipans,** and cemented horizons. The latter pans were formed, mostly in the geologic past, as soluble cements, including carbonates, sulfates, silica, and iron and aluminum forms, moved downward and precipitated in layers of soil. These pans can frequently be ripped and opened by tillage, allowing water and root penetration. Deep chiseling is an energy-expensive operation but may be justified by the greater production that results (see Fig. 14-7).

In Clovis, California, a 2.1-m (7-ft)-shank chisel was used to break deep silica-cemented hardpans common in the San Joaquin soil series. The 2.1-m ripper requires four D-9 caterpillar tractors to pull and push it along (see Fig. 14-8).

Shaping the Soil   Surface flow irrigation or ridges to allow drainage of excess water are often essential. About 70% of irrigated land is surface irrigated, and most of this is by furrow irrigation. Tillage is needed to form the ridges and furrows and to smooth the land surface to avoid "hills and valleys" in the field, over which surface flow irrigation would otherwise be difficult or impossible.

FIGURE 14-7   Tillage pans can be broken or shattered by some type of chisel or sweep set at a depth of about 30 cm (1 ft). Note that the greatest compaction occurs in this soil at and just below the plow depth.   (Courtesy of Caterpillar Tractor Co.)

FIGURE 14-8 A 2.1-m (7-ft.) tall ripper chisel used near Clovis, California, to rip deep hardpans in the San Joaquin soil series. The large ripper requires enormous power (four D-9 Caterpillar tractors) to pull it through such deep and hardened soil masses. Note the man standing to the right of the chisel. Smaller rippers are used for shallower depths to increase root and water penetration. (Courtesy USDA— Soil Conservation Service; photo by G. Kennedy.)

Incorporation of Lime and Fertilizers Lime, to be most efficient, needs to be mixed with the soil. If it is added on the soil surface, it will move down slowly; the soil acidity at deeper depths will be corrected slowly. The "over-liming" at the soil surface could also favor some fungi diseases. Many management schemes involve broadcast followed by incorporation of fertilizer additions. All of these require tillage for incorporation.

Control of Insects and Disease The timely plowing under of crop residues is an effective means of controlling certain insects. The Hessian fly, which is a serious infestation in many wheat fields, can be controlled by plowing under infested wheat stubble and volunteer wheat. The wheat jointworm is held in check by destroying all volunteer grain. Plowing under corn stalks reduces the next year's crop of European corn borers. Timely cultivation aids in reducing grasshopper infestations by drying out their eggs. The cotton boll weevil and the pink boll worm in cotton are held in check by the early destruction and plowing under of cotton stalks.

Improving Water Relations Tillage destroys soil crusts, loosens soil, and generally increases water intake when water is next applied after tillage. Tillage should be minimal and the land left as rough as feasible to still have good seedling establishment. Crop residues help keep cultivated soil porous. In Minnesota, when residues were partially incorporated to 15 cm (6 in.) with a chisel cultivator in the fall, they provided eight times more infiltration before runoff started and four times more infiltration during runoff than occurred on spring-plowed, -disked, and -harrowed land.

The higher water storage in the root zone resulting from deep tillage permits irrigation before planting and fewer irrigations during crop growth. Tillage on a clay loam to a depth of 16 in. (41 cm) was still beneficial after 3 years.

When no crop is grown for a year for the purposes of increasing the total soil-stored water, it is called **fallow.** Weeds and volunteer growth are controlled by two to four weed-control tillage passes (cutting blades slide along below the soil surface) during fallow. Water storage is often low, but the storage of even an inch or two of water is essential for increasing yields in areas of limited rainfall. In the northern Great Plains in the United States 1 in. (2.5 cm) of water produces, on the average, an extra 2.4 bushels per acre of wheat (161 kg/ha). A gain of 2–6 in. (5–15 cm) of soil water can increase grain yields appreciably. Water gains from fallow vary considerably. Except in low-rainfall areas of less than 8–12 in. (20–30 cm), where a profitable yield is impossible without fallow, the value of allowing land to lie idle a full crop year is now being questioned. Even if water buildup is low, a fallow year is sometimes necessary to save enough moisture in the *fall* to germinate the seed.

Aerating Soil and Speeding Warming   Compacted soil may have inadequate air exchange. This lowered aeration can be growth limiting in some clayey soils, but it is unlikely to be a problem in sandy soils. Any soil cover reduces the amount of heat radiation reaching the soil. This, plus keeping the soil wetter, slows warming of the soil surface in a cool spring. Tillage that exposes soil to the sun, especially as *drained* ridges, will speed warming of the soil (where seed will be planted).

Ease of Planting   Anyone who has attempted to operate a planter in soil covered by bulky plant residues has experienced the frustration of constantly removing accumulated trash from the equipment and trying to achieve uniform planting depths. Tillage to incorporate portions of the land's residue greatly facilitates planting. For example, if a no-till planter goes over a pile of residue 7–10 cm (2.7–4 in.) thick, it is possible for the seed to be left up in the residue (not the soil), where the seed dries out and dies. There will be no crop there.

14:3.2 Reasons for Reduced Tillage

The reasons for going to reduced-tillage systems are primarily to reduce erosion of soil, save time in operations, and save on costs of fuel. Fuel costs are about 35% of the energy used in production with conventional tillage. Fertilizers are about 20–25%, machinery is about 20%, and other chemicals are about 5–8% of production energy costs. Reducing tillage operations should reduce costs for pesticides and nitrogen fertilizers. The total reduction in tillage costs with reduced tillage are about 5–10% lower than with conventional moldboard or disk-plow tillage.

Erosion of soil has finally been recognized by decision makers as the bankrupting policy of soil management that it is. Topsoil must be kept in place. Pressure from national and state soil conservation groups and controls by the U.S. Environmental Protection Agency (EPA) to prevent erosion of soil sediments into waters have encouraged the use of reduced tillage. Reduced tillage lessens erosion. The U.S. Department of Agriculture started in 1988 to implement a plan to allow government payments only to those farmers (1) who have a farm plan to limit soil erosion and (2) who follow the plan to reduce erosion so that it does not exceed the rate of soil formation. This allowable erosion is known as the "T" (tolerance) rate.

Other benefits or factors contributing to the increased use of reduced tillage are the following:

1. More area can be planted in a short time; important if weather is bad
2. No-tillage soil dries less than with conventional tillage
3. Double-cropping (two crops per growing year) is possible with no-till by reducing time needed for seedbed preparation in many temperate areas usually producing one crop per year
4. Steeper land can be farmed because of better erosion control
5. Usually more water is stored in the root zone
6. Equipment requirements are lower
7. Chemicals are now available for weed and disease control
8. The lower costs permit the farming of marginal lands
9. Less tillage reduces soil compaction

As the world's population mushrooms, it would seem logical that food products would be used entirely for food but other needs already compete for some crop land and food products. For example, producing cotton uses land that could grow food, and some corn is used to produce ethyl alcohol, which in turn is used as a petroleum substitute in some gasoline. The diminishing oil reserves will cause an increased use of ethyl alcohol in fuel.

Now another food staple—wheat—is being considered for plastics. Cornstarch is already being used for plastics, but the cornstarch mol-

ecule is large and limits the film thinness that is possible. Wheat starch comes in two sizes. The smaller wheat starch diameter (6 microns) is less than half the size of cornstarch (15 microns). Plastic made from these starches is biodegradable and thus very desirable.

One of the problems with using wheat starch is sorting the two sizes of starch grains. One method employs varied air-stream velocities. Another method uses a liquid cyclone, much like the old-time cream separators, which used centripetal force to separate the heavier cream particles form the lighter milk particles.

Adapted from: Anonymous, *Agricultural Research,* **32** (1990), no. 6, p. 15.

14:3.3 The Goal for the Future

The goal of agriculture is to develop a *permanent agriculture*. That goal has not changed over the years, but the concepts we have to accomplish it may now be somewhat different from what they were previously. For the foreseeable future, agriculture must supply the food and much of the fiber to an increasing world population (Detail 14-1). This production should be as economical as possible, for both the producers and the consumers. While doing this, the soil must not be permitted to be further degraded or destroyed. To accomplish this goal, the following important actions are needed:

1. Reduce erosion to low, replaceable rates of loss. This will maintain what soil we have, even perhaps improve it.

2. Minimize the costs of production by lowering energy consumption: Use less fuel and less equipment, use chemicals at their highest efficiency but lowest essential amounts.

3. Select management systems that do not cause land degradation or uncontrollable ecological upsets.

4. Select management systems that will accomplish items 1, 2, and 3 but will still be highly productive.

14:4 Conventional Versus No-till Farming

There are two objectives in evaluating conventional versus no-till systems. The first objective is to minimize as many of the disadvantages of each system as possible, while taking advantage of the benefits of each. In most instances, any system selected for a crop, soil, and climate will result in some *intermediate compromise—a reduced tillage system of some kind.* The second objective is to recognize the reasons for and benefits of certain practices so that better equipment can be designed, certain practices can be modified, and cost-benefit concerns can be included in management plans.

No-till farming is the practice of doing no more tillage than opening a slit or very narrow strip of soil just enough to plant the seed (Fig. 14-9). The entire field is not plowed or

(a)

(b)

FIGURE 14-9 (a) No-till seeding of corn in Indiana into a wheat cover crop grown to protect the soil and retain leachable nutrients. A fluted coulter cuts a narrow strip of soil, which is the total seedbed preparation. Fertilizer is banded in the same operation. (b) Soybeans growing in grain stubble from no-till planting. (*Source:* USDA—Soil Conservation Service; photos by Bob Steele and J. B. McDonald.)

disked. It also means that surface plant residues remain standing, lime is not incorporated into the soil, and the only fertilizer incorporated is by banding it during planting.

The reasons for going to reduced tillage are also its advantages. These are primarily (1) control of erosion, (2) reduced costs in energy and labor, (3) less time needed for land preparation and planting, (4) the ability to use steeper land with less hazard of erosion, and (5) the possibility of, or the greater ease in, double-cropping.

No-till and reduced tillage have numerous disadvantages. Some disadvantages of no-till and reduced tillage are given in the following list:

1. *Reduced tillage requires better management,* particularly more advanced planning. More planning regarding timing, amounts and types of weed control chemicals, lime addition, and fertilization methods are essential. One successful farmer said, "I now use my head more than my rump . . . I spend more time thinking about what I'm doing than sitting on a tractor seat."[3]

2. Minimal seedbed preparation techniques adapt well to large-seeded crops, such as corn and soybeans, but not to fine-seeded ones. Tomatoes and peppers may have lower yields and plant survival than with conventional tillage.

3. Less plant residues are incorporated into the soil, which means less "active" soil humus. However, the organic matter *content* of no-till soils may stay higher than with conventional tillage because of slower organic-matter decomposition.

---

[3] J. Howard Turner and Robert J. Rice, eds., *Fundamentals of No-Till Farming,* American Association for Vocational Instructional Materials and Ortho, Chevron Chemical Co., San Francisco, Calif., 1983, p. 42.

4. Soluble fertilizer salt residues will concentrate more in the top 10 cm than where plowing is done.

5. Weed problems are accentuated because of increased water and nutrient content at the soil surface and more weed seed production, unless herbicides are constantly employed. The amounts of herbicides used must be large enough to kill the most resistant weed species because there will be no control by cultivation.

6. Herbicide buildup and residual time in soil is usually greater and can cause damage to sensitive, subsequent crops.

7. No-till plant roots are shallower.

8. Plant emergence for most crops is reduced, and stand density is decreased unless extra seed is added. (No-till requires 10–15% more seed than used in conventional tillage planting.)

9. Plant residues from previous seasons on top of the soil keep soil wetter and cooler during spring germination periods. In Minnesota early growth was only 60% of the normal (no mulch) because of the 6.7-Mg/ha (3-t/a) crop residue cover that existed with conservation or no-till.

10. Pest damage from birds, field mice, and insects is greater.

11. Reduced tillage may favor crop diseases, especially fungal diseases, by leaving moisture-holding residues on the soil surface in an organic material easily attacked by fungi.

12. Stronger, heavier planters are usually required to rip and plant in a nontilled soil or even in a recently growing plant cover (Fig. 14-10).

FIGURE 14-10 This heavy no-till drill is seeding dryland wheat on the Clarkston watershed in northern Utah. This "Yielder-type" research model drill was built by Yielder Drill, Inc., of Spokane, Washington. It weighs about 8864 kg (19,500 lb), nearly 10 times more than conventional drills of similar width. It has large dual disks to allow deep placement of fertilizer up to 20 cm (8 in.) deep. It can also surface band pesticides. (Courtesy of V. P. Rasmussen, Utah State University.)

The question for each land user becomes one of adapting some form of reduced tillage, even no-till, to their soils, to their preferences, to their abilities, to their climate, and to their cropping interests.

14:5 Understanding Reduced Tillage

Nature has no provision to till the soil, except slowly by insects and other macroorganisms. The natural vegetation of the dense forests and grasslands do not require tillage. People initiated tillage because they wanted to grow large areas of a single cultivar, to produce blemish-free foods, and to produce maximum yields. People also wanted to irrigate dry lands and control weeds. Tillage is one way people have tried to improve on nature's methods. Although there are some benefits from tillage, there are also some damages. Those facts bring up many questions: How much tillage is good? What happens if tillage is done or not done? The following sections provide some insights into the benefits and hazards of tillage.

14:5.1 The Soil's Physical Properties

Tillage loosens soil; no-till allows soil consolidation and gradual compaction during years of wetting and drying. The no-till soil often has a greater number of small pores, and soil protection and aggregation are better at the surface because of the crop residues there.[4]

Less tillage favors more earthworm activities and burrowing-mixing actions by other soil macroanimals; these activities improve the soil by increasing its water infiltration rate.[5] Freeze-thaw cycles over the winter ameliorate some surface compaction, but compaction may gradually develop under reduced tillage. Periodic chisel plowing or ripping in the top 30–40 cm (12–15.5 in.) of reduced tillage may be beneficial to the soil (Fig. 14-11).

The inclusion of deep-rooted crops in a rotation will help soil permeability as old roots decompose, leaving vertical channels.

14:5.2 Weed Control [6,7]

Weed control is a major problem. Weeds include any unwanted plants, among them lost seeds from the previous crop and volunteer grain. If farmers do not control weeds, their crops fail. Reduced tillage puts the burden of weed control on herbicides.

The importance given to weed control by farmers is illustrated by a 1983 survey in which the three most important reported reasons for not wanting to use conservation tillage were (1) inadequate weed control, (2) herbicide costs, and (3) a lack of proper equipment. In a 1984 survey in Indiana the reporting farmers claimed they had *satisfactory weed control* 87% of the time with conventional tillage, 65% of the time using reduced tillage, and only 42% of the time using no-till systems. These surveys emphasize the importance of weed control. Even more critical is that the increased costs of herbicides may eat up any savings on fuel and labor. One survey on corn in 17 states reported that herbicides in no-till cost 34% more than in conventional tillage. Also, the high organic matter left on the surface absorbs some herbicide and requires that more be added to get to the weeds (Fig. 14-12).

[4] A Systems Approach to Conservation Tillage, pp. 100–105
[5] Ronald E. Phillips and Shirley H. Phillips, *No-Tillage Agriculture,* Van Nostrand Reinhold, New York, 1984, pp. 206–215.
[6] Ibid., pp. 152–170.
[7] William C. Koskinen and Chester G. McWhorter, "Weed Control in Conservation Tillage," *Journal of Soil and Water Conservation,* **41** (1986), pp. 365–370.

(a)   (b)

FIGURE 14-11 Compaction of soils into pans hinders water and root penetration. Tillage is needed to open paths through the pan. Photo (a) shows increased growth in corn as a result of ribbon tillage (narrow but deep tillage, right) compared to no-till (left). Ribbon tillage is narrow but deep strip tillage. The plow pans and other shallow pans can be broken by chisels or deep fluted coulters. Photo (b) shows soybeans.   (Courtesy of Albert C. Trouse, Jr., Tilth International, Auburn, Ala.)

With reduced tillage, control of many increased populations of perennial weeds will require increased herbicide use. Large populations of rhizomatous grasses (quackgrass, Johnson grass, Bermuda grass), nutsedge, and hemp dogbane may develop in reduced-tillage systems. In contrast, large-seeded broadleaf weeds (cocklebur, velvetleaf) usually are less of a problem and require less herbicide.

Some recommended practices to aid in weed control when going to a reduced tillage system are the following:

1. Employ a judicious use of cultivation if it is feasible
2. Spray lower rates of herbicide application on young weeds, which is often cheaper than the heavier application needs of preemergence herbicides
3. Calibrate the spray equipment frequently to minimize excess applications
4. Consider fallow as a possible system in marginal-rainfall dryland grain areas
5. Rotate tillage level to an *increased* tillage for a year after each few years of no-till or reduced tillage

Control of weeds is critical. If cultivation is necessary to control weeds, it had better be done.

FIGURE 14-12 No-tillage (zero tillage) is shown with soybeans being planted in last year's corn field with no prior tillage. The large amount of residues would absorb large amounts of any preemergence herbicide, increasing the amounts needed to accomplish good weed control. Often, postemergence, rather than preemergence, herbicides are applied as a more economical method of weed control. (*Source:* USDA—Soil Conservation Service.)

14:5.3 Insect Control [8,9,10]

Pest problems, both soil insects and above-ground insects, will increase in reduced-tillage systems. Army ants, armyworms, corn borer, and many other insects will increase rapidly when not disturbed by tillage. Fortunately, better pesticides are being developed for some of these. Plant residues keep the soil surface moist and make conditions favorable for slugs, deer-mice, weevils, and many other pests. Some of these are only minor problems under conventional tillage. Also, birds and rodents may become more severe pests because of seeds left on the soil surface. Rotation crops must be carefully selected. Many insects that damage corn, for example, are commonly associated with many grass weeds or ryegrass crops. Some of these pests are the armyworm, lined stalk borer, hop vine borer, and potato stem borer.

---

[8] Philips and Philips, *No-Tillage Agriculture,* pp. 171–189.
[9] D'Itri, *A Systems Approach to Conservation Tillage,* pp. 137–143.
[10] Turner and Rice, *Fundamentals of No-Till Farming,* pp. 97–108.

Soil insects are more difficult to control than above-ground plant pests. In one survey in Ohio in no-till corn, the black cutworm had attacked 15% of the plants in no-till but only 1% of the plants in conventional tillage.

Although insect control in reduced tillage is considered to be manageable, it does require more attention to frequent monitoring of the fields. Some suggestions for aiding in insect and disease control when using reduced tillage are the following:

1. Select varieties having quick emergence, tolerance to cold, disease-resistant varieties, ability to perform well under high population, etc.
2. Rotate crops to remove hosts for certain problem insects
3. Monitor the crop frequently and carefully for evidence of approaching problems (eggs, young hatches)
4. Carefully select the pesticides that are proven effective, and be prompt but correct in their use and timing

14:5.4 Fertilizer and Lime Applications[11,12]

It is desirable to place phosphorus and potassium into the soil (band or broadcast and plow in) because they are not very mobile in soil. Lime is mixed as intimately with the soil as is practical because it is not very soluble or mobile. Using no-till limits the preferred methods of addition that can be selected. Nitrogen in the nitrate form is quite mobile in the soil but, volatilization losses of both ammonia gas from surface-applied urea and gases from denitrification in overly wet soil may be greater than in conventional tillage.

In soils where reduced tillage is used, most of the lime and fertilizers can be concentrated near the surface. Recently, greater effort has been made to band fertilizer a few centimeters into the soil. The shaded soil surface and higher soil-water content in reduced tillage causes higher root densities to develop in the soil's top 5 cm (2 in.). As much as 10 times more roots were measured in no-till soil surfaces in climates where conventional tillage allowed extensive drying. Nitrification in this shallow soil layer can strongly acidify the soil surface. All of these factors suggest that yields could be reduced if short droughts occurred. Dryness would hinder the mass of roots in the shallow soil from absorbing nutrients.

Particularly, reduced tillage may affect phosphorus uptake. First, the surface residues keep the surface soil cooler and wetter. If the temperature is lower than about 10°C (50°F) for corn, phosphorus uptake will be reduced. Second, the higher water content in reduced-tillage soils will *increase* phosphorus uptake by increasing the amount of diffusion, the main mechanism for the movement of phosphorus to roots. Third, fertilizer placement in bands usually prolongs availability. Surface application might be considered a *horizontal surface band,* but in a lower-accessibility location for roots, unless the soil surface is shaded and kept moist.

Recommendations for fertilization and liming in reduced-tillage systems will probably include most of the following items:

1. If persistent drought is likely, any economical method to place fertilizer deeper than the top 4–5 cm will probably help.

2. With adequate irrigation or rainfall, surface applications of both lime and fertilizers, including phosphorus and potassium, seem to be about as effective as banding or broadcast-plus-incorporation.

---

[11] Philips and Philips, *No-Tillage Agriculture,* pp. 87–126.
[12] D'Itri, *A Systems Approach to Conservation Tillage,* pp. 89–98.

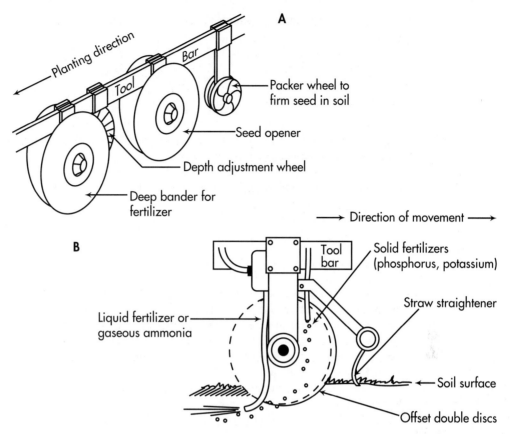

FIGURE 14-13 Diagram (a) of a combined fertilizer-planter using double discs to cut through the surface mulch, band fertilizer, plant seed, and firm the soil onto the seed. Diagram (b) shows how liquid or solid fertilizers, or both, can be added with a double-disc assembly. Notice the straw straightener to minimize plugging by residues. A narrow rotary plow could be attached in front of these discs to "strip till." A planter double-disc could also be added (Diagram a). (*Source:* Modified by Raymond W. Miller from "Go for Yield," a publication of YielderTM (manufacturer of a no-till drill), S. 4305 University Road, Spokane, Wash. 99206.)

3. Lesser fertilizer efficiency may be observed in these conditions: (a) for phosphorus, if the temperature is less than about 15°C (59°F), (b) for potassium, if the soil is compacted, wet, and cold, and (c) for nitrogen, if conditions for ammonia loss (urea on basic soils, gaseous ammonia knifed into wet soils but the chisel slits do not close) and leaching losses (sandy, high rainfall) are favorable.

4. Surface lime application seems to be more effective than first anticipated, affecting pH even to depths of 30 cm (1 ft).

5. With *reduced tillage,* rather than no-till, the opportunity exists to knife-in fertilizer bands or to do some tillage every several years (Fig. 14-13). This may help to move plant nutrients and lime deeper into the soil. On a long-term basis, fertilizing and liming in reduced-tillage systems may not be as troublesome as initially expected.

6. Use crop rotations for long-term fertility in reduced-tillage systems. Legumes will distribute nitrogen in the soil profile. Winter annuals (small grains) or legumes usually provide the winter cover crop to protect the soil from erosion.

14:5.5 Problems from Allelopathy [13]

Reduced tillage, especially no-till, provides conditions favoring the production of organic toxins. Layers of residues on the soil remain wet, which can produce allelopathic materials. Wheat residues are well known to produce some toxins that reduce growth of the following wheat crop during early growth stages. The cause of toxin production is the wetness of heavy residues, even anaerobic conditions in some thick accumulations. Corn residues can be allelopathic to the following crop of corn, too. Clearing residues from near the seed (removing most residue from the immediate seed row) eliminates the major allelopathic effects. Most problems are in early growth stages.

14:5.6 Problems of Special Crops [14]

Certain crops pose special problems for reduced-tillage systems. Sugar beets must be planted at precise depths because they are small seeds: potato planters will need special adaptation. Harvesting root crops cause some tillage as underground parts are dug. Also the soil compaction or aeration can affect the shape and quality of the root or tuber. In no-till, sugar beets have an increased branching of roots. With potatoes, tubers expand; without "hilling-over," the parts that are exposed to the sun turn green and are toxic to humans. In conventional tillage, these tubers would be covered by tillage (hilling). With both cotton and tobacco, reduced tillage has made weed control marginal and difficult. Many other crops might have unique difficulties, such as fungal diseases on the wet plant residues when growing cucumbers and strawberries. Each of these problems must be viewed as unique, and compromises must be made accordingly (Fig. 13-12).

The more drastic the rotation, the greater will be the impact on weeds. This concept also applies somewhat to other pests: insects, diseases, and macro-animalia. The selection of plants in the rotation is a very important choice to be made.

14:5.7 Preparation of Soil for Irrigation

Reduced tillage is difficult in arid climates where surface irrigation is planned (Fig. 14-14). Although sprinkler and drip irrigation are possible with reduced tillage, surface flow in furrows is nearly impossible. The surface residues plug the furrows so that the water erodes the ridges. Also, furrows must be remade or rebuilt for each crop. "Border irrigation" could be used, which allows wide, uniformly graded flow runs. Plant residues would still hinder flow rates down the field, causing excess deep percolation at the upper ends of the fields.

Because these arid areas are less likely to erode (they have few heavy rains), reduced tillage and greater residue incorporation into the soil in the spring prior to planting have a low erosion hazard, unless the slope is great. Economics, local climate, and crop choices will determine whether using sprinkler irrigation or using soil tillage to allow furrow preparation will be an acceptable choice for a particular area.

14:5.8 Sandy and Organic Soils and Wind Erosion

Soils subject to serious wind erosion need careful management to maintain a protective residue cover. In areas with windy periods, sandy and organic soils have the most severe erosion hazard (Fig 14-15). Erosion by wind is increased by clean tillage; the protective organic mulch is often removed by tillage. Where surface flow irrigation is not used, the most effec-

---

[13] Jane Paul, "Resisting Residue?" *Agrichemical Age,* **31** (1987), no. 10, pp. 6–8.

[14] M. A. Sprague and G. B. Triplett, No-Tillage and Surface Tillage Agriculture: The Tillage Revolution, Wiley-Interscience, (1986), pp. 149–182.

FIGURE 14-14 Furrow irrigation of carrots with siphon tubes (spiles) in the Lompoc Valley, California. The need to make ridges and furrows encourages clean tillage as a convenience. Some row crops (sugar beets, onions, carrots, celery, tomatoes, and others) are difficult to plant, and furrows are hard to make in trash-covered soils. Fortunately, many areas needing furrow irrigation have little soil erosion hazard from rain during the cropping season. Alternatives to ridge furrow needs are to use drip and/or sprinkler systems, both of which are more expensive. (*Source:* USDA—Soil Conservation Service.)

tive tillage involves zone tillage. **Zone tillage** is tilling a strip of the land only 15–25 cm (6–10 in.) wide at normal row spacings (70–100 cm or 27–39 in., apart). The crop is planted and band-fertilized in the narrow tilled strip.

The technique is successful for many crops, including potatoes and corn.[15] In organic soils (mucks) zone tillage is effective for onions and carrots, particularly because their small seeds require a good seedbed for successful planting. The registration of good postemergence herbicides for use on vegetables has made it possible to grow rye and barley as winter cover crops. These crops are killed by sprays just before the zone tillage and planting of tomatoes is done. The cover crops can be used without tillage for asparagus; the cover crop is killed in the asparagus strip by postemergence herbicides.

14:6 Tillage and the Environment[16]

Reduced tillage will reduce soil sediments that move into the air (dust) and waters. Thus, reduced tillage seems desirable. However, reduced tillage also increases the amounts of pesticides applied. There is a tendency to use the more persistent types for better long-term weed control; these are more likely to be persistent long enough to move into waters. It is typical

---

[15] Judy Ferguson, "Zone Tillage," *Ag Consultant*, **45** (1989), no. 2, p. 17.
[16] D'Itri, *A Systems Approach to Conservation Tillage*, pp. 299–313.

(a)

(b)

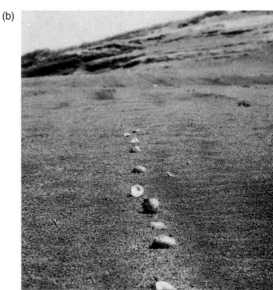

FIGURE 14-15 These sand dunes (a) near Pasco, Washington, appear to be anything but inviting as profitable farmland. In (b) potatoes are uncovered by wind erosion of the sandy soil. However, 5 years after enough leveling (some 3-m, 10-ft, cuts) to permit the use of center-pivot sprinklers, potatoes were producing 60,480 kg/ha (54,000 lb/a), wheat yielded 5376 kg/ha (80 bu/a), and alfalfa produced 17.9 Mg/ha = megagrams (8t/a). (*Source:* Jerry Schleicher, "Sand Dune Potatoes," *Irrigation Age,* **9** [Sept. 1975], no. 9, pp. 25–26.)

to require 15–40% more pesticides in reduced tillage compared to needs in conventional tillage. Higher pesticide additions increase solubilized pesticides. Any eroded soil sediment has higher amounts of both pesticides and phosphorus adsorbed to it. The reduction in fossil fuels burned while using reduced tillage will probably be insignificant in its effect on the environmental air pollution. Crop rotations would reduce the amounts of pesticides required and would be a logical consideration to help reduce environmental pollution from croplands.

14:7 What Tillage Is Best?

To suggest that there is a "best" tillage is to ignore that there are special needs of each climate, each soil, each crop, each rotation, each social and economic society, and each farmer's preferences. Each system will be tailored to its own needs and peculiarities. The following items are a partial summary of the considerations to ponder when planning a tillage system for a field or a farm:

1. If soil erosion is not a problem, serious thought should be given to the weed, insect, and disease problems before converting to reduced tillage or no-till. These items are easier to control using conventional tillage.

2. Use reduced tillage or no-till if soil erosion control is a critical requirement. Reduced tillage is much more adaptable to more systems than is no-till. It allows for some tillage, such as an overall cultivation or cultivation of narrow strips.

   Some of the unique reduced-tillage suggestions for different situations include:

   - Use the paraplow to loosen soil, but avoid inverting the soil.
   - Use zone tillage to prepare a good seedbed, but retain a protective cover on most of the soil
   - Use ridge-till to avoid water-logging and to warm the seeding zone earlier (Fig. 14-16)

3. Reduced tillage requires better management skills and planning than does conventional tillage. Reduced tillage will reduce soil preparation and planting time, *but it may not reduce total costs*. If you are using reduced tillage primarily to reduce costs, do some careful bookkeeping to be sure you are ahead economically.

4. Different types and/or toughness of planters and tillers are needed in reduced tillage. Will the changeover be too great a financial burden? Conservation tillage planters must be able to (1) cut and clear residue in the seed row, (2) penetrate the soil uniformly, (3) have uniform predetermined depth control, and (4) have uniform seed placement and density.

5. The least adaptable soils to reduced tillage are clayey soils in cold climates where crop residues cause slow drying and warming. In such areas, the least amount of residue still capable of controlling erosion should be used.

FIGURE 14-16 An example of ridge-till tillage illustrating the sequential operations from (a) early-spring preplant conditions through (b) planting, (c) precultivation, and finally (d) midsummer postcultivation. (*Source:* Modified and redrawn by Raymond W. Miller from "Conservation Systems for Row Crops," University of Nebraska publication EC76-714, as published in "Ridge-Till Row Crop Production," Cooperative Extension Service Publication C-662, Kansas State University [1985].)

(a) **Early spring** after corn stalks are chopped (optional).

(b) **Late spring** after ridge is scalped off by 25- to 35-cm-wide sweeps to lay bare the ridge top that is now about one-third of the row spacing width. Plant the seed, band the fertilizer, and apply pre-emergence herbicide.

(c) **Summer** after considerable residue decomposition and befor cultivation.

(d) **Summer** after the one or two cultivations to control weeds and to rebuild ridges.

For good corn growth, a residue cover of less than 10% is necessary to affect temperature and allelopathic effects. A temperature increase of 1°C has been reported to cause a difference of 20% in growth. Other crops like soybeans are less sensitive to temperature changes. According to one scientist at the University of Minnesota,[17] "you can probably plant right into 30 percent or 40 percent cover with a conventional planter. If you're worried about growth for corn, you want to keep the row area down to 20 percent or less cover." Fluted coulters open a 7–10 cm-wide band relatively clear of residue for the seed row.

6. Preparation of a good seedbed will be needed for some small-seeded crops, such as tomatoes, lettuce, carrots, sugar beets, alfalfa, and clovers. A fine seedbed for small seeds requires a tilled soil, preferable zone (strip) tillage. Removal of a strip of mulch residue along the planting row and careful depth control during planting with paired discs may be developed as a suitable alternative. If one of the paired discs runs slightly in advance of the second one, cutting of the residue is better.

7. Fertilizer and lime additions as strips on the soil surface (horizontal bands) are nearly as effective as are the normal banding and broadcast-incorporation. In cold periods (<15°C, or <59°F), both potassium and phosphorus uptake may be reduced. This reduction may be temporary and may not alter yields appreciably. The greatest nutrient deficiency will probably be in rain-fed crops having a prolonged drought because most nutrients and roots are in the top 5–8 cm (2–3 in.) of soil. On basic soils urea applications may have large losses as ammonia, unless the urea is watered-in with irrigation. Injection of ammonia gas into wet clayey soils may have large losses of ammonia if slits do not close adequately.

8. Crop rotations, although they may appear to produce less income, may be as good or better financially over a long time. Usually, rotations reduce the weed and insect problems, distribute the work load, and allow more options in tillage, fertility, and crop residue management.

9. The most useful tillage operations depend on the most critical needs. Where weeds are a problem, one or two cultivations are very effective. Where drying and warming are problems, light disking to incorporate residues partially and leave some soil barren of cover is critical. Where surface flow furrow irrigation is planned, ridge furrow construction is the most essential tillage operation.

10. For sustained agriculture on the land, reduced tillage should have as its priorities, in order of decreasing importance: (a) erosion control, (b) weed control, (c) pest control, (d) preparation for surface irrigation, (e) minimizing environmental pollution, (f) reducing required land preparation time, and (g) reducing fossil-fuel energy needs (future fuel costs will increase rapidly).

11. In the 1983 survey reported earlier, the three most important reasons farmers gave for opposing *conservation* tillage were (a) inadequate weed control, (b) herbicide costs, and (c) lack of the proper equipment. Do you feel competent to solve these three problems, as the beginning problems? Reduced tillage may be better, but it is not usually easier.

---

[17] Jane Paul, "Resisting Residue?" *Agrochemical Age,* **31** (1987), no. 10, pp. 6–8, quoting extension soil scientist, John Moncrief.

For conservation tillage to work, crop residue must be managed; there lies the problem.

—Jane Paul

So many people who get into conservation tillage try to go out and do no-till as early as the guy who is doing it conventionally. He tries to plant to early, . . . when it's too cool and wet; it's a problem.

—Jim Lake

Questions

1. Describe these tillage implements (a) bedder, (b) chisel, (c) coulter, (d) cultivator, (e) disk, (f) harrow, (g) rotary plow, and (h) stubble mulch "sweep."
2. Define (a) *conservation tillage,* (b) *reduced tillage,* (c) *conventional tillage,* and (d) *minimum tillage.*
3. To what extent is no-till actually zero tillage?
4. Discuss the purposes of tillage.
5. Discuss the reasons for doing reduced tillage.
6. Compare conventional tillage with no-tillage, by evaluating these items: (a) the required ability of the farmer, (b) the ease of weed control, and (c) the problems of lime and fertilizer additions.
7. What advantages does reduced tillage have over no-till systems?
8. Are no-till or reduced-tillage systems more economical than conventional tillage systems? Discuss.
9. With reduced tillage, there are certain to be some kinds of savings (but they are not always dollar savings). What are several of these savings?
10. In what types of situations would savings in labor be particularly important? Explain each.
11. (a) How does reduced tillage affect physical properties of soils? (b) How is soil permeability maintained under reduced tillage?
12. What is probably the major problem that becomes more severe when going from conventional to reduced tillage?
13. What are the recommendations and precautions for weed control in reduced tillage?
14. What recommendations and precautions are given for controlling insects in reduced-tillage systems?
15. (a) How effective are fertilizer and lime on no-till systems? (b) How effective will these be in reduced-tillage systems? Explain.
16. What particular problems are greatly increased when using reduced tillage on clayey soils in cold areas?
17. When considering reduced tillage, what particular problems occur if the land is to be (a) irrigated by surface flow in furrows or (b) planted with small-seeded crops (such as sugar beets and carrots)?
18. How does tillage or the lack of it influence pollution of the environment?
19. To what extent can one define "the best tillage system"? Explain.
20. Give a 50-word summary of why reduced tillage should be used on *most* cropped land.

15

Soil Erosion and Sediment Control

We travel together, passengers on a little spaceship, dependent on its vulnerable resources of air, water, and soil . . . preserved from annihilation only by the care, the work, and the love we give our fragile craft.

—Adlai E. Stevenson

15:1 Preview and Important Facts

PREVIEW

Do we treat our soils like dirt?[1] With the enormous losses of soil from many construction sites, wild lands, and farms, we certainly believe many persons should answer Yes! Soil is not always treated as a treasure and essential *nonrenewable resource,* which, for all intents and purposes, it is for life on earth. Quick money and needed paper and wood cause us to abuse forests, whose soils can then erode. Highways and homes cause more clearing of vegetation from the land, concentrating runoff to wear more gullies and erode steep slopes.

Soil erosion is the removal of soil by water and/or wind. Erosion is slight from soil well covered by dense grasses or forests but is enormous from steep, poorly covered soils that are exposed to heavy rainfall or strong winds. Well-aggregated soils resist erosion but pulverized silts and very fine sands are the most easily eroded.

On an average, each hectare in the United States is losing soil and producing an average annual sediment loss of 6.7 Mg/ha (about 3 t/a). This is an average of about 0.02 inch loss per acre. However, some areas lose much more than this (Fig. 15-1). Croplands are the source of nearly 50% of the sediment and erosion; these lands erode about 30% more than the overall erosion average for all lands. Extending "clean" cultivation to sloping lands and construction have accelerated erosion losses (Fig. 15-2). For example, in the last 100 years more than 40% of the original topsoil has been lost from the dryland, rolling hills of the Palouse loess wheatlands of eastern Washington and adjacent areas in Idaho and Oregon (an erosion rate of 2.5 cm, or 1 in., each 15 years). Nature required about 800 years to form the

---

[1] Boyd Gibbons, "Do We Treat Our Soils Like Dirt?" *National Geographic,* (Sept. 1984), pp. 350–389.

FIGURE 15-1 Erosion by water in 1961 in this Kentucky land was ignored until it became severe. It was probably further ignored because of the high cost to control the chaotic condition. Government financial aid plus regulatory laws to control erosion are needed to minimize this kind of waste of our soil resources. (Courtesy of USDA—Soil Conservation Service.)

FIGURE 15-2 Construction areas, such as this highway, often have limited vegetative cover. Steep slopes easily slip and/or erode. One storm saturated this West Virginia clayey soil and caused the slip. The repair cost in 1970 was $418,000. Prevention may have been possible by (1) a better routing of the highway, (2) better water disposal, and/or (3) more deeply rooted vegetation. (Courtesy USDA—Soil Conservation Service.)

erosion rate of 2.5 cm, or 1 in., each 15 years). Nature required about 800 years to form the amount of topsoil lost in only about 15 years in this situation.

Several calculations have emphasized the serious cost in soil for the amount of food harvested:

1. An average of 545 kg of topsoil was lost in the Palouse wheatlands of eastern Washington for every 27 kg (1 bushel) of wheat produced. This is a ratio of 20 kg of soil lost per kilogram of wheat produced.

2. In the cornbelt, a bushel of corn (25 kg) was produced for each 1.5 "bushel of soil" eroded.[2]

Such high soil losses are long-term bankruptcy policies of soil management.

It is impossible to stop all erosion and sedimentation, but both of these problems can be greatly reduced. It will not always be easy. Since 1935 over $30 billion has been put into erosion control efforts, yet erosion is still a major agronomic and economic loss.

In 1972 the U.S. Congress passed P.L. 92-500, the Federal Water Pollution Control Act (FWPCA). Attention was focused on eroded soil as an extensive water pollutant. The Clean Water Act of 1977 amended Section 208 of the PWPCA; this amendment required states to develop plans to control "non-point sources" of pollution, such as eroded soil coming off many lands.

In 1981 a renewed emphasis to control erosion was begun. Money and expertise were "targeted" on the nation's areas having the most severe erosion. By 1986 some improvements were already evident.[3] Associated with this renewed effort were increasingly strict environmental laws directed against pollution of waters. Numerous states have also passed laws against persons permitting "excess"erosion. The control of soil erosion is past due, but it is better now than later.

IMPORTANT FACTS TO KNOW

1. (a) The relative erosion under different climates
   (b) The permissible rates that allow equal rates of soil formation to replace the soil lost
2. The high quality of the eroded topsoil lost
3. The general conditions and actions that increase erosion by water
4. The importance of soil cover in reducing erosion
5. The Universal Soil Loss Equation (USLE) and the relative importance of each factor
6. The characteristics of a soil that alter its erodibility (the K factor)
7. The importance, in controlling erosion, of both (a) the kind of cover and (b) the amount of cover
8. The Revised Universal Soil Loss Equation (RUSLE) and how it differs from the USLE.
9. The Erosion Tolerance Equation and the approximate allowable erosion usually established for the T value
10. The methods used to reduce detachment processes
11. The methods used to reduce soil transport processes
12. The factors affecting soil erosion by wind (those factors in the equation)

---

[2] Nani G. Bhowmik, et al., "Conceptual Models of Erosion and Sedimentation in Illinois: Vol. 1. Project Summary," Illinois Scientific Surveys Joint Report 1, Illinois State Geological Survey and Illinois Natural History Survey, Champaign, 1984.
[3] James Nielson, "Conservation Targeting: Success or Failure?" *Journal of Soil and Water Conservation,* **41** (Mar.-Apr. 1986), no. 2.

13. The particle sizes most likely to be eroded by wind
14. The various ways to control soil erosion by wind

15:2 Nature of Water Erosion

Natural or **geologic soil erosion** does not occur at a constant or consistent rate. Semiarid and arid soils, which lack protective plant covers, may erode naturally at rates averaging 10–50 times greater than those for humid-climate soils. In Indiana, where rainfall is adequate to support extensive plant ground covers, soil erosion loss averages about 1 metric ton per hectare (0.5 t/a) annually, a rate that does not seriously deplete soil productivity and that is about the same erosion rate as that for undisturbed or well-managed forest lands. This rate of erosion is a loss of 2.5 cm (1 in.) of surface soil every 300 years. These losses are compensated by new soil formation, which for Indiana glacial till is about equal to that lost from natural erosion. New soil forming from bedrock will take more than 10 times longer, so natural erosion of soils forming on bedrock is a more serious loss than from deep sediment. This is also true for soils in cold and arid climates, as in Alaska, because both conditions retard soil formation.

The natural progress of soil erosion can be increased horrendously by human activities, such as overcultivating depleted soils until the protective ground covers are gone and accelerated erosion takes place. Severe erosion exceeds 200 Mg/ha/yr (about 90 t/a/yr).

Soil erosion, by whatever cause, destroys human-made structures, fills reservoirs, lakes, and rivers with wasted soil sediment, and badly damages the land. Whether it is called mud, silt, or sediment, it is all soil material that should have been kept in place: on top of the land, where it can support plant growth, and plants can in turn stabilize the soil. Erosion sediment is the richest part of the soil, the nutritive topsoil containing most of the organic matter. The cost of dredging the several billion tons of sediment from rivers and harbors *each year* is about 15 times more than the cost of holding the soil on the land from where it eroded. More than 12,300 hectare-meters (1 million acre-feet) of sediment settles annually in reservoirs, reducing water storage by the same volume (Fig. 15-3). The water that cannot be stored because of lowered reservoir capacity due to silting could irrigate 100,000 hectares (approximately 250,000 acres) of alfalfa in the dry areas of the western United States.

15:3 Causes of Water Erosion of Soil

Water erosion of soil starts when raindrops strike bare soil peds and clods, separating particles and causing the finer particles to move with the flowing water as suspended sediments. This soupy, muddy water moves downhill, scouring channels along the way. Each subsequent rain erodes additional amounts of soil until erosion has transformed an area into gullies, rills, and eroded land that often has reduced productivity.

People bare the soil when they remove protective plant covers by plowing, cultivating, burning crop residues, overgrazing ranges and pastures, overcutting forests, and causing drastic soil disturbance by using heavy machinery in road and building construction and surface mining or by using off-road vehicles in easily erodible areas.[4,5] Soil disturbance is especially disastrous in arid and arctic regions, where water deficit and cold weather, respectively, slow the reestablishment of protective vegetation.

---

[4] Warren E. Rickard, Jr., and Jerry Brown, "Effects of Vehicles on Arctic Tundra," *Environmental Conservation,* **1** (1974), no. 1, pp. 55–62.
[5] U.S. Environmental Protection Agency, *Logging Roads and Protection of Water Quality,* EPA-910/9-75-007, 1975, 311 pp.

(a)

(b)

FIGURE 15-3 (a) The city of Ballinger, Texas, used the water stored behind this dam as the city water supply from 1920 to 1952. (b) By the early 1970s soil erosion sediments filled the lake to a height of more than 35 ft (10.7 m), destroying the ability of the dam to hold water. Soil sediments are the principal cause of water pollution. (*Source:* Environmental Protection Agency.)

FIGURE 15-4 Ice lense in permafrost in Alaska (between arrows). Such ice lenses are common in soils of northern and central Alaska, but recede with continuous field crop production.   (Courtesy of Mark P. Kinney, USDA —Soil Conservation Service, Fairbanks, Alaska.)

FIGURE 15-5 This cave-in pit in a farmer's field in the permafrost region of Alaska was caused by melting of ice lenses and subsequent vertical erosion. Note the large size of the pit by comparing it with the man indicated by the arrow.   (*Source:* USDA—Soil Conservation Service.)

Most soils in permafrost regions have 15–30 cm (6–12 in.) of organic matter on the surface, which protects against warming in the sun. Many of these soils have ice lenses. Any fire, land clearing, or construction activity that exposes bare soil to the sun causes the ice in the soil to melt (Fig. 15-4). Following the melting, slopes fail and soil caves in, channeling water that causes erosion and often produces cave-in pits or cuts gullies (Fig. 15-5).

Erosion of soil sediments by water and wind results from two physical processes: **detachment** and **transport.** Detaching forces include glacial ice, tillage, wind, flowing water, and crushing by vehicles, animal hooves, and people's feet. Transporting forces include glacial ice, gravity, strong wind, and flowing water.

15:4 Classification of Water Erosion

Erosion by water is classified as **raindrop splash** erosion, **surface flow** or **sheet** erosion, and **channelized-flow** erosion.

15:4.1 Raindrop Splash Erosion

Raindrops fall with an approximate speed of 914 cm/s (30 ft/s). When raindrops strike bare soil, they beat it into flowing mud, which splashes as far as 61 cm (2 ft) high and 152 cm (5 ft) away (Fig. 15-6).

The soils most readily detached by raindrop splash erosion are fine sands and silt. Coarser particles are not shifted about as much because of their greater volume and weight. Most soils of finer texture, such as clays and clay loams, are not readily detached because of the strong forces of cohesion that keep them aggregated.

Clays can be dispersed by repeated freezing-thawing actions, by high exchangeable sodium, by tillage when they are very wet, and by allowing destruction of most of the soil's organic matter.

During heavy rain, bare soil aggregates are disrupted, splashed, shifted about, and packed together more closely. As muddy water flows down through natural openings of the soil into the profile below, the soil sediment plugs the pore openings. The result is a nonaggregated soil surface, which, upon drying, forms a crust that is only slowly permeable to air and water. Water that should move downward into the soil profile during subsequent storms or irrigation now flows over the wet clogged surface, carrying soil particles with it to pollute surface waters.

15:4.2 Surface Flow Erosion

Runoff water is responsible for much soil erosion, moving the soil particles by surface creep, saltation (vaultation, or leaping), and suspension. **Surface creep** means movement of wet and supersaturated soil downhill by a rolling or dragging action. **Vaultation** results when turbulent water causes soil particles to hop or skip in water as they move downward. Smaller soil particles that never touch the soil surface as they are moved along in the waterflow are carried by **suspension.** Uniform cutting off of the soil surface is **sheet** or **laminar** erosion.

15:4.3 Channelized Flow Erosion

As water moves over the surface of the soil, some of it concentrates in low places to cut deeper depressions or channels. Continued flow develops minor channels called **rills;** later, major rills and large **gullies** may be formed by the scouring action of increasing volumes of channeled muddy water carrying enormous amounts of sediment that will be deposited somewhere downstream.

15:5 Factors Affecting Erosion by Water

In 1965 the **Universal Soil Loss Equation (USLE)** was proposed for estimating sheet and rill erosion sediment losses from cultivated fields in the United States east of the Rocky

FIGURE 15-6 The impact of falling raindrops on bare soil (a) beats the soil into flowing mud, and soil erosion has begun degradation of the soil. (b) This farm in Ripley, Mississippi, should have been left in trees. (*Source:* USDA—Soil Conservation Service.)

(a)

(b)

Mountains. The equation has since been adapted for use in other cultivated areas of the United States, in Europe, and in tropical western Africa. It also is useful for predicting erosion on rangelands and forestlands in the United States and for most situations in the tropics.[6]

---

[6] G. R. Foster, W. C. Moldenhauer, and W. H. Wischmeier, "Transferability of U.S. Technology for Prediction and Control of Erosion in the Tropics," in *Soil Erosion and Conservation in the Tropics,* ASA Special Publication 43, American Society of Agronomy and Soil Science Society of America, Madison, Wis., 1982.

In fall 1993 a revision of the USLE was distributed to Soil and Water Conservation Service (SWCS) offices as SWCS RUSLE Version 1.01. The major changes are in the values given for erosion as modified by vegetative cover and better calculations of the slope factors (Detail 15-1). Most of the general procedure and values for many of the factors in the equation are unchanged. The RUSLE is computerized to allow ease of calculation.

The expected soil loss is determined from the product of six factors: the rainfall, the erodibility of the selected soil, the length and steepness (gradient) of the ground slope, the crop grown in the soil, and the land practices used. The **Universal Soil Loss Equation** is defined as[7,8]

$$A = R \cdot K \cdot L \cdot S \cdot C \cdot P$$

where A = *erosion soil loss* in tons per acre per year
 R = *rainfall* factor
 K = *soil erodibility* factor
 L = *slope length* factor
 S = *slope gradient* factor (percentage steepness)
 C = *vegetative cover and management* factor
 P = *practices* used for erosion control (terraces, contouring)

The many possible numerical values for each factor require extensive knowledge and preparation. A practicing conservationist normally works in a small area, maybe a county, and often will need only one or two rainfall factors (R), values for only a few soils (K), and only a few plant cover systems (C). The remaining data can be tabulated quite easily for the small area, and few or no extensive computations need to be done in the field. The RUSLE computer version has been released and is being used by the Soil Conservation Service.

The reference soil erosion conditions used, for comparison with the soil being studied and its conditions, is the erosion from a soil plot that

- is 22.1 m (72.5 ft) long, with
- a uniform 9% slope, in
- fallow (barren, no plant cover), which is
- tilled up- and down-slope

The rainfall (R) and soil erodibility (K) are established with numerical values determined for the developed equation. Almost all identified phases of soil series in the United States have had K valaues determined (Table 15-1). For an example of calculation of or erosion, see Calculation 15-1. A modified (Revised) USLE called RUSLE has been computerized and is just beginning to be used. The individual factors are discussed in the following sections.

15:5.1 Rainfall Factor (R)

The **rainfall factor (R)** is a product of the kinetic energy (falling force) of a rainfall times its *maximum 30-minute intensity* of fall (Fig. 15-7). When snowmelt is included in an area precipitation pattern, the symbol R_s is often used as the snowmelt contribution (and added to

---

[7] W. H. Wischmeier and D. D. Smith, "Predicting Rainfall-Erosion Losses: A Guide to Conservation Planning," USDA Agriculture Handbook 537, 1978.
[8] F. R. Troeh, J. A. Hobbs, and R. L. Donahue, "Soil and Water Conservation," 2nd ed., Prentice-Hall, Englewood Cliffs, N.J., 1991, p. 530.

The **Universal Soil Loss Equation (USLE)** has been under considerable attack by many persons who feel some portions of the equation are in error. Scientists working with the equation knew there was a need for some changes. The *revised equation* is called the **RUSLE.**

The sudden concern about the accuracy of the USLE is related to the recent use of the equation for *regulatory* purposes. During development, the USLE was intended to be as accurate as possible, of course, but when it began to be used to regulate who was or was not in conservation compliance, many landowners felt that some farm practices that were said to be needed to control erosion were not adequately evaluated by the USLE.

The 1977 Clean Water Act amended Section 208 of the Federal Water Pollution Control Act (P.L. 92-500) of 1972. Current provisions of the law include **conservation compliance,** which requires land owners to have an *approved conservation plan* (including control of erosion) in order to be eligible for federal program benefits. The Environmental Protection Agency (EPA) is the agency charged with controlling water pollution, but it has limited experience in erosion control and few personnel available for this concern throughout the United States. Consequently, the EPA encouraged a linkage between the Soil Conservation Service and the Soil and Water Conservation Districts to control soil erosion.[a] Conservation plans developed were to be fully implemented by the end of 1994.[b]

This is where the USLE enters into the situation. A *conservation compliance plan* requires control of excess erosion; the only extensive method for estimating erosion and predicting what is needed for a certain level of erosion control is the USLE. Various owners feel that they have been required to expend money and effort beyond what they felt was needed to control erosion on their soils. They challenged the accuracy of the USLE, and those using the equation recognized that some problems existed that needed revision.

The new computer model of the Revised USLE is part of the Agricultural Research Service (ARS) Water Erosion Prediction Project (WEPP). A wind erosion model, called Wind Erosion Prediction System (WEPS), is also being developed. In general, these are some of the early conclusions about the USLE that are at least partially corrected in the new RUSLE:

- The RUSLE gives more generous credit to surface residue's ability to reduce erosion[c]
- The RUSLE gives credit for residues incorporated in the soil *near* the soil surface[c]
- The new data in WEPP will apply to a much broader range of conditions than did the USLE[d]
- The WEPP program will be more accurate than the USLE[d]
- Numerous changes in the method of calculating the *LS* factors are expected to result in a better fit of the *LS* portion[e]

---

[a] Jim Riggle and Karl Berger, " 'Tolerable' Soil Loss," *American Farmland,* (Fall 1992), pp. 10–13.

[b] Douglas Osterman, Frederick Steiner, Theresa Hicks, Ray Ledgerwood, and Kelsey Gray, "Coordinated Resource Management and Planning: The Case of the Missouri Flat Creek Watershed," *Journal of Soil and Water Conservation,* **44** (1989), no. 5, pp. 403–406.

[c] Charlene Finck, "SCS Under Fire," *Farm Journal,* (March 1993), pp. 16–17.

[d] W. Doral Kemper, "Farmers Look to Science for Anti-Erosion Plan," *Agricultural Research,* (April 1988), pp. 7–10.

[e] Ian D. Moore and John P. Wilson, "Length-Slope Factors for the Revised Universal Soil Loss Equation: Simplified Method of Estimation," *Journal of Soil and Water Conservation,* **47** (1992), no. 5, pp. 423–428.

rainfall to equal total R). Obviously, fluctuations in duration and intensities of storms from year to year complicate getting "average" R values for any location. The R values shown in Fig. 15-7 are for 22-year averages; a given year may vary from one-half to more than double the average value. If all other factors are equal, the most erosive rainfall patterns are in the Southern states, particularly coastal Louisiana, Mississippi, Alabama, Georgia, and Florida.

Table 15-1 Selected Values for Soil Erosion Hazard *K* for Selected Phases of Soil Series[a]

Phases of Soil Series and Location	*K*	Phases of Soil Series and Location	*K*
Albia gravelley loam (N.Y.)	0.03	Austin clay (Tex.)	0.29
Freehold loamy sand (N.J.)	0.08	Mansic clay loam (Kans.)	0.32
Tifton loamy sand (Ga.)	0.10	Marshall silt loam (Iowa)	0.33
Molokai clay (Hawaii)	0.20	Shelby loam (Ind.)	0.41
Boswell fine sandy loam (Miss.)	0.25	Keene silt loam (Ohio)	0.48
Cecil sandy loam (N.C.)	0.28	Dunkirk silt loam (N.Y.)	0.69

Source: USDA—Soil Conservation Service, *Engineering Field Manual,* 1975, pp. 2–3 to 2–29.

[a] Generalized *K* factors in four runoff potential groups known as hydrologic soil groups have been developed for the 8000 principal soil series in the United States.

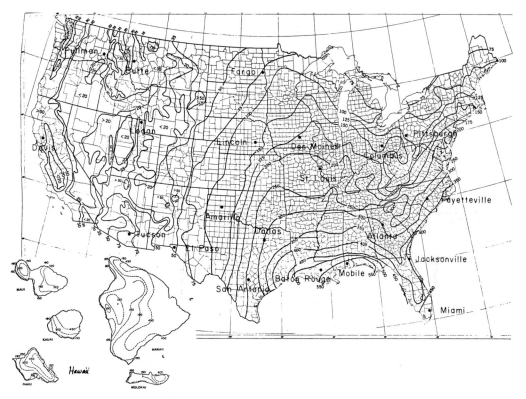

FIGURE 15-7 Average annual values for *R,* the rainfall erosion index. Individual values for small areas may differ considerably from these average values because elevation or mountains alter the amount and intensity of rainfall. Elevation differences can occur without being on a mountain. Death Valley—below sea level, Our Valleys in Utah—4,000 ft., Altaplano of Bolvia—10,000–12,000 ft. Mountains cause air lift-cooling-rain. So elevated plains can be different in effect from mountains themselves. (*Source:* W. H. Wischmeier and D. D. Smith, *Predicting Rainfall Erosion Loses: A Guide to Conservation Planning,* Agriculture Handbook 537, USDA, Washington, D.C., 1978.)

The soil erodibility factor (*K*) is estimated from four soil properties: texture, organic-matter content, soil structure, and permeability data. When compared with measured *K* values, about 95% of tested soils varied less than 0.04 *K* value from calculated values by this technique. To calculate a soil *K* value, in this example of a soil with 65% silt plus very fine sand, 5% other sands, 2.8% organic matter, structure 2, and permeability 4, follow the dashed line on the figure starting at the left at 65% (silt + very fine sand).

The *K* value is 0.31. With the appropriate data for other soils, enter the scale at the left and proceed to lines for sand, organic matter, structure, and permeability, in that order. These data indicate the relative values of these four determinant soil characteristics in allocating the erodibility factor.

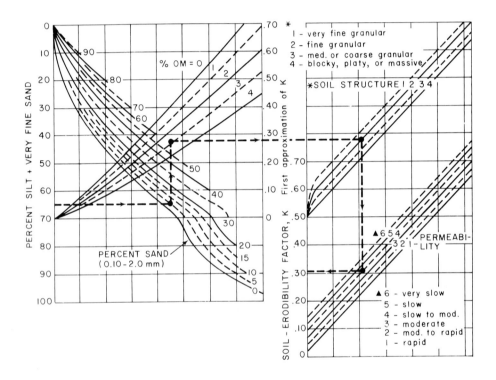

Arid lands (western United States) have lower *R* values but have extensive erosion because of other factors, such as poor plant cover or steep slopes.

15:5.2 Soil Erodibility Factor (K)

Soil erodibility (*K*) is the ease with which a soil can be eroded. Values range from 1.0 (most easily eroded) to 0.01 (almost nonerosive). Soils high in silt and very fine sand are more easily eroded than other soils. Organic matter, larger structural aggregates, and rapid soil permeability all lessen the soil *K* factor (Table 15-1 and Calculation 15-1).

15:5.3 Length-Slope Factor (LS)

The **length-slope value (LS)** is the ratio of soil loss from the slope in question to the soil loss from the reference, which is a slope 22.1 m (72.5 ft) long and of a uniform 9% grade (Table 15-2). The reference slope has an LS factor of 1.00.

Longer slopes increase erosion because water accumulates and increases in speed, collecting more cutting sediment and doing proportionally more damage. Doubling slope steepness (percentage grade) usually more than doubles the erosion; on long slopes it may triple the erosion. Doubling slope length (with the same grade) increases erosion about 20–40%. *Convex* slopes (dome-shaped) have more erosion than indicated by calculations using the average slope; *concave* (saucer-shaped) slopes have less erosion than calculations using average slopes. Changes in plant cover between the comparative slopes complicate these calculations, and recognition must be made of the difference.

The new RUSLE equation has major changes in the manner that LS values are calculated. Some of the changes are as follows[9]:

- The USLE assumed that runoff was uniform over the catchment rather than that some runoff was channeled into rills and gullies. Rill erosion is a major component in the RUSLE.
- The amount of runoff water was calculated as the excess applied minus infiltration in the USLE, but it did not consider that long rains would saturate a soil, intake would slow down, and runoff would be greater. It is runoff water that causes erosion.
- Sediment deposition, as at the bottom of concave slopes, was not considered. The USLE applied only to those areas with net erosion. The RUSLE considers areas of net sedimentation.
- The USLE was not designed to handle converging and diverging terrain. The RUSLE does handle these.

Table 15-2 Combined LS Factor (Slope Length and Steepness) for the Soil Erosion Equation When Compared to the Standard Reference of a Slope 72.6 ft Long with Steepness of 9% (= 1.0)

Slope Percentage	Slope Length (ft)[a]							
	25	50	100	200	400	600	800	1000
0.5	0.073	0.083	0.096	0.110	0.126	0.137	0.145	0.152
2	0.13	0.16	0.20	0.25	0.31	0.35	0.38	0.40
4	0.23	0.30	0.40	0.53	0.70	0.82	0.92	1.01
8	0.50	0.70	0.99	1.41	1.98	2.43	2.81	3.14
12	0.90	1.28	1.80	2.65	3.61	4.42	5.11	5.71
16	1.42	2.01	2.84	4.01	5.68	6.95	8.03	8.98
20	2.04	2.88	4.08	5.77	8.16	10.0	11.5	12.9

[a] Feet times 0.3048 equals meters.

[9] Ian D. Moore and John P. Wilson, "Length-Slope Factors for the Revised Universal Soil Loss Equation: Simplified Method of Estimation," *Journal of Soil and Water Conservation,* **47** (1992), no. 5, pp. 423–428.

15:5.4 Cover and Management Factor (C)

The **cover and management factor** (*C*) considers the type and density of vegetative cover on the soil and all related management practices, such as time between operations (delay in planting after plowing, and so forth), weed control, tillage, watering, fertilization, and so on. This *C* factor is very complicated because of the wide range of possibilities in cover material, management, and the manner in which crop residues can be left on soil.

Tables 15-3, 15-4, and 15-5 list a few selected values from a wide range of the research-verified factors. Each value is the ratio of erosion under specified cover and management to the amount of erosion from continuous fallow (barren). If the value is 0.33, for example, the cover and management factor reduces erosion to one-third that which would occur on soil under continuous fallow. Lower factors mean less erosion.

The best cover for minimum erosion is dense forest or tall, thick grass with dead residue ground cover 4–5 cm (1.5-2 in.) thick. Plant material in contact with the surface protects the soil from raindrop splash and erosive flowing water. In contrast, continuous cotton cropping would result in 40–60 times more erosion than thick grass residue cover.

15:5.5 Practice Factor (P)

The **practice factor** (*P*) recognizes the influence of contour planting, strip cropping, terracing, and combinations. Table 15-6 lists several of these practices and their effect on erosion.

Table 15-3 The *C* Factor, Which is the Ratio of Soil Loss from Areas with Protective Cover and Various Management Practices to the Corresponding Loss from Continuous Fallow

| | C Values as Functions of the Stage of Growth[a] | | |
| | | (50–75%) | |
Vegetative Condition	Seedbed	Plant Cover	Mature Crop
1. *Corn after corn, grain sorghum, cotton:*			
Conventional till (plowing), plant in 3400 lb of residue			
(75–99 bushel) left	0.60	0.41	0.20
No-till, plant in crop residues, 3400 lb of residues in spring	0.08	0.08	0.06
No-till, plant in 2600 lb of residue	0.21	0.18	0.11
Strip-till one-fourth of row spacing, on contour, 3400 lb			
of residue	0.16	0.12	0.10
2. *Corn after winter cover grain:*			
No-till, plant in 3000 lb of residue	0.11	0.11	0.07
Plow, conventional seedbed, 3000 lb of residue	0.60	0.41	0.20
Strip-till on contour, 3000 lb of residue	0.15	0.15	0.09
3. *Corn after soybeans:*			
Conventional spring till (plow), 3400 lb of residue	0.78	0.51	0.25
No-till in 3400 lb of residue	0.25	0.19	0.11
4. *Beans after corn,* conventional till, 3400 lb of residue	0.64	0.41	0.18
5. *Grain after summer fallow* in 1500 lb of grain residues	0.20	0.12	0.05
6. *Potatoes, rows with slope*	0.64	0.36	0.16

Source: Selected data from W. H. Wischmeier and D. D. Smith, *Predicting Rainfall Erosion Losses: A Guide to Conservation Planning,* Agriculture Handbook 537, USDA, Washington, D.C., 1978.

[a] To obtain a single *C* value for a cropping system, the time for each growth stage (some of which are not in this table)

Table 15-4 The *C* Factor, Which Is the Ratio of Soil Loss from Areas with Protective Cover to the Corresponding Loss from Continuous Fallow

Vegetative Condition	C Value
Cotton after cotton, seedbed period, 80% soil cover (plant mulch)	0.64
Cotton after cotton, 35–60% of full development, expect 80% final soil cover, conventional tillage	0.46
Corn mulch, 40% of soil actually covered, no-till	0.21
Corn mulch, 90% of soil actually covered, no-till	0.03
Soybean mulch, 40% of soil actually covered, no till	0.26
Grain stubble, 40% of soil actually covered	0.30
Undisturbed forest, 90–100% duff cover, 75–100% area has canopy	0.001–0.0001
Undisturbed forest, 40–70% with 5 cm duff, 20–40% with canopy	0.003–0.009
Pasture or range, 10% grass cover	0.20
Pasture or range, 40% grass cover	0.10
Pasture or range, 80% grass cover	0.013
Pasture or range, 95+% grass cover	0.003
Brush, 25% plus 40% grass cover	0.09
Brush, 75% plus 40% grass cover	0.08
Brush, 75% plus 40% covered by weeds and broadleaf plants	0.12
Trees, 25% plus 20% covered by weeds and broadleaf plants	0.23
Trees, 75% plus 20% covered by weeds and broadleaf plants	0.20
Trees, 25% with no ground cover	0.42

Source: Selected data from W. H. Wischmeier and D. D. Smith, *Predicting Rainfall Erosion Losses: A Guide to Conservation Planning,* Agriculture Handbook 537, USDA, Washington, D.C., 1978.

The total benefit of terracing is not evident in the table because, after terracing, each terrace is treated as a separate slope, which is less steep than the original slope. These values are put into the *LS* factor. The greatest erosion control by choosing the optimum practice would be to have level terraces (zero slope) having excellent protective cover on connecting slopes.

15:5.6 Calculation of Erosion by Water

To calculate erosion, determine the value for each of the six factors in the USLE and multiply each factor by the next to obtain the product of the six values (Calculation 15-2). Notice that the values given in tables in this chapter are in the English system and represent erosion loss in *tons per acre.*[10] Conversions can be performed with new models and data to work in metric units. The calculation of erosion is simple; validating the numerical values to be used for each factor and situation is the difficult part.

15:6 Erosion Tolerance (*T*)

Landowners, in order to benefit from any federal agricultural-benefit programs, must have in operation by January 1, 1995, a **Conservation Compliance Plan** for their land. Control of erosion is one requirement in a conservation compliance plan (Detail 15-1). **Highly erodible land (HEL)** must have erosion reduced to a pre-determined tolerance level. The **erosion tolerance level,** *T, is usually 4–5 tons/acre per year.* Some experts in erosion control believe that the value of *T* for some soils should be as low as 1–2 tons/acre. One definition

---

[10] All units in this section are in U.S. units. To avoid confusion, no conversion to metric equivalents will be made here.

Table 15-5 The *C* factor, Which Is the Ratio of Soil Loss from Areas with Protective Cover to the Corresponding Loss from Continuous Barren Ground (Fallow), for Rangelands, Pastures, and Idle Lands

Vegetation Canopy	Canopy Cover[a]	Type	Percentage Ground Cover					
			0	20	40	60	80	95-100
No appreciable canopy	0	G[b]	0.45	0.20	0.10	0.042	0.013	0.003
		W[c]	.45	.24	.15	.090	.043	.011
Tall weeds or short	25	G	.36	.17	.09	.038	.012	.003
brush (average 0.5 m		W	.36	.20	.13	.082	.041	.011
canopy height)	50	G	.26	.13	.07	.035	.012	.003
		W	.26	.16	.11	.075	.039	.011
	75	G	.17	.10	.06	.031	.011	.003
		W	.17	.12	.09	.067	.038	.011
Appreciable brush,	25	G	.40	.18	.09	.040	.013	.003
bushes (average 2 m		W	.40	.22	.14	.085	.042	.011
canopy height)	50	G	.34	.16	.085	.038	.012	.003
		W	.34	.19	.13	.081	.041	.011
	75	G	.28	.14	.08	.036	.012	.003
		W	.28	.17	.12	.077	.040	.011
Trees, but few low	25	G	.42	.19	.10	.041	.013	.003
brush (average 4 m		W	.42	.23	.14	.087	.042	.011
canopy height)	50	G	.39	.18	.09	.040	.013	.003
		W	.39	.21	.14	.085	.042	.011
	75	G	.36	.17	.10	.039	.012	.003
		W	.36	.20	.13	.083	.041	.011

Source: USDA—Soil Conservation Service.
[a] Portion of the soil surface hidden from a bird's-eye view (vertical projections)
[b] G = grass, grasslike plants, decaying duff, or litter at least 5 cm (2 in.) deep
[c] W = surface cover, mostly broadleaf plants (weeds) with little lateral network of roots near the surface

of *T* is the *maximum rate of annual soil loss that will permit crop productivity to be obtained economically and indefinitely.* The tolerance value is supposed to approach the rate of natural replacement of the soil. Obviously, some soils forming from hard bedrock may "produce" less than 0.5 ton/acre of new soil annually.

The equation for erosion tolerance, *T,* simply replaces *T* for *A* in the USLE:

$$T = R \cdot K \cdot L \cdot S \cdot C \cdot P$$

The **erodibility index, *EI*,** is

$$EI = \frac{R \cdot K \cdot L \cdot S}{T}$$

When *EI* = 8 or greater, the land is considered to be highly erodible land. In order to be in conservation compliance, the plan for soil erosion control must be activated to bring all erosion eventually down to *T* or less (Detail 15-2).

Because the rainfall factor (*R*) and the soil of a field (*K*) are not changeable, the factors that can be modified (*L,S,C,P*) are grouped to determine alternatives to reduce erosion. Rearranging the soil-loss equation (where *T* = *A*),

$$\frac{T}{R \cdot K} = L \cdot S \cdot C \cdot P$$

Table 15-6 The *P* Factor, Which Is the Ratio of the Erosion Resultant from the Practice Described to That Which Would Occur with Up-and-Down-Slope Cultivation

Vegetative Condition	P Value
A 1–2% slope, contoured, in 400-ft lengths	0.60
A 3–5% slope, contoured, in 300-ft lengths	0.50
A 6–8% slope, contoured, in 200-ft lengths	0.50
A 9–12% slope, contoured, in 120-ft lengths	0.60
A 21–25% slope, contoured, in 50-ft lengths	0.90
A 1–2% slope, contour strip cropping (row crop—grain), 130-ft strip width, 800-ft slope length (along direction of strip)	0.60
A 6–8% slope, contour strip cropping (row crop—grain), 100-ft strip width, 400-ft strip length	0.50
A 13–16% slope, contour strip cropping (row crop—grain), 80-ft strip width, 160-ft strip length	0.70
A 21–25% slope, contour strip cropping (row crop—grain), 50-ft strip width, 100-ft strip length	0.90
A 6–8% slope, contour strip cropping with a 4-year rotation of 1 row crop, 1 small grain, 2 meadow	0.25

Source: Selected data from W. H. Wischmeier and D. D. Smith, *Predicting Rainfall Erosion Losses: A Guide to Conservation Planning,* Agriculture Handbook 537, USDA, Washington, D.C., 1978.

A landowner's alternatives in controlling erosion are some combination of altering the slope features (*LS*), cover/management/cropping (*C*), and erosion control practices (*P*). The value *T/RK* is constant for a given field. An example calculation is given in Calculation 15-3.

The soil loss calculated by the USLE is for soil that actually leaves the field or slope, not in-field movement of sediment. The equation is so complex that it should not be used "cookbook" style, but rather by scientists well trained in its use. Some cautions on the use of the USLE include knowing that the *R* factor, which is a long-time average, is different from the average in any one year. It is important to correctly select factor values, to anticipate poorer-than-normal yields some years, and to correctly estimate such items as slope lengths and vegetative cover weight.[11]

15:7 Water Erosion Control Techniques

Soil erosion is lessened by reducing either soil detachment or soil sediment transport or both. The various methods of reducing detachment are much the same, whether protecting against erosion caused by wind or by water. Methods for controlling transport are vastly different.

Aggregated soils (soil peds or clods) may be disintegrated by the direct impact of falling raindrops. In the process of disintegration, sand, silt, clay, and humus in the aggregate are separated. Clay and humus particles may be splashed as far as about 150 cm (5 ft) from the point of impact; the larger silt and sand particles are moved shorter distances. Being less dense, humus floats away with surface flow; the water also suspends some clay and silt as a muddy

---

[11] W. H. Wischmeier, "Use and Misuse of the Universal Soil Loss Equation," *Journal of Soil and Water Conservation,* **31** (1976), no. 1, pp. 5–9.

Problem What maximum erosion loss might normally be expected of a Shelby loam in central Indiana on a 380-ft-long, 4% slope adequately cultivated on the contour with continuous corn (80 bushel) using conventional plant and till methods?

Solution The factors needed are

Shelby loam (*K*)	0.41 (Table 15-1)
Rainfall (*R*)	180 (Fig. 15-7)
Slope (*LS*)	0.70 (Table 15-2 for 400 ft)
Erosion control practice (*P*)	0.5 (Table 15-6)
Crop system, corn (*C*)	0.40 (Table 15-3) [(0.6 + 0.41 + 0.20)/3]

Using the equation $A = R \cdot K \cdot L \cdot S \cdot C \cdot P$ and substituting the values

$$A\text{(tons/a)} = 180 \times 0.41 \times 0.70 \times 0.40 \times 0.5$$

$$= 10.3 \text{ t/a/yr lost as eroded soil}$$

Notice that if continuous pasture with 95% cover (0.003 in Table 15-5) were substituted in the equation for continuous corn (0.40), the calculated loss would be reduced from 10.3 t/a to 0.08 t/a/yr.

---

runoff. This is soil detachment and surface erosion; the cause is raindrop impact on exposed soil and runoff flow. The soil productivity is lessened, and sediment pollutes surface waters.

15:7.1 Controlling Soil Detachment (Cover and Management Factor C)

Soil detachment can be controlled by cropping or other vegetative cover practices (*C*) that keep the soil covered as much as possible. When plants, either living or dead (grasses, leaves, mulches), cover the soil surface, the energy of falling raindrops is dissipated by the springy vegetation. As raindrops fall, the vegetation absorbs the energy; then the water gently slides off, to be absorbed into the soil.

Maintaining some protective cover requires deliberate action. During winter months, when cool-area lands might be left barren, planting a cover drop of cool-season grasses, legumes, or small grains protects the soil surface. Dryland areas, which are used to grow winter wheat, are often left barren (farrow) in alternate years to store some extra water for the next growing year. Fallow soil is susceptible to erosion. To protect it, the grain stubble is not plowed into the soil, but rather is left standing or is only partly incorporated by special equipment.

Stubble mulches help to control soil erosion. Crop residues on or near the surface reduce the impact of the falling raindrop, help to hold winter snows, protect topsoil from winds, improve soil structure, and increase the infiltration of rainwater.

Leaving a surface mulch of crop residues is an effective way to slow water runoff and lessen raindrop destruction of soil aggregates. **Conservation tillage** maintains at least 30% of the soil surface covered by residues (Calculation 15-4).

Long-term, economical control of soil erosion involves changes in management: tillage, drainage ways, irrigation, alteration of the crops grown, use of cover crops, leaving plant residues on soil, land grading, and other practices. Landowners must have conservation compliance plans filed by 1990 and in action by 1995 or lose the chance to be involved in federal benefit programs. To meet compliance, erodible land must have erosion controlled; the cost for this control has raised questions about the accuracy of the USLE, which is being used to assess suitable planning for erosion control. The present requirement is voluntary. However, because erosion control is essential to minimize water pollution, it is only a matter of time before regulations will replace voluntary action as non-point-source pollution is tackled.

A few states have anticipated strong regulations, as well as the importance of retaining our soils, and they have set erosion control goals at what they consider to be a reasonable rate. Illinois has 24 million acres (nearly 10 million hectares) that need to meet the "T-by-2000" guideline. Illinois set up a time table for all this land:

- Bring losses to "4T" between 1983 and end of 1987
- Bring losses to "2T" between 1988 and end of 1993
- Bring losses to "1.5T" between 1994 and end of 1999
- Meet T-value losses after January 1, 2000

There are, as yet, no civil or criminal penalties for noncompliance. However, violators may be pressured and shamed by local media exposure and be required to travel to the state capital for public hearings.

There are some serious reservations about the accuracy of T values for water erosion; there is even greater concern when the wind erosion equation is used for regulatory purposes.

[a]Jim Riggle and Karl Berger, "'Tolerable' Soil Loss," *American Farmland,* (Fall 1992), pp. 10–13.
Charles M. Benbrook, "First Principles: The Definition of Highly Erodible Land and Tolerable Soil Loss," *Journal of Soil and Water Conservation,* **43** (1988), no. 1, pp. 35–38.

A 1991 survey of 1941 Iowa farmers listed 64% that used reduced tillage, 7% in no-till, and 22% in conventional tillage.[12] Although 93% had an approved conservation plan, only 64% indicated that they understood the plan. Interestingly, 58% supported the concept of conservation compliance, but 26% opposed it. To implement their plans, 25% had to change tillage systems, 32% needed to change rotations, and 44% needed to install terraces, waterways, or filter strips.

The practice of using subsurface tillage implements that leave much of the crop residues standing on the surface of the soil is **stubble mulch farming,** an effective wind erosion control technique on semiarid wheat land (see Chapters 7 and 14).

The need to control erosion on problem areas—highway banks in arid regions, gully control in the Southern states, acid mine spoils—has accelerated the search for especially suitable plants.[13] A Cape Cod beach grass is now a widely used dune stabilizer; emerald crown vetch in the corn belt stabilizes road banks where soil is too shallow for grass. Pink Lady winterberry, a shrub used for windbreaks and for wildlife food, came from a stone wall

[12] Kelly O'Brien-Wray, "Helping Producers with Conservation Compliance," Solutions, **36** (1992), no. 3, pp. 18–22.
[13] Robert S. MacLauchlan, "The Search for `Workhorse' Plants," Soil Conservation, **42** (1977), no. 12, pp. 5–10.

Problem If the soil in Calculation 15-2 has a soil loss tolerance (maximum) of only 2.5 t/a per year, what *cropping practice* would be permitted if no changes in slope factors or erosion control practices were wanted?

Solution Substitute the known values of *T, R, K, L, S,* and *P* in the equation, and solve for the value for the cropping system, *C*:

$$\frac{T}{R \cdot K} = \frac{2.5}{(180)\,(0.41)} = L \cdot S \cdot C \cdot P = (0.70)\,(C)(0.5)$$

$$0.034 = 0.35C$$

So *C* must be 0.034/0.35 = 0.097 or lower. In Tables 15-3, 15-4, and 15-5 many pastures with more than 40% cover (probably 45–50%) and probably some no-till rotations that include pastures would adequately control erosion.

in Beijing, China; Tegmar intermediate wheat grass, brought from Turkey, is used in diversions and waterways in the West, and Nortran tufted hairgrass developed from breeding materials is used in Iceland and Alaska.

15:7.2 Controlling Soil Transport (Special Practice P)

Soil transportation is hindered by slowing the eroding water, reducing the steepness of slope, and erecting barriers such as brush dams, terraces, contour cultivation, or contour strip cropping (special practice *P*) (Figs. 15-8, 15-9, and 15-10). Success in achieving adequate solutions depends on the cost, whether equipment can be used, and the extent of the owner's desire to make long-range plans for use of the land.

In areas of high-intensity rainfall, the slope steepness for runoff water can be reduced by reshaping the land by **terraces.** Water is directed to follow the gentler slopes of the terraces rather than the steeper natural slopes. Figure 15-8 diagrams cross sections of various types of terraces.

Terracing is usually recommended only for intensively used, eroding cropland. Land that has grass or other dense perennial cover seldom needs terracing for water runoff erosion control. Terraces are costly to construct and need annual maintenance. They are feasible where arable land is in short supply or valuable crops can be grown. Although well-designed and adequately maintained terraces are very effective in reducing soil transport, they are not included in estimating the *P* factor in the Universal Soil Loss Equation. Their effectiveness results from reducing the effective slope length *L* and gradient *S* in the equation. Terraces are considered as permanent changes and new *LS* values are determined for terraced fields.

Contour cultivation is tilling and planting at right angles to the natural slope of the land. On terraced fields contour tillage should be parallel to the terraces. Contour tillage alone successfully controls erosion during low-intensity rainfall on moderate slopes of 2–8%. Ridges formed during contour tillage are effective in reducing erosion. Obviously, contour tillage combined with terracing and/or contour strip cropping is more effective than contour tillage alone.

Contour strip cropping is the practice of planting on the contour strips of intensively cultivated crops alternating with strips of sod-forming crops. Erosion sediments from the clean-tilled strips are filtered out and retained on the sodlike-crop strips. The greater the pro-

Calculation 15-4 Measuring Residue Percentage [a]

When residues are left on the soil, how is the percentage coverage determined? To be in *conservation tillage,* the soil must have at least 30% cover.

Problem What is the percentage cover of a field covered by plant residues?

Solution Use a 100-ft tape or cord marked at each 1-ft interval. Then follow these steps:

1. Select an area representative of the whole field.

2. Anchor one end of the tape or cord and stretch it diagonally across several rows and fix the other end of tape or cord.

3. Count the number of residues that are *under* the tape or cord at each of the 1-ft. interval marks. For corn residues, count only those with at least a 1/4-inch diameter. Read the "contact points" only on one side of the tape or cord. Repeat three times at different locations in the field and average the counts (see illustration below).

4. The number of contact points equals the percentage cover. If you have 31 contact points at the 1-ft-interval marks, the coverage is 31%.

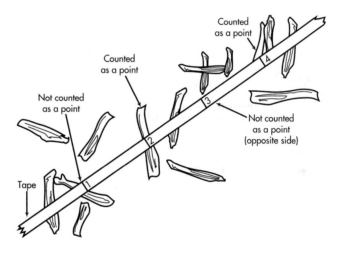

[a] Kelly O'Brien-Wray, "Helping Producers with Conservation Compliance," *Solutions,* **36** (1992), no. 3, pp. 18–22.

portion of sod-crop strips to cultivated strips, the less the erosion. If the sod strips occupied one-half of the field, the P value would be half that of contour cultivation alone (Fig. 15-10).

The width of each strip should be a multiple of the width of the machinery used to plant and harvest the crops. Other factors that determine the width of crop strips are: rainfall intensity (R), soil erodibility (K), slope length (L), slope gradient (S), the kind of crop (C), and whether the field is terraced (P). For a 5% slope typical strip widths are 10 m (33 ft) for a sod crop and 25 m (82 ft) for a cultivated crop.

15:8 Nature of Wind Erosion

Wind erosion is accelerated when soil is dry, weakly aggregated, bare and smooth, and the winds are strong. Winds segregate dry humus, clay, silt, and sands—the least dense being

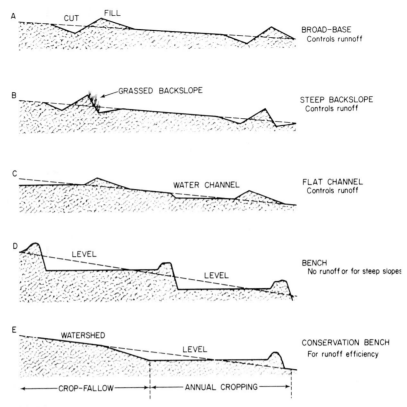

FIGURE 15-8 Cross sections of several kinds of terraces. In the broadbase terraces (a), all of the surface area is planted (steepness is accentuated in all drawings). Thus, (a), (b), and (c) are quite similar in use. The bench terrace (d) is much more expensive if it is on steep slopes. However, gentle slopes, as in the plains of Texas, easily can be made into terraces leveled in both directions to hinder all runoff losses. The conservation bench terrace (e) attempts to take advantage of areas with runoff to produce good annual yields on some nearly level portions of sloping soil areas. (Courtesy of Raymond W. Miller, Utah State University.)

carried the farthest. Even moderate wind velocities can keep individual particles of humus, clay, and silt (particles <0.05 mm in diameter) in **suspension** simultaneously. Very fine, fine, and medium sands (0.05–0.5 mm in diameter) are moved by wind in a succession of bounces known as **saltation.** Coarse sand (0.5–1.0 mm in diameter) is not usually airborne but rather is rolled along the soil surface. This kind of erosion is called **surface creep.** Very coarse sand (1–2 mm in diameter), gravels, peds, and clods are too large and/or too dense to be rolled by the wind, so wind-eroded soils have surfaces covered with coarse fragments larger than 1 mm in diameter. This kind of arid soil surface is known as **desert pavement.**

Eroding winds deplete soil productivity and change the soil texture toward coarseness.[14]

---

[14] Frederick R. Troeh, J. A. Hobbs, and Roy L. Donahue, *Soil and Water Conservation for Productivity and Environmental Protection,* 2nd ed., Prentice-Hall, Englewood Cliffs, N.J., 1991, p. 541.

(a)

(b)

FIGURE 15-9 Terraces are usually intended to reduce erosion and to control water runoff. Terraces can be essential, but they are also expensive. (a) Thai farmers learning to construct bench terraces in China; (b) Ethiopians making bench terraces in rocky hills;— (continued on next page) (By permission, Food and Agricultural Organization, United Nations. Photo a, by F. Botts; photo b by D. Craig.)

15:9 Extent and Causes of Wind Erosion

Prevention of wind erosion is the dominant management factor on an estimated 29 million hectares (69 million acres) in the United States. Of this total area 22 million hectares (54 million acres) are used for cropland, 3.6 million hectares (9 million acres) are used for range grazing, and the balance is in other uses.

(c)

(d)

FIGURE 15-9, cont'd (c) bench terraces in East Java being prepared for rice paddy; (d) bench terraces in the deep loess hills in northern Shaanxi Province, China, to try to stop excessive erosion into the Yellow River. (By permission, Food and Agricultural Organization, United Nations, Photos c and d by F. Botts.)

There are an estimated 5 billion tons of soil eroded from U.S. croplands yearly, and it is estimated that one-third is stripped off by wind erosion. Wind erosion is most severe in areas of arid and semiarid climates, which comprise one-third of the land surface of the world, excluding polar deserts. Arid climate Aridisols comprise 18.76% of the world soils. The principal deserts of the world comprise 11.6% of the land surface, excluding Antarctica. The largest desert, the Sahara in northern Africa, has an area of 9 million square kilometers (3.6 million square miles), 6% of the world's land surface.

FIGURE 15-10 Contour strip cropping of hay and corn near North Springville, Wisconsin. The grassed diversion ditch (downslope at extreme right center) carries away runoff water. True contour strips would be of unequal widths, making cultivation awkward. These strips of equal width are a compromise between true contours and cultivation conveniences. (*Source:* USDA—Soil Conservation Service.)

Most deserts of the world are formed because of two climatic and topographic factors:

1. **Tropical deserts,** or high-pressure deserts, are formed at about 20°–30° north or south of the equator. High-pressure air masses become warmer and relatively drier as they descend toward the earth's surface. The Sahara is the best-known example of a tropical desert (Fig. 15-11).

2. **Topographic deserts** develop on the leeward (eastern) side of high mountains where air masses are descending and are becoming warmer and drier. The Mojave in the southwestern United States just east of the Sierra Nevada is a good example of a topographic desert.

Although wind erosion in desert areas is very severe, few people live there; for this reason the social impact is not great. The most severe wind erosion with social impact results from people working the soil and leaving it bare in semiarid areas. Very serious wind erosion has occurred and continues to occur during dry seasons in northern Africa, the USSR, the Middle East, China, India, Pakistan, Australia, Argentina, Peru, western Canada, and in the Great Plains of the United States (Fig. 15-12). In all countries it seems to be economically sensible, during seasons of favorable rainfall, to expand cultivation in "marginally" suitable climates and soils; during dry years, however, crops fail and strong winds cause serious erosion damage.

FIGURE 15-11 Example of a tropical desert in central Asia that is threatening productive farm land. Such conditions require immediate and often costly measures. Dune encroachment has threatened Cairo, Egypt, for decades. (*Source:* By permission, from *Food, Fiber and the Arid Lands,* by William G. McGinnies, Bram J. Goldman, and Patricia Paylore, eds. University of Arizona Press, Tucson, ©1971.)

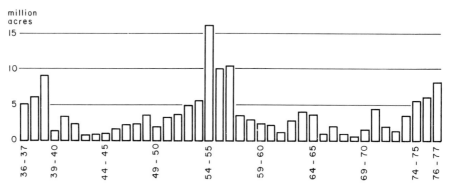

FIGURE 15-12 Land damage by wind erosion in the Great Plains in the United States since 1936. In 1976–1977 nearly 3.3 million hectares (8 million acres) were damaged by wind erosion; Colorado had severe wind erosion on more than 1 million hectares (2.52 million acres), and Texas had damage on more than 887,000 hectares (2.19 million acres). Most damage occurred from November through May; 80% was on cropland. (*Source: Soil Conservation,* **42** [1977], no. 12, p. .

Wind erosion is also severe in some humid regions. Strong winds may develop shifting dunes from humid-region beach sands along oceans throughout the world. This hazard is especially serious in the United States along the Atlantic Ocean, the Gulf coast, and the Great Lakes.[15,16] Drained and bare Histosols (peats and mucks) are subject to wind erosion also.

15:10 Factors Affecting Erosion by Wind

Erosion by wind increases where soil is less cohesive, loose particles are smaller, land cover is less, and wind speeds are higher. Soils low in clay but high in fine sands and coarse silts are usually weakly structured. These soil particles are easily detached and transported if wind speeds are above about 20 km/h (13 mi/hr). Wet soils, being denser, are less easily borne by winds.

Scientists are able to predict the seriousness of wind erosion before it happens. This makes it possible to control wind erosion by suitable management practices. The calculation of erosion by wind using the **wind erodibility equation** is very involved, as is the Universal Soil Loss Equation for erosion by water. For this reason, only the concept of how the equation works and an example using some typical values are given here for the area of northwest Texas. For all areas, many tables and figures are required.[17]

The amount of erosion, E, by wind is expressed by the potential wind erodibility equation.[18,19]

$$E = f(I, K, C, L, V)$$

where E = *total erosion* loss in tons per acre per year (times 2.242 = metric tons/hectare).

f = indicates that erosion is a *function* of the various items.

I = soil *erodibility index* based on texture and aggregation. I ranges from 0 (as stony) to over 300 (very fine nonaggregated sands). Use 70 to 100 for the U.S. southern Great Plains (Table 15-7).

Table 15-7 Some Typical Soil Erodibility Values (I) for Various Soils

Percentage of Dry Soil Larger than 0.84 mm after Sieving When Dry	Erodibility (t/a)
1	310
5	180
10	134
15	117
20	98
30	74
50	38
80	2

---

[15] W. W. Woodhouse, Jr., "Dune Building and Stabilization with Vegetation," Special Report 3, U.S. Army Corps of Engineers, 1978.

[16] E. L. Skidmore and N. P. Woodruff, *Wind Erosion Forces in the United States and Their Use in Predicting Soil Loss,* Agricultural Handbook 346, USDA, Washington, D.C., 1968.

[17] Ibid.

[18] Lee Wilson, "Application of the Wind Erosion Equation in Air Pollution Surveys," *Journal of Soil and Water Conservation,* **30** (1975), pp. 215–219.

[19] E. L. Skidmore, P. S. Fisher, and N. P. Woodruff, "Wind Erosion Equation: Computer Solution and Application," *Soil Science Society of America Proceedings,* **34** (1970), pp. 931–935.

Table 15-8 Selected Climatic Factor Values (*C*) for Various Locations

Location	Climatic Factor for:			
	February	April	July	October
Albuquerque, N.M.	1.20	2.20	1.30	1.00
Denver, Colo.	0.70	0.70	0.40	0.30
Dodge City, Kans.	0.80	1.50	0.70	0.70
El Paso, Tex.	2.00	3.00	1.50	1.00
Ft. Worth, Tex.	0.20	0.25	0.10	0.10
Laramie, Wyo.	0.80	0.65	0.30	0.30
Mandan, N.D.	0.30	0.50	0.20	0.30
Omaha, Neb.	0.20	0.30	0.10	0.20
Pocatello, Ida.	0.20	0.50	0.30	0.20
Wichita, Kans.	0.30	0.35	0.20	0.30
Yakima, Wash.	0.20	0.50	0.30	0.20

$K =$ *surface roughness;* it varies form 1.0 for smooth soil surfaces to 0.5 if rough surface has at least a 4-in. vertical microrelief variation.

$C =$ *climate factor* (wind speed and effective soil moisture). For west Texas use 1 to 2 (100–200%); the lower values for early fall, higher values for midspring (Table 15-8).

$L =$ effect of *field size* (length). This value ranges from 0 (small protected areas) to 1 (wide open area for many hundreds of feet). Without effective barriers, use values of 0.8 to 1.0, depending on openness. Normally, this is taken from a special complex table.

$V =$ equivalent quantity of vegetative cover, calculated from tables according to the kind of cover (stubble, cut residues) left on the field. Roughly, the *equivalent cover* in thousands of pounds per acre is

Grain sorghum, flat	1.5 times actual weight
Grain sorghum, 12 in. tall	2.3 times actual weight
Grain sorghum, 20 in. tall	3 times actual weight
Wheat stubble, standing	6 times actual weight
Desert range vegetation	7 to 8 times actual weight

The *calculation* is in two steps, as follows:

- Step 1: partial estimate $E_A = I \times K \times C \times L$.
- Step 2: final estimated $E = (E_A)$ (vegetative cover factor). The vegetative cover factor reduces erosion as more cover exists, and the graph in Fig. 15-13 is used to get E from E_A (Calculation 15-5).

The complexity of the tables and the determination of values for L and V in the complete procedure have prompted the development of computer programmed solutions. The Wind Erosion Prediction System (WEPS) is well along as an up-to-date computer model that

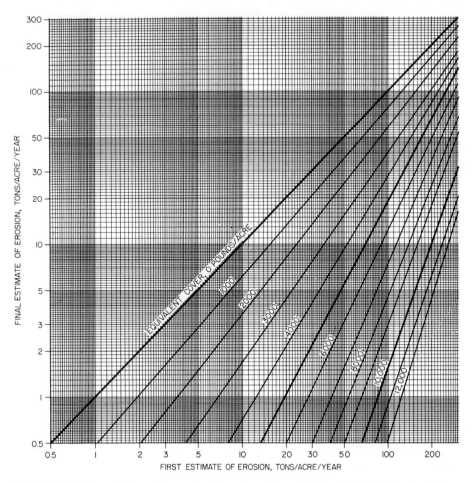

FIGURE 15-13 Completion of the calculation for soil erosion by wind. Find E_A, the *first estimate of erosion*, along the horizontal scale. Read upward until the line for "equivalent cover" that the field has is intersected. Follow that point horizontally left to the vertical scale. That intersect value is the *final value, E,* calculated for erosion. (Note: tons/acre/year × 2.242 = Mg/ha/yr.)

takes into account the (1) wind speed and direction, (2) soil characteristics, (3) crop growth, (4) types of tillage, and (5) erosion mechanics.[20]

15:11 Wind Erosion Control Techniques

Some factors that influence wind erosion cannot be controlled by the land manager. Not readily controllable are the soil erodibility index (I) and the climate (C). However, the soil's erodibility index can be altered slowly by changes in soil humus levels and tillage practices.

[20] W. Doral Kemper, "Farmers Look to Science for Anti-Erosion Plan," *Agricultural Research,* (April 1988), pp. 7–10.

Calculation 15-5 Sample Wind Erosion Calculation

Problem For a stubble sorghum field in early October in northwest Texas, calculate the potential erosion under these conditions: Stubble averages about 12 in. tall and has an actual weight of 1800 lb/a. The field is large, smooth, and without windbreaks. The soil is a loam.

Solution All values except those of Step 3 are estimates. Several graphs and tables would normally be available for obtaining more-exact values.

1. Estimate from previous discussion in the text the following values for the equation $E = f(I, K, C, L, V)$:

$I = 85$ (a loam soil)

$K = 0.95$ (a smooth terrace, but with some terrace ridges)

$C = 1.2$ (an estimated value for lower windspeeds)

$L = 1.0$ (wide-open, level-terrace land)

$V =$ See step 3

2. E_A (the partial estimate) is determined as

$$E_A = I \times K \times C \times L$$

$$= (85)(0.95)(1.2)(1.0)$$

$$= 96.9 \text{ t/a/yr}$$

3. Calculate the effective cover as 2.3 times the actual weight of 1800 lb/a because it is 12 in. tall. This is 4140 lb/a. Then using the graph of Fig. 15-13, find 96.9 along the "first estimate" (E_A) axis and follow it upward to where it cuts the "extrapolated" 4140 lb/a line and follow the point left to the E_T (total) scale, which reads about 22. The expected soil erosion is 22 t/a/yr (49.3 Mg/ha).

Notice how great the influence of quantity of cover has on the expected erosion. For example, if the equivalent cover was only 500 lb/a, the estimated erosion would be about 85 t/a/yr, not 22.

Also, the wind effects of climate can be altered by using windbreaks or irrigation to wet the soil. Soil surface roughness (K), field length (L), and vegetative cover (V) can be controlled by the land manager.

When soils are fine or medium textured (not sandy), soil surface roughness can be increased by establishing ridges in a field by tillage implements. Research has established that soil ridges from only 5–10 cm (2–4 in.) in height are most effective. Lower ridges were not effective either in reducing wind velocity or in trapping soil. Higher ridges increased wind velocity over the ridge tops and increased erosion.

Field length (L) can be shortened to reduce wind erosion by establishing porous-to-wind mechanical or vegetative windbreaks at suitable intervals at right angles to the most erosive winds. Porosity of about 40–50% seems to be ideal; and the higher the barrier, the more effective it is.[21,22]

Vegetative cover (V) is the most important factor in wind erosion under the control of the land manager. The erosion-resisting value of vegetation (Fig. 15-14) depends on the amount of plant material, its coarseness or fineness, its height when standing, and whether it is living or dead, standing or flattened. Because the original research was conducted on flattened wheat straw, the relative value of V in the wind erodibility equation is based on

---

[21] L. J. Hagen, "Windbreak Design for Optimum Wind Erosion Control," *Proceedings of the Symposium on the Great Plains*, Denver, Colo., April 20–22, 1976.
[22] Woodhouse, "Dune Building and Stabililzation with Vegetation."

FIGURE 15-14 Field windbreak protecting a corn crop in North Dakota. The species planted are buckthorn, boxelder, cottonwood, elm, ash, red cedar, and plum. (*Source:* USDA—Soil Conservation Service.)

this condition. Various graphs and tables have been developed to convert the principal kinds and conditions of plant materials to their equivalent of flattened wheat straw to obtain their *V* value for use in the equation.[23] Examples of wind erosion control practices are shown in Fig. 15-15.

> *We abuse land because we regard it as a commodity belonging to us. When we see land as a community to which we belong, we may begin to use it with love and respect.*
>
> **—Aldo Leopold**

> *Liberty exists in proportion to wholesome restraint.*
>
> **—Daniel Webster**

---

[23] Troeh, Hobbs, and Donahue, *Soil and Water Conservation,* p. 541.

(a)

(b)

FIGURE 15-15 (a) This special Calkins sweep plow was designed to be used in the fall, after the harvest of grain sorghum or wheat, to provide subsurface tillage that leaves approximately 90% of the stubble on the soil as a mulch to reduce wind erosion. Texas Panhandle. (b) A stubble mulch established by the Calkins sweep plow holds winter snows, reduces soil detachment by water or wind, improves soil structure, increases the infiltration of water, and reduces wind erosion. (*Source:* USDA—Soil Conservation Service.)

Questions

1. List some typical values (t/a or Mg/ha) per year for these erosion rates: (1) common maximum rates permitted by law (the tolerance *T* values), (2) the typical value for a high loss of about 2.5 cm (1 in.) of soil in 15 years, and (3) some very low erosion losses occurring under some well-stocked forests. (Assume that a hectare–15 cm weighs 2.0 million kg [calculate from typical values for the USLE].)
2. Explain the damages caused by allowing rainfall to hit bare soil.
3. Itemize a number of activities that increase detachment of soil.
4. Discuss the damages of erosion from the points of view of (a) quality of the soil lost and (b) rates of loss versus the rates of soil formation.
5. Define (a) *sheet* or *laminar erosion,* (b) *rill erosion,* and (c) *gully erosion.*
6. Write the Universal Soil Loss Equation (USLE), and define each term in it.
7. List the condition for each factor in the USLE that will allow (a) very large amounts of erosion and (b) very low amounts of erosion.
8. What is the RUSLE, and how does it differ from the USLE?

9. Define and briefly discuss these terms: (a) *highly erodible land* (HEL); (b) *erosivity index,* EI; (c) *conservation compliance,* and (d) *conservation tillage.*
10. Calculate the expected yearly erosion by water for (a) a field near Des Moines, Iowa, with $K = 0.32$, $LS = 1.41$, with a soybean mulch with 40% soil cover, on a 1–2% slope contour-planted in 390-ft lengths and (b) the same system if the P factor is a 3–5% slope contour-planted in 300-ft lengths.
11. What are the four characteristics about a soil that are used in estimating its K value?
12. Which causes the greatest increase in soil erosion by water: doubling the slope percentage or doubling the slope length?
13. (a)What vegetative cover has generally allowed the least soil erosion? (b) Explain why this is expected.
14. (a) Give the "reference condition" used to compare the relative amount of erosion by each factor in the equation. (b) As examples, give the meaning of these values: (1) an LS value of 8.2. (2) a C factor of 0.3, and (3) a P factor of 0.8.
15. In the erosion tolerance equation: (a) What does T mean? (b) Explain why T/RK is a "constant." (c) How and why is this equation used? (d) What are the typical values used for $T?$
16. Discuss the factor P in the USLE.
17. What are the factors in the Wind Erodibility Equation?
18. (a) In wind erodibility, what soil factors (I) influence the soil's erodibility index? (b) Describe a soil of high erodibility index.
19. Discuss the importance of soil cover on the wind erodibility.
20. Briefly discuss methods to control wind erosion.
21. What are the "units" of A, T, and E in the equations for erosion?

16

Water Resources and Irrigation

I have made water flow in dry channels and have given an unfailing supply to the people. I have changed desert plains into well-watered land.

—Hammurabi

16:1 Preview and Important Facts

PREVIEW

The planet we live on extends over 500,000,000 square kilometers, (158,000,000 square miles) and about three-fourths of it is covered by water. As the saying goes, "water, water everywhere, . . ." but not always enough in the required place or of suitable quality. Water is essential to plant and animal life; it is our best solvent, it carries off our wastes, and it modifies our climate. The mushrooming population and the extensive water pollution of the 1960s forced public awareness of the fragile nature of water and its limited distribution. The massive California water system and the other large structures developed to move water hundreds of miles emphasize the uneven distribution of water on the land area and the magnitude of the costs and labor required to translocate it. People must have water, and, to live in the affluent U.S. manner, they need large amounts of water for urban and industrial uses, as well as for agriculture. Where the water is, how much exists and of what quality, and potential water demands are important problems of water management.

The importance of water has many monuments and evidences around the world (Figs. 16-1 and 16-2). Raised waterways are found throughout the lands conquered by ancient Rome. In many arid Middle Eastern countries, underground "springs" (*ganats*) and canals were dug. Water flowed by gravity long distances. The use of underground canals eliminated the need to "lift" the water to the soil surface and reduced evaporation losses during transport (see Fig. 16-4).

Surface waters make up about 75% of the U.S. water being used, although they are only a fraction of the water supplies. Subterranean **groundwaters** are extensive, but they are expensive to use because of pumping costs. The porous strata called **aquifers** that contain these waters occur at various depths underground and may have very slow to very fast water flow and recharge rates. Surface waters and groundwaters also vary in composition.

FIGURE 16-1 Roman aqueduct (by the painter Zemo Diemer, the original now in the German Museum in Munich). Many remnants of the aqueducts remain to furnish information on the shape, size, and extent of the Roman systems. (*Source:* By permission, from *Food, Fiber and the Arid Lands,* by William G. McGinnies, Bram J. Goldman, and Patricia Paylore, eds., University of Arizona Press, Tucson, ©1971; and from *Arid Lands in Perspective,* William G. McGinnies and Bram J. Goldman, eds., University of Arizona Press, Tucson, ©1969.)

During a growing season plants use about 20–100 cm (8–40 in.) of water, with 40–75 cm (16–30 in.) being average for most crops. This water use is larger where crops are dense and grow longer in hot, dry climates. Plants vary in their rooting depths, in their susceptibility to damage when water is available slowly, and in the growth stage, during which water is most critical.

Irrigation waters vary in quality. The primary concerns are the salt content and the proportion of cations that are sodium. High salt or high relative sodium ion contents are undesirable.

Irrigation water can be added in many ways. Various furrow flow techniques are the cheapest and most used. Sprinklers are increasing in use but are expensive. Drip techniques are most efficient in water use but are expensive. The scientific approach to irrigation—when to irrigate and how much to add—has been extensively studied. With the increasing demands on a finite yearly supply of water, many changes in techniques, supplies, and water law are expected in the next few decades.

IMPORTANT FACTS TO KNOW

1. The definitions of *aquifer, spring, saltwater intrusion boundary, recharge area,* and *artesian well*
2. The hydrologic cycle and composition of aquifers
3. The extent to which over-use of surface waters and groundwaters is occurring
4. The major parameters determining the quality of water for irrigation use
5. The meaning of *SAR* and its relationship to the adjusted SAR
6. The approximate water requirements of most crops and how climate affects these requirement values
7. The soil depth from which most water for the plant is extracted

FIGURE 16-2 Ancient civilizations developed unique ways to get water from below ground. Underground tunnels *(ganats)* tapped the water-carrying strata. Access holes to the soil surface needed to be frequent (a), and sometimes the tunnels needed reinforcement (b). Near Shiraz, Iran. (Courtesy of Bruce Anderson, Utah State University.)

(a)

(b)

8. How to determine the time to irrigate and the amount of water to add
9. The advantages and disadvantages of irrigating with surface flow in furrows, basin and border strips, sprinkler systems, and drip or trickle systems
10. Some of the modifications needed to irrigate low-permeability soils, to use surge flow, and to use less than optimum quantities of water

16:2 Water Resources

The earth is estimated to have about 1359 million cubic kilometers (327 million cubic miles) of water with 97.22% of it contained in oceans, 2.15% in glaciers and icebergs, and about 0.03% of it circulated annually by precipitation, transpiration, and evaporation. This 0.03% is the critical amount that includes snow and rain, flowing surface water, underground recharge water, and atmospheric evaporative water vapor—most of the agricultural, industrial, and culinary water used daily (Fig. 16-3). This natural water circulation system is called the **hydrologic cycle.**

16:2.1 Surface Waters

About 75% of the water used in the United States is surface water because using it is usually cheaper and easier than to use groundwater. The composition of surface waters varies with the

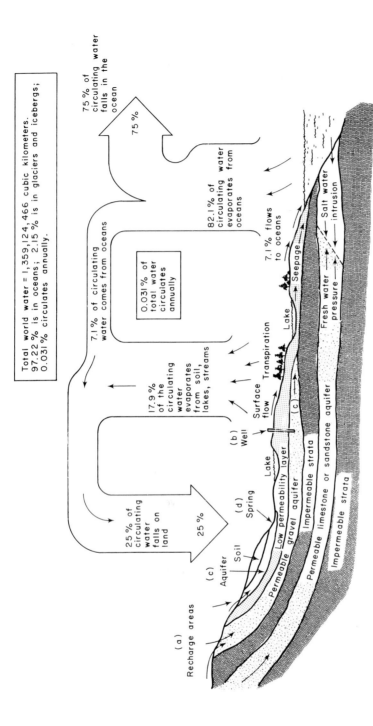

Total world water = 1,359,124,466 cubic kilometers. 97.22% is in oceans; 2.15% is in glaciers and icebergs; 0.031% circulates annually.

FIGURE 16-3 Hydrologic cycle, showing relative water distribution. Only about 0.03% of the total earth water recycles each year. (Percentage numbers are the portion of recycled water moving in that form.) Groundwaters are recharged from this circulated water. Salt water intrusion is caused by pumping out so much fresh water from underground sources that nearby ocean pressure fills the aquifer layer with salt water. (a) Recharge areas are land surfaces where porous strata are open to the soil surface. (b) Wells may flow or require pumping. If the level of recharge area water is higher than the well top, water can flow by gravity. This is called artesian well flow. (c) Aquifers often are coarse sediment laid down from former geologic erosion. These tend to become thinner or "pinch out" farther from the deposition front. (d) A spring is an exposed opening in an aquifer. (Courtesy of Raymond W. Miller, Utah State University.)

terrain through which the waters flow because the dissolving minerals that enter the water solution vary. The kinds and amounts of materials that are dissolved in surface waters exhibit extreme variations. Table 16-1 illustrates this variation in composition of rainwaters, river waters, and groundwaters with selected examples of the concentrations of the most common waterborne ions. The composition of a water is often critical to the uses that can be made of it.

16:2.2 Groundwaters

Much of the recent increase in irrigated land area has been possible because of a greater use of **groundwaters,** those waters in underground reservoirs in the deeper soil and substratum. These underground reservoirs are usually found in porous rock formations called **aquifers** (Fig. 16-3). Most aquifers extend from a few kilometers to 20–30 km (12–18 mi) long but may connect for hundreds of kilometers. These porous strata can be sands, gravels, porous sandstones, and channels in partly dissolved limestones.

How important are groundwaters in comparison to surface waters? Total groundwater stored is about 25 times more than the amount of water in all the world's lakes, rivers, and streams combined.[1,2] Groundwater is a vast and important resource. In the United States approximately 15 quadrillion gallons (56 quadrillion liters) of water are stored within 0.5 mi (0.8 km) of the land surface. Groundwater supplies about 25% of all fresh water used. Fifty percent of U.S. citizens obtain all or part of their drinking water from groundwater; and 95% of rural households depend on it totally. Commercially, groundwater is employed extensively in agricultural practices, particularly for irrigation, and in various industries.[3]

Groundwaters generally have a *more constant temperature, less sediment, less dissolved material,* and are *more omnipresent* than surface waters. With these groundwater advantages, why are surface waters so much more extensively used? Cost is the primary reason. Drilling wells and pumping to lift water are much more expensive than are the surface reservoirs and/or simple gravity-flow diversion canals needed for surface waters. Ancient and modern users have found unique ways to tap groundwater to avoid the problem of lifting it. *Horizontal wells* called *ganats* (or *khanats* or *kharezes*) have been used in Iraq, Iran, Afghanistan, and other countries for thousands of years (Figs. 16-2 and 16-4).

A serious problem in using groundwaters is the rate of recharge of the aquifer. Some aquifers are quite porous, and percolating recharge water can readily move hundreds of meters in a few days to refill the underground reservoir. Retaining adequate water in these aquifers depends upon the rate of water recharge of these reservoirs and the rate of water use from each aquifer.

In contrast to rapid-recharge aquifers, some aquifers allow very slow water intake and flow; water may flow only a few hundred meters per year. The Carrizo-Wilcox sandstone aquifer in Texas permits water movement at the rate of only 2–16 m (6.6–52.5 ft) per year.[4] If recharge areas were, say, 160 km (96 mi) away, natural recharge would take 3000–4000 years. If an aquifer layer is tapped by too many wells, the water is removed faster than recharge can refill it—a condition called *mining*. The water table drops, and the flow from wells decreases or stops entirely. As aquifer water—which helps to support the land above—is removed and not replaced, the surface land may sink (*subside*).

---

[1] Robert A. Weimer, "Prevent Groundwater Contamination before It's Too Late," *Water and Wastes Engineering,* **17** (1980), no. 2, pp. 30–33, 63.

[2] Brana Label, *EPA Groundwater Research Programs,* EPA/600/S8–86/004, Environmental Protection Agency, Washington, D.C., 1986.

[3] Ibid.

[4] Keith Young, *Geology: The Paradox of Earth and Man,* Houghton Mifflin, Boston, 1975, pp. 120–121.

Table 16-1 Principal Ions in Various Rainwaters, River Waters, and Groundwaters

Location of Sample	Cations				Anions			
	Na^+	K^+	Ca^{2+}	Mg^{2+}	Cl^-	SO_4^{2-}	NO_3^-	HCO_3^-
Rainfall[a]	ppm	ppm	ppm	ppm	ppm	ppm	ppm	ppm
Cape Hatteras, North Carolina, on coast	4.5	0.2	0.4	N.D.[b]	6.5	0.9	1.0	N.D.
Brownsville, Texas 1 mile from coast	22.3	1.0	6.5	N.D.	22.0	5.3	1.8	N.D.
Ely, Nevada	0.7	0.1	3.8	N.D.	0.3	1.0	0.8	N.D.
Columbia, Missouri	0.3	0.3	2.2	N.D.	0.1	1.2	3.8	N.D.
Rivers								
Savannah, Georgia[a]	3.5	1.0	4.0	1.3	2.5	3.0	0.3	22.0
Mississippi, at Minneapolis[a]	——10——		40	14	2.0	18	2.0	190.0
Colorado, at Grand Canyon[a]	120	6.0	94	34	90	320	5.0	200
California Aqueduct, Los Angeles (1965)[c]	40	5.0	26	6.0	19	23	0.2	N.D.
Groundwaters								
Saline Utah wells[c]	685	N.D.	226	139	1940	33	N.D.	67
Pennsylvania Appalachian Plateau, 4 m deep[d]	10	3.5	33	8.0	6.5	50	0	N.D.
Pennsylvania strip mine contaminated, 6 m deep[d]	18	3.6	160	230	6.0	950	0.2	N.D.

[a] A. N. Stahler and A. H. Stahler, *Introduction to Environmental Science,* Hamilton Publishing Co., Santa Barbara, Calif., 1974, pp. 340–341.
[b] N.D. means *not determined.*
[c] M. W. Kellogg Co., *Saline Water Conversion Engineering Data Book,* U.S. Department of Interior, Washington, D.C., 1965.
[d] Joseph R. J. Studlich, "Recharge of Wells from Spoil Banks," *Ground Water,* **16** (1978), no. 3, pp. 204–205.

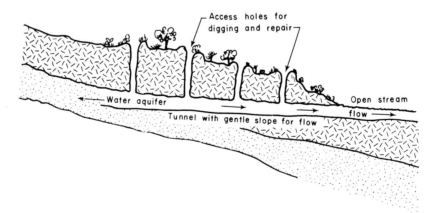

FIGURE 16-4 Cross section of a *ganat,* an ancient well made by digging slightly sloping tunnels into aquifers to obtain irrigation water. These are still in use. (Courtesy of Raymond W. Miller, Utah State University.)

To solve the problem of inadequate water distribution, water is often transported great distances in expensive canals. Israeli scientists in a water-short country discovered that some of their aquifers were continuous from north to south. Pumping water in the north from Lake Tiberius (Galilee) into depleted wells replenished the aquifer's water, which flowed as far as 242 km (150 mi) south, where it could be pumped out for use.[a]

Many other areas are now using deep wells as water disposal sinks. In Orange County, California, treated sewage effluent pumped into deep wells helps keep salt water intrusion from moving farther inland and also recharges the aquifer for use by others.

[a]Michael Overman, *Water,* Doubleday, New York, 1979, pp. 11–14.

As groundwater use increases, a balance of extraction and recharge is essential (Detail 16-1). For example, groundwater supplies about 40% of Arizona's water, but it is being pumped out almost five times faster than it is being recharged.[5] In south-central Kansas farmers above the Equus Beds aquifer north of Wichita were limited by 1979 state regulations to 450 mm (17.6 in.) of groundwater per year, about two-thirds of optimum needs.[6] In western Kansas water resources are in even worse condition.

Examples of the groundwater problem in the United States are given in Table 16-2 and Fig. 16-5. The western 60% of the country, except for upper Colorado, is using water faster than recharge rates. Overuse ranges from 8.5% to 77.2% annually. Subregions in the Arkansas-White-Red and the Texas-Gulf regions have more than 60% overdraft.

16:3 Water Supplies for Irrigation

From its beginning, irrigation agriculture has expanded in the western United States to encompass 12 million hectares (30 million acres). Although this represents only 3.5% of all farmland in these Western states, the crops harvested from irrigated land represent 35% of the value of all crops harvested in that region.

In other parts of the United States irrigation is expanding rapidly. Irrigation is concentrated in southern California, the high plains of Texas and Nebraska, the rice belts of Texas, Louisiana, and Arkansas, and the vegetable farms in central and southern Florida, but all 50 states practice irrigation to some extent.

Recent work in North Carolina indicates the value of irrigation in humid regions. In an area with an average annual precipitation of 102 cm (40 in.), as many as 50 days of drought may occur one year in five. Although the average precipitation seems to be adequate for crop production, during some years irregularities in rainfall may make irrigation profitable here.

A similar study was made of "drought days" in the lower Mississippi Valley where average annual precipitation varies from approximately 127 cm to 152 cm (50–60 in.) a year. If properly distributed throughout the growing season, this amount of rainfall would be sufficient to grow all crops except sugarcane and flooded rice. Because of distribution irregularities, however, many areas have drought days half the time, whereas only a few areas have drought days less than one-third of the time. As a consequence, irrigation is used to achieve maximum crop production on all but the most water-retentive soils in the lower Mississippi Valley. Even though the average annual precipitation in the Mississippi Delta is 127 cm

[5] Anonymous, "Like Having Your Dad Die," *Time,* Mar. 7, 1977, p. 80.
[6] Ron Larsen, "Water Control Imperative in Kansas," *Irrigation Age,* **13** (1979), no. 5, pp. 38–41.

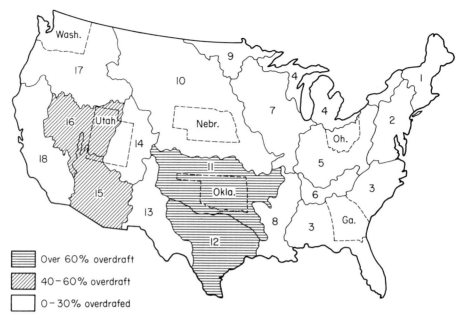

FIGURE 16-5 The 21 Water Resource Regions, which are subdivided into 106 subregions. The three not shown are Alaska, Hawaii, and the Caribbean.   (*Source:* U.S. Water Resources Council, *The Nation's Water Resources, Vol. 1: Summary,* U.S. Government Printing Office, Washington, D.C., 1978, 87 pp.)

Table 16-2 Groundwater Withdrawals in 1975 from the 21 U.S. Water Resource Regions and the 106 Subregions

Water Resources Region[a]	Overdraft[b] in Region (%)	Number of Subregions	Subregions with Overdraft
1. New England	0	6	0
2. Mid-Atlantic	1.2	6	3
3. South Atlantic Gulf	6.2	9	8
4. Great Lakes	2.2	8	1
5. Ohio	0	7	0
6. Tennessee	0	2	0
7. Upper Mississippi	0	5	0
8. Lower Mississippi	8.5	3	3
9. Souris-Red-Rainy	0	1	0
10. Missouri	24.6	11	10
11. Arkansas-White-Red	61.7	7	7
12. Texas-Gulf	77.2	5	5
13. Rio Grande	28.1	5	4
14. Upper Colorado	0	3	0
15. Lower Colorado	48.2	3	3
16. Great Basin	41.5	4	4
17. Pacific Northwest	8.5	7	6
18. California	11.5	7	5
19. Alaska	0	1	0
20. Hawaii	0	4	0
21. Caribbean (Puerto Rico)	5.1	2	1

Source: U.S. Water Resources Council, *The Nation's Water Resources, Vol. 1: Summary,* U.S. Government Printing Office, Washington, D.C., 1978, p. 18.

[a] Number before region is the number in the map (Fig. 16-5).

[b] Overdraft is a greater use over several years than is being replaced by recharge.

Table 16-3 Values of Total Cropland, Harvested Cropland, Irrigated Cropland, and Crops for 1978

State	Total Cropland[a] (ha)[b]	Harvested Cropland (ha)	Irrigated Cropland		Value of Crops ($millions)
			(ha)	Percentage of Total Cropland	
Western states					
Arizona	646,241	454,038	499,453	77.3	669
California	4,755,321	3,606,867	3,469,185	73.0	5,917
Colorado	4,352,329	2,384,266	1,394,936	32.1	575
Idaho	2,695,474	1,981,156	1,421,891	52.8	800
Montana	6,637,568	3,566,490	844,696	12.7	490
Nevada	349,287	241,444	363,960	100+[a]	60
New Mexico	935,278	494,086	366,695	39.2	190
Oregon	2,153,161	1,332,123	780,069	36.2	694
Utah	827,361	476,825	480,270	58.0	105
Washington	3,415,823	2,061,879	689,087	20.2	1,309
Great Plains states					
Kansas	12,161,128	7,731,055	1,055,285	8.7	1,449
Nebraska	9,068,305	6,655,838	2,243,202	24.7	1,725
North Dakota	11,672,221	7,740,318	58,183	0.5	1,296
Oklahoma	5,930,853	3,523,996	250,383	4.2	625
South Dakota	7,635,515	5,643,145	139,769	1.8	568
Texas	16,278,851	8,426,494	2,841,248	17.5	2,711
Wyoming	1,119,718	733,097	684,605	61.1	91
Midwestern states					
Illinois	10,269,800	9,241,483	53,167	0.5	3,915
Indiana	5,537,939	4,822,950	30,773	0.6	1,938
Iowa	11,443,188	9,631,606	40,961	0.4	3,150
Kentucky	3,873,187	1,865,227	6,280	0.2	978
Michigan	3,507,249	2,813,968	91,882	2.6	1,014
Minnesota	9,185,664	7,774,201	110,170	1.2	2,082
Missouri	8,356,167	5,098,195	139,085	1.7	1,404
Ohio	4,953,642	4,165,799	10,708	0.2	1,611
West Virgina	641,557	245,538	688	0.1	47
Wisconsin	4,959,764	4,037,353	95,366	1.9	677
Southern states					
Alabama	2,281,713	1,390,709	23,977	1.1	510
Arkansas	4,281,738	3,097,541	683,827	16.0	1,194
Florida	1,829,472	1,121,145	810,396	44.3	2,187
Georgia	2,827,366	1,923,022	187,706	6.6	925
Louisiana	2,646,894	1,992,460	276,855	10.5	848
Mississippi	3,441,837	2,432,185	128,888	3.7	962
North Carolina	2,526,265	1,849,906	38,249	1.5	1,705
South Carolina	1,409,043	1,040,712	13,570	1.0	551
Tennessee	3,241,836	1,812,407	6,428	0.2	674
Virginia	1,932,645	1,096,811	18,013	0.9	529
New England and Atlantic coastal states					
Connecticut	101,373	72,257	2,823	2.8	85
Delaware	214,747	200,782	13,663	6.4	102
Maine	278,703	197,588	2,986	1.1	110
Maryland	768,486	618,685	11,685	1.5	281
Massachusetts	126,120	126,120	6,880	5.5	101

Continued.

[a] Cropland does not include woodlands, noncultivated pasturelands and rangelands, or land in house lots, roads, etc. The data for Nevada appear inconsistent (irrigated cropland exceeds total cropland); perhaps it includes irrigated rangelands for pasture.
[b] Hectares times 2.471 equals acres.

Table 16-3, cont'd Values of Total Cropland, Harvested Cropland, Irrigated Cropland, and Crops for 1978

State	Total Cropland[a] (ha)[b]	Harvested Cropland (ha)	Irrigated Cropland		Value of Crops ($millions)
			(ha)	Percentage of Total Cropland	
New Hampshire	80,306	55,435	731	0.9	23
New Jersey	305,487	250,594	31,238	10.2	254
New York	2,504,124	1,815,230	23,754	0.9	548
Pennsylvania	2,354,879	1,758,504	6,130	0.3	648
Rhode Island	14,831	10,179	1,209	8.2	16
Vermont	349,284	235,459	658	0.2	17
Alaska	11,879	8,315	372	3.1	5
Hawaii	134,931	64,243	64,510	47.8	331

Source: Arranged and calculated by Raymond W. Miller from *1978 Census of Agriculture,* U.S. Department of Commerce, Bureau of the Census, Washington, D.C., for individual states.

[a] Cropland does not include woodlands, noncultivated pasturelands and rangelands, or land in house lots, roads, etc. The data for Nevada appear inconsistent (irrigated cropland exceeds total cropland); perhaps it includes irrigated rangelands for pasture.

[b] Hectares times 2.471 equals acres.

(50 in.), growing season droughts occur (with an 80% probability). Corn will need 29.2 cm (11.4 in.) of irrigation water for maximum yield.[7]

An indication of the growth of irrigation in the United States can be seen in Table 16-3 and the extent of irrigation worldwide in Table 16-4. In the United States about 10% of cultivated land is irrigated, and of that two-thirds is by surface flow irrigation methods.

16:4 Irrigation Water Quality

Water quality is determined according to the purpose for which it will be used. For irrigation waters, the usual criteria include salinity, sodicity (sodium content), and element toxicities. (These are discussed in detail later.) Many criteria important in assessing water quality for other uses—taste, color, odor, turbidity, temperature, hardness, pH, BOD or COD, nutrient content (N and P), and pathogenic organisms—are sometimes, but not usually, important for irrigation water.

Turbidity is water opaqueness caused by the presence of suspended solids of clays, silts, sands, and organic materials. These materials may fill up irrigation canals or reservoirs, seal up the surface pores of soils, grind away turbine blades of electrical generators, and clog sprinkler and trickle irrigation systems. Sediment problems can be *reduced* by the use of settling basins and filters, but this is expensive.

Water temperature is of limited concern in irrigation except where it may be cold enough to reduce growth, such as for flooded rice production. Usually, the soil heat and incoming solar radiation modify the water's initial temperature enough that it does not greatly affect plant growth.

---

[7] Jonathan W. Pote and Charles L. Wax, "Climatological Aspects of Irrigation Design Criteria in Mississippi," *Technical Bulletin 138,* Mississippi State University, Starkville, 1986.

Table 16-4 Major Irrigating Countries in the World according to Total Amount of Irrigated Area and Percentage of Cultivated Area That Is Irrigated (Data Mostly for Years 1968–1971)

Country	Irrigated Area (millions of acres)[a]	Cultivated Area (millions of acres)[a]	Percentage of Cultivated Area Irrigated	Ranking Based on Percentage of Area Irrigated[b]
China	187.7	272.4	68.9	2
India	67.9	406.6	16.7	23
United States	39.0	475.0	8.2	—
Pakistan	30.9	47.4	65.2	3
USSR	27.4	575.0	4.8	—
Indonesia	16.8	44.5	37.8	8
Iran	13.0	41.2	31.6	12
Mexico	10.4	58.8	17.7	21
Iraq	9.1	25.1	36.3	10
Egypt	7.0	7.0	100.0	1
Japan	7.0	13.6	51.5	5
Italy	6.0	30.6	19.6	18
Spain	6.0	50.9	11.8	—
Thailand	4.5	28.2	16.0	25
Argentina	3.8	64.2	5.9	—
Turkey	3.8	67.6	5.6	—
Australia	3.6	110.2	3.3	—
Peru	2.7	7.2	37.5	9
Taiwan	1.2	2.1	57.1	4
Albania	0.6	1.4	42.9	7
Israel	0.4	1.0	40.0	6

Source: "World Food Situation and Prospects to 1985," Foreign Agricultural Economic Report 98, Economic Research Service, USDA, Washington, D.C., 1974, pp. 70–71.

[a] To convert to hectares, multiply acres by 0.405.

[b] Where no ranking is given, the nation does not rank in the top 25 in percentage of cultivated area irrigated.

BOD (biological oxygen demand) or **COD** (chemical oxygen demand) are measures of how much of the oxygen dissolved in water will be used as the organic material and certain chemicals in the water are decomposed (oxidized). A high BOD or COD is caused by organic materials such as algae, plant residues, or manures in the water (Fig. 16-6).

The same materials are usually high in nitrogen and phosphorus, which are nutrients for the growth of algae. If these manures and organic materials are present in sufficient concentrations, the contaminated water (especially if it is static or slow moving) will soon be covered and filled with excessive growth of green algal slimes.

Sludge and manure effluents have high BODs and CODs and are undesirable for aquatic life (fish, algae, protozoa). Such waters would also be more rapidly depleted of oxygen when added to poorly drained soils. High BOD waters when used on soil can cause poor aeration (inadequate oxygen) conditions faster than would low-BOD waters.

Pathogenic organisms are disease bearing. It is becoming more common to find some pathogenic organisms in natural freshwater sources in the United States; they are almost certain to be present in inadequately treated sewage effluents. If sewage effluents are to be used for irrigation (or disposal) on agricultural land, they should first be certified by the Public Health Service to be free of viable pathogenic organisms.

FIGURE 16-6 A small stream ponded near areas of animal habitation where manure washed into the stream results in the growth of algae (the light-colored floating material). This algal-growth condition indicates an undesirably high level of nutrients (especially of nitrogen and phosphorus) called *eutrophication.* Large amounts of decomposable organic wastes, including the dead algae, result in high BOD and depleted oxygen levels in the water. (Courtesy of Raymond W. Miller, Utah State University.)

16:4.1 Salinity

Salinity, or total soluble salts (TSS), is one of the most critical criteria for irrigation water quality. Salts affect plants by increasing the osmotic pressure of water, making the plant exert more energy to absorb soil water. Salt concentration of a few tenths of a percent by weight can affect or hinder plant growth. Salt contents are measured by electrical conductivity (EC) in siemens meter^{-1} (S m^{-1}) or decisiemens meter^{-1} (dS m^{-1}) for soil solutions or millisiemens meter^{-1} (mS m^{-1}) for waters. Previously these measurements were reported as mmhos/cm and micromhos/cm, respectively.

A typical water classification is given in Fig. 16-7. The arbitrarily selected boundaries are only approximate; other classification systems have been proposed.

Whether a given water is usable or unsuitable for plants depends on the plant grown, the amount of leaching permitted during each irrigation, and how dry the soil is allowed to get before the next irrigation. If the soil is allowed to dry, salts will move to the surface with evaporating water instead of moving deeper into the soil by leaching. The more leaching that occurs and the wetter the soil is kept, the higher is the salt content in soil that can be tolerated by plants.

A second system of classifying irrigation water is given in Table 16-5. Notice that the salinity boundaries show the C1 and C2 waters from Fig. 16-7 as "no problem." Most of the C4-category water and higher salt in Figure 16-7 are categorized as a severe problem (3 mmhos/cm is the same as 3000 micrommhos and equals 300 mS m^{-1}).

16:4.2 Sodium Hazard (Sodicity)

High concentrations of sodium are undesirable in water because sodium adsorbs onto the soil cation exchange sites, causing soil aggregates to break down (disperse), sealing the pores of

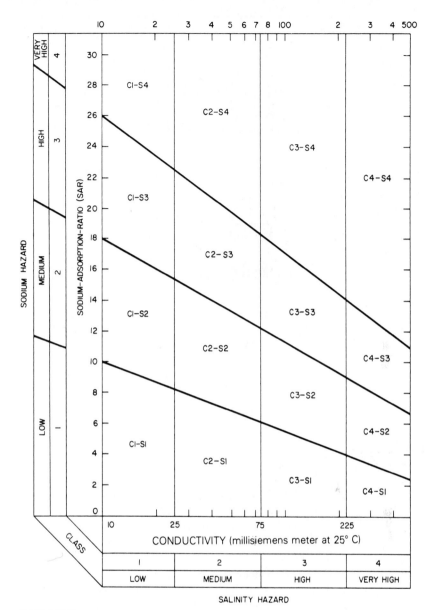

FIGURE 16-7 One of several classifications of waters to be used for irrigation. The best waters are toward the lower left corner; the poorest waters are toward the upper right. (*Source:* Modified from L. A. Richards, ed., *Diagnosis and Improvement of Saline and Alkali Soils,* Agriculture Handbook 60, USDA, Washington, D.C., 1954, p. 80.)

the soil, and making it less permeable to water flow. The tendency for sodium to increase its proportion on the cation exchange sites at the expense of other types of cations is estimated by the ratio of sodium content to the content of calcium plus magnesium in the water. This is called the sodium adsorption ratio (SAR). A small SAR value indicates a desirably low sodium content.

The Food and Agriculture Organization (FAO) guideline (Table 16-5) refers to the sodicity problem as *permeability*. The guide includes salt concentrations because, at very low salt levels, the soil particle flocculation (which occurs with any high salt concentration) is

Water Constituent	Intensity of Problem[a]		
	No Problem	Moderate	Severe
Salinity (decisiemens meter $^{-1}$)	<0.70	0.70–3.0	>3.0
Permeability (rate of infiltration affected)			
Salinity (decisiemens meter $^{-1}$)	>0.5	0.5–0.2	<0.2
Adjusted SAR; soils are:			
Dominantly montmorillonite	<6	6–9	>9
Dominantly illite-vermiculite	<8	8–16	>16
Dominantly kaolinite-sesquioxides	<16	16–24	>24
Specific ion toxicity			
Sodium (as adjusted SAR) (sprinkler)	<3	3–9	>9
Chloride (mmol/L) (sprinkler)	<3	>3	>10
Boron (mmol/L)[b] as B	<0.70	0.70–30	>3.0
Miscellaneous			
NO_3^-—N or NH_4^+—N (mmol/L)	<5	5–30	>30
HCO_3^- (mmol/L) as damage by			
overhead sprinkler	<1.5	1.5–8.5	>8.5
pH	6.5–8.4		0–5, 9.5+

Source: Modified from R. S. Ayres and D. W. Westcott, "Water Quality for Agriculture," Irrigation and Drainage Paper 29, FAO, Rome, 1976; rev. 1986.

[a] Based on the assumptions that the soils are sandy loam to clay loams, have good drainage, are in arid to semiarid climates, that irrigation is sprinkler or surface, that root depths are normal for deep soil, and that the guidelines are only approximate.

[b] Assumes molecular weight = mole$_c$ weight (one charge) because it is slightly ionized or nonionized.

lost and permeability decreases. Waters that are very low in salt (<0.2 dS m^{-1}) may accentuate poor permeability.

The FAO guidelines recognize also that the problem of sodium is most severe with montmorillonitic soils and least with kaolinitic and sesquioxide (metal oxide) clays that have slight swelling. The adjusted SAR is a value corrected to account for the removal of calcium and magnesium by their precipitation with bicarbonate and carbonate ions in the water added, giving higher values for adjusted SAR than for SAR and a truer picture of the sodicity of the soil (Calculation 16-1). Notice in Table 16-5 that "severe" SAR values in montmorillonitic soils are those about 9 and larger. This is lower than the SAR of 13 for soil paste extracts used to define a sodic soil in Chapter 11.

16:4.3 Toxicities

Boron is the most commonly encountered element found in toxic concentrations in water. Because it is quite soluble, boron is found in water where drainage and geologic strata supply boron source minerals. The problem of boron levels for plants is accentuated because the range between nutritionally deficient and toxic levels of boron is relatively narrow. Boron cannot be precipitated or otherwise easily removed from water. The only known remedy is to dilute high-boron water with low-boron water or to grow boron-tolerant crops.

Chloride and bicarbonate may cause problems. Many plants (avocado, tobacco, berries) are sensitive to high chloride concentrations and sometimes to high sodium levels in their leaves. Bicarbonates and carbonates promote precipitation of calcium as calcium carbonate (lime) during drying periods, resulting in a higher SAR in the water (higher sodium hazard) because of the lowered calcium content.

The SAR is defined, in $mmol_c/L$, as

$$SAR = \frac{Na^+}{\sqrt{(Ca^{2+} + Mg^{2+})/2}}$$

The SAR indicates the tendency for the soil to become higher in exchangeable sodium; higher SAR values mean higher exchangeable sodium percentages and lower soil permeability. If the water contains bicarbonate (HCO_3^-) and carbonate (CO_3^{2-}) ions, these will precipitate with calcium and magnesium, which increases the SAR. The formula is the *adjusted SAR*, and it is defined as

$$adj\ SAR = \frac{Na^+}{\sqrt{(Ca^{2+} + Mg^{2+})/2}}[1 + (8.4 - pHc)]$$

and

$$pHc = (pK_2' - pKc') + p(HCO_3^- + CO_3^{2-}] + p[Ca^{2+} + Mg^{2+}]$$

$(pK_2' - pKc')$ is essentially the sum of Ca + Mg + Na ion concentrations, $p(HCO_3^- + CO_3^{2-})$ is the bicarbonate + carbonate concentration, and p(Ca + Mg) is the concentration of the Ca plus Mg ions only. At the present time there is concern that this calculation needs correction because the adjusted SAR may sometimes be too different from the SAR.

Problem Calculate the adj SAR for water with these ion contents: 7 $mmol_c/L$ of Ca^{2+}, 2 $mmol_c/L$ of Mg^{2+}, 5 $mmol_c/L$ of Na^+, 4 $mmol_c/L$ of HCO_3^-, and a total cation concentration of 14 $mmol_c/L$.

Ion Concentrations ($mmol_c/L$)	$(pK_2' - pKc')$[a] (Ca + Mg + Na)	p(Ca + Mg) (Ca + Mg)	$p(HCO_3^-) + CO_3^{2-}$ ($HCO_3 + CO_3$)
0.5	2.11	3.60	3.30
1.0	2.13	3.30	3.00
4.0	2.20	2.70	2.40
8.0	2.25	2.40	2.10
10	2.27	2.30	2.00
20	2.35	2.00	1.70
30	2.40	1.82	1.52
40	2.44	1.70	1.40
50	2.47	1.60	1.30

[a] Essentially, the total cation content in the water (Ca + Mg + Na ions).

Solution Interpolating to get values between those listed in the table, $(pK_2' - pKc')$ at 14 $mmol_c/L$ total ions = about 2.30. The p(Ca + Mg) (7 + 2 $mmol_c/L$) = 2.35 and $p(HCO_3)$ (4 $mmol_c/L$) = 2.40. The pHc = 2.30 + 2.40 + 2.35 = 7.05:

$$adj\ SAR = \frac{5}{\sqrt{(7 + 2)/2}}[1 + (8.4 - 7.05)]$$

$$= 2.36(1 + 1.35)$$

$$= 5.55 \quad \text{rather than the uncorrected SAR of 2.36}$$

In localized areas, other elements toxic to plants or animals, such as lithium (California) and selenium (Wyoming), may contaminate water. Nitrates can also be toxic to mammals if present in water at concentrations of more than 15 parts per million of nitrate-nitrogen.

16:5 Water Needs of Plants

Most commercial crops cannot store water to carry them through a dry period; plants need a continuous water supply. Any serious reduction in available water reduces plant growth to some extent. The growth reduction is more pronounced as (1) the time of dryness increases, (2) the rate of transpiration is greater, and (3) the rate of water movement from drying soil areas to root surfaces is slower (because thinner water films in soils are held with greater force). To irrigate efficiently, it is necessary to know the amount of water needed by the chosen crop and the method of applying it that will provide the best results.

Consumptive use is increased by conditions that increase evaporation: warm days, dry air, wind if the atmosphere is dry, and maximum plant-available water in the soil. Some typical consumptive use (evapotranspiration, *ET*) values are given in Table 16-6; note that total water needs will vary from year to year as climatic conditions vary (Table 16-7). **Daily consumptive use** ranges from low values of about 2 mm (a tenth of an inch) to maxima of about 1–1.5 cm (0.4–0.6 in.). For example, peak water use in midsummer in California is generalized as follows[8]:

Area	Water Use per Day by Plants
Coastal fog belt	0.25–0.38 cm (0.10–0.15 in.)
Coastal valley	0.50–0.64 cm (0.20–0.25 in.)
Interior valleys	0.64–0.76 cm (0.25–0.30 in.)
Desert areas	0.64–1.02 cm (0.25–0.40 in.)

Some additional values and conditions are given in Chapter 6.

16:6 Amount and Frequency of Irrigation

The objective of irrigation is to add the amount of water needed when the plant requires it. A water shortage reduces vegetative growth, forces premature "seed" production, or both. Some factors to be considered when planning the amount and frequency of irrigation are

1. The depth and distribution of plant roots
2. The amount of water retained within the rooting depth
3. The minimum water potential to be maintained in the root zone
4. The rate of water use by the plant (consumptive use)
5. Whether adequate irrigation water is available to add when it is needed

---

[8] L. N. Brown and L. J. Booher, "Irrigation on Steep Land," California Agricultural Experiment Station—Extension Service Circular 561, 1972.

Table 16-6 Selected Total Evapotranspiration Values for Crop Season or Crop Year Growth When the Crop Is Well Watered[a]

Crop	Location	Crop Duration	Evapotranspiration in.	Evapotranspiration mm
Alfalfa	North Dakota	143 days (summer)	23.4	594
	Nevada	124 days (summer)	39.9	1013
Grass	Canada	—	22.8	579
	Davis, Calif.	12 months	51.8	1316
Barley	Wyoming	May–Aug.	15.2	386
	Mesa, Ariz.	Dec.–May	25.3	643
Beans	South Dakota	105 days	16.4	417
	Davis, Calif.	92 days	15.9	404
Corn	Ohio	124 days	18.5	470
	Bushland, Tex.	122 days	24.3	617
Potatoes	Alberta, Can.	—	19.9	505
	Phoenix, Ariz.	Feb.–June	24.3	617
Rice, flooded	Davis, Calif.	150 days	36.2	919
Sorghum	Kansas	—	21.7	551
	Mesa, Ariz.	July–Nov.	25.4	645
Wheat, hard	South Dakota	—	16.3	414
	Bushland, Tex.	Oct.–June	28.3	719
Sugar beets	Montana	Apr.–Sept.	22.5	571
	Kansas	Apr.–Nov.	36.5	927
Safflower	Southern Idaho	Apr.–Sept.	25.0	635
Soybeans	South Dakota	—	15.7	399
Cotton	Arvin, Calif.	12 months	35.9	912
	Mesa, Ariz.	Apr.–Nov.	41.2	1046
Cabbage, late	Mesa, Ariz.	Sept.–Mar.	24.9	632
Lettuce	Mesa, Ariz.	Sept.–Dec.	8.5	216
Peas, green	Alberta, Can.	—	13.4	340
Tomatoes	Alberta, Can.	—	14.4	366
	Davis, Calif.	May–Oct.	26.8	689
Apples	Wenatchee, Wash.	Apr.–Nov.	41.7	1059
Oranges	Phoenix, Ariz.	12 months	39.1	993
Turf	Reno, Nev.	112 days	21.8	554

Source: Selected and modified from a tabulation by Marvin E. Jensen, ed., *Consumptive Use of Water and Irrigation Water Requirements,* American Society of Civil Engineers, New York, 1973.

[a]Notice that values for a given crop usually increase from cooler Northern areas to warmer Southern regions. Notice also that in the table the crop durations and times of the year vary considerably.

16:6.1 Plant Root Systems

Rooting depths for any given plant are altered by shallow hardpans, dense soil, and clayey soil texture, as well as by the watering and tillage methods. The discussion that follows assumes deep, medium-textured soil without physical limitations for roots—more like ideal conditions.

The effective rooting depth for numerous plants is given by showing the suggested irrigation depth for some crops (Fig. 16-8). To understand the importance of keeping the upper soil layers moist, a rule of thumb for water uptake by plants is

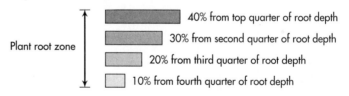

Plant root zone

40% from top quarter of root depth

30% from second quarter of root depth

20% from third quarter of root depth

10% from fourth quarter of root depth

Table 16-7 Variations in Total Water Required (Evapotranspiration, *ET*) for Various Crops in Three Successive Years in Ankara, Turkey

Crop	1967–1968 ET (in.)[a]	1967–1968 Days	1968–1969 ET (in.)[a]	1968–1969 Days	1969–1970 ET (in.)[a]	1969–1970 Days
Wheat	22.5	289	23.0	279	28.6	264
Alfalfa	28.4	183	29.0	168	41.7	199
Sugar beets	26.8	172	33.3	163	25.0	165
Potatoes	18.8	164	20.9	131	36.7	145
Beans	—	—	16.2	121	22.6	146
Honeydew melons	13.4	153	—	—	—	—
	1961		*1962*		*1963*	
Tomatoes	25.0	104	25.4	152	19.8	102
Cotton (in Tarsus)	21.0	169	18.8	184	21.0	136

Source: Modified from O. Beyce, "Water Requirements of Various Crops in Arid and Semi-arid Zones of Turkey," in *CENTO Seminar on Agricultural Aspects of Arid and Semi-arid Zones,* Tehran, Iran, Sept. 19–23, 1971, pp. 191–211.
[a] To convert to centimeters, multiply inches by 2.54.

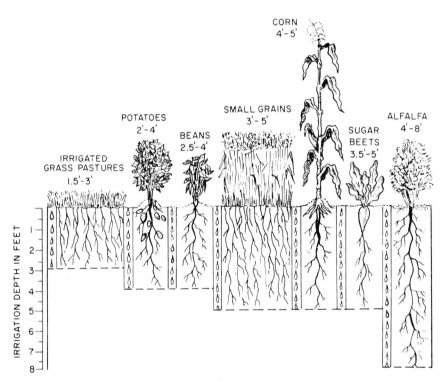

FIGURE 16-8 Normal irrigation depth for common farm crops varies from 1½ ft for some grass pastures to as much as 8 ft for alfalfa. Some not shown are carrots and peas (2–3 ft), cantaloupe (3–5 ft), grape (5 ft), and tomatoes (3–5 ft). Feet x 30.5 = cm. (*Source: Irrigation on Western Farms,* USDA Agricultural Information Bulletin 199, 1959.)

Obviously, most roots and the uptake of most water occur in shallow soil depths. Allowing the surface 30–60 cm (1–2 ft) of soil to dry will greatly limit the plant uptake of water and nutrients.

16:6.2 Minimum Water Potential

Although plants can readily use soil water held with water potentials approaching –1500 kPa (–15 bars, permanent wilting point), most of the plant-available water is held in high potentials, from –33 to –100 kPa (–⅓ to – 1 bar). Irrigation is recommended when about 50% of plant-available water has been used in the zone of maximum root activity, probably within the 15–60-cm (6–24-in.) depth (Fig. 16-9). Except in ripening seed crops or sugar crops, the moisture content in soil should never approach the permanent wilting point unless rapid maturing is needed.

In Arizona, one of the high-yielding, cotton-producing states, cotton yields were increased 25% by irrigating four times in July rather than the usual two times.[9] The increase is claimed to be because of less stress to the plant during its critical boll-producing stage of growth. Approximately the same amount of total water for July was used, but only one-fourth was added per irrigation in each of the four irrigations. Also, the frequent irrigations keep the surface soil, where most roots and nutrients are located, wet more of the time. Cotton plants, under water stress, shed some fruits. Water stress may also cause some root die-back during boll loading, again reducing water uptake.

16:6.3 Calculating When to Irrigate

The objective of irrigation is to keep adequate water available to crops or, if water is in short supply, to use what is available most effectively. The most critical time to keep water available for fruits, nuts, grains, or cotton is the several weeks following flowering. If the desired yield is vegetative growth rather than fruit, the critical period may be the latter half of the growth period prior to flowering, when most *size* growth occurs.

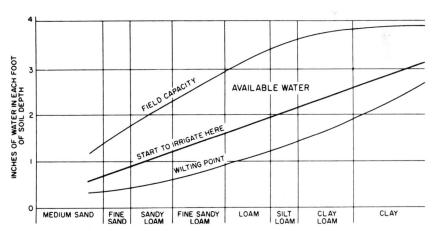

FIGURE 16-9 The amount and frequency of irrigation are determined partly by soil texture. A silt loam soil has the largest available water capacity and, therefore, should require irrigation water less frequently than a sand or a clay soil. Feet × 30.5 = cm. (*Source:* "Water," *Yearbook of Agriculture,* USDA, Washington, D.C., 1955.)

---

[9] Dennis Senft, "Extra Irrigations Protect Cotton," *Agricultural Research,* **40** (1992), no. 2, p. 9.

The method of calculating CU or ET from climatic data is too complex for a short discussion in this text, but Table 16-8 indicates how the calculated data can vary if the equation used is not "calibrated" with measured data for the area. For example, the Thornthwaite equation in Table 16-8 gave values that overestimated ET in Copenhagen, Denmark, by 36% and underestimated it in Kimberly, Idaho, by 58%.

In comparison to estimating CU or ET from climatic-data formulas, the use of the weather station class-A pan evaporation to estimate ET is quite accurate and simple.[10] The assumption is that all effects of climate on ET (temperature, wind, relative humidity, day length, etc.) act similarly on water in a pan and on water in crops. After measuring loss from a pan, the next requirement is to correlate the pan loss to actual measured losses for that crop, then use the pan water loss relationship to figure ET values for the crop. Two examples are given in Calculations 16-2 and 16-3.

Irrigation timing by most commercial agriculturalists is planned, based on (1) "an art" learned over the years, (2) the use of simple measurements, or (3) careful measurements estimating consumptive use.

The first method, the art of irrigation learned by experience, can be effective in determining watering timing, but extra water beyond minimum need is usually applied "to be sure." The irrigator looks at early signs of dryness: darkened foliage color; slight temporary wilts in the hot afternoon; and the wetness of the soil down to 30–40 cm (12–16 in.).

The second method, the use of simple measurements in the field, determines irrigation need and timing by the measurement of soil water with tensiometers or gravimetric soil water measurements. When the moisture readings of instruments reach a predetermined value (e.g., −50 kPa or other water potential value), irrigation is needed. The gauge of a moisture sensor can be wired to trip an automatic sprinkler or other device for automatic watering systems. Tensiometers have been successfully used for many situations, but they are most useful in crops where water content is kept relatively high.

The third method, called the **water budget method,** requires consumptive use data; it is the more scientific approach to detailed scheduling. Both the time for irrigation and the es-

Table 16-8 Estimated Potential *CU* or *ET* Values as a Percentage of Actual Measured *ET* for Various Locations by Several Widely Used Equations[a]

Location	Equation for Estimating CU or ET[b]				
	Th	*Pen*	*J-H*	*B-C*	*C-H*
		(calculated ET as % of measured ET)			
Aspendale, Austria	64	133	69	75	91
Copenhagen, Denmark	136	130	66	135	113
Ruzizi, Zaire	66	87	108	82	69
Brawley, California	60	106	102	85	82
Kimberly, Idaho	42	86	76	56	66
Coshocton, Ohio	65	89	80	81	81

Source: Selected data from M. E. Jensen, ed., *Consumptive Use of Water and Irrigation Water Requirements,* American Society of Civil Engineers, New York, 1974.

[a] The table illustrates that such equations should not be used unless it is established that they are suitable to the area where data are wanted.

[b] Abbreviations for equation names are: Th = Thornthwaite, Pen = Penman, J-H = Jensen-Haise, B-C = Blaney-Criddle, C-H = Christiansen-Hargreaves.

---

[10] G. W. Bloemen, "A High-Accuracy Recording Pan-Evaporimeter and Some of Its Possibilities," *Journal of Hydrology,* **39** (1978), pp. 159–173.

Calculation 16-2 Calculation of Irrigation Frequency[a]

Problem If a deciduous orchard is to be irrigated when half of the available water in the top 76 cm (30 in.) is used up, how often must the orchard be irrigated during August when the class-A pan is evaporating 5.6 cm (2.2 in.) of water per week? The clay loam soil holds 5.0 cm (2.0 in.) of plant-available water in the top foot of soil and 4.3 cm (1.7 in.) of plant-available water in each additional foot of subsoil.

Solution

1. First, the total plant-available water in the top 76 cm (30 in.), when the soil is wetted, is 5.0 cm (in the top 30 cm) plus 4.32 cm (in each 30 cm of soil). Thus, 5.0 cm for the top 30 cm plus 6.62 cm in the next 46 cm equals 11.6 cm (4.52 in.).

2. Only half of the plant-available water can be used before irrigating again. Half of 11.6 cm is 5.8 cm (2.26 in.).

3. The last problem to solve is how fast the water is used. The class-A pan loses 5.6 cm (2.18 in.) per week or 0.80 cm (0.312 in.) per day.

   Referring to Table 6-4, deciduous orchards in August are seen to have a consumptive use estimated to be 65% as high as the class-A pan evaporation. So the daily use is

   (pan loss)(percentage by plant)
   = 0.80 cm (0.65)

   = 0.52 cm used by the orchard per day
   (= 0.20 in.)

4. The final step is to see how long the 5.8 cm (2.26 in.) in the soil that can be used (see step 2) will last if 0.52 cm (0.20 in.) per day is used.

$$\frac{5.8 \text{ cm total water}}{0.52 \text{ cm water per day}} = 11.2 \text{ days}$$

Irrigation should be about every 11 or 12 days.

The methods actually used by laboratories and consulting specialists are more complex—some are computerized, and all take into account any added rainfall when consumptive use (*CU*) is calculated.

---

[a]Further reading: E. C. Stegman, "Microcomputer Applications to Irrigation System Management," *North Dakota Farm Research,* **43** (Nov.–Dec. 1985).

---

timated amount of water to add are calculated by measuring (1) the amount of water the soil holds and (2) how much of that water has been used. When the soil reaches a predetermined dryness, irrigation is recommended. The major difficulty is in determining the consumptive use (*CU*) or evapotranspiration (*ET*). (These are almost the same amount.) Two procedures are used:

1. *Measuring the water lost from a free water surface:* such as loss from a weather station class-A pan.

2. *Calculating CU or ET values from climate:* solar radiation, average temperatures, and other factors such as relative humidity and wind. Many equations are used. No one formula is suitable for all climatic areas[11] (see Table 16-8).

---

[11] R. J. Hanks and G. L. Ashcroft, *Applied Soil Physics,* Unit 4: "Soil–Plant–Atmosphere Relations," Springer-Verlag, New York, 1980, pp. 99–124.

The use by most farmers of scientific methods to determine timing and quantity of irrigation requires that these methods be convenient. Complex equipment (expensive computers, weather stations etc.) or formulas that require frequent recalculation to avoid errors are not useful for fieldwork. However, local areas may be able to establish and use the simple systems to be described here.

Having no summer rainfall, the San Joaquin Valley of California has a yearly summer climate that seldom deviates more than about 10% from the average summer climate. For each crop and planting date a certain soil can be programmed for irrigation in the man-ner indicated in the graph.[a] The program can be set up from a master plan created months ahead. These data are required for the initial plan: (1) class-A pan evaporation data for the area, (2) the crop coefficients (see Table 6-4), (3) the crop *ET* value, (4) the soil's depth and water-holding capacity, (5) the crop sensitivity to water stress, which determines allowable depletions during each irrigation cycle, and (6) any modifications based on the cultural practices used. The result is a graph (such as that shown) for each crop, each soil, and each planting date, but all available in this convenient form before planting is ever started.

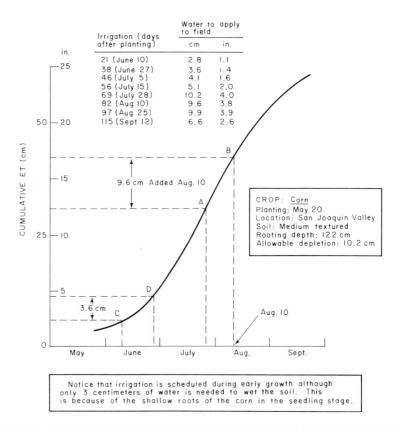

Irrigation (days after planting)	Water to apply to field	
	cm	in.
21 (June 10)	2.8	1.1
38 (June 27)	3.6	1.4
46 (July 5)	4.1	1.6
56 (July 15)	5.1	2.0
69 (July 28)	10.2	4.0
82 (Aug 10)	9.6	3.8
97 (Aug 25)	9.9	3.9
115 (Sept 12)	6.6	2.6

CROP: Corn
Planting: May 20
Location: San Joaquin Valley
Soil: Medium textured
Rooting depth: 122 cm
Allowable depletion: 10.2 cm

Notice that irrigation is scheduled during early growth although only 3 centimeters of water is needed to wet the soil. This is because of the shallow roots of the corn in the seedling stage.

[a] Redrawn by Raymond Miller from Elias Fereres, Patricia M. Kitlas, Richard E. Goldfien, William O. Pruit, and Robert M. Hagan, "Simplified but Scientific Irrigation Scheduling," *California Agriculture,* **35** (1981), nos. 5–6, pp. 19–21.

16:6.4 Water Use Efficiency

Water use efficiency (WUE) is defined in many ways. Those who are interested in reducing losses during transportation may compare "water transferred to the field" with "water diverted to the canal system." Another comparison is the "water added to the field" divided by the "water used by the plants." We will define water use efficiency as *the amount of growth produced divided by the amount of water used by the plant*. This relationship can be given in several ways:

1. Dry weight of produce per volume of water used, such as kg of grain per hectare–cm of water added (or in lb/acre–in.). Clover in Davis, California, had efficiencies that ranged from 400 to 640 lb/acre–in. of water in the first four cuttings (January through April) and 300 lb/acre–in. in the last two cuttings in warmer May and June.[12] Irrigated alfalfa in various states has been reported as 390 lb/acre–in. in Idaho, 410 lb/acre–in. in Texas, 520 lb/acre–in. in Utah, and 610 lb/acre–in. in California.

2. **Transpiration ratio (*TR*),** which is the kilograms of water used per kilogram of dry yield produced. This is the inverse of the above system. *Notice how important it is to report the units of the WUE relationship used. TR* values of 400–1000 are common. For example, a *TR* of 560 means that 560 kg of water were used per kg of dry grain produced. An efficiency of 400 lb/acre–in. of water equals a *TR* of 566. A 610 lb/acre–in. equals a *TR* of 371.

Recent trends in determining irrigation frequency are based partly on the crop's moisture condition—that is, petiole moisture content or leaf moisture potential. A study on cotton grown on Miller clay, an alluvial soil in Texas, showed higher water use efficiencies (WUE) when irrigation timing was based on plant moisture (82% WUE) rather than on soil water status (30% WUE). The plant measurements include the atmospheric effects as well as soil conditions. **Water use efficiencies** are calculated as yield divided by water used. Water use efficiency is greater under slight water stress with grain sorghum, wheat, and cotton but not with corn (i.e., limited irrigation of corn is not recommended).[13–15]

Increasing water demands and/or scant water supply usually encourage more efficient water use. In Israel, where water requirements are carefully measured, a maximum quantity of water is allotted on the basis of soil properties and the crop to be grown.[16] Since 1965, crop yields in Israel have doubled while water use per hectare dropped 20%. Numerous computerized scheduling models for water application have been developed to increase WUE (Detail 16-2).

16:7 Methods of Applying Water

Once the seasonal and daily water use is determined, the next question is how to apply that required amount of water most effectively. Irrigation water can be applied by methods

---

[12] William A. Williams, Walter L. Groves, Kenneth G. Cassman, Paul R. Miller, and Craig D. Thomson, "Water-Efficient Clover Fixes Soil Nitrogen, Provides Winter Forage Crop," *California Agriculture,* **45** (1991), no. 4, pp. 30–32.

[13] Harold V. Eck, "Effects of Water Deficits on Yield, Yield Components, and Water Use Efficiency on Irrigated Corn," *Agronomy Journal,* **78** (1986), pp. 1035–1040.

[14] B. A. Stewart, J. T. Musick, and D. A. Dusek, "Yield and Water Use Efficiency on Grain Sorghum in a Limited Irrigation Dryland Farming System," *Agronomy Journal,* **75** (1983), pp. 629–634.

[15] M. E. Jenson, *Design and Operation of Farm Irrigation Systems,* American Society of Agricultural Engineers, St. Joseph, Mich., 1983.

[16] Ron Ross, "Israel . . . Where Irrigation Is Art," *Irrigation Age,* **13** (1978), no. 1, pp. 6–9.

People have always wanted to take the guesswork out of irrigation and other required activities. Irrigation scheduling models attempt to do that. With increasing data available from weather stations and field water measurements, computers enable us to make sophisticated calculations on water loss by evapotranspiration and on when to add water. The following models are some of those available. (Listing these models does not imply promotion of them by the authors or publishers nor criticism of models not listed.)

1. **SCHED:** This model is developed to run on IBM-PC or IBM-compatible computers. It was developed in Colorado by ARS engineers. It uses some special weather information, which is usually available from radio and newspapers. Contact Harold Duke or Dale Heermann, USDA-ARS Irrigation and Drainage Research Unit, Colorado State University, Ft. Collins, CO 80523.

2. **Soil-Water-Plant-Atmosphere-Irrigation-Salinity models.** 1991.

Department of Plants, Soils, and Biometeorology, Utah State University, Logan, UT 84322-4820.

a. **SOWATET:** R. J. Hanks, M. N. Nimah, and J. K. Cui. A general-purpose water flow model that takes into account infiltration, redistribution, and root uptake of water and subsequent loss of water to the atmosphere.

b. **SOWATSAL:** R. J. Hanks and J. K. Cui. This model provides, in addition to the processes in SOWATET, for the flow of noninteracting salt with soil water, root uptake of pure water leaving salt behind, and salt flow to or from a water table.

c. **PLSALPIS:** R. J. Hanks, J. H. Schick, J. K. Cui, and W. R. Mace. This model simulates water and salt flow. The model predicts the effects of salt on plant root uptake and subsequent yield, assuming osmotic effects only. The model assumes that relative yield is directly related to relative transpiration.

ranging from haphazard flooding to enormous center-pivot mobile sprinklers that cover 53 hectares (130 acres) in one circular sweep (Fig. 16-10). The use of sprinklers and, more recently, trickle (drip) irrigation has made it possible to irrigate almost all arable soils, even those of rolling hills' deep and steep slopes. However, making irrigation possible does not necessarily make it economical, practical, or even desirable.

16:7.1 Border Strip and Check Basin Irrigation

The **border strip** method of irrigation is illustrated in Figs. 16-11 and 16-12. In **border irrigation,** soil ridges keep the water flowing down a strip of land. Border irrigation works well if the soil surface is level in the direction perpendicular to water flow and the slope in the direction of water flow is gentle. Percolation losses will occur in the intake end of the field, and runoff is usually appreciable. Loss of 20–45% of applied water is common in border-irrigated fields when runoff water is not reused.

Basin irrigation, creating soil-ridged basins to hold water, can be quite efficient and has been widely used for pastures and for orchards where each tree is within its own check or basin (Fig. 16-13). Basin irrigation is not suitable for highly permeable soils (sands, organic soils), for irregularly sloped land, for crops harmed by temporary flooding (tomatoes, beans, corn in early stages, lettuce), or for very slowly permeable soil (clays).

Basin tillage is the practice of placing mounds of soil at intervals across the irrigation furrow to hold water so it runs in, not off. The use of this technique on cotton fields in Texas has increased yields and decreased runoff for three consecutive years. In an 8-day period

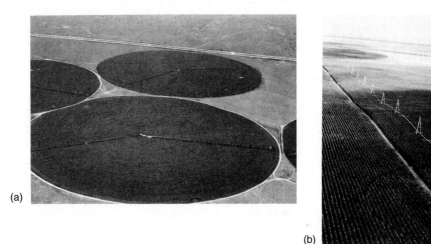

(a)

(b)

(c)

FIGURE 16-10 A typical center-pivot sprinkler system is shown in these photos, ranging from a high aerial view (a), to a nearer view (b), to a close-up of the mobile unit (c). One unit covers 53 ha (131 a), nearly a ¼ mi². Water is supplied by a well at the center pivot. An automatic fertilizer injection system is sometimes located at the center-pivot area. (*Source:* USDA — Agricultural Research Service, Ft. Collins, Colo.; photos by Dale F. Heermann.)

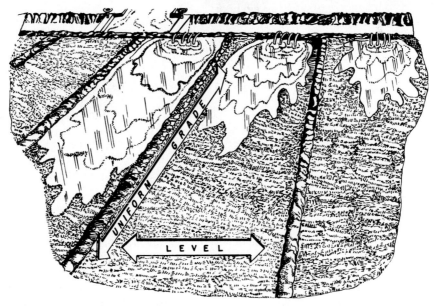

FIGURE 16-11 The field has been prepared for border irrigation by building small levees around each leveled area; then the areas are flooded to irrigate them. (*Source:* USDA.)

FIGURE 16-12 Border irrigation of pecan trees near Fabens, Texas. This type of irrigation requires large streams of water, a soil that is not permeable too rapidly, and a nearly level land surface perpendicular to the direction of water flow. (*Source:* U.S. Department of Interior — Bureau of Reclamation; photo by H. L. Personius.)

FIGURE 16-13 Where a field slopes in more than one direction, basin flooding can still be done satisfactorily by making the ridges of the check basins (dikes) on the land contour. Each tree or several trees are enclosed by a small dike. As each dike is filled with water, it is cut to allow water to flow into the next lower dike. Stanislaus County, California. (*Source:* USDA — Soil Conservation Service.)

100 mm (4 in.) of rain fell. In the basin-tilled field 80 mm (3.2 in.) of this rainwater infiltrated; in the regularly tilled fields only 27 mm (1.1 in.) stayed on the field.[17]

16:7.2 Furrow Irrigation

Furrow irrigation, including small, close furrows termed **corrugations,** is the oldest form of irrigation. In this method water flows by gravity from a main ditch and down each furrow. The crop is usually planted atop the ridges before water is applied. About 42% of all irrigated land is furrow irrigated.

Field crops such as corn and cotton have a furrow to carry water between all planted rows. Crops that are planted in double rows or beds (wide enough for two or more crop rows between furrows) are irrigated by directing the water between the beds. Crops planted in a wide spacing—such as berries, grapes, and orchards—usually have two furrows for irrigation between adjacent rows of plants.

One problem associated with all surface flow methods of irrigation is *deep percolation* and *runoff water loss.* Deep percolation occurs at the head of the field (see Section 16:8.7). To reduce the deep percolation, the full length of the furrow needs to be wet quickly so that "soaking" will be nearly the same duration over the full length of the field. Water runs down the furrow faster if the furrow is smooth rather than rough. Some irrigators in California have dragged "torpedoes" (25-cm-diameter steel cylinders, coned at the front end, filled with cement, and about a meter long) down non-wheel-track rows to smooth the furrow.[18] The most furrow smoothing was in the cloddy, clayey soil. If the soil is moist, torpedoes "slicken" the furrow more than if the soil is dry; the slicking might even partially seal the soil. Water advance rates increased about 15–30% in the studied fields.

In general, soil erosion is excessive when the furrow method of irrigation is used on rows that have a slope of more than 2%; ideally, the slope of the furrows should be less than 0.25%. However, on erosive slopes in Idaho, Kentucky bluegrass was established in each furrow. This reduced erosion to a minimum (Fig. 16-14).

The goal of the irrigator should be to obtain the maximum flow of water down each furrow without causing excessive erosion. In this way, water will soak into the soil at a fairly uniform rate all along the furrow instead of wetting the soil at the upper ends of the rows deeply and wetting the lower ends to shallow depths. Water should soak two to three times as long as the time required to initially wet the entire row length. (For example, if it takes 1 hour to wet to the lower end of a row, the water should be run another hour for sandy loam, or up to 3 more hours for soil as fine as clay loam.) To reduce water loss by runoff, the amount of water running down the furrows should be reduced when the water reaches the lower end of the furrow.

There are several common methods of controlling the distribution of water in surface irrigation:

1. Large lateral ditches across the field with smaller equalizing ditches leading directly to each furrow or border
2. Large lateral ditches with siphon tubes leading to each furrow (Fig. 16-15)
3. Field lateral ditch with spiles (small straight pipes) leading directly through the bank to each row or border

---

[17] Anonymous, "Basin Tillage Increases Yields," *Agricultural Research,* **29** (1979), no. 2, pp. 6–7.
[18] Lawrence J. Schwanki, Blaine R. Hanson, and Anthanosios Panoras, "Furrow Torpedoes Improve Water Irrigation Advance," *California Agriculture,* **46** (1992), no. 6, pp. 15–17.

FIGURE 16-14 Erosion in furrows during furrow irrigation was controlled by planting and maintaining Kentucky bluegrass in them. Beans are on the ridges, but corn, wheat, and barley were also grown satisfactorily. However, sugar beet yields with grass in the furrows were not satisfactory (Idaho). (Courtesy of John Cary USDA — Agricultural Research Service, Kimberly, Idaho; used with permission.)

FIGURE 16-15 Furrow irrigation of cotton, showing plastic siphon tubes in use for moving water from the main ditch to each furrow. Notice the temporary dam in the lower right to keep the water level where siphons are set to a nearly uniform elevation for more equal water flows in each furrow. (Courtesy of Drue W. Dunn, Oklahoma Extension Service.)

4. Irrigation pipe with large openings (gates) emptying into each furrow (Fig. 16-16)
5. Buried pipe to carry the water to the field, with risers (vertical pipe outlets) emptying into each furrow or series of furrows

Cablegation is a surface-flow-controlled pipe system that is placed on grade and delivers water out plastic distribution tubes. Timing the opening and closing of outlets is predetermined. The operation is done with a fishing reellike cable that slowly opens new out-

FIGURE 16-16 Irrigation water is transported by pipe to the field where large openings in the pipe (gates) occur at each furrow. Note the canvas (or plastic) sleeves that lead the water into the furrows (at arrows) without causing excessive erosion. (Courtesy of Atto C. Wilke, Texas Agricultural Experiment Station, Lubbock.)

lets at one end as it slowly reduces and then closes flow on the longest-flowing outlets on the other end as they complete irrigation. The number of furrows irrigated and the flow times and rates are adjustable. Cablegation is more efficient than siphon tubes and is usable in some situations not suitable to surge flow.

16:7.3 Sprinkler Methods

The supplying of **sprinkler systems** is a rapidly expanding multimillion dollar agribusiness. The overhead application of water in simulated rainfall (sprinkling) makes it possible to irrigate both normal lands and the unlevel and sandy lands (neither of which is readily irrigated otherwise except by drip irrigation) (Figs. 16-17 and 16-18).[19]

Sprinkler systems have decided advantages. More-uniform application of water is usually possible than with surface irrigation (except in high winds). In California some crops are given a light sprinkler irrigation just after planting to germinate the seed; then surface irrigation is used the rest of the season. With sprinklers, the amount of water added can be calculated and applied with great precision and usually at the rate desired; watering ditches are eliminated, the physical labor is usually less, and fertilizer can be applied simultaneously through the system. For example, on the sandhills of Nebraska, fertilizer is applied in small regular increments as the crop needs it to prevent large leaching losses. Yields are around 8467 kg/ha (135 bu/a) of corn on these soils by this method.

There are concerns, other than costs, in using sprinkler systems. The non-uniform application of water and the chemicals in the water can cause problems (Detail 16-3). Wetting foliage and wetting the soil surface increase the hazard from many fungi and bacterial diseases. For example, significant reduction of the fungus *Botryosphaeria dothidea* (panicle

[19] J. W. Cary, "Irrigating Row Crops from Sod Furrows to Reduce Erosion," *Soil Science Society of America Journal,* **50** (1986), pp. 1299–1302.

FIGURE 16-17 Portable or solid-set systems, similar to the one shown, are the least expensive sprinkler systems. After each area is wetted, portable systems must be moved by hand to a new location. This system near Morganfield, Kentucky, applies 2.5 cm (1 in.) of water per hour, pumped from a 3.0-ha (7.5 acre) pond containing water accumulated from a 21.9-ha (54-acre) watershed. The pond, at upper left, is also used for recreation. Grassed waterways reduce the amount of sediment carried into the pond. (*Source:* USDA — Soil Conservation Service.)

FIGURE 16-18 (a) Many automatic-drive sprinkler systems are available. This one, photographed from the outer end, is a center-pivot system near Klamath Falls, Oregon, irrigating wheat. The entire circle of 51.8 ha (128 acres) can be irrigated in either 11 or 22 hours. The system has 14 tower stations, each with its own drive system. The spray is a heavy mist. This system now costs over $70,000. (b) The extent of center-pivot systems is indicated in this space satellite photo over Texas. Each circle is a 56.7-ha (140-acre) area. Such scenes are also common over Nebraska and neighboring states. (*Sources:* (a) USDA — Soil Conservation Service; (b) Brantwood Publications, cover of *Irrigation Journal* [July–Aug. 1974]. Price updated.)

(a)

(b)

There are better and worse ways of applying water by sprinkler systems. More recently, **low-energy precision applicators (LEPA)** have become one way to increase application uniformity. In Texas, studies of center pivots using nozzles that shoot water with pressures near 60 psi (pounds per square inch) had as high as 15–20% of the water that *didn't reach weighing lysimeters* (3 ft × 3 ft) in the field. With nozzles at 30 psi, the loss was 10–15%. With LEPA systems (6 psi), the loss was only about 4%.[a]

Generally, **emission uniformities** over 90% are considered excellent, 80–90% are good, 70–80% are fair, and less than 70% are poor. A study of 112 systems in the San Joaquin Valley of California showed 62% were good to excellent, with a mean value of all systems of 80.3%. A very large portion of distribution uniformity problems were caused by *poor pressure regulation* and to *plugged nozzles.*[b]

Where fertilizer or pesticide is applied in the water, drift (movement away from the target area) can be reduced by equipment adjustments. Some of these alterations are[c]

1. Select nozzle type to produce large drops. Large drops drift less.

2. Use lowest pressure feasible. High pressures form small droplets.

3. Lower the boom height. Wind speed, thus drift, increases with height above ground.

4. Spray when wind speeds are low (less than 10 mph). Wind blows the spray off target.

5. Increase nozzle size. Larger nozzles form larger droplets at a given pressure.

6. Do not spray when air is completely calm or an inversion exists. Spray can slowly move downwind before it falls.

7. Use a drift control additive, if needed. These increase the droplet size produced.

---

[a] Don Comis, "Lower Water Pressure, Less Water Loss," *Agricultural Research,* **40** (1992), no. 5, p. 23.
[b] Dale Handley, Henry J. Vaux, Jr., and Nigel Pickering, "Evaluating Low-Volume Irrigation Systems for Emission Uniformity," *California Agriculture,* **37** (1983), nos. 1/2, pp. 10–12.
[c] Anonymous, "Seven Ways to Reduce Drift," *Solutions,* **37** (1993), no. 1, pp. 36–37.

---

and shoot blight of pistachio) was observed when sprinkler irrigation times were reduced from 24 to 12 hours in Sacramento Valley and from 48 to 24 hours in the San Joaquin Valley of California.[20] The disease has optimum development at about 80–85°C. Spores germinate within 2 hours after a period of wetness but need about 12 hours to penetrate the petioles and leaves. The orchards had previous losses of 76–99% using 24- to 46-hour-long irrigations. Shorter irrigations reduced fruit losses as much as to one-third the loss in longer irrigation sites.

Sprinkler systems come in all shapes, sizes, and kinds. Small, portable, rotating heads, such as are used in home gardening, are often employed agriculturally. *Solid-set* stationary systems and *mobile* sprinkler lines come in many sizes. **Center-pivot** or **lateral-move line** systems are usually installed on large acreages. Another innovation is large **sprinkler "guns"** that deliver up to 4.54 m³/min (1200 gal/min), sprinkling a circle of 139 m (620 ft) diameter.[21] These guns have large nozzle openings and do not easily plug with debris in the water; sewage effluents can be spread easily in this way (Fig. 16-19).

---

[20] Themis J. Michailides, David P. Morgan, Joseph A. Grant, and William H. Olson, "Shorter Sprinkler Irrigations Reduce Botryosphaeria Blight of Pistachio," *California Agriculture,* **46** (1992), no. 6, pp. 28–32.
[21] Bob Rupar, "The Big Gun Boom," *Irrigation Journal,* **25** (1975), no. 1, pp. 16–18.

FIGURE 16-19 Large sprinkler guns, such as this single unit in western Washington, have large nozzle openings and do not easily become plugged. They are used to spread manure slurries, as shown here, containing 5–8% solids, to apply other effluents, or to irrigate normally. Because the water falls with considerable force, sprinkler guns are best used on established crop cover, such as sugarcane and pastures, rather than on barren or newly planted land. (Courtesy of Darrell Turner, Washington State University.)

Sprinkler systems have two major disadvantages: their high cost and plugging of the nozzles by debris in the water. Costs for solid-set systems range from about $1000 upward per hectare (about $400 per acre) for installation.

16:7.4 Drip (Trickle) Irrigation[22–24]

The newest method of watering is **drip (trickle) irrigation.** As the name suggests, drip irrigation is the frequent, slow application of dripping water to soil through small outlets (**emitters**) located along small plastic delivery lines, 1.3–2.5 cm (0.5–1.0 in.) in diameter. The application rate is so slow—often less than 3.7 liters (1 gal) per hour per emitter—that surface water flow is almost nil. Water movement is by saturated and unsaturated (capillary) flow, and seldom is all the surface of the field wetted. The small emitters can be placed at any frequency wanted. For example, a large tree in an orchard may have four to eight emitters spaced around it; a grapevine may have one or two emitters.

Why drip irrigation? Costs are only 50–75% of some sprinkler systems; the system uses smaller lines (slower water flow) at low water pressures; and once the system is in-

[22] "Drip/Trickle Irrigation in Action," *Proceedings of the Third International Drip/Trickle Irrigation Congress,* Fresno, Calif., Nov. 18–21, 1985, American Society of Agricultural Engineers, St. Joseph, Mich., 1986.

[23] P. Tscheschke, J. F. Alfaro, J. Keller, and R. J. Hanks, "Trickle Irrigation Soil Water Potential as Influenced by Management of Highly Saline Water," *Soil Science,* **117** (1974), pp. 226–231.

[24] J. Ben-Asher, "Trickle Irrigation Timing and Its Effect on Plant and Soil Water Status," *Agricultural Water Management,* **2** (1979), pp. 225–232.

stalled, the labor cost is low. Drip irrigation is adaptable to very steep hills, although differential hydraulic pressure in lines at the top and bottom of the slope must be considered. Of increasing importance, drip systems conserve water because of lower distribution and evaporation losses; savings of 20–50% are expected. Water of higher salt content can also be used because of the nearly constant, high soil moisture maintained in part of the root zone, and salt is constantly being moved away from the plant roots to the outer part of the wetted soil volume. Drip systems are currently used on orchards, in vineyards, on sugarcane in Hawaii, and for numerous other crops. Drip irrigation is popular in water-short Israel.

Drip irrigation is not without its problems. The small emitters plug easily, so careful filtering at the water source is essential. Without water added to promote leaching, salts can accumulate, although this occurs at the periphery of the wetting area. Also, the plant root zone is often less deep and extensive with drip irrigation, and a water failure can quickly cause water stress problems to the plant. Drip water lines may require special cultivation and harvesting arrangements. In sugarcane harvesting (18–24 months after planting), the plastic lines are considered expendable and are "wasted" (destroyed during harvest).

Drip irrigation is usually a water application to the soil surface, but subsurface application lines and emitters are also used. Drip irrigation allows us to wet only the root zone desired. For example, if plants are widely spaced (such as grapes or orchards), only the soil needed to be wet is drip irrigated. For many crops (tomatoes, melons, cucumbers, strawberries), buried lines allow wetting the root zone soil without wetting all the surface soil. Wet surface soil causes rot and spotting of fruits lying on the wet ground. Underground lines and computer-controlled linkage for irrigation scheduling allow a yield of 100 tons of ripe tomatoes per acre (224 Mg/ha) compared to the area average of about 26 tons/acre (58 Mg/ha).[25]

A simple modification of drip irrigation on gently sloping land is **bubbler irrigation,** a method using low water pressure and open, standing outlet tubes. The rate of water flow is determined by line size, water pressure, and the height of each standing outlet tube (Fig. 16-20). The large flow opening (which may be a fully open tube end) reduces the problem of clogged emitters. Setting flow by elevation adjustment can be tedious and frustrating if the area is unlevel and the line system has many outlets.

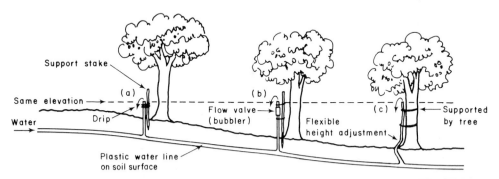

FIGURE 16-20 Diagram of a bubbler irrigation system, a variation of drip irrigation. The open outlet system controls waterflow by (a) its pipe elevation, (b) with a valve, or (c) by both. The riser lines can be flexible and attached to the grape trellis or tree (as in c) to simplify connections. A flexible line with extra length allows vertical adjustment.   (Courtesy of Raymond W. Miller, Utah State University.)

---

[25] Marcia Wood, "Underground Drip Irrigation Yields Record Tomato Harvest," *Agricultural Research,* **36** (1988), no. 7, p. 14.

The special problems of drip irrigation have prompted proposals of criteria for water quality in such systems. One quality scheme proposed is given in abridged form in Table 16-9. This classifies water based on three categories of constituents:

1. *Physical, suspended solid material* (column 2 of Table 16-9) such as sands, silts, clays, humus, other suspended organic materials, and small organisms.

2. *Chemical precipitation,* such as carbonates, gypsum, hydroxides of iron and manganese, fertilizer phosphates, dissolved salts and substances (columns 3 and 4 of Table 16-9) that might precipitate, thereby clogging emitters.

3. *Biological,* the microorganisms (column 5 of Table 16-9) whose actions produce slimes, filaments, and chemical deposits that plug emitters.

Rating these three categories (averaging the two columns for chemical precipitation) in the order *physical-chemical-biological,* a 0-0-0 water should be excellent for use in a drip system, whereas a 10-10-10 water would be almost unusable. If the rating is between 0 and 10 after adding the three numbers, few problems are anticipated. A rating of 10–20 indicates some problems will occur, and 20–30 indicates severe problems. Above 10, filtration and/or other remedial measures would be required.

16:7.5 Subirrigation

Some areas have unique soil properties that permit irrigation by raising the localized water table. In this method, called **subirrigation,** water is applied to an area via open ditches or by buried pipes and flows into the water table to raise it to the root zone level. Subirrigation is effective on land with sandy, peat, or muck soils from a root depth to 1.5–3 m (5–10 ft), a level soil surface, and an impermeable layer beneath the root zone to hold the water. A typi-

Table 16-9 Proposed Scheme for Classifying Waters as to Their Suitability for Use in Drip (Trickle) Irrigation

Numerical Rating[a]	Suspended Solids[b] (max. mg/L)	Dissolved Solids[c] (max. mg/L)	Iron and/or Manganese[c] (max. mg/L)	Bacteria Population[d] (no./mL)
(Best)				
0	<10	<100	<0.1	<100
1	20	200	0.2	1,000
2	30	300	0.3	2,000
4	50	500	0.5	4,000
6	90	800	0.7	10,000
8	120	1,200	0.9	30,000
10	>160	>1,600	>1.1	>50,000
(Poorest)				

Source: These data have been selected and modified from D. A. Bucks, F. S. Nakayama, and R. G. Gilbert, "Trickle Irrigation Water Quality and Preventive Maintenance," *Agricultural Water Management,* **2** (1979), pp. 149–162.

[a] The complete proposal has 11 rating classes from 0 to 10; only seven are listed here.

[b] These are sands, silts, clays, organic substances of many kinds, bacteria, phytoplankton, and zooplankton.

[c] Dissolved solids are mostly soluble salts. Columns 3 and 4 make up the "chemical" portion of water quality and affect the plugging of emitters by precipitation. If water pH is 7.5 or higher, increase suspended solids rating by 2.

[d] Action of microorganisms increases potential to plug emitters. If snails are common, increase the rating by 4.

cal subirrigation area in northern Utah has river terraces of loamy sands 2.1–2.4 m (7–8 ft) deep over impermeable clay deposits from ancient Lake Bonneville (Great Salt Lake). Tile pipes with controllable gates (closures) allow early-spring drainage. As the water table becomes sufficiently low, the gates in the drains are closed to restrict further water drainage loss and, later in the season, to allow subirrigation of the crop.

Subirrigation is commonly practiced in tile-drained or tube-drained peat or muck soils by closing the outlets and raising the water table to the bottom of the rhizosphere.

16:8 Special Irrigation Techniques

Modifications of irrigation techniques are employed in special circumstances. Recently, innovations in furrow irrigation, automated water delivery, and sprinkler system improvements have been subjects of investigation. Special techniques may not decrease consumptive water use for a given yield, but they may improve efficiency by reducing the leaching of nutrients, water losses by drainage, and so on.

16:8.1 Alternate-Row Irrigation

Grain sorghum in the Great Plains of Texas, corn in Nebraska, and cotton in California and Texas have been irrigated by applying water to **alternate rows** during a given irrigation and then watering the missed rows in the next irrigation period. Water savings are large here (nearly 50%), and yields are normal or only slightly reduced. Alternate-row watering allows more rapid coverage of a field during an irrigation period, saves water, and requires less labor. An alternate-row system also leaves some soil dry enough to absorb any rainfall that comes, which reduces erosion and runoff loss and maximizes use of rainfall. Time intervals between irrigations must be shorter because only half as much water is added to the area at a given irrigation.

16:8.2 Mini-Watersheds or Catchment Basins

Where water is deficient but wells or surface water is too expensive or is unavailable, wide spacing of crops (hills of corn, widely spaced fruit trees or vines), each fed by runoff water, is sometimes possible (Fig. 16-21). Normally, the plants must be quite tolerant of drought but able to produce better yields when rainfall is more plentiful than during drier periods. Numerous modifications of these **mini-watershed** patterns are possible. **Contour furrows**—ditches or furrows dug along the contour at various distances apart—are used on arid rangelands to concentrate enough water in and near furrows to reestablish grasses. These increase yields of better grasses on parts of a range.

All of these processes *concentrate sparse water supplies onto parts of the area.* This concentrated water allows production of desired plants that generally grow unsatisfactorily, or not at all, without water concentration.

16:8.3 Timing for Limited Irrigation

Plants have critical periods of water demand, depending upon the stage of plant development. Water use efficiency can be improved by increasing the water supply during critical plant demand and reducing it proportionally during minimum demand periods.

In Texas, on the southern Great Plains, grain sorghum can best use water from a single 10-cm (4-in.) furrow irrigation if the water is added at the heading or milk stage of the grain. When two 10-cm (4-in.) irrigations were replaced by two 5-cm (2-in.) irrigations, yields were lower but water use efficiency was higher (more yield per unit of water added). Sprinkler systems would provide better addition of small quantities of water than does surface

FIGURE 16-21 Pomegranate trees in Israel grown in *negarins* (mini-watersheds). A small watershed area for each tree supplies runoff water to a small catchment basin in the center of which the tree is planted. In this area of Israel, with an annual rainfall of 150–200 mm (6–8 in.), orchards could not grow without some method of increasing the water available to them. The distance between trees indicates the watershed area for each tree. (By permission, from *Food, Fiber and the Arid Lands,* by William G. McGinnies, Bram J. Goldmon, and Patricia Paylore, eds., University of Arizona Press, Tucson, 1969.)

irrigation and would reduce the critical timing problem at various stages of plant growth. In water shortage areas the timing of crop irrigation is crucial.

Early morning, evening, or night irrigation helps reduce the excessive loss of water by evaporation that occurs during hotter daytime hours, but often this is impractical. Yet in the Sinai Desert daily drip irrigation on sand dunes is actually most efficient when done during the day because the greatest water loss (up to 70% of the water applied) is due to internal deep drainage. If water is applied during active growth periods, more is absorbed by the plants and less is lost by drainage.[26]

16:8.4 Irrigating Clay Soils

Soils with high clay percentages (more than 40%) have special problems. Clayey soils in the United States usually contain montmorillonite, illite, or kaolinite. Of these, montmorillonitic clay soils are the greatest problem because of large volume changes that occur when the clays are wetted and dried, their great stickiness and plasticity, and their exceptionally high water retention capacities.

The three major problems of clay soils are (1) inadequate aeration when wet, (2) slow water infiltration, and (3) a limited moisture range that is suitable for tilling. The Sharkey clay, an alluvial soil of the Mississippi River delta area, exemplifies two of these problems. Its pores are all very small and drain no water when fully wetted—all pore space is filled with water having water potentials lower than −33 kPa.[27] When dried and cracked, this clay readily absorbs several inches of rain, but after it becomes wetted, additional rain runs off the surface. Only open-ditch drains are helpful in draining these clays.

Treat clayey soils as shallow soils. Because poor aeration reduces rooting depths, *frequent but shallow irrigations* are a practical approach to irrigation of many clayey soils. When cracks in dry clayey soils are not wide, wetting and quick swelling soon close those cracks and reduce water intake. One clay studied decreased from a very rapid infiltration rate of 25 cm/hr (9.8 in./hr) to a slow rate of 0.6 cm/hr (0.23 in./hr) while absorbing only 2 cm

---

[26] Ben-Asher, "Trickle Irrigation Timing and Its Effect on Plant and Soil Water Status."
[27] W. M. Broadfoot, "The Fame of Sharkey Clay," *Forests and People,* **12** (1962), no. 1, pp. 30, 40.

(0.8 in.) of water, a reduction to less than $\frac{1}{40}$ of the maximum rate. Because of the greater soil surfaces exposed to the wind and sun, cracking of clayey soils increases evaporation losses 12–30% more than if there were no cracks.

The rate of water infiltration is a major problem when irrigating clay soils. Where deep and extensive cracks occur, wetting to capacity (up to 15–20% of the soil volume) is relatively easy. But if the clay does not crack extensively or is not allowed to dry enough, water intake is slow. A surface mulch of organic residues and soil tillage help some, but usually not enough. Methods for holding water on the soil longer include lengthening the rows (longer fields or serpentine rows), using check furrows (small dams across the furrow), slow rate of sprinkling, and drip irrigation.

Slow infiltration rates in the range of 0.76 cm/hr (0.3 in./hr) or less after the initial few minutes require 10–20 hours to wet clay soils to normal rooting depth. Maximum row lengths for nearly level clays are about 244–366 m (800–1200 ft). Where field lengths are rigidly established, irrigation furrows can be lengthened by a serpentine scheme, as shown in Fig. 16-22. **Serpentine** schemes are made by cutting passageways across the furrow direction to connect several channels so that the water flows back and forth like a snake's

INTAKE   TOP OF FIELD

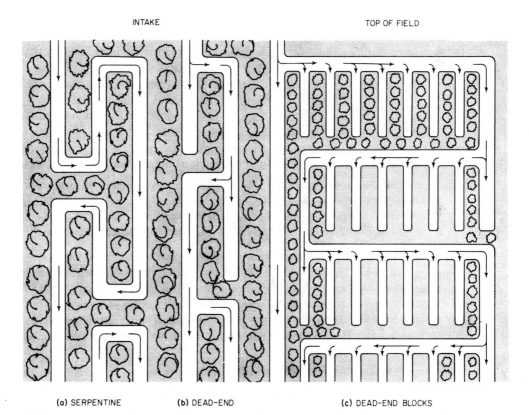

(a) SERPENTINE   (b) DEAD-END   (c) DEAD-END BLOCKS

FIGURE 16-22 Serpentine channels increase infiltration on clayey soils by maximizing stream flow, minimizing erosion and water runoff, and increasing water-soil contact time by means of longer furrows. (a) A typical serpentine involves three meander furrows. (b) The dead-end serpentine keeps a large soil area in contact with water as flow occurs in only half of it. (c) Modifications are possible for special uses, such as onion sets or other intensive crops. Serpentines require more work than simple irrigation furrows and should be used only where the benefits exceed disadvantages.   (Courtesy of Raymond W. Miller, Utah State University.)

(serpent's) trail, greatly increasing the effective length of the furrow. The many patterns have been given names such as *dead-end serpentine* or *block serpentine* and *meander furrows*. Longer furrows allow larger volumes of water to be added per furrow, which increases the soil-water contact surface and, thus, total water intake.

16:8.5 Irrigating with Salty Water

In some world areas water shortages are relieved by using brackish (slightly to moderately salty) water for irrigation, although most salty waters have the problems of high sodium (undesirable SAR), high chloride (toxic on foliage of many plants, especially citrus and avocados), and the salt itself (which affects osmotic relationships). Countries such as Israel have developed extensive canal and reservoir systems where both low-salt and salty waters are mixed to obtain usable water. These irrigation techniques are selected: first, to hold the soil near field capacity to keep salt as dilute as possible; second, to avoid application techniques that wet the foliage damaging the leaves; and third, to leach accumulated salts periodically. To accomplish these objectives, three general rules follow:

1. Apply water at or below soil surface. Sprinklers should be used only if they avoid wilting foliage (as sprinkling before plant emergence or below-canopy to avoid salt-burn damage).

2. Keep water additions almost continuous, but at or below field capacity so that most flow is unsaturated. This maintains adequate aeration.

3. Enough water should be added to keep salts moving downward, thus avoiding salt buildup in the root zone.

These requirements are difficult to meet and are best satisfied by some form of drip irrigation. Due to the need for high moisture levels and because of the high sodium (SAR) problem, sands seem to be most adaptable to irrigation with salty water. Where the soil includes considerable clay and silt, the use of salty water is less likely to be suitable.

16:8.6 Rice Production: From Paddy to Sprinkler?

Growing rice without flooding (paddy) is done in many countries, even though paddy rice has usually produced higher yields and has fewer weed problems. But paddy rice also uses more water, requires flat or terraced land, uses aerial fertilizer application in the United States, and sometimes requires aerial seeding.

However, rice grown under sprinkler irrigation in Texas and Arkansas has been successful, with lower total costs, higher yields, and fewer problems than paddy production.

Some advantages of using sprinkler irrigation on rice rather than flooding follow:

1. Soils that slope too much or have too high a water percolation rate can be used for rice
2. Without dikes, 10–12% more land in each field can grow crops; nondiked land is also easier on equipment and is cheaper to manage
3. The sprinkler system and the land can be used for crops other than rice

In the first years of sprinkler rice production in Texas and Arkansas, water requirements as low as 20–30% of that required for paddy production have been noted (300–400 mm for sprinkler vs. about 1200 mm for paddy), with yields of 4500–5400 kg/ha (4008–4810 lb/a).

Fertilizers and herbicides are added in the sprinkler water. As yet, not all of the problems are solved, and extensive use of sprinkler rice cultivation is only a potential.

16:8.7 Surge Flow Surface Irrigation[28]

The low water efficiency of surface flow irrigation has encouraged more use of sprinkler and drip systems. Now, a technique of automated surface flow, called **surge flow,** may help those farmers who prefer surface furrow irrigation. Surge flow delivers water intermittently. Larger flows for short on-off periods have long been known to wet a longer furrow distance more quickly than the same amount of water added continuously at a smaller flow rate. For example, applying 1000 gallons of water during 100 minutes produced these results:

Flow Rate (gal/min)	Time Flowing	Furrow Distance Wetted (ft)
10	Continuously	240
15	⅔ of time	350
20	½ of time	490
30	⅓ of time	600

Surge intervals may be from 1–10 minutes or more (1 minute on, 1 minute off).

The advantage of surge flow is in minimizing the deep percolation of water at the head of the furrow in order to wet the soil at the end (Fig. 16-23). Although it is impossible to get uniform wetting from top to bottom, the surge flow method greatly increases efficient water use. The automated system involves gated pipe valves mechanically adjustable for flow rate but opening and closing pneumatically (or by other mechanisms) for surge flow. Air pressure to run the pneumatic valves is produced with a small compressor.

16:8.8 Flexible Irrigation-Dryland System (IDS)[29]

When natural rainfall limits plant-available water, crop yields are greater if planting densities are reduced and fertilization is lower than optimum for the crop when adequate water is available. In areas of Texas that need irrigation but have less than optimum amounts of plant-available water, a new technique of mixed irrigation and semidryland farming is being tried, which is described here as the irrigation-dryland system (IDS).

A 575-m (1900-ft) grain sorghum field was divided into three parts as follows:

Field Portion	Irrigations	Seeding Rate (kg ha⁻¹)	N Application (kg N ha⁻¹)
Upper half	Five of 37 mm each	6.7	168
Third one-fourth	Tailwater only	3.4	84
Bottom one-fourth	Dryland or tailwater	1.7	84

After planting, check dams were placed in each irrigation furrow at about 3-m (10-ft) intervals. As the furrows were irrigated, each dam along the furrow would break (wash away)

---

[28] A. Alvin Bishop, "Surge Flow," *Crops and Soils Magazine,* **33** (1980), no. 2, pp. 13–16.
[29] B. A. Stewart, D. A. Dusek, and J. T. Musick, "New Land Management Technique Proves Water-Efficient," *Irrigation Age,* **15** (1980), no. 1, pp. 52–53.

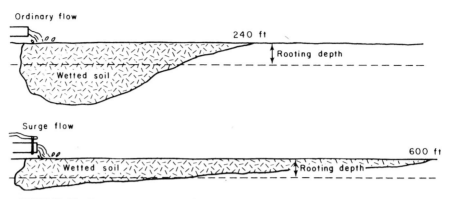

FIGURE 16-23 The same quantity of water applied intermittently, but in larger flow rates, wets farther down the irrigation row but lessens deep percolation loss at the upper end compared with continuous flow of an equal amount of water for the same time period. On susceptible soils, the larger heads of water may increase on-field erosion. This method of intermittently adding water is called *surge flow.* (Courtesy of Raymond W. Miller, Utah State University.)

as water built up against it. Although the tailwater would seldom wet the bottom half of the field, the portion wetted would infiltrate enough water to be effective. Only alternate furrows were irrigated. The nonirrigated furrows trapped all rainfall. No runoff occurred from the field. The benefits in terms of yields and water efficiency of one study are given below:

	Applied Water (mm)	Seasonal Rainfall (mm)	Total Runoff (mm)	Grain Yield $(kg \cdot ha^{-1})$	Water Use Efficiency $(kg \cdot ha\text{-}cm^{-1})$
Control, with full irrigation	592	220	210	9300	310
The IDS field	188	220	0	7300	520
Control, all dryland	—	220	—	3410	—

The encouraging aspects of this IDS system are the absence of runoff water and the higher water efficiency (higher grain yield per unit of added water).

As our case is new so must we think and act anew.

—Abraham Lincoln

There is no medicine like hope, no incentive so great, and no tonic so powerful as expectation of something better tomorrow.

—O. S. Marden

Questions

1. (a) What is an aquifer? (b) What are the materials from which aquifers form?
2. (a) What causes an artesian well to flow? (b) Why might a flowing well later cease to flow naturally but can be pumped?
3. (a) Why would some aquifers produce lower flow from wells and recharge more slowly than other aquifers? (b) Define "overuse" of groundwater.
4. (a) What causes saltwater intrusion? (b) What can be done to minimize it?
5. What are the advantages and disadvantages of groundwater compared to surface waters as sources of irrigation water?
6. What areas of the United States are overusing groundwaters?
7. (a) What is the sodium hazard? (b) How is it indicated? (c) What is measured?
8. What electrical conductivity values for irrigation waters are (a) moderate and (b) severe?
9. (a) How much water do "average" field crops require? (b) What factors increase or decrease the amount of water needed by a given crop?
10. (a) From which soil depth is most water absorbed by plant roots? (b) As the upper soil layers dry, from where must the next water and nutrients be taken?
11. In water budget irrigation scheduling: (a) How is the rate and amount of water loss measured? (b) How is the amount of water to be added determined?
12. Discuss briefly how dry a soil should be allowed to get before irrigating it. Consider (a) rooting depth, (b) the different soil depths, and (c) the nature of the crop grown.
13. Define and describe each of these irrigation techniques: (a) furrow, (b) basin, (c) border, (d) drip, and (e) sprinkler.
14. What are the disadvantages of furrow irrigation, and how is the extent of these disadvantages reduced?
15. List the disadvantages and advantages in using (a) drip systems and (b) sprinkler systems.
16. Discuss the filtering requirement for drip irrigation systems.
17. When irrigating sloping lands, sands, or shallow rolling-land soils, discuss the relative values of surface flow, drip, and sprinkler systems.
18. If only a small amount of irrigation water is available, when should it be added?
19. What are some surface flow techniques used to increase wetting depth of poorly permeable (clayey) soils?
20. To what extent can rice be grown without paddy (ponded water) conditions?
21. (a) What is surge flow? (b) What are its general principles of operation? (c) Why is it of interest?

Land Drainage

If the land is wet it should be drained with trough-shaped ditches dug three feet wide at the surface and one foot at the bottom, and four feet deep. Bind these ditches with rock. If you have no rock, then fill them with green willow poles braced crosswise. If you have no poles, fill them with faggots [bundles of twigs or sticks tied together]. Then dig lateral trenches three feet deep and four feet wide in such a way that the water will flow from the trenches into the ditches.

—**Cato (234–149 B.C.E)**

17:1 Preview and Important Facts

PREVIEW

Artificial drainage is needed and used on more than 10% of the world's cropland and on about one-third of the cropland in Canada and the United States. In humid areas it enhances plant growth by increasing the oxygen supply for plant roots and permits timely planting and harvesting of crops. Some high-salt, arid soils that are irrigated must also be artificially drained to permit control of salt concentration in the root zone. Not all wetlands should be drained; some may be of more value as wetlands than as cropland. The extent of worldwide drainage utilization is tabulated in Table 17-1 which shows that more than 10% of all croplands in the world and 25% of those in the United States and Canada are artificially drained.

The numbers of drainage systems are increasing rapidly in such areas as the Indus Valley of Pakistan; the Mekong Valley of China, Laos, Cambodia, and Vietnam; the Niger basin in Mali, Guinea, and Sierra Leone; the Congo basin in Zaire; and the Amazon Valley of Brazil.

Drainage is the removal of excess gravitational water from soils by natural or artificial means. Unless otherwise specified, drainage refers to artificial (anthropic) systems for removal of water in soils.

The need to remove excess salt in arid and semiarid irrigated areas may require the establishment of an adequate drainage system simultaneously with the irrigation system. The principal difference between drainage practices in humid and arid regions is that in humid regions the drains are normally established at depths of 75–120 cm (2.5–4 ft), whereas in arid

513

Table 17-1 Estimates of Croplands Artificially Drained[a]

Area	Total Cropland (10^3 acres)	Cropland Artificially Drained (10^3 acres)	Cropland Drained (%)
World	3,627,497	383,492	10.6
Asia	1,178,020	78,756	6.7
Africa	522,090	5,925	1.1
North and Central America	674,531	167,195	24.8
South America	220,858	19,276	8.7
Europe	353,988	87,844	24.8
Oceana	116,599	2,234	1.9
USSR	561,411	22,335	4.0
Canada	108,148	36,741	34.0
Mexico	63,692	3,385	5.3
United States	467,718	147,766	31.6

Source: Calculated from *Food and Agriculture Production Yearbook, 1974* and P. P. Nosenko and I. S. Zonn, *Land Drainage in the World,* ICID Biennial Bulletin, International Commission of Irrigation and Drainage, 1976.
[a] Acres × 0.405 = hectares.

regions 180-cm (6-ft) depths are common. Greater depths of drains in arid regions are necessary to leach surplus salts below the root zone.

IMPORTANT FACTS TO KNOW

1. Soil and vegetation clues that indicate a soil has poor drainage and thus would produce higher yields if it were drained
2. Numerous benefits of drainage, particularly the crops that are adaptable, rate of soil warming, and depth that plants root
3. Some of the undesirable results from drainage
4. The "drainage" characteristics and nature of paddy rice culture
5. The various ways to aid soil drainage and the major advantages and disadvantages of each of these: (a) surface smoothing, (b) open ditch, (c) subsurface tubes, and (d) mole drains
6. The shapes, sizes, and reasons for "beds" used in poorly drained soils
7. How water enters drainage lines and how the entry zones are protected from sediment plugging
8. The special natures of sump-and-pump systems and of vertical drainage systems
9. The need for drainage in irrigated soils that can accumulate salts

▰▰▰ 17:2 Soils That Need Drainage

Much can be learned about the internal drainage of a soil by digging into it. The light-gray colors of gley indicate long periods of continuous saturation. Periodic aeration usually produces orange-red mottles caused by oxidation of iron to various iron oxides (such as "rust"). This coloring develops in the most easily oxidized parts (cracks and other large channels like old root channels). The rest of the soil may be gray.

Except for rice, almost all crop plants of economic importance grow best when soil pores contain air that is easily exchangeable, because some space is required for the release of CO_2 from plant root respiration and bacterial respiration and for the entry of O_2 from the atmosphere. The purpose of artificial drainage is to increase aeration to the growing plant roots.

17:2.1 Reasons for Poor Drainage

A soil may need artificial drainage because of a high water table that should be lowered or because of excess surface water that cannot move off the surface or downward into the soil fast enough to keep from "suffocating" plant roots. If the condition persists, it causes an oxygen (O_2) deficiency. The results are (1) a reduction in root respiration and hence in growth, (2) increased resistance to water and nutrient movement inside the root, and (3) formation of substances (such as manganese cations) of a kind or in concentrations that are toxic to plants (Detail 17-1).

Usually, poor internal drainage is caused by shallow depths to bedrock or by low-permeability clayey layers (Fig. 17-1). Depressions with clayey bottoms will pond water, eventually forming peat and muck soils after thousands of years. Most organic soils need drainage; they formed because of poor drainage. Clayey soils in flat relief in humid climates are usually poorly drained. Soils may also have poor drainage conditions simply because they accumulate more water (high rainfall or collection of runoff water) than they can dissipate by their slow natural drainage. After heavy rains poorly drained soils take longer to lose surface water (Fig. 17-1). Such soils may have poor-drainage *indicator plants,* such as marsh grasses, sedges, and willow trees. Surface salt accumulations during dry periods and light-gray gley soil are other clues to poor drainage.

17:2.2 Redox Potentials and Poor Drainage

There are various degrees of poor drainage. Poorly drained soil conditions are usually indicated by the **redox potential,** the relative electron concentration in the soil. In well-aerated soils, the free gaseous oxygen accepts the electrons produced during decomposition of organic matter (Fig. 17-2). If the free oxygen is used up, *nitrate* becomes the electron acceptor and denitrification occurs producing N_2, N_2O, and NO gases. Manganese as MnO_2 can also be reduced, releasing soluble Mn^{2+}. As the nitrate and manganese dioxide are "used up," ferric iron can be reduced to ferrous (Fe^{2+}). Eventually, in very poorly aerated conditions, sulfate or sulfur will be reduced to sulfide. Even carbon dioxide can be reduced to methane gas (reduced carbon) and be given off in very poorly aerated conditions. Thus, the development of maximum anaerobic conditions occurs with (1) exclusion of air (waterlogging), (2) warm temperatures for organic-matter decomposition, and (3) large amounts of easily oxidized organic material.

17:2.3 The Anaerobic Crop: Paddy Rice

Rice is the major food grain for about 60% of the world's population. It is grown in the tropics and subtropics. Only about 1% is grown in the United States.

Rice is an aquatic plant and the only major food crop that can germinate and grow throughout its life under continuous shallow water. It can also be grown as an upland crop like wheat or corn. However, rice is grown usually in soil covered with a layer of water 5–10 cm (2–4 in.) deep (called "paddy") because yields are much greater than when grown on drained soils (see Fig. 13-11).

All rice in the United States is grown with flood irrigation and about 80% of the rice in Asia is so grown. Thirteen percent of Asian rice is grown on dryland (rainfed, no levees) and 7% is rainfed, but with levees around each field to retain rainwater. One reason for paddy culture is that it provides good weed control in naturally wet soils. Rotating rice with non-paddy crops helps in weed control of both crops.

Most rice soils throughout the world have low hydraulic conductivity either naturally or artificially. To reduce hydraulic conductivity, the soil is puddled. Puddling consists of repeated tillage during soil saturation to achieve an artificial tillage pan and "soupy" mud above it. The objectives of puddling are to reduce percolation losses of standing water and dissolved plant nutrients, to control weeds, and to facilitate hand transplanting of rice seedlings where this practice is followed.

Some symptoms of waterlogging include drooping leaves (apparent wilting and epinasty—leaves curve downward at margins), decreased stem growth rate, leaf abscission (prepares to drop), leaf chlorosis (pale coloring), adventitious (secondary) root formation, decreased root growth, death of smaller roots, absence of fruits, and reduced yields. The overriding effect of soil flooding is the limited diffusion of oxygen to roots. The plants that are tolerant of waterlogging have a good flow of oxygen from shoots to roots inside the plant, but the majority of plants require most of their oxygen from soil air around the roots. This oxygen flow internally in waterlogged-tolerant plants apparently occurs through larger air spaces within stems and roots than in those parts in plants not tolerant to waterlogging. Plants that adapt to waterlogged conditions during growth do so by forming larger internal air spaces, even at the expense of some cell destruction and the dissolution of some cellulose in cell walls.

The exact effects of waterlogging damage are rapid but still not clearly understood. The following changes and effects are those generally proposed:

1. A lack of adequate O_2 to roots initiates changes of the amino acid methionine to *S*-adenosylmethionine (SAM). SAM is converted by an enzyme to 1-aminocyclopropane-1-carboxylic acid (ACC). The ACC can then be converted to ethylene *but only in the presence of* O_2. Ethylene ($H_2C=CH_2$) in high concentrations is known to alter plant growth.

   In anaerobic (inadequate free oxygen) conditions growing plants that are not tolerant of waterlogging (poor O_2 transfer to roots through stems) have the ACC produced in roots translocated to stems and petioles where O_2 is available. The ACC is then quickly changed to ethylene. The presence of high ethylene concentrations in petioles causes rapid expansion of cells on the upward side, causing the edges of leaves to droop (**epinasty**). This feature can be seen in less than a day. Other stimulators converting SAM to ACC are increased amounts of the plant hormone indoleacetic acid (IAA), wounds to the plant, and senescence (old age).

2. Apparent wilting of leaves, although not necessarily a general loss of leaf turgor, may be a combination of reduced water permeability to roots (low oxygen lowers roots' abilities to absorb water) and a water loss from leaves (which is unchanged) that exceeds intake. Rapid closure of stomata allows leaves to regain turgor but hinders growth by hindering CO_2 uptake into the leaf.

3. Contents of certain growth hormones, such as gibberellic acid (GA) and abscisic acid (ABA), are reduced in the transport parts of the plant, perhaps another "chemical signal" forcing the stomata to close. Reduced transport of IAA from leaves may cause the epinastic response by petioles. Increases in concentrations of growth substances (auxins) because of reduced transport from roots to the stem may be responsible for adventitious (new secondary) root formation.

4. The plant system can also transport upward some toxins produced in roots by anaerobic conditions.

5. Anaerobic energy transformations are poor; thus, growth rates are slowed because of limited energy.

6. Some toxic substances are produced, including hydrogen sulfide, butyric acid, and volatile fatty acid components of carbohydrate decomposition.

Sources: (1) K. A. Smith and P. D. Robertson, "Effect of Ethylene on Root Extension of Cereals," *Nature,* **234** (1971), no. 5325, pp. 148–149. (2) Makoto Kawase, "Anatomical and Morphological Adaptation of Plants to Waterlogging," *HortScience,* **16** (1981), no. 1, pp. 30–34. (3) Kent J. Bradford and Shang Fa Yang, "Physiological Responses of Plants to Waterlogging," *HortScience,* **16** (1981), no. 1, pp. 25–30.

FIGURE 17-1 This Coker clay soil (Vertisol) near Vilas, Oregon, has poor internal and surface drainage. All Vertisols have very slow infiltration when saturated. During periods of high rainfall or surface flooding, this home site is a sticky mess. (*Source:* USDA—Soil Conservation Service.)

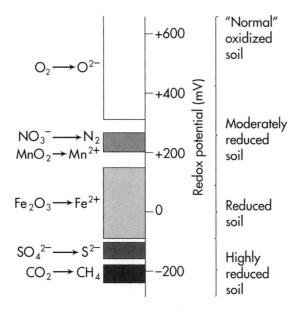

FIGURE 17-2 Example of the range in redox potentials in waterlogged soils and the location in the redox range where the various electron acceptors are active. (Courtesy of Raymond W. Miller, from data of W. H. Patrick, Jr., "The Role of Inorganic Redox Systems in Controlling Reduction in Paddy Soils," *Proceedings of the Symposium on Paddy Soil,* Institute of Soil Science, Academia Sinica, Science Press, Beijing, and Springer-Verlag, New York, 1981.)

Rice can transport oxygen from leaves to roots internally in contrast to most crops that depend on oxygen transport through the soil pores to roots.[1]

In the United States rice is grown on about 1.2 million hectares (3 million acres), with an average yield of about 5044 kg/ha (4,500 lb/a). The six rice-producing states' rank in acreage is Arkansas > Louisiana > California > Texas > Mississippi > Missouri. Throughout the tropical and subtropical world rice is grown in 89 countries.[2]

---

[1] P. K. Sharma and S. K. DeDatta, "Puddling Influence on Soil, Rice Development, and Yield," *Soil Science Society of America Journal,* **49** (1985), pp. 1451–1457.

[2] S. K. DeDatta and S. S. Hundal, "Effects of Organic Matter Management on Land Preparation and Structural Regeneration in Rice-Based Cropping Systems," in *Organic Matter and Rice,* International Rice Research Institute, Los Baños, The Philippines, 1984, pp. 399–416.

Rice paddies are laid out in large beds surrounded with a levee called a *bund.* Seed may be planted with a small-grain drill and each bed then wetted enough to germinate them. When the plants grow up to a good stand (five-leaf stage), more water is added to form a depth of 5–10 cm (2–4 in.). This depth is maintained until near maturity. Fields are then drained 2–3 weeks prior to harvest to facilitate combine harvesting. In some places barely sprouted rice seeds are sown from an aircraft. Rice seeds have the ability to sprout and grow through shallow planting in soil or through shallow water on the soil surface but *not through soil and water.* Most paddy outside the United States uses seedling transplants.

Rice tolerates a soil pH of 4.5–7.0; above pH 7.0 Zn deficiency may occur. When flooded, the pH of such soils shifts toward neutrality. Even somewhat saline soils can be used for wetland rice because water ponding dilutes the salt and thus the plasmolytic effect (death of plant cells by desiccation).[3-6] Reclamation of polders from the sea often uses paddy rice as the first reclamation crop, if the climate is adapted.

Fertilizer N is necessary to maximize yields of rice. Under lowland rice culture ammonium-nitrogen is more efficient than nitrate sources because of the hazard of denitrification of the nitrate.

Urea is the cheapest source of solid N fertilizer. When topdressed into floodwaters, urea increases the pH of the water, which causes volatilization loss as gaseous NH_3. Ammonia loss was measured by N isotope (^{15}N) techniques in the Philippines. During the 30 days following the application of 80 kg N/ha (71 lb N/a) of urea, 55% was lost by ammonia volatization.[7] Slow-release urea fertilizers include supergranules, sulfur-coated urea, urea mixed into large mud balls, and the use of nitrification inhibitors.

When flooded rice is followed by wheat or soybeans, phosphorus deficiency is common. Here are the reasons[8]:

- During flooding, phosphorus and iron become reduced and are more soluble and more available to rice than in aerobic soil
- During the months after drainage of paddy, iron becomes reoxidized and the larger amounts of solubilized iron combine with soil solution (available) phosphorus to make phosphorus and iron less available to the following upland crop

17:3 Drainage Benefits

Some of the more obvious benefits of artificial drainage are as follows:

1. Wet soils are usually the most fertile soils (high clay and high organic matter). Drainage permits them to be used for more-productive purposes.

2. Drained soils warm earlier in the spring. In temperate climates wet soils warm more slowly in the spring than drained soils because water requires four to five times more heat to raise a unit weight of it 1° than is needed for the same weight of dry soil minerals. Plant growth and other chemical reactions are slowed approximately 25% for

[3] Frans R. Moorman and Nico van Breemen, *Rice: Soil, Water, Land,* International Rice Research Institute, Los Baños, The Philippines, 1978, pp. 185.

[4] International Rice Research Institute, *Field Problems of Tropical Rice,* IRRI, Los Baños, The Philippines, 1983, 172 pp.

[5] International Rice Research Institute, *Wetlands Soils: Characterization, Classification, and Utilization,* IRRI, Los Baños, The Philippines, 1985, pp. 559

[6] D. S. Mikkelson and D. M. Brandon, "Zinc Deficiency in California Rice," *California Agriculture,* **38** (1984).

[7] S. K. DeData, et al., "Comparison of Total N Loss and Ammonia Volatilization in Lowland Rice Using Simple Techniques," *Agronomy Abstracts,* (1986), p. 197.

[8] R. N. Sah, D. S. Mikkelsen, and A. A. Hafez, "Phosphorus Behavior in Flooded-Drained Soils. II. Iron Transformation and Phosphorus Sorption," *Soil Science Society of America Journal,* **53** (1989), pp. 1723–1729.

each 4.7°C (10°F) decrease in soil temperature. Plant uptake of phosphorus, potassium, and most other nutrients is increased by drainage (aeration).

3. Proper drainage makes the entire field more uniform in soil moisture (elimination of wet spots) and thus results in earlier, more predictable, and more efficient tractor tillage, planting, and harvesting operations.

4. Removal of surplus water increases aerobic (well-aerated) microbial activity by permitting air (with its 20% oxygen) to replace water in more soil pore spaces. The resulting larger population of aerobic microorganisms produces more organic-matter decomposition, and thus more potential plant nutrients are made available.

5. Drainage decreases the potential losses of nitrogen from the soil by microbial denitrification (which occurs under anaerobic conditions).

6. Drainage reduces the buildup of toxic substances in the soil, such as soluble salts, ethylene gas, methane gas, butyric acid, sulfides, and excessive ferrous and manganous ions.

7. Drained land is adapted to a wider variety of more-valuable crops. Wet soils can produce well only with the limited crop types that can tolerate "wet roots." Water ponded in small surface depressions, even when the profile is not too wet, can kill many crops, such as seedling corn.

8. Drainage permits a deeper penetration by plant roots, thereby increasing the amounts of nutrients available to growing plants and resulting in greater crop yields. Deeper roots also make plants more drought resistant.

9. Drained land is more capable of supporting buildings and roadbeds.

10. Septic systems perform best in well-drained soils; so do sanitary landfills.

11. Controlled drainage allows use of lands for high value crops but with lessened erosion (Fig. 17-3).

12. Mosquito and fly populations are reduced by adequate drainage of wetlands.

13. The market value of drained land is greater than that of comparable ponded ground. Drainage also preserves the land for arable use.

14. Adequate drainage is essential for reclaiming saline, saline-sodic, and sodic soils.

15. Drainage systems reduce heaving (pushing out of the ground) of plants by ice crystals forming around the plant crowns.

▬▬ 17:4 Drainage Hazards

A distinction should be made between *wetlands* and *wet soils* in need of drainage. **Wetlands** are grounds too wet to drain economically for growing upland agricultural crops (or have greater value as animal and bird habitats); **wet soils** are suitable for producing agricultural crops more productively when drained. The dividing line between wetlands and wet soils, however, is often shifted by public policy, technology, and economics. Before the United States was settled, the wetlands area was estimated at 51 million hectares (126 million acres). This area had shrunk to 28 million hectares (69 million acres) by 1971 because water tables had been lowered by artificial drainage. Restrictions to draining wetlands now exist.[9]

---

[9] U.S. Geological Survey and U.S. Environmental Protection Agency, *Environmental Statistics, 1978,* U.S. Government Printing Office, Washington, D.C., 1979, p. 39.

FIGURE 17-3 Controlled drainage and contour planting in contour blocks help conserve soil planted to pineapple. Notice the diversion terraces and sodded waterways (upper center). The high rainfall in this Maui Pineapple Co. field in Hawaii requires controlled surface drainage to avoid severe erosion. (Courtesy USDA—Soil Conservation Service; photo by Arnold Nowotny.)

Swampbuster is the term given to a part of the Food Security Act of 1985, which denies federal subsidies to persons draining and farming wetlands outside strict regulations. This is an attempt to control and reduce conversion of wetlands to farmlands, thereby helping to retain wetland habitats.[10] This law concerns conversion of wetlands to croplands, not the drainage of land already cropped to produce better yields. The law deals only with land on which drainage was initiated after the law was enacted.

Some wet soils should *not* be drained, as in the following cases:

1. Land where drainage would adversely affect the environment (e.g., by reducing the number and variety of aquatic animals, such as beaver, muskrat, and waterfowl). Income from leasing, hunting, and trapping rights before establishing drainage systems may even exceed probable crop income after drainage. Environmental protection of some areas may also override the economic benefits of draining the land for crop production.

2. Land where drainage would lower the ambient (surrounding) water table and thereby decrease the level and volume of water flowing into nearby streams, ponds, springs, and shallow wells.

3. Land where excessive drainage would adversely affect capillary water rise in deep sandy soils (Psamments) and deep organic soils (Histosols). Such soils may lack sufficient water after drainage because of a slow and low capillary rise from the watertable. This hazard is less on medium- and fine-textured soils, for the soil pores are smaller and more continuous, both of which aid capillary action.

---

[10] *Federal Register,* Sept. 17, 1987.

4. Wet soils that contain excess amounts of iron disulfide (FeS_2, pyrite, fool's gold) should not be drained. On being drained, the iron disulfide oxidizes to ferrous sulfate and *sulfuric acid,* thereby lowering the pH, sometimes even to a toxic pH 2.0.

5. Land where drainage, particularly for soils low in fertility, may cost more than the increased value of crops grown on the soil.

17:5 Drainage System Selection

The two chief types of drainage systems are *surface* and *subsurface.* **Surface drainage** is achieved primarily by constructing gently sloping open ditches for water collection and by smoothing the soil surface and creating enough slope to facilitate runoff toward the drainage ditches. The most prevalent kind of **subsurface drainage** consists of burying conduits (sections of tile or porous plastic tubing) on specified grades (specified slopes).

There are no technical restrictions on the use of surface drains; they can be used on all kinds of soils. Subsurface drains, however, require a soil profile that is sufficiently porous for water to percolate through it to the buried drain tile or porous tubing. Many of the finest-textured clays have such small pores between the particles that they are too slowly permeable to be drained adequately by any *subsurface* drainage system. Such soils must be drained by a *surface* drainage technique or an open-ditch system.

17:5.1 Workability of Subsoil Drainage

It is often difficult to determine whether a soil will drain rapidly enough to permit the use of some form of subsurface drainage. With such a large potential investment, it is usually worthwhile to make some field and laboratory determinations to find which drainage system is best suited to any particular soil. Pore spaces that are nearly full of water at field capacity do not transmit much water down through the pores to drain lines. The soil pore air space through which gravitational water moves is the total pore space minus the pore space occupied by water held at field capacity (–33 kPa matric potential).[11] This value is known as the **drainage capacity.** As a percentage, drainage capacity is determined in this way:

percent drainage capacity = (% soil volume that is pore space)

– (% water-filled pore space at field capacity, on a volume basis)

$$DC_p = E_p - P_v$$

In Mississippi, for example, the surface soil of Memphis silt loam (a Typic Hapludalf developed from loess) has a total pore space of 59% and a field capacity of 20% (Table 17-2). The drainage capacity is, therefore, 59% – 20% = 39% a readily drainable soil.

Soils with drainage capacities greater than about 10% can be drained with subsurface systems. From Table 17-2, the Bosket sandy loam (a Mollic Hapludalf), for instance, with a drainage capacity of 12%, will drain fairly readily. On the other hand, Sharkey clay, a Vertic Haplaquept—for its moisture retention at –33 kPa (field capacity) is greater than the total dry soil pore space—*will not drain through tile drains.* (Sharkey clay expands greatly when wet; so the total pore space wet is greater than when it is dry. For this reason, the total *dry* pore

---

[11] Percentage moisture times the ratio of bulk soil density to water density equals volume percentage of water. This gives the percentage of the soil *volume* containing water at that moisture content.

Table 17-2 Total Pore Space, Pore Volume at $-1/3$ Matric Potential, and Drainage Capacity of Three Mississippi Soils

Soil Type	Depth (cm)[a]	Total pore Space (%)	Pore Volume at $-1/3$ bar of Matric Potential (%)	Drainage Capacity (%)
Memphis silt loam	0–15	59	20	39
Bosket sandy loam	0–20	51	39	12
Sharkey clay	0–15	51	81	0

Source: W. M. Broadfoot and W. A. Raney, "Properties Affecting Water Relations and Management of 14 Mississippi Soils," Mississippi Agricultural Experiment Station Bulletin 521, 1954.
[a] Centimeters times 0.4 = inches.

space percentage is exceeded by the water-filled pore space percentage.) This soil can be drained only by surface ditches because water will not flow through the soil into tile or tube drains at a rate sufficient to remove the surplus water.

17:5.2 Drainage System: Which One?

Some of the more important relationships to be considered when choosing a drainage system are given in Table 17-3. The major factors are soil impermeability (the cause of the poor drainage) and the topographic features of the area.

17:6 Surface Drainage Systems

Surface drainage systems are best adapted to drain flat or nearly flat soils that are (1) slowly permeable, (2) shallow over rock or fine clay, (3) have surface depressions that trap water, (4) receive runoff or seepage from uplands, (5) require the removal of excess irrigation water, and/or (6) require lowering of the water table.[12] The three principal surface drainage systems are *open ditch, smoothing,* and *drainage-by-beds* (bedding).

17:6.1 Open Ditches

Open ditches must be designed to conform to topography, natural drainways, land use, and soil characteristics. Such ditches are usually parallel and designed as a grid or in a herringbone pattern. Spacing between the parallel ditches is determined by the amount and intensity of the rainfall in relation to the infiltration capacity of the soil. The depth and width of the ditches, as well as the spacing, must be determined by using all available local data. Data on rainfall intensity are readily available from the U.S. Weather Service. Farm field drains are normally designed to carry a maximum volume of water from the maximum expected 1-hour rainfall once in 5 years. When larger areas are drained, they may be designed to direct the surface flow adequately from a maximum 1-hour storm once in 20 years (the largest predicted storm for 1 hour that might be expected once in 20 years).

Field ditches should be at least 30 cm (1 ft) deep and, if machinery is to cross them, have side slopes of 8 : 1 (8 linear units horizontal to 1 vertical). If not crossed with machinery, it is cheaper to build them at the angle of repose (natural stability) of the respective soil. The angle of repose for loam, for example, is 2 : 1; for clay, 1.5 : 1; and for peat, muck, or sand, 1 : 1.

---

[12] USDA—Soil Conservation Service, *National Handbook of Conservation Practices,* USDA, Washington, D.C., 1977, p. 607-A-1.

Table 17-3 Soil and Topographic Parameters Important in Selecting the Type of Drainage System to Install on a Poorly Drained Area

Parameter	Suggested Drainage System
1. Sloping areas	Interceptor lines in low areas or seepage spots
2. Closed basins, level areas	Outlet problems, may need to pump
3. Deep, permeable sands	Any system is adequate
4. Deep, impermeable clays	Careful irrigation management; mole drains, surface and open ditch drains
5. Shallow, permeable soil over impermeable layers	Tube or tile drains just above impermeable layer; careful irrigation management
6. Deep (9–12 ft, 2.8–3.7 m thick), impermeable soils over coarse sands, gravels	Sump or well[a] drainage; surface and open-ditch drains in humid areas
7. Water table fluctuates with irrigation	Tube drain system on a grid and careful irrigation control
8. Water table fluctuates with rainfall	Surface drainage better; also, consider tube or tile drains
9. Ponded water in fields	Surface grading (sloping) and surface drains

Source: Selected, modified, and tabulated from W. W. Donnan and G. O. Schwab, "Current Drainage Methods in the USA," in *Drainage for Agriculture,* J. Van Schilfgaarde, ed., No. 17 in the Agronomy Series, American Society of Agronomy, Madison, Wis., 1974, pp. 93–114.
[a] Shallow wells or sumps with the water collected are pumped to surface drainage lines.

Seeding an adapted grass on the bottom and sides of the drain ditches helps to stabilize them. In most regions timothy and reed canarygrass are recommended for drain ditches. Under no circumstances should Bermuda grass or quackgrass be seeded in drain ditches on croplands because these grasses will spread to the fields. An alternative technique, known as the **open-W drainage system,** is to establish side slopes of 8 : 1 and move the soil so excavated toward the middles between each two drains (making a hump resulting in a W-shaped soil cross section, with the ditches forming the low points). When dry enough, crops can then be planted *in the ditches* at the same time as between them. Rows should run parallel to the ditches to avoid crossing them with tillage implements and thereby carrying soil into them. This modification is similar to the drainage-by-beds system, described in the next section.

Open ditches do not satisfactorily drain any field if the soil surface is not uniformly smooth and sloping toward the ditches. The field should be plowed or disked and then leveled by a landplane or leveler. Laser-beam-guided equipment can greatly aid land leveling. Greater leveling efficiency is achieved if the technique is followed as shown in Fig. 17-4.

17:6.2 Drainage by Smoothing

Smoothing (eliminating minor ridges and depressions) of the field without altering the general topography) and **grading** (making a uniform slope) usually need 2 years to complete. After the initial work the first year, a second smoothing is done the second year, once the soil has settled. The quality of the smoothing is best viewed immediately after a storm. If puddles persist—even small ones, a half meter (couple of feet) across—the smoothing is not adequate. Yearly repairs are desirable. Rapid "ditchers" are available for making nonpermanent shallow-surface field ditches.

Maintenance of open ditches should be done at least once a year after crop harvest, or more often if the need arises. Grasses and other weeds often grow rampant. Sediment must be cleaned out, willow sprouts and water weeds removed, and, in some areas, tumbleweeds disposed of. Often a suitable herbicide may be used on growing plants, followed by burning to dispose of the plant residues.

(a)

Ditch   Fill   Cut   Ditch

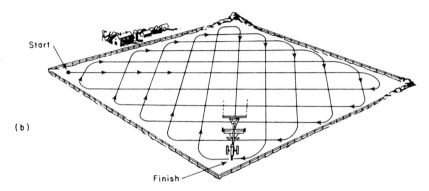

Start

(b)

Finish

FIGURE 17-4 Surface drainage system showing (a) a system of field ditches with the land smoothed to facilitate surface flow toward the ditches and (b) the recommended pattern for operation of a leveler for smoothing or grading. (*Sources:* (a) U.S. Department of Agriculture; (b) U.S. Department of the Interior.)

17:6.3 Drainage-by-Beds

Fine-textured, fairly level fields in areas of high rainfall are usually drained most advantageously by the construction of a precision land-forming drainage-by-beds system. Also known as **bedding,** the system of **drainage-by-beds** is similar to the open-W drainage system. One difference is that the drainage-by-beds system usually has a higher crown and is narrower between drains, as depicted in Fig. 17-5. The more water to be disposed of, the narrower must be the beds and the higher the crown. The bed widths should be in multiples of the width of the tillage and planting implements available. Maintenance of bedding drainage includes plowing a back furrow at the crest of the ridge and plowing "uphill," leaving the final furrows (dead furrows) at the drains.

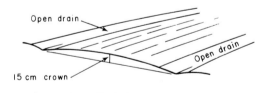

Open drain

Open drain

15 cm crown

Open - W drainage system

FIGURE 17-5 Comparison of the open-W drainage system of open-ditch technique (above) and the drainage-by-beds system (below). (*Source:* Adapted from USDA—Soil Conservation Service.)

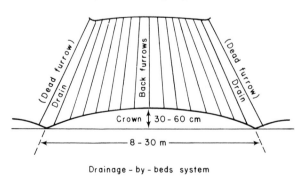

(Dead furrow)
Drain

Back furrows

(Dead furrow)
Drain

Crown ⏐ 30 - 60 cm

8 - 30 m

Drainage - by - beds system

17:7 Subsurface Drainage Systems

The principal subsurface drainage systems include tile drainage, tube drainage, mole drainage, sump-and-pump drainage, and special vertical techniques, such as relief wells, pumped wells, and inverted or recharge wells. Subsurface drainage systems have become increasingly popular in recent years because prices of productive crop-land make installing drainage systems on wet, less expensive land more feasible than acquiring better land. Crops can be planted over subsurface drainage systems but not over most surface drainage ditches.

Tile and tube drainage have the same technical layout parameters; they differ only in the kinds of pipe used. **Tiles** are short sections of pipe made from fired clay or concrete and may be 30–60 cm (1–2 ft) long and 10–30 cm (4–12 in.) or more in diameter. **Porous plastic tubes** are now made in similar diameters but are a hundred meters (305 ft) or more long per section.

To inhibit the entrance of soil particles into openings in the sides of tile or plastic drain pipe, gravel is usually poured around the tile and pipe when laid or before backfilling. In some locations gravel or chat (ground stone) is scarce. In these areas Drainguard®, a nylon substitute for gravel, has been tried and proved satisfactory (Detail 17-2).

17:7.1 Tile Drainage System

Satisfactory layout of a tile drainage system requires considerable planning and a great deal of experience. Technical assistance in layout and installation of tile drainage systems can be obtained from local Soil Conservation Service offices.

A tile drainage system can operate satisfactorily for a century or more if properly planned, adequately constructed, and carefully maintained. Concrete and fired-clay tile are both satisfactory drain materials, but in strongly acid soils the concrete is dissolved by the acid. The depth and spacing of the lines of tile vary with the crops grown and the kind of soil. Soils with slow downward movement of water should have shallower placings of the lines of tile, and the lines should be laid closer together. Alfalfa and orchards need drainage with a depth of tile of about 1.2 m (4 ft). Corn needs intermediate depths, and grasses and small grains can get along adequately with tile lines placed about 60 cm (2 ft) deep. Horizontal spacings may vary from 12 to 92 m (40–300 ft) between tile lines, depending on soil drainage capacity, which is related to soil texture and soil structure. Clayey soils need close spacings.

In low-permeability clay and clay loam soils, tile depth should not exceed 1 m (3 ft), and the horizontal spacing no more than 21 m (about 70 ft). Tile lines in silt loam soil can be placed 1.2 m (4 ft) deep and 30 m (about 100 ft) apart. The respective maximum spacing and depth suggested for sandy and organic soils are 92 m (300 ft) and 1.4 m (4.5 ft), respectively. In irrigated arid soils it is common for tile to be laid 1.8 m (6 ft) or deeper.

Outlets for tile lines should be screened to prevent rodents and insects from plugging them. They should also be encased in cement, with a suitable apron to prevent undercutting by flowing water; the last 3 m (10 ft) of tile back from the outlet should be cemented at the joints. Other tile in the lines is placed end to end to permit water to seep in between the sections of tile (Fig. 17-6).

Trees and shrubs near tile lines should be removed so that the roots cannot grow into cracks between the joints of tile. A small sink hole in the soil above a tile line indicates that one of the sections has been broken or displaced. It should be repaired before the whole tile system is ruined by being plugged with soil. Sometimes an outlet will erode and render the entire drainage system useless. Timely maintenance pays big dividends in extending the useful life of a tile drainage system.

Although gravel is the customary material used in the United States as an envelope filter around newly laid subsurface drains, cheaper materials have been effective in other countries. These materials include peat, lime, gypsum, and small-grain straws.

Peat has proven satisfactory as a filter material in Belgium, the Netherlands, and Great Britain. Small-grain straws have also been used as drain filters in Great Britain. Calcium oxide (CaO) has been used successfully in Hungary for three different soil conditions. On acid soils CaO precipitates soluble iron, which reduces its buildup inside the drain pipe. Calcium oxide flocculates clay soils, making them more permeable to water and also less likely to wash into the drain tubes. On sodic soils CaO is mixed with gypsum ($CaSO_4$) to replace soil sodium ions (Na^+) with calcium ions (Ca^{2+}), which increases the movement of water through the soil particles (better hydraulic conductivity).

Source: Food and Agriculture Organization of the United Nations, "Drainage Materials," Irrigation and Drainage Paper 9, FAO, Rome, 1972, 122 pp.

17:7.2 Plastic Tube Drainage System

A major disadvantage of tile drainage is its cost, particularly in labor. Plastic tube drains are less expensive than are tile drains. The major advantages of plastic tube lines are savings in installation labor, near-foolproof alignment, and lower material cost.

Corrugated plastic drain tubing, in use in the United States since about 1967, has become increasingly popular; in 1975 about 152 million meters (463 million feet) of corrugated plastic tubing was installed in the United States and Canada.

The diameter of the early tubing was 10 cm (4 in.), but in recent years diameters up to 46 cm (18 in.) have been produced. Repeated field tests have shown that all diameter sizes have performed satisfactorily when installed in sand, loam, clay, and muck soils according to manufacturer's directions.[13]

The perforated, corrugated, polyethylene and polyvinyl chloride 10-cm (4-in.)-diameter tubing comes in 76-m (250-ft) lengths weighing 31.8 kg (70 lb) and with easy-to-connect couplers to form an endless tube (Fig. 17-7). Laser-guided machines with a single operator can easily lay over 1.6 km (1 mi) of tube line daily with a vertical-alignment accuracy to within 3 mm ($^1/_8$ in.), eliminating the problems of poor drainage flow resulting from humps or dips in the tubing. Although installation is rapid, the laser-guided laying equipment is expensive; so plastic tubing drainage is best used on large fields needing extensive drains over long distances. Clay or concrete tile is a more rigid material than plastic and can still be used economically on short lines and special small drainage systems. Rodents sometimes cause failure of plastic drainage tubing by chewing holes in it.

17:7.3 Mole Drainage System

Mole drains are most useful when temporary drainage is needed, such as for salty land that needs rapid reclamation but that may not require permanent improved drainage once the excess salt is removed. Generally, mole drains can be cheaply and rapidly installed or even redone when needed every few years.

---

[13] Carroll J. W. Drablos, Paul N. Walker, and James L. Searborough, "Field Evaluation of Corrugated Plastic Drain Tubing," in *Third National Drainage Symposium Proceedings,* American Society of Agricultural Engineers, St. Joseph, Mich., 1976, pp. 69–74.

(a)

(b)

(c)

FIGURE 17-6 A tile drainage system will be satisfactory for a century or more if properly planned, adequately constructed, and carefully maintained. (a) A tile ditching machine seen in operation. West Virginia. (b) A tile line that has been laid but not yet covered. Oregon. (c) A line of tile that has not yet been completed but is already draining the field into an open ditch. The outlet will be screened to keep out rodents. Washington. (*Source:* USDA.)

FIGURE 17-7 (a) A laser-guided plastic drainage tube layer ensures great accuracy in the control of the line elevation and rapidity of installation. This installation near Florence, South Carolina, can lay drains 6 ft (1.8 m) deep at rates of 3000 ft (915 m) per hour, and can work at a 40-acre (16-ha) field with only one setting of the laser plane command post shown in the left foreground. (b) A trench-cutting, laser-guided layer with following disks, which partially cover the tube to fix it in place. Careful covering is essential to minimize crushing the tube. Installation cost in 1989 was about $2.50–3.00 per foot plus gravel and pipe. The tube often used has a nylon web welded around the plastic tubing that eliminates the need for gravel. (c) A less rapid trench-cutting tube layer, but with a less costly machine, in operation near Lewiston, Utah. The trailing box screens soil or gravel added to cover the tube first so large rocks or clods do not jar tile out of alignment. (*Sources:* (a) USDA—Agricultural Research Service, from J. L. Fouss; (b) United Enterprises, Orem, Utah; (c) USDA—Soil Conservation Service.)

(a)

(b)

(c)

Mole drains are shallow, short-lived drainage channels created by pulling a torpedo-shaped object (the "mole") through fine-textured soil about 51–61 cm (20–24 in.) deep (Fig. 17-8). Even in the first year some channels will partly fill with soil. To compensate for this loss of drainage capacity, the channels can be dug only a few feet apart, more closely than is needed originally. Thin, perforated, plastic liners are available for an additional cost, but they extend the mole drain channel life considerably.

17:7.4 Sump-and-Pump Drainage System

Sump-and-pump drainage is used for areas in which water removal is not possible by gravity flow. Water is collected in a low-lying area and then removed by pumping. The sump is a low place in which water collects; water may flow into the sump from surface or subsurface sources or both. To avoid collapse of the sides of the sump depression, it should be lined with sand, then gravel, and on top a layer of coarse gravel or broken rock. The sump motor and pump should be mounted rigidly above the sump depression and a float valve installed to start and stop the motor and pump.

FIGURE 17-8 Organic soils usually require drainage because of high water tables. A mole drainage machine, in operation in Florida, is making a cut from the main drainage ditch and continuing the mole drainage line across the field. The mole drain *plow* has a knife-edged shank that eases the pull by cutting through the soil, and the torpedo-shaped mole tool has a trailing chain and ball to smooth the tunnel a second time. (*Source:* USDA—Soil Conservation Service.)

With increasing energy costs, every practical effort should be made to avoid sump-and-pump drainage by shaping the field to permit gravity flow of drainage water. For polder lands reclaimed from the sea, the surface level of the land is below sea level and pumping is necessary to remove drainage water.

17:7.5 Vertical Drainage Systems

The principal kinds of vertical drainage systems are vertical subsurface drainage inlets, inverted wells, pumped wells, and relief wells.

Vertical Subsurface Drainage Inlets   An increasing use of herbicides to kill weeds on terraced cropland causes runoff of sediments with attendant herbicides that often kill grasses used in stabilizing grassed waterways. Furthermore, land now used for grassed waterways could be producing crops under a different drainage system. **Vertical subsurface drainage,** an alternative technique for moving runoff water from higher-level to lower-level terraces, consists of installing vertical pipes with screened tops into which surplus water enters. The water then moves through a subsurface drainage system to a lower terrace or a protected outlet.

Inverted Wells   In this **vertical well drainage system** excess water is directed *into* a hole in the ground (the reverse of the usual; hence the name "inverted" wells). Inverted shallow wells near Fresno, California, are being used in soil underlaid with hardpan, which ponds the surface water. The wet fields become hatching grounds for mosquitoes when water stands for as long as a week. A screw auger drills holes 23 cm (9 in.) in diameter and 2.7–3.4 m (9–11 ft) deep where needed to drain the wet spots. The bore holes, filled with coarse gravel and a few inches of soil on top, drain the soil in less than 24 hours, causing mosquito eggs to dry and die. An added benefit of the drainage is that the soil can be cropped earlier.

First law of ecology: In nature you can never do just one thing, so always expect the unexpected; or there are numerous effects, often unpredictable, to everything we do.

The danger is not what nature will do with man, but what man will do with nature.

—Evan Esar

Questions

1. What evidence can be seen that might indicate that a soil has very poor drainage? (Consider native vegetation, appearance in rainy weather, and soil profile colors.)
2. (a) What is gley? (b) What are mottles? (c) What are the colors of gleyed and of mottled soils?
3. State how drainage affects these items: (a) rate of warming, (b) the variety of adaptable crops, (c) toxic organic substances produced, (d) the depth of the root zone, and (e) salt accumulation or removal.
4. Recent "swampbuster" laws restrict indiscriminant drainage of wetlands or swamps. What is the justification given?
5. What are some kinds of "toxins" produced in poorly drained soils?
6. What is meant by *paddy?*
7. Describe the changes that occur in rice paddy soils as they are flooded: (a) pH changes, and (b) aeration.
8. Why is the soil puddled for paddy?
9. Because all plants need oxygen for root respiration, where do rice roots in paddy get oxygen?
10. Ironically, some poorly drained soils may, after drainage, become droughty. Explain.
11. Why might some clayey soils not be easily drainable by subsurface drainage?
12. List advantages and disadvantages of (a) open-ditch drains, (b) tile drains, and (c) mole drains.
13. (a) Describe how surface-smoothing drainage is done and is checked for quality. (b) Why and when might it be used in preference to other methods?
14. (a) Define *drainage-by-beds.* (b) What are the advantages of beds?

15. (a) Describe the formation of mole drains. (b) Are they as permanent and deep as tile drains? Explain.
16. (a) What is the composition of tile in tile drains? (b) How does water seep into the tile lines? (c) What problems can occur to tile lines?
17. What is the purpose of lasers in establishing drainage systems?
18. Explain why a person may use a sump-and-pump system. (Remember, sea bottoms—polders—may be drained this way.)
19. When might vertical-well drainage be practical? Explain.
20. (a) Define *subirrigation*. (b) How might some small, unique areas adapt drainage systems for subirrigation?
21. (a) Why is soil drainage sometimes needed in *arid* climatic areas? (b) Explain how it is used.

18

Soils and Environmental Pollution

The far more difficult problems [in controlling affronts to law and decency] in a complex society are those where there is no evil intent, no deliberate intrusion on the security of others, not even outrageous stupidity, just normal, law-abiding, enlightened folks going about their daily business.

—Lawrence W. Libby

18:1 Preview and Important Facts

PREVIEW

Soils are nature's dispose-all, its sewage treatment plant, its water purifier, and, at times, also a pollutant. Soils are valuable as cleansers of the earth's environments. The soil is a **physical filter** (sieving action), a **chemical filter** (adsorption and precipitation), and a **biological filter** (decomposition of organic materials), as well as the receptacle for all things buried and disposed of beneath and on the surface.

Although the soil is the most universal and extensive substance that cleans waters and recycles wastes, it is not infinite in capacity. High land costs near most large cities increase the expense of waste disposal. Many toxins added to soils can build up to concentrations that become serious threats to plant and animal health. Some toxic substances become residual in the soils; perhaps centuries, even millennia, will pass before they again permit normal use of the soil. Even harmful organic substances that will decompose eventually to nonharmful recycled elements of carbon, oxygen, hydrogen, phosphorus, nitrogen, and sulfur are dangerous until that decomposition is well along. Materials accumulate when they are added in larger amounts than can be accommodated by their decomposition rates (Fig. 18-1). Materials that are toxic to soil microorganisms further slow recycling.

Pollution is *adding something to air, soil, or water that makes it less desirable for people's use.* Probably all activities have the potential to pollute with **noise, wastes, toxins, tastes, odors, carcinogens,** and other undesirable products. Agriculture, forestry, and homeowners, among many other sources, add pesticides, nutrients, and soil sediment to air, soil, and water. Industry, businesses, and individuals pollute with solid wastes, odors, gasoline

FIGURE 18-1 Oil-waste land in Jim Hogg County, Texas, is wasteland caused by excessive pollution by oil, saltwater, and oil-drilling muds left in the early days of oil well drilling. Modern drilling techniques control waste disposal and do relatively little damage. Many of these areas will improve gradually, but most will require the help of irrigation, tillage, and lots of time.   (*Source:* USDA — Soil Conservation Service, Temple, Tex.)

and natural gas fumes, cosmetics, heavy metals, solvents, and hundreds of organic and inorganic hazards.

Pollution has always been a fact of peoples' use of resources. But nature is able to handle "reasonable" amounts of most natural substances. We have just become more consumptive and more numerous, discarding large amounts of "wastes." We have manufactured hard-to-decompose materials—metal tools, homes, and machines; resistant synthetics and plastics; and countless varieties of glass, ceramics, and heavy metals. We have invented poisons and toxins to control unwanted weeds and other pests. We are overdue in facing the havoc we have wrought. The question now is whether we can afford the reclamation without major changes in the lifestyles to which most people in developed countries have become accustomed.

IMPORTANT FACTS TO KNOW

1. The definitions for *bioaccumulate, pollution, eutrophication, half-life, heavy metals, PANs, pesticides, carcinogenic, teratogenic, radioactive*
2. The seriousness of nutrient pollution of waters and what alternatives are available
3. The legal bases on which the EPA controls many activities on lands and waters
4. The meaning of *nonpoint sources of pollution*
5. The meaning of and need for *BMPs*
6. The problem of selenium contamination and its relation to salt accumulation
7. The general problem of heavy-metal contamination
8. The reason we will continue to use pesticides for at least several decades more and the results if pesticides were to be banned from use

9. The need and methods to reuse wastewaters
10. The health hazards from infection and from heavy-metal bioaccumulation when applying sewage sludges and various other solid wastes to cropped soils
11. The nature and longevity of the methane explosion hazard from organics buried in landfills
12. The extent to which the soluble-salt hazard is increasing worldwide and whether the problem is likely to lessen or worsen
13. The extent of radioactive fallout onto soils and the alternatives available to "clean up" the polluted soils
14. The extent and kinds of damages caused by soil erosional sediments
15. The relation between population, "civilization," and *prime farmlands* and the solutions currently being tried in order to continue to use these lands for agriculture

18:2 Definitions

In considering sensitive topics, such as pollution, the meanings for the terms used must be clear and exact. Some of these terms are given in the following list:

Acceptable daily intake (ADI) The amount of material, including a "safety factor," that is set as safe to ingest.

Acute toxicity Pronounced and immediate toxicity. Symptoms are exhibited by the organism in a short period of time.

Antagonism An effect of substance A to reduce the detrimental effects of substance B below the effect substance B would have if it were alone in the system.

Bioaccumulate (biomagnify) This term indicates an accumulation of a substance in increased concentrations over time by an organism. Biomagnify usually means increasing the concentration upward in the food chain.

Biological oxygen demand (BOD) The amount of dissolved oxygen in water that will be consumed as the organic matter present is decomposed. High BOD means low water quality and probably the development of anaerobic waters. It usually results when waters have received organic wastes.

Cancer potency factor A "negligible" risk, or potentially causing no more than *one additional cancer per million people.*

Carcinogenic The ability to cause cancer in animals and, by implication, in humans.

Chemical oxygen demand (COD) The amount of dissolved oxygen in water that will be consumed, as all oxidizable substances (organic and inorganic) are oxidized. A high COD is undesirable.

Chlorinated hydrocarbon Any of a number of organic pesticides with chloride groups on carbon atoms. Nearly all chlorinated hydrocarbons have long half-lives in the environment, are of low water solubility, and are nonselective in their action. The best known is DDT (**d**ichloro**d**iphenyl**t**richloroethane). Others include chlordane, aldrin, heptachlor, and dieldrin.

Chronic toxicity Low-level toxicity. Distinct effects or symptoms may be very slow to develop or not be evident at all; they may appear months or years after the initial exposure.

Contamination (pollution)   Accumulation of any substance that makes the soil, air, or water less desirable for people's use. The contaminant may be any substance from simple nontoxic substances, such as soluble salts or phosphorus, to toxic and poisonous substances.

Eutrophic (eutrophication)   Nutrient enrichment of water. It is usually most affected by phosphorus; less so by nitrogen. Increased algal growth eventually loads the water with dead algae, which, during microbial decomposition, results in consumption of the water's dissolved oxygen, causing anaerobic water.

Food chain (food web)   The transfer of energy and material through a number of organisms, one being food for the next (small fish eaten by larger fish eaten by a bird, then eaten by a hunter). The food web is the complex intermeshing of many food chains.

Half-life   The time required for half of a substance to be destroyed, made inactive, or to be lost. Used for loss of activity of pesticides, of other materials, or of radiation from radioactive substances.

Heavy metals   High-atomic-weight metals such as cadmium, lead, mercury, nickel, and chromium. These and several others are, in large amounts, toxic to animal life and bioaccumulate.

Mutation   A random but inheritable gene change, which can occur naturally but may be caused by increased exposure to radioactive radiation and/or certain chemicals. Usually, a mutation is a physical or health defect.

Ozone shield   A concentration of ozone gas (O_3) in the upper atmosphere, which reacts with and reduces much of the harmful ultraviolet radiation from the sun that reaches the earth. The shield is believed to be reduced or destroyed by certain chemicals (e.g., chlorofluorocarbons) in polluted air.

PANs   A group of chemicals (**p**erox**y**a**c**et**y**l**n**itrate**s**) of various origins present in photochemical smog. They are extremely toxic to plants and irritating to eyes and membranes of the nose and throat.

Pesticide   Any chemical designed to kill pests (weeds, insects, mites, nematodes, fungi, rodents, algae, bacteria). The term now includes many natural chemicals (juvenile hormones or attractants) to trap or confuse insects, hindering them from finding mates (as pheromones).

Pollution (pollutant)   *See* Contamination.

Radioactive substance   Substances of, or containing, unstable chemical isotopes of elements and that give off radiation (alpha, beta, and gamma rays and others) at some specific half-life rate. These are destructive to biological tissues and can cause cancer or mutations in some instances.

Reference dose (RFD)   Level of daily exposure that, over a 70-year human life, is believed to have no negative effect.

Resistance, developed   Organisms, through natural selection, develop populations that are no longer killed by a particular pesticide. These may be genetic changes to a few organisms. These "changed" organisms breed new populations which are resistant to the pesticide.

Risk-benefit analysis An evaluation of pesticides when considering their registration and changes in residues.

Synergism Two factors interact in some manner to cause a greater effect than expected by summing the effects of the two substances separately, when each acts alone.

Teratogenic The substance causes tissue deformations, when eaten by the mother at certain stages during gestation, usually in early stages of the development of the fetus.

Threshold level The maximum level of a substance or condition that can be tolerated without ill effects. Often these values are difficult to determine unequivocally.

18:3 Water and Air Quality Regulations

National water and air quality control is embedded in numerous laws. An extensive law was passed in 1972 and has had periodic updates. Several laws passed in 1987 include the Federal Water Pollution Control Act, the Safe Drinking Water Act, the Solid Waste Disposal Act, the Comprehensive Environmental Response Act, the Compensation and Liability Act, the Resource Conservation and Recovery Act, and the Federal Insecticide, Fungicide, and Rodenticide Act. These many acts empower and require the **Environmental Protection Agency (EPA)** to regulate water and air quality. The EPA must control management of land to the extent necessary to limit pollution of water and air. For the past two decades considerable control has been exerted on **point sources** of pollution (pipelines, smokestacks, hauled wastes, and disposal canals). In Section 319 of the 1987 Water Quality Act, Congress extended the EPA's obligation to set up methods to control **nonpoint sources** of pollution (underground leaching, general erosion, or evolution of gases from large areas, each difficult to pinpoint as a pollution source) (Fig. 18-2).

The problem with controlling nonpoint sources of pollution is whom to blame because sources of pollution are widespread and may come from many land users in small amounts. This problem has forced the EPA to require land users to practice what we might label "reasonable responsibility." Unfortunately, freedom of action in the United States is difficult to

FIGURE 18-2 Blowing soil is a pollutant in air, as well as in water or on lands where the windblown soil materials fall out. This storm in Floydada, Texas, had winds 50–75 mph and buried a road adjacent to this unprotected cotton field. It is a *nonpoint source of pollution* because miles of fields had wind erosion at the same time. (Courtesy of USDA — Soil Conservation Service; photo by Earl Spendlove.)

Statewide and nationally, best management practices (BMPs) are being devised for all imaginable activities that might contribute to pollution of air and water. The two schemes listed below are from Arizona and apply generally to the entire state. The BMPs are to be followed to reduce pollution by (a) nitrogen from fertilizer applications and (b) various pollutants from animal-feeding operations (manures, washings).

1. BMPs for application of nitrogen fertilizers to crops
   a. Application shall be limited to the amount necessary to meet projected crop needs (eliminates freedom to add extra, "insurance," fertilizer; requires knowledge of crop and soil needs)
   b. Applications shall be timed to be as efficient as possible
   c. Method of addition shall be to add nitrogen to the area of maximum crop uptake
   d. Irrigation shall be managed to minimize nitrogen loss by leaching and runoff
   e. Tillage practices that maximize water and nitrogen uptake shall be used

2. BMPs for animal feeding operations
   a. Minimize leaching and runoff losses of nitrogen by harvest, stockpiling, and disposing of manures as economically as is feasible
   b. Have water control facilities (ponds, etc.) to control and dispose of nitrogen-contaminated water in case of a 25-year, 24-hour storm event equivalent as economically as is feasible
   c. Close facilities, as necessary, in an economically feasible manner, to minimize nitrogen pollutant discharges

Source: Modified from Brian E. Munson and Carrol Russell, "Environmental Regulation of Agriculture in Arizona," *Journal of Soil and Water Conservation,* **45** (1990), no. 2, pp. 249–253.

stifle. The most direct way to deter actions that cause contamination or pollution is to declare those actions illegal. Consequently, state and federal agencies establish standards and then declare as illegal any actions that violate those standards. Sources of nonpoint pollution are reduced by making certain actions illegal. The list of permissible actions are termed **best management practices (BMPs).** Detail 18-1 lists some state-developed BMPs for application of nitrogen fertilizers and for managing animal-feeding operations in Arizona. What BMPs can be installed to control erosion in Grand Canyon National Park (Fig. 18-3)?

The essential purpose of water and air quality control policies are to change certain of peoples' behaviors. Commonly, people do change if there is evidence that a failure to change could be painful. Establishing certain actions as illegal and imposing penalties for failure to curb illegal actions are effective, even though they are unpleasant and restrictive of our freedoms.

Liability for damages done to an individual, group, city, state, or federal area is quite straightforward when the action causing the problem is a *deliberate act of defiance,* or is done knowing it is illegal. The most common predicament for liability is likely to be because of *negligence,* in which the defendant caused a problem because she/he was not sufficiently careful or informed. Proving *causation*—that the action was the actual cause of the claimed damage—may be the most difficult aspect of liability. *Blameless contamination* denotes limited situations in which contamination occurs even *when normal and approved actions are followed.* Blameless contamination does not include such things as *spills, accidents, use at wrong time of year, application of chemicals by a person untrained in their use,* and *actions contrary to labels or BMPs.*

Lt. Ives's report, April 18, 1857: "Our reconnoitering parties have now been out in all directions, and everywhere have been headed off by impassable obstacles . . . The region last explored is, of course, altogether valueless. It can be approached only from the south, and after entering it there is nothing to do but to leave. Ours has been the first, and will doubtless be the last, party of whites to visit this profitless locality. It seems intended by nature that the Colorado river, along the greater portion of its lonely and majestic way, shall be forever unvisited and undisturbed."

FIGURE 18-3 What BMPs could fit the Grand Canyon National Park, which ranges from 4 to 18 miles wide and is 280 miles long? We look at it now as useful — not worthless, as did Lt. Ives in 1857. We hope to control, to the degree feasible, further erosion and degradation, mostly to protect the Colorado River water users downstream. (Courtesy Raymond W. Miller, Utah State University.)

One of the frightening aspects of pollution regulation for those of us who consider ourselves environment-friendly and law-abiding persons is that future legislation may impose cleanup and other liability for contamination that occurred from activities *that were entirely legal at the time they occurred.* There is a need for care, common sense, and a concern for others as we protect our own environment.

18:4 Plant Nutrients

Nitrogen and phosphorus, which are added in agricultural operations, are the major nutrients of concern in pollution. Some of the expected effects of various forms of nitrogen are shown in Table 18-1. Phosphorus exerts its greatest influence by increasing eutrophication.

Table 18-1 Some of the Potentially Adverse Environmental and Health Effects Caused by Forms of Nitrogen

Effect	Causative Agent
Environmental quality	
Eutrophication	Nitrogen sources in surface waters
Corrosive damage	HNO_3 in rainfall (acid rain)
Ozone layer depletion	Nitrous oxides from fuels, denitrification, and industrial stack emissions
Human health	
Methemoglobinemia in infants and elderly; also livestock	Excess NO_3^- and NO_2^- in water and food
Respiratory illness	PANS and other nitrogen oxides
Cancer	Nitrosamines from NO_2^- and secondary amines in food

Source: Modified and abbreviated from D. R. Keeney, "Nitrogen Management for Maximum Efficiency and Minimum Pollution," in *Nitrogen in Agricultural Soils,* F. J. Stevenson, ed., American Society of Agronomy, Madison, Wis., 1982, pp. 605–649.

18:4.1 Eutrophication

Fertilizers increase algal growth in surface waters (and mosquitoes in shallow water) into which they are washed. **Eutrophication** is defined as *increased water fertility,* which causes accelerated algae and water-plant growth. Often, phosphorus (which is usually scarce in water because it has low solubility) is the major cause of eutrophication, but nitrogen and other nutrients also contribute to the problem. Eventually, the considerable masses of dead organic material accumulated undergo decomposition. Because decomposition uses up oxygen from the water, these large masses of decomposing tissue deplete the gaseous oxygen (O_2) in water, causing other aquatic life to die and near-anaerobic conditions to develop. Although the green algal growth is unsightly, the anaerobic (low O_2) condition of the water body is the most deleterious effect of eutrophication.

It is likely that most surface waters already have adequate nitrates in them, which would encourage growth of water plants and N_2-fixation by algae. The major mechanism to limit algal growth is to minimize *phosphate* additions. Phosphorus is not a mobile nutrient in soils; it too easily forms insoluble substances with many common cations. Because of its insoluble nature, phosphorus is often the limiting nutrient for algae and other plants growing in water. The usual ways phosphate enters waters are (1) from municipal sewage with its high phosphate detergents, (2) from certain industrial wastes, and (3) "piggyback" as phosphate adsorbed onto soils particles eroded into waters.

Phosphorus is immobile in soils, so what is the source causing water contamination and are phosphorus fertilizers partly responsible? The major source of phosphorus contamination in surface waters comes from direct dumping of wastes (sewage, animal wastes, industrial wastes) and from eroded suspended sediment from urban and agricultural lands. Eroding soil and other phosphorus-carrying solids continually supply phosphorus to waters. In Minnesota more than 95% of measured phosphorus loss from soils was carried away in eroded sediment. In Canada, of the phosphates applied in late fall on top of frozen ground—a nearly optimum condition for maximum spring runoff losses as soluble phosphorus—only 10% (4.7 kg/ha, or 4.2 lb/a) was dissolved and carried off the field.[1] Certainly, BMPs to minimize erosion are necessary to minimize pollution of waters by phosphorus.

In another example, a concentration of 0.01 mg/L of inorganic phosphorus enhances algae growth in waters, and Canadian water quality criteria suggest a maximum of 0.15 mg/L of phosphorus as acceptable. Interestingly, one study from 6 years of data showed that natural phosphorus losses from *unfertilized* fields *exceeded* the quality criteria guidelines, and this was in semiarid dryland wheat soils (little water for leaching, not a highly fertile soil). Unnecessary pollution should be avoided by wise use of fertilizers, but *zero* contamination is impossible, unrealistic, and catastrophic to the production of food supplies.

Fertilizer use, including organic nutrient sources, does increase nutrient contamination in runoff waters and groundwaters. Generally, the portion of added nutrients lost is a small percentage of the total, less than 1% in one study in Louisiana. Losses from naturally fertile or fertilized soils can be much higher if extensive leaching and erosion occur. Unfortunately, even low-percentage losses may cause serious pollution.

18:4.2 Nitrogen in Groundwater

Nearly all nitrogen fertilizers (except organic and other slow-release fertilizers) are very soluble in water and the final oxidized form, nitrate, moves readily in the water. Heavy rain or

---

[1] W. Nicholaichuk and D. W. L. Read, "Nutrient Runoff from Fertilized and Unfertilized Fields in Western Canada," *Journal of Environmental Quality,* **7** (1978), no. 4, pp. 542–544.

irrigation following nitrogen application can wash the fertilizer off the soil into surface flow and carry urea and nitrates deep into the drainage waters; sandy soils are particularly susceptible to this.

Fertilizers are not the only source of soluble nitrogen in soil water. Nitrogen also comes from decomposing organic matter. Approximately 37% of the U.S. total soluble nitrogen comes from soil humus, 22% from human and animal manures, 18% originally from fixation by soil bacteria and algae, and 9% from rainfall (Fig. 18-4). *Only 13% comes from fertilizers* added to soils. Nitrogen in rain and snow comes from dissolved nitrogen oxides. Sometimes the source of soluble nitrogen moving into waters is unusual. For example, in deep geologic substratum (27 m) in Nebraska, nitrate concentrations were high (25–85 ppm), and it seemed the nitrates would in time be washed into groundwaters. Investigation proved that the source was not fertilizer but deeply buried ammonium and decomposing organic matter deposited 35,000 years ago, before being covered by loess.[2]

The most important factors that influence the amount of nitrate movement to groundwater or surface waters are (1) the amount of nitrate dissolved in the soil solution, (2) the rate of its use by plants, (3) the rate of immobilization into soil microorganisms or newly synthesized soil organic matter, (4) the amount of water available for runoff and leaching through the soil, and (5) soil permeability. Soils heavily fertilized above levels recommended by BMPs and soils naturally high in fertility are potential sources of nitrate contamination in groundwaters or runoff waters. If much of the soluble nitrate in soil solution is quickly absorbed by a rapidly growing plant or by multiplying populations of soil microorganisms, less is available to become leached. Nitrates are soluble ions but are not leached unless enough water exists to leach through the soil and the soil is permeable enough to allow leaching. Some example losses of nitrogen and concentrations of nitrate that could be leached are indicated in Table 18-2. Any soil with a large amount of readily decomposable organic material can, in favorable

FIGURE 18-4 In Venezuela thousands of acres of brush and trees were pushed down by these land clearers, the dead vegetation pushed into piled rows and burned. The ashes were readily washed into groundwater, streams, and lakes. Ashes from natural wildfire burns would have the same fate. (Courtesy Raymond W. Miller, Utah State University.)

[2] J. S. Boyce, J. Muir, E. D. Seim, and R. A. Olson, "Scientists Trace Ancient Nitrogen in Deep Nebraska Soils," *Farm, Ranch, and Home Quarterly,* **22** (1976), no. 4, pp. 2–4.

Table 18-2 Loss of Nitrates from Various Sites by Leaching or Runoff

Site Description and Time Interval	Annual Nitrogen Loss (kg N/ha)
1. Ontario, Canada, tile-drained Humaquepts,[a] in drainage water	
Clay, corn-soybeans-vegetables, 110 kg N/ha added, 85 kg N recommended	64
Clay, soybeans-wheat-barley-corn, 35 kg N added, 50 kg N recommended/ha	16
Sand, corn, 150 kg N added, 150 kg N recommended/ha	4
Sand, corn, 200 kg N added, 150 kg N recommended/ha, high humus	49
2. Ontario, Canada, tile-drained organic Borosaprists,[b] in drainage water	
Well decomposed, onions, 70 kg N added yearly, site 1/ha	96
Well decomposed, onions, 70 kg N added yearly, site 2/ha	188
Well decomposed, sandy loam at 50 cm, onions-carrots, 55 kg N added/ha	202
3. New Hampshire, hardwood forest	
Clearcut area, no vegetation removed, herbicide-inhibited regrowth	97
Control area not cut or sprayed	2
4. Alsea River, Oregon, coniferous forest	
Clearcut area with slash burned	15
Control area, no vegetation removed	4

	Nitrate Concentrations (mg/L)
5. Intact forest sites,[c] on soil water	
Indiana, maple and beech, no root uptake of N	91.0
Indiana, maple and beech, control (normal forest cover)	0.7
New Mexico, ponderosa pine, no root uptake of N	2.7
New Mexico, ponderosa pine, control (normal forest cover)	0.04
Oregon, western hemlock, no root uptake of N	33.0
Oregon, western hemlock, control (normal forest cover)	1.1
Washington, red alder, no root uptake of N[d]	70.0
Washington, red alder, control (normal forest cover)	17.0

Sources: (1) P. M. Vitousek, J. R. Gosz, C. C. Grier, J. M. Melillo, W. A. Reiners, and R. L. Todd, "Nitrate Losses from Disturbed Ecosystems," *Science,* **204** May 4, 1979, pp. 469–474. (2) M. H. Miller, "Contribution of Nitrogen and Phosphorus to Subsurface Drainage Water from Intensively Cropped Mineral and Organic Soils in Ontario," *Journal of Environmental Quality,* **8** (1979), no. 1, pp. 42–48.

[a] Normally poorly drained soil high in humus.

[b] An organic, well-decomposed, cold-area soil.

[c] Control plots did have normal plants with root uptake; other plots were isolated portions where tree roots, etc., were cut by a trench to avoid major root activity.

climates and with adequate leachings, supply considerable nitrate to waters. These can be rangeland, forest, urban, or cultivated soils.

Four methods can help to minimize nitrate losses where serious leaching is expected:

1. Nitrification inhibitors, such as N-Serve, that retard oxidation of NH_4^+ to NO_3^- can be applied
2. Organic and other slow-release or slowly soluble fertilizers, where they are suitable, can be added at appropriate rates
3. Applications of predetermined minimum amounts of fertilizer can be carefully timed (split applications)

4.   Flashboard partitions can be installed for adjustable flow restriction or other controls in drainage lines to regulate water table levels and to permit more denitrification of nitrate before it moves deeply into subsoils or into drainage ways

18:4.3 HNO₃ and H₂SO₄ in Acidic Rain and Fog

Acidic rain, defined as rainfall (or melting snow) with enough dissolved acids to have a pH < 5.7, is found in many areas. The most severe conditions are near large industrial valleys. The major contributing acid is sulfuric acid (H_2SO_4), but nitric acid (HNO_3) adds to that acidity and is an oxidizer. Because these acids are in the air, they can be carried hundreds of miles from their origin.

The effect of acid is to dissolve carbonates (e.g., limestone and marble buildings), corrode metals, and kill much aquatic life. Some lakes have become almost devoid of fish and of most other large organisms. The acidic water increases the solubility of aluminum, which may already be toxic. On oxygenated fish gills the soluble soil aluminum may form a gelatinous aluminum hydroxide coating, which suffocates fish. In Norway and Sweden, fish have died in over 6500 lakes and in seven Atlantic salmon-containing rivers. Similar fates have affected 1200 lakes in Ontario, Canada, and over 200 lakes in the Adirondacks of the United States. A similar problem exists in the Alps.

Acid fog in southern California has been more damaging to plants than acid rain.[3] The lower amount of water in the fog than in rains allows acidity to reach values of pH 2–3. A single 2-hour exposure to fog at pH 2.8 causes low injury to most plants. Marketability of cauliflower, spinach, and lettuce is reduced by pH 2.4–2.6. Reduced crop yields require pH values more acidic than pH 2. Injuries to plants from acidic fog are still less than the damages from ozone.

High acidity of water also increases soluble aluminum (Al), even to levels toxic to plants. Although high Al in drinking water increases with acidity, an association of Al-Alzheimer's disease-Parkinsonism-dementia is suspected by some medical researchers but is not yet proven.[4] Soluble aluminum is toxic to gill-breathing animals (fish) by causing loss of plasma- and hemolymph ions, leading to osmoregulatory failure.[5] Aluminum reduces the activities of gill enzymes important in the active uptake of ions.

The nitric and sulfuric acids in the atmosphere come from various oxides of nitrogen accumulated in the atmosphere from denitrification, volatilized ammonia that is later oxidized in air, burning of fossil fuels, and the burning of vegetation at hot temperatures (especially in forest fires). Any high-temperature burning or heating of organic materials of plant origin, which is 1–4% nitrogen on a dry-weight basis, may evolve some of the nitrogen as oxides, which then oxidize to nitric acid.

18:4.4 Methemoglobinemia

Nitrate becomes toxic to any animal with a disrupted digestive tract in which the conditions cause microbes to reduce nitrate ion to nitrite ion in large amounts; this effect is called **methemoglobinemia.** The nitrite is absorbed into the bloodstream, where it oxidizes *oxyhemoglobin* (the oxygen carrier) to *methemoglobin* (which cannot carry oxygen). The young

[3] Robert C. Musselman, Patrick M. McCool, and Jerry L. Sterrett, "Acid Fog Injures California Creps," *California Agriculture,* **42** (1988), no. 4, pp. 6–8.

[4] Trond Peder Flaten, "Geographical Associations between Aluminum in Drinking Water and Death Rates with Dementia (including Alzheimer's Disease), Parkinson's Disease and Myotrophic Lateral Sclerosis in Norway," *Environmental and Geochemical Health,* **12** (1990), nos. 1/2, pp. 152–167.

[5] B. O. Rosseland, T. D. Eldhuset, and M. Staurnes, "Environmental Effects of Aluminum," *Environmental and Geochemical Health* **12** (1990), nos. 1/2, pp. 17–27.

mammal "suffocates." The blue coloring ("blue baby") is called *cyanosis.* When 70% of the hemoglobin is changed, death may occur. In mammals 0–3 months old, the stomach's lower acidity and certain bacteria it contains (later killed by stomach acids) causes a high rate of conversion of nitrate to nitrite. In older mammals and adults the stomach acids and rapid absorption and excretion of unused nitrates make this nitrate "poisoning" unlikely.[6] It can be a problem in sick and elderly adult mammals.

Although deaths of infants from methemoglobinemia are rare, other possible effects of reduced oxygen transport may be extensive. Death is not the only possible damage.

The nitrate content permitted by public health regulations for drinking water is 45 parts per million (ppm) nitrate (or 10 ppm nitrate-nitrogen). Many wells have been tested that approach or exceed this value; some are near the 100 ppm nitrate level. Obviously, there are some concerns with nitrate concentrations.[7,8]

Nitrate contents may be hazardous to grown animals. The bacteria that reduce nitrate to nitrite are found in adult ruminants (cows, sheep), as well as in infant pigs and chickens, and in the secum and colon of the horse. Thus, the following conditions could cause damage to animals from methemoglobinemia, to some degree, and should be given attention:

1. The mammal infant about 0–3 months old is susceptible. Concentrations as low as 65 ppm nitrate have caused the disease.

2. Cattle and sheep (rumen) and horse (secum and colon) can develop the condition. It is especially common in cattle.

3. High-nitrate vegetables or other foods (particularly spinach) stored where microbial growth is expected may magnify the problem to those eating the food.

4. Damp forage materials containing high nitrate contents add to the hazard to those eating the forage.

5. Ensiled forage crops contain denitrifying bacteria. These produce nitrites and brown gases of NO, NO_2, and N_2O_4. These gases may accumulate in chutes or attached enclosures and can be lethal when inhaled. High nitrate contents produce the highest concentrations.

6. Grazing of newly greened pastures in spring or after heavy frost.

The *acute* dosages of nitrate for cattle are about 50 mg of nitrate-nitrogen taken in per kilogram of animal weight (25 g, or 1.5 oz, per large cow).[9] *Chronic* (lower level) intakes of 0.5 ppm *nitrite* or 100–150 ppm *nitrate* may cause degeneration of vascular tissues of the brain, lungs, heart, liver, and kidneys.

Comparisons of analytical values of nitrate found in plants at the turn of the century with those of recent years give no evidence that plants today have higher nitrate contents as a result of using chemical fertilizers. However, total nitrate in the environment has increased.

A number of N-nitroso compounds are formed by reduction of nitrate to nitrite forming nitrous acid (HNO_2), which reacts with a variety of nitrogenous compounds. Some of

---

[6] Gary W. Hergert, "Consequences of Nitrate in Groundwater," *Solutions,* **30** (1986), no. 5, pp. 24–31.

[7] Nebraska Cooperative Extension Service, "Living with Nitrate," Paper EC 81-2400, 1981.

[8] W. L. Magette, R. A. Weismüller, J. S. Angle, and R. B. Brinsfield, "A Nitrate Groundwater Standard for the 1990 Farm Bill," *Journal of Soil and Water Conservation,* **44** (1989), no. 5, pp. 491–494.

[9] G. B. Garner, W. H. Pfauder, G. E. Smith, and A. A. Case, "Nature and History of the Nitrate Problem," in Science and Technology Guide, Columbia Extension Division, University of Missouri, Columbia, 1979, p. 9800.

these compounds are highly toxic and some are animal carcinogens. The amount of nitrite formed in the *saliva* of the mouth is 5 times greater than the amount in most dietary sources and 12 times greater than nitrites in meat *curing* actions. The small hazard reduction by eliminating nitrite in meat curing would be overwhelmed by the increased hazard of the bacterium *Clostridium botulinum* (controlled by nitrite) that produces the toxin causing botulism.[10]

18:5 Pesticides

Pests are any noxious, destructive, or troublesome organisms; **pesticides** are chemicals to kill pest organisms. The "-cide" ending means "to kill." Although pests are a part of ecosystems, they interfere with people's effort to do various things. Various kinds of pesticides have been used for a long time. The Greek poet Homer, for example, wrote of pesticides 1000 years before Christ. He referred to "pest-averting sulfur with its properties of divine and purifying fumigation." In the 1930s sulfur was still powdered onto the human body to "clean up" some fungus skin effects. It was also volatilized in closed greenhouses for fumigation even in the middle 1950s.

18:5.1 DDT—A Magic Substance[11,12]

The story of **DDT** is a classic example of a "magic substance" capable of unbelievably good pest control. Eventually, it was also shown to be a problem. DDT was developed by a chemist in the 1880s, but not as a pesticide. In 1938 Paul Müller tried it as a pesticide (among many other chemicals). It killed almost all insects easily and quickly. The chemical had *low toxicity to people and animals, was inexpensive, and was long-lasting.* Its discovery brought Müller the Nobel prize in 1948. DDT was used during World War II to delouse people, to control mosquitoes for malaria control, on thatched-roof homes to control the chagas beetle, and for long-term general insect control. Millions of people were saved from typhus and malarial deaths alone by DDT.

DDT is still an excellent insecticide, but it has two major disadvantages: (1) its half-life in the environment is too long (10–25 years) and (2) it is bioaccumulated in the fat of animals. Thus, in 1962, Rachael Carson published a book called *Silent Spring* that emphasized the long-term, nonselective killing of insects by DDT in the environment. Birds ate the dead insects and *biomagnified* the DDT into their bodies. Eventually, the accumulated DDT would in some manner cause a bird population decrease, maybe to zero—thus, a silent spring. DDT was banned from most uses in the United States in the early 1970s.

18:5.2 Pesticides Today

Hundreds of pesticides have been synthesized, with a lesser number being used extensively as suitable pesticides. There are now about 600 commercially important pesticides and more than 1500 registered for sale. Newer ones will be developed more slowly because of the high cost of testing required before they can be licensed for sale and use. Cost is about $6 to $10 million plus more than 4 years of testing and review for each pesticide brought to the market.

---

[10] Charles A. Black, "Reducing American Exposure to Nitrate, Nitrite, and Nitroso Compounds: The National Network to Prevent Birth Defects Proposal," Council for Agricultural Science and Technology, Ames, Iowa, 1989, p. 13.

[11] Orie L. Loucks, "The Trial of DDT in Wisconsin," in *Patient Earth,* J. Harte and R. H. Socolaw eds., Holt, Rinehart and Winston, New York, 1971, pp. 88–107.

[12] Bernard J. Nebel, *Environmental Science,* 3rd ed., Prentice Hall, Englewood Cliffs, NJ, 1990, pp. 404–414.

What criteria define a "good pesticide"? Many of the highly regarded pesticides are quite toxic to animals, so it is not low toxicity. Acceptable pesticides must have at least the following characteristics:

1. *The pesticide must be short-lived in the environment.* It must not exist long enough to bioaccumulate and biomagnify in the food chains. The residues must be gone from foodstuffs. Usually, it must be dissipated within a week or two of application time. To allow workers into fields, it is desirable for many pesticides to be gone within a few days.

2. *The pesticide must not be carcinogenic, teratogenic, or cause mutations.* There is no easy or clear-cut way to determine these characteristics with certainty. Much disagreement is voiced about many decisions related to these properties for pesticides.

3. *The pesticide must be effective yet able to be safely handled.* The chemical must be retained on the sprayed or treated area, must not be overly volatile so that it moves extensively into the air, and must permit safe handling by the applicator when reasonable care is used.

To these characteristics could be added others. The chemicals are preferably low-cost, easily washed from the body and equipment, of low caustic nature to skin and eyes, have low corrosiveness to equipment, and exist as liquids at normal working temperatures.

Pesticides currently used range from those of low toxicity to animals (many herbicides) to those that are very toxic to animals (some of the organic phosphates). Most pesticides are extremely toxic but short-lived, although some herbicides may remain active in the soil for more than a year (triazines).

The negligible-risk level for *noncarcinogenic substances* is usually set as shown in Fig. 18-5. The no-observable-effect level (NOEL) is reduced from 0.1 to 0.001 of the NOEL, usually to the 0.01 level of the NOEL. The safety factor is based on the concept that people are

FIGURE 18-5 Approximate concept for setting "safe" dosage levels (*acceptable daily intake,* ADI) of a noncarcinogenic pollutant. The *no-observable-effect level* (NOEL) was observed in animal studies. (*Source:* Modified from Gary A. Beall, Christine M. Bruhn, Arthur L. Craigmill, and Carl K. Winter, "Pesticides and Your Food: How Safe Is 'Safe,'" *California Agriculture,* **45** [1991], no. 4, pp. 4–8).

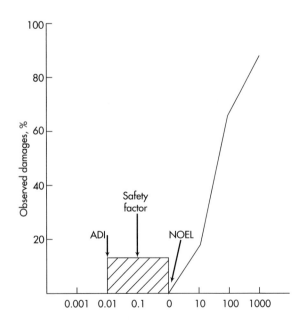

10 times more sensitive than are animals and that the most sensitive persons are 10 times more sensitive than the average person. California's Proposition 65 states that a cancer-causing substance posing a negligible risk of exposure at regulatory levels causes no more than *one additional cancer per million people.* The **Delaney clause** (passed in 1958 in the Federal Food, Drug, and Cosmetic Act) *inhibits addition of any amount of any known carcinogen to foods.* A new reaffirmation of *zero addition,* by court ruling, may trigger new pesticide cancellations.[13]

18:5.3 Pesticide Problems and the Extent of Pollution

Toxicity to animals is the major concern about pesticides. "Safe levels" of several of these materials are given in drinking-water standards and food standards. Unreasonable standard limits are sometimes established, such as "no measurable amount." Analysts continue to perfect analytical methods that can detect smaller and smaller quantities. Also, some materials may be present naturally (e.g., formed during fires).

The grower faces an increasingly serious problem: *resistance.* Some organisms have built up resistance to certain frequently used pesticides, by chemical changes or by natural selection of the more resistant organisms over time. Development of resistance forces growers to use larger pesticide applications. One example was the use of the excellent triazine herbicide, which keeps most weeds out of corn fields. Unfortunately, many weeds (pigweed, ragweed, lambsquarters) developed resistance.[14] The major way to fight resistance, currently, is to change pesticides for a while. Just for this reason of resistance, new pesticides must continually be sought.

Agriculture accounts for more than 67% of all pesticides used in the United States.[15] Numerous spills and data from monitoring of surface waters and groundwaters point unequivocally to agriculture as an important pollutor. Part of today's pollution was put in the leaching cycle three decades ago; some is from present overuse, careless use, spills, and waste and wash waters.

Although *less than 5% of pesticides added are lost into groundwater,* the desired limit is zero contamination.[16] The uncertainty about the hazards of many pesticides in such low concentrations has kept the permissible levels low. As a result, many wells in various states have been closed, some for over a decade.[17] Groundwaters cleanse themselves, but often very slowly (Detail 18-2).

How great is the problem of pesticide contamination today? It is difficult to say, but one author put it boldly: "Groundwater pollution is probably the major environmental problem facing agriculture over the next 10 years."[18] With over half of the people in the United States using groundwater for drinking, this statement may well be accurate, despite the increased effort to reduce pollution. *Many pollutants may already be en-route to groundwaters from past applications and disposals of pesticides.*

---

[13] Melnicoe Stinmann, "Delaney Clause Ruling May Trigger Pesticide Cancellations," *California Agriculture,* **48** (1994), no. 1, pp. 30–36.

[14] "The Trouble with Triazines Is Resistance," *Agrichemical Age,* Aug. 1986, p. 32.

[15] U.S. Department of Agriculture, *1983 Handbook of Agricultural Charts,* Agricultural Handbook 619, USDA, Washington, D.C., 1983.

[16] G. R. Hallberg, "Agrichemicals and Water Quality," in *Proceedings of the Colloquium on Agrichemical Management and Water Quality,* National Academy Press, Washington, D.C., 1986.

[17] G. R. Hallberg, "From Hoes to Herbicides: Agriculture and Groundwater Quality," *Journal of Soil and Water Conservation,* **41** (1986), pp. 357–364.

[18] Ken Cook, "The Big Seep," *Journal of Soil and Water Conservation,* **41** (1986), pp. 235–237.

Everything that leaches into subsoils with waters has a good chance to end up in groundwater (aquifers). Half of all Americans and 95% of rural Americans get their household water from groundwater. We are finding pollutants from waste disposal that occurred decades ago now showing up in some groundwaters. Although we should be concerned about any hazard, pesticides have received the most attention. Twenty-two municipal wells were closed in 1989 in Fresno, California, because of contamination with the nematocide *dibromochlorpropane* (DBCP)—which had not been used in the area since 1977!

Pesticides are not the only, nor even the major, contaminants in groundwaters; they have just had high "visibility" by testing. Imagine the variety of substances from these general sources of groundwater contamination:

1. *Septic tanks, cesspools, and privies.* Nearly one-fourth of U.S. homes use septic systems that dump relatively untreated wastes into the ground: human wastes, nitrates, detergents, household chemicals, and viruses.

2. *Surface impoundments* used by industries, municipalities, businesses, and farms. Supposedly, sealing liners cover the bottoms, but some are punctured and leak. Some states now require double liners, forcing the costly removal and safe disposal of the sediment already in such impoundments. The University of California expects their cost to remove and replace one-liner impoundments on nine experiment stations to be greater than $1.5 million.

3. *Agricultural activities* dispense pesticides, fertilizers, salts, fuels, and solvents from both agricultural lands and from urban homes and gardens.

4. *Landfills* may have almost anything in them. About 500 hazardous-waste facilities and 16,000 other landfills exist nationwide.

5. *Underground storage tanks,* some 5 to 6 million of them, store gasoline, fuel oil, solvents, and numerous other chemicals.

6. *Abandoned wells,* if not properly sealed, can be conduits from surface contamination to groundwater.

7. *Accidents and illegal dumping* can cause major pollution in localized areas. A cleanup cost in Massachusetts—after 2000 gallons of gasoline had leaked from an underground tank near a municipal well—cost more than $3 million. Small spills and accidents happen daily and can gradually move into groundwaters.

8. *Highway deicing salts.*

Sources: Brenda Simmonds and Dennis Brosten, "DBCP Haunts California Wells," *Agrichemical Age,* **34** (1990), no. 1, pp. 11–12; Anonymous, "Citizens Guide to Ground-Water Protection," U.S. Environmental Protection Agency, *Social Issues Resources Series,* **5** (April 1990), Article 9, pp. 1–10; Dennis Brosten, "University and Applicators Grappling with Rinsate Morass," *Agrichemical Age,* **31** (1987), no. 7, pp. 8, 15.

18:5.4 Are There Alternatives to Pesticides?

A popular saying several decades ago about judging others went something like this: "Don't judge a man harshly until you have walked in his moccasins for a while." We are quick to set laws for someone else and to criticize others. Pollution control laws greatly influence both the health and the livelihood of many people. Laws need to be made with reason, logic, and fairness and based on the most accurate data possible. Many proponents of stopping the use of pesticides may be the first to select unblemished fruits in the market, to buy sprays to control home mosquitoes, and to complain about increased food prices. The world existed without pesticides up until the 1940s. Obviously, we could exist again without them. But all al-

FIGURE 18-6 Without pesticides, insects like the Japanese beetle (top left), Mediterranean fruit fly (top right), and gypsy moth (larva and adult), which infest more than 300 species of trees, vegetables, flowers, and fruits, would have many opportunities to chew their way through our crops. (*Source:* Photos courtesy of USDA — *Agricultural Research, **27,*** No. 9; and California Department of Food and Agriculture; photos by D. Zadig and M. Pendrak.)

ternatives have a cost. The cost without pesticides would be higher food costs, more "blemished" products, less control of nuisance insects (mosquitoes, cockroaches, weevils, locusts, etc.), more people working on the farms, and lower crop yields for a hungry world (see Fig 18-6).

Benefits from the banning of organic pesticides are harder to document unequivocably. Clearly, numerous accidental deaths and lesser poisonings would be eliminated. Damage of a carcinogenic and teratogenic nature, attributed to some pesticides, would be lessened. People would be forced to adopt farming practices that are more self-sustaining and compatible with nature. Undoubtedly, greater effort would be expended toward developing suitable cultural and biological methods of pest control, which would be helpful in self-sustaining agriculture.

There are alternatives—or perhaps more correctly termed, partial substitutes—to the use of these chemical pesticides. The use of natural predators (**biological control**) is increasing. The cottony-cushion scale insect, which devastated citrus trees, is controlled by two Australian insects: a small parasite fly and a predatory ladybird beetle. Another pest, the sugarcane leafhopper, threatened to devastate the Hawaiian sugarcane industry. A total of five insects parasitizing the leafhopper eggs finally brought the problem under control. Other control techniques involve disease-causing bacteria that kill certain pest larvae and larvae that feed on certain weeds, thus partially controlling the larvae. But biological controls are usually slow (may take many days or weeks for populations to build up) and incomplete (may have only 70–90% control). Pesticides, in contrast, may kill 95–99% in a few hours or a day or two. Unfortunately, only a few of these biological-control systems are known and developed.

Biological control is slow to be developed and has a lag time when it is needed. When new pests, such as the Russian wheat aphid, infests suddenly into an area it had not infested before, it causes severe damages. Pesticides suitable to control the aphid developed may not be developed for several years.[19] Estimated crop damage from this aphid in 1987 in the ten Great Plains states was $53.8 million. Texas alone lost $100 million in 1988 and expected a loss of $140 million the next year. No dependable biological control is known yet.

Cultural control is the control of pests by management and physical activities. We may chop and bury old plant residues to kill eggs laid on/in them. We eliminate weeds in adjacent areas before they go to seed (to reduce weed seeds). We select crops that have fewer pest problems in our area. We rotate crops to remove the *host plant* for several years, causing the pest to die out or be reduced to low numbers. We bury residues by tillage; no-till and reduced-tillage practices have more pest problems. We are then faced with the question: Do we reduce soil erosion, use less fuel, and add more pesticides, or do we use the methods employing extensive tillage that require less pesticides? This is an example of more compromises that need good scientific data on which to make rational decisions.

Other ways to reduce pesticide uses include (1) **breeding disease-resistant plant varieties,** (2) **male sterilization by irradiation,** and (3) some use of **natural chemicals.** One example of natural chemicals is the use of **juvenile hormones,** which keep the larval stage from becoming an adult. **Sex attractants** (included as **pheromones**) have been used to (1) attract insects to traps and (2) confuse males so that they cannot find mates, thus reducing viable eggs.

A unique nonpesticide control for some crops is **vacuum cleaning** the plants.[20] Special vacuums to suck pests off strawberry plants and leaf lettuce have worked well. Vacuuming two or three times a week usually has controlled the major problem—lygus bugs—on strawberries and cut pesticide expenses by 73%. But vacuums cost over $20,000 each, remove only 30–40% of adult lygus bugs per pass, and miss a lot of immature bugs that are still very damaging. The use of vacuums is being tried on grapes. Vacuuming will be useful for flying insects, but it is probably less valuable for nonflying types, for compact crops, such as head lettuce, and for very open crops, such as cotton.

All of these possible ways to control pests work best if they are used in the correct *combination,* called **integrated pest management (IPM).** Smaller amounts of chemical pesticides are needed with IPM; and immediate control of infestations is possible. Unfortunately, suitable biological and cultural controls are available for only a relatively few pests. *Chemical pesticides are likely to continue to be used in the next few decades.* We can reduce the amounts of the chemicals needed by IPM and by careful applications. As one scientist put it, IPM should mean *Integrated Pharm* (aceutical) *Management,* the careful, comprehensive, knowledgeable use of the best techniques for all practices in agriculture.

---

[19] Vernon M. Stern and Steve B. Orloff, "Controlling Russian Wheat Aphid in California," *California Agriculture,* **45** (1991), no. 1, pp. 6–8.

[20] Richard Steven Street, "Is Vacuum Pest Control for Real?" *Agrichemical Age,* **34** (1990), no. 2, pp. 22–23, 26.

▌ *18:6* Wastewaters Added to Soils

Wastewaters that must be cleaned up include domestic-sewage effluents as well as an almost unlimited variety of liquid industrial wastes. Many industrial wastewaters have been dumped into municipal sewage lines routinely. Tighter Environmental Protection Agency (EPA) restrictions now require that such liquids be treated extensively prior to disposal into municipal treatment plants or back into groundwater. Although irrigating with wastewaters partially cleans water by percolation through the soil, not all contaminants in the water are removed. Soluble salts and organic chemicals may continue to flow with the water to groundwater or surface waters.

In general, the disposal of wastewaters on land or into waters is permitted only if it does not cause (1) extensive groundwater pollution, (2) a direct public health hazard, (3) an accumulation in the soil or water of hazardous substances than can get into the food chain, (4) an accumulation of pollutants such as odors into the atmosphere, and (5) other aesthetic losses, within limits. By July 1987 industries needed to comply with EPA standards for control of all pollutants by using the **best available technology (BAT)** to remove offending substances. The guide used to regulate wastewater disposal is that the water must not exceed the allowable contamination limits for **drinking-water standards.** These standards designate allowable maximum levels of many metals, anions, toxic organic substances (such as PCBs and cyanide), pesticides, strong-tasting phenols, coliform bacteria, radioactivity, total dissolved solids, and even maximum water temperature and pH.

Wash waters are one of the most common wastewaters. Wash waters in large quantities are used in vegetable and fruit processing plants. Dade County, Florida, is an example of a sensitive area.[21] Over 1.8 million people there obtain drinking water from a shallow, porous, limestone aquifer that is easily polluted (according to past experiences). The county also produces vegetables, fruits, and ornamentals worth over $350 million a year on over 32,000 hectares (80,000 acres). Composition of the effluent of 18 of 38 packing operations—particularly those of limes, mangoes, tomatoes, and potatoes—exceeded water quality standards. The obvious result is to alter processing and/or treatment of effluents to improve water quality.

Water, the universal dissolver and cleaner, must periodically be cleaned itself before it becomes reusable. Whether or not it can be allowed to be cleansed by percolation through soil depends on whether it is actually "cleansed" by the soil and whether it leaves undesirable residues in the soil through which it filters (Detail 18-3).

▌ *18:7* Solid Wastes Applied to Soils

Solid wastes are animal manures, sewage sludges, food-processing wastes, municipal leaf and garbage composts, and many other residues. These materials are being added in increasing amounts to soils. The organic matter is usually beneficial to soil productivity. The alternative disposal method—landfills—is less desirable because of limited available land and increasing costs. With more disposal of wastes onto cropped land, this question needs an answer: To what extent do these wastes added to soils pose environmental pollution problems?

---

[21] Dennis F. Howard, Sue M. Alspack, and Nancy D. Stevens, "Wastewater Disposal at Fruit and Vegetable Packing Facilities in Dade County, Florida," *Journal of Soil and Water Conservation,* **45** (1990), no. 2, pp. 274–275.

In the Phoenix, Arizona, area the Salt River formation is a thick geologic deposit of sand and gravel that can absorb water at a rate of about 60 m (197 ft) of water yearly, allowing it to seep to groundwater. Wastewaters are collected into standing infiltration ponds built on this formation. During percolation from these ponds, more than 60% of the nitrogen and essentially all bacteria, viruses, and suspended solids are removed. Besides recharging groundwaters, which are clean enough for repumping, the cost is much less than would be needed to clean these waters by ordinary tertiary (final-stage) sewage treatment. This method is one-half to one-third as expensive and uses only one-fifth as much energy as additional sewage treatment would require.

Source: Paul Dean, "Recycling Waste Water—From City to Farm," *Agricultural Research,* **28** (1979), no. 6, pp. 4–7.

18:7.1 Sewage Sludges

For centuries people in Asia have recycled human waste through the soil to utilize the nutrients and organic materials it contains. In Western developed countries such use has not been popular, largely due to concern about spreading human diseases and the aesthetics of the practice. Animal manures are beneficial to crops; human wastes could be as useful. Municipal sewage systems have frequently buried residual sewage solid wastes (called **sewage sludge**) in garbage dumps or sanitary landfills. This is expensive and wasteful because the sludge, which contains organic material, nitrogen, phosphorus, micronutrients, and other substances, can be applied to land to improve plant growth (Fig. 18-7).

FIGURE 18-7 Chisel injectors are used effectively to apply slurries of sludge and animal manure. This method conserves ammonium nitrogen and removes much of the undesirable appearance and smell associated with surface applications. (Courtesy of Big Wheels, Inc., Paxton, Ill.)

Table 18-3 Average Composition of Sewage Sludges from More Than 200 Municipalities in Eight States

Component	Minimum	Maximum	Median[a]
	Concentration on a Dry-Weight Basis		
	(percent)		
Organic carbon[b]	6.5	48.0	30.4
Total nitrogen	0.1	17.6	3.3
Total phosphorus	0.1	14.3	2.3
Total sulfur	0.6	1.5	1.1
Calcium	0.1	25.0	3.9
Sodium	0.01	3.1	0.2
Potassium	0.02	2.6	0.3
	(parts per million)		
Zinc	101	27,800	1,740
Copper	84	10,400	850
Nickel	2	3,515	82
Chromium	10	99,000	890
Cadmium	3	3,410	260
Lead	13	19,730	500
Mercury	1	10,600	5
Arsenic	6	230	10

Source: Selected data from L. E. Sommers, "Chemical Composition of Sewage Sludges and Analysis of Their Potential Use as Fertilizers," *Journal of Environmental Quality,* **6** (1977), pp. 225–232.

[a] Median is the middle number of all numbers ranked sequentially.

[b] Organic carbon times 1.5–1.8 estimates the organic-matter content. Most sewage sludges are between 40% and 60% organic matter, although some will be even higher.

Table 18-3 lists some of the components typically contained in sewage sludges. Although only about 2% of all U.S. cropland is needed to accommodate all sewage sludge currently produced, high-population areas such as New Jersey may require as much as 50% of its cropland on which to dispose of its sludge, if this is the only method of disposal.[22]

Sludges do have faults. They may contain enough live viruses and viable intestinal worm eggs to require careful handling for several months. They should not be used on crops that are grazed (grasses, clovers) or those eaten fresh by humans (such as lettuce, carrots, and radishes). High concentrations of soluble salts may be troublesome also. Heavy metals—such as cadmium, zinc, chromium, copper, lead, cobalt, nickel, and mercury—accumulate by adsorption in the soil to which sludge containing them (Table 18-3) is applied and remain for centuries. These metals may be adsorbed by plants grown in contaminated soil, then be accumulated in animals eating those plants, perhaps reaching chronic toxic levels. Most heavy metals become quite insoluble in soils of about pH 6 or more basic. Cadmium, being more highly soluble than other heavy metals, is a frequently found contaminant, as are nickel, copper, molybdenum, and zinc to a lesser extent. Federal regulations for sludge use require that the soil pH be kept at 6.5 or higher to reduce the solubility of heavy metals. *Annual sludge applications must not exceed 0.5 kg Cd/ha annually.*[23] *Accumulative totals* on

[22] Council for Agricultural Science and Technology, "Application of Sewage Sludge to Cropland: Appraisal of Potential Hazards of the Heavy Metals to Plants and Animals," CAST Report 6A, EPA 430/9-76-013, 1976, p. 1.

[23] Environmental Protection Agency, *Federal Register,* **44** (Sept. 13, 1979), no. 179, pp. 53461–53462.

a soil must not exceed 5 kg Cd/ha for soils with less than 5 $mmol_c$/kg cation exchange capacity, (CEC, sandy soils), or 20 kg Cd/ha for soils with a CEC above 15 $mmol_c$/kg (clayey or high humus soils). PCBs (polychlorinated biphenyls, which have long half-lives, similar to DDT) must not exceed 10 mg/kg in sludge.

The crops most likely to be high in heavy metals when grown on sludge-treated soil are leafy vegetables such as Swiss chard and spinach. The least likely foods are grains, fruits, and other seed or fruit products. As pollution controls reduce indiscriminant dumping into sewer lines, sewage sludge should become a product with lower amounts of potential hazards. If serious contaminants are avoided, sludge could be used more extensively on agricultural lands and with greater safety.

With the ban on ocean dumping (which went into effect January 1, 1992), solid wastes must be recycled, burned, or buried. Recycling allows sludge to be used as you would manures. The city of Holyoke, Massachusetts, had to pay $108–$125 per ton to dispose of wastes into distant landfills.[24] Negotiation reduced the cost to $87.50 per wet ton. A private operator is now *composting* the sludge with wood chips for $72.50 a wet ton. The city of about 45,000 people produces over 60 tons of sludge per day (32% solids). The State of Washington in 1992 passed the nation's first *biosolids law,* establishing a policy for the beneficial use of treated sludge.

18:7.2 Animal Manures

Today's animal manures are not the same as yesterday's. The push to increase weight on beef cattle and to encourage appetite in many animals has promoted the practice of increasing salt in animal diets. Manures may have from several percent to more than 10% soluble salts by dry weight. Heavy application of manures to soils without periodic leaching could cause a salt hazard to plants in a few years. Leaching these salts into groundwaters to free the rhizosphere of salts may pollute groundwaters. Eutrophication may also result from simultaneously leaching nitrates or eroding portions of the manure into surface waters (Calculation 18-1).

Some poultry and swine are fed enough disease-control medicines to leave significant amounts of antibiotics, copper, and some other metals in their manures. These chemicals accumulate in soils. Animal disease organisms in manures are also of concern—along with odors and aesthetics—in dealing with manure disposal.

18:7.3 Municipal Garbage, Composts, and Sanitary Landfills

The major technical problems of garbage disposal (other than the sheer volume of waste) are the toxic chemical cleaners, pesticides, solvents, and medicines contained therein, the leaching by water of garbage solubles, the volatilization of solvents, and the gases formed by anaerobic decomposition of organic wastes.

Most municipal wastes are disposed of by burial in **sanitary landfills** (soil-covered trenches or holes filled with garbage), although some cities are composting it or using it as fuel in power plants. The danger of pollution of groundwater by leaching and of the air as volatile gases escape has caused a tightening of regulations for land suitable for sanitary landfills (Fig. 18-8). Some EPA requirements for suitable landfills are summarized and briefly discussed in the following list.[25,26]

---

[24] Robert Spencer, "Sludge Composting Takes Town out of Landfill," *Biocycle,* **33** (1992), no. 1, pp. 52–54.

[25] Environmental Protection Agency, "Solid Waste Disposal Facilities," *Federal Register,* **43** (Feb. 6, 1978), no. 25, pp. 4952–4955.

[26] Environmental Protection Agency, "Landfill Disposal of Solid Waste: Proposed Guidelines," *Federal Register,* **44** (Mar. 26, 1979), no. 59, pp. 18138–18148.

The estimated amount of manure that can safely be added to soil without causing pollution is dependent upon (1) the nitrogen content of the manure, (2) the climate (warmer areas have more rapid manure decomposition), (3) the amount of residual manure left from previous additions, (4) storage and application methods (which affect nitrogen losses before manure is added to soils), and (5) the nitrogen requirement of the crop to be grown. Manure may best be used to supply only part of the total nitrogen requirement if the amount of nitrogen needed is large.

As an example, consider an operation in south-central Washington State that has closeby manures available for use. To calculate the weight of the available manure needed to add the amount of nitrogen suggested for the crop grown (fertilizer guide nitrogen = *FGN*), the following data could be used:

1. Nitrogen content (*NC*) in fresh manure, kg N/1000 kg manure (of 80% water):

Dairy cow	10 kg	Sheep feeder	17 kg
Beef feeder	9 kg	Layer hen	43 kg
Swine feeder	16 kg	Broiler chicken	52 kg

2. Fraction of nitrogen remaining (*NR*) from fresh manure after losses during handling and storage:

Anaerobic lagoon, oxidation ditch, liquid spreading	0.16%
Deep-pit storage, liquid spreading	0.34%
Open-stockpile storage, solid spreading	0.67%[a]
Fresh manure incorporated within 1–4 days, warm, dry soil	0.65%
Fresh manure incorporated within 1–4 days, warm, wet soil	0.85%

3. Fraction of initial manure nitrogen made plant available (*A*) each year after application:

		Years after Application			
Type of Manure	*Application Year*	*1*	*2*	*3*	*4*
Dairy, fresh	0.50	0.15	0.05	0.04	0.04
Beef, feedlot, piled	0.35	0.10	0.05	0.03	0.02
Dairy, liquid-manure tank	0.45	0.10	0.06	0.04	0.04
Poultry, fresh	0.75	0.05	0.05	0.05	0.03

4. Fraction of nitrogen left after denitrification losses (*D*):

Excessively drained soil	1.00
Well-drained soil	0.85
Poorly drained soil	0.70
Very poorly drained soil	0.60

Problem (a) How much stockpiled beef manure should be added per year to supply 100 kg of N per hectare (FGN) to corn on moderately well-drained soil?

(b) How much beef manure is needed each year after the fourth year in continuous corn cropping needing 100 kg N/ha yearly?

(c) How much fresh broiler poultry manure, rather than beef manure, would be needed each year after the fourth successive year of its use for 100 kg of N released per year? Assume wet soil.

Continued.

Solution   (a) Weight of manure to supply:

$$100 \text{ kg N/ha} = \frac{(FGN)}{(NC)(NR)(A)(D)}$$

$$= \frac{100}{(9)(0.67)(0.35)(0.85)}$$

$$= 55.7 \text{ Mg/ha (24.8 t/a)}$$

(b) The fifth year would have 0.35 from the year of addition +0.10 from manure added 2 years previous +0.05 from third year previous +0.03 from fourth year previous +0.02 from fifth year previous = 0.55 for *A:*

$$\text{metric tons} = \frac{100}{(9)(0.67)(0.55)(0.85)} = 35.5 \text{ Mg/ha (15.8 t/a)}$$

(c) Fresh poultry has 0.03 nitrogen fraction (*A*) for fifth year, so *A* = 0.75 + 0.05 + 0.05 + 0.03 = 0.93:

$$\text{weight of poultry manure} = \frac{100}{(52)(0.85)(0.93)(0.85)} + 2.86 \text{ Mg/ha (1.3 t/a)}$$

This 3.03 metric tons of broiler manure is considerably less than the 35.5 metric tons of beef manure with its lower N content and lower percentage availability.

---

[a] 40% of original, but concentrated by drying and decomposition.

1. *Site selection and general information*
    (a) Avoid environmentally sensitive areas (for example, floodplains, permafrost areas, critical habitats of endangered species, and recharge zones that supply the sole source of culinary water).
    (b) Avoid active earth faults and **karst terrain** (limestone areas with caves and sinkholes).
    (c) Avoid sites traversed by pipes or conduits (such as sewer pipes, storm drains, and water lines).
    (d) Evaluate on-site soil, which should have low to moderate permeability and support vehicular traffic in bad weather.
    (e) Consider special site problems, such as the nuisance of attracted birds if the site is near an airport.
    (f) Know the kinds and amounts of wastes expected.

2. *Soil, geology, and hydrology*
    (a) Determine the water balance for the site (rainfall, water moving through soil, soil permeability, water table).
    (b) Evaluate trade-offs between environmental impact, economic considerations, and future use alternatives of the site. (Would the filled site make a good location for a park, or for farming?)

FIGURE 18-8 Disposal of solid wastes is most often done by burial in soil. These photos of a Salt Lake City sanitary landfill serving about 1 million people show wide trenches to be filled with garbage and covered by soil; the trenches are 30.5 m (100 ft) wide, 2.1 m (7 ft) deep, and 1.6 km (1 mile) long. No-burning laws increase the volume of garbage buried. Twelve trenches were filled in 8 years in Salt Lake City. (*Source:* USDA — Soil Conservation Service; photo by D. C. Schuhart.)

3. *Leachate control*
   (a) The bottom of the landfill should be at least 1.5 m (5 ft) above the seasonal high level of the groundwater table. Water on landfill areas should not have direct-flow connection to other standing or flowing surface waters.
   (b) Determine the water–soil strata (hydrogeologic) conditions to predict possible contamination of groundwater. Without deep permeable soil or deep groundwater, a slowdown barrier or a complete bottom-and-side sealing barrier

FIGURE 18-9 This small landfill on Lakeland sand in South Carolina would seem to be on soil much too permeable and in danger of polluting the groundwater. Fortunately, just below the trench bottom is soil of only moderate permeability. The soil beneath the landfill is important if pollution is to be avoided. (*Source:* USDA — Soil Conservation Service.)

may be needed for the landfill area (Fig. 18-9).

Lining materials can be clays, soil cements, crushed limestone,[27] and artificial materials such as asphaltic and plastic membranes. If little substratum flow is wanted, the lining material should have a permeability of 1×10^{-7} cm/s (about 0.1 ft/yr) or less. Natural soil liners should be at least 30 cm (1 ft) thick; synthetic membranes should be 20 mils or more thick.

4. *Gas Control.* Control methane gas concentration in the atmosphere (produced by anaerobically decomposing wastes) to no more than 5% at the property boundary or 1.25% in buildings on site. Compacted clays, asphaltic and plastic liners, and cement help to retain gases within the landfill. The addition of perforated pipe in gravel-filled trenches with or without air pressure or suction can be used to remove unwanted gases before high concentrations build up within the landfill (see Detail 18-4).

5. *Runoff control.* Locate the landfill site so as to avoid accumulations of runoff water from adjacent lands. Surface grading and an impermeable soil surface can reduce leaching; ditches and dikes can be constructed to move and hold runoff water.

6. *Monitoring.* In conformance with laws passed in 1975, monitors continuously sample and analyze groundwaters from deep bore holes and check for gases and for pollution in groundwaters sampled from nearby wells.

Composting nontoxic organic wastes is a recent practice of some municipalities that helps to dispose of some garbage volume usefully. Leaves that are collected in

---

[27] Juan Artiola and Wallace H. Fuller, "Limestone Liner for Landfill Leachates Containing Beryllium, Cadmium, Iron, Nickel, and Zinc," *Soil Science,* **129** (1980), no. 3, pp. 167–179.

Near Commerce City, Colorado, a 150-cm (5-ft)-diameter water conduit was constructed in 1977 to carry water to the eastern part of Denver. During inspection and cleaning to begin use of the conduit, a welder requested two workmen to install a fan at a distant manhole. At the manhole a worker lit a match and touched off an explosion that sent flames 12 m (39 ft) into the air. This burning created a vacuum that pulled more air (oxygen) into the conduit. Four explosions during 90 seconds rocked the area and knocked the welder about 10 m (33 ft). Four of the firemen who came to fight the fire were sent to the hospital for treatment of carbon monoxide inhala-

tion; one workman was dead on arrival, and another died 3 weeks later.

The exploding gas was **methane** produced by anaerobic bacteria decomposing organic matter in a nearby landfill. The gas diffused laterally through the soil and entered the eastern end of the closed conduit. Tests of gas from the landfill indicated 31.8–52.5% methane by volume (lower explosive limits are near 5% for methane); the eastern end of the conduit (near the landfill) contained 6.1–14.4% methane. Large gas concentrations occured in the gravelly soil at distances of 120 m (394 ft) from the landfill.

Source: J. W. Martyny et al., "Landfill-Associated Methane Gas a Threat to Public Safety," *Journal of Environmental Health,* **41** (1979), no. 4, pp. 194–197.

large amounts without toxic wastes mixed with them are frequently used. Such compost is sometimes given away just for hauling it. The problem in using composts rather than burying the materials is that composting costs more to make than people want to pay.

Graven County, North Carolina, started a pay-as-you-go program to increase recycling of municipal waste at the grass roots—the home. Each household must attach a sticker costing $1.25 to each 33-gallon bag of garbage.[28] A 90-gallon trash cart can be used with three stickers attached. Curbside recycling was made available and doubled in a short while. To encourage recycling, a household is sent only four stickers a month. Any additional stickers must be purchased at specified outlets, mainly grocery stores.

18:7.4 Food-Processing Wastes

Food-processing wastes are as varied as the foods processed: peapods; tomato and potato peels; soybean, peanut, and cottonseed pulp after extracting oils; sugarcane pulp; waste from cheese making; and any chemicals used in food processing.

These products (except some chemical processing solutions) are organic and can be composted, added to soil, burned, used in animal feeds, or buried. Food wastes contain considerable amounts of nitrogen and phosphorus; their disposal can result in large nitrate concentrations in the soil (similar to the disposal of fresh animal manure).

The most serious environmental threat posed by food-processing waste disposal is that of water pollution by nitrogen. If such materials are dumped or eroded into surface waters, they also reduce oxygen in the water because of high chemical oxygen demand (COD) values and cause eutrophication from nitrogen and phosphorus added by the wastes. **Chemical oxygen**

---

[28] Robert Bracken, "North Carolina County Institutes Sticker System," *Biocycle,* **33** (1992), no. 2, pp. 35–37.

demand (COD) is a measure of decomposable material and other oxygen consumers in the wastewater or slurry. **Biological oxygen demand (BOD)** is a measure of the amount of oxygen required from water for biological decomposition of organic material. It is an indication of the oxygen stress that organic pollutants will have on living aquatic organisms. BODs are determined on a more restrictive basis than CODs and so are of smaller values than CODs for the same water samples. Organic materials (high CODs) cause microbes decomposing them to use the water's oxygen and make the system anaerobic, an undesirable condition. Treatment chemicals—such as sodium hydroxide (lye) in potatoes, waste syrups, salts, and cleanup detergents—can become disposal problems under rigid disposal regulations.

In one study of potato wastes in wastewater, the sodium hydroxide used to peel potatoes was beneficial as a feed. The waste solids were collected by filtration and used as 20–25% of the total feed ration for cattle. In studying other disposal alternatives, the potato waste slurry, when added to soil at rates as high as 435 kg N/ha (388 lb/a), was expected to cause excessive percolation of nitrate into the soil substratum and groundwaters. The temporary anaerobic condition formed by the ponded slurry caused large nitrate losses, but by denitrification; insignificant losses occurred in percolation water (1 ppm nitrate) because of the lowered nitrate content after denitrification.

18:7.5 Industrial Sludges and Solid Wastes

Industrial sludges are even more difficult than industrial solid wastes to dispose of tidily. Compositions of industrial sludges vary enormously; two common ones are boiler scale (calcium carbonates) and flue gas sludge. *Flue gas desulfurization sludge* (FGDS) is generated when lime [$Ca(OH)_2$] or limestone slurries are used to trap sulfur oxides from escaping gases in coal-fired power plants. The waste contains fly ash (burned coal ash), calcium salts, and volatile elements such as mercury, arsenic, selenium, lead, and cadmium. Disposal restrictions to avoid heavy-metal contamination are similar to those discussed previously for sewage sludges.

Phosphogypsum, a by-product of phosphoric acid production, has been used as a non-lime source for peanuts in the southeastern United States. In 1989 the EPA banned the material because it was claimed to have unsafe levels of *radionuclides* that could produce radioactive radon gas.[29] The material was cheap but was low in calcium (16%) and caked, making spreading it physically hard and even hazardous. It was also difficult to know exactly how much calcium was being applied. The more expensive landplaster (21–25% calcium) is easier to spread and is of more constant composition than the phosphogypsum.

Oil sludges at oil terminals are difficult wastes to dispose of. Large quantities are produced as "empty" oil tankers are filled with ballast water on the return trips. The oil left in the tankers mixes with this ballast water and is accumulated at oil terminals when it is removed to fill tankers again with oil. At the Valdez oil terminal in Alaska, the oil sludge has been mixed with soil (7.0–7.5% sludge), limed, fertilized with nitrogen, phosphorus, and potassium to aid breakdown, and planted to various grasses.[30] Annual grasses are most used to allow yearly liming, fertilization, and cultivation of the mixture.

Heavy metals (mercury, lead, cadmium, chromium, nickel, zinc, selenium, and many others) are elements that are very immobile in soils and last forever. Contamination in soils comes from additions in sludges, phosphate fertilizers, atmospheric fallout (from ore smelting), from electroplating wastes, paint wastes, fumes from burning gasoline, natural gas, coal, and many other sources. Almost all heavy-metal tailings ponds have potential problems of (1) developing strong acidity as their pyrites oxidize and (2) accumulating soluble heavy metals.

---

[29] Anonymous, "Farewell, Phosphogypsum," *Agrichemical Age,* **34** (1990), no. 4, pp. 17, 18, 30.
[30] William W. Mitchell and G. Allen Mitchell, "Land Farming of Oil Sludge at Valdez Oil Terminal," *Agroborealis,* **22** (1990), no. 1, pp. 18–21.

In mid-Wales many abandoned metal mines are sources of pollution. Oxidizing pyrites form sulfuric acids that produce drainage waters with pH 2.6 and heavily contaminated with aluminum, zinc, cadmium, and nickel.[31] The large quantities of aluminum (200–411 micrograms/L) were in the area drainage. EPA has suggested a 4-day aluminum concentration average should not exceed 87 micrograms/L to avoid damage to fish. Zinc levels in river waters near drainage entryways were three times higher than the recommended limits. Even drainage from acidic peaty soils have high aluminum contents.

Cadmium, because it is readily absorbed by plants, is a major concern in foods, as was discussed in the section on sewage sludges.

Lead also is a hazard when direct ingestion of lead-containing materials is common (children eating old paint chips, contaminated soil, and contaminated dust on fruits, berries, and vegetables). After comparing lead contents in 500-year-old frozen mummies from Greenland with the current contents in people in Denmark (1979) and in people in U.S. metropolitan areas, these conclusions were reached.[32]

- U.S. metropolitan people had about seven times more lead than the Danish people
- The 500-year-old mummies had about $\frac{1}{30}$ the lead content of present-day Danish people
- Present-day exposures are about 10- to 1000-fold larger than those centuries ago
- Some medieval samples had high lead levels, perhaps due to use of lead ceramic glaze, pewterware, lead water pipes, lead therapeutic agents, and lead to preserve certain beverages.

18:7.5 Sugarcane Field Trash[33]

Restrictions on burning (which have been recommended to reduce air pollution) seriously concern sugarcane growers, who customarily burn tons of cane leaf in the field before harvest. Such burning reduces weeds, removes hazardous snakes in tropical areas, reduces insect pests, and helps clear the field for harvesting (Fig. 18-10). One study in Florida indicates what growers may face if cane burning is not permitted.

FIGURE 18-10 Preharvest burning of sugarcane converts large amounts of trash leaves and weeds to ashes and partially burned waste. It also kills or chases out snakes! Burning releases particulates and gases into the air. Burning of straw and range grasses are common in many countries. (Courtesy Raymond W. Miller, Utah State University.)

[31] Ronald Fuge, Isan M. S. Laidlaw, William T. Perkins, and Kerry P. Rogers, "The Influence of Acidic Mine and Spoil Drainage on Water Quality in the Mid-Wales Area," *Environmental Geochemical Health* **13** (1991), no. 2, pp. 70–75.

[32] Philippe Grandjean and Poul J. Jorgensen, "Retention of Lead and Cadmium in Prehistoric and Modern Human Teeth," *Environmental Research,* **53** (1990), pp. 6–15.

[33] J. R. Orsenigo, "A Harvest Comparison of Green and Burned Sugarcane," Florida Agricultural Experiment Station Technical Bulletin 794, 1978.

1. Hand-cutting green cane would increase the labor requirement at least 50%. Cutting quality would also decrease.

2. The extra bulk would require double, and perhaps triple, the number of cane loaders in operation to keep an adequate flow of cane to the processors.

3. Transport vehicles, or number of trips, would need to be increased 50%.

4. The volume of trash at mills would increase threefold. Excess trash increases the cost of processing and lessens sugar recovery. Some of the actual figures determined for these comparisons are:

Operation	Burned	Unburned
Hand-cutting green cane (man-hours/ton)	0.66	1.00
Loader capacity (ton/hour)	149.9	62.0
Field wagon capacity (ton/wagon load)	3.95	2.62
Mill transport units (ton/unit)	21.60	15.50
Trash in cane (%)	4.73	13.04

18:8 Soluble Salts

Salt accumulation has been a perpetual problem of civilizations in arid and semiarid regions. The United Nations Food and Agricultural Organization (FAO) states that half of the irrigated farms in the world are damaged by salt (Fig. 18-11). Scientists, in their attempts to increase water efficiency (less water per unit of crop yield), have sometimes increased salt problems when leaching is too little.

All natural waters contain dissolved mineral substances called *soluble salts*. Some rainwaters, far from coastal salt sprays, may be very low in salts. As water flows over and through soils, it picks up salt loads. If the water has rapid evaporation as it flows on the surface (such as the Colorado River in the western United States), the salt concentration increases as water is evaporated. The erosion of salts and return-flow waters with salts in them all add to the "load" of salt. Deicer salts, salty wastes dumped in streams and lakes, and sea spray are all sources of salts.

18:8.1 Damages from Soluble Salts

The environmental concern with soluble salts is the cost damages (crop losses, metal corrosion, and costly cleansing activities needed by industry). Salt washed from one field ends up in groundwaters or rivers to be used by someone else. The major question related to soluble salts is how to best manage the salt with the least cost and damage to everyone.

An example of costs from salts is given by the complicated United States–Mexico agreement, signed in 1973, concerning the Colorado River water flowing into Mexico.[34,35] The agreement states that the United States will keep the salt content in the water behind Imperial Dam (the last U.S. dam on the Colorado before it enters Mexico) to 879 ppm salt.

---

[34] M. B. Holburt, "The 1973 Agreement on Colorado River Salinity between the United States and Mexico," in *Irrigation Return Flow Quality Management,* J. P. Law, Jr., and G. V. Skogerboe, eds., Proceedings of National Conference, U.S. EPA and Colorado State University, Fort Collins, 1977, pp. 325–333.
[35] Fred Pearce, "Banishing the Salt of the Earth," *New Scientist,* **114** (1987), no. 1564, pp. 53–56.

FIGURE 18-11 Salt accumulation (white crusts) in Isphahan, Iran. The salt in the surface is caused by irrigating a soil in an arid climate where most water loss was by evapotranspiration and few salts were removed by leaching. (Courtesy Bruce Anderson, Utah State University.)

At places along the river, this value is exceeded at times. From the Imperial Dam, water goes either to Mexico (12% of the Colorado River flow) or to the Welton-Mohawk Irrigation District in southwestern Arizona. The drainage from the irrigation district is high in salt and must be cleaned. The agreement furnishes Mexico with salinity of less than 1000 ppm.

To clean up the salty water of the lower Colorado River, an enormous desalting plant was planned. Some of its details are given in the following items:

1. About 350 million liters (about 95 million gallons) of water per day need to be desalinized.

2. The expected completion date was 1991. For 15 years the saline drainage of the Welton-Mohawk District has been channeled to drain into the sea in the Gulf of California.

3. The desalinization plant cost is $300 million.

4. The desalination process used in the plant is *reverse osmosis,* which will cost about $264 per thousand cubic meters. This cost is estimated as about 10 times more than the water is worth to farmers as irrigation water. Most of this cost will be borne by the taxpayer.

5. The total expenditures for all costs through 1976 were about $317 million. This included land reclamation in Mexico, canal repair, desalinization plant, and other costs.

Some of the individual elements in soluble salts can also be problems. As total salts accumulate, hazardous concentrations of boron, selenium, molybdenum, and/or arsenic may increase. Often these hazards occur where regional soils and rocks contain high levels of these metals, and they are dissolved into drainage water. Soluble contents in solution phase in ponds in California are as follows:

Selenium	1.0 mg/L
Arsenic	5.0 mg/L
Boron	70.0 mg/L
Molybdenum	350.0 mg/L

Selenium has become a severe problem in some evaporation ponds and other salt-accumulation areas (Detail 18-5).[36]

18:8.2 Solution to the Soluble-Salt Problem

No simple solution to environmental accumulation of soluble salts is known. People can reclaim the soil by washing out the salt, but the salt then goes into groundwaters or surface waters. The only uncontested disposal sites are the oceans, which are already salty, and a few "salt basins" that are near, and convenient, to some areas. One such area is the originally dry Salton Sea of California, now used to collect drainage (Detail 18-5). In the future the actions *that will not be allowed* are indiscriminant leaching of salts into groundwaters and the wash-

---

[36] Kenneth K. Tanji, Colin G. H. Ong, Randy A. Dahlgren, and Mitchell J. Herbel, "Salt Deposits in Evaporation Ponds: An Environmental Hazard?" *California Agriculture,* **46** (1992), no. 6, pp. 18–21.

Detail 18-5 Toxicity of Selenium

Selenium (Se) is an element not needed by plants but required by animals in small amounts to avoid "white-muscle disease," which weakens the heart. Too much selenium causes severe liver and kidney damage, odoriferous body and breath, and eventually death.

Selenium is chemically similar to sulfur and has four soil forms: (1) selenide (Se^{2-}), (2) elemental Se^{o}, (3) selenites (SeO_3^{2-}), and (4) selenates (SeO_4^{2-}). Selenates are quite soluble, just as are sulfates, so they move and accumulate with other soluble salts.

The San Joaquin Valley in California has poor drainage and accumulates salts from the irrigation and other drainage waters. The Salton Sea, flooded long ago by ocean waters, continues to accumulate salts from added waters. In 1992 about 150,000 eared grebes died at the Salton Sea; selenium is suspected as part of the problem. In the San Joaquin Valley health problems with nesting birds were noted in 1983. In 1984 about 16,000 birds died of what is believed by some to be Selenium poisoning, although avian cholera was the cause listed by local wildlife

specialists. Selenium is blamed for deformation and death of ruddy ducks, mallards, grebes, killdeer, coots, and other birds.

Studies on the selenium hazard in these areas are currently in progress to seek ways to reduce soluble selenium in the high-selenium soils of the area. High-uptake plants (mustard, milk vetch, prince's plume) are possible extractors to lower soil selenium.

Sources: Gary S. Banuelos, "Selenium-Loving Plants Cleanse the Soil," *Agricultural Research,* **37** (1989), no. 5, pp. 8–9; Gary Banuelos and Gerrit Schrale, "Plants That Remove Selenium from Soils," *California Agriculture,* **43** (1989), no. 3, pp. 19–20; Robert H. Boyle, "The Killing Fields," *Sports Illustrated,* March 22, 1993, pp. 62–69.

ing of salty irrigation wastewaters into surface waters. A number of states already have some regulations on this salt problem.

If "dumping" is more restricted, how are the leaching to be regulated and the sources of contaminating salts to be identified? The more careful use of soil as a receptor of the salt will be necessary (salt precipitation). Careful irrigation to avoid excess water application will allow precipitation of calcium, magnesium, bicarbonates, silicates, and some sulfates as carbonates, silicates, and sulfates during drying cycles. Many of these precipitated salts do not redissolve very easily, especially the carbonates and silicates. In some waters with low sodium but a high proportion of calcium and magnesium, as much as 60–80% of the soluble salts may be precipitated. There is a catch to this seeming "light at the end of the tunnel." The salt *not precipitated* is high in sodium and chloride ions. The sodium and potassium cause soil dispersion and are highly corrosive to metals. The chloride is toxic to plants in high concentrations. Thus, periodically, even these soils will need leaching and removal of the exchangeable sodium in the soil. Careful addition of water will, however, lessen salt moving into downstream waters and the total amount of salt in return-flow waters.

Most dissolved inorganic chemicals in natural waters are soluble salts. They are found in all soil solutions. In high concentrations, salts are unwanted because they reduce or

FIGURE 18-12 What can be done with the pollutant called "soluble salts"? Washing (leaching) salts may pollute groundwaters; not leaching them leaves a soil that will not grow plants — even plants that have some salt tolerance, like the cotton in this photo. This soil near Bakersfield, California, is typical of what can happen without careful control when salty waters (wastewaters, sewage effluents, brackish waters), and even "normal" waters are used. (*Source: Agricultural Research,* **27** [1978], no. 11, p. 14.]

hinder plant growth, speed corrosion of metals, make drinking water unpalatable, and interfere in many other uses of water (Fig. 18-12). California has already established regulations that wastewaters must have total dissolved solids (TDS) reduced to 500 ppm before they can be discharged. In the near future more controls likely will be placed on irrigation runoff water disposal. Salt pollution from agricultural runoff water is largely **nonpoint pollution** (which means the pollution does not always derive from one source, or point, but from a combination of sources—for example, lateral seepage flow and salt carried from the fields in irrigation wastewater). As mentioned earlier, individual states are to develop best management practices for on-farm improvements to control salt contents in waters leaving the farm. These studies have been in progress for several years.

18:9 Toxic Elements Natural in Soils[37]

In 1856 a doctor at Fort Randall (now central South Dakota) observed a fatal disease that he linked with pastures in the area. The same problem (losing portions of horses' hooves and hair, listlessness, and even death) was noted in 1275 by Marco Polo. Although early reports are not wholly reliable or specific, we now know various plants that, in certain kinds of soils, are toxic to animals.

The economic loss from animal consumption of poisonous rangeland plants in the 17 Western states of the United States is estimated at about $340 million annually.[38] These dam-

---

[37] Gary S. Banuelos, "Selenium-Loving Plants Cleanse the Soil," *Agricultural Research,* **37** (1989), no. 5, pp. 8–9.
[38] Julie Corliss, "Toxic Encounters with Range Plants," *Agricultural Research,* **39** (1991), no. 12, pp. 4–7.

ages vary from reduced production to animal death. Here are some examples of individual losses:

1. Sheep kills from grazing halogeton (*Halogeton glomeratus*) totaled 1300, 1200, 800, and 600 in individual losses.

2. Losses from locoweed (*Astragalus* spp. and *Oxytropis* spp.) poisoning included 6000 sheep killed in eastern Utah and losses of $125,000 by one rancher in 1964.

3. Grazing of larkspur (*Delphinum* spp.) on mountain ranges caused the death of 103 mature cattle in one Forest Service allotment. U.S. ranchers in the intermountain West lose more cattle to larkspur than to any other poisonous plant.

The more than 200 range plants containing toxic substances usually cause problems because of the innate toxic nature of the plant and/or management of livestock; only a few plants are toxic because of excess or deficient absorption of soil elements. The more common toxic range plants are locoweed, halogeton, saltbush (*Atriplex nutallii*), goldenweed (*Oonopsis* spp.), larkspur, lupine, prince's plume (*Stanleya pinnata*), and woody aster (*Xylorrhiza* spp.) (Fig. 18-13).

Selenium is a soil-supplied essential nutritive element whose concentration is critically important. Too little selenium in forage foods can cause white-muscle disease in animals; too much selenium causes blind staggers, alkali disease, or even death. Plants that accumulate very high selenium contents of 1000 to 10,000 mg/kg include milkvetch (*Astragalus*), woody aster (*Machaeranthera*), mustard (*Brassica*), and prince's plume (*Stanleya*).

 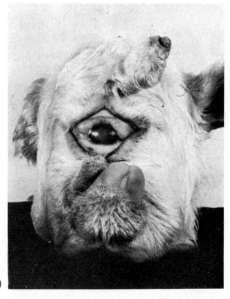

(a) (b)

FIGURE 18-13 Deformation caused by eating poisonous range plants during pregnancy. (a) Deformed lamb from ewe fed locoweed *(Astragalus pubentissimus).* (b) Malformed head of a lamb from ewe fed false hellebore *(Veratrum californicum)* on day 14 of gestation.   (Courtesy USDA Poisonous Plant Research Laboratory, Logan, Utah; photo by Lynn F. James.)

Legend

⬤ General location of relatively broad areas of high (>10 to 100 ppm)
    molybdenum forage plants associated with incidence of molybdenosis

✕ Localized areas, principally isolated alluvial fans, swales and
   depressions, of high molybdenum plants

▒ Areas where general background levels of molybdenum are moderately
   high in soils and plants

⋀ Principal mountain ranges

⌇ Principal rivers

FIGURE 18-14 Areas of molybdenosis and other potential problem areas having high molybdenum in soils and plants. (*Source:* Modified from Joe Kubota, "The Poisoned Cattle of Willow Creek," *Soil Conservation,* **40** [1975], no. 9, pp. 18–21).

Molybdenosis, caused by excess molybdenum, is an animal disease that hinders utilization of copper and so results in poor growth. A common treatment is to feed supplemental copper. Areas that may cause molybdenosis in the West are *wet* alluvial soils formed from granite. Problem areas are shown in Fig. 18-14. The Florida Everglades is the only area east of the Rockies where molybdenosis is known to be a problem.

Here are general guidelines for dealing with toxic range plants[39]:

1. Do not drive or unload hungry animals into areas low in good range plants but high in poisonous plants. Dumping trucked sheep in overgrazed unloading areas where halogeton is the common ground cover has caused heavy losses.

[39] William C. Krueger and Lee A. Sharp, "Management Approaches to Reduce Livestock Losses from Poisonous Plants on Rangeland," *Journal of Range Management,* **31** (1978), pp. 347–351.

2. Maintain free access to salt to limit grazing of salty poisonous plants (halogeton).

3. Animals low in phosphorus tend to have depraved appetites and eat abnormally. Provide known needed minerals.

4. Fence animals out of extremely toxic plant areas.

5. Maintain good range feed and avoid putting new arrivals on bad areas of range until they become accustomed to low levels of the scattered toxic plants.

▬▬ *18:10* Radionuclides

Radioactive elements differ from nonradioactive *isotopes* (elements with similar properties) by the emission at some time of a high-energy particle. These emissions are **radiation** and include, among others, gamma rays, beta rays, alpha rays, and neutrons. Bombardment of a living body by enough radiation changes atoms of compounds into different elements, altering their action, even killing cells or portions of cells. Exposure to high radiation levels causes nausea, diarrhea, vomiting, hemorrhages, leukemia, sterilization, or death. At lower dosage levels, cell membranes are damaged or destroyed and leukemia (a low count of white blood cells) is common.

Radioactive elements have specific rates at which they emit radiation (**decay**). This rate is measured as the time needed for a mass of the element to reach a radiation rate that is half of what it was at time zero, when it began to radiate; this is called the **half-life** of the element. At the end of the two half-life periods, a mass of radioactive material would have one-fourth (one-half of one-half) as much radiation as it had at the initial starting time of measurement (Fig. 18-15). The half-lives of some common radioactive elements are

^{238}uranium	4,510,000,000 years
^{14}carbon	5,730 years
^{137}cesium	30.2 years
^{90}strontium	28.1 years
^{131}iodine	8.0 days

Radioactivity is of great concern because of its invisibility, its insidious damages, and the fact that there is no known way to reduce or stop the radiation process. Shielding the source with energy-absorbing heavy metals (e.g., lead) and/or increasing the distance between source and victim (a thickness of solid materials and even air) are the only known protections. Atomic bomb testing has covered all areas of the earth with radioactive strontium, cesium, and/or iodine. The radioactive element ^{90}strontium, taken up in pastures, becomes substituted into bones and milk in place of calcium; ^{137}cesium accumulates in muscle tissue. Claims in Arizona, Utah, and Nevada of increased leukemia deaths (2.4 times more than normal) during and since the Nevada atomic bomb tests[40] are generally accepted as valid claims by the U.S. government.

Where radioactive fallout has occurred, there is no known way to eliminate the hazard, except by scraping and carrying off contaminated soil to which the cations have adsorbed. Radioactive elements in the soil are absorbed into plants, as are other elements and become a part of whatever consumes the plants. Wildlife grazing on forages on radioactive tailing piles or on plants containing fallout elements and later eaten for food is one way that isolated

---

[40] "A Fallout of Nuclear Fear," *Time,* **113,** Mar. 12, 1979, p. 84.

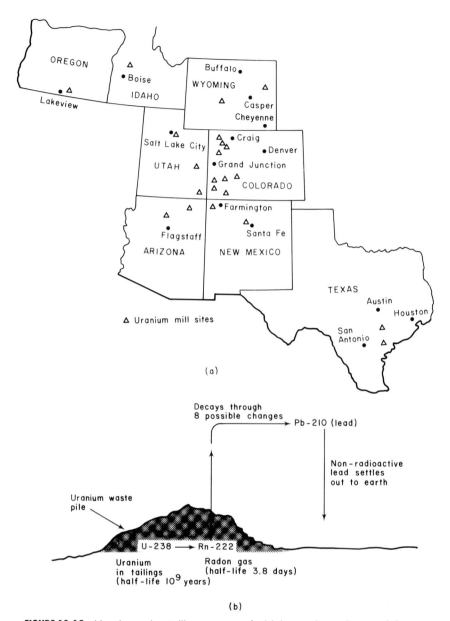

△ Uranium mill sites

(a)

Decays through
8 possible changes → Pb-210 (lead)

Non-radioactive
lead settles
out to earth

Uranium waste
pile

U-238 → Rn-222

Uranium
in tailings
(half-life 10^9 years)

Radon gas
(half-life 3.8 days)

(b)

FIGURE 18-15 Uranium mine tailings, some of which are shown in map (a), contain unextracted radioactive uranium-238. Left as spoil piles, these tailings now cause concern. Some of the sandy waste has found its way into cement, brick mortar, and as fill under sidewalks and buildings. In (b) the general decay of uranium (4.5 billion year half-life) to radon gas and several other elements shows the radiation hazard. In some areas large amounts of sandy uranium tailings were used as filler under buildings and sidewalks or even in the cement. Major, expensive excavations have been undertaken to remove these tailings, where it is possible. This action is necessary to reduce radon gas pollution in the area and in the buildings. (Courtesy of Raymond W. Miller, Utah State University.)

radioactivity may affect people. The inability to mitigate contamination by radioactive substances except by the passage of time makes many people strongly oppose development of nuclear energy systems and weapons testing.

18:11 Soil Sediments Are Pollutants

Eroded soil becomes a serious pollutant, both as a physical problem and because of chemicals it carries adsorbed to the particles' surfaces. The control of soil erosion will be beneficial to solving several pollution problems.

18:11.1 The Sediment Problem

Soil itself becomes a spectacular pollutant when large amounts slide down to cover homes or roadways, when receded floodwaters leave behind mucky, muddy messes, or reservoirs and harbors fill with silt. Pollution by sediment modifies the environment in the following ways:

1. Suspended sediment is usually eroded topsoil, the most fertile portion of soil (Fig. 18-16). The eroded soil is deteriorated, and the carried topsoil could deposit in places where the fertility is a liability, such as in bodies of water where eutrophication would be increased by the nutrients, especially phosphorus. In contrast, the great deltas of major rivers are formed from deposited sediments and can be beneficial as productive cropland.

2. Water reservoirs can be filled by sediments, decreasing their storage capacities. Tarbela Dam reservoir in Pakistan, one of the world's largest, has a silt load about 16 times larger than predicted by the dam's engineers. The life expectancy of the reservoir is now only 50 years. In India the Kosi Canal has been so heavily silted in its 20-year existence because of overgrazing its watershed that it now provides water for only one-seventh its original irrigation area (shrunk from 570,000 ha to 81,000 ha).[41]

3. Suspended solids reduce sunlight penetration into water, thereby reducing production of microscopic-sized organisms, which begin the aquatic food chain. Suspended soils also cover lake bottom plants and fish eggs, to the detriment of both, and interfere with the gill action of fish and generally suffocate them.

4. Sediment on land can cover good soil with poorer or even rocky debris.

Removing unwanted sediment—from agricultural land, residences, roadways, harbors, and streams—is toilsome and expensive.

Federal regulations for water pollution control, recognition of the importance of soil as a nonrenewable resource, and the need to minimize other damages from soil erosion have increased restrictions on land management. Typical of these new land management restrictions is the North Carolina state Sedimentation Pollution Control Act of 1973 to control erosion from all land-disturbing actions, particularly commercial development, including road construction. Some of its requirements follow[42]:

---

[41] Sid-Gautam, "Dam Building No Longer Means 'Instant Progress,'" *Water and Sewage Works,* **125** (Aug. 1978), pp. 30–32.

[42] Joseph A. Phillips and Jesse L. Hicks, "Sediment Control: The North Carolina Law," *Journal of Soil and Water Conservation,* **31** (1976), no. 2, pp. 76–77.

FIGURE 18-16   All of this California orchard's lost soil ended up somewhere downslope or in the river. In areas with intensive cropping, this kind of erosion can and should be better controlled.   (*Source:* USDA — Soil Conservation Service.)

1.   Any land disturbance near a lake or natural watercourse must have a wide enough buffer zone that the 25% of the buffer zone nearest the activity collects all of the visible siltation. The law excludes constructions on, over, or under the water.

2.   The angles of graded slopes are limited to those on which vegetative cover can be maintained; graded slopes must be planted within 30 working days after any phase of grading is completed.

3.   Any land area of more than 1 acre (0.4 ha) that has been disturbed must be planted to ground cover within 30 working days on portions not to be further worked.

Minimum tillage is a management system that reduces soil erosion. It maintains crop residues on the soil surface and thus reduces the dispersing action of rainfall and decreasing the speed of water runoff. However, the increased need for herbicides and other pesticides has caused some people to wonder if minimum tillage *is* reducing pollution. An increase in one pollutant (pesticides) is substituting for a decrease in a second pollutant (soil).

The 1985 Food Security Act (PL-99-198) has these three conservation provisions that should help to reduce erosion[43]:

1.   *Conservation Reserve Program.* Persons seeding highly erodible land to grass for 10 years will receive a per-acre payment.

---

[43] Clayton W. Ogg, "Erodible Land and State Water Quality Programs: A Linkage," *Journal of Soil and Water Conservation,* **41** (1986), pp. 371–373.

2. *Swampbuster provision.* This act eliminates any federal incentives for all farm benefits to anyone converting wet lands to crop production without specified conservation treatments.

3. *Sodbuster provision.* This is similar to the swampbuster provision, but this act refers to putting certain *fragile* (highly erosive) grasslands into crop production.

It is estimated that only a small fraction of the cropland has excessive erosion of more than 33.6 Mg/ha (about 15 t/a). Other associated estimates are these[44]:

- 3.5% of cropland, which is losing over 56 Mg/ha (25 t/a), accounts for 32% of the total soil loss.
- 7% of cropland, which is losing over 33.6 Mg/ha (15 t/a), accounts for 44% of the total soil loss (includes the estimate above).

In 1979 a national erosion estimate anticipated that at least 3.6 billion metric tons of soil would be lost annually by erosion, enough to cover 850,000 ha (over 2 million acres) with sediment nearly 30 cm (1 ft) deep. The major part of this erosion is natural and nearly impossible to remedy. Rangelands, which are enormous contributors to sediment pollution, are usually in low rainfall areas and are too arid to maintain good plant cover, so wind carries away the barren soil and occasional flash floods wash away more (Fig. 18-17).

It is not just erosion by croplands that is of concern. Wind erosion causes many kinds of damage. Dune stabilization has been worked at for decades to protect cities (Cairo, Egypt, for example), and facilities in many areas are forced to close due to blowing sands. In China, storms originating in inner Mongolia sweep into Beijing (old Peking) and across 8 million hectares of farmland and pastures. To protect against this wind, an enormous shelter belt 7000 km (4375 mi) long was begun in 1978. By 1985 the first phase of 6 million hectares of barren land were planted from Heilongjiang province to Xinjiang Uyger region. About 51% of the seedlings survived in the area having 400 mm (15.6 in.) of rainfall yearly. Peasants who plant will "own the trees" to bequeath to their children. The shelter belts have reduced water transpiration on croplands and yields are up by one-fifth. Chinese agronomists set a net benefit on the "Green Wall of China" at $630 million a year already. It is expected that the total "wall" will take several generations to plant.

18:11.2 Adsorbed Chemicals

Eroded sediment can carry appreciable amounts of phosphorus and pesticides and large quantities of nitrogen and organic matter to surface waters. One report suggests that 80% of phosphorus and 73% of nitrogen loadings of surface waters nationally are from eroded soil. Cropland erosion accounts for about one-third of these (27% of phosphorus and 24% of nitrogen) and costs of losses are $2.2 billion per year.[45] These authors state, however, that their most notable finding was: "cropland erosion control achieves acceptable phosphorus concentrations in only a few regions." There are sources other than croplands providing phosphorus. If the phosphorus pollutant has many sources, cropland controls alone seem

---

[44] Clayton W. Ogg and Harry B. Pionke, "Water Quality and the New Farm Policy Initiative," *Journal of Soil and Water Conservation,* **41** (1986), pp. 85–88.

[45] Leonard P. Glanessi, Henry M. Peskin, Pierre Crosson, and Cyndi Puffer, "Nonpoint-Source Pollution: Are Cropland Controls the Answer?" *Journal of Soil and Water Conservation,* **41** (1986), pp. 215–218.

FIGURE 18-17 These dunes in North Africa are widespread, and they are promoted by arid years, overgrazing, and cutting vegetation for fuel. Containment followed by reclamation is slow, precarious, and expensive. (By permission of Food and Agriculture Organization, Rome.)

inadequate to bring about major improvements in water quality. The quality is already adversely affected by other nonpoint-source pollution.

In 1977 the Great-Lakes-Pollution-from-Land-Use-Activities Reference Group (PLUARG), composed of U.S. and Canadian specialists, believed the Great Lakes were being appreciably polluted by runoff of phosphorus from agricultural and *urban* areas. They estimated that 41% of the total phosphorus load came from fine-textured sediments washed into the lakes.

18:11.3 Controlling Sediment-Caused Pollution

In 1976 the National Commission on Water Quality concluded that despite intense efforts to treat **point sources** of pollution (sources clearly identifiable, such as factory discharge lines), 92% of the total solids suspended in water would continue to come from uncontrolled sources. These uncontrolled sources would also furnish 79% of the total nitrogen, 53% of the total phosphorus, and 37% of the biological oxygen demand (BOD) material. This depressing estimate emphasizes the magnitude of costly and/or uncontrollable pollution sources and the continual need to reduce water pollution, including that from suspended solids resulting from agricultural activity, natural erosion, or anthropic (man-caused) disturbance.

The Federal Water Pollution Control Act of 1972, amended in 1977 by the Clean Water Act and the Resource Conservation Act, was the first giant step toward national regulation of the pollution problem. Congress specified 1985 as the date by which the United States was to "restore and maintain chemical, physical, and biological integrity to the Nation's waters." There is general unanimity that more must be done to halt water pollution, but more dispar-

FIGURE 18-18 Without adequate regulation, some people build homes in places where they should not be built, as in this Layton, Utah, site. Then they clear the steep slopes nearby. With this lack of realism, pollution of drainage water is only one of the hazards here. (Courtesy USDA — Soil Conservation Service.)

ity of opinion exists about the cost of the final phase of regulation versus the probable benefit (Fig. 18-18).

Certainly the role of eroding cropland in causing pollution is important. Erosion control is a sensible approach in looking at conserved soil for the well-being of future generations. Soil erosion does come from many noncultivated crop areas, construction sites, and mine areas. The costs to control erosion and the social benefits of that control all need to be a part of the planning and legislative regulations.

18:12 Impacts on Air Quality

Agriculture produces pollutants both to the air and to water. Particulates (dust) are added by wind erosion of unprotected soil, by tillage, and by burning of many fuels and residues. Fuels release oxides of sulfur and nitrogen. Volatile pesticides are carried great distances in wind. Those persons affected by asthma testify to the many pollens and spores of various types suspended in the atmosphere. Anaerobic decomposition (wetlands, ponds, silage, manure piles) provides methane, sulfur oxides, ammonia, and numerous odoriferous organic gases. Large amounts of heat and water vapor join these other pollutants. Even noise may be of concern in very limited sites.

Some fertilizer nitrogen produces nitrous oxide by denitrification. The EPA estimates global nitrous oxide emissions are in the range of 11–17 million tons (about 10–15.5 million metric tons).[46] Only about 3–5% of these emissions come from nitrogen fertilizers. Major nitrous oxide sources are burning fossil fuels and biomass and release of nitrous oxide from tropical and subtropical forest soils.

18:12.1 Ozone Layer Depletion

Ozone (O_3) in the stratosphere, at 10–15 km above the earth, screens out as much as 99% of the ultraviolet (**UV**) rays that could otherwise be absorbed preferentially in proteins and nucleic acids. These would cause tissue damage and possibly genetic mutations. The UV that does get through the ozone layer may cause sunburn and skin cancer. Actually, the UV rays form the ozone by splitting O_2 molecules into active O* atoms. These active atoms

---

[46] Dennis Brosten and Brenda Simmonds, "Do Fertilizers Affect the Atmosphere?" *Agrichemical Age,* **38** (1990), no. 2, pp. 6–7.

recombine with the O_2 molecules to form ozone (O_3). This process is an equilibrium between ozone destruction and formation. Chlorine gas atoms are effective catalysts for ozone breakdown. Frequently, the chlorine has been supplied by leakage of Freon gas (a chlorofluorocarbon) from refrigeration units and by various chlorofluorocarbons used in aerosol containers and as solvents for cleaning special electronic equipment. In 1974 the United States alone sprayed 230 million kilograms of these gases into the atmosphere. These chemicals are now banned in aerosols in the United States.[47]

Other chemicals also destroy ozone. Carbon tetrachloride (a common grease solvent) and nitric oxide (NO) destroy ozone. The effect of NO supplies on ozone destruction is about 10% as important as those of chlorofluorocarbons. It is estimated that by the year 2100 about 1.5–3.5% of the ozone layer will be destroyed by NO. Some scientists say this is too high an estimate. Whatever the facts are, the NO produced in various agricultural activities adds to this hazard of somewhat unknown seriousness.

At the earth's surface ozone is not desirable. Almost 90% of crops lost to pollutants are lost because of ozone and sulfur dioxide; ozone is about 10 times more toxic than sulfur dioxide.[48] It is estimated that reducing ozone by 25% would boost farmers' incomes $1.9 billion annually. Ozone destroys chloroplasts, weakens cell walls, and allows leaching of nutrients from plants. A stressed plant has higher sugar content, which increases damage from insects it attracts.

18:12.2 Methane

Methane (CH_4) is given off from rice paddies and swamp areas. About one-fourth of the 500 million tons of methane released yearly into the atmosphere comes from flooded rice fields (Fig. 18-19 and Detail 18-6).[49] A methane molecule traps heat about *30 times more effectively* than a carbon dioxide molecule (methane may cause 15% as much global warming as does carbon dioxide). Because about 80% of the methane from a rice paddy escapes from the root area *up through* plants, plant selection (breeding) is one approach to reducing methane loss from paddy.

18:12.3 Carbon Monoxide

Carbon monoxide (CO) occurs from incompletely oxidized carbon; burning or decomposition of any carbon source with limited access to oxygen can produce carbon monoxide. Landfills produce considerable CO. Carbon monoxide has about 200 times the affinity for hemoglobin as does oxygen, so CO causes loss of oxygen transport to body cells. Small amounts cause persons to become nondiscriminatory—more accidents. High doses cause headaches, dizziness, and eventually death. Fortunately, soil bacteria (autotrophs) convert CO to CO_2. This is another good reason for parks and lawns in dense cities.

18:12.4 Burning

Burning vegetation can produce carbon particulates, CO_2, CO, heat, and fiberlike silica particles from burning rice straw.[50] It is the silica particles that are of concern; it is feared that they may act in the same harmful way asbestos fibers, but little is known about them. Rice

[47] B. J. Nebel, *Environmental Science,* 2nd ed., Prentice-Hall, Englewood Cliffs, N.J., 1987, p. 388.
[48] Sandy Miller Hays, "Wanted: Breathing Room for Crops," *Agricultural Research,* **37** (1989), no. 7, pp. 4–6.
[49] Anonymous, "IRRI Studies Role of Ricefield Methane in Global Climate Change," *The IRRI Reporter,* Dec. 1991, pp. 1–2.
[50] Bryan M. Jenkins, Scott Q. Turn, and Robert B. Williams, "Survey Documents Open Burning in the San Joaquin Valley," *California Agriculture,* **45** (1991), no. 4, pp. 12–16.

When rice soils are flooded, O_2 becomes in short supply and aerobic bacteria are replaced by facultative anaerobes, which in a few days are replaced by obligate anaerobes.

In aerobic biological oxidation, free molecular O_2 is the ultimate electron acceptor. Facultative anaerobes use nitrate, manganic oxide, ferric oxide, carbonate, and other compounds with high oxidation levels as the electron acceptors. The most striking difference between aerobic and anaerobic decomposition lies in the nature of the end products. In normal, well-drained soils the main end products of organic-matter decomposition are CO_2, H_2O, nitrate, sulfate, and resistant humus; in submerged anaerobic soils the end products are CO_2, H_2O, H_2 CH_4, NH_3, H_2S, mercaptans, and partially humified residues. The successive microbial changes are accompanied by a lowering of the redox potential (from $+600$ to -300 mV) and a change in pH to near neutral. The pattern of changes and the accompanying formation of gases are related to soil properties, environmental factors, and management.

The range in the percentages of the gases in submerged rice soils is: N_2 is 35–9810%; CH_4 is 4–55%; CO_2 is 2–10%; H_2 is 0–9%; and O_2 is as low as about 1–6%.

Source: Heinz-Ulrich Neue and Hans-Wilhelm Scharpenseel, "Gaseous Products of the Decomposition of Organic Matter in Submerged Soils," in *Organic Matter and Rice,* International Rice Research Institute, Los Baños, The Philippines, 1984.

FIGURE 18-19 Large areas of paddy rice produce large amounts of anaerobic gases, including methane, carbon monoxide, nitrogen oxides plus dinitrogen, and dimethyl sulfide or hydrogen sulfide.   (By permission of Food and Agriculture Organization, Rome.)

is known to need considerable silica in growth (probably for straw strength), and much of this silica appears to become particulate fibers of silica when the straw burns.

18:13 The Ultimate Pollutant: People

The definition of pollution—the degradation of a substance or system for people's use—would seem not to include people as a pollutant. A little inspection dispels that myth. People destroy land as they build on it, pollute waters as they wash and swim in them, pollute air with gasoline fumes as they travel long distances, overload treatment systems, dump wastes into landfills, degrade inner cities, make water shortages, and cause many social stresses to each other (Fig. 18-20).

18:13.1 The Nature of the Problem

Every ecosystem has a limit to the numbers of organisms that it can support and/or live among without causing a catastrophe. All pollutants are not the result of people's ingenious discoveries; some are simply the result of too much natural "wastes," such as sewage, garbage, burning, dust, heat dumped into waters, etc. Fewer people would need fewer factories, fewer acres for production, less gas and oil, and fewer other luxuries. Most people are in favor of environmental integrity, but few suggest any major project to control our major pollutant: people (Fig. 18-21).

The reasons for a lack of progress in population control are many (1) people do not agree on how many people constitute pollution, (2) population control is unacceptable to many persons on religious grounds, (3) population control measures would likely be taken by nations, with some ignoring controls, and (4) a lack of knowledge is coupled with an apathetic attitude in most people. Many persons fear racial genocide, or they dislike infringements of their personal choices. Too many people are willing to face a problem only when a solution becomes crucial. The exact outcome of continued population growth is uncertain.

FIGURE 18-20 People damage land in many ways, including mixing material that is best left unmixed. These long piles of overburden to get at coal seams is destructive to many soils and hydraulic patterns. The size of this operation can be scaled by looking at the pickup truck (lower right, at the arrow). (Courtesy USDA — *Agricultural Research.*)

One thing is a certainty, however: *The more nature is bent abnormally by more and more people, the more catastrophic will be the results, whenever we temporarily lose control.*

18:13.2 Protecting Land from People

The American Farmland Trust stated that unless California's agricultural problems were addressed in the next 10–20 years, the state's farming industry would likely decline. The four major problems are these:

1. Agricultural-land conversion to nonagricultural uses
2. Soil erosion
3. Increasing salinity of soil and water
4. Diminishing water supplies and diversion to nonagricultural uses

Over 17,000 ha yearly are converted to urban uses, of which over 80% were irrigated croplands.

How can farmland be protected from people pollution (use)? So far, only limited controls are available. Many states have erosion control legislation, which forces the land owner (farmer, contractor, businessperson of other trades) to establish controls, if the erosion exceeds a defined value (usually 4–10 Mg/ha [2–5 t/a]). Poor management practices allowing erosion are not tolerated. In addition, the **Environmental Protection Agency (EPA)** has strict regulations on allowing pollution of water and air by excessive fertilizer and pesticide use and uncontrolled burning practices.

A number of states have varying regulations to restrict unlimited land conversion to nonagricultural uses. New York bought "development rights" to thousands of acres of farmland on Long Island. It was thought that the land as farmland had high economic and social values for (1) producing vegetables near a large market, (2) a way for city dwellers to quickly get into "the country," and (3) to control somewhat the urban sprawl.

The criteria for the best farmlands are approximately those given by the Soil Conservation Service (SCS).

Prime Farmlands These lands have the best combination of physical and chemical characteristics for producing food, feed, forage, fiber, and oilseed crops, and they are available for these uses (not already covered with urban development, roads, etc.). The lands must meet *all of these criteria* (condensed and approximated by the authors):

1. Good temperature for most crops. Average summer temperatures over 15°C (59°F).

2. Sufficient natural rainfall or irrigation water 7 out of 10 years for the commonly grown crops.

3. A soil depth of at least 1 m (40 in.).

4. A pH in the top 1 m of between 4.5 and 8.4.

5. No shallow water table to interfere with normal plant growth.

6. The soil can be managed to keep within the top 1 m of soil (a) a low salt content (<4 dS/m) and (b) a low exchangeable-sodium percentage (<15%).

7. Soils are not flooded in the growing season more often than once in 2 years.

8. The soil is not highly erodible. The K (erodibility factor) times the percentage slope is less than 2.0 and, for wind erosion, I (erodibility) times climate factor does not exceed 60.

9. Permeability is not a limiting factor in using the soil. The permeability is >0.15 cm/h in the top 50 cm (20 in.) of soil.

10. The soil is not rocky. It must have less than 10% of the surface with rocks larger than 7.6 cm (3 in.) in diameter.

Unique Farmlands Unique farmlands are lands *other than prime farmlands* that are used to produce specific crops (such as citrus, tree nuts, olives, cranberries, and certain other fruits and vegetables). Some of the requirements for land to be classified as unique lands are these:

1. The land must be used for a specific high-value food or fiber crop (with unique soil and/or climatic requirements).

2. The land has adequate moisture available for the specific crop.

3. The land area combines the favorable factors of soil qualities, temperature, humidity, aboveground air drainage (frost control), steepness of slope, aspect (direction of slope), or other attributes (such as nearness to market) that favor the growth and/or distribution of the specific crop.

Source: Raymond W. Miller, with data from "Part 657—Prime and Unique Farmland," *Federal Register,* **7CFR** (Jan. 1, 1986), Ch. VI, p. 657.1.

Most land control measures have attempted to pinpoint the saving of the best farmland for agricultural use. This best land is defined as **prime farmland** or **unique farmland.** Details of these lands are given in Detail 18-7.

As an example, Ontario, Canada, proposed in 1988 new and stronger guidelines for the preservation of prime farmland. These new regulations replace its 1978 guidelines. The regulations are designed to minimize conflicting land uses within agricultural areas and preserve prime agricultural land. These are the methods by which these lands will be preserved:

1. Direct any new nonagricultural development to urban areas or onto marginal land.

2. Require documented evidence to justify any nonagricultural development that is proposed to go onto prime farmland.

3. Designate certain kinds of uses that are incompatible for any development on prime farmland. These could include many industrial uses, recreation facilities, conventional residential development, and mobile-home parks.

These examples and other laws already in use in many U.S. states emphasize the trend to protect agricultural lands. These lands are needed for food production for local people, to protect local economies, to maintain open space, and to increase exports to balance U.S. foreign-trade deficits.

As important as technology, politics, law, and ethics are to the pollution question, all such approaches are bound to have disappointing results, for they ignore the primary fact that pollution is primarily an economic problem, which must be understood in economic terms.

—Larry E. Ruff

The reason we have water pollution is not basically the paper or pulp mills. It is, rather, the social side of humans—our unwillingness to support reform government, to place into office the best qualified candidates, to keep in office the best talent, and to see to it that legislation both evolves from and inspires wise social planning with a human orientation.

—Stewart L. Udall

Questions

1. (a) In what ways can soil act to remove pollutants from waters? (b) Are all pollutants "filtered" out by soil? Explain.
2. (a) What is meant by *pollution?* (b) Must a pollutant be toxic or carcinogenic? Explain.
3. Explain what BMPs are and discuss why they are used.
4. By what authorization does the EPA control land use?
5. (a) What is meant by eutrophication? (b) Explain why phosphates have greater influence than nitrates on eutrophication.
6. (a) How is biomagnification related to half-life? (b) Explain why half-life of many hazardous substances is very important.
7. Probably the major pollution by nitrogen is in eutrophication and methemoglobinemia. Explain each of these and indicate the relative hazard of each compared to other hazards.

8. What are the suggested approaches to minimize pollution from the use of fertilizers?
9. (a) To what extent was DDT more "poisonous" to animals than other currently accepted pesticides? (b) What are the criteria a pesticide must meet to be acceptable for use today?
10. (a) To what extent is agriculture responsible for pollution with pesticides? (b) What are the alternatives to using pesticides? (c) What is required to make pesticide usage drop sharply in the next few decades?
11. Discuss the problem of selenium in soils and wastewater impoundments.
12. (a) What are some of the common heavy metals? (b) Why are they of concern?
13. (a) What undesirable by-products are added to soils in manures and sewage sludges that may make their use hazardous? (b) Explain the hazardous nature of each "pollutant" mentioned.
14. (a) Since adding soluble salts in natural waters is usually a natural process, why are soluble salts treated as pollutants? (b) How serious is the salt problem?
15. (a) What hazardous gas is produced when organic substances are buried in landfills? (b) What is the longevity of this gas production? (c) Is there a solution to the problem? Explain.
16. How is BOD (or COD) related to undesirable water conditions and waste disposal?
17. Is there an easy solution to the soluble-salt problem? Explain.
18. (a) What are the kinds and sources of radioactive pollutants most likely to occur on crops and in soils? (b) What is the nature of the hazard?
19. (a) Even though pollution by radioactive substances is quite small, why is it given such serious attention? (b) How can such pollution in soil be corrected?
20. (a) How do soil sediments fit the definition of a pollutant? (b) What are the damages from soil sediments?
21. How has minimum tillage and no-till reduced pollution by soil sediments?
22. Although air pollutants have been given less attention than water pollutants, agriculture does produce some air pollutants. List and briefly discuss three of these.
23. How are some soil resources protected from people and their detrimental actions?
24. Define (a) *prime farmlands,* and (b) *unique farmlands.*
25. What activities are used in various localities or countries to protect prime farmlands and unique farmlands?

19

Soil Surveys, Interpretations, and Land-Use Planning

Our duty to the whole, including the unborn generations, bids us restrain an unprincipled present-day minority from wasting the heritage of these unborn generations.

—Theodore Roosevelt

19:1 Preview and Important Facts

PREVIEW

The first soil surveys were simple and limited. They were intended to answer practical agronomic questions of soil differences and limitations important in improving and expanding crop production. Was a new soil area suitable for crops that had never been planted there? How much additional fertilization did it need? What were the problems of water, salts, or acidity? And could other crops be grown more profitably?

Soil surveys expanded in detail and concept with the increase in scientific knowledge and demand for more useful information. Today's surveys include information to make scientific interpretations about using each soil mapping unit. A **soil mapping unit** is an area of soil that is delineated from adjacent soil on a map; those differences may be slope, erosion, and other features as well as differences in the soil profile itself. The information of the survey helps in engineering construction, locating sources of sand and gravel, forestry management, urban development, game management, recreation development, predicting erosion hazards, irrigation, urban housing, taxation, and land use planning, as well as the traditional agricultural guidelines for increasing production on farms and ranches.

The National Cooperative Soil Survey was organized in 1952 to coordinate and simplify soil survey information. It is part of the U.S. Department of Agriculture (USDA) within the Soil Conservation Service (SCS), now called the Natural Resources Conservation Service (NRCS). In most states the agricultural university serves as the statewide cooperating agency. The U.S. Forest Service and the U.S. Bureau of Land Management (BLM) cooperate closely in surveying areas within their respective jurisdictions.

Progress continues in making soil survey maps and reports. Remote and wild areas (as in most of Alaska today) usually have only an *exploratory* soil survey made for most of the land; *reconnaissance* surveys are available for areas of intermediate intensive use; and *detailed* soil surveys are made of the areas with greatest intensive or potentially intensive use.

IMPORTANT FACTS TO KNOW

1. The organization responsible for U.S. soil surveys
2. The general method used to prepare a soil survey
3. The measurements and descriptions included in soil surveys
4. How quickly the data from soil surveys will change
5. The definition and equation to calculate SPI
6. The definition and use of *corrective measures* (CM) and of *continuing limitations* (CL)
7. The determination and use of the *yield* or *performance standard* (P)
8. The LESA approach to land use planning
9. The variety of types of evaluations of soils in a soil survey
10. The uses made of soil survey report data
11. The definition of *benchmark soils* and some reasons why they are used
12. The general properties of prime farmlands and unique farmlands and the importance of this information
13. The land capability class and subclass
14. How soils are rated as having slight, moderate, or severe limitations for each particular major use
15. The approaches available for governmental control of land use

19:2 Making a Soil Survey[1,2]

The soil survey of an area begins with numerous conferences, the collection of aerial photographs, and some initial field reviews to prepare a partial legend of properties and descriptions of some of the most extensive soils of the area. The **legend** is prepared during the initial field reviews from soil pits dug in the most extensive landforms where large areas of similar soil are expected to occur. Then the detailed soil survey is ready to be done.

The surveyor needs to be a self-starter, adaptive, competent, and good at public relations. As surveyors work, they encounter rain, locked gates, unfriendly landowners, dogs, nuisance insects, and rocky soils. At times, they work hard physically, and they constantly need to interpret correctly the soil profiles that they dig and probe, always looking for the boundary where "different" soils meet each other. Short travel is by foot, but maps and materials usually require a car nearby. In unique situations, such as in mapping mountain terrain, even helicopters might be used to reach high or distant locations. Helicopters have been used in many locations, including Vermont, Pennsylvania, New Mexico, and Alaska.[3] Various kinds of satellite maps have been used to assist in the mapping.[4]

---

[1] K. C. Thomas, "Computer Assisted Writing—Its Application in Soil Survey Manuscripts," *Agronomy Abstracts* (1985), p. 199.

[2] Soil survey reports are available for about one-fourth of the 3059 counties in the United States and may be obtained from the county agricultural agent, the state land grant university, the Soil Conservation Service, or U.S. congress members.

[3] John B. Carey and C. Wesley Keetch, "Helicopter Mapping of Soils in San Juan County, New Mexico," *Journal of Soil and Water Conservation,* Mar.–Apr. 1979, pp. 99–102.

[4] R. N. Fernandez, C. R. Valenzuela, and M. F. Baumgardner, "Soil Differentiation Using Digital Analyses of Lansat MSS Data in Tippecanoe Co., Indiana.," *Agronomy Abstracts* (1985), p. 191.

Working alone, a surveyor carries an aerial photograph of the area, digging tools, a hand level (the Abney level) for measuring slope percentages, a pH kit, Munsell color chip book, and 10% hydrochloric acid to identify the presence of lime. The surveyor carries the legend describing the profile characteristics of most of the extensive soils of the area. By frequent borings (with an auger) or pits (using shovels) and observations and notations of soil color, horizon thicknesses, texture by feel, pH, soil structure, and other features, the surveyor will determine the soil mapping units and soil boundaries where it changes to another soil mapping unit. In Alaska power ice augers are used on permafrost layers. Backhoes have been used for pits. The boundaries of each mapping unit are drawn on the map and the area labeled with a mapping code symbol from a legend.

Soil map units are identified within each boundary by a symbol consisting of numbers and letters keyed to a legend. A mapping unit in Illinois—for example, HiE3—may consist of a geographic (series) name, such as Hickory (Hi); a textural phase of series, such as loam; a slope class, such as E (e.g., 18–30%); and an erosion class, such as 3 (severely eroded).

All soils within the mapped unit *called* Hickory soil series may not actually *be* Hickory soil series. Natural soil variations develop different series, but the areas may be too small to be practical to separate them out on the scale of map selected. Thus, a map unit (a natural body or area of soil) is not identical to a taxonomic unit (a conceptual body—the Hickory series, for example). Soils distributed in an area differ because of the soil-forming factors (Fig. 19-1).

19:3 Soil Survey Report

Soil surveys have traditionally been made on a county basis, but a survey may be a valley or part of a valley in mountain areas. Typically, each survey has about the same contents. Table 19-1 lists typical soil survey report contents and how the total report is distributed among the various items. About half of the volume of each report is maps with delineations of soils. About 50–80% of the text is details of the soil profiles, including their classification.

Mapping units, delineated on the maps, are named for the most extensive soil series within each unit, but each unit is often a natural mixture of two to five soil series. These mixtures are referred to as **soil associations.** The most likely landscapes to have extensive intermixing of series are small rolling hills or ridges-and-depressions landscapes. In these areas there will usually be one series on the ridges or hilltops (eroded, shallow), perhaps another series along the slope, and at least a third series in the bottoms (deep deposits or wetter than tops).

Each *detailed soil survey map* consists of many folded double-sheet maps that are prepared on an aerial photographic base on which mapping units that are *phases of series* are drawn. One scale used is 1:20,000 (3.16 in./mi, or 5 cm/km). Soil surveys are made at several detailed levels. On a scale of 1:12,000 the smallest area that can be conveniently outlined (delineated) is about 0.6 ha (1.5 a) or a bit smaller. With maps at 1:12,000 to 1:32,680, areas from 0.6 ha to 4 ha would be the smallest areas to delineate. The soil map has sufficient detail to be suitable for making most decisions on land use. Additional uses for urban planning, erosion control, and engineering are common.

19:3.1 The Value of Detailed Soil Survey Reports

Soil surveys are long-time inventories because soils are not easily changed by common soil management. It is true that careless persons may cause or permit rapid destruction of a few soils. Soils shallow over bedrock may be allowed to erode away. Organic soils can be burned. Permanent toxins, such as heavy metals, may be dumped onto soils. However, most soil properties recorded in soil surveys change slowly or not at all. These include (1) the land relief, (2) soil texture and coarse fragments, (3) general organic matter contents, (4) soil lime

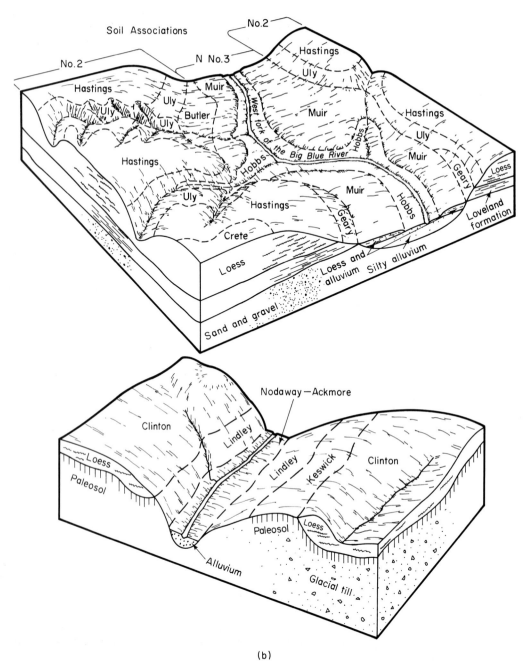

FIGURE 19-1 Soils differ from each other. In a given climatic area, these differences are greatly determined by (1) parent material and (2) extent of soil formation. In (a) the different soils follow boundaries between different parent materials (*Hastings* from loess, *Muir* from mixed loess and alluvium on a river terrace, and *Hobbs* on the bottom floodplain) and in the slopes of the material, thus the erosion occuring (*Uly* is on steeper loess slopes than is *Hastings*). In (b), even the portions of an ancient buried soil (paleosol), now exposed, will be more developed and more weathered (the *Keswick* series) than is the younger glacial till soil (*Lindley*) forming on substratum when the paleosol cover has been eroded off. (*Sources:* U.S. Soil Conservation Service, *Surveys of Fillmore County, Nebraska,* 1986; *Survey of Washington County, Iowa,* 1987.)

Table 19-1 Contents of Soil Survey Reports in the United States and the Distribution of the Total Report Length among the Various Contents

Soil Survey Contents Report[d]	Montana Report[a]	Indiana Report[b]	Florida Report[c]	Kansas
Index to map units, pages	1	1	1	1
Summary of tables, pages	1	1	2	2
Foreword, pages	1	1	1	1
Introduction, pages	5	4	10	3
General soil map units, pages	8	7	13	6
Detailed soil map units, pages	60	39	65	23
Prime farmland, pages	1	1	1	1
Use and management of the soils, pages	13	12	13	13
Soil properties, pages	4	4	7	4
Classification of soils, pages	19	14	40	13
Formation of the soils, pages	2	2	3	3
References, number	6	6	24	12
Glossary, number of items defined	159	177	163	110
Number of tables	14	19	21	18
Number of photographs	4	7	14	11
Drawings, diagrams	2	4	3	4
Total text, pages	193	149	257	119
Total double-page maps	254	64	119	72

Source: Raymond W. Miller, Utah State University.
[a]Roosevelt and Daniels Counties, Montana, 1985.
[b]Wayne County, Indiana, 1987.
[c]Alachua County, Florida, 1985.
[d]Jewell County, Kansas, 1984.

contents, (5) geologic origin, (6) natural fertility, (7) soil depth, (8) the tendency to accumulate soluble salts, (9) soil structure, and (10) the soils' engineering properties. Many other, unchanging, site factors tell much about the area. Some of these are (1) climate (the frost-free periods, average annual rainfall, average annual temperatures, windiness), (2) natural vegetation, (3) adaptable crops, (4) general productivity of the soils, and (5) land use problems, such as poor drainage or shallow depth to bedrock.

Thus, soil surveys are nearly a "permanent to slightly changing record" of a very important world resource. Nevertheless, some areas have been resurveyed after several decades. This is usually done because many kinds of information were not gathered in the early surveys and are now needed. Certainly, management practices, crops grown, and yields might be expected to change somewhat over several decades. As soil science progresses, more-detailed or more-accurate information will be sought for use in planning and management.

19:3.2 Uses of Detailed Soil Survey Reports

Engineers and builders can use the soil and the corresponding interpretations to locate and design scientifically road and building sites for efficient land use. Soil and plant scientists, foresters, watershed scientists, range scientists, engineers, and horticulturists can establish field research plots on the most extensive of the suitable soil series and phases of soil series so that research results can be applied with equal success on all similar soil mapping units throughout the United States and the world.

Soil survey reports provide an enormous amount of detail to help researchers, county agents, and others to locate good or problem areas on which to work. The survey report provides details on drainage, parent materials, soil depths, and other details. Such details would be time consuming for each researcher to search out alone. Unfortunately, a given soil series is not so uniform that the series in all locations react alike.

▇▇▇ 19:4 Benchmark Soils[5,6]

Benchmark soils are those few soils considered to be of *great importance* and of *extensive area,* and/or *occupying a key position in the USDA system of soil taxonomy.* At this time there are 78 nationally designated benchmark soils. Detailed analyses and extensive characterization information are collected for these soils. The hope is that results of studies on these soils can be better extrapolated so that the results for those studies, when done on other soils, can be anticipated. It is true that 78 soils do not sound like just a few soils. However, when compared to the 15,000 series described in the United States alone, it is less than 0.5% of all series.

▇▇▇ 19:5 Prime and Unique Farmlands and Farmlands Critical to the State

Soil surveys emphasize the importance of prime farmland with this statement in each soil survey (emphasis added by this book's authors):

> Prime farmland is one of several kinds of important farmland defined by the U.S. Department of Agriculture. It is of major importance in meeting the nation's short- and long-range needs for food and fiber. The acreage of high-quality farmland is limited, and the U.S. Department of Agriculture recognizes that government at local, state, and federal levels, as well as individuals, must encourage and facilitate the wise use of our nation's prime farmland.
>
> **Prime farmland soils,** as defined by the U.S. Department of Agriculture, are soils that are best suited to producing food, feed, forage, fiber, and oilseed crops. Such soils have properties that are favorable for the economic production of sustained high yields of crops. The soils need only to be treated and managed using acceptable farming methods. The moisture supply, of course, must be adequate, and the growing season has to be sufficiently long. Prime farmland soils produce the highest yields with minimal inputs of energy and economic resources, and farming these soils results in the least damage to the environment.
>
> Prime farmland soils may presently be in use as cropland, pasture, or woodland, or they may be in other uses. They either are used for producing food or fiber or available for these uses. Urban or built-up land and water areas cannot be considered prime farmland.
>
> Soils that have a high water table, are subject to flooding, or are droughty may qualify as prime farmland soils if the limitations or hazards are overcome by drainage, flood control, or irrigation.

Many states have low percentages of their land as prime farmland because of their cold areas with short growing seasons or their mountainous regions. Other states have large percentages of prime lands. As an example, Jewell County, Kansas, is 67% prime farmland, and Wayne County, Indiana, is nearly 74% prime farmland. These are not the highest percentages; some counties in Iowa are over 90% prime farmland.

Some criteria that soils must meet to qualify for prime farmlands, for unique farmlands, and for "lands critical to the state" are given in Detail 19-1.

---

[5] National Soils Handbook 430-VI, USDA—Soil Conservation Service, Washington, D.C., 1983, pp. 604-1 to 604-3.

[6] James A. Silva, ed., "Soil-Based Agrotechnology Transfer," in *Benchmark Soils Project,* Department of Agronomy and Soil Science, Hawaii Institute of Tropical Agriculture and Human Resources, University of Hawaii, Honolulu, 1985, p. 269.

Detail 19-1 Prime and Unique Farmlands, and Lands Critical to the State

The term **prime land** means the "best land." It can be determined by a standard, detailed soil survey. The definition of prime land will differ for each different kind of use; there is unlikely to be general agreement as to what those properties are, even for a given single use. For farmland use it is proposed that **prime farmland** meet all these requirements[a]:

1. Adequate natural rainfall or adequate and good-quality irrigation water to meet normal needs 7 out of 10 years.

2. Mean annual summer temperatures warmer than 15°C (59°F) at a depth of 50 cm (20 in.).

3. Lack of excessive moisture: (a) no flooding more often than once in 2 years; (b) water table below rooting zone.

4. Soil not excessively acid or basic, not saline nor sodic. The pH between 5.5 and 8.6.

5. Permeability of at least 0.39 in./hr (1.0 cm/hr) in the upper 20 in. (51 cm) of soil.

6. Gravel, cobbles, or stones not excessive enough to interfere seriously with power machinery.

7. Soil deep enough to any restricting layer to permit adequate moisture storage and unhampered root growth.

8. Soil not excessively erodible. The universal soil loss equation K-factor times slope percentage is 5 or less.

Although prime lands are very important, **unique lands,** those particularly suited to some single use, are equally valuable. Some such unique lands are those suited to the production of special crops like citrus (no frost, sunny), cranberries (acid, marshy, organic soils), and artichokes (foggy, cool climates). Because each criterion is different, the following suggestions are criteria for all unique lands:

1. Adequate water for the crop.

2. Adequate season and temperature for a good harvest.

3. A location that has a unique combination of soil qualities, temperature, humidity, air drainage, steepness of slope, aspect (direction of slope), or other attributes (such as nearness to market) that favor the growth and/or distribution of the specific crop.

Lands critical to the state do not qualify for either prime or unique lands. However, their extensiveness or locations may make them particularly valuable to the state. These soils are identified and their conservation is encouraged.

---

[a] William M. Johnson, "Classification and Mapping of Prime and Unique Farmlands," in *Perspectives on Prime Lands,* USDA, Washington, D.C., 1975, pp. 189–198. Also, the *Federal Register,* Jan. 31, 1978.

▄▄ 19:6 Land Evaluation for Land-Use Planning[7–9]

Land evaluation for land use planning can be done for urban areas, rural areas, or both. The effects of land use on (1) quality of the environment and (2) environmental sustainability of agricultural production systems are not major issues. The problems of concern include pollution with nitrate, phosphate, and biocides; erosion of land; declining soil fertility; low-input farming; exploitation of timber and range resources; and information for engineering

---

[7] C. A. Van Diepen, et al., "Land Evaluation," *Advances in Soil Science,* **15** (1991), pp. 145–203.

[8] R. W. Dunford, R. D. Roe, F. R. Steiner, W. R. Wagner, and L. E. Wright, "Implementing LESA in Whitman County, Washington," *Journal of Soil and Water Conservation,* **38** (1983), pp. 87–89.

[9] J. Bouma, "Using Soil Survey Data for Quantitative Land Evaluation," *Advances in Soil Science,* **9** (1989), pp. 177–213.

uses. Land use problems may have agronomic, economic, political, and social dimensions. There will be conflicting uses and multipurpose uses.

The USDA Soil Conservation Service uses a system called **Land Evaluation and Site Assessment (LESA)** for purposes of *guiding the conversion of farmland to urban uses.* One major aim is to preserve the best agricultural land for growing plants.

The land evaluation approach involves three procedures: First, land capability classifications make preliminary subdivisions and characterizations of the areas. Second, the current category of land use is documented. (Is the land prime farmland or of lower quality?) Third, each soil is given a rating according to capability for use and is sorted into about ten agricultural groups based on each soil's quality.

The site assessment portion of LESA considers the location of the farmland (Fig. 19-2). How far is it from present urban areas and services? What are the zoning regulations now in effect? What are adjacent land uses? How large is the farmland? Do its physical properties match up with adopted plans of use (Fig. 19-3)? Has the site any unique properties? When sites are given ratings, it is preferred that the sites with the lowest ratings be put to nonagricultural use.

The Canadian Land Evaluation Group developed a more mathematical program under the name *integral land evaluation.*[10] Its prototype model (LEM2) provides information on the flexibility of land used and the feasibility of land use options. Other models exist (Detail 19-2).

19:6.1 Land Capability Classification

The USDA Soil Conservation Service uses a uniform system of two levels of soil management—**land capability classes** and **land capability subclasses**—for all soil mapping units in the United States. A third level, **land capability units,** is also common in some surveys.

Level 1 Soil management **land capability classes** are numbered from one to eight. Classes I through IV can be used for cultivation; Classes V through VIII cannot be cultivated in their present state under normal management (Fig. 19-4).
Class I soils can be used continuously for intensive crop production with minimum attention other than good farming practices.
Class II soils have more limitations than Class I soils for intensive crop production, such as moderately steep slopes (2–5%).

FIGURE 19-2 Land use planning information will readily identify areas of unusual soils. This barren lava bed in Hawaiian "*Aa* rock" was covered only with a thin leaf mat prior to its very limited use for farming. However, in its location (warm humid Hawaii), where land is scarce, and with crops that can adapt, this "soil" is better than it might be at some other site. These papaya trees do well in this "soil." Macadamia nuts and coffee are also produced on these Typic Tropofolists (soils with thin leaf mats). (Courtesy of USDA—Soil Conservation Service; photo by R. P. Yonce.)

[10] B. Smit, M. Brklacich, J. Dumanski, K. B. MacDonald, and M. H. Miller, "Integral Land Evaluation and Its Application to Policy," *Canadian Journal of Soil Science,* **64** (1984), no. 4, pp. 467–479.

FIGURE 19-3 Evaluating land allows potential users to foresee problems for each use. These "dream acres" near Canton, Ohio, have a problem of temporary ponding, for which home owners and urban developers need to plan. (Courtesy USDA— Soil Conservation Service; photo by Marion F. Bureau.)

Class III soils have severe limitations and require more special conservation practices than Class II soils to keep them continuously productive. For examples, they have shallow soil, steep slopes of about 6–10%, or shallow water tables.

Class IV soils have severe limitations and need a greater intensity of conservation practices for cultivated crops than Class III soils. Most of the time these soils should be in "permanent" crops, such as pastures.

Class V soils are not likely to erode but have other limitations, such as boulders or wetness, which are impractical to correct and thus cannot be cultivated. They should be used for pasture, range, woodland, or wildlife habitat.

Class VI soils are suitable for the same uses as Class V soils, but they have a greater need for good management to maintain production because of such limitations as steep slopes or shallow soils.

Class VII soils have very severe limitations and require extreme care to protect the soil, even with low intensity use for grazing, wildlife, or timber.

Class VIII soils have such severe limitations (steep slopes, rock lands, swamps, delicate plant cover) that they can be wisely used *only* for wildlife, recreation, watersheds, and esthetic appreciation.

Level 2   Soil management **land capability subclasses** are soil groups within the eight classes that explain the reasons for the limitations of intensive crop production. Subclasses are designated by lowercase letters that follow the Roman numeral of the soil class. The soil capability subclasses recognized are:

> e   **erosion hazard** is the main limitation (Fig. 19-5).
> w   **wetness**
> s   **shallow, droughty, stony,** or **permafrost**
> c   **climate,** too cold or too dry

Detail 19-2 Land Evaluation Computer Models

Many computer models are in various stages of development and are adaptable to various uses. (Those listed below are for convenience and information of the readers but do not imply any recommendation or endorsement from the authors or publishers, nor does the omission of models indicate any lack of support or recommendation.)

LUPLAN: A model developed by CSIRO in Australia. It is related to the LESA method but is more comprehensive; it is not exclusively oriented to farmland protection. LUPLAN calculates a score for each land use on each mapping unit. The *land use* with the highest score is the preferred use for that soil.

CRIES: This Comprehensive Resource Inventory and Evaluation System (CRIES) was developed at Michigan State University for application in developing countries, focusing on evaluation of alternative land use options. It considers derived public and private benefits.

Sources: J. R. Ive, J. R. Davis, and K. D. Cocks, "LUPLAN: A Computer Package to Support Inventory, Evaluation and Allocation of Land Resources," *Soil Surv. Land Eval,* **5** (1985), no. a, pp. 77–87; G. Schultink, "The CRIES Resource Information System: Computer-Aided Resource Evaluation for Development Planning and Policy Analysis," *Soil Surveys and Land Evaluation,* **7** (1987), pp. 47–62; D. Rossiter, "ALES: A Microcomputer Program to Assist in Land Evaluation," pp. 113–116 in J. Bouma and A. K. Bregt, eds., *Land Qualities in Space and Time*, Proceedings of the ISSS Symposium, Wageningen, Pudoc, Wageningen, The Netherlands, 1989.

FIGURE 19-4 Land capability classes used by the U.S. Department of Agriculture's Soil Conservation Service. (*Source:* USDA—Soil Conservation Service.)

FIGURE 19-5 This spinach field in New Jersey has been mapped as Collington sandy loam, 2–4% slopes, and classified in the subgroup Typic Hapludults and in the fine-loamy, mixed, mesic family. The capability classification is IIe-9, meaning soil Class II with Subclass e—erosion hazard. (*Source:* USDA—Soil Conservation Service.)

Level 3   Soil management **land capability units** are subdivisions of the subclasses; soils in one *unit* are enough alike to be suited to the same crops and pasture plants, to require similar management, and to have similar productivity and other responses to management. Usually Arabic numerals are assigned locally or by states to indicate certain features. Letters may also be used. Fewer soil surveys now publish details of how capability units are separated, but most surveys list soils at either the *subclass* or *unit* level.

A soil may be designated as a IIc land capability subclass. If this soil had a hardpan at a shallow depth in California, the land capability unit would be written IIc-8.

19:6.2 Soil Potential Ratings[11]

The USDA Soil Conservation Service develops **soil potential ratings,** which are classes of soils that indicate the relative quality of a soil for a particular use compared with other soils in a given area. The ratings are developed for planning purposes but are not intended as recommendations for soil use. These ratings supplement land capability classes, woodland suitability groups, range sites, soil limitation ratings, or other interpretations and technical guides.

---

[11] U.S. Department of Agriculture, Soil Conservation Staff, *National Soils Handbook,* USDA—Soil Conservation Service, Washington, D.C., 1983.

Each **soil potential index (SPI)** is defined to equal *performance* or *yield* (*P*) (established locally) minus the sum of (1) costs of *corrective measures* (*CM*) and (2) costs from *continuing limitations* (*CL*),

$$SPI = P - (CM + CL)$$

Use the same units and time scale for all factors. The value of *P* can be chosen to be above the "average" soil of the area, yet below the *P* for the best soil of an area.

Costs of **corrective measures** (CM) can involve building terraces or drainage systems, prorated to an annual cost; increasing septic tank or drain fields; or control barriers to stop erosion (Fig. 19-6).

Continuing limitations (CL) can include three types of costs (1) lower yields, inconvenience, discomfort, probability of periodic failure, and limitations because of field size or location, (2) regular maintenance costs, such as pumping, irrigation, removal of septic tank solids, and pollution control devices or repair, and (3) offsite damages from sediment or other pollution as part of corrective measures.

Large amounts of data are needed to evaluate a soil for various potential uses. Two very simplified examples are given in Details 19-3 and 19-4.

19:6.3 California Storie Index[12,13]

A unique suitability soil-rating system, the **Storie Index,** was developed for use in southern California but is adaptable to many other arid or semiarid regions. The four soil characteristics used in the soil–plant rating scale are

- *Factor A:* profile characteristics that influence the depth and quality of the root zone
- *Factor B:* textural class of the surface soil as it relates to infiltration, permeability, water capacity, and ease of tillage
- *Factor C:* slope as a soil–plant limitation related to irrigation potential

FIGURE 19-6 This builder needed to know the soil potential index (SPI) for this land. These houses in El Dorado Estates in New Mexico are on salty and sodium-affected soil. Many remain vacant, probably partly because septic systems failed in the low permeability soil and landscaping plants and gardens did poorly in the salty soil. (Courtesy USDA—Soil Conservation Service; photo by D. S. Pease.)

[12] Roy H. Bowman, et al., *Soil Survey of the San Diego Area, California, Part II,* USDA—Soil Conservation Service, in cooperation with the University of California Agricultural Experiment Station, U.S. Marine Corps, 1973, p. 92.

[13] K. Koreleski, "Adaptations of the Storie Index for Land Evaluation in Poland," *Soil Surveys and Land Evaluation,* **8** (1988), pp. 23–29.

Detail 19-3 Soil Potential Rating for Dwellings Without Basements

		Cost	Index[a]
P	Index of performance, set locally and arbitrarily for a given area—say, a 1200 sq ft dwelling on a ¼ acre lot (40 yr value)	$85,000	850 (21.2/yr)
CM	Cost of corrective measures, which could include		
	Drainage of footing and slab floor	$600–800	7
	Excavation and grading of 8–15% slope	$1000–1400	12
	Reinforced slab floor, moderate shrink-swell	$1500–2000	17
	Areawide surface drainage needed, per lot	$100–200	2
	Importing topsoil for lawn and garden, as needed	$1000–1400	12
	Prorated for 40 years for value/year		50/40
CL	Costs of continuing limitations, which could include		
	Maintenance of off-lot drainage	$25–50	0.5
	Repair of drainage system for footing and slab	$50–100	1.0
	Replacement of eroded-topsoil loss	$25–50	0.25
	Flooding insurance	$40	0.4

$$\text{SPI}_{dwelling} = 21.2/yr - (1.25 + 2.15) = 17.8$$

[a]Index is 1% of dollar value (= dollars/100)

- *Factor X:* other characteristics that limit plant growth, including poor drainage, excessive erosion, excessive salts, high exchangeable sodium, low inherent fertility, and/or high acidity

Factors A, B, C, and X are individually assigned a rating, with a maximum of 100%. For example, a specific field mapping unit, according to the Storie Index, might be rated as follows:

- *Factor A:* profile is permeable and deep, rating of 100% (= excellent)
- *Factor B:* texture of surface is a loam with high water capacity, rating of 100% (= excellent)
- *Factor C:* slope is nearly level, rating of 100% (= excellent)
- *Factor X:* total soluble salts is 10 dS m^{-1} (classified as a saline soil), rating of 10% (= poor) (see Fig. 19-7)

The composite soil index rating is determined by multiplying the four separate ratings as decimal fractions: $1.0 \times 1.0 \times 1.0 \times 0.1 = 0.1$. Changed back to a percentage, the 0.1 becomes 10% and, according to Table 19-2, has a Storie numerical index rating of "5, poor" for intensive agriculture.

Once the soils of a county have been mapped, laboratory analyses on collected samples are made to characterize the soils more scientifically. Using field observations, field notes, and laboratory data, each soil mapping unit is interpreted as to its relative suitability or degree of limitation for the anticipated major land uses. These interpretations are explained for cropland, pastureland, and rangeland; woodlands and windbreaks; engineering and environment; soil hydrology; and general soil fertility.

			Cost	Index[a]
P	Index of yield, set locally at 130 bu/a, which is a bit above the average for area ($420)		130 bu	4.20
CM	Cost of corrective measures, which could include			
	Equipment costs to convert to no-till, prorated for 10 years use to cost per year, per acre for 300 acres		$60,000	0.20
	Install center-pivot sprinkler, prorated for 10 years use for 130 acres		$80,000	0.06
CL	Cost of continuing limitations, which could include			
	Yearly maintenance of equipment per acre		$30	0.3
	Lower yields than average, only 110 bushels per acre, a loss of 20 bushels = $65		$65	0.65
	Maintenance of runoff sedimentation control ponds required, per acre		$15	0.15

$$SPI_{corn} = 4.20 - (0.19 + 1.10) = 2.91$$

[a] Index is 1% of dollar value (= dollars/100). For corn yield, index could be bushels per acre for all factors or converted to value in dollars, as done here.

19:7 Interpretations for Various Soil Uses

There are many approaches to evaluating land for various uses. Some of them have been briefly discussed earlier. As we consider the specific kinds of uses that can be made of soils, we recognize that they are used for homes, roads, large and small buildings, sources of gravel, septic drain fields, playgrounds, golf courses, ponds, and roads, to name a few. The numerous tables in each soil survey report will list, in part, these data (1) areas and locations of all described soils, (2) land capability classification, (3) soil physical and chemical properties, (4) crop and range productivity, (5) nonagricultural-use potentials, (6) climatic data, (7) the soil's mechanical properties (engineering data) and suitability for building sites, (8) construction materials, (9) information for sanitation facilities, and (10) information for recreational development.

FIGURE 19-7 A saline soil with a Storie Index rating of 5 (poor for intensive agriculture) is suited only to the most salt-tolerant of crops, such as barley, sugar beets, rape, upland cotton, or safflower. (*Source:* USDA—Soil Conservation Service.)

Table 19-2 Composite Storie Index Rating Scale for Intensive Agriculture

Composite Soil Rating (Storie Index)	Soil Grade for Intensive Agriculture	
	Numerical	Relative
80–100%	1	Excellent
60–80%	2	Very good
40–60%	3	Good
20–40%	4	Moderately good
10–20%	5	Poor
Less than 10%	6	Very poor

For each kind of use, soil can be rated as having (1) slight, (2) moderate, or (3) severe limitations because of its texture, depth, slope, position, and/or other factors. The soil factors that are important to consider will vary with the use intended. A soil may have many kinds of limitations. Physical problems include shallow to bedrock, low permeability, and slippage. Chemical problems include strong acidity, high corrosiveness, and high salts. Developing the soil may require high costs to use it (too rocky to dig easily, needs leveling, needs drainage, needs dikes to protect against flooding). In Tables 19-3, 19-4, 19-5, 19-6, 19-7, 19-8, and 19-9 are shown eight of about 34 tables used by soil conservationists to aid in the evaluation of soil areas for various engineering uses. A rating of "severe" indicates that the soil will require major soil reclamation, special design, or intensive maintenance to use for the item listed. "Very severe" indicates that the properties of the soil cause either or both great difficulty in use for the item listed or high expense (Fig. 19-8).

As an illustration of how soils are rated, five contrasting soils, each from a different order, are classified in Table 19-10. Some general features about the five soils used will help in understanding Table 19-10.

1. **Aridisol: Mohave series,** *Typic Haplargid.* Sandy clay loam, 0–5% slopes, deep, well-drained, neutral to alkaline, developed on alluvium of mixed origin. Mean annual precipitation is 380 mm (15 in.); mean annual temperature is 4.4°C (40°F). Growing season ranges from 179 to 200 days, depending on elevation. It is a benchmark soil but not prime farmland. The name is from Mohave County, Arizona (Mohave Desert).

2. **Mollisol: Marshall series,** *Typic Hapludoll.* Silty clay loam, 0–2% slopes, well-drained, deep, slightly acidic, and developed from upland loess. Mean annual precipitation is 400–800 mm (16–32 in.). Average annual temperature is 10.6°C (51°F); the average growing season is 141 days. The name "Marshall" is from Marshall County, Iowa. The soil is a benchmark soil and is prime farmland.

3. **Spodosol: Berryland series,** *Typic Haplaquod.* It is loamy sand, 0–2% slopes, very deep, poorly drained, very strongly acidic, and developed on glacial outwash plains. Mean annual precipitation is 1170 mm (46 in.); mean annual temperature is 9.4°C (49°F). The average growing season is 155 days. It is not considered as a benchmark soil nor as prime farmland because of poor drainage, low inherent fertility, and a strongly cemented high-iron *B* horizon.

4. **Ultisol: Ruston series,** *Typic Paleudult.* A fine sandy loam, 2–5% slopes, deep and well-drained, and acidic. It developed from sandy marine sediments. Mean annual precipitation is 1420 mm (55 in.); mean annual temperature is 18°C (65°F). The mean growing season is 221 days. "Ruston" is from the town of Ruston, Louisiana. Ruston soil is prime farmland but is not a benchmark soil.

Table 19-3 Evaluating the Use of Soils for Septic Tank Absorption Fields and Sanitary Landfill Area

Septic Tank Absorption Fields

| | Limits | | | Restrictive |
Property	Slight	Moderate	Severe	Feature
Surface texture	—	—	Ice	Permafrost
Total subsidence (in.)	—	—	>24	Subsides
Flooding	None	Rare	Common	Flooding
Depth to bedrock (in.)	>72	40–72	<40	Depth to rock
Depth to cemented pan (in.)	>72	40–72	<40	Cemented pan
Depth to high water table (ft)	—	—	+	Ponding
	>6	4–6	<4	Wetness
Permeability (in./hr)				
(24–60 in.)	2–6	0.6–2	<0.6	Percs slowly
(24–40 in.)	—	—	>6.0	Poor filter
Slope (%)	<8	8–15	>15	Slope
Fraction >3 in.[a] (wt %)	<25	25–50	>50	Large stones
Downslope movement	—	—	II/	Slippage
Formation of pits	—	—	III/	Pitting

Sanitary Landfill (Area)

| | Limits | | | Restrictive |
Property	Slight	Moderate	Severe	Feature
Surface texture	—	—	Ice	Permafrost
Flooding	None	Rare	Common	Flooding
Depth to bedrock[b] (in.)	>60	40–60	<40	Depth to rock
Depth to cemented[b] pan (in.)	>60	40–60	<40	Cemented pan
Permeability[b] (in/hr) (20–40 in.)	—	—	>2.0	Seepage
Depth to high water table (ft)	—	—	+	Ponding
Apparent	>5	3.5–5	<3.5	Wetness
Perched	>3	1.5–3	<1.5	Wetness
Slope (%)	<8	8–15	<15	Slope
Downslope movement	—	—	Occurs	Slippage
Formation of pits	—	—	Occurs	Pitting
Differential settling	—	—	Occurs	Unstable fill

Source: National Soils Handbook, USDA—Soil Conservation Service, 1983, pp. 603–61, 603–68.
[a] Weighted average to 40 in.
[b] Disregard (1) in all Aridisols except Salorthids and Aquic subgroups, (2) all Aridic subgroups, and (3) all Torri great groups of Entisols except Aquic subgroups.

5. **Vertisol: Houston Black series,** *Udic Pellustert.* It is clay, 0–1% slopes, deep, moderately well-drained, alkaline, and developed from marine chalk (impure calcite). Mean annual precipitation is 810 mm (32 in.); mean annual temperature is 19°C (66°F). The average growing season is 202 days. The name "Houston Black" came from a county name in Texas and its deep black color. It is prime farmland and is a benchmark soil.

In Table 19-10, notice particularly these items: (1) differences exist in crop potential in different soils, (2) the soils that are good croplands often are poor (low stength or high swell-shrink) as construction materials (see "local roads"), and (3) poorly permeable soils may be good for water reservoirs (see "sewage lagoons/ponds") but poor for septic absorption fields, irrigation, and playgrounds.

Table 19-4 Evaluating the Use of Soils for Local Roads and Streets

Property	Limits			Restrictive Feature
	Slight	Moderate	Severe	
Surface texture	—	—	Ice	Permafrost
Total subsidence (in.)	—	—	>12	Subsides
Depth to bedrock (in.)	>40	20–40	<20	Depth to rock
Depth to cemented pan (in.)	>40	20–40	<20	Cemented pan
Shrink-swell[a]	Low	Moderate	High very high	Shrink-swell
Depth to high water table (ft.)	>2.5	1.0–2.5	<1.0	Wetness
Slope percentage	<8	8–15	>15	Slope
Flooding	None	Rare	Common	Flooding
Potential frost action	Low	Moderate	High	Frost action
Fraction >3 in. (wt %)	<25	25–50	>50	Large stones
Downslope movement	—	—	Occurs	Slippage
Formation of pits	—	—	Occurs	Pitting
Differential settling	—	—	Occurs	Unstable fill

Source: National Soils Handbook, USDA—Soil Conservation Service, 1983, pp. 603–681.
[a] Thickest layer 10–40 in.

Table 19-5 Evaluating the Use of Soils for Dwellings (homes) with Basements

Property	Limits			Restrictive Feature
	Slight	Moderate	Severe	
USDA texture	—	—	Ice	Permafrost
Total subsidence (in.)	—	—	>12	Subsides
Flooding	None	—	Rare, common	Flooding
Depth to high water table (ft)	—	—	0.5	Ponding
	>6	2.5–6	<2.5	Wetness
Depth to bedrock (in.)				Depth to rock
hard	>60	40–60	<40	
soft	>40	20–40	<20	
Depth to cemented pan (in.)				Cemented pan
thick	>60	40–60	<40	
thin	>40	20–40	<20	
Slope (%)	<8	8–15	>15	Slope
Shrink-swell[a]	Low	Moderate	High, very high	Shrink-swell
Unified (bottom level)	—	—	OL, OH, PT	Low strength
Fraction >3 in.[b] (wt %)	<25	25–50	>50	Large stones
Downslope movement	—	—	Occurs	Slippage
Formation of pits	—	—	Occurs	Pitting
Differential settling	—	—	Occurs	Unstable fill

Source: National Soils Handbook, USDA—Soil Conservation Service, 1983, pp. 603-681.

Table 19-6 Evaluating the Use of Soils for Lawns, Landscaping, and Golf Fairways

Property	Limits			Restrictive Feature
	Slight	Moderate	Severe	
USDA texture	—	—	Ice	Permafrost
Flooding	None	Occasionally	Frequently	Flooding
Slope percentage	<8	8–15	>15	Slope
Depth to bedrock (in.)	>40	20–40	<20	Depth to rock
Depth to cemented pan (in.)	>40	20–40	<20	Cemented pan
Surface texture[a]	—	—	SiC, C, SC	Too clayey
Surface texture	—	—	Muck, peat	Excess humus
Surface texture	—	LCoS, S	CoS	Too sandy
Salinity, dS m^{-1}	<4	4–8	>8	Excess salt
Sodium adsorption ratio	—	—	>12	Excess sodium
Soil reaction (pH)	—	—	>3.6	Too acidic
Coarse fragments[b] (wt %)	<25	25–50	>50	Small stones
Fraction >3 in.[b] (wt %)	<5	5–30	>30	Large stones
Depth to high water table (ft)	>2	1–2	<1	Wetness

Source: National Soils Handbook, USDA—Soil Conservation Service, 1983, pp. 603–681.
[a]Thickest layer 0–40 in.; Co = coarse, C = clay, L = loam(y), S = sand(y), Si = silt(y).
[b]Fraction all passes No. 10 sieve.

Table 19-7 Evaluating the Use of Soils for Topsoil

Property	Limits			Restrictive Feature
	Slight	Moderate	Severe	
Surface texture	—	—	Ice	Permafrost
Depth to bedrock (in.)	>40	20–40	<20	Depth to rock
Depth to cemented pan (in.)	>40	20–40	<20	Cemented pan
Depth to bulk density >1.8 Mg/cm^3	>40	20–40	<20	Too compacted
Surface texture[a]	—	LCOS, LS, LFS, LvFS	CoS, S FS, vFS	Too sandy
Surface texture[a]	—	SCL, CL, SiCL	SiC, C, SC	Too clayey
Fraction >3 in.[b] Surface (wt. %)	<5	5–30	>30	Large stones
Coarse fragments[b] Surface (wt. %)	<25	25–50	>50	Small stones
Salinity, dS m^{-1}	<4	4–8	>8	Excess salt
Layer thickness (in.)	>40	20–40	<20	Twin layer
Depth to high water table (ft)	—	—	<1	Wetness
Sodium adsorption ratio (0–40 in.)	—	—	>12	Excess sodium
Soil reaction (pH)	—	—	<3.6	Too acidic
Slope percentage	<8	8–15	>15	Slope

Source: National Soils Handbook, USDA—Soil Conservation Service, 1983, pp. 603–689.
[a]Thickest layer 0–40 in.; Co = coarse, L = loam(y), S = sand(y), Si = silt(y), C = clay, F = fine, vF = very fine.
[b]Sum > 3 in., all passing No. 10 sieve.

Table 19-8 Evaluating the Soil for Drainage

Property	Limits	Restrictive Feature
Surface texture	Ice	Permafrost
Depth to high water table (ft)	>3	Deep to water
Permeability (in./hr)	<0.2	Percs slowly
Depth to bedrock (in.) or cemented pan	<40	Depth to rock (cemented pan)
Flooding	Common	Flooding
Total subsidence	Any	Subsides
Fraction >3 in. (wt %)	>25	Large stones
Potential frost action	High	Frost action
Slope percentage	>3	Slope
Texture, thickest layer 10–60 in.[a]	CoS, S, FS, vFS, LCoS, LS, LFS, LvFS, SG, G	Cutbanks cave
Salinity (dS m^{-1})	>8	Excess salt
Sodium adsorption ratio	>12	Excess sodium
Soil reaction (pH)	<3.6	Too acidic
Downslope movement	Occurs	Slippage
Complex landscape	Occurs	Complex slopes
Availability of outlets	Difficult to find	Poor outlets

Source: National Soils Handbook, USDA—Soil Conservation Service, 1983, 603:p.101.
[a] Co = coarse, S = sand(y), F = fine, vF = very fine, L = loam(y), G = gravel.

Table 19-9 Evaluating the Use of Soils for Irrigation

Property	Limits	Restrictive Feature
Surface texture	Ice	Permafrost
Slope percentage	>3	Slope
Fraction >3 in. (wt. %)	>25	Large stones
Depth to high water table (ft)	<3	Wetness
Available water capacity (in./in.)	<0.10	Droughty
Surface texture[a]	CoS, S, FS, vFS, LCoS, LS, LFS, LvFS	Fast intake
Surface texture[a]	SiC, C, SC	Slow intake
Wind erodibility group	1, 2, 3	Soil blowing
Permeability (in./hr)	<0.2	Percs slowly
Depth to bedrock (in.) or cemented pan	<40	Depth to rock (cemented pan)
Bulk density (Mg/m³)	>1,7	Rooting depth
Erosion factor (surface K)	>0.35	Erodes easily
Flooding	Common	Flooding
Sodium adsorption ratio	>12	Excess sodium
Salinity (dS m^{-1})	>4	Excess salt
Soil reaction (pH)	<3.6	Too acidic
Complex landscape	Exists	Complex slope
Formation of pits	Occurs	Pitting

Source: National Soils Handbook, USDA—Soil Conservation Service, 1983, 603:p.103.
[a]Co = coarse, S = sand(y), F = fine, vF = very fine, L = loam(y), Si = silt(y), C = clay.

FIGURE 19-8 Consider what you need to know about this Berks County, Pennsylvania, landscape, if it were to have constructed on it (1) additional roads, (2) some houses with septic tanks, (3) a small park, and (4) a school and playground. At the least, you would need good information on slope, internal and external drainage, depth to water table and hardpans, ability of soil to support small buildings and a roadbed, and what is needed to control erosion. A soil survey and land evaluation would provide that information. (Courtesy USDA—Soil Conservation Service; photo by Ledbetter.)

Any single soil will likely be good for some uses but poor for others. A permeable, well-drained soil may be good for crops but poor for landfills, lagoons, and ponds. A clayey soil without good, deep drainage will be good for lagoons and ponds but may be difficult to use for crops, unstable for roads and small buildings, and poor topsoil. Soil evaluations (land assessments) not only indicate the best uses for each soil, but also indicate the degree to which the good and poor characteristics of each soil will affect its use for a given purpose.

▪▬▬▬ 19:8 Controls in Land-Use Planning[14]

Land-use planning utilizes information, foresight, and a wise compromise of the possibilities to designate the best use of land for its manifold purposes, now and in the future. Difficulties arise because private and public rights can clash and there are not enough appropriate soils in the right locations to satisfy all demands. Unfortunately, *master land use plans* usually are formulated to benefit urban populations. The concept of reserving prime land areas for agricultural, wildlife, or recreational use because they are particularly valuable for those purposes is not as common as it needs to be. Only recently have many of these kinds of land zoning decisions been made. A soil survey is the *only* means of establishing a scientific basis for planning the most appropriate uses of every acre or hectare of land.

---

[14] Gerald W. Olson and Arthur S. Lieberman, eds., *Proceedings of the International Symposium on Geographic Information Systems for Conservation and Development Planning*, Cornell University, Ithaca, N.Y., Apr. 4–6, 1984.

Table 19-10 Interpretive Uses of Five Soil Series in Five Soil Orders under Good Management

Proposed Use	Soil Order and Soil Series (State)				
	Aridisols, Mohave (New Mexico)	Mollisols, Marshall (Iowa)	Spodosols, Berryland (Massachusetts)	Ultisols, Ruston (Mississippi)	Vertisols, Houston Black (Texas)
Crop production potential					
Sorghum, corn (bu/a)	Not adapted	109 (corn)	Not adapted	65 (corn)	90 (sorghum)
Oats, wheat (bu/a)		62 (oats)		30 (wheat)	45 (wheat) 90 (oats)
Soybeans (bu/a)	Not adapted	41	Not adapted	25	Not adapted
Improved Pasture/Hay (t/a)	Not adapted	7.6	Not adapted	12.0	8.0
Engineering uses					
Dwellings with basements	Fair (too clayey)	Fair (shrink-swell clay)	Very poor (wetness)	Good	Very poor (shrink-swell clay)
Local roads	Poor (low strength)	Poor (low strength)	Poor (wetness)	Fair (low strength)	Very poor (poor bearing)
Septic absorption fields	Severe (slow percolation)	Slight	Very poor (wetness)	Moderate (slow percolation)	Very poor (slow percolation)
Landfills	Fair	Fair	Very poor (wetness)	Moderate (too clayey)	Poor
Sewage lagoons/ponds	Moderate (slopes)	Moderate (seepage)	Poor	Moderate (slow perc)	Fair
Topsoil	Fair (too clayey)	Fair (slopes)	Poor (sandy, wet)	Fair (small stones)	Good
Irrigation	Moderate (slopes)	Favorable	Poor	Fair (sloping)	Poor (too clayey) Poor (slow percolation)
Playgrounds	Moderate (slopes)	Favorable	Poor (too wet)	Fair (slopes)	Moderate (too clayey)
Habitat potential for					
Openland wildlife	Very poor	Good	Poor	Good	Fair
Woodland wildlife	Not adapted	Good	Poor	Good	Poor (few trees)
Wetland wildlife	Not adapted	Very poor	Good	Very poor	Poor
Corrosion risk for					
Uncoated steel	High (too salty)	Moderate	High (acidic)	Moderate	High
Concrete	Low hazard	Moderate	High (acidic)	Moderate	Low hazard

Source: Ellis Knox, National Leader for Soil Research, USDA—Soil Conservation Service, Washington, D.C.

19:8.1 Conventional Controls on Land Use

Even though there is no federal land use planning law, prior to 1977 state laws were passed by Vermont, New York, New Jersey, Florida, and Oregon.[15] Other states are considering similar laws. Conventional controls on land use include zoning laws, differential taxation, easements and contracts, and public purchase.

Zoning laws passed by local, regional, or national government agencies serve to guide the orderly development and use of land. Laws may soon be utilized to protect valuable, irreplaceable lands from other uses. Hawaii, for example, has zoned the state by major land use classes: agricultural, urban, conservation, and residential, with soil survey information as the scientific base.

Zoning to enforce a county land use plan was established in Black Hawk County, Iowa.[16] Worried about the rapid loss of Class I agricultural land to urban developments, officials in the county used a county soil survey as the scientific physical basis for passing a specific land use zoning ordinance to restrict land capability Class I land for agricultural purposes.

The prime objective of **differential taxation** is the preservation of land for specified uses, such as agricultural, timber, recreational, urban, and suburban. Taxes are assessed on "current" values (using soil surveys as the scientific physical basis) rather than on the principle of "highest and best," "potential future use," or "politics." Differential-taxation laws usually permit productive timber lands, as interpreted partly from soil survey reports, to be taxed primarily at time of harvest rather than on an annual basis, as is the practice for most other land uses. Agricultural land is taxed on the basis of its use for farming and ranching and not at the inflationary rate of its potential as a subdivided residential or commercial area. These "differential" taxes may need to be paid later when a change in use occurs.

Easements and **contracts** are used by some governmental bodies to implement land use planning by mandatory action. A municipality, county, state, or national government may demand an easement or a contract to purchase (valid for a period of perhaps 10 years) that specifies the future use of the land for public purposes, such as a park. (Soil surveys are used to interpret potential recreational areas.) An initial payment is made to the owner, and tax reductions are offered. Such easements and contracts can be broken by the public agency but not by the owners.

Public purchase is the most direct method of implementing a land use plan. Most federal and state departments have funds set aside for the purchase of lands for national or state forests, parks, public hunting areas, flood control dams and backwater areas, and for other land use purposes to promote the general welfare. The sale of private land to a public body may be voluntary or involuntary; the right of eminent domain prevails. **Eminent domain** holds that the public good (for a satisfactory and nonarbitrary reason) prevails over the private good and so condemnation proceedings result in fair-market price compensation to private landowners for forced sales of their property.

19:8.2 Newer Trends in Controlling Land Uses

Will there be enough first-class agricultural soils to sustain the growing world population? More than 400,000 hectares (1 million acres) of prime agricultural soils and three times this hectarage of all soils used for agriculture are being diverted to other uses in the United States each year. Concern over the magnitude of these losses has resulted in laws being passed by 48 of the 50 states to retain prime agricultural soils for agricultural use.

---

[15] John B. Mitchell, "Land Use Policies in Selected States," Ohio State University Bulletin 623, 1977.

[16] Louise A. Lex and Louis Lex, Jr., "Land Use Control: The Black Hawk County Experience," *Soil Conservation*, June 1975, pp. 16–18.

Nonurban land use planning has had limited effect in reserving agricultural lands for agricultural use. Most plans *encouraged* compliance but did not provide any enforcement capability. The basis of zoning laws dating from 1926 required that regulations imposed on the use of private property must be for purposes of *promoting health, safety, morals, or the general welfare* of the community (people). With the passage of the Clean Air Acts and the Clean Water Acts in the 1970s, additional control of land was possible if that control was essential to hinder or reduce polluting of water or air. These laws and increasing public concern about the vanishing agricultural lands have resulted in *preferential tax assessment, buying development rights, transferable development rights,* and altering *enforced planning* as methods to achieve the land use decided by legal authorities for the common good.

Preferential tax assessments are made to fit the current use of land instead of potential uses that usually are assessed at higher rates. The additional accumulated taxes it might have had if valued as its potential selling price are collected when the land use is changed (e.g., subdidivided for a housing development). A lower tax while in agricultural production allows the owner to continue to farm the land without the additional burden of high land taxes that might make an agricultural use unprofitable or that might make subdividing too attractive.[17,18]

Buying development rights is a mechanism by which governments control land use. The owner sells development rights to the government but retains ownership of the land in its present use and can sell it for that continued use. The private landowner is recompensed for lost potential profit and the public retains land in agricultural or other use. Suffolk County in Long Island, New York,[19] has begun long-range plans to buy development rights on 12,000 ha (at $13,000/ha or over $5000/a prior to 1978) of farmland to keep a rural area for city residents to visit. This technique is costly to governments.

Transferable development credits (TDCs) is a technique of allotting a specified amount of development per area. If a developer wanted a higher-than-normal-density development, he or she could purchase development credits from other landowners who were not interested in developing their own properties but who could not later change the use once they sold development credits. This process allows more efficient urban planning and reimburses farmers for their lost development potential.[20]

Enforced planning grew out of a concern about present land use. Increasing numbers of states require local governments *to establish and enforce land use plans.* Such plans include identifying the most valuable agricultural lands and housing-growth areas. The object is to plan for growth by saving those lands best suited for certain uses and reserve them for the "highest and best" purposes and to designate them as prime land.

Although these laws embody the many variations of the methods of preservation already mentioned, the most common technique has been differential taxation.

Honest differences of view and honest debates are not disunity. They are the vital process of policymaking among free men.

—Herbert Hoover

[17] Richard Barrows and Douglas Yanggen, "The Wisconsin Farmland Preservation Program," *Journal of Soil and Water Conservation*, **33** (1978), no. 5, pp. 209–212.
[18] K. R. Olson and G. W. Olson, "Use of Agronomic Data and Enterprise Budgets in Land Assessment Evaluations," *Journal of Soil and Water Conservation*, **40** (1985), no. 5, pp. 455–458.
[19] John V. N. Klein, "Preserving Farmland on Long Island," *Environmental Comment*, no. 5, Jan. 1978, pp. 11–13.
[20] Peter J. Pizor, "New Jersey's TDC Experience," *Environmental Comment*, Apr. 1978, p. 11.

Man must go back to nature for information.

—Thomas Paine

Questions

1. What organization is responsible for making soil surveys of the United States?
2. (a) What equipment is used by the soil surveyor? (b) What profile features will the surveyor describe and measure in the survey?
3. What kinds of information are tabulated in survey reports?
4. What three categories constitute the bulk of each soil survey report?
5. How "permanent" is information in a soil survey report? Discuss.
6. What are the major uses made of soil survey reports?
7. What is the LESA approach to land use planning?
8. (a) What is a *soil potential index* (SPI)? (b) Is there only one SPI for each soil? Explain.
9. Describe what each of the factors is in this equation:
$$SPI = P - (CM + CL)$$
10. Could any building contractors or city and county planners find useful information in a soil survey report? Explain.
11. Define (a) a *benchmark soil,* and (b) *land critical to the state.*
12. What are some of the important soil properties that may result in slight, moderate, or severe ratings for (a) drainage of soils, (b) local roads and streets, and (c) lawns, landscaping, and golf fairways?
13. In general terms, list several of the properties that soils must have to qualify as prime farmland. List information for (a) depth, (b) pH, (c) erosion hazard, (d) flooding permitted, (e) temperature, and (f) available water.
14. How does unique farmland differ from prime farmland?
15. Define (a) *land capability class III land,* (b) *class V land,* (c) *subclass IIs land,* and (d) *subclass IVw land.*
16. Do the subclass notations—e, s, c, w—always mean the same about a soil? Explain, particularly for "s."
17. List five nonagricultural uses made of soil survey data.
18. What are three general requirements of a soil that are important in its use for (a) septic system drainage fields, (b) topsoil, and (c) small buildings? Explain the importance of each criterion (requirement) listed.
19. What might be a *severe* restriction in a soil that was considered as a source of topsoil?
20. Controlling land use is of increasing interest and concern. What are the mechanisms of eminent domain and preferential tax assessments used for control?
21. How does prime farmland relate to land use control and planning?

Glossary*

A

A horizon. *See* Soil horizon.

AB horizon. *See* Soil horizon.

Abiotic. The nonliving factors in the biosphere: rainfall, minerals, temperature, wind, and others.

Absorption, active. Movement of ions and water into the plant root as a result of metabolic processes by the root, frequently *against* an activity gradient. *See also* Adsorption.

Absorption, passive. Movement of water into roots resulting from "pulling" forces on the water column in the plant as water is lost to the atmosphere through the leaves by transpiration (a kind of *wick* action).

AC horizon. *See* Soil horizon.

Accelerated erosion. *See* Erosion.

Acidic cations. Hydrogen ions or cations in water that undergo hydrolysis to form an acidic solution, as do Al^{3+} and Fe^{3+}.

Acidic soil. Soil with a pH lower than 7.0.

Acidic sulfate soils. Soils strongly acidic (pH 3.5) or potentially so and with relatively large amounts of sulfides (or jarosite) when anaerobic or sulfates when aerobic.

Acre. An area of 43,560 ft² or 0.405 ha.

Sources:
(1) *Glossary of Soil Science Terms,* Soil Science Society of America, 1979, 1987. Selections used with permission.
(2) *Resource Conservation Glossary,* Soil Conservation Society of America, 1976. The complete glossary is available from 7515 Northeast Ankeny Road, Ankeny, IA 50021. These selections have been used with permission.
(3) *Soil Series of the United States, Puerto Rico, and the Virgin Islands: Their Taxonomic Classification.* USDA—Soil Conservation Service, 1972.
(4) Soil Survey Staff, Soil Conservation Service, *Soil Taxonomy: A Basic System of Soil Classification for Making and Interpreting Soil Surveys,* Agriculture Handbook 436, USDA, Washington, D.C., 1975.
(5) Robert L. Bates and Julia A. Jackson, eds., *Glossary of Geology,* 2nd ed., 1980. American Geologic Institute, Falls Church, Va.
(6) Soil Survey Staff, "Soil Families and Their Included Series," and "Classification of Soil Series," USDA—Soil Conservation Service, Washington, D.C., 1988.

Actinomycetes. A nontaxonomic term applied to a group of filamentous bacteria with characteristics intermediate between simple bacteria and the true fungi.

Activation energy. The energy required before the reaction of concern will proceed. Often, heating will supply the needed energy. Enzymes or other catalysts lower the activation energy of a reaction.

Adhesion. Molecular attraction that holds the surfaces of two dissimilar substances in contact, such as water and rock particles. *See also* Cohesion; H bond.

Adsorption. The bonding, usually temporary, of ions or compounds to the surfaces of a solid, such as a calcium ion held on the surface of a clay crystal.

Aeration, soil. The process by which air in the soil is replaced by air from the atmosphere. Poorly aerated soils usually contain a much higher percentage of carbon dioxide and a correspondingly lower percentage of oxygen than the atmosphere above the soil.

Aerobic. 1. Having molecular oxygen as a part of the environment. 2. Growing only in the presence of molecular oxygen, as aerobic organisms. 3. Occurring only in the presence of molecular oxygen (said of certain chemical or biochemical processes, such as aerobic decomposition).

Aflatoxins. Potent carcinogens in nature that are antibodies produced by fungal growth on many grains and nuts.

Aggregation, soil. The cementing or binding together of several soil particles into a secondary unit, aggregate, or granule.

Agronomy. A specialization of agriculture concerned with field crop production and soil management. The scientific management of land.

Albic horizon. A strongly leached, light-colored mineral horizon, designated by letter *E. See* Chapter 2.

Alfisols. *See* Soil classification: Order.

Alkali. (obsolete) *See* Saline-sodic soil; Sodic soil.

Alkaline soil. Any soil having a pH greater than 7.0. A *basic* soil. *See also* Reaction, soil.

Allelopathy. The action of some substance produced by or in one plant species that reduces the growth of another plant species. Mostly these are toxic organic materials produced either by the plant or during decomposition of plant residues.

Allophane. Amorphous (noncrystalline) clay-sized aluminosilicate.

Alluvial fan. Fan-shaped alluvium deposited at the mouth of ravines or canyons as the slope lessens and water-carried debris settles out of the slowed water.

Alluvium. Eroded soil sediments deposited from flowing water.

Aluminosilicate. Minerals dominantly of aluminum (Al), silicon (Si), and oxygen (O).

Amendment, soil. Any substance (gypsum, lime, manures, sewage sludge, sawdust, composts, etc.) added to the soil to improve plant growth.

Amensalism. Production by one organism of a substance that is inhibitory to other organisms. Similar in use to the term *allelopathy.*

Ammonia volatilization. Loss of nitrogen to the atmosphere as gaseous ammonia (NH_3), increasing with pH above 8–8.5.

Ammonification. The biochemical process whereby ammoniacal nitrogen is released from nitrogen-containing organic compounds.

Ammonium fixation. Adsorption of ammonium ions (NH_4^+) by the soil mineral fraction into interlayer positions that cannot be replaced by a neutral potassium salt solution (e.g., 1 N KCl).

Amorphous materials. Lacking adequate crystal orientation to diffract x-rays. Without regular crystalline form in molecular sizes.

Anaerobic. 1. The absence of molecular oxygen. 2. Growing in the absence of molecular oxygen (such as anaerobic bacteria). 3. Occurring in the absence of molecular oxygen (as a biochemical process).

Anaerobic respiration. Metabolic transfer of electrons to ions other than oxygen, predominantly to nitrate, manganese oxide, ferric ion, sulfate, and carbon forms.

Andisol. Soils with weakly developed horizons; formed in volcanic ejecta parent materials.

Animalia. One of the five kingdoms categorizing all living organisms. *See* Chapter 7.

Anion. (an′-eye-on) Negatively charged ion; ion that during electrolysis is attracted to the anode (positively charged electrode).

Anion exchange capacity. The sum total of exchangeable anions that a soil can adsorb.

Anthropogenic, anthropic. (soils) Changes in soils caused by people, such as plowing, fertilizing, and construction.

Aquic conditions. A soil water regime, mostly too wet (reducing conditions, waterlogged) for parts of the year. *See* Chapter 3.

Aquifer. A geologic formation that transmits water underground, usually sands, gravel, and fractured, porous, cavernous, and vesicular rock.

Arable land. Land suitable for the production of cultivated crops.

Argillan. *See* Clay film.

Argillic horizon. A diagnostic horizon of clay accumulation often designated as *Bt*. (*See* Chapter 2).

Arid. A term applied to regions or climates that lack sufficient moisture for crop production without irrigation. The upper annual limit for cool regions is 250 mm (10 in.) or less and for tropical regions as much as 380–510 mm (15–20 in.). *See also* Semiarid.

Aridic. A soil water regime with long dry periods. *See* Chapter 3.

Aridisols. *See* Soil classification: Order.

Aspect. (forestry) The direction that a slope faces.

Autotrophic. Capable of utilizing carbon dioxide and/or carbonates as a sole or major source of carbon and of obtaining energy for carbon reduction and biosynthetic processes from radiant energy (photoautotroph) or oxidation of inorganic substances (chemoautotroph).

Available nutrient. That portion of any nutrient in the soil that can be absorbed readily by growing plants. Same meaning as *plant-available*.

Available water. The portion of water in a soil that can be readily absorbed by plant roots. Considered by most workers to be that water held in the soil with a water potential of –33 tPa up to approximately –1500 kPa. Same as plant-available. *See also* Field (water) capacity; Permanent wilting point; Soil water potential.

B

BA horizon. *See* Soil horizon.

Bacteroid. "Clump" of bacteria cells, particularly the swollen, vacuolated mass of bacteria in *nodules* of legumes.

Badland. A highly eroded area with little vegetation, often with narrow ravines and sharp ridges. Common in arid regions with deep alluvium deposits.

Bajada. (ba-ha′da) Coalescing adjacent alluvial fans; a bajada is almost a continuous slope of alluvium along the side of a mountain.

Banding. Applying fertilizer or other amendment into the soil (7–15 cm, or 2.7–6 in., deep) in a thin narrow strip (band), as beside or beneath a planted row of seeds or plants.

Bar. (obsolete) A unit of pressure equal to 1 million dynes/cm^2 or 100 kilopascals. *See* Pascal.

Base saturation percentage. (basic cation saturation percentage) The percentage of the adsorption complex of a soil saturated with basic cations (cations other than hydrogen and aluminum).

B horizon. *See* Soil horizon.

BC horizon. *See* Soil horizon.

BE horizon. *See* Soil horizon.

B/E horizon. *See* Soil horizon.

Bedder. A tillage tool to form ridges or narrow beds. *See* Chapter 14.

Bentonite. A highly plastic clay consisting of the minerals montmorillonite and beidellite (smectites), which swell extensively when wet. Also called *volcanic clay, soap clay,* and *amargosite.*

Best management practices (BMP). The management practices determined to permit the least pollution or other undesirable effect. *See* Chapters 14 and 18.

Bioaccumulation. Accumulation of a substance in an organism over time because excretion is slower than intake rates.

Biochemical oxygen demand (BOD or B.O.D.). The amount of oxygen required for bacteria to decompose the organic matter in the solution. *See also* Chemical oxygen demand.

Biological magnification. (biomagnification) The accumulation of increasing amounts of a substance in organisms as one progresses up the food chain (e.g., low in plankton, higher in small fish, large fish, and highest in fish-eating birds).

Biomass. Total amount of living organisms and their residues in a volume or mass of the environment.

Biosphere. The environment in which living organisms live: the earth's crust, vegetation, and atmosphere near the earth.

BMP. *See* Best management practices.

Brackish. Slightly salty. Water with a content of 1.5–3% salts; seawater has over 3% salts.

Buffering. 1. The capacity of the soil solids and liquids to resist appreciable change in pH of the soil solution. 2. The ability to maintain the approximate concentration desired of any ion in the soil solution.

Bulk density, soil. The mass (weight) of dry soil per unit bulk volume.

C

C horizon. *See* Soil horizon.

Cablegation. A surface-flow furrow irrigation system that uses a long cable to control opening and closing of irrigation pipe gates.

Calcarious soil. Soil with sufficient carbonates (mostly calcium) to effervesce visibly with cold 0.1 *N* hydrochloric acid.

Calcic horizon. A diagnostic mineral horizon of carbonate accumulation. Indicated by letter *k*. *See* Chapter 2.

Caliche. A pedologic layer near the surface, cemented by secondary carbonates of calcium or magnesium precipitated from the soil solution. *See also* Hardpan; Duripan and Petrocalcic diagnostic horizons in Chapter 2.

Cambic horizon. A weakly developed diagnostic subsoil horizon. Indicated by the letter *w See* Chapter 2.

Capillary water. Plant-available water held in the "capillary" or small pores of a soil, usually with attraction forces exceeding the pull of a 60-cm (2-ft) column of water.

Carbon:nitrogen ratio. The ratio of the weight of organic carbon to the weight of total nitrogen in the soil or in organic material.

Carbonation. A chemical weathering process combining the action of acidity and formation of soluble bicarbonates from carbon dioxide dissolved in water to form carbonic acid.

Carboxyl group. The "acid" group of soil organic matter (—COOH), which is the most active group in soil humus in adsorbing cations.

"Cat clays." Strongly acidic clays resulting from aeration of anaerobic soils containing many sulfide minerals. The sulfides are oxidized to sulfuric acid in aerated soil. In the Netherlands called *katteklei* (cat's coat).

Catena. A sequence of different soils formed under similar soil-forming factors except for the effect of their relief positions which altered erosion and drainage of each different soil.

Cation (cat′-eye-on) exchange. Replacement by a cation in solution for an adsorbed cation on negatively charged sites of a solid.

Cation exchange capacity (CEC). The sum total of exchangeable cations that a soil can adsorb, expressed in centimoles$_c$ per kg of soil or colloid.

CB horizon. *See* Soil horizon.

Cemented (indurated). Having a hard, brittle consistency, even when wet, because the particles are cemented together. *See* Hardpan, Indurated.

Chelation. (key-lay′-shun) The formation of strong bonds between metals and organic compounds. Some chelates are insoluble, such as in soil humus.

Chemical oxygen demand (COD or C.O.D.). A measure of the oxygen-consuming capacity of inorganic and organic matter present in water or wastewater. *See also* Biological oxygen demand.

Chemigation. Application of pesticides and fertilizers through mixing them into irrigation water, usually in sprinkler or drip systems. *See also* Fertigation.

Chiseling. Breaking or loosening the soil, without inversion, with a chisel cultivator or chisel plow, usually at a deeper-than-normal plow depth.

Chlorite. A nonexpanding clay mineral having a silica tetrahedral, an alumina octahedral, a silica tetrahedral, and a magnesium hydroxide (brucite) octahedral layer; has a 2:2 or 2:1:1 crystal structure.

Chlorosis. A loss of normal green color of the plant.

Chroma. The relative purity, strength, or saturation of a color.

Chromatography. The movement of substances in gas or liquid through a stationary matrix, usually solid, at different rates because of differential "attraction" of the substances to the stationary matrix.

Clay. (soils) 1. A mineral soil separate consisting of particles less than 0.002 mm in equivalent diameter. 2. A soil textural class. 3. A fine-grained soil that has a high plasticity index in relation to the liquid limits (engineering). 4. A specific mineral structure. *See* Clay minerals in Chapter 5.

Clay films. Coatings of clay on the surfaces of soil peds and mineral grains and in soil pores. Also called *clay skins, cutans, argillans,* or *tonhautchen.*

Claypan. A dense, compact layer in the subsoil having a much higher clay content than the overlying material. It usually impedes the movement of water and air and the growth of plant roots. Compare with Hardpan. *See also* B horizons, Table G-2 and in Chapter 2.

Cleavage of minerals. The smoothness of regular breaking surfaces of minerals; "perfect" is very smooth; "conchoidal" is convex and not smooth.

Clod. A compact, coherent mass of soil ranging in size from 5 to 10 mm (0.2–0.4 in.) to as much as 200–250 mm (8–10 in.); produced artificially by digging or tillage. *See also* Ped.

Coarse fragments in soils. Includes rock fragments between 2 mm and 25 cm (0.08–9.8 in.) in diameter. *See* Chapter 4.

Coarse texture. Texture is sands, loamy sands, and sandy loams.

Cobblestone. Rounded rock and mineral particles of 8–25 cm (3–10 in.) diameter.

COD or C.O.D. (chemical oxygen demand). The oxygen-consuming capacity of all the substances in water.

Cohesion. Force holding a solid or liquid together because of attraction between like molecules. *See also* Adhesion.

Colloid. A substance that, when suspended in water, diffuses not at all or very slowly through a semipermeable membrane; a substance in a state of fine subdivision with particles from 1 micrometer to 1 nanometer.

Colluvium. A deposit of rock fragments and soil material accumulated at the base of steep slopes as a result of gravitational action. *See also* Creep; Earth flow.

Color. *See* Munsell color notation system.

Columnar structure. Vertically oriented, round-topped structural prisms. *See* Chapter 4.

Compost. Organic residues or a mixture of organic residues and soil that have been piled and allowed to undergo biological decomposition.

Concretion. (soil) A local concentration of a chemical compound, such as calcium carbonate or iron oxide, in the form of an aggregate or nodule.

Conductivity, hydraulic. *See* Hydraulic conductivity.

Conservation. The protection, improvement, and use of natural resources according to principles that will ensure their highest economic, social, and psychological benefits to people in perpetuity.

Conservation tillage. 1. Tillage sequence that reduces loss of soil or water relative to conventional tillage. 2. Retention of residues to cover over 30% of soil surface. *See* Chapter 14.

Consistence. (soil) 1. The resistance of a material to deformation or rupture. 2. The degree of cohesion or adhesion of the soil mass. Terms used for describing consistency of soil materials at various soil moisture contents and degrees of cementation are as follows.

Wet (stickiness). Stickiness is the quality of adhesion to other objects. For field evaluation of stickiness, soil material is pressed between thumb and forefinger and its relative adherence is noted.
Nonsticky. Practically no soil material adheres to thumb or finger.
Slightly sticky. Slight adherence to thumb and finger.
Sticky. Soil adheres to both thumb and finger and tends to stretch somewhat.
Very sticky. Soil material adheres strongly to fingers.
Wet (plasticity). Plasticity is the ability to change shape continuously under the influence of an applied stress and retain the impressed shape on removal of the stress. For field determination of plasticity, roll the soil material between thumb and forefinger and observe whether a thin rod ("wire") of soil can be formed.
Nonplastic. No "wire" is formable.
Slightly plastic. "Wire" is formable, but soil mass is easily deformable.
Plastic. "Wire" is formable; moderate pressure is required for deformation.
Very plastic. "Wire" is formable; much pressure is required for deformation.
Moist. Consistency when soil is moist is determined at a moisture content approximately midway between air-dry and field capacity.
Loose. Noncoherent.
Very friable. Soil material crushes under very gentle pressure; coheres when pressed together.
Friable. Soil material crushes easily under gentle pressure.
Firm. Soil material crushes under moderate pressure between thumb and forefinger; resistance is distinctly noticeable.
Very firm. Soil barely crushable between thumb and forefinger.
Extremely firm. Soil cannot be crushed between thumb and forefinger.
Dry. The consistency of soil material when dry is characterized by rigidity, brittleness, maximum resistance to pressure, and tendency to crush to a powder.
Loose. Noncoherent.

Soft. Soil mass is very weakly coherent; breaks easily to powder.
Slightly hard. Easily broken between thumb and forefinger.
Hard. Moderately resistant to pressure; can be broken in the hands without difficulty but is barely breakable between thumb and forefinger.
Very hard. Can be broken in the hands only with difficulty.
Extremely hard. Cannot be broken in the hand.
Cemented. Cementation of soil material refers to a brittle, hard consistency.
Weakly cemented. Is brittle and hard; can be broken by the fingers.
Strongly cemented. Cannot be broken in the hand; easily broken with a hammer.
Indurated. Very strongly cemented; brittle, does not soften under prolonged wetting; for breakage, a sharp blow with a hammer is required.

Consumptive use. The total quantity of water transpired by vegetation plus that evaporated from soil plus that in the plant material. Almost equal to evapotranspiration. *See also* Evapotranspiration.

Contamination. Any substance added or accumulating in air, water, or soil that makes the air, water, or soil less desirable for people's use. Same as *pollution*.

Contour. 1. An *imaginary* line on the surface of the earth connecting points of the same elevation. 2. A *true* line drawn on a map connecting points of the same elevation.

Contour stripcropping. Layout of crops in comparatively narrow strips in which the farming operations are performed approximately on the contour. Usually, strips of grass, close-growing crops, or fallow are alternated with those in cultivated crops.

Control section. A defined depth of the soil profile used for a specific determination, e.g., water control section, texture control section, or temperature control section. *See* Chapter 3.

Coulter. A disk used in tillage to cut surface litter or open a slit to facilitate the passage of following chisels, knives, or chemical drops in reduced tillage. *See* Chapter 14.

Creep. (soils) Slow mass movement of soil and soil material down relatively steep slopes, primarily under the influence of gravity but facilitated by saturation with water and by alternate freezing and thawing. *See also* Colluvium; Earth flow.

Crust. A surface layer on soils, ranging in thickness from a few millimeters to perhaps as much as 25 mm (1 in.), that is much more compact, hard, and brittle when dry than the material immediately beneath it.

Cryic. An annual soil temperature regime (0–8°C) in quite cold environments. *See* Chapter 3.

Cyanobacteria. New name for N_2-fixing microbes; previously called *blue-green algae.*

D

Darcy's law. A law describing the saturated flow of water through a porous media.

Deflocculation. *See* Dispersion, soil.

Degraded soil. Soil that is less productive than previously, usually because of damages from erosion, loss of humus, loss of fertility, or accumulated salts or other pollutants.

Delta. An alluvial deposit formed where a stream or river drops its sediment load on entering a body of more quiet water.

Denitrification. The biochemical reduction of nitrate or nitrite to gaseous nitrogen, either as molecular nitrogen or as an oxide of nitrogen.

Desert pavement. The layer of gravel or stones left on the land surface in desert regions after the removal of the fine material by wind and water erosion.

Desertification. The decline in productivity and vegetative cover of arid and semiarid soils caused by natural or person-made (anthropogenic) stresses.

Diagnostic horizons. Horizons with combinations of specific substances that are indicative of certain kinds of soil development. *See* Chapter 2.

Diffuse double layer. A theoretical zone of the soil solution near the soil particles' negatively charged surface, only a few ionic

diameters thick, where cations are strongly adsorbed.

Diffusion. The movement of an ion in water mostly by its own kinetic motion. Very slow, usually in nanometers per hour (10^{-4} cm/h).

Dinitrogen fixation. Conversion of gaseous dinitrogen (N_2) in the air to organic nitrogenous substances by certain bacteria, algae, and actinomycetes.

Disk. A round, convex-shaped, tillage tool used for shallow tillage. *See* Chapter 14.

Dispersion, soil. The breaking down of soil aggregates into individual particles. Generally, the more easily dispersed the soil, the more erodible it is. Favored by high exchangeable sodium. *See also* Flocculate.

Drainage class. The description of the ease with which a soil drains off excess water by percolation. Terms such as *well drained, imperfectly drained,* or *very poorly drained* are used.

Dry weight. (soils) The equilibrium weight of the solid soil particles after the water has been vaporized by heating to 105°C (221°F).

Dryland farming. Rainfed farming in arid and semiarid regions without the use of irrigation.

Duripan. *See* Diagnostic horizons.

E

E horizon. A strongly leached layer, often referred to as an *albic horizon. See also* Soil horizons.

Earth flow. The process of soil moving downslope because of the pull of gravity, usually lubricated by water. *See also* Colluvium; Creep.

EB horizon. *See* Soil horizons.

EC (electrical conductivity). Measured in Siemens/meter. EC_e = value from a saturated soil paste extract.

Ecology. The totality of relationships among organisms and their ambient (surrounding) environment.

Edaphology. The science that deals with the influence of soils on living things, particularly plants, including human use of land for plant growth.

Efflorescence. Accumulation of dried soluble salts on the soil surface, left as the water carrying the salts has evaporated.

Effluent. 1. The discharge or outflow of water from ground or subsurface storage. 2. The fluids discharged from domestic, industrial, and/or municipal waste collection systems or treatment facilities.

Eh. The electrical potential between a half-cell oxidation or reduction reaction and the H electrode.

Eluviation. The downward removal of soil material in suspension (or in solution) from a layer of soil. *See also* Leaching, Illuvial.

Entisols. *See* Soil classification: Order.

Enzyme. A protein mass with surface chemical groups arranged so that atoms around a certain kind of bond can attach to the enzyme, allowing the enzyme to act as a catalyst to reduce the activation energy for a chemical reaction (such as splitting of a bond).

Eolian (aeolian) soil material. Soil material accumulated through wind action.

Epipedon. A term used with diagnostic horizons meaning that the horizon formed while at or near the soil surface. *See* Chapter 2.

Erosion. The wearing away of the land surface by water, wind, ice, or other geological agents, including such processes as gravitational creep.

Accelerated erosion. Erosion much more rapid than normal, natural, or geologic, primarily as a result of the influence of the activities of humans or other animals or natural catastrophies, such as fires and earthquakes.

Geological erosion. The normal erosion caused by geological processes acting over long geologic periods. Synonym: *natural erosion.*

Gully erosion. Erosion whereby water accumulates in narrow channels and over short periods removes the soil from this narrow area (gully) to depths ranging from 30 cm to 30 m (1–100 ft) or more.

Rill erosion. An erosion process in which numerous small channels several centimeters deep are formed. *See also* Rill.

Saltation. Bouncing or jumping action of particles falling from wind or water or impacted by other falling particles in wind or water flow.

Sheet erosion. The removal of a fairly uniform layer of soil from the land surface.

Splash erosion. The spattering of small soil particles caused by the impact of raindrops on wet soils.

Surface creep Rolling or slow movement of particles too large to be carried in wind or flowing water (also gravity "flow").

Erosion pavement. *See* Desert pavement.

Essential element. (plant nutrition) A chemical element required for the normal growth and reproduction of plants, people, or other animals.

Eucaryote. (you-car′-ry-ot) Organism cell type that has the genetic material enclosed inside a nucleus. *See also* Procaryote.

Eutrophication. Enrichment of waters with nutrients, primarily phosphorus, causing abundant aquatic plant growth.

Evaporites. Residue of gypsum plus all salts more soluble than gypsum, which solidify as water evaporates.

Evapotranspiration (ET). Water transpired by vegetation plus that evaporated from the soil. Approximate synonym: consumptive use.

Exchangeable cation percentage (ECP). The percentage of the cation adsorption complex of a soil occupied by a particular cation. It is epressed as follows:

$$ECP = \left[\frac{exchangeable\ cation}{cation\ exchange\ capacity} \right] (100)$$

Exchangeable ion. Any ion held through electrical attraction to a charged surface; can be displaced by other ions from surrounding solution.

Exchangeable sodium percentage (ESP). The percentage of the cation exchange capacity occupied by sodium.

Extrusive rocks. Volcanic igneous rocks formed from molten magma that cooled rapidly (has glassy nature of small crystals) by exposure above the earth's crust.

Exudation. The excretion or natural elimination of substances from the plant.

F

Fallow. Allowing cropland to lie idle during the growing season.

Family, soil taxonomy. *See* Soil classification: Family; Chapter 3.

Feldspars. Minerals made up of silicates of aluminum, with potassium, calcium, or sodium ions; very low solubility.

Ferrihydrite. An amorphous or poorly crystalline iron hydrous oxide mineral.

Fertigation. A term coined for application of fertilizers in irrigation waters, usually through sprinkler systems. *See also* Chemigation.

Fertilizer. Any material, except lime, added to soil to supply one or more essential elements.

Fertilizer analysis. The actual composition of a fertilizer as determined in a chemical laboratory by standard methods. *See also* Fertilizer grade.

Fertilizer grade. The guaranteed minimum analysis in whole numbers, in percent, of the nitrogen, phosphorus, and potassium in a fertilizer material. For example, a fertilizer with a grade of 20-10-5 is guaranteed to contain 20% *total* nitrogen (N), 10% *available* phosphoric acid (P_2O_5), and 5% *water-soluble* potash (K_2O). *See also* Fertilizer analysis; Chapter 13.

Fertilizer, starter. A small application of fertilizer applied with or near the seed for accelerating early growth of the crop.

Fibric material. Organic material, only slightly decomposed. *See* Histosols in Chapter 3.

Field (water) capacity. The amount of water remaining in a soil after the soil layer has been saturated and the free (drainable) water has been allowed to drain away (a day or two). Estimated at −33 kPa water potential.

Fixation. The processes by which chemical elements are converted from a soluble or exchangeable form to a much less soluble or to a nonexchangeable form.

Flocculate. To aggregate, or clump together, small soil particles into small clumps or granules that usually settle out of suspension quickly. *See also* Dispersion, soil.

Floodplain. Nearly level land situated on either side of a channel that is subject to overflow flooding.

Flux. The time rate of flow of a quantity (substance) across a given area cross section.

Foliar diagnosis. An estimate of a plant's nutrient deficiencies or sufficiencies by analytical measurement of selected parts of the plants.

Folic. A thin organic "soil" of leaves over solid rock.

Formative element. A syllable indicating a soil property that is added to a soil order name's "root" to make a suborder name (or to a suborder name to make a great group name). *See* Chapter 3.

Fragipan. A hard, dense, brittle-when-dry soil pan. *See* Chapter 2.

Fragmental. Stones, cobbles, gravel and very coarse sand with too little fine sand, silt, and clay to fill soil interstices larger than 1 mm.

Friable. Soils that crumble easily. *See also* Consistence.

Frigid. A cool soil temperature regime (0–8°C) with warm summers. *See* Chapter 3.

Fritted trace elements. Sintered silicates (glass fragments) having guaranteed analyses of micronutrients with slow release characteristics. *See* Micronutrient.

Frost heaving. The lifting of the surface soil by growing ice crystals in the underlying soil. Such action often pushes plants out of the ground.

Fulvic acid. An indefinite term for the mixture of organic substances remaining soluble after a soil extract, using dilute alkali, has been acidified.

Fungi. One of five kingdoms into which all organisms are fitted. *See* Chapter 7.

G

Ganat. (also Qanat) Gently sloping underground tunnels into shallow aquifers to obtain gravity flow streams.

Gated pipe. Portable pipe with small gates (slide openings) installed along one side for uniformly distributing irrigation water to corrugations or furrows.

Geological erosion. *See* Erosion.

Geophagy. The eating of soil by animals and people. *See also* Pica.

Gilgai. The microrelief of clayey (montmorillonitic) soils produced by expansion and contraction with wetting and drying. A succession of microbasins and microknolls. Diagnostic for Vertisols.

Glacial till. Unstratified, nonsorted materials deposited from melting glaciers and consisting of clays, silts, sands, gravels, and boulders.

Gley. Some layer of mineral soil developed under conditions of poor drainage (poor aeration), resulting in reduction of iron and other elements and in gray colors and mottles (blobs of variously colored soils).

Goethite. A yellowish-brown iron oxide mineral, a common cause of yellowish-brown soil color.

Granule. A natural soil ped or aggregate.

Gravitational potential. The amount of work an infinitesimal amount of pure free water can do at the site of the soil solution as a result of the force of gravity.

Great soil group. A category of the U.S. Soil Taxonomy system. Soils are placed according to soil moisture, temperature, base saturation status, and expression of horizons. *See* Chapter 3. *See also* Soil classification.

Green-manure crop. Any crop grown for the purpose of being turned under while green or soon after maturity for soil improvement.

Groundwater. Subsurface water in the zone of saturation. *See* Chapter 16.

Gully. Large eroded channels. *See also* Erosion.

Gypsic horizon. A diagnostic horizon having an accumulation of gypsum. *See* Chapter 2.

Gypsum requirement. The quantity of gypsum or its equivalent required to reduce the exchangeable-sodium percentage of a soil to an acceptable level.

H

H bond. (hydrogen bond) The bond of a hydrogen ion, already strongly attached to one electronegative ion (such as oxygen, nitrogen, sulfur), less strongly to a second electronegative ion. H bonds are common between water molecules, between water and mineral oxygens, and between organic-substance hydrogens, and mineral oxygens.

Half-life. The time for half of a substance to be destroyed or inactivated or, for radioactive substances, to lose half of its radiation.

Halophyte. Any organism that grows in salty environments and readily absorbs salts.

Hardpan. A hardened soil layer in the lower *A* or a deeper horizon caused by cementation of soil particles with organic matter, silica, sesquioxides, or calcium carbonate. *See also* Caliche; Diagnostic horizons in Chapter 2.

Heavy metals. Those metals of high atomic weight having densities greater than 5.0 Mg/m^3. Many are toxic when accumulated into animal bodies; some heavy metals are arsenic, cadmium, chromium, copper, beryllium, lead, manganese, mercury, nickel, and zinc, among many others.

Heavy soil. (obsolete scientifically) An inexact term still widely used to indicate clayey soils (hard to till).

Hectare. (ha) A metric unit equalling 10,000 square meters or 2.471 acres.

Hematite. A red iron oxide (Fe_2O_3, "rust") that contributes red coloring to soils.

Hemic material. Organic residues in organic soils (Histosols) having an intermediate stage of decomposition. *See* Histosols in Chapter 3.

Heterotroph. An organism capable of deriving energy for life processes from the oxidation of organic compounds.

Histic horizon. A thin diagnostic epipedon of organic material formed by periods of excess wetness (anerobic, waterlogged).

Histosols. *See* Organic soils; Soil classification: Order.

Horizon. *See* Soil horizon.

Hue. One of the three variables of color, the "rainbow" color of light reflected from each soil. *See also* Chroma; Munsell color notation system; Value, color.

Humic acid. A mixture of dark-colored organic materials of indefinite composition extracted from soil with dilute alkali and precipitated upon acidification.

Humin. Soil organic matter insoluble in dilute alkali solution.

Humus. The fraction of the soil organic matter remaining, usually amorphous and dark colored, after the major portion of added residues have decomposed.

Hydraulic conductivity. (K) The proportionality factor in Darcy's law, indicating the soil's ability to transmit flowing water.

Hydrogen bond. *See* H bond.

Hydrologic cycle. The circuit of water movement from the atmosphere to the earth and return to the atmosphere through various stages or processes, as precipitation, runoff, percolation, storage, evaporation, and transpiration.

Hydrolysis. That chemical reaction involving double displacement in which hydrogen of water combines with the anion of the mineral and hydroxyl of water combines with the cation of the mineral to form an acid and a base.

Hydrous mica. (illite) (obsolete) A hydrous aluminosilicate clay mineral with structurally mixed mica and smectite or vermiculite, similar to montmorillonite but containing potassium between the crystal layers. Sometimes referred to as *illite* or *mica* (*see* Chapter 5). No term given to replace these names. Still widely used.

Hydroxyl. Oxygen with one hydrogen forming OH⁻, the anion of bases.

Hyperthermic. Continuously hot temperature regime (15–22°C). *See* Chapter 3.

I

Igneous rock. Formed by solidification from a molten or partially molten state. Synonym: primary rock. Example: granite.

Illite. (obsolete) *See* Hydrous mica.

Illuvial horizon. A soil layer or horizon in which material carried from an overlying layer has been precipitated from solution or deposited from suspension. The layer of accumulation. Contrast to eluviation. *See* B horizons, Table G-2.

Immobilization. The transfer of an element from the soluble inorganic into the organic form of microbial or plant tissues.

Imogolite. Slightly crystalline allophane.

Impervious soil. A soil through which water, air, or roots cannot penetrate.

Inceptisols. *See* Soil classification: Order.

Indurated. (soil) Soil material cemented into a hard mass that will not soften on wetting. *See also* Caliche; Fragipan; Hardpan.

Infiltration. Entry of water downward into the soil surface.

Inoculation. The process of introducing cultures of microorganisms into soils or culture media, such as by adding *Rhizobia* bacteria coated on legume seed.

Integrated pest management (IPM). The use of many different techniques in combination to control pests, such as the combined uses of resistant plant varieties, natural predators of the pest, specific chemical pesticides, good preventive measures, and good management practices, such as crop rotations.

Internal soil drainage. The downward movement of water through the soil profile. *See also* Percolation, soil water.

Intrusive rocks. Rocks formed from molten magma that cooled slowly (has large crystals) because it cooled beneath the earth's crust. Also called *plutonic igneous rocks.*

Ions. Atoms or groups of atoms that are electrically charged as a result of the loss of electrons (cations) or the gain of electrons (anions).

Iron-pan. An indurated soil horizon in which iron oxides are the principal cementing agent along with aluminum oxides. If from plinthite, it is called *ironstone.*

Ironstone. *See* Iron-pan.

Irrigation application efficiency. Percentage of irrigation water applied to an area that is stored in the soil for crop use.

Irrigation methods. The manner in which water is artificially applied to an area. *See* Chapter 16.

Isomorphous substitution. The replacement of one atom by another of similar size in a crystal lattice during crystal growth without changing the crystal structure.

K

Kandic horizon. A diagnostic argillic horizon having mostly low activity (1:1 crystal) clays, such as Kaolinite. *See* Chapter 2.

Kaolinite. Hydrous aluminosilicate clay mineral of the 1:1 crystal structure group—that is, consisting of one silicon tetrahedral sheet and one aluminum oxide-hydroxide octahedral sheet.

L

Labile pool. The total sum of a nutrient that readily solubilizes or exchanges to become available to plants during a season.

Lacustrine deposit. Sediments deposited in fresh (nonsaline) lake water and later exposed either by lowering of the water level or by the elevation of the land.

Land capability class. One of eight classes of land in the land capability classification of the USDA—Soil Conservation Service, distinguished according to the risk of land damage or the difficulty of land use. *See* Chapter 19.

Land capability subclass. The four kinds of limitations recognized at the subclass level are risks of erosion, designated by the symbol (e); wetness, drainage, or overflow (w); other root zone limitations (s), and climatic limitations (c).

Land capability unit. A group of soils that are nearly alike in suitability for plant growth and that respond similarly to the same kinds of soil management.

Land evaluation and site assessment (LESA). An evaluation of land for guiding the conversion of farmland to urban areas and to protect prime farmland from other uses.

Landform. A discernible natural landscape, such as a floodplain, stream terrace, plateau, or alluvial fan.

Latent heat. The amount of heat involved per unit of mass to undergo a phase change, such as heat of melting.

Layer. (mineralogy) A combination of sheets in minerals in a 1:1, 2:1, or 2:2 combination. *See* Chapter 5.

Leaching. The downward removal of materials in solution from the soil. *See* Eluviation.

Leaching requirement. The extra fraction of the amount of water needed to wet the soil that must be added to keep soil salinity below a predetermined tolerance concentration.

Legume. A member of the legume or pulse family *Leguminosae:* peas, beans, peanuts, clovers, alfalfas, sweetclovers, vetches, lespedezas, and kudzu. Most are nitrogen-fixing plants.

Legume inoculation. *See* Inoculation.

LESA. *See* Land evaluation and site assessment.

Lichen. A symbiotic relationship of a fungus and an alga whereby the fungus supplies water and dissolved nutrients and the alga photosynthesizes carbohydrates and fixes nitrogen. Lichens colonize bare minerals, rocks, and large trees, a first step in rock weathering and soil formation.

Ligand. An organic molecule that can bond to metals through two or more bonds. The ligand-metal is called a *chelate. See also* Chelation.

Light soil. (obsolete scientifically) Indicates sandy textures (easy to till).

Lime. 1. *Chemistry:* calcium oxide (CaO); 2. *Agriculture:* a variety of acid-neutralizing materials; most are the oxide, hydroxide, or carbonate of calcium, or of calcium and magnesium (ground limestone, marl, oyster shells [carbonates], wood ashes).

Lime (calcium) requirement. The amount of agricultural limestone required to raise the pH of the surface soil to about pH 6–6.5.

Limestone. A sedimentary rock composed of over 50% calcium carbonate ($CaCO_3$).

Liquid lime. Lime materials that have been pulverized and added to soil as a liquid suspension.

Lithic contact. A boundary between soil and continuous, coherent underlying material that has a hardness of 3 or more (Mohs' scale). *See also* Paralithic contact.

Loess. Material transported and deposited by wind; predominantly silt sized. *See also* Eolian (aeolian) soil material.

Longwave radiation. Radiation of infrared and radio wavelengths emitted from the earth's surface. *See also* Shortwave radiation.

Lysimeter. Container of soil to measure the water movement, gains, or losses through that block of soil, usually undisturbed or in situ.

M

Macronutrient. A chemical element needed in amounts usually >1 part per 500 in the plant for plant growth. Examples: C, N, O, K, Ca, Mg, S, H. *See also* Essential element; Micronutrient.

Mapping unit. *See* Soil mapping unit.

Marl. An earthy, unconsolidated deposit formed in freshwater lakes, consisting chiefly of calcium carbonate mixed with clay or other impurities.

Mass flow. Movement of nutrients with the overall flow of water to plant roots.

Mass water percentage. The mass water ratio times 100.

Mass water ratio. The water content expressed as the weight of water in a soil divided by the oven-dry weight of soil. When expressed as a percentage, it is the mass water percentage or simply water percentage.

Massive. Lack of soil structure in coherent materials; structureless but holds together. *See* Chapter 4.

Matric potential. The amount of work an infinitesimal quantity of water in the soil can do as it moves from the soil to a pool of free water of the same composition and at the same location. This work is less than zero, or *negative* work, thus reported in negative values. Matric potential nearly equals water potential in nonsalty soils. *See also* Soil water potential.

Melanic horizon. Deep, black surface horizon, over 10% organic carbon, formed in volcanic material (Andisols). *See* Diagnostic horizons in Chapter 2.

Mesic. A soil temperature regime of intermediate range (8–15°C). *See* Chapter 3.

Metal oxides. *See* Sesquioxides.

Metamorphic rock. Igneous or sedimentary rock that has changed because of high temperature, high pressure, and the chemical environment while deep in the crust of the earth. Examples: marble, slate, gneiss.

Methemoglobinemia. "Blue baby disease," caused by high nitrate intake in very young mammals, which when nitrate is reduced to nitrite can reduce the blood's ability to carry oxygen. *See* Chapter 18.

Micronutrient. A chemical element necessary in only small amounts (usually less than several parts per million in the plant) for the growth of plants. Examples: boron, copper, iron, and zinc. *See also* Fritted trace elements, Macronutrient.

Mineralization. The conversion of an element from an organic combination to an inorganic form as a result of microbial decomposition.

Mohs' scale of hardness. Relative hardness of minerals ranging from a rating of 1 for the softest (talc) to 10 for the hardest (diamond).

Moisture suction. *See* Matric potential; Soil water potential.

Moisture tension (or pressure). (obsolete) *See* Soil water potential.

Mole drain. Drains formed by pulling a bullet-shaped cylinder through the soil at a depth of 30–91 cm (1–3 ft).

Mollic horizon. A diagnostic epipedon of dark color, of moderate pH, and quite deep. *See* Chapter 2.

Mollisols. *See* Soil classification: Order.

Monera. (moan-ee′-ra) One of five kingdoms for organisms and having procaryotic cells. Includes bacteria, actinomycetes, and cyanobacteria. *See also* Animalia; Fungi; Plantae; Protista.

Montmorillonite. A hydrous aluminosilicate clay mineral with 2:1 expanding crystal structure—that is, with two silicon tetrahedral sheets enclosing an aluminum octahedral sheet. Considerable expansion may be caused by water.

Mor. Forest surface humus in which the *Oa* horizon has little mixing into mineral soil beneath it. *See also* Mull.

Moraine. An accumulation of glacial drift formed chiefly by the direct deposition from glacial ice. Examples are ground, lateral, recessional, and terminal moraines.

Morphology. The physical nature of the soil as exhibited by horizon differences and such physical properties as texture, porosity, and color.

Mottles. (soils) Irregular soil mass spots of various colors. A common cause of mottling is impeded drainage.

Muck. Organic soil whose organic material is too decomposed to be recognizable.

Mulch. A natural or artificial layer of crop residues, leaves, sand, plastic, or paper on the soil surface.

Mull. Forest surface humus with or without *Oe* and without an *Oa*. Organic matter is intimately mixed with mineral soil so the transition to an *A* is gradual. *See also* Mor.

Munsell color notation system. A system that specifies the relative degrees of the three simple variables of color: hue, value, and chroma. For example: 10YR 6/4 is a color with hue 10YR, value 6, and chroma 4. *See also* Chroma; Hue; Value, color.

Mycelia. A cottony mass of individual hyphae (filaments or elongated "threads") typical of growth structures of many fungi.

Mycorrhiza. (my-core-rise′-a) Literally "fungus root." The association, usually symbiotic, of specific fungi with the roots of specific higher plants.

N

Natric horizon. An argillic horizon with >15% exchangeable sodium. *See* Chapter 2.

Nematode. Small (0.5–1.5 mm long) nonsegmented worms. Some are parasitic on plant roots; others parasatize insects.

Newton. (N) Unit of force required to accelerate a mass of 1 kg 1 meter per second per second. It equals 100,000 dynes. *See also* Pascal.

Nitrate reduction. Reduction process converting nitrate to ammonium for use by plants and microorganisms.

Nitrate toxicity of forage. Forage containing more than 6000 ppm of nitrate (NO_3^-) may be toxic to cattle. Causes of high nitrate include drought, cool weather, cloudy weather, and heavier-than-recommended applications of manures, sludges, and nitrogen fertilizers.

Nitrification. The biological oxidation of ammonium salts to nitrites and the further oxidation of nitrites to nitrates.

Nitrogen fixation. *See* Dinitrogen fixation.

Nitrogenase. The enzyme involved in biological dinitrogen fixation.

No-tillage. (no-till) A method of growing crops that involves no seedbed preparation other than opening a small slit or punching a hole into the soil in order to place the seed at the intended depth.

Nodule. A "growth" developed on the roots of plants in response to the stimulus of root nodule bacteria or actinomycetes.

Nonpoint (pollution) source. Coming from a general area (as in sheet erosion) but unable to identify to a ditch, field, or home lot. *See also* Point (pollution) source.

O

O horizon. Surface organic layers. *See also* Soil horizon.

Ochric horizon. A diagnostic epipedon of light color, low humus, or shallow depth. *See* Chapter 2.

Order, soil. *See* Soil classification.

Organic phosphorus. Phosphorus present as a part of organic compounds, such as glycerophosphoric acid, inositol phosphoric acid, and cytidylic acid.

Organic soils. Histosols 1. Saturated with water for prolonged periods unless artificially drained and having at least 12% or 18% organic carbon by weight, depending on the mineral fraction and the kind of organic materials. 2. Never saturated with water for more than a few days and having 20% or more organic carbon by weight. *See also* Soil classification: Order.

Fibrists. The least decomposed of all the organic soils; high amounts of fiber are well preserved and readily identifiable as to botanical origin; a bulk density of less than 0.1 Mg/m³ (6.2 lb/ft³). *See also* Peat.

Folists. Freely drained Histosols that consist primarily of leaf litter that rests on rock or on gravel, stones, and boulders whose voids are filled with organic matter; contain 20% or more organic carbon.

Hemists. Histosols that are intermediate in degree of decomposition between the less decomposed Fibrists and the more decomposed Saprists; saturated with water for 6 months or more in a year. *See also* Muck; Peat.

Saprists. The most highly decomposed of the Histosols; a bulk density of more than 0.2 Mg/m³ (12.5 lb/ft³); saturated for 6 months or more in a year. *See also* Muck.

Osmotic potential. (solute potential) The amount of work an infinitesimal quantity of water will do in moving from a pool of free water the same composition as the soil water to a pool of pure water at the same location. The effect of dissolved substances. Usually very small.

Outwash. Stratified glacial drift, often high in sands and gravels, that is deposited by meltwaters near glaciers.

Oven-dry soil. Soil dried at about 105–110°C (221–230°F).

Oxic horizon. A diagnostic horizon common to Oxisols. It is thick, low in weatherable minerals, and contains low CEC clays. *See* Chapter 2.

Oxidation. (1) Combination with oxygen. (2) Removal of electrons from an atom, ion, or molecule during a reaction. *See also* Reduction.

Oxisols. *See* Soil classification: Order.

P

Paleosols. Soils formed under (ancient) climates different from those that now exist.

Pan. A layer in soils that is strongly compacted, indurated, or very high in clay content. *See also* Caliche; Claypan; Hardpan; Diagnostic horizons in Chapter 2.

Pan, pressure or induced. A subsurface soil layer having a high bulk density and a lower total porosity than the soil directly above or below it as a result of pressure. Frequently referred to as *plow pan, plowsole, tillage pan,* or *traffic pan.*

Paralithic contact. Soil contact at a hard layer, which can be dug by a spade, though with difficulty. *See also* Lithic contact.

Parent material. (soils) The unconsolidated, chemically weathered mineral or organic matter from which the *A* and *B* horizons (solum) of soils may have developed by pedogenic processes.

Particle density. The mass per unit volume of the soil particles. *See also* Bulk density, soil.

Particle-size classes for family groupings. *See* Chapter 3.

Pascal. A unit of pressure equal to 1 newton per square meter.

Peat. Undecomposed or only slightly decomposed organic matter accumulated under conditions of excessive moisture. *See also* Fibric material; Muck; Organic soils.

Ped. A unit of soil structure; an aggregate, such as prism, block, or granule, formed by natural processes. *See also* Clod.

Pediment. A uniformly sloped (graded) rock surface cut by erosion, such as a mountain slope, and with only a shallow cover of alluvium.

Pedogenesis. Caused by the natural processes of soil development. Synonyms: *soil genesis, soil development,* and *soil formation.*

Pedon. The smallest volume that can be called "a soil." It has three dimensions. It extends downward to the depth of plant roots or to the lower limit of the genetic soil horizons. Its lateral cross section is roughly hexagonal and ranges from 1 to 10 m² in size, depending on the variability in the horizons.

Peneplain. A "near" plain; the alluvium-covered low areas that have almost formed a plain near the level of water drainage.

Perc test. A rate-of-percolation test for site evaluation for drainage, waste disposal, septic drain fields, and so on.

Percolation, soil water. The downward movement of water through soil, especially the downward flow of water in saturated or nearly saturated soil.

Pergelic. A soil temperature regime (0°C) averaging subzero temperature (permafrost).

Permafrost. A permanently frozen layer.

Permanent wilting point. The largest water content in soil at which plants will wilt and not recover when placed in a humid chamber. It is estimated at about –1.5 MPa matric potential.

Permeability, soil. The quality of a soil layer that enables water or air to move through it.

Petrocalcic horizon. A diagnostic horizon with carbonate accumulation and hard cementation. *See* Chapter 2.

Petrogypsic horizon. A diagnostic horizon with gypsum accumulation and hard cementation. *See* Chapter 2.

pH, soil. A numerical measure of the acidity or hydrogen ion activity of a soil. Exactly, the negative logarithm of the hydrogen-ion activity of a soil. *See also* Reaction, soil.

pH-dependent charge. The portion of the cation or anion exchange capacity that varies with pH.

Phase, soil. A subdivision of a soil taxon, usually a soil series or other unit of classification, based on characteristics that affect the use and management of the soil; phases include degree of slope, degree of erosion, content of stones, and texture of the surface. *See also* Soil mapping unit.

Phosphorus or potassium fixation. *See* Fixation.

Pica. An unnatural appetite of animals and people that results in eating soil, bark, bones, and/or hair. *See also* Geophagy.

Plane of atoms. A flat (planar) array of one atomic thickness. Example: in soil mineralogy, a plane of basal oxygen atoms within a tetrahedral sheet.

Plant available nutrient. *See* Available nutrient.

Plantae. One of five kingdoms into which all organisms are fitted. Includes such plants as algae, mosses, grasses, and trees.

Platy structure. Soil aggregates developed along the horizontal direction; flaky. *See also* Soil structure in Chapter 4.

Playa. A shallow basin on a plain where water gathers and is evaporated.

Plinthite. A nonindurated mixture of iron and aluminum oxides, clay, quartz, and other diluents that commonly occurs as red soil mottles, usually arranged in platy, polygonal, or reticulate patterns. Plinthite changes irreversibly to ironstone hardpans or irregular ironstone aggregates on exposure to repeated cycles of wetting and drying.

Plow pan. *See* Pan, pressure or induced.

Plutonic rocks. *See* Intrusive rocks.

Point (pollution) source. Identifiable source of pollution, such as a smokestack or discharge from a pipe or channel. *See also* Nonpoint (pollution) source.

Polder. A low-lying land area normally under water but reclaimed from the water and protected from resubmersion by dikes, as in the Netherlands.

Pollution. *See* Contamination.

Polypedon. Two or more contiguous pedons, all of which are within the defined limits of a single soil series. *See also* Soil mapping unit.

Pore space. Total space not occupied by soil particles in a bulk volume of soil.

Potassium fixation. *See* Fixation.

Pressure potential. The amount of work an infinitesimal amount of soil water can do in moving from a pool of pure water under the pressures common to that soil position to a pool of pure water at the same location and at normal atmospheric pressure.

Primary mineral. A mineral that has not been chemically altered since it crystallized from molten magma.

Prime agricultural lands. Soils capable of the highest production levels with the least hazard of erosion or damage. *See* Chapters 18 and 19.

Prion. Virallike proteins without a protective coat. Carries genetic material. *See also* Viroid; Virus.

Prismatic structure. A soil structure type with a long vertical axis that is prism shaped. *See* Soil structure in Chapter 4.

Procaryote. (pro-car′-ry-ot) Organism cell type without nuclear materials enclosed in a nuclear membrane (bacteria, cyanobacteria, actinomycetes). *See also* Eucaryote.

Profile, soil. A vertical section of the soil through all its horizons and extending into the parent material.

Protista. One of five kingdoms into which all organisms are fitted. Protista includes mostly one-celled organisms with eucaryotic cells: amoeba, sporozoans, slime molds, ciliates, and flagellates.

Puddling. The act of destroying natural soil structure by intensive tillage when soil is saturated with water.

Pyroclastics. Volcanic materials explosively or aerially ejected from a volcanic vent.

Pyrophyllite. An aluminosilicate mineral of 2:1 layer structure and with no isomorphous substitution.

Q

Qanat. *See* Ganat.

Quick test, soil. Simple, routine analysis on soils, usually to measure pH, soluble salts, and nutritional status.

R

R horizon. Solid-rock horizon, usually found below parent material, but may be at great depth. *See also* Soil horizon.

Radiation. The process of emitting, from atoms, energy as waves and particles through space.

Reaction, soil. The degree of acidity or alkalinity (basicity) of a soil, usually expressed as a pH value. Descriptive terms commonly associated with certain ranges in pH are extremely acid, less than 4.5; very strongly acid, 4.5–5.0; strongly acid, 5.1–5.5; medium acid, 5.6–6.0; slightly acid, 6.1–6.5; neutral, 6.6–7.3; mildly alkaline, 7.4–7.8; moderately alkaline, 7.9–8.4; strongly alkaline, 8.5–9.0; and very strongly alkaline, more than pH 9.0 *See also* pH, soil.

Redox. 1. A term for the overall reactions in which one substance is oxidized while another is reduced by the electron transfers. 2. The electron density of the media. Redox is measured in units of millivolts.

Reduced tillage. Less tillage than in "conventional" tillage. *See* Chapter 14.

Reduction. Atoms or ions that gain electrons. Often associated with very wet, waterlogged soil conditions. *See also* Oxidation.

Regolith. The layer or mantle of loose, noncohesive or cohesive rock material that nearly everywhere forms the surface of the land and rests on bedrock.

Relief. The difference between the high and low areas of a landscape. Approximate synonyms: *topography, earth surface contour, elevation differences.*

Residual material. Unconsolidated and partly weathered mineral materials derived from rock "in place."

Revised universal soil loss equation (RUSLE). An update of the universal soil loss equation (USLE). *See* Chapter 15.

Rhizobia. Collective name for bacteria of the genus *Rhizobium,* which are capable of symbiotic nitrogen fixation with legume plant roots.

Rhizosphere. The zone of soil immediately adjacent to plant roots in which microbial numbers and kinds may be much different than in the bulk soil in general.

Rill. A small, eroded ditch, usually only a few inches deep and hence no great obstacle to tillage operations. *See* Erosion.

Root nodule. A swelling formed on the roots of leguminous plants, caused by the symbiotic nitrogen-fixing bacteria *Rhizobium.*

Root of soil order. Syllable of the order name prior to the vowel connecting to -*sol*. Examples: *od* from Spodosol, *ert* from Vertisol, and *ept* from Inceptisol.

S

Salic horizon. A mineral horizon with enough accumulated soluble salts to affect plant growth. *See* Chapter 2.

Salination. The process whereby soluble salts accumulate in soil. It replaces the obsolete term *salinization.*

Saline seep. Accumulation of salt by local movement of salts into subsoil at a higher location, movement downslope at a shallow depth, and accumulation into the surface soil in downslope areas as salt moves up and water is evaporated.

Saline-sodic soil. 1. A soil containing an exchangeable-sodium percentage greater than 15 (or SAR greater than 13) and a conductivity of the saturation extract greater than 4 dS m^{-1} (25°C). Growth of most crop plants is reduced.

Saline soil. A nonsodic soil containing sufficient soluble salts to impair its productivity; conductivity of the saturation paste extract is greater than 4 dS m^{-1} (25°C).

Salinization. (obsolete) *See* Salination.

Salt balance. The quantities of dissolved salts removed by drainage water minus the quantities of dissolved salts carried to that area in irrigation water.

Salt-affected soil. Soil that has been adversely modified for the growth of most crop plants by the presence of soluble salts, exchangeable sodium, or both. *See also* Saline-sodic soil; Saline soil; Sodic soil.

Saltation. Synonym: vaultation. *See* Erosion.

Sapric materials. *See* Organic soils.

SAR. *See* Sodium adsorption ratio.

Saturated flow. Movement of water through soil by gravity flow, as in irrigation or during a rainstorm.

Saturation paste extract. The extract from a saturated soil paste. Field capacity times 2 approximates saturation percentage.

Secondary mineral. A mineral formed by the precipitation of the soluble weathered products of a primary mineral.

Seed inoculation. *See* Inoculation.

Self-mulching soil. A soil with a high shrink-swell potential in the surface layer so that portions of its surface yearly fall deeply into cracks (self-mixing). *See also* Gilgai; Soil classification: Vertisols.

Semiarid. Regions or climates where moisture is normally greater than under arid conditions but still definitely limits the growth of most crops. The upper limit of average annual precipitation in the cool semiarid regions is as low as 38 cm (15 in.), whereas in tropical regions it is as high as 114–127 cm (45–50 in.). *See also* Arid.

Separate. *See* Soil separates.

Septage. The anaerobic residues from septic tanks.

Sequestrene. *See* Chelation; Ligand.

Series, soil. *See* Soil classification.

Sesquioxides. (metal oxides) A term for minerals containing 1.5 atoms of oxygen per atom of the metal, particularly Al_2O_3 and Fe_2O_3. Often TiO_2 is included, although it does not strictly fit the meaning of *sesqui* (=1.5 times).

Sewage sludge. *See* Sludge.

Sheet erosion. *See* Erosion.

Sheet of atoms. (mineralogy) A flat array of more than one atomic thickness and composed of one or more levels of linked coordination polyhedra. A sheet is thicker than a plane and thinner than a layer. Examples: tetrahedral sheet, octahedral sheet.

Shortwave radiation. High energy radiation (ultraviolet, visible, and near-infrared). Sun's light. It readily passes through air. *See also* Longwave radiation.

Shrink-swell potential. Susceptibility to volume change due to loss or gain in moisture content.

Siderophore. A nonporphyrin metabolite, secreted by certain microorganisms, that forms highly stable bonds with iron, thereby mobilizing iron in soil. The two major types are catecholate and hydroxamate.

Siemen. A unit of electrical inductance; in SI metric units, electrical conductance is measured in siemens per meter (S m^{-1}). Decisiemens per meter (dS m^{-1}) is equivalent to millimhos per centimeter (mmhos/cm).

Siliceous. *See* Soil mineralogy classes for family groupings.

Silt. 1. A soil separate consisting of particles between 0.05 and 0.002 mm in equivalent diameter. *See also* Soil separates. 2. A soil textural class. *See also* Soil texture.

Single grain. (obsolete when called "structure") A term for a lack of structure in which individual particles do not cohere together. Example: coarse sands.

Sink. 1. Sunken depression in the land surface. 2. A material or reaction that acts as an infinite reservoir or removal mechanism. Example: The ocean is a large reservoir (a "sink") to absorb carbon dioxide.

Skeletal. A textural term used in soil "families" to indicate that the soil has greater than 35% coarse fragments (larger than sand).

Slick spots. Small areas in a field that appear wet longer and are slick due to a high content of clay with high exchangeable sodium (low permeability).

Slickensides. Polished and grooved clayey surfaces produced by one soil mass sliding past another. Common in Vertisols. *See also* Clay films.

Slip. The downslope movement of a soil mass under wet or saturated conditions; a microlandslide that produces microrelief in soils. *See also* Creep.

Sludge. A general term for solid wastes, usually collected by sedimentation from water. Common sludges are sewage sludge, food-processing sludges, boiler sludge, electroplating sludge, and sugar-processing sludge.

Smectite. A group of minerals of 2:1 layer silicates having high cation exchange capacities and variable interlayer spacing. Typified by montmorillonite.

Sodic soil. A soil that contains an exchangeable sodium percentage of 15 or more or a saturation extract SAR of 13 or more.

Sodium adsorption ratio (SAR). A value representing the relative hazard of irrigation water because of a high sodium content relative to its calcium plus magnesium content.

$$SAR = \frac{Na^+}{\sqrt{(Ca^{2+} + Mg^{2+})/2}}$$

The ions are in millimoles$_c$ per liter.

Soil. 1. The unconsolidated mineral and organic material on the immediate surface of the earth that serves as a natural medium for the growth of land plants. 2. The unconsolidated mineral matter on the surface of the earth that has been subjected to and influenced by genetic and environmental factors of parent material, climate, macro- and microorganisms, and topography, all acting over a period of time and producing a product—soil—that differs from the material from which it is derived in many physical, chemical, biological, and morphological properties and characteristics.

Soil alkalinity (soil basicity). *See* Reaction, soil.

Soil amendment. *See* Amendment, soil.

Soil association. A mapping unit in which two or more defined taxonomic units occurring together are combined because the scale of the map or the purpose for which it is being made does not require delineation of the individual soils.

Soil bulk density. *See* bulk density, soil.

Soil classification. The systematic arrangement of soils into classes in one or more categories or levels of classification for a specific objective. The relationship between the orders of the present system and approximate equivalents of the 1938 system used in the United States are shown in Table G-1. *See* Chapter 3.

Order. The category at the highest level of generalization in the soil classification system. The properties selected to distinguish the orders are reflections of the degree of horizon development and the kinds of horizons present. *See* Chapter 3.

Suborder. This category narrows the ranges in soil moisture and temperature regimes, kinds of horizons, and composition, according to which is most important. Moisture and/or temperature or soil properties associated with them are used to define suborders of Alfisols, Mollisols, Oxisols, Ultisols, and Vertisols. Kinds of horizons are used for Aridisols, composition for Histosols and Spodosols, and combinations for Entisols and Inceptisols. *See* Chapter 3.

Great group. The classes in this category contain soils that have the same kinds of horizons in the same sequence and have similar moisture and temperature regimes. Exceptions to the horizon sequences are made for horizons near the surface that may get mixed or lost by erosion if plowed. Formative elements used to formulate great group names are given in Chapter 3.

Subgroup. The great groups are subdivided into subgroups that show the central properties of the great group, intergrade subgroups that show properties of more than one great group, and other subgroups for soils with atypical properties that are not characteristic of any great group.

Table G-1 Comparison of the Present United States Soil Classification System Adopted in 1965 with the Approximate Equivalents in Use before 1965

Soil Order (Adopted in 1965)	Approximate Equivalents (In Use before 1965)
Alfisols	Gray Brown Podzolic, Gray Wooded soils, Noncalcic Brown soils, Degraded Chernozem, and associated Planosols and some Half-Bog soils
Andisols	Volcanic materials formed into weakly developed soils, soils with *and-* and *vitr-* prefixes, and selected soils from most categories
Aridisols	Desert, Reddish Desert, Sierozem, Solonchak, some Brown and Reddish Brown soils, and associated Solonetz soils
Entisols	Azonal soils and some Low-Humic Gley soils
Histosols	Bog soils
Inceptisols	Ando, Sols Bruns Acides, some Brown Forest, Low-Humic Gley, and Humic Gley soils
Mollisols	Chestnut, Chernozem, Brunizem (Prairie), Rendzina, some Brown, Brown Forest, and associated Solonetz and Humic Gley soils
Oxisols	Laterite soils, Latosols
Spodosols	Podzols, Brown Podzolic soils, and Ground-Water Podzols
Ultisols	Red-Yellow Podzolic soils, Reddish Brown Lateritic soils of the U.S., and associated Planosols and Half-Bog soils
Vertisols	Grumusols, Rendzinas

Source: Soil Survey Staff, *Soil Taxonomy: A Basic System of Soil Classification for Making and Interpreting Soil Surveys,* Agriculture Handbook 436, USDA—Soil Conservation Service, Washington, D.C., 1975, pp. 433–435. Andisols inserted by authors of this textbook in 1988.

Family. Families are defined largely on the basis of physical and mineralogical properties of importance to plant growth.

Series. The soil series is a group of soils having horizons similar in differentiating characteristics and arrangement in the soil profile, except for texture of the surface, slope, gravel, stones, and erosion.

Soil complex. Much like a soil association, but the different soils occur in a regular pattern and as smaller, discrete units.

Soil creep. *See* Creep.

Soil development. 1. The breakdown of rocks to unconsolidated materials called *soil.* 2. The changes in the soil profile brought about by natural processes of leaching, translocation of colloids, accumulation of organic materials, and continued mineral and rock weathering.

Soil-formation factors. The variables (parent material, climate, organisms, topography, and time) active in and responsible for the formation of soil.

Soil genesis. Formation of the soil with special reference to the processes or soil-forming factors responsible for the development of the solum or true soil from the unconsolidated parent material. Synonyms: *pedogenesis, soil formation.*

Soil horizon. A layer of soil or soil material approximately parallel to the land surface and differing from adjacent genetically related layers in physical, chemical, and biological properties or characteristics, such as color, structure, texture, consistency, amount of organic matter, and degree of acidity or alkalinity. Tables G-2 and G-3 list the designation and description of the master soil letter horizons and subscript horizons. *See also* Diagnostic horizons in Chapter 2.

Soil individual. *See* Pedon; Polypedon.

Soil interpretations. Predictions of soil behavior under specific uses or management and based on inferences from the soils' characteristics as reported in a soil survey report. *See* Chapter 19.

Soil loss tolerance (*T*). In conservation farming, the maximum average annual soil loss in tons per acre per year that should be permitted on a given soil. In general, the rate of soil formation should equal or exceed soil erosion loss.

Soil mapping unit. A kind of soil, a combination of kinds of soil, or miscellaneous land type or types that can be shown at the scale of mapping for the defined purposes and objectives of the survey. The soil mapping legend lists all mapping units for the survey of an area. *See also* Phase, soil.

Soil mineralogy classes for family groupings. The family category includes mineralogy classes for specific control sections that are similar to those used for particle-size classes for family groupings. *See* Chapter 3.

Soil moisture regimes. Moisture regimes indicate the time periods during which the soil "control section" (loam soil between about 10 and 30 cm [4–12 in.] deep) is either moist or dry (wilting percentage). In addition to details below, *see also* aquic, udic, ustic, xeric, aridic, and torric moisture regimes. To simplify this tabulation, the data given are not precise in all particulars. All temperatures are at the 50-cm (20-in.) depth in soils; "parts" refers to parts of the control section (CS).

Aquic conditions. Must have reducing conditions for at least long enough to exhibit one or more of (1) redoximorphic features (wetness-caused mottles), (2) redox concentrations of Fe and Mn, (3) redox depletions of Fe and Mn (leaves low-chroma wetness mottles), and (4) reduced matrix that changes color when exposed to oxygen (air).

Aridic regime. A dry and arid soil environment. Dry in all parts more than half the time above 5°C. Never moist in any part as long as 90 consecutive days when above 8°C.

Udic regime. Wet but aerated soil; moist year around. Not dry in any part as long as 90 cumulative days. If colder than 22°C, the control section is not dry in all parts 45 consecutive days in the 4 months after the summer solstice (June 22) in 6 of 10 years.

Ustic regime. Between aridic and udic; moist summers. Above mean annual temperature of 22°C, CS is dry in some parts 90 or more days. Moist in some part for more than 180 cumulative days *or* continuously moist in some part for 90 consecutive days. Below MAT of 22°C, soil is dry in some parts 90 or more accumulative days. Not dry in all parts more than half the time that it is above 5°C. Not dry in all parts 45 consecutive days in the 4 months after the summer solstice (June 22) in 6 of 10 years.

Xeric regime. Mediterranean climate: moist winters, dry summers. Control zone is dry 45 or more consecutive days in the 4 months after the summer solstice (June 22) in 6 of 10 years. Control zone is moist 45 or more consecutive days in the 4 months after the winter solstice (December 22) in 6 of 10 years. Moist in some parts more than half the time (accumulative) when soil is above 5°C. or during 90 consecutive days when soil is above 8°C. Mean annual temperature is below 22°C. *See* Chapter 3.

Soil moisture suction. *See* Soil water potential.

Soil monolith. A soil profile removed from the soil and mounted for display, often 10–20 cm (4–8 in.) wide, 5–10 cm (2–4 in.) thick, profile depth (1.5 m), and stabilized with transparent resin (vinyl).

Soil morphology. The constitution of the soil, including the texture, structure, consistency, color, and other physical, chemical, and biological properties of the various soil horizons that make up the soil profile.

Soil order. The most inclusive category of the U.S. soil classification system. *See also* Soil classification; Chapter 3.

Soil phase. *See* Phase, soil.

Soil potential index (SPI). Numerical ratings of soils for each of various uses, using the equation:

$$SPI = P - (CM + CL)$$

where P = performance or yield, CM is costs of corrective measures and CL is costs of continuing limitations.

Soil reaction. *See* Reaction, soil.

Soil science. That science dealing with soils as a natural resource on the surface of the earth, including soil formation,

Table G-2 Designation and Description of Master Soil Horizons and Layers (*See also* Table G-3), and Horizons by Name

Horizon Designation		
Current	*Old*	*Description*
O	O	Organic horizons of minerals soils. Horizon (i) formed or forming in the upper part of mineral soils above the mineral part, (ii) dominated by fresh or partly decomposed organic materials, and (iii) containing more than 30% organic matter if the mineral fraction is more than 50% clay, or more than 20% organic matter if the mineral fraction has no clay. Intermediate clay content requires proportional organic-matter content.
Oi	O1	Organic horizons in which essentially the original form of most vegetative matter is visible to the naked eye. The Oi corresponds to the L (litter) and Oe or Oa to fermentation layers in forest soils designations and to the horizon formerly called Aoo or O1.
Oa or Oe	O2	Organic horizons in which the original form of most plant or animal matter cannot be recognized with the naked eye. The Oa corresponds to the H (humification) and Oe to F (fermentation) layers in forest soils designations and to the horizon formerly called Ao or O2.
A	A	Mineral horizons consisting of (i) horizons of organic-matter accumulation formed or forming at or adjacent to the surface, (ii) horizons that have lost clay, iron, or aluminum with resultant concentration of quartz or other resistant minerals of sand or silt size, or (iii) horizons dominated by (i) or (ii) but transitional to an underlying B or C.
A	A1	Mineral horizons, formed or forming at or adjacent to the surface, in which the feature emphasized is an accumulation of humified organic matter intimately associated with the mineral fraction.
E	A2	Mineral horizons in which the feature emphasized is loss of clay, iron, or aluminum, with resultant concentration of quartz or other resistant minerals in sand and silt sizes.
AB or EB	A3	A transitional horizon between A or E and B, dominated by properties characteristic of an overlying A or E but having some subordinate properties of an underlying B.
AB	AB	A horizon transitional between A and B, having an upper part dominated by properties of A and a lower part dominated by properties of B; the two parts cannot be conveniently separated into A and B.
E/B	A & B	Horizons that would qualify for E except for included parts constituting less than 50% of the volume that would qualify as B.
AC	AC	A horizon transitional between A and C, having subordinate properties of both A and C but not dominated by properties characteristic of either A or C.
B	B	Horizons in which the dominant feature or features are one or more of the following: (i) an illuvial concentration of silicate clay, iron, aluminum, or humus, alone or in combination, (ii) a residual concentration of sesquioxides or silicate clays, alone or mixed, that has formed by means other than solution and removal of carbonates or more soluble salts, (iii) coatings of sesquioxides adequate to give conspicuously darker, stronger, or redder colors than overlying and underlying horizons in the same sequum but without apparent illuviation of iron and not genetically related to B horizons that meet requirements of (i) or (ii) in the same sequum, or (iv) an alteration of material from its original condition in sequums lacking conditions defined in (i), (ii), and (iii) that obliterates original rock structure; that forms silicate clays, liberates oxides, or both;

Continued.

Horizon Designation		
Current	Old	Description
		and that forms granular, blocky, or prismatic structure if textures are such that volume changes accompany changes in moisture. A sequum is an E horizon and its related B horizon.
BA or BE	B1	A transitional horizon between B and A or between B and E in which the horizon is dominated by properties of an underlying B but has some subordinate properties of an overlying A or E.
B/E	B & A	Any horizon qualifying as B in more than 50% of its volume, including parts that qualify as E.
B or Bw	B2	That part of the B horizon where the properties on which the B is based are clearly expressed characteristics, indicating that the horizon is related to an adjacent overlying A or an adjacent underlying C or R.
BC or CB	B3	A transitional horizon between B and C in which the properties diagnostic of an overlying B are clearly expressed but are associated with clearly expressed properties characteristic of C.
C	C	A mineral horizon or layer, excluding bedrock, that is relatively little affected by pedogenic processes, and lacking properties diagnostic of A or B but including materials modified by (i) weathering outside the zone of major biological activity, (ii) reversible cementation, development of brittleness, development of high bulk density, and other properties characteristic of fragipans, (iii) gleying, (iv) accumulation of calcium or magnesium carbonate or more soluble salts, (v) cementation by accumulations, as calcium or magnesium carbonate or more soluble salts, or (vi) cementation by alkali-soluble siliceous material or by iron and silica.
R	R	Underlying consolidated bedrock, such as granite, sandstone, or limestone. If presumed to be like the parent rock from which the adjacent overlying layer or horizon was formed, the symbol R is used alone. If presumed to be unlike the overlying material, the R is preceded by an arabic numeral denoting lithologic discontinuity, such as 2R.

Sources: Modified from (1) Soil Survey Staff, *Soil Taxonomy: A Basic System of Soil Classification for Making and Interpreting Soil Surveys,* Agriculture Handbook 436, USDA—Soil Conservation Service, Washington, D.C., 1975, pp. 461–462; (2) Soil Survey Service directive 430-V-SSM, May 1981; (3) USDA—Soil Conservation Service draft of *Soil Survey Manual,* Chapter 4, pp. 4-39 to 4-51, 1981.

classification and mapping, and the physical, chemical, biological, and fertility properties of soils per se; and these properties in relation to their management for the growth of plants and to cleanse the environment.

Soil separates. Mineral particles, less than 2.0 mm in equivalent diameter, ranging between specified size limits. The names and size limits of separates recognized by the National Cooperative Soil Survey in the United States are found in Chapter 4.

Soil series. *See* Soil classification.

Soil structure. The combination or arrangement of primary soil particles into secondary particles, units, or peds. *See* Chapter 4.

Soil subgroup. *See* Soil classification; Chapter 3.

Soil suborder. *See* Soil classification; Chapter 3.

Table G-3 Suffixes to Master Letter Horizons to Indicate Subordinate Distinctions within Master Horizons. Current and Old Symbols

Current	Old	Description of Symbol	Example
a	—	Highly decomposed organic matter	Oa
b	b	Buried genetic soil horizon. Must be by at least 50 cm (20 in.) of sediment	Ab
c	cn	Accumulation of concretions or other hand nodules enriched in sesquioxides	Bc
e	—	Intermediately decomposed organic matter	Oe
f	f	Frozen soil; mostly for permanently frozen (permafrost)	Af
g	g	Strong gleying. Intense reducing (anaerobic) conditions due to stagnant water. Base colors approaching neutral (grays), with or without mottles (brighter orange and yellow colors of oxidized iron)	Cg
h	h	Illuvial humus, usually coating on sands or as pellets. Mostly used with B horizons	Bh
i	—	Slightly decomposed organic matter (litter)	Oi
k	ca	Accumulation of carbonates, usually calcium and magnesium salts	Bk
m	m	Strong cementation by various materials	Cm
n	sa	Accumulation of sodium salts (not just any salts)	Bn
o	—	Residual accumulation of sesquioxides	Bo
p	p	Plowing or other disturbance of soil	Ap
q	si	Accumulation of silica, usually cementation	Bq
r	r	Weathered or soft bedrock	Cr
s	ir	Illuvial accumulation of sesquioxides, often as coatings on sands, as pellets, or as cementing material	Bs
t	t	Accumulation of clay, usually large amounts	Bt
v	—	Plinthite present	B2v
w	—	Color or structural B (weakly developed)	Bw
x	x	Fragipan character; the horizon has firmness, is brittle, and has a high density	B3x
y	cs	Accumulation of gypsum	C1
z	sa	Accumulation of salts more soluble than gypsum	B3

Sources: Modified from (1) Soil Survey Staff, *Soil Taxonomy: A Basic System of Soil Classification for Making and Interpreting Soil Surveys*, Agriculture Handbook 436, USDA—Soil Conservation Service, Washington, D.C., 1975, pp. 461–462; (2) Soil Survey directive 430-V-SSM, May 1981; (3) USDA–Soil Conservation Service draft of *Soil Survey Manual*, Chapter 4, pp. 4–39 to 4–51, 1981.

Soil survey. The systematic examination, description, classification, and mapping of soils of an area.

Soil test. A chemical, physical, or microbiological operation that estimates a property of the soil.

Soil texture. The relative proportions of the various soil size separates. *See* Chapter 4.

Soil variant. A kind of soil that differs enough from recognized series to justify a new series name but is so limited in area that creation of a new series is not justified at the given time.

Soil water potential. The amount of work (usually given in KiloPascals) an infinitesimal quantity of soil water can do in moving from the soil to a pool of pure, free water at the same location and at normal atmospheric pressure. It is mostly matric potential.

Solifluction. The slow downhill flowage or creep of saturated soil and other loose materials. *See also* Creep.

Solum. (plural: sola) The *O, A, E,* and *B* horizons of a soil profile, above the parent material.

Solute potential. *See* Osmotic potential.

Sombric horizon. An illuvial humus subsurface horizon, darker than the horizon above it but without characteristics of a spodic. *See* Diagnostic horizons in Chapter 2.

Specific gravity. The relative weight of a given volume of any kind of matter (volume occupied by solid phase, pore space excluded) compared with an equal volume of distilled water at a specified temperature. The average specific gravity for soil is about 2.65 Mg/m^3. *See also* Bulk density, soil.

SPI. *See* Soil potential index.

Spodic horizon. A diagnostic *B* horizon accumulating amorphous alumina and/or humus colloids, with or without iron, designated *Bs*. *See* Diagnostic horizons in Chapter 2.

Spodosols. *See* Soil classification: Order.

Spoilbank. A pile of soil, subsoil, rock, or other material excavated from a drainage ditch, pond, road cut, or surface mining.

Starter fertilizer. *See* Fertilizer, starter.

Stokes' law. Equation to calculate settling rates: fall of sands, silts, and clay when they are suspended in water; used to determine soil texture.

Strip cropping. The practice of growing crops for controlling erosion that requires different types of tillage, such as row and sod, in alternate strips along contours or at right angles to the prevailing direction of erosive winds.

Strip tillage. Tillage operations performed in isolated bands separated by bands of soil essentially undisturbed by the particular tillage equipment.

Structural charge. The negative charge on a clay mineral caused by isomorphous substitution within the layer.

Structure, soil. *See* Soil structure.

Stubble mulch. The stubble of crops or crop residues left essentially in place on the land as a surface cover during fallow.

Subgroup, soil. *See* Soil classification: Subgroup.

Subsidence. A lowering of the land elevation caused by solution and collapse of underlying soluble deposits, reduction of fluid pressures within an aquifer or petroleum reservoir, or decomposition of organic soils (Histosols).

Subsoiling. The tillage of subsurface soil without inversion, to break up dense soil layers that restrict water movement and root penetration.

Substrate. 1. An underlying layer. 2. Substance or nutrient on which an organism grows. 3. Substances acted upon by enzymes and changed to other compounds in the reaction.

Substratum. Any layer lying beneath the solum, which is the *A* and *B* horizons.

Suction. *See* Soil water potential.

Sulfuric horizon. A diagnostic horizon with both a pH < 3.5 and yellow jarosite (sulfate) mottles.

Summer fallow. *See* Fallow.

Sweep. In tillage, an implement with a point and flat or curved "wings" going back from the point at 30–60° angles. *See* Chapter 14.

Symbiosis. The living together in intimate association of two dissimilar organisms, the cohabitation being mutually beneficial, such as *Rhizobia* legume bacteria with the host leguminous plant.

Synergism. The action of two or more substances, organs, or organisms to achieve an effect which is greater than the sum of the two effects when acting separately.

T

Talus. Fragments of rock and other soil material accumulated by gravity at the foot of cliffs or steep slopes.

Taxon. In the context of soil survey, a class at any categorical level in the U.S. system of soil taxonomy.

Tensiometer. Instrument used for measuring the water potential (suction or negative pressure) of soil water.

Tension, water. *See* Water tension.

Terrace. 1. A natural level plain bordering a river, lake, or sea. 2. A raised, level strip of earth usually constructed on or nearly on a contour designed to make the land suitable for tillage and to prevent accelerated erosion.

Texture. *See* Soil texture.

Thermic. A soil temperature regime averaging between 15° and 22°C. *See* Chapter 3.

Thermophilic organisms. Organisms that grow readily at temperatures of about 113°F (45°C).

Tile drain. Lines of concrete or ceramic (clay) pipe placed in the subsoil to collect and drain water from the soil to an outlet.

Till. 1. Unstratified glacial drift deposited directly by the melting ice and consisting of clay, sand, gravel, and boulders intermingled in any proportion. 2. To plow and prepare for seeding; to seed or cultivate the soil.

Tillage pan. *See* Pan, pressure or induced.

Tilth. The physical condition of soil as related to its ease of tillage, fitness as a seedbed, and impedance to seedling emergence and root penetration. Tilth is a result of tillage.

Top-dressing. An application of fertilizer to a soil after the crop stand has been established.

Toposequence. A sequence of soils that differ primarily because of differences in topography as a soil-forming factor.

Torric. A water-deficient soil water regime defined like aridic but used in a different category level of the U.S. soil taxonomy than is "aridic."

Tortuosity. The winding, twisting, and intermeshing of soil pores.

Trace elements. *See* Micronutrient.

Transpiration. The process by which water vapor is released from plants to the atmosphere primarily through the leaf stomata.

U

Udic. *See* Soil moisture regimes.

Ultisols. *See* Soil classification: Order.

Umbric horizon. A dark-colored, high-organic-matter, diagnostic epipedon similar to mollic but more acidic. *See* Diagnostic horizons in Chapter 2.

Unique agricultural lands. Soils uniquely suited to the high production of certain desired crops. The uniqueness may be aided by a lack of frost, a needed wetness or acidity, a peculiar climate, or a particular location relative to markets.

Universal soil loss equation. *See* Revised universal soil loss equation.

Unsaturated flow. The movement of water in a soil that is not filled to capacity with water. Water moves because of water-potential differences toward areas of greater water potentials (drier soil).

Ustic. *See* Soil moisture regimes.

V

Value, color. The relative lightness or darkness of color. *See also* Chroma; Hue; Munsell color notation system.

Vermiculite. A clay similar to hydrous mica and having 2:1 layers of 2 tetrahedral sheets to 1 octahedral sheet. Vermiculite has the layers held together by hydrated cations and has less swelling than montmorillonite.

Vertical mulching. A subsoiling operation in which a vertical band of mulching material is placed into a vertical slit in the soil immediately behind a special chisel.

Vertisols. *See* Soil classification: Order.

Virgin soil. A soil essentially undisturbed from its natural environment; never cultivated.

Viroid. A virallike RNA protein without a protective coating. *See also* Prion; Virus.

Virus. Any of numerous submicroscopic pathogens consisting of a single nucleic acid and one or two protein coats, capable of being replicated only inside a living cell. *See also* Prion; Viroid.

Void ratio. The ratio of the volume of soil pore space (voids) to the solid-particle volume.

Volcanic ash. Fine particles of rock (ash) blown into the air from a volcano and settled on land, often in layers several meters thick.

Volume water percentage. Volume water ratio multiplied by 100.

Volume water ratio. The ratio of the volume of water in a soil to the total bulk volume of the soil, in decimal form.

W

Water content. *See* Soil water potential.

Water potential. *See* Soil water potential.

Water retention curve. A graph showing the soil–water content versus applied tension, suction, or water potential.

Water rights. The legal rights to the use of water. They consist of riparian rights and those acquired by appropriation and prescription. Riparian rights are those by virtue of ownership of the banks on the stream, lake, or ocean. Appropriated rights are those acquired by an individual to the exclusive use of water, based strictly on priority of appropriation and application of the water to beneficial use. Prescribed rights are those to which legal title is acquired by long possession.

Water suction. *See* Water tension.

Water table. The upper surface of groundwater; that level below which the soil is saturated with water.

Water table, perched. The surface of a local zone of saturation held above the main body of groundwater by an impermeable layer, usually clay or rock, and separated from the main body of groundwater by an unsaturated zone.

Water tension. The equivalent negative pressure in the soil water. It is equal to pressure needed to bring the water into hydraulic equilibrium.

Water use efficiency. Crop production per unit of water used, irrespective of water source, expressed in units of water depth per unit weight of crop; for example, 25 cm (10 in.) of water used per ton of alfalfa hay produced.

Waterlogged. Saturated with water, usually developing anaerobic conditions. Most or all pore space filled with water; natural or induced by humans.

Weathering. The group of processes (such as chemical action of air and rainwater and the biological action of plants and animals) whereby rocks and minerals change in character, disintegrate, decompose, and synthesize new compounds and clay minerals.

Wilting point. *See* Permanent wilting point.

Wind strip cropping. The production of crops in relatively narrow strips placed perpendicular to the direction of prevailing erosive winds.

X

Xeric. *See* Soil moisture regimes.

Xerophytes. Plants that tolerate and grow in very dry soils.

Index

determination of, 107

example values of, 105, 108

use in calculations of pore space, 108–109

use in calculations of water contents, 170–172

Butte, 28–29

C

Cadmium, 13. *See also* Heavy metals

Calcic horizon, 40–41

Calcite, 5., 11

Calcium, 13., 178
  amendments (carriers), 244, 246
  and agricultural lime, 244–249
  carbonate, 247
  deficiency, 366
  exchangeable, 147
  nutrition in plants, 13

Calcium carbonate equivalent, 247

Caliche *See* Calcic horizon; Petrocalcic horizon; Duripan horizon

Cambic horizon, 40–41

Cancer–causing substances. *See* Carcinogens

Capability classification, USDA. *See* Land capability classification

Capillary water. *See* Soil water

Carbon, 13, 142, 217
  as a nutrient, 13, 142
  in organic compounds, 142, 217, 220, 222

Carbonation, as a weathering process, 10

Carbon dioxide, 245
  from burning and combustion, 220
  and soil acidity, 245

Carbonic acid, 245

Carbon monoxide, 576

Carbon:nitrogen ratio, 220–223

Carboxyl, active (functional) group, 142

Carcinogens, 533, 535, 546

Catalyst, 218

Cat clays, 301, 302

Cation exchange, 127, 141–150
  available nutrients from, 143
  cause of, 142–143
  definition of, 142
  examples of, 142
  factors influencing, 142–143
  importance of, 143–145

Cation exchange capacity, 127, 141–150
  calculation of, 148–149
  definition of, 127, 146
  estimation of, 149–152
  influence of soil texture on, 146–147
  means of expressing, 146
  of humus, 147
  of metal oxides, 147
  of representative mineral soils, 146–147
  of silicate clays, 147

Cations, 130, 141
  adsorbed by colloids, 130
  chromatographic movement of, in soils, 144
  hydrated size, 135
  in soil solution, 130

CB horizon, 36

CEC. *See* Cation exchange capacity

Centimoles per kilogram (cmol$_c$/kg), 146

Chalcedony, 6

Chalk, 246

Chelates, 224, 306. *See also* Ligands
  and animal manures, 232
  and soil humus, 224
  for applying micronutrient metals, 305–309, 311–313, 393
  natural, 224, 232

Chemical analyses. *See* Soil diagnosis; Plant analyses

Chemical oxygen demand (COD), 535, 560

Chemical weathering, 10

Chemoautrophs, 207–208

Chemigation, 392. *See also* Fertigation

Chert, 6

Chiseling, 414

Chisel plow, 414

Chloride, 13, 314
  as a plant nutrient, 13
  burning by, on foliage, 314
  for reducing disease, 414

Chlorinated hydrocarbons, 535

Chlorite clay. *See* Clays, chlorite

Chlorosis, 365–367

Chroma (color notation), 114

Chromatographic ion movement in soil, 144

Chronic toxicity, 535, 544

Cinders *See* Volcanic ash; Pumice

Class "A" evaporation pan, 186, 189

Classes, USDA Capability Classification. *See* Soil classification

Clays, 127–140
  and cation exchange, 132–135, 141–152
  allophane, 132, 133, 137, 140, 141
  amorphous, 132, 137
  chemical and mineralogical properties of, 132–152
  chlorite, 132–133, 139, 140
  crystalline, 132
  ferrihydrite, 132
  hydrous micas (illites), 11, 132, 139–141
  imogolite, 132
  inherited, 11, 129
  isomorphous substitution and, 133–135
  kandites, 137
  kaolinite, 11, 132–133, 137, 140–141
  metal oxides (sesquioxides), 132–133, 140–141
  montmorillonite (smectites), 11, 132–133, 137–138, 141
  nature of, 129–141
  neoformed, 129
  non-silicates, 132, 139, 141
  origin of (formation of), 128–129
  reactions of, 142–154
  sesquioxides (metal oxides), 132, 139, 141
  silicates, 132, 134–139, 141
  smectites. *See* montmorillonite
  source of negative charge in, 133–134
  structures of, 131, 134–137
  transformed, 129
  vermiculite, 11, 132–133, 139, 141

Clay films, 39, 43

Clay minerals. *See* Clays

Clay soils, 97–98, 506. *See also* Vertisols
  aeration of, 506
  drainage of, 506–508
  irrigation of, 506–508
  management of, 506–508. *See also* Vertisols
  textural class, 98
  tillage of, 506–508. *See also* Vertisols

Climate and soil formation, 19–20

rate of uptake, 262–263
recommendations for, 352, 370
soil balance of, 266, 274
symbiotic fixation of, 209–211, 267
volatilization of, 270–273
Nitrogen cycle, 266, 274
Nitrogenase, 209
Nitropyrin (N-Serve), 270
Nitrosomonas, 208
Nonpoint pollution. *See* Pollution,
nonpoint
Nonsilicate clays, 132–133, 140–141
Nonsymbiotic N2 fixation, 212, 267
Nonsoils, 83, 88
No-till farming, 413, 416, 422–425
N-Serve. *See* Nitrapyrin
Nutrient balance, 377–378
Nutrient immobilization, 262
Nutrients, plant
availability and deficiency, 378
chromatographic movement, 144
content in plants, 378
critical (threshold) range of,
363–364
deficiency symptoms, 363,
365–367
diffusion, 263
essential, 12–13
in fertilzers, 380–382
ion forms in soils and plants, 13,
262
mechanism of uptake of, 262–264
mobility of, 382–383
movement of ions to root, 263
removed by crops of, 378
visual deficiency symptoms of,
363, 365–367

O

Obsidian, 8
Ochric epipedon, 37–39
Octahedra sheets, 131
O horizon, 36
Onyx, 6
Opal, 6
Order, soil. *See* Soil orders
Organic amendments, 196, 229–237
Organic colloids. *See* Humus
Organic compounds. *See* Humus
Organic farming, 237–239
Organic fertilizers, 225
Organic matter, 196. *See also* Humus
and plant nutrition, 221, 225, 233,
235

and soil aggregates, 224
and soil conservation, 450–452,
454–455, 464–467
benefits to soil, 224–226
composition of, 217–218, 221–222
contribution to soil CEC by,
146–147, 224
decomposition of, 196–197, 218–222
detrimental effects in soil of,
227–229
energy concepts and, 224
example contents in soils, 230
factors affecting decomposition of,
223–224
management of, 229
optimum levels of, 229
soil formation and, 20, 22, 24
Organic matter decomposition,
220–222
factors affecting, 223–224
products of, 200
Organic soil materials, 9, 27,
224–239
Organic soils, 3, 9. *See* Histosols
Organisms. *See* Animalia; Fungi;
Monera; Plantae; Protista
Osmotic effect of salts, 319, 320
Osmotic potential, 163. *See* Solute
potential
Outwash plains, 32
Oven dry soil, 169
Oxic horizon, 40–41
Oxidation, 111–112. *See also* Redox
reactions
as a weathering process, 10
of ammonium, iron, manganese,
and sulfur, 268–269, 395–396,
516–517
Oxidation numbers. *See* Valence
Oxidation-reduction potential,
395–396, 515–518
Oxides. *See* Metal oxides;
Ferrihydrite; Hematite; Quartz;
Gibbsite
Oxisols, 46–48, 79–82
classification of, 79
color profile of, *insert,* 76–77
management of, 80–81
properties of, 79–80
world location map of, 82
Oxygen, 12, 13, 110
and aeration in soil, 110–112
in redox reactions, 268–269,
395–396, 515–518
need by roots for respiration, 111

Oxygen diffusion rate, 111
Ozone shield of earth, 536, 575–576

P

Paddy soils, 394–397, 515
aeration in, 395–396
and phosphorus availability, 397
fertilization with nitrogen in,
395–396
pH change in, 395
toxins produced in, 395–396, 577
Paleosols, 25–27
PANs in air pollution, 536
Paraplow, 416
Parent materials, 18, 27
alluvial, 27, 29–30
classification of, 27
colluvial, 27, 34–35
eolian, 27, 30–31
flood plains, 27, 29
glacial, 27, 31–34
lacustrine, 27, 30
loess, 27, 30–31
marine, 27, 30
residual, 27–28
soil formation and, 19, 24
terraces, 27, 29
transported, 27, 29–35
Particle density, 102–103
Pascals (Pa), 162
Passive absorption of water, 181–182
Peat, 3, 27. *See also* Organic soil;
Histosols
Ped, 100
Pediment, 28–29
Pedon, 40
Peneplain, 29
People, a pollutant, 578–581
Percolation of water, 176–180
Permanent wilting point, 165–168.
See also Wilting point
Permeability. *See* Infiltration
Pest control, 545–550
biological, 549–550
cultural, 550
integrated pest management
(IPM), 550
natural chemicals (pheromones,
hormones), 550
Pesticides, 536, 545–550
acceptable, 546
and environment, 545–550
definition of, 545
negligible risk, 546

Water use efficiency, 184–190, 493
Weathering, 9–11, 17
    chemical, 10
    climate and, 10–11
    physical, 10
Wetlands, 519–520
Wilting point, 165–172
Wind erodibility equation, 459–460
Wind erosion, 459–467
    control of, 465–468
    predicting, 463–465

processes of, 460–463
Windbreaks, 467

X

Xeric water regime, 56

Y

Yeasts, 199, 203–204

Z

Zero tillage (no-till). *See* No-till
    farming
Zinc, 19, 308–310
    amendments (carriers) of, 309–311
    availability of, 308–309
    chelates, 309
    deficiency symptoms, 308–309

FACTORS FOR CONVERTING NON-SI UNITS TO SI UNITS

Non-SI Units	By	SI Units
(Multiply this) ⟶	By	(To obtain this)
(To obtain this)	By ⟵	(Divide this)

AREA		
acre, a	0.404 685	hectare, ha (10^4 m^2)
acre, a	4 046.85	square meter, m^2

LENGTH		
Angstrom unit, Å	0.1	nanometer, nm (10^{-9} m)
foot, ft.	0.304 8	meter, m
inch, in.	25.40	millimeter, mm (10^{-3} m)
micron, u	1.00	micrometer, um (10^{-6} m)
mile	1.609 34	kilometer, km (10^3 m)

MASS		
ounce, oz	28.349 5	gram, g (10^{-3} kg)
pound, lb	453.592	gram, g (10^{-3} kg)
quintal (metric), q	100	kilogram, kg
tonne, metric ton, Mg	1000	kilogram, kg
ton (2000 lb), t	907.185	kilogram, kg

VOLUME		
bushel, bu (U.S.)	35.238 1	liter, L (10^{-3} m^3)
cubic feet, ft^3	0.028 317	cubic meter, m^3
cubic feet, ft^3	28.316 5	liter, L (10^{-3} m^3)
cubic inch, in.^{3}	0.000 016 4	cubic meter, m^3
gallon, gal.	3.785 31	liter, L (10^{-3} m^3)
ounce, fluid, oz.	0.029 573	liter, L (10^{-3} m^3)

PRESSURE		
atmosphere, atm	14.70	lb/in.^{2}, psi
bar	14.50	lb/in.^{2}, psi
bar	0.986 9	atmosphere, atm
bar	100.0	kilopascal, kPa (10^3 Pa)
kg (weight) cm^{-2}	14.22	lb/in.^{2}, psi
pounds per square foot, lb ft^{-2}	47.88	pascals, Pa

OTHER CONVERSIONS		
bushel	32	lb of oats
bushel	45	lb of rice
bushel	56	lb of corn, sorghum
bushel	60	lb of wheat, soybeans, potatoes
cubic feet	0.028	cubic meters
cubic feet	28.3	liters
cubic inches	1.64×10^{-5}	cubic meters
dyne	1×10^{-5}	newton, N
erg	1×10^{-7}	joule, J
foot-pound	1.356	joule, J
gallon per acre, gal./a	9.35	liter per hectare, L ha^{-1}
gram per cubic centimeter	1.00	megagram per cubic meter, Mg m^{-3}
hectare-30 cm of soil	4×10^6 kg	kg per hectare-30 cm of soil
milliequivalents per 100 g soil	10.0	mmol$_c$ kg^{-1} soil
milliequivalents per 100 g soil	1.00	cmol$_c$ kg^{-1} soil
ounce (fluid)	2.96×10^{-2}	liter, L
pounds per acre, lb/a	1.121	kg per hectare, kg ha^{-1}
pounds per cubic ft, lb ft^{-3}	16.02	kilogram per cubic meter, kg m^{-3}
square feet	0.092 9	square meters
square miles	2.54	square kilometers
temperature (°C)	1.80	+ 32 = temperature, °F
temperature (°F) −32°	0.555 5	temperature, °C
temperature kelvin (°K) − 273.15	1.0	temperature, °C
ton (2000 lb) per acre, t/a	2.242	metric tons (1000 kg) per hectare, Mg ha^{-1}
ton (metric), Mg	1000	kilogram, kg